THOMAS JOHNSON

RADNOR J. PAQUETTE, M.S., Michigan College of Mining and Technology, has been Professor of Civil Engineering at Georgia Institute of Technology for many years. He is now Professor Emeritus at that institution. He is the co-author, with the late Leo J. Ritter, Jr., of *Highway Engineering,* published by The Ronald Press Company.

NORMAN ASHFORD, Ph.D., Georgia Institute of Technology, served as Assistant Professor of Urban and Regional Planning at Florida State University. He is presently Director of the Institute of Transportation for the Florida State University System. He is past Chairman of the Air Centers Committee of the American Society of Civil Engineers.

PAUL H. WRIGHT, Ph.D., Georgia Institute of Technology, is Professor of Civil Engineering at that institution. He is actively involved in research in highway and traffic engineering and planning.

TRANSPORTATION ENGINEERING

PLANNING AND DESIGN

RADNOR J. PAQUETTE
GEORGIA INSTITUTE OF TECHNOLOGY

NORMAN ASHFORD
FLORIDA STATE UNIVERSITY SYSTEM

PAUL H. WRIGHT
GEORGIA INSTITUTE OF TECHNOLOGY

JOHN WILEY & SONS

New York • Chichester • Brisbane • Toronto

ISBN 0 471 06670-2

Library of Congress Catalog Card Number: 79–190209

PRINTED IN THE UNITED STATES OF AMERICA

10 9 8 7 6

Preface

This book is designed as a basic text for transportation courses in civil engineering and city planning. It will also prove valuable as a professional reference for the engineering consultant confronted by transportation problems.

The authors have presented the historical development of all modes of transportation with special emphasis on the roles played by the federal and state governmental agencies which regulate them. Descriptive material is also included on the economics and operating characteristics of the various transportation modes. This material is included in order to encourage the student to view the problems of transportation in a broad context—as a complex system for moving people and goods.

In the planning and design of land transportation facilities, special treatment is given to urban transportation emphasizing the latest state of the art in the use of mathematical planning models and computer techniques in forecasting future urban traffic. A unified discussion of the design of streets, highways, and railroads is given showing the similarities and differences that exist in railroad and highway design.

Specific planning procedures and design criteria for air transportation facilities are given. Airport planning, site selection, layout and design are discussed along with such topics as terminal layout, runway orientation, airport configuration, runway and taxiway dimensioning, and airport lighting.

Criteria for water transportation facilities, including up-to-date material on the design for the coastal environment, the design of harbors, and the design of port facilities, are discussed. Separate chapters on pipelines and belt conveyor systems are given describing the important role these predominantly land transportation modes play in the movement of freight.

Finally, the authors take a look at probable future developments in transportation. Futuristic vehicles and control systems, including those now in the research and development phase are described, and the probable effect that these vehicles may have on the planning and design of transportation facilities is explored.

For the reader who wishes to go deeper into the transportation problem, an excellent list of references is given at the end of each chapter.

Since many agencies and organizations are active in the transportation field, publications of many of these groups have been freely consulted in the preparation of this volume. The authors are indebted to many agencies and individuals who provided photographs and source material for the book. Special thanks are extended to Mr. Theron Schnure who wrote the air transportation section of Chapter 5.

RADNOR J. PAQUETTE
NORMAN ASHFORD
PAUL H. WRIGHT

April, 1972

Contents

Part IV

DESIGN OF AIR TRANSPORTATION FACILITIES

Part V

DESIGN OF WATER TRANSPORTATION FACILITIES

Part VI

CONCLUSION

I

INTRODUCTION TO TRANSPORTATION SYSTEMS

1

Governmental Activity in Transportation

1–1. Introduction. Transportation has profound and enduring effects on a nation and its people. It is, in the words of a recent Congressional committee report, "unquestionably the most important industry in the world" [1]. The presence or lack of adequate transportation facilities shapes the boundaries of national, state, and local governments. Cities are located to take advantage of favorable transportation. Wars may be won or lost because of the mobility or lack of mobility of a nation's troops and weapons. The adequacy of the transportation system may determine a nation's foreign policy and the extent of its political control over its people. In the lives of people, transportation affects our economic well-being, our cultural development, and our social and recreational habits and customs.

In a free society, the interest and concern of the government in transportation matters, as in other matters, is a reflection of the moods and attitudes of the people. In turn, government interest and concern is manifest in legislation, court actions, and executive orders. Because of the vital importance of transportation to the well-being of the nation, it is not surprising that the federal and state governments have through the years seen fit to enact laws promoting the development of transportation facilities and the regulation of various carriers. In the paragraphs that follow, the dominant role the government has played in the development of the transportation system of the United States will be described.

WATER TRANSPORTATION

1–2. Canal Building Prior to the Civil War. When studying the history of transportation in the United States, it is important to remember that formidable physical obstacles deterred the development of land transportation facilities. In addition to mountainous terrain, the early settlers were confronted with vast expanses of primeval forestland which was practically impenetrable. It is, therefore, understandable that the earliest settlers took advantage of existing lakes, streams, and rivers and that the first federal aid to transportation was for water transportation facilities.

Although federal aid was granted for improvement of harbors as early as 1789, practically all of the impetus for the development of water transportation prior to the Civil War came from the various states. Examples of canals constructed by states or with state aid are listed as follows:

1817–1825 The Erie Canal, the first large canal project undertaken in the nation, was constructed by the State of New York. This route was 364 miles long and cost about $7 million. Later, in 1918, at a cost of $176 million, the state built the New York Barge Canal by deepening and widening the main routes of the old Erie Canal.

1826–1834 The State of Pennsylvania constructed a canal-railroad route which extended from Philadelphia to Pittsburgh on the Ohio River. This project cost $10 million.

1828–1836 The State of Ohio built the Ohio Canal extending from Cleveland to Portsmouth on the Ohio River and the Miami and Erie Canal from Toledo to Cincinnati.

1828–1850 A joint project of Virginia and Maryland, the Chesapeake and Ohio Canal extended from Washington, D.C. to Cumberland, Maryland. This canal operated until 1924, and a portion of it has been restored as a historical park.

1–3. Later Government Interest in Inland Waterways. While the federal government began early to maintain river channels and to provide coastal surveys and lighthouses and other navigation aids, the amounts of federal aid to water transportation were relatively small until about 1890.

The Rivers and Harbors Act of 1902 created the Board of Engineers for Rivers and Harbors in the War Department, forerunner of the U.S. Army Corps of Engineers. This board was given the responsibility of investigating the feasibility of proposals for improvements of waterways.

In 1907, President Theodore Roosevelt appointed the Inland Waterways Commission to study the status of waterways and carriers on inland waterways. This Commission recommended in 1908 that more suitable provision should be made for improving the inland waterways of the country.

Congress passed the Panama Canal Act in 1912 in anticipation of the completion of the Panama Canal. This act provided that without specific approval of the Interstate Commerce Commission, railroads could not own, operate, control, or have interest in common carriers using the Panama Canal. In cases where water lines were competitive with rail lines, the act made it illegal for other water carriers to be controlled by railroads unless specifically approved by the ICC. The increased activity in inland water transportation which subsequently occurred is attributed in part to the passage of this act.

The Transportation Act of 1920 stated the intent of Congress to promote, encourage, and develop water transportation. Four years later, Congress incorporated the Inland Waterways Corporation to demonstrate the service capabilities of water transportation. Historians generally date the beginnings of modern inland water carrier operations from this period [2].

The Transportation Act of 1940 established a system of regulation for domestic water carriers. It sought to control the route and service of water carriers and to prevent undesirable duplication of transportation services, undue competition, and the entry of incompetent carriers. This act established Part III of the Interstate Commerce Act extending its coverage to inland water transportation and charged the ICC with its implementation. Part III does not apply to private water carriers, and only about 10 per cent of the for-hire carriers are regulated under this act.

A government report [1] estimated that about $7.1 billion in federal funds were provided for navigation during the period 1917–1960. The majority of these expenditures was made in support of domestic navigation. This aid constituted about 16 per cent of the total funds provided for all transportation modes.

1–4. Legislation Relating to Ocean Transportation. Events during the decade of World War I prompted increased interest in ocean transportation and several acts were passed assisting the merchant marine. The Seamans Act of 1915 provided requirements of a minimum space per man, a separate berth for each seaman, and proper drainage, lighting, and ventilation of seamans' quarters.

The Shipping Board Act of 1916 created the U.S. Shipping Board, a five-member independent agency, the precursor of the Maritime Commission. The purpose of this act was to prohibit monopolistic, unjust, and discriminatory practices in ocean transportation. The act applied to common carriers operating over regular ocean and coastwise routes and on the Great Lakes.

In 1917, the Congress appropriated $500 million for a government ship building program for World War I. Although the program was too late to aid significantly the war effort, vessels continued to be launched under the program until 1921.

The Congress passed the Merchant Marine Act of 1920 in an attempt to establish and temporarily maintain ocean steamship lines operating over the major shipping routes from U.S. ports. The government intended to sell these lines, if necessary, at very low prices, to American companies for continued operation. The results of this act were disappointing.

Direct subsidies for ocean transportation were provided by the Merchant Marine Act of 1928. This act provided for a revolving construction loan fund of $250 million for ship building and permitted low interest loans up to three-fourths of the cost of vessels to be built or reconditioned. Loans with interest rates as low as one-fourth of 1 per cent were made under this act. Substantial mail subsidies were also provided by this law.

The functions of the Shipping Board were transferred to the Shipping Board Bureau in the Department of Commerce in 1933. These functions were subsequently transferred to the U.S. Maritime Commission by the Merchant Marine Act of 1936.

The purpose of the 1936 act was to permit the transportation of a substantial percentage of American foreign commerce in American flag ships. Two kinds of subsidies were provided by this legislation:

1. Construction subsidies for the difference between the cost of building a ship in a foreign yard and the cost of building it in the United States.
2. Operating subsidies to allow for the difference in wages between American flag ships.

By 1960, about $1.3 billion had been granted in subsidies to the merchant marine, approximately three-fourths of which was for ship construction and one-fourth for operating subsidies [1].

1–5. Recent Governmental Activity Relating to Water Transportation. In 1962, President Kennedy in his message to Congress on transportation recommended that a user tax of two cents per gallon should be applied to all fuels used in transportation on inland waterways. In his final budget message of January 15, 1969, President Johnson recommended ". . . a charge of two cents per gallon, increasing to ten cents over the next five years, on the now untaxed fuel used by vessels on inland waterways." In his January, 1971 budget message to Congress, President Nixon stated:

> Federal funding in inland waterways is estimated to be in excess of $250 million annually. Unlike highway and airport/airway

> users, waterways users bear no portion of the costs of the facilities and services provided for them. Therefore, in order to assure a more equitable distribution of the tax burden among transportation users and to move toward a more efficient allocation of transportation resources, legislation providing for the recovery of a portion of the costs of inland waterways will be submitted to the Congress.

Federal expenditures for water transportation in 1968 amounted to more than $1.2 billion. This amount included $545 million for the U.S. Coast Guard, $314 million for the merchant marine, and $391 million for inland waterways, intracoastal waterways, Great Lakes, and coastal harbors.

Increased concern for the U.S. Merchant Marine and the nation's shipyards was reflected in the passage of the Merchant Marine Act of 1970. This act, which is perhaps the most significant piece of legislation relating to water transportation since the 1936 act, promises to end the nation's decline as a shipping power. The bill will provide $4 billion during the decade of the seventies in ocean transportation subsidies. Of this total, $2.2 billion will be used to subsidize the construction of ships in U.S. shipyards, and the remaining $1.8 billion will provide operating subsidies for the owners of U.S. merchant ships.

RAIL TRANSPORTATION

1–6. Railroad Legislation Prior to 1834. The first significant legislative act to provide assistance to railroads was the General Survey Bill of 1824. Haney [3] observes:

> In the early years of the century civil engineers were very scarce relatively to the demand for them. The settlement of our vast public domain and numerous roads and canals constantly required the service of surveyors and engineers, and on some of the larger work foreigners were employed. The period was one in which new lines of communication and transportation were rapidly projected—one of births and expansions. The work of plans, surveys and estimates was in great demand and of high relative importance at this stage, and it was natural that a government which was desirous of increasing the facilities for its commerce, and which maintained a corps of engineers in its employ, should have begun with aid to these initial steps toward the construction of such facilities. The first assistance ever granted by Congress for railway purposes was given in the shape of a survey by government engineers.

> The survey bill of 1824 authorized the President to cause the necessary surveys, plans and estimates to be made for the routes of "such roads and canals as he may deem of national importance, in a commercial or military point of view, or necessary for the transportation of the public mail;" and a limited number of civil engineers and officers of the corps of engineers was placed at his disposal. Thirty thousand dollars annually was the amount appropriated to cover expenditures.

Although railways were not specifically mentioned in this bill, by 1826 railways were being surveyed by government engineers under this bill. Before the repeal of this act in 1838, sixty-one railway surveys had been made.

The second important law promoting railroad transportation was the Railroad Iron Bill of 1832. At this time the nation's iron industry was primitive and its output was relatively small. The railroads were commencing a period of vigorous growth, and the supply of rails had to be purchased from foreign sources. Earlier legislation had placed protective tariffs on imported iron to assist the growth and development of the iron industry in this country. After several years of debate, the Congress passed the Railroad Iron Bill of 1832 which removed the tariffs on iron imported for use as rails.

1–7. Federal Land Grants. There is little question that the most important legislative acts promoting railroad transportation were the land grant acts. These acts generally provided grants of federal lands for rights-of-way and additional sections of land which the railroads could sell or use in some other way to raise capital for railroad construction. The purpose of these grants was to encourage the development of the western portion of the nation.

As the Association of American Railroads [4] points out, less than 8 per cent—18,737 miles—of railroad mileage was involved in either federal or state land grants at the peak of expansion in the United States. Furthermore, the land grants imposed certain obligations on the railroads such as the transportation of mail and other federal property and the transportation troops over the land-grant portions without charge or at reduced rates. These obligations remained with the railroads until 1940, when the Transportation Act of 1940 canceled the rate concessions applicable to the transportation of mail and federal property. The concessions applicable to the transportation of troops were dropped in 1945.

The first railroad land grant was made to the Tallahassee Railroad in 1834. Several subsequent requests by railroads for land grants resulted in a number of grants being given in the years that immediately followed.

One of the largest land grants to railroads was made in 1850. This act provided alternate sections of land to the states of Illinois, Mississippi, and Alabama for railroads. Under this act, more than 3,750,000 acres of land were granted for the Illinois Central, Mobile and Ohio, and Mobile and Chicago railroads.

By 1852, the demand for land grants was so great that a general law was passed providing for right-of-way through public lands. The act provided for a right-of-way of 100 feet in width for all rail and plank road or macadam turnpike companies then chartered or to be chartered within ten years. These companies were also given the right to take construction materials from nearby public lands, and additional land was provided for watering places, workshops, and depots.

On July 1, 1862, enabling legislation for the construction of the first transcontinental railroad was signed by President Lincoln. The con-

Fig. 1–1. The driving of the golden spike at Promontory, Utah on May 10, 1869, signaling the completion of the first chain of railroads to span the American continent. (Courtesy Union Pacific Railroad and the Association of American Railroads.)

struction of this line, which culminated in the celebrated driving of the golden spike at Promontory, Utah in 1869, played a prominent part in the development of the West.

1–8. Railroad Taxation and Regulation. During the Civil War, the Revenue Bill of 1862 was passed, levying a tax on gross earnings from railroad passenger services and a tax on interest and dividend payments. Another wartime measure, the Revenue Bill of 1864, placed a tax on gross railway receipts including both passenger and freight earnings.

There had been earlier complaints about aid to the railroads for transportation of the mails and, by 1850, the mail service relation had ceased to be one of aid and had become one of regulation. Opposition to government land grant policies, long present, grew stronger as railroads engaged in practices which were not in the public interest. There were frequent cases of overbuilding, and the economic struggles that followed often resulted in price discrimination against the shippers. Collusion among the carriers and other abusive practices were not uncommon. By about 1870, opposition to further land grants became so strong as to bring about the end of the land grant policy.

The first positive regulation of interstate commerce came in 1873 with the passage of a bill to prevent cruelty to animals in transportation. This act, which applied to the interstate transportation of livestock, provided that no railway should confine animals in cars, boats, or other vessels for a period longer than twenty-eight consecutive hours without unloading them for at least five consecutive hours for rest, water, and feeding.

The pace of railroad construction increased and, by 1880, the railway system in the United States consisted of over 93,000 miles. The increase in railway mileage was accompanied by growing public fear of and hostility toward the railroad companies.

In 1887, the Congress passed the Interstate Commerce Act, a comprehensive measure which applied to railroads engaged in interstate or foreign commerce. This act:

1. Required "just and reasonable" rates.
2. Made it unlawful for a carrier to charge one person more than another person for like transportation services.
3. Prohibited "undue or unreasonable preference or advantage" to any person, place, or kind of traffic.
4. Made it unlawful for a carrier to charge more for transportation over a short distance than over a longer distance over the same line and in the same direction.
5. Prohibited the pooling of freights of different and competing railroads.

6. Required carriers to keep printed schedules of rates and fares for public inspection.
7. Established an Interstate Commerce Commission to administer the law.

In the years that followed, three laws were passed which strengthened the Interstate Commerce Act, and increased and broadened the powers of the Interstate Commerce Commission. The Elkins Act of 1903 strengthened the preferential treatment section of the Interstate Commerce Act by making both the carrier and shipper guilty of violations.

The Hepburn Act of 1906:

1. Extended the ICC's authority over related rail activities such as private car lines, sleeping car companies, and terminal facilities.
2. Required accounting reports as prescribed by the ICC.
3. Allowed ICC to prescribe *maximum rates.*
4. Strengthened ICC's enforcement procedures by providing for requests for court orders to comply.

The Mann-Elkins Act of 1910 authorized the ICC to suspend proposed rail rate changes, either upon complaint or upon its own initiative, for a period of 120 days, during which time a study of the merits of the proposal would be made.

1–9. Railroad Legislation After World War I. At the beginning of World War I, the federal government took control of and operated the railroads during the war and until March, 1920. The Transportation Act of 1920 returned the carriers to private operation.

> The provisions of the Transportation Act of 1920 grew out of the economic status of the railroads at the termination of the war period and the necessity for a transition from federal control of the railroads to private management and operation. The major issue facing Congress then was to prevent a complete breakdown of the railroad transportation system because of the precarious financial condition of the "weak roads" [1].

The basic purpose of this act was to establish a stronger economy for the railroad industry by encouraging consolidation of rail carriers into a limited number of balanced systems under the supervision of the ICC. This act provided that carriers could earn a fair return on their investments. The Commission was given permission to establish minimum rates in order to prevent rate wars and discrimination, and the power to prescribe exact transportation rates. The act also gave the ICC control over the issuance or purchase of securities by railroads.

During the years since 1920, the railroads have experienced relative declines in both passenger and freight traffic. The railroads' share of passenger traffic by common carrier declined from 74.9 per cent in 1929 to only 9.2 per cent in 1968 [6]. More importantly, the absolute rail passenger traffic has dropped precipitously since the end of World War II, as more and more intercity trips have been made by automobiles and airplanes. The magnitude of railroad problems with passenger transportation is shown by the cumulative passenger traffic deficit which increased from $139,776,000 in 1946 to $723,670,000 in 1957 [1]. In 1968, the railroads reported a $468-million deficit in passenger service [5], while experiencing a 14-per-cent decline in revenue passenger miles from the previous year [6].

The railroads' share of total intercity freight declined from 76.5 per cent in 1926 and 66.6 per cent in 1946 to 41 per cent in 1968 [6]. The trend in terms of ton-miles of freight, except for the period of the Great Depression of the thirties, was upward until 1956. It was reported in 1961:

> There has been a downward trend of railroad freight traffic revenue since 1956 and it took place during a period of expanding economic activity, except for the mild recessions of 1949 and 1957–58, and during a period of expanding trade for other modes [1].

This subject is covered in somewhat greater depth in Section 2–6.

The Transportation Act of 1958 sought to give relief to the railroads in two important ways:

1. It provided a more liberal policy in train service abandonment.
2. It provided that "the rates of a carrier shall not be held up to a particular level to protect the traffic of any other mode of transportation . . ."

President Kennedy alluded to some of the economic problems of the railroads in his comprehensive message to Congress in 1962 on national transportation policy. He stated:

> But pressing problems are burdening our national transportation system, jeopardizing the progress and security on which we depend. A chaotic patchwork of inconsistent and often obsolete legislation and regulation has evolved from a history of specific actions addressed to specific problems of specific industries at specific times. This patchwork does not fully reflect either the dramatic changes in technology of the past half-century or the parallel changes in the structure of competition.
>
> The regulatory commissions are required to make thousands of detailed decisions based on out-of-date standards. The management of the various modes of transportation is subjected to

excessive, cumbersome and time-consuming regulatory supervision that shackles and distorts managerial initiative. Some parts of the transportation industry are restrained unnecessarily; others are promoted or taxed unevenly and inconsistently.

Some carriers are required to provide, at a loss, services for which there is little demand. Some carriers are required to charge rates which are high in relation to cost in order to shelter competing carriers. Some carriers are prevented from making full use of their capacity by restrictions on freedom to solicit business or adjust rates. Restraints on cost-reducing rivalry in rate-making often cause competition to take the form of cost-increasing rivalry—such as excessive promotion and traffic solicitation, or excessive frequency of service. Some carriers are subject to rate regulation on the transportation of particular commodities while other carriers, competing for the same traffic, are exempt. Some carriers benefit from public facilities provided for their use, while others do not; and of those enjoying the use of public facilities, some bear a large part of the cost, while others bear little or none.

No simple Federal solution can end the problems of any particular company or mode of transportation. On the contrary, I am convinced that less Federal regulation and subsidization is in the long run a prime prerequisite of a healthy inter-city transportation network. The constructive efforts of State and local governments as well as the transportation industry will also be needed to revitalize our transportation services [7].

In 1970, a bill was passed which was expected to bring financial relief to railroads operating unprofitable passenger service. This legislation authorized the creation of the National Railroad Passenger Corporation, informally termed "Amtrak." It is intended that this corporation will take over the management of a national passenger train network in the hope that improved service will attract riders back to passenger trains.

The bill required that the new network and schedules of service be initially designated by the Secretary of Transportation. The Amtrak corporation is authorized to contract with railroad companies to provide the right-of-way and service. Railroad employees would operate the trains under the management of Amtrak.

The legislation includes a complex financial plan wherein each railroad that elects to participate will contribute cash or equipment to the new corporation in return for which the railroad will receive common stock in the company or a tax deduction. The law provided a direct grant of $40 million and loan guarantees of up to $100 million for the improvement of roadbeds and the purchase of new equipment.

By May 1, 1971, eighteen railroads had signed contracts with the National Railroad Passenger Corporation, and on that date the corporation began operating its intercity passenger network.

HIGHWAY TRANSPORTATION

1–10. Colonial Roads. Up to the time of the Revolutionary War, the public roads consisted mainly of horse paths and were suitable for carts and wagons only near the large centers of population. Gradually, these paths, which often followed the Indian trails or traces, were widened into wagon roads. Because of the formidable, heavily wooded terrain, the major emphasis in colonial times was on water transportation, and the roads were crude and localized.

In the colonies, the settlers instituted the statute labor system for the building and maintenance of roadways. Under this system, which had been used in England since 1555, several days were set aside each year in which the men were required to work on the roads. This practice was called "working out the road tax." It was supplemented by donations, bridge tolls, proceeds from public land sales, and fines from those who failed to perform the road work. Although the statute labor system was notoriously inefficient, it continued to be used in the United States until the beginning of the twentieth century.

1–11. Road Building in the Early Days of the Republic. Prior to the Revolutionary War, the majority of long distance travel was either on foot or by horseback.

> Stage coaches had come into such general use, by 1785, that an urgent need arose for surfaced roads travelable at all seasons of the year. Impoverished by contributions of men, money and supplies during the War of Independence and by a ten-year postwar competition with British merchants, New England townships were unable to raise the necessary taxes for road improvement. The financial condition of the States was no better. The answer was found in turnpike companies chartered by the State and financed and operated by private citizens [8].

One of the most important of these roads was the Philadelphia and Lancaster Turnpike Road which was completed in 1795. This 62¼-mile stone and gravel road was built at a cost of $465,000, all from private sources. Other toll roads built during this period included:

1. The Northwestern Turnpike, which extended from Winchester, Virginia to Parkersburg, Virginia. It was completed in 1838 at a cost of $400,000.
2. The Maysville Turnpike, a 64-mile section extending from Lexington, Kentucky to Maysville, Kentucky on the Ohio River. This

stone-surfaced road was completed in 1835 at a cost of $426,000, one-half of which was provided by the state.

3. The Shenandoah Valley Turnpike, a 92-mile macadam road which extended from Staunton, Virginia to Winchester, Virginia. The $385,000 road was completed in 1840.

Although the individual states frequently purchased stock in turnpike companies, no federal aid was provided for these roads. Indeed, the role of the federal government in roadbuilding during this period was rather limited, consisting principally of the construction of several primitive military roads and The National Pike.

1–12. The National Pike. The National Pike, or Cumberland Road as it was also called, was the first important road to be built with federal funds. This road, which extended from Cumberland, Maryland to Vandalia, Illinois, was built over a 33-year period ending in 1841. The legislation which authorized the construction of the National Pike provided for compacts or agreements between the federal government and the states of Ohio, Indiana, and Illinois. Congress provided that a "two per cent fund" be established from the sale of public lands and that money from this fund would be used to reimburse the federal government for costs of the construction of the National Pike. This plan of reimbursement was never realized, and about $6.7 million was spent by the federal government in the construction of this road.

The steam railroads, which came into prominence in the 1840's, contributed to the financial failure of a large number of turnpike companies and decreased federal interest in road building. As a consequence, the responsibility for building and maintaining wagon roads and bridges was returned to the local governments for the remainder of the nineteenth century.

1–13. The Bicycle Craze. During the latter years of the nineteenth century, great interest was manifest in bicycles, first in "ordinary bicycles" which had large front wheels, and later in "safety bicycles" which resembled bicycles of the present day. Bicycle clubs were formed, and, in 1880, a federation of local bicycle clubs was organized and named the League of American Wheelmen. Largely through the efforts of this organization, there grew a strong movement for the construction of good roads. This movement led to the establishment of the Office of Roads Inquiry in the U.S. Department of Agriculture in 1893, forerunner of the Bureau of Public Roads. Several state laws were also enacted at this time providing financial aid for local road improvements.

1–14. Federal Aid to Highways. Although the federal government has been involved in road construction since the early days of the republic, the current procedure whereby the states and the federal government share road building costs had its beginning in 1916. Ritter and Paquette [9] describe the development of the federal-aid program as follows:

> The modern era of federal aid for highways began with the passage of the Federal Road Act of 1916, which authorized the expenditure of $75 million over a period of five years for improvement of rural highways. The funds appropriated under this Act were apportioned to the individual states on the basis of area, population, and mileage of rural roads, the apportioning ratios being based upon the ratio of the amount of each of these items within the individual state to similar totals for the country as a whole. States were required to match the federal funds on a 50–50 basis, with sliding scale in the public land states of the West.
>
> The Federal-Aid Highway Act of 1921 extended the principle of federal aid in highway construction and strengthened it in two important respects. First, it required that each state designate a connected system of interstate and intrastate routes, not to exceed 7 per cent of the total rural mileage then existing within the state, and it further directed that federal funds be expended upon this designated system. Second, the Act placed the responsibility for maintenance of these routes upon the individual states.
>
> During this period many states, experiencing difficulty in securing the necessary funds to match the federal appropriations, looked for new sources of revenue. This fact led to the enactment of gasoline taxes by several states in 1919, and other states rapidly followed suit.
>
> During the ensuing years, appropriations for federal aid were steadily increased, while the basis of participation remained practically the same. Funds were provided (as now) for the improvement of roads in national parks, national forests, Indian reservations, and other public lands. The principle of federal aid was broadened to allow the use of federal funds for the improvement of extensions of the federal-aid system into and through urban areas and for the construction of secondary roads.
>
> Another important step taken during this period was the authorization by Congress, in 1934, of the expenditure of not more than 1.5 per cent of the annual federal funds by the states in making highway planning surveys and other important investigations. The first planning survey was inaugurated in 1935, and work of this nature was being conducted by all the states by 1940.
>
> World War II focused attention upon the role of highways in national defense and funds were provided for the construction of access roads to military establishments and for the performance of various other activities geared to the war effort. Normal highway development ceased during the war years.
>
> The Federal-Aid Highway Act of 1944 provided funds for highway improvements in the postwar years. Basically, the Act provided

$500 million for each of three years. Funds were earmarked as follows: $225 million for projects on the federal-aid system; $150 million for projects on the principal secondary and farm-to-market roads; and $125 million for projects on the federal-aid system in urban areas. The law also required the designation of two new highway systems. One is the National System of Interstate Highways, which will be discussed later in this chapter, and the other is one composed of the principal secondary routes (Federal-Aid Secondary System).

The Federal Aid Highway Act of 1956 must be regarded as one of the most important laws ever passed regarding transportation. The act provided nearly $25 billion[1] for the construction of the National System of Interstate and Defense Highways, the system envisioned by the 1944 highway act. The law specified that the funds for this system would be made available on a 90–10 matching basis, the larger share being provided by the federal government. When completed, this system will connect all of the major cities of the United States with controlled access freeways constructed to approved modern design standards. According to the provisions of this act, the federal share of the costs of this system will be paid from highway user taxes levied by the federal government and placed in a special trust fund called the Highway Trust Fund. The act also provided that several important studies be made including a study of equitable tax allocations. The importance of the 1956 act on highway development may be seen in Fig. 1–2, which shows the annual highway appropriations from 1948 to 1969.

Prior to 1958, biennial highway appropriations were made by amending and supplementing the Federal Aid Road Act of 1916. On August 27, 1958, Congress passed a law which repealed all obsolete provisions of Federal Aid Highway Acts and codified existing provisions into Title 23–Highways of the United States Code. Subsequent appropriations made by the Congress have been made "in accordance with Title 23 of the United States Code." This law and other federal laws and regulations relating to highways are given in a booklet published by the Bureau of Public Roads [10].

The growing problems of urban transportation were recognized by the Federal Aid Highway Act of 1962. This act gave greater emphasis to planning and research and required cooperative and comprehensive planning of urban transportation facilities by state and local governments. Specifically the act:

[1] By January, 1972, some $47.19 billion had been spent or authorized for the Federal Aid Interstate Program. Including the ten-per-cent state shares, the cost through 1974 was expected to exceed $56 billion.

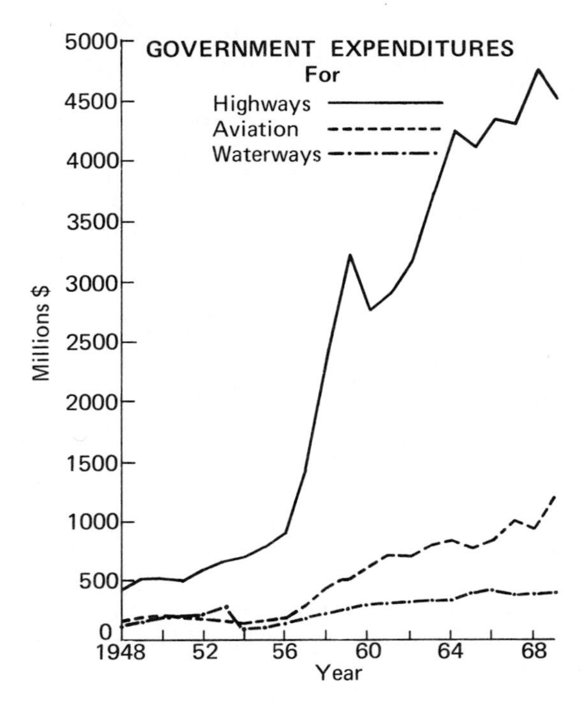

Fig. 1–2. Recent government expenditures for highways, aviation, and waterways.

1. Permitted more extensive use of federal aid secondary funds in urban areas.

2. Required that, after July 1, 1965, the Secretary of Commerce (Bureau of Public Roads) not approve any program for highway projects in an urban area of more than 50,000 population unless such projects are based on a continuing, comprehensive transportation planning process carried on cooperatively by the states and local communities.

3. Recognized the importance of urban mass transportation by requiring that urban highway systems be "an integral part of a soundly based, balanced transportation system for the area involved."

4. Required that the 1.5 per cent planning and research funds be used only for these purposes, or otherwise lapse. (The Hayden-Cartwright Act, passed in 1934, was a permissive act. It authorized, but did not require, the use of up to 1.5 per cent of the annual federal highway funds for planning and research purposes.) The 1962 act provided that an additional one-half of one per cent of federal aid highway funds be available for expenditure for planning and research upon request by the State Highway Department.

5. Provided relocation assistance from construction funds for those required to move and relocate because of highway construction. Maximum relocation payments of $200 for individuals or families and $3,000 for business concerns and nonprofit organizations were authorized by this act.

1–15. Highway Safety Legislation. Long a matter of national concern, the traffic safety problem worsened after World War II as gains in motor vehicle registrations were accompanied by similar increases in traffic crashes. Higher vehicle speeds were reflected by larger economic losses due to traffic accidents and increases in traffic deaths. Highway engineers could take some satisfaction in the fact that the traffic safety problem was not growing in proportion to the growth in travel. While the number of traffic fatalities had steadily increased over the years, the trend of the fatality rate in terms of deaths per vehicle mile had been downward. However, the troublesome highway safety problem reached epidemic proportions in the 1960's as the number of fatalities climbed higher than 50,000 per year. Reflecting an aroused public concern, the Congress passed two important safety measures: the National Traffic and Motor Vehicle Safety Act of 1966 and the Highway Safety Act of 1966. Administration of these acts was to be accomplished by a new federal agency, the National Traffic Safety Agency, later renamed and reorganized as the National Highway Safety Bureau in the Department of Transportation.

The National Traffic and Motor Vehicle Safety Act of 1966 provided for the establishment of federal motor vehicle safety standards and prohibited the manufacture or import of any motor vehicle which fails to conform to the established standards. Violators of these provisions may be penalized up to $1,000 for each violation, up to a maximum of $400,000. This act requires that motor vehicle manufacturers notify buyers and dealers of safety defects and provides for the correction of these defects. The act provided for research of the adequacy of safety standards and inspection requirements which apply to used motor vehicles and studies of the relationship between motor vehicle performance characteristics and accidents, injuries, and deaths. It was required that motor vehicle tires be labeled, giving the number of plies and maximum permissible load, and that new vehicles be equipped with tires which could safely support maximum loads. Under this act, a national register has been established identifying individuals reported by state and local authorities who have been denied a driver's license or who have had their driver's license withdrawn.

The Highway Safety Act of 1966 required each state to have a highway safety program designed to reduce traffic accidents and deaths, injuries, and property damages resulting from these accidents. Administration of these programs is typically accomplished by a Traffic Safety Coordinator on the Governor's staff who oversees programs carried out by various state and local agencies. According to this act, the various programs were to be in accordance with uniform standards promulgated by the federal government. The uniform standards [11] promulgated in 1967 by the Secretary of Transportation provided for programs in thirteen problem areas in traffic safety:

1. Periodic motor vehicle inspection
2. Motor vehicle registration
3. Motorcycle safety
4. Driver education
5. Driver licensing
6. Codes and laws
7. Traffic courts
8. Alcohol in relation to traffic safety
9. Identification and locations of accidents
10. Traffic records
11. Emergency medical services
12. Highway design, construction, and maintenance
13. Traffic control devices

In 1968, the Secretary promulgated three additional standards providing for programs in pedestrian safety, police traffic services, and debris hazard control and cleanup.

The Highway Safety Act of 1970 changed the name of the National Highway Safety Bureau to the National Highway Traffic Safety Administration. Further, the act gave the Federal Highway Administration the responsibility of administering the highway safety programs related to highway design, construction, and maintenance, traffic control devices, identification, and surveillance of accident locations, and highway related aspects of pedestrian safety.

1–16. Recent Federal Aid Highway Acts. The federal-aid highway acts of 1968 and 1970 demonstrated the government's increasing concern for transportation problems in urban areas and the social, economic, and environmental consequences of highway construction. The 1968 legislation contained three landmark provisions:

1. It authorized the use of federal aid urban funds for fringe parking projects to reduce urban congestion.

2. It established TOPICS, Traffic Operation Programs to Increase Capacity and Safety of urban facilities.

3. It provided for more liberal compensation to families and businesses displaced by highway projects. For example, it makes possible the payment to resident home owners of up to $5,000 above

fair market value where this is necessary to enable them to purchase fully comparable replacement property. Similarly, renters unable to afford comparable replacement property may be paid up to $1,500 in rent subsidies or as part of a down payment on a home. The bill also provides for the reimbursement of property owners for all expenses incidental to the sale of their property and the payment of certain displaced businesses up to $5,000 for loss of neighborhood patronage and good will.

The 1970 Federal Aid Highway Act provided for the establishment of a federal-aid *urban* highway system and made $100 million available for expenditure on the new system during each of the fiscal years ending June 30, 1972 and June 30, 1973. The act specified that the urban system would serve major centers of activity, the highest traffic volume corridors, and the longest urban trips. Each route in the proposed system will connect with another route on a federal-aid system.

The 1970 act further emphasized the government's interest in encouraging the use of buses and other urban highway mass transportation systems. It provided that under certain conditions federal highway funds may be used to help finance the construction of exclusive or preferential bus lanes, traffic control devices, bus passenger loading areas and shelters, and fringe and transportation corridor parking facilities which serve mass transportation passengers.

The act required the Secretary of Transportation to promulgate guidelines designed to assure that possible adverse economic, social, and environmental effects are properly considered during the planning and development of federal-aid highway projects. The intent of this requirement is to insure that decisions on federally funded projects be made in the best overall public interest, taking into consideration any adverse effects due to air, noise, and water pollution, the displacement of citizens, farms, and businesses, and the disruption or destruction of natural resources, public facilities, community cohesion, and aesthetic values.

One of the sections of the 1970 legislation illustrates the important role of highways in carrying out national goals relating to land use and population distribution. This section authorized demonstration grants to states for the construction and improvement of *economic growth center development highways* on the federal-aid primary system. The purposes of these highways would be to revitalize the economy of rural areas and smaller communities, to enhance and disperse industrial growth, to encourage more balanced population patterns, and to check and possibly reverse the current migration from rural areas and smaller communities. The law provides payments of up to 100 per cent of the cost of engineering and economic surveys and an increased federal share of the cost of construction of these facilities. An amount of $50 million was provided for each of the fiscal years ending June 30, 1972 and June 30, 1973.

The act authorized the establishment of a National Highway Institute for the development and administration of training programs for federal, state, and local highway employees engaged in federal-aid highway work. The Secretary of Transportation was authorized to deduct up to $5 million per fiscal year from federal-aid appropriations for use in carrying out training programs. In addition, the act authorized the States (subject to the approval of the Secretary) to expend up to one-half of one percent of all federal-aid highway funds for payment of up to 70 per cent of the cost of tuition and direct educational expenses of state and local highway department employees. This latter training could be carried out through grants and contracts with public and private agencies, institutions, and individuals.

The Federal Aid Highway Act of 1970 included the following important additional provisions:

1. It authorized the establishment of an emergency fund to be spent for the repair or reconstruction of roadways seriously damaged by widespread natural disasters or catastrophes or of dangerous bridges which have been closed to traffic because of structural deficiencies or physical deterioration.
2. It extended the time of completion of the Interstate System from 18 to 20 years making the new date of completion June 30, 1976. The act authorized appropriations for the Interstate System of $4 billion for each of the fiscal years ending June 30, 1974, June 30, 1975, and June 30, 1976.
3. The act authorized slightly over $1.7 billion for fiscal years 1972 and 1973 for the various highway systems and programs. Of this amount, $1.1 billion was specified for the federal aid primary, the Federal aid secondary and their extension within urban areas, the so-called ABC system.
4. It provided that beginning after June 30, 1973, the federal share of funds for the ABC system and the new Federal aid urban system will be increased from 50 per cent to 70 per cent.

PUBLIC MASS TRANSPORTATION

1–17. The Government and Mass Transportation. Public transportation has deep and enduring problems. In 1969, President Nixon stated:

> In the last 30 years urban transportation systems have experienced a cycle of increasing costs, decreasing funds for replacements, cutbacks in service and decrease in passengers.
>
> Transit fares have almost tripled since 1945; the number of passengers has decreased to one third the level of that year.

> Transit industry profits before taxes have declined from $313 million in 1945 to $25 million in 1967. In recent years 235 bus and subway companies have gone out of business. The remaining transit companies have progressively deteriorated. Today they give their riders fewer runs, older cars, and less service.
>
> Local governments, faced with demands for many pressing public services and with an inadequate financial base, have been unable to provide sufficient assistance.
>
> This is not a problem peculiar to our largest cities alone. Indeed, many of our small and medium-sized communities have seen their bus transportation systems simply close down [12].

Until recently, the federal government has shown almost total indifference to the problems of urban mass transit. Prior to 1961, there was practically no federal commitment to assist or promote mass transit systems.

One of the earliest legislative references to mass transportation was in the Federal Aid Highway Act of 1962, which required that urban highway systems be "an integral part of a soundly based *balanced* transportation system." President Kennedy's transportation message of that same year to Congress reflected concern with mass transportation and requested a capital grant authorization of $500 million to be made available over a three-year period. Congress, by means of the Urban Mass Transportation Act of 1964, appropriated $375 million for mass transit assistance and gave the responsibility for the administration of these funds to the Housing and Home Finance Agency (later to become the Department of Housing and Urban Development). In 1968, this responsibility was shifted to the Department of Transportation when the Urban Mass Transportation Administration was formed within that Department.

The Urban Mass Transportation Administration has reported that programs for the assistance and promotion of urban mass transportation totaled approximately $134 million in fiscal year 1968. Financial aid under this program included assistance of the following types:

1. *Capital Improvement Grants*—for new transit system equipment and modernization of transit facilities.
2. *Demonstration, Research, and Development Projects*—for studies, tests, and demonstrations of new ideas, methods, systems, and equipment for improved transportation planning and operations.
3. *Technical Studies Grants*—for engineering plans and designs of urban mass transportation systems, and for other technical studies.
4. *Managerial Training Grants*—for fellowship awards for advanced training of personnel employed in managerial, technical, and professional positions in the urban mass transportation field.

5. *University Research and Training Grants*—for nonprofit educational institutions combining comprehensive research and research-training in urban transportation problems.

Government concern regarding the need for improved intercity mass transportation facilities was shown by the passage of the High Speed Ground Transportation Act of 1965. This act appropriated $20 million for fiscal year 1966 and $35 million for each of the two following fiscal years for research and development in high speed ground transportation systems.

The Office of High Speed Ground Transportation in the Department of Transportation is conducting two conventional rail demonstrations: one using advanced electric powered equipment to run between Washington, D.C. and New York City, and the other using jet powered trains for the New York to Boston run. This office is also sponsoring research in innovative mass transportation systems including tracked air cushion vehicles and tube vehicle systems.

Despite increased government aid for mass transit facilities, it became clear in the later 1960's that much more aid would be required to reverse the downward trends in mass transit usage. President Nixon reflected the deep concern of the federal government about public transportation in his message to Congress on August 7, 1969:

> We cannot meet future needs by concentrating development on just one means of transportation. We must have a truly balanced system. Only when automobile transportation is complemented by adequate public transportation can we meet those needs.
>
> I propose that we provide $10 billion out of the general fund over a 12-year period to help in developing and improving public transportation in local communities. To establish this program, I am requesting contract authorization totaling $3.1 billion for the first five years starting with a first year authorization of $300 million and rising to $1 billion annually by 1975 [12].

The essence of the President's request was granted by Congress with the passage of the Urban Mass Transportation Assistance Act of 1970. This legislation, which amended the 1964 act, stated a federal intention to provide $10 billion for urban mass transportation during the period 1970–82. A total of $3.1 billion was authorized by the law, increasing amounts being specified for each of the fiscal years 1971–75.

The act contains safeguards to protect the environment and requires that public hearings be held on environmental as well as social and economic impacts of urban mass transportation projects. The act has a provision that special consideration be given to the mass transportation needs of the elderly and the handicapped.

Although the act does not provide assistance for operating costs of mass transit system, it requires the Secretary of Transportation to conduct a study of the desirability of providing federal grants to assist local systems in meeting their operating costs.

PIPELINE TRANSPORTATION

1–18. The Government and the Pipeline Industry. Pipeline companies are, for the most part, private corporations, and the federal government has given little direct assistance for the promotion of pipeline transportation. Indeed, the sole instance of direct assistance to pipeline transportation occurred during World War II when the federal government financed the construction of two pipelines from Texas to New Jersey because of the threat of German submarines to coastal shipping. One of these pipelines was a 24-inch line, 1,341 miles long, and the other was a 20-inch line 1,475 miles long.

On occasion, pipelines have indirectly benefitted from government actions which adversely affected competitive transportation modes. For example, in 1890, the government aided the cause of pipeline transportation by outlawing the payment of rebates to the railroad industry.

Interstate oil pipelines were made common carriers subject to the Interstate Commerce Act by the Hepburn Amendment of June 29, 1906. Litigation followed, and, in 1914, the U.S. Supreme Court ruled in favor of the ICC except in the case of the Uncle Sam Oil Company which was simply transporting oil from its own wells to its own refinery [13]. The decision in favor of the Uncle Sam Oil Company is the legal basis for present-day privately owned products pipeline systems that transport products from the owners' refineries to their own distribution terminals.

Pipelines that transport natural gas are subject to regulation by the Federal Power Commission.

AIR TRANSPORTATION

1–19. Early Government Promotion of Air Transportation. There was incredible public apathy and governmental indifference to air transportation in the years immediately following the Wright brothers' first flight.[2] However, by 1908, the Army had expressed interest in the airplane and contracted with the Wrights to build an airplane for military use. The plane was delivered the following year.

[2] On three occasions in 1905 the Wright brothers offered their invention and scientific knowledge to the War Department, which rejected their offers without even bothering to investigate [14].

Less than a decade after the Wrights' premiere flight, the government was becoming keenly interested in airplanes for the transportation of mail. In 1910, a law was passed "to definitely determine whether aerial navigation may be utilized for the safe and rapid transmission of the mails" [15]. The first official air mail flights were made the following year on Long Island, covering a distance of about ten miles. Thirty-one additional air mail flights were made in 1912. The Post Office cooperated in these experimental flights, but did not share the expenses.

In 1916, Congress appropriated $50,000 for air mail service. However, because of World War I, nothing was accomplished with this appropriation. In 1918, a sum of $100,000 was appropriated for the establishment of an experimental air mail route, as well as the purchase, operation, and maintenance of airplanes. In May, 1918, an experimental route was established between Washington, D.C. and New York City with a stopover in Philadelphia. For about three months, the air mail service was performed by six army pilots using army training planes. In August, 1918, the Post Office bought four planes, hired six civilian pilots, and took charge of the air mail flights. The Post Office extended the air mail service to additional cities in 1919, and in 1920 transcontinental air mail service was achieved with the mail being transported by airplanes during daylight hours and by trains at night. During this period pilots often got lost and there were frequent crashes.

In 1925 the Kelly Act (Air Mail Act of 1925) was passed. This act provided for the transportation of mail by private air mail contractors, and by the beginning of 1926, twelve contract air mail routes had been established. No subsidy was provided by this act as it stipulated that not more than 80 per cent of the air mail postage revenue could be paid to the air carrier, with at least 20 per cent going to the Post Office for ground handling expenses. Under this act, payment was based on the number of letters, which meant that the letters had to be counted. The act was amended in 1926 to provide payments to the air carriers on the basis of weight and distance rather than number of letters. A second amendment to the Kelly Act, passed in 1928, opened the door for air mail subsidies to the carriers as it removed the provision in the original act which assured the government against losses.

1–20. The Air Commerce Act of 1926. In the 1920's, air transport was a risky business. The pilots were without radios and had no navigational aids nor system of aerial guide posts. In 1926, the Congress passed a law which marked the beginning of a recognition by the federal government of its leading role in the promotion and regulation of air transportation. Entitled the Air Commerce Act of 1926, this law gave the federal government the responsibility of the operation and maintenance of the airway system and other aids to air navigation, as well as the responsibility to provide safety in air commerce through regulation. The safety regulations, which were to be administered by the Department of Commerce, included registration and licensing of aircraft and the certification and medical examination of pilots. Thus, the scope of this act was much more comprehensive than the earlier airmail acts. Its purpose was to foster the growth and development of the air industry.

1–21. Airmail Acts of the Thirties. Two additional airmail acts were passed in the early thirties: the Airmail Act of 1930 (McNary-Watres Act) and the Airmail Act of 1934 (Black-McKellar Act).

The McNary-Watres Act gave the Postmaster General unprecedented control over the airline industry, providing that airmail contracts could be granted without competitive bidding. The forceful administration of this act by Postmaster General Walter Brown contributed to the merger of several of the smaller airline companies and their placing greater emphasis on passenger transportation. This act also changed the method of compensating airlines for the transportation of mail, basing payments on space rather than weight.

The Black-McKellar Act placed the control of the air industry under three government agencies: the Post Office, the Interstate Commerce Commission, and the Department of Commerce. The Post Office retained its responsibility for awarding airmail contracts and enforcing airmail regulations. However, this act required that airmail contracts be made on the basis of competitive bidding. The Interstate Commerce Commission was given the responsibility of establishing and reviewing the rates of pay to the air carriers for the transportation of mail. The Department of Commerce was responsible for air safety, including the maintenance, development, and operation of the airway system.

The Airmail Act of 1934 also created a Federal Aviation Commission to study aviation policy. The findings of this commission and the difficulties encountered in regulating and promoting air transportation by three government agencies set the stage for the passage of the Civil Aeronautics Act of 1938.

1–22. The Civil Aeronautics Act of 1938. This was a comprehensive act which provided for the development and regulation of air transportation. This act repealed or amended all major existing legislation having to do with aviation. It established three agencies for the regulation of air transportation. By amendment to the act in 1940, these agencies were reorganized and the various regulatory functions were placed under the control of:

1. The Civil Aeronautics Board, an independent organization, which was to exercise judicial and legislative authority over civil aviation and provide economic regulation of the industry. The CAB was

also made responsible for the investigation of aircraft accidents.
2. The Civil Aeronautics Administration, a part of the Department of Commerce, which was given the responsibility for safety regulation and the operation of the airway system.

1–23. The Federal Airport Act of 1946. This act must be regarded as one of the most important laws ever passed promoting air transportation. It provided $520 million to aid in the development of a comprehensive national system of airports. In addition, the act provided federal aid to airport sponsors for the construction of operational facilities (e.g., runways) on a 50–50 matching basis. Although the original bill provided funds for only a seven-year period, other bills have since been passed extending the federal aid program.

1–24. The Federal Aviation Act of 1958. The Federal Aviation Act of 1958 was largely a reenactment of the 1938 bill. Its principal effects were to remove the Civil Aeronautics Administration from the Department of Commerce and to give it increased authority in air safety. It was renamed the Federal Aviation Agency and under the new organizational structure, the Federal Aviation Administrator reported directly to the president. The Federal Aviation Agency later became the Federal Aviation Administration, a part of the Department of Transportation, established in 1966.

In addition to outlining in general the organizations, powers, and responsibilities of the Federal Aviation Agency (FAA) and the Civil Aeronautics Board (CAB), this act:

1. Provided for the economic regulation of air carriers and the issuance of certificates of public convenience and necessity by the CAB. The act authorized the CAB to control the changing and publishing of fares and related charges and the consolidations and mergers of air carrier corporations and to fix rates of compensation to the carriers for transportation of mail.
2. Granted the FAA broad powers relating to safety regulation of civil aeronautics, including the issuance of licenses or certificates for aircraft and airmen.
3. Assigned the CAB the responsibility for investigation of aircraft accidents.
4. Authorized the Postmaster General to make rules and regulations, consistent with the provisions of the act, for the safe and expeditious transportation of mail by aircraft.

1–25. The Airport and Airway Development Act of 1970. Recognizing the inadequacy of the nation's airport and airway system in meeting the current and projected growth in aviation, the Congress enacted Public

Law 91–258. This law consisted of two parts: Title I, the Airport and Airway Development Act of 1970; and Title II, Airport and Airway Revenue Act of 1970.

Title I included the following provisions:

1. Authorized the expenditure of $280 million[3] for each of the fiscal years 1971 through 1975 for airport development and an additional $250 million annually for acquiring, establishing, and improving air navigation facilities. Generally, the airport development funds are to be provided on a 50–50 matching basis; however, the federal share of costs of certain lighting systems may be as much as 82 per cent. The act does not allow federal aid for automobile parking lots or airport buildings except "such of those buildings or parts of buildings intended to house facilities or activities directly related to the safety of persons at the airport."
2. Authorized an amount of up to $15 million annually for airport planning studies. Under this provision, two-thirds of the cost of these studies may be paid with federal funds.
3. Required that public hearings be held to consider the economic, social, and environmental effects of the airport location and its consistency with urban planning goals and objectives.
4. Directed the Secretary of Transportation to formulate and recommend to the Congress within one year after enactment of the law a national transportation policy considering the development of all modes of transportation. An annual report on the implementation of the national policy was also required.
5. Directed the Secretary to prepare and publish a national airport system plan for the development of public airports in the United States. The Secretary was directed to publish this plan within two years after enactment of the legislation and, thereafter, to review and revise it as necessary.
6. Established an Aviation Advisory Commission of nine members to formulate recommendations concerning the long-range needs of aviation.
7. Stated that "no airport development project involving the location of an airport, an airport runway, or a runway extension may be approved by the Secretary unless the public agency sponsoring the project certifies to the Secretary that there has been afforded the opportunity for public hearings for the purpose of considering the economic, social, and environmental effects of the airport location and its consistency with the goals and objectives of such urban planning as has been carried out by the community."

[3] The act specified that $250 million would be for the purpose of developing airports served by air carriers and $30 million would be for general aviation airports.

Title II of the 1970 legislation provided for taxes on aircraft fuel, air fares, and the transportation of property. This act also established the Airport and Airway Trust Fund and provided for the transfer from the general treasury to this fund amounts equivalent to the taxes imposed under the act. It was specified that the amounts in the trust fund were to be used to meet the obligations incurred under Title I of the act and under the Federal Aviation Act of 1958 as well as certain administrative expenses of the Department of Transportation.

THE DEPARTMENT OF TRANSPORTATION

1–26. The Department of Transportation Act. The need to have one federal agency responsible for the overall supervision of transportation activities in the nation has long been recognized. The establishment of such an agency was proposed in 1805 by President Thomas Jefferson's Secretary of the Treasury Albert Gallatin. The creation of a national Department of Transportation has often been proposed since that time.

Earlier in this chapter, we have noted that, at various times during the nation's history, Congress has passed laws which regulated or promoted one transportation mode with little regard for its effect on other carriers. Transport policies for one mode have frequently been at cross-purposes with policies for other modes, and there has been little effort to coordinate government transport programs or to study transportation problems as an integrated system. In recognition of the vital need for better coordination of national transportation matters, the Congress passed the Department of Transportation Act (Public Law 89–670) in 1966. The act was signed by President Lyndon Johnson on October 16 of that year, and on January 5, 1967, Alan S. Boyd took office as the first Secretary of Transportation and a member of the President's Cabinet.

1–27. Mission and Organization of the Department of Transportation. In the 1966 act, the Congress listed the following purposes in establishing the Department of Transportation (D.O.T.):

1. to develop national transportation policies and programs conducive to the provision of fast, safe, efficient, and convenient transportation at the lowest cost consistent therewith and with other national objectives, including the efficient utilization and conservation of the Nation's resources;
2. to assure the coordinated, effective administration of the transportation programs of the Federal Government;
3. to facilitate the development and improvement of coordinated transportation service, to be provided by private enterprise to the maximum extent feasible;

4. to encourage cooperation of Federal, State, and local governments, carriers, labor, and other interested parties toward the achievement of national transportation objectives;

5. to stimulate technological advances in transportation;

6. to provide general leadership in identification and solution of transportation problems;

7. to develop and recommend to the President and the Congress for the approval of national transportation policies and programs to accomplish these objectives with full and appropriate consideration of the needs of the public, users, carriers, industry, labor, and the national defense.

Establishment of D.O.T. brought more than 30 offices throughout the Federal Government under one administrative umbrella. Figure 1–3, which shows the genealogy of D.O.T., dramatizes the diverse and uncoordinated manner in which transportation matters were handled prior to the passage of this act. Figure 1–4 shows an organization chart for D.O.T.

The Department includes seven operating administrations:

1. United States Coast Guard
2. Federal Aviation Administration
3. Federal Highway Administration
4. Federal Railroad Administration
5. Urban Mass Transportation Administration
6. St. Lawrence Seaway Development Corporation
7. National Highway Traffic Safety Administration

The U.S. Coast Guard is the nation's maritime law enforcement agency. It maintains an extensive program to prevent maritime disasters and to provide effective search and rescue operations when off-shore incidents occur. It is responsible for the establishment and maintenance of lighthouses, lightships, buoys, and other navigation aids for ocean transportation. It enforces standards for the construction and manning of all U.S. merchant marine vessels and carries out a boating and safety education and enforcement program for pleasure boats on domestic waters. An armed force, the Coast Guard polices the nation's coasts and, in time of war, supplements the operations of the U.S. Navy.

The Department of Transportation Act transferred to the Secretary of Transportation the functions, powers, and duties of the Federal Aviation Agency held under the Federal Aviation Act of 1958. The act specified, however, that certain functions, powers, and duties relating to aviation safety would be the sole responsibility of the Federal Aviation Administrator.

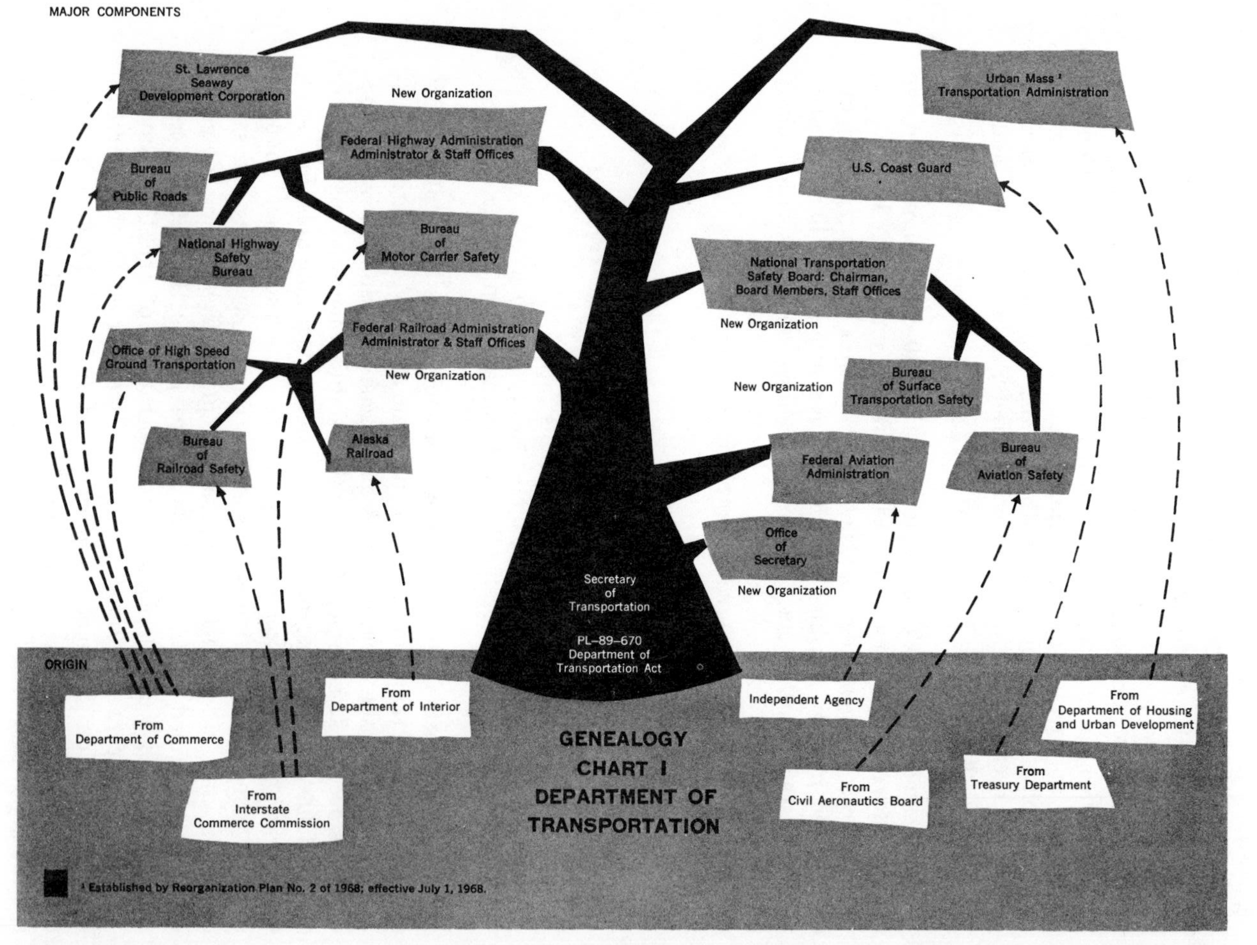

Fig. 1–3. Genealogy chart for the Department of Transportation. (Courtesy U.S. Department of Transportation.)

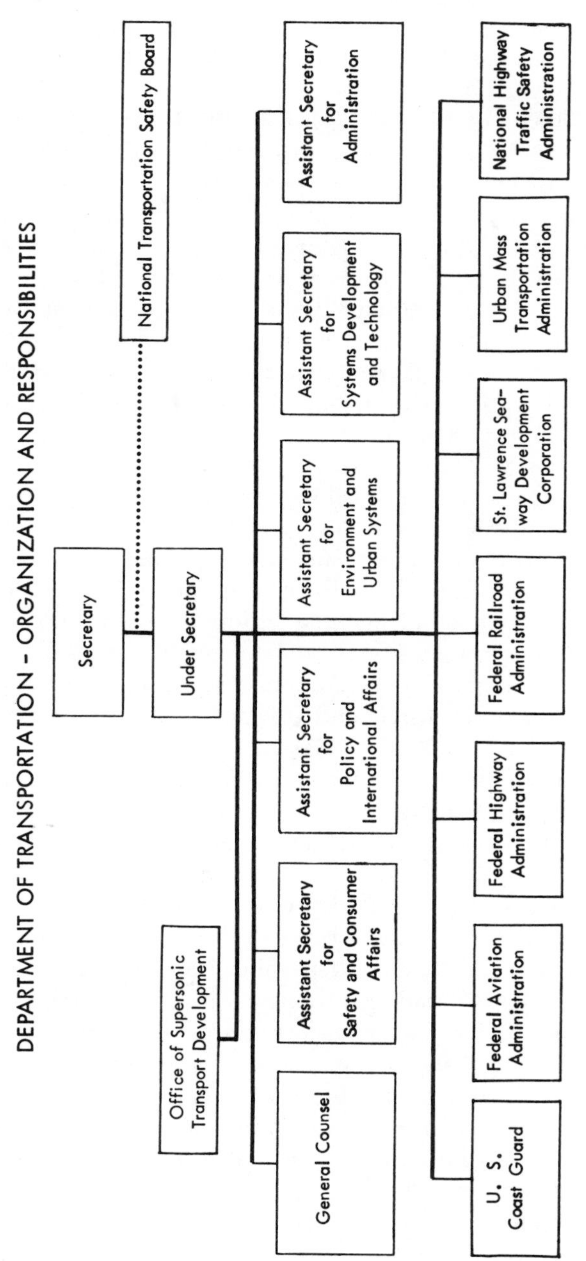

Fig. 1–4. Organization chart for the Department of Transportation. (Adapted from *Transportation Facts and Trends,* Transportation Association of America, 7th edition, April, 1970.)

The act created a National Transportation Safety Board, an autonomous five-man board appointed by the President, and charged it with reducing accidents in all modes of transportation. The functions, powers, and duties of the Civil Aeronautics Board regarding the investigation of aircraft accidents were transferred to the National Transportation Safety Board.

Three operating bureaus were placed under the newly created Federal Highway Administration: the Bureau of Public Roads, the National Highway Safety Bureau, and the Bureau of Motor Carrier Safety. The name "Bureau of Public Roads" was dropped in 1970 and the organization was merged with the Federal Highway Administration. The National Highway Safety Bureau became the National Highway Traffic Safety Administration with its administrator reporting directly to the Secretary of Transportation.

In cooperation with state and local governments, the Federal Highway Administration has the responsibility of completing the construction of the Interstate System, improving other federal-aid highway systems, conducting highway planning and research, and implementing certain programs in traffic safety. Its Bureau of Motor Carrier Safety investigates motor carrier accidents and enforces federal safety regulations relating to the operation of buses and commercial trucks.

The National Traffic Safety Administration has the responsibility for implementing comprehensive and diverse safety programs as specified by the safety acts of 1966 and 1970.

The Federal Railroad Administration is responsible for three on-going programs: the Alaska Railroad, the Bureau of Railroad Safety, and the Office of High Speed Ground Transportation. The Alaska Railroad, which is owned by the United States, has a 483-mile line which provides common carrier service between Seward and Fairbanks. The Bureau of Railroad Safety investigates railroad accidents and studies all aspects of railroad safety in order to determine whether regulatory action can contribute to safer railroad operations. The Office of High Speed Ground Transportation, which was transferred from the Department of Commerce, is primarily concerned with developing means of moving large numbers of people at high speeds between dense centers of population.

The Urban Mass Transportation Administration, previously discussed in Section 1–17, was not originally included in the Department of Transportation but was transferred from the Department of Housing and Urban Development on July 1, 1968.

The St. Lawrence Seaway Development Corporation, in cooperation with Canada's St. Lawrence Seaway Authority, operates the St. Lawrence Seaway, which, by means of canals and locks, connects the mid-continent of North American to the Atlantic Ocean. The Corporation collects tolls, and earnings in excess of operating costs are used to pay the U.S. Treasury for a $131-million construction loan.

PROBLEMS

1. Prepare a report which outlines the various public statements regarding transportation which have been made by presidents of the United States.
2. Contrast the national transportation policy of the United States with that of one or more foreign countries.
3. Make a list of the various ways in which your state and federal government regulates:
   a. railroad transportation
   b. truck transportation
   c. water transportation
   d. air transportation
   e. pipelines.
4. (a) Prepare a report on the procedures used by the Interstate Commerce Commission to establish freight rates for motor carriers. (b) What organization in your state government is responsible for the regulation of intrastate transportation? Briefly describe its functions.
5. What efforts have been made in your state to coordinate the transportation activities of the various modes. If a state Department of Transportation has been organized, obtain and present an organization chart for that department.
6. Determine what efforts are being made in your state to carry out highway safety programs under the Highway Safety Act of 1966.

REFERENCES

1. *National Transportation Policy,* Committee on Interstate and Foreign Commerce (Doyle Report), January, 1961.
2. Carr, Braxton B., "Inland Water Transportation Resources," *U.S. Transportation, Resources, Performance, and Problems,* National Academy of Sciences–National Research Council, Publication 841–S, 1961.
3. Haney, Lewis H., *A Congressional History of Railways in the United States,* August M. Kelley Publishers, New York (1908, reprinted in 1968).
4. Association of American Railroads, "Land Grants, Public Service Role Put in Focus," News Release, September 17, 1969.
5. Association of American Railroads, *Statistics of Railroads of Class I in the United States, Years 1958 to 1968,* September, 1969.
6. Association of American Railroads, *Yearbook of Railroad Facts, 1969 Edition,* April, 1969.
7. President John F. Kennedy's Message on Transportation to the Congress of the United States, April 4, 1962.
8. American Association of State Highway Officials, *Historic American Highways,* 1953.
9. Ritter, L. J., Jr., and Paquette, R. J., *Highway Engineering,* Third Edition, The Ronald Press Company, New York (1967).
10. Bureau of Public Roads, *Federal Laws, Regulations, and Other Materials Relating to Highways,* August, 1960.

11. *National Uniform Standards for State Highway Safety Programs,* Letter from the Secretary of Transportation, June 27, 1967, House Document No. 138.
12. President Richard M. Nixon's Message to the Congress of the United States, August 7, 1969.
13. BURKE, J. L., "The Oil Pipelines Place in the Transportation Industry," *ICC Practitioners' Journal,* Washington, D. C., April, 1964.
14. *The American Heritage History of Flight,* The American Heritage Publishing Company, New York (1962).
15. KANE, ROBERT M., and VOSE, ALLAN D., *Air Transportation,* Second Edition, Kendall/Hunt Publishing Company, Dubuque, Iowa (1969). Quotation reprinted by permission of the authors and publisher.

OTHER REFERENCES

FAIR, M. L., and WILLIAMS, E. W., *Economics of Transportation,* Harper & Row Publishers, New York (1958).

NORTON, HUGH S., *National Transportation Policy: Formation and Implementation,* McCutchan Publishing Company, Berkeley, Calif. (1966).

2

The Transportation System of the United States

An inherent need of highly developed and industrialized nations is a sophisticated and widespread transportation system. In developed nations there must be an easy mobility of persons and goods. As the degree of industrialization of an economy increases, there is a change in preponderance from *basic* production industries, sometimes called primary industries, to the service industries of a *secondary, tertiary,* and *quaternary* character. Primary industries have a great need for freight transportation. Industries in the service area display a need for both freight and extensive personal mobility. The United States, whose economy depends greatly on nonprimary industries, has found a need to provide the world's most sophisticated and widespread transportation system which is expanding under the constant pressure of increasing demand. Because of the unprecedented need of nationwide mobility, even in the previously remote areas of the country, there is a requirement not only for various modes of transport but also for increasingly sophisticated interfaces between the modes of subsystems.

2–1. Subsystems and Modes. The overall transportation system is required to provide for both passenger movements and freight movements. Both subsystems of movement are able to supply various levels of service (measured by such criteria as convenience, speed, safety, and availability), at varying levels of price. The user of transportation service is seldom in the position of having only one service available. Monopoly conditions, prevalent in the days of the early railroads, were broken by the introduction of motor vehicle and air transportation.

Modern technology has evolved a great variety of transporting devices either in the design or development stage. Five modes however, account for all but a minute fractional percentage of all ton mileage of freight and passenger mileage of person travel:

1. motor vehicles
2. railroads
3. air transport
4. water transport
5. pipelines.

2–2. Highway System Classification. The highway system of the United States is publicly owned. The elements of the system are customarily classified according to the jurisdiction or level of governmental unit which has the overriding influence in the construction and operation of the system. In Section 2–7, the various subsystems of highways are discussed in detail. These subsystems are:

1. the national system of interstate and defense highways
2. the federal aid system—primary, secondary, and urban extensions
3. the state primary and secondary systems
4. local and county roads
5. city streets
6. toll roads.

Roads are also classified by *function*. Used mainly for planning purposes, the rural and urban classifications are slightly different:

Rural:
- Arterials:
  - Principal—Interstate
  - Principal—Other
  - Minor
- Collectors:
  - Major
  - Minor
- Local Roads

Urban:
- Arterials:
  - Principal—Interstate
  - Principal—Other Freeway and Expressway
  - Principal—Other
  - Minor
- Collector Streets
- Local Streets

2–3. Railroad System Classification. Railroad systems are customarily designated by two separate system classifications, *revenues* and *use*.

Revenue classification, as set down by the Interstate Commerce Commission, divides the railroads into:

Class I railroads
Class II railroads
switching and terminal companies.

Class I railroads are those companies whose average annual operating revenues are $5 million or more. Class II railroads are those companies with operating revenues less than $5 million. Switching and terminal companies provide no line haul service, their facilities connect to Class I and Class II facilities to provide switching and terminal service. In urban areas these services eliminate a duplication of switching facilities and also enable railroad service to be provided at one terminal location. In the United States there are 77 Class I, 294 Class II, and 187 switching and terminal companies. Even among Class I companies there is domination by a few large roads. In 1966, six companies, for example, accounted for 27 per cent of all track mileage, 41 per cent of freight ton mileage, and 43 per cent of the operating revenues of all 77 Class I carriers. Some observers have felt that future development of the railroad system will lead to widespread mergers, resulting in all rail service being provided by a few giant companies with regional monopoly. Two of these six companies, the New York Central and the Pennsylvania Railroads, merged in 1968 with Supreme Court approval. This giant company accounts for approximately 12 per cent of all ton mileage carried by Class I railroads. Despite its size, the Penn Central filed bankruptcy proceedings in 1970.

Use Classification. Railroads may also be classified by use. The various classes are *line-haul, switching, belt-line,* and *terminal* companies. Line-haul companies concentrate essentially on providing intercity movement of freight and passengers. Belt-line and switching companies connect to intercity facilities to provide special access to warehousing, wharves, and industrial areas. Terminal companies operate terminals and approach trackage; in this way, more than one line-haul company has use of terminal facilities, avoiding a duplication of plant.

2–4. The Airport System Classification. All airports, both public and private, which contribute significantly to the national transportation system are included in the National Airport Plan. Criteria for inclusion in the plan include scheduled airline passenger service, a substantial degree of non-local aviation activity, lack of satisfactory surface transportation, or evidence that air transportation is an essential ingredient of the local and national economy. The Federal Aviation Administration classifies airports according to function.

PASSENGER CARRIER AIRPORTS

1. *Trunk* airports—used or proposed for use for servicing trunk line carriers alone, or some mix of trunk and local service carriers. These trunk airports include all the major air hubs of the country.

2. *Local Service* airports—used by local service carriers, Intra-Alaskan and Intra-Hawaiian lines only. No airports which service trunk lines are included in this category.

GENERAL AVIATION AIRPORTS

1. *Basic Utility* airports—designed to accommodate about 95 per cent of the general aviation fleet. They are divided into *Stage I* and *Stage II* subclassifications with runway lengths of 2,200 feet and 2,700 feet respectively, under standard conditions.

A Basic Utility Stage I airport is defined by the FAA as one which accommodates about 75 per cent of the propeller aircraft under 12,500 pounds. It primarily serves areas with low air activity such as small communities and remote recreational areas.

A Basic Utility Stage II airport accommodates about 95 per cent of propeller aircraft under 12,500 pounds. It generally serves locations that have a medium size population with a diversity of usage and potential for growth in air activity.

2. *General Utility* airports—accommodate all current general aviation aircraft except transport types. Under standard conditions the runway length is 3,000 feet minimum.

3. *Larger than General Utility* airports—generally designed specifically to accommodate large transport aircraft and general aviation aircraft with large runway demands. Under standard conditions, runway length requirements are in the range of 5,000–9,000 feet.

4. *Exceptions*—designed for special usage. Design criteria are sometimes used which are less than Stage I Basic Utility requirements. These exceptions bear a special classification under the Federal Aviation Administration.

5. *Heliport* classification—given to airports whose helipad and related taxiways, aprons and lighting provide service for vertical take-off aircraft.

6. *Seaplane Bases*—facilities constructed for the use of aircraft with water landing capabilities.[1]

[1] In 1971, the F.A.A. announced a new National Airport Functional Classification System. The new system consists of three subsystems of airports: *primary, secondary,* and *feeder.* These subsystems are differentiated by the level of public service, i.e., the number of emplaning passengers accommodated by the airports. Each subsystem is further classified into three levels of aeronautical operational density (*high, medium,* and *low*), depending on the number of aircraft operations [8].

2–5. Pipeline System Classification. Pipelines are most commonly classified according to type of commodity carried and function. The two principal classes of commodity carried are crude oil and its refined products and natural gas. A classification scheme for all pipelines would be of the following form:

OIL PIPELINES
  Gathering Lines
  Trunk Lines
    Crude oil
    Products
GAS PIPELINES
  Field Lines
  Trunk Lines

For further discussion of pipeline classification, the reader is referred to Section 2–17.

2–6. Transportation Demand by Mode. The rising population and increasing industrialization of the United States have been accompanied by an increasing demand for transportation. As an industry, transportation amounts to a significant proportion of the economy. Economists estimate that the annual bill for transportation is close to 20 per cent of the Gross National Product and the industry employs approximately 13 per cent of the labor force. Approximately 20 per cent of all federal taxes are collected from the transportation sphere. Figure 2–1 indicates the national trends in transportation demand since 1939. A clear relationship can be discerned between transportation demand and such indexes as total population, industrial production, and the Gross National Product.

While overall demand for transportation, both passenger and freight, has been increasing rapidly with marked acceleration since World War II, the modal shares of this total demand have not been constant. Each mode has shown significant secular trends which tend to reflect governmental regulatory policy and technological development.

Rails. In the last three decades, rail intercity freight traffic has accounted for a decreasing share of the total ton mileage. The rate of decline in recent years has slowed until the share is almost constant, averaging slightly above 42 per cent since 1960. Although the market share has declined, the absolute ton mileage has shown steady increase since the late 1950's. The principal benefactor from the rail's relative decline in the freight transportation market has been truck transport. The steadily increasing standards of the intercity road network has promoted intercity truck freight from 53 billion ton miles in 1939 to 404 billion ton miles in 1969. This represents an increase of 660 per cent

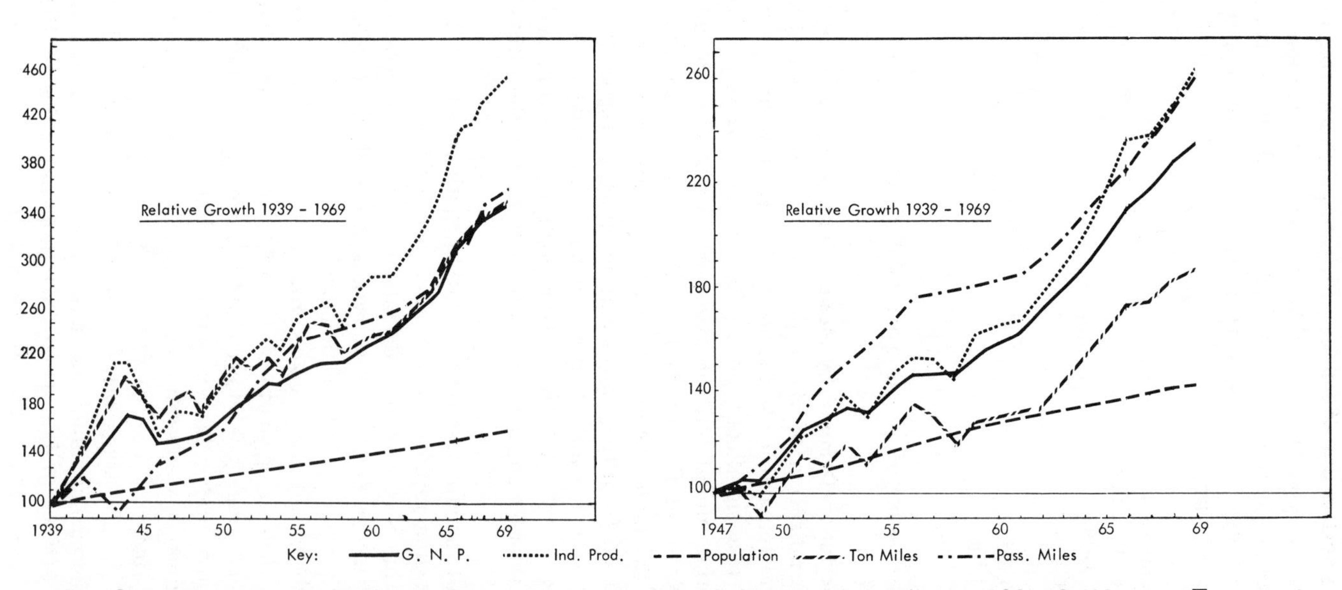

Fig. 2–1. Relative growth of transportation—economic and demographic indices, 1939–69. (Source: *Transportation Facts and Figures.* 1970 edition, Transportation Association of America.)

TABLE 2-1

Intercity Freight by Modes
(Billions of ton miles)

Year	Rail	Truck	Oil Pipeline	Inland Waterways	Air
1939	334	53	56	96	.01
1945	691	67	127	143	.09
1950	597	173	129	164	.30
1955	631	223	203	217	.49
1960	579	285	229	220	.89
1965	709	359	306	262	1.91
1969	780	404	411	302	3.20

Source: *Transportation Facts and Trends,* Transportation Association of America, April, 1970.

compared with an increase of 130 per cent for rail freight. Table 2–1 shows intercity freight ton mileage for all modes for the period 1939–69.

In contrast with freight trends, rail passenger traffic has shown a long term secular decline both in market share and in absolute figures. Since the end of World War I, except for a brief period during gasoline rationing in World War II, intercity passenger mileage has steadily decreased. In 1939, 62 per cent of all intercity passengers carried by common carrier traveled by rail. By 1969, that percentage had declined to 20.7 per cent. Rail movements in this year accounted for 1 per cent of all intercity passenger mileage. The degree to which air travel has provided competition to the railroads is clearly indicated by the dramatic rise of air passengers in the last 30 years. In 1939, air passengers amounted to 1.7 million persons or 0.5 per cent of all intercity travelers. While railroads showed a significant long term decline, air travelers reached 142.3 million in 1969 or 33.5 per cent of all intercity common carrier movements. Rail passenger traffic suffers stiff competition not only from air carriers but also from the auto mode. The greatest amount of intercity travel is not made by any form of common carrier. Throughout the 1950's and first half of the 1960's private auto accounted for over 89 per cent of intercity passenger mileage. This figure has remained relatively constant over the last three decades with the exception of the war years, declining to 86.5 per cent in 1969. As a result of the decline of demand for passenger service, railroads have operated passenger trains at a deficit. In 1969 this deficit was estimated to be $200 million by the American Association of Railroads [1]. The heavy financial burden placed upon the railroads in providing this uneconomic service has resulted in numerous pleas for discontinuance of service. The Interstate Commerce Commission has heard and has granted discontinuances, where it found that passenger service was no longer required by public convenience and necessity and its continuance constituted an unjust and undue burden on interstate commerce.

Highways. The United States is the world's most mobile country. The right of personal mobility is almost universally accepted on a nationwide basis. This mobility is due mainly to the widespread provision of a superb road system combined with high per capita car ownership. In 1968 there was one motor vehicle for every 2.1 persons in the United States. Vehicle registrations followed the long term increase shown in Fig 2–2, reaching a figure of just over 101 million vehicles,

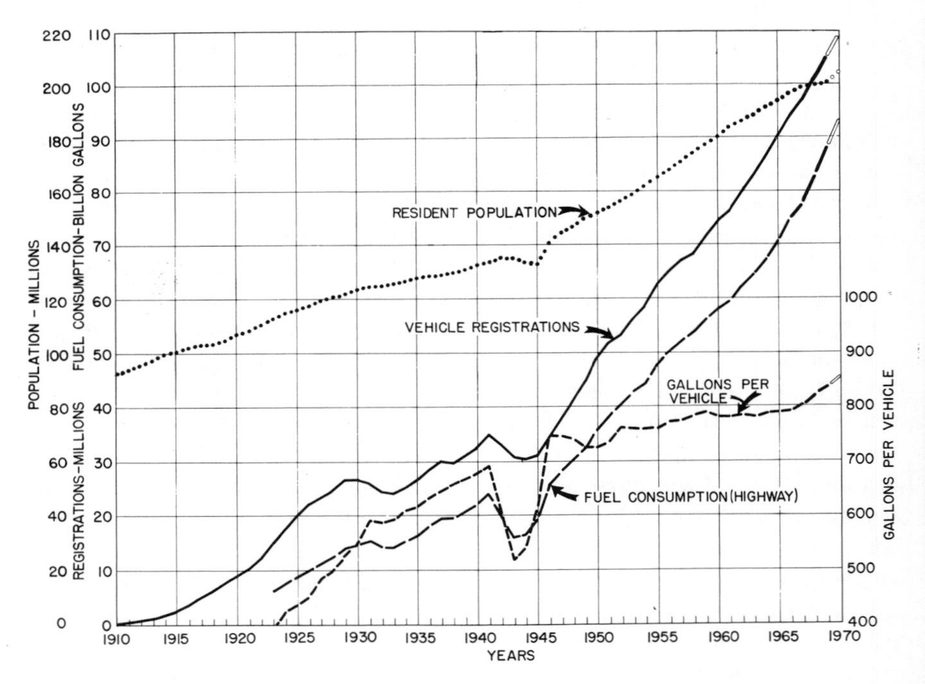

Fig. 2–2. Total population, motor vehicle registrations, and motor fuel consumption. (Source: *Highway Statistics 1969,* Federal Highway Administration.)

made up of 83 million autos, 350,000 buses, and 16 million trucks. The total amount of vehicle travel in this year was 1,016 billion vehicle miles, an average of 9,847 miles per vehicle. The degree of personal mobility is underscored by the fact that of a population of approximately 200 million persons over 105 million are licensed drivers [2].

Fifty per cent of the mileage traveled each year is estimated to occur in the urban areas. The amount of urban travel carried out by highway systems varies from city to city depending on availability of rail transit systems. Where no rail transit exists, urban mobility is totally dependent upon the highway system. Where rail transit does exist there is considerable diversion of downtown oriented trips but a much smaller effect on a citywide basis. For example, in Chicago, in 1956 the rail systems accounted for only 7.3 per cent of all trips in the urban area. In comparison, 26 per cent of all trips into the downtown area were made by rail, the remaining 74 per cent being made by bus transit or automobile.

Similarly, intercity passenger and freight movements are heavily dependent on highway facilities. Rural road travel amounted to 502 billion vehicle miles in 1967. Of all intercity passenger miles traveled, private auto constantly accounts for approximately 87 per cent of the total, and intercity bus accounts for an additional 2½ per cent. The remaining amount of travel is divided between air and rail with an increasing share being taken by the former.

The relative importance of highway trucking on intercity freight movements can be inferred from reference to Table 2–1, which shows that freight ton mileage moved on the highways accounts for approximately 22 per cent of all intercity ton mileage.

Inland Waterways. Since water transportation is slow and the modal interface relatively inconvenient, water transportation is primarily a question of freight movement. The type of freight moved tends to be limited to those classes which are bulky and of low cost per unit volume. Bituminous coal and lignite alone amounted to 23 per cent of all tonnage moved by water in 1967. Eight classes of commodities accounted for 81 per cent of net tonnage in the same year. A breakdown of principal commodities transported is given in Table 19–1. The market share of total ton mileage moved over the last three decades shows a very small long term decline, falling from 18 per cent in 1939 to 16 per cent in 1967. The actual ton mileage shows a steady growth, amounting to a tripling of ton miles in this period, as shown in Table 2–1. Without a dramatic breakthrough in technology, the economics of water transportation are likely to remain at approximately the same level. Under these conditions, this mode can be expected to retain approximately the same share of the total freight market with an increasing annual total ton mileage closely following the increase in the Gross National Product.

Air Transport. Air transportation is the fastest growing mode, both in the areas of passenger and freight transport. Figure 2–3 shows the rapidly increasing rate of air passenger miles in comparison with rela-

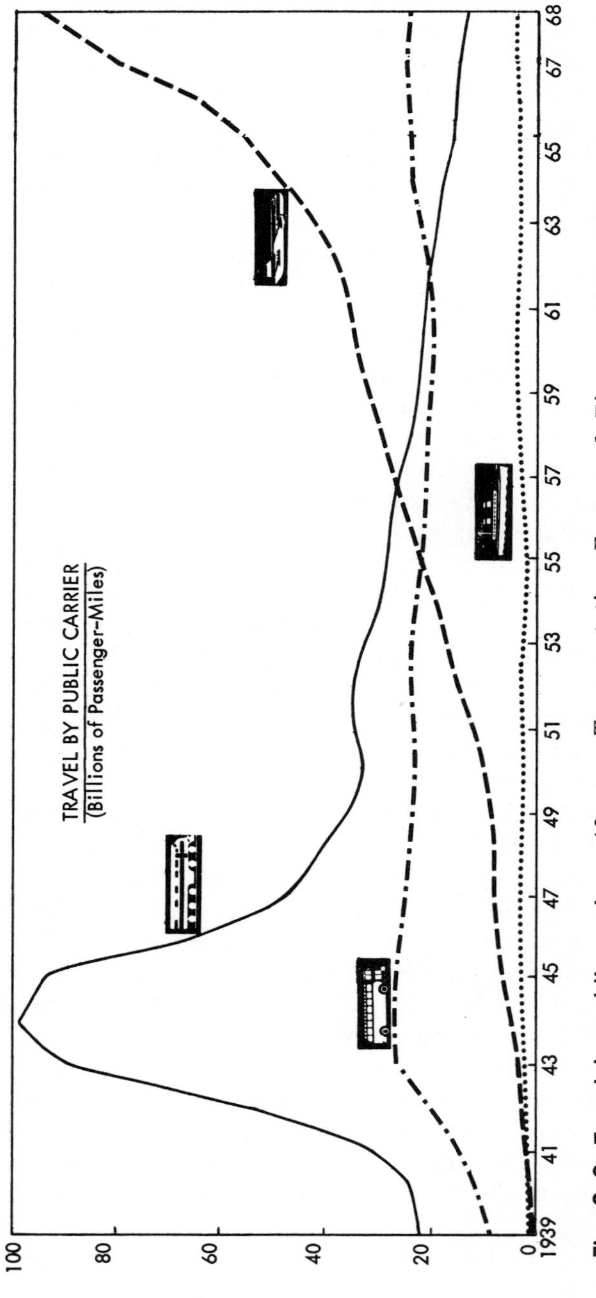

Fig. 2–3. Travel by public carrier. (Source: *Transportation Facts and Figures,* 1970 Edition, Transportation Association of America.)

tively stable or declining figures for other public passenger carriers. The increase in air travel is expected to continue at a similar rate to that observed in the early 1960's. As a result, air travel has been projected to increase from 29.2 billion passenger miles in 1960 to 230 billion passenger miles in 1980 [3]. Aircraft manufacturers have a somewhat more buoyant forecast of 500 billion passenger miles in 1985 [4]. The rapid increase in air travel is caused by many underlying factors including increasing gross national product, increasing affluence of the middle-income groups, a greater preponderance of nonprimary industry in the economic system, and improved technology. The introduction of the jumbo jets in late 1969 promoted an expansion of the air passenger travel market that, it is hoped, will continue into the seventies. This equipment, known as "jumbo jet service," flies approximately 400 passengers on transcontinental and intercontinental routes at cruising speeds of 625 mph.

In addition to the important passenger carrier sector needs, there is increasing demand in the areas of *general aviation.* General aviation is the term used for all flying done by aircraft other than military aircraft or by passenger and cargo carriers. From 1958 to 1967, general aviation aircraft handlings by the F.A.A. at air route traffic control centers have increased almost fivefold, from 535,517 to about 2½ million [5]. The level of general aviation activity is expected to increase at a continuing rate as businesses use private aircraft and more middle-income individuals use aircraft for recreational purposes.

Other areas of demand for air transports are *express and freight* and *mail.* The trunk and local passenger/cargo carriers carry most of this material. The remainder is carried by a relatively small number of all-cargo carriers. Air freight is still a transportation mode in its relative infancy. Improved technology has resulted in a steady lowering of costs which has encouraged a rapid growth rate. If mail and express are included, the amount of intercity air freight has increased from 0.01 billion ton miles in 1939 to 3.20 billion ton miles in 1969. While this rate of increase is greater than for any other mode, it must be born in mind that in 1966 air freight amounted to only 0.12 per cent of total intercity freight.

Pipelines. In the terminology of transportation, normally pipeline transportation refers only to the oil and oil-products lines, since the natural gas lines are not in competition with other carriers, and gas transporters are regulated by the Federal Power Commission rather than by the Interstate Commerce Commission.

As a mode of freight transportation, the pipelines are often overlooked. The total intercity mileage of oil pipeline in 1967 was 209,478 miles compared with 122,500 miles in 1939 [6]. During the same period the ton mileage of oil moved showed a steady increase from 56 billion ton miles to 361 billion ton miles [7]. The 1967 figure represented almost 21 per cent of all intercity freight ton mileage. In this year more ton miles of freight were moved by the nation's oil pipelines than were carried on the Great Lakes and the Inland Waterways.

THE PHYSICAL SYSTEMS

2–7. National System of Interstate and Defense Highways. In 1956, the Congress approved the construction of the 41,000-mile National System of Interstate and Defense Highways. This system, extending from coast to coast, was to provide a high level of service to interstate travel by the provision of nationwide freeway facilities. When completed, all of the Interstate routes will be at least four-lane divided highways, growing to six and eight lanes in and near the large metropolitan areas.[2]

The extent of the proposed system is shown in Fig. 2–4. The original cost as authorized by the 1956 act was slightly in excess of $27 billion. This same act provided for federal participation in construction funding (including right-of-way acquisition) on a 90–10 matching basis. Beginning July 1, 1956, federal funding was authorized to the amount of $24.825 billion, matched by expenditures of the individual states amounting to approximately $2.6 billion. All maintenance costs are borne by the states. Inflationary trends in construction and right-of-way costs coupled with increased design standards have inflated the estimated cost to over $60 billion. The initial target date for completion of the system, 1975, was abandoned and completion of the total system is likely sometime late in the 1970's.

The function of the system is to provide high level-of-service roads both within and between urban areas. Urban areas account for 7,500 miles of the total system. The Interstate constitutes only 1½ per cent of the total road and street mileage, yet it is estimated that it will carry over 20 per cent of all vehicle mileage. In comparison with other road systems, the Interstate will be remarkably efficient. The degree of coverage that this limited system will provide on a nationwide basis can be seen from the fact that 42 state capitals and 90 per cent of all cities with population in excess of 50,000 are located on it.

The Interstate System represents a massive public involvement in transportation. It constitutes the world's most expensive public works

[2] The current Interstate "traveled way" includes certain relatively low-standard segments of roadway which will later be improved or replaced. Over half of the current system is divided highway with four or more lanes and full control of access. In areas of low density population, two-lane undivided sections account for approximately 30 per cent of the mileage.

program. Originally the financing was to be carried out on a "pay-as-you-go" basis. Revenues from federal gasoline and other user taxes were placed in an earmarked Highway Trust Fund from which federal expenditures on the system would be drawn. Because insufficient revenue was available for construction, this feature of the 1956 act was suspended in 1960.

2–8. The Federal Aid Primary System. The main highway system interconnecting the cities and towns of the United States is the Federal Aid Primary System. It is composed of 252,651 miles of traveled way, of which 30,390 miles are in the urban areas. The Interstate System is included within the Primary System. All but approximately 500 miles of this system is surfaced. Approximately 83 per cent of the rural mileage and 34 per cent of urban mileage is comprised of two-lane facilities. Federal highway funds are available for construction and improvement of the Federal Primary System. For those sections not on the Interstate system, federal funding is on a 50–50 matching basis with the states, including right-of-way purchase. Maintenance of these facilities falls totally on state and local authorities.

2–9. The Federal Aid Secondary System. The Federal Aid Secondary System is a feeder system under state, county, or local authority control. It comprises approximately 633,000 miles of road, of which 95 per cent of the rural mileage and 96 per cent of the urban mileage is surfaced. As with the Federal Aid Primary System, federal highway funds are available for construction and improvement. These funds, while on a 50–50 federal–state matching basis, are not available for the purchase of right-of-way. Maintenance of the system is carried out by state and local authorities.

2–10. Urban Extensions. Prior to 1944, only token amounts of federal or state funds were expended for road construction within urban areas. To relieve the plight of the urban areas, 25 per cent of federal funds was earmarked, starting in 1944, for use inside urban areas. Arterial routes within the urban areas connecting to the Federal Primary and Secondary systems became known as the urban extensions. The total system of Primary, Secondary, and Urban Extensions is sometimes called the ABC system.

2–11. State Systems. Within each state, state highway departments have designated State Primary and State Secondary systems. These systems overlap with the Federal Primary and Secondary systems as shown by Table 2–2. Local and county authorities bear different fiscal responsibilities on these two systems, with less state contribution to the secondary system. In 1969, state highways accounted for over 742,000 miles of road. Total state expenditures on highways in 1967 amounted

Fig. 2–4. The national system of interstate and national

defense highways. (Source: Federal Highway Administration.)

TABLE 2-2

Total Road and Street Mileage in the United States—1967

State or Local Road System	Interstate	Federal Aid Primary	Federal Aid Secondary	Total Federal Aid Systems	Total Non-Federal Aid Systems	Total
State Primary	38,893	245,825	197,313	443,138	34,731	477,869
State Secondary	131	1,349	74,022	75,371	47,227	122,598
County Roads Under State Control	51	205	45,713	45,918	96,061	141,979
Total State Highways	39,075	247,379	317,048	564,427	178,019	742,446
County Roads	10	629	294,921	295,550	1,433,313	1,728,863
Local	2	86	5,689	5,775	586,115	591,890
City Streets	149	1,504	15,387	16,891	435,798	452,689
Other	180	915	149	1,064	184,838	185,902
Toll Facilities	2,016	2,138	5	2,143	981	3,124
Total Existing Mileage	41,432	252,651	633,199	885,850	2,819,064	3,704,914

Source: Federal Highway Administration

to $13.2 billion. Federal expenditures by the Federal Highway Administration amounted to $4.1 billion.

2–12. Local and Other Systems. County, town, township, and other local systems comprise a total of 2.32 million miles of road of which nearly 75 per cent is under county control. In addition, slightly over 450,000 miles of road fall under city street classification. Maintenance, improvement, and construction of these facilities are the responsibility of the local authorities.

In addition, there are approximately 186,000 miles of road not overlapping state, county, or other local systems. Generally these roads are national and state park, forest, and reservation roads. The vast majority of this system is not under the Federal Aid System.

2–13. Toll Roads. As seen in Table 2–2, there are over 3,000 miles of toll road in operation in the United States. Over two thousand miles are now under the Federal Aid Systems. Construction of these facilities stretched from the late 1930's to the middle 1950's. Toll facilities were conceived to solve the traffic problems of special areas, both urban and rural. For this purpose, the states created special local or statewide authorities empowered to build and operate toll facilities. Financing was carried out by revenue bonds paid off from toll charges levied against the users. The New York State Thruway and the Sunshine State Parkway in Florida are examples of statewide toll facilities. Local bodies such as the Jacksonville Expressway Authority are responsible for the provision of extensive intraurban toll networks.

The construction of toll roads, especially interurban facilities, fell into disfavor with the passage of the 1956 Federal Highway Act which authorized construction of competitive free facilities. In some areas, toll facilities have been incorporated into the Interstate System. It is anticipated that these facilities will become toll free when the bonded indebtedness is repaid.

2–14. Railroads. Virtually every community of reasonable size in the United States is either served by or within close reach of rail service. Figure 2–5 shows the extent of the system of Class I railroads. This represented a net capitalized investment of $27.6 billion in 1969 in plant and equipment. In 1969 Class I railroads operated 207,500 miles of line, 27,040 locomotives, 1,794,655 freight cars, and 12,800 passenger cars. With minor exceptions, operations are carried out on company-owned and -maintained rights-of-way.

In addition, in 1969 railroads owned and operated over 50,000 service vehicles (ballast cars, snow removing cars, boarding outfit cars, etc.), and such floating equipment as tugboats, car ferries, lighters, and barges. These amounted to 990 vessels for all railroad companies. Stationary

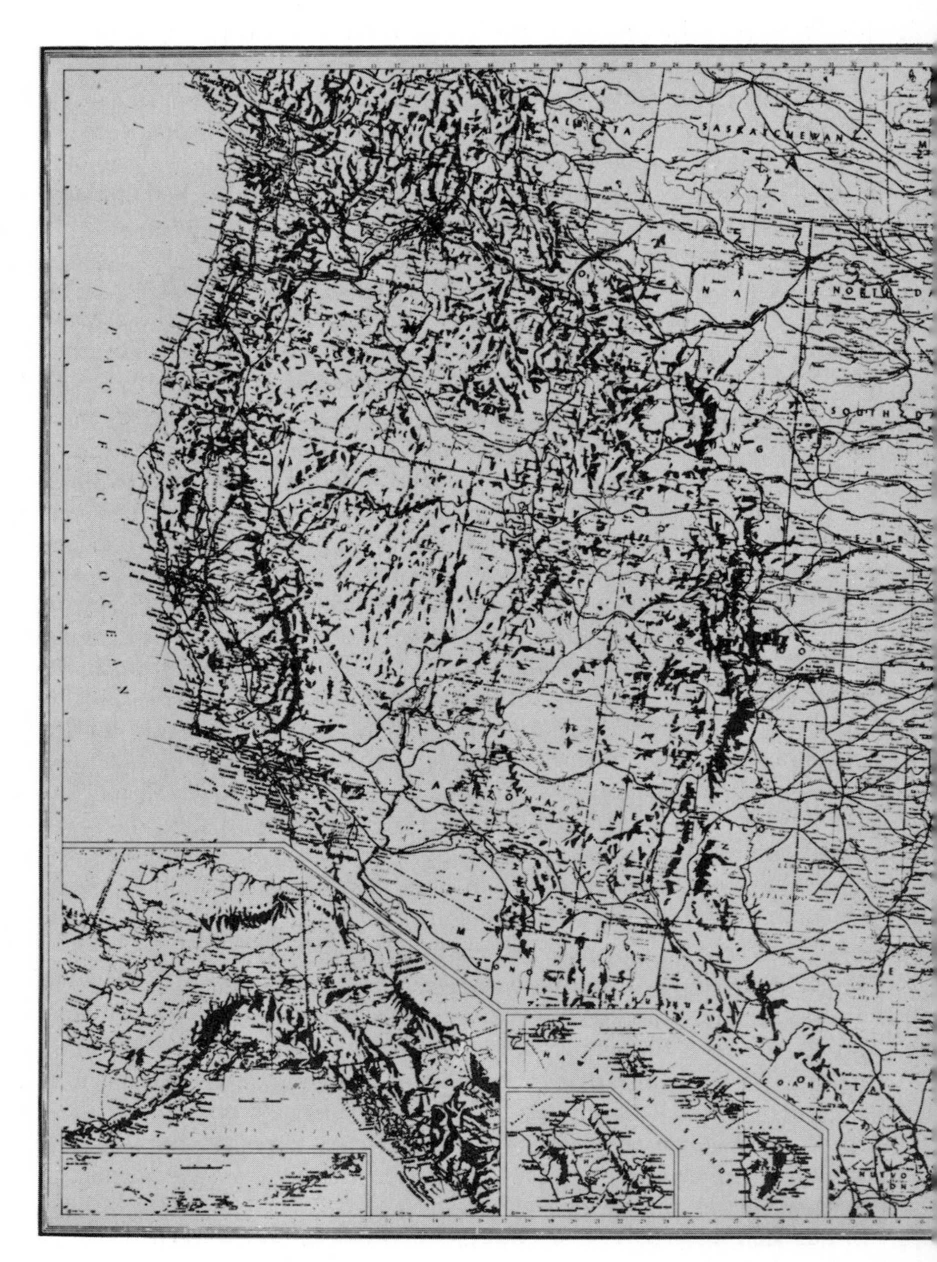

Fig. 2–5. American railroads Class I—Map of the United

States. (Source: Association of American Railroads.)

plants both owned and operated by the companies include tunnels, bridges, stations, and buildings, water and fuel stations, grain elevators, warehouses, wharves, communications systems, systems powerplants, transportation systems, shops, and yards. Operating revenues amounted to approximately $11.5 billion with income payments to employees amounting to $5.4 billion. Employment in Class I railroads was 578,302 in 1969.

The railroad system is totally privately owned. In general, no public monies are allocated to the construction, operation, or maintenance of the system.[3]

2–15. The Airways. The National Airspace System has been established by the federal government for the purpose of providing a safe environment for air travel through the airspace above the United States. Two systems of en route guidance exist. Most important is the *very high frequency* (VHF) multidirectional system with an airway mileage of 133,177 miles, direct, at low altitudes, and 95,944 miles at jet altitudes. The low-altitude system operates from 700 feet up to 18,000 feet while the high-altitude jet route mileage extends from 18,000 feet to 45,000 feet. The VHF system, often referred to as the VOR/VORTAC system, was introduced in 1952 phasing out the older *low and medium frequency* (L/MF) four-course system which constituted the airways prior to that time. By 1967, the L/MF system had declined from a peak of 74,424 miles to only 388 miles. This older system, known as the *colored* airways system, consisted of routes in an east–west direction designated red and green, and north–south routes designated amber and blue. The four-course system was marked by visual ground markings and radio guides. Currently the airspace is also under Air Positive Control (APC). All airspace from 24,000 to 60,000 feet is covered by APC. Operations within this airspace must be under the control of the Federal Aviation Administration using Instrument Flight Rules. In the busy north and northeast sections of the United States, the ceiling has been lowered to 18,000 feet.

Ground facilities represent an important element of the air transportation system. In 1967, there were 10,126 airport facilities in the United States, of which 852 were heliports and seaplane bases only. Scheduled airline service is carried out in 826 of these airports. It is apparent that small general aviation airports constitute the large majority of the facilities. During the period 1947–67, 2,184 airports received federal aid, 613 of which were facilities with air carrier service, while 1,571 served general aviation alone.

[3] As already noted in Section 1–9, public financing of limited rail passenger traffic was initiated by the National Railroad Passenger Corporation (Amtrak), which commenced operations in May, 1971.

The air vehicle, the remaining major element of the air transportation system, is chiefly privately owned. While the air carrier fleet constitutes only 2,595 airplanes, there were approximately 114,000 airplanes engaging in general aviation activities in 1967. There were, at the same time, 617,931 registered pilots, of whom approximately 21,000 were in the employment of the air carriers. The amount of flying time generated by general aviation amounted to 22.2 million hours, averaging approximately 200 hours per aircraft.

In addition to the provision of air traffic control in the interests of safety, federal aid to the air transportation system is available in the form of matching construction and planning grants for the airports themselves. To be eligible for federal aid, an airport facility must be publicly owned and must be included in the National Airport Plan. User charges in part repay this federal assistance. Airlines contribute to user charges in the form of landing fees, and passengers are required to pay federal excise taxes. There is no doubt, however, that the air transportation system currently is heavily subsidized by government. The actual amount of subsidy is subject to dispute among the carriers of various modes.

2–16. The Inland Waterways. The extent of the inland waterway system of the United States is shown in Fig. 2–6. In total, there are approximately 29,000 miles of waterway on the system with 15,000 miles having navigable depths in excess of 9 feet. As the map shows, the degree of coverage of the system is unevenly spread throughout the country with very few waterways west of the Mississippi River where few rivers of navigable depth are available. The total system can be considered in three major divisions, the inland rivers and canals, the coastal waterways, and the Great Lakes system.

The inland rivers and canals are natural, improved, and manmade waterways which thread deep into the eastern portion of the United States. The extent of the system is limited by the number of rivers which have navigable depths of 9 feet. Three rivers form the most important sections of the inland waterways: the Mississippi (which connects to the Great Lakes at Chicago by means of the Illinois Waterway), the Ohio, and the Tennessee. While limited in extent, the contribution of the inland waterways to freight transportation is considerable. The Mississippi River alone accounts for approximately 60 per cent of the tonnage carried in domestic water transportation.

The coastal system includes the protected intracoastal waterways of the Atlantic and Gulf Coasts and the Pacific Coast Waterways in California, Washington, and Oregon. In addition to movements on these *intracoastal* waterways, there are also deep-water movements of ocean going vessels on the Atlantic, Gulf, and Pacific Coasts. Because of

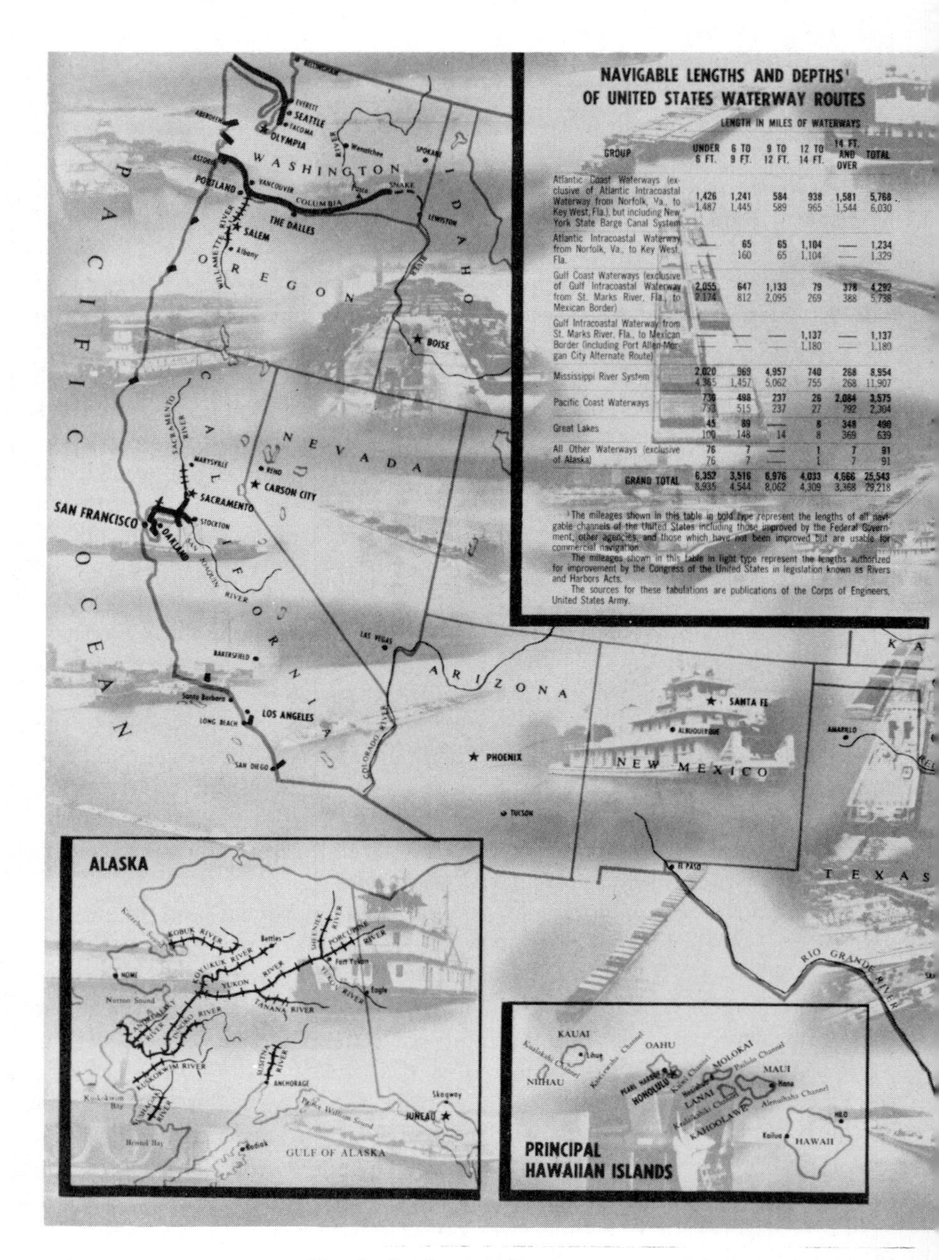

NAVIGABLE LENGTHS AND DEPTHS[1] OF UNITED STATES WATERWAY ROUTES

GROUP	LENGTH IN MILES OF WATERWAYS: UNDER 6 FT.	6 TO 9 FT.	9 TO 12 FT.	12 TO 14 FT.	14 FT. AND OVER	TOTAL
Atlantic Coast Waterways (exclusive of Atlantic Intracoastal Waterway from Norfolk, Va., to Key West, Fla.), but including New York State Barge Canal System	**1,426** 1,487	**1,241** 1,445	**584** 589	**938** 965	**1,581** 1,544	**5,768** 6,030
Atlantic Intracoastal Waterway from Norfolk, Va., to Key West, Fla.	— —	**65** 160	**65** 65	**1,104** 1,104	— —	**1,234** 1,329
Gulf Coast Waterways (exclusive of Gulf Intracoastal Waterway from St. Marks River, Fla., to Mexican Border)	**2,055** 2,174	**647** 812	**1,133** 2,095	**79** 269	**378** 388	**4,292** 5,738
Gulf Intracoastal Waterway from St. Marks River, Fla., to Mexican Border (including Port Allen-Morgan City Alternate Route)	— —	— —	— —	**1,137** 1,180	— —	**1,137** 1,180
Mississippi River System	**2,020** 4,365	**969** 1,457	**4,957** 5,062	**740** 755	**268** 268	**8,954** 11,907
Pacific Coast Waterways	**736** 733	**498** 515	**237** 237	**26** 27	**2,084** 792	**3,575** 2,304
Great Lakes	**45** 100	**89** 148	— 14	**8** 8	**348** 369	**498** 639
All Other Waterways (exclusive of Alaska)	**76** 76	**7** 7	— —	**1** 1	**7** 7	**91** 91
GRAND TOTAL	**6,352** 8,935	**3,516** 4,544	**6,976** 8,062	**4,033** 4,309	**4,666** 3,368	**25,543** 29,218

[1] The mileages shown in this table in bold type represent the lengths of all navigable channels of the United States including those improved by the Federal Government, other agencies, and those which have not been improved but are usable for commercial navigation.

The mileages shown in this table in light type represent the lengths authorized for improvement by the Congress of the United States in legislation known as Rivers and Harbors Acts.

The sources for these tabulations are publications of the Corps of Engineers, United States Army.

Fig. 2–6. The inland waterways of the United States.

(Source: American Waterways Operators, Inc.)

the maritime nature of these vessels and their operation, they are controlled and regulated in interstate commerce by the United States Maritime Commission, rather than by the Interstate Commerce Commission.

The Great Lakes system includes Lake Superior, Lake Michigan, Lake Huron, Lake Erie, Lake Ontario, and Lake St Clair, and connecting waterways such as the Soo Canal, the St. Mary's River, and the Welland Canal. Connecting systems such as the St. Lawrence Seaway system, the New York State Barge Canal, and the Hudson River are usually grouped within the Great Lakes system. Exclusive of Seaway traffic, the Great Lakes system handles approximately 20 per cent of water borne freight reported to the I.C.C.

The Great Lakes and Mississippi systems together provide an extensive network of navigable waters linking the eastern seaboard to states as far west as Iowa and Nebraska, and also connecting the natural resource areas of the Great Lakes region with the Gulf Coast and the Port of New Orleans.

The St. Lawrence Seaway system, opened by President Eisenhower and Queen Elizabeth II in 1959, was completed over a period of 5 years at a total cost of over $1 billion. A joint project of the Canadian and United States governments, the seaway permits ocean going ships to travel 2,430 miles into the heart of the continents to such ports as Chicago, Duluth, and Port Arthur. While not usually considered part of the inland waterways system because of the ocean going nature of the vessels for which it was designed, the waterways of the Seaway are also used by domestic vessels.

There are approximately 17,000 vessels operating on the inland waterways of the United States excluding the Great Lakes. In comparing the Mississippi River system to the Atlantic, Gulf, and Pacific Coasts, it can be seen that there are comparable numbers of self-propelled vessels. The Mississippi, however, has a vast preponderance of non-self-propelled vessels, as indicated in Table 2–3. The high ratio of non-self-propelled vessels to self-propelled is indicative of the larger scale of tows which are common on the Mississippi system, as shown in Chapter 4, Fig. 4–19.

The waterways are constructed and maintained by the Corps of Engineers, since navigable waterways come under the jurisdiction of the federal government. Operation of the waterways is also freely supplied by the federal government since the United States Coast Guard assumes responsibility for the supply and operation of navigation aids. Private operations do not contribute to the construction of the waterways, and there are no user taxes on water transportation which correspond to the taxes associated with highways. The carriers of other modes, there-

TABLE 2-3

Number of Vessels on Inland Waterways of the United States

ypes of Vessels	Mississippi River System	Atlantic, Gulf, and Pacific Coasts	Great Lakes	Total
elf-Propelled				
Towboats and Tugs	2,245	1,995	155	4,395
on-Self-Propelled				
Dry Cargo Barges and Scows	12,153	3,505	172	15,830
Tank Barges	2,267	491	23	2,781

ource: 1967 Inland Waterborne Commerce Statistics, The American Waterways Operators, Inc., as of January 1, 1968.

fore, strenuously assert that water transportation is heavily subsidized by government policy.

2–17. Pipeline Systems. Two principal pipeline systems serve in the transportation of the United States. These are the *oil* and *natural gas* pipelines.

Oil pipeline operations which engage in interstate common carriage are controlled by the Interstate Commerce Commission. The lines are classified as *gathering* lines, *crude oil* lines, and *products* lines. Crude oil and products lines are termed trunk lines to differentiate their line-haul function from that of the gathering lines whose purpose is to bring the crude oil in from the fields to the primary pumping station at the beginning of the trunkline. Products lines carry gasoline, fuel oils, and kerosene from the refineries. A description of the operation of oil pipeline systems is given in Chapter 14.

In 1968, there were 169,307 miles of oil pipeline, of which 61,807 were crude oil lines, 53,431 were products lines, and 46,886 were gathering lines. Figure 2–7 shows the location of these lines. A large proportion of pipelines originates in the oil-producing states of Illinois, Kansas, Ohio, Pennsylvania, Texas, and Wyoming. Texas alone accounts for 31 per cent of all pipeline mileage in the nation, and 42 per cent of all gathering lines [7]. The orientation of the pipelines connects the producing states with the industrial areas of the north and northeast and with the Gulf ports.

Gas pipelines are under the control of the Federal Power Commission. In 1967, the total amount of transmission pipeline reported by Class A and B companies was 166,739 miles. In addition, there were almost 50,000 miles of field lines used to connect the fields to the primary

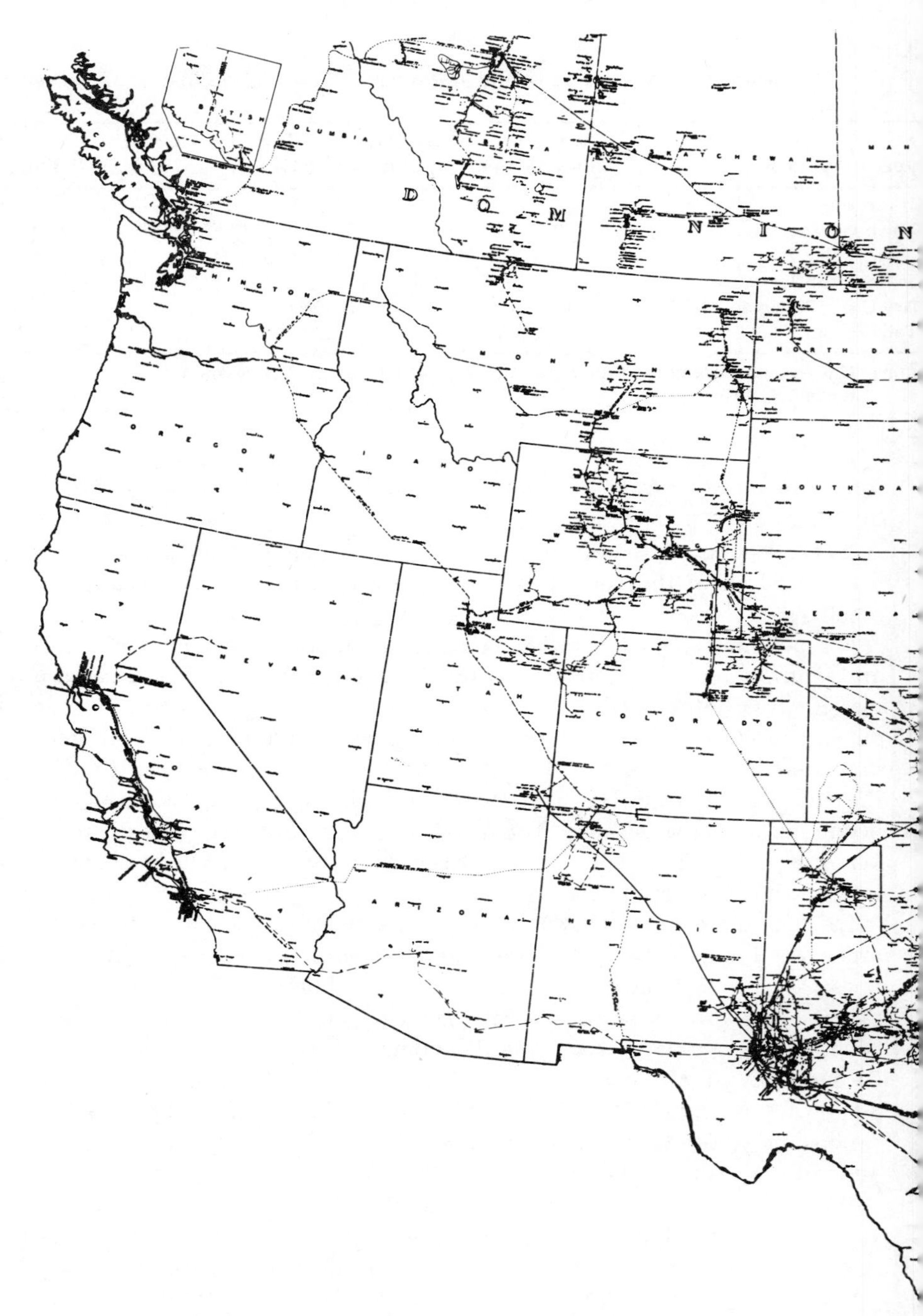

Fig. 2–7. Oil pipeline systems of the United States. (Source:

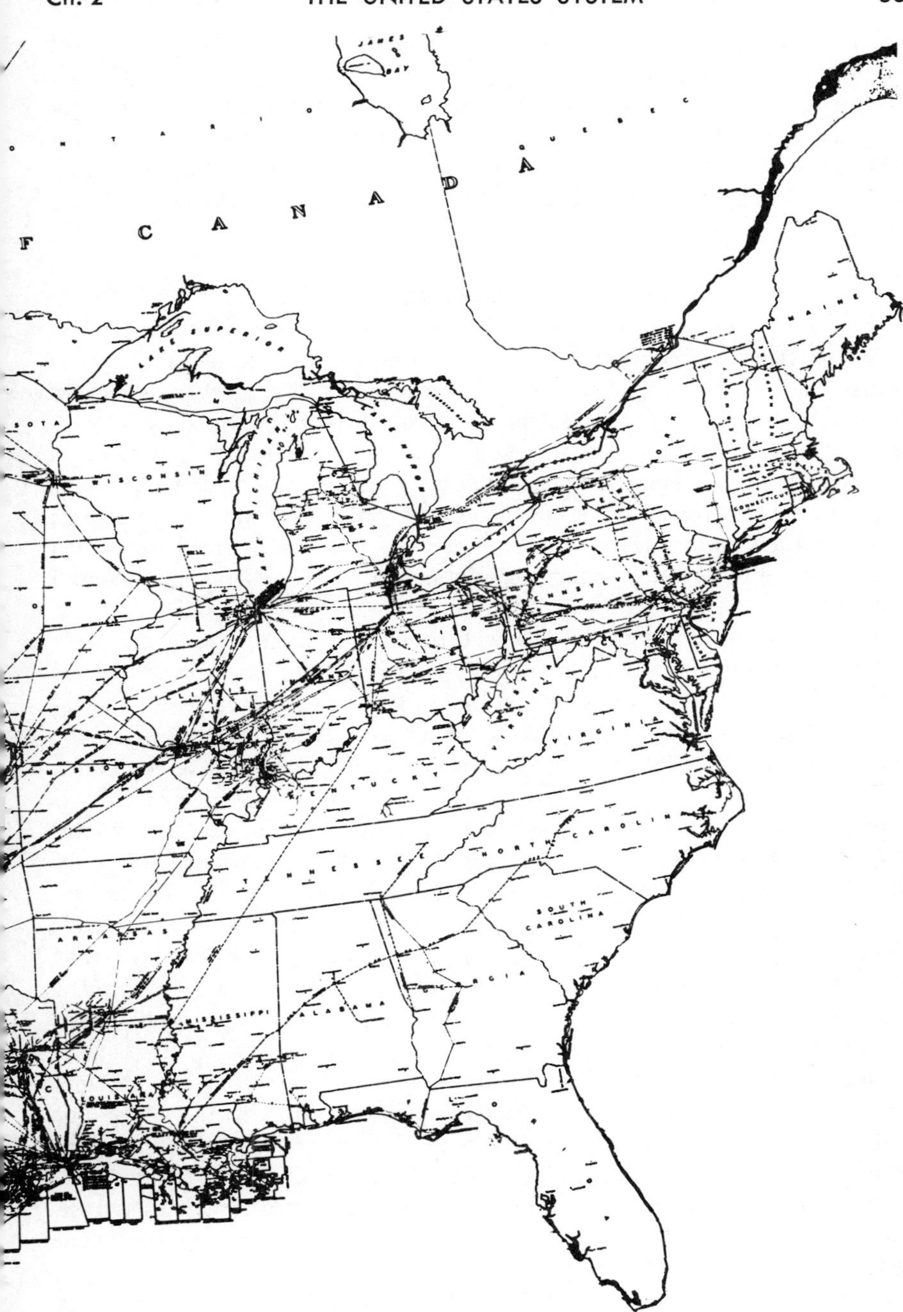

Oil and Gas Journal, Petroleum Publishing Company.)

trunk pumping stations. This system conveyed 14.5 trillion cubic feet of gas to the consumer.

The total pipeline mileage of interstate and intrastate facilities for oil, gas, and oil products, including gathering lines, amounted to over 400,000 miles. It is of interest to recall that pipelines are the only mode of transportation which has had no direct government aid.

THE CARRIERS

2–18. The Motor Carriers—Trucks. In comparison with other modes of transportation, the highway truck carriers offer the most varieties of forms of carriage. The most basic division of the carriers is between those carriers that are *for hire* and those that are *not for hire,* often called *private* carriers. Private carriers are not in the business of hauling the goods of others, but are concerned only with moving their own goods in their own trucks. For-hire carriers may be further subdivided into three categories: common carriers, contract carriers, and exempt carriers.

The *common* carrier is the one who presents a standing offer to carry the public's goods within the capability of the service and schedules which he publishes. Such carriers are subject to extensive federal legislation when operating in interstate commerce, and are subject to the control of state regulatory agencies when engaged in intrastate commerce only.

Opposed to the concept of common carriage, available to the general public, is the concept of *contract* carriage, where the carrier hauls only under specific contract, and does not hold himself out to hire to the general public. While the classification does not affect the type of goods that a carrier is capable of carrying, in general it may be stated that regulations are considerably less stringent for contract carriers than for common carriers. In the latter instance, regulations have been drawn up for protection of the general public, whereas contract carriage is frequently negotiated between business firms. Since contract carriers are not subject to common carriage regulation, freight rates are not subject to regulation by state and federal agencies Contract carriers carefully protect their classification, since failure to observe the requirements for contract carriage could result both in a loss of contract carriage classification and litigation against the carrier for engaging in illegal common carriage.

Exempt carriers are exempt from governmental control either from the nature of the services they provide or from the nature of the goods carried. For example, private carriers are exempt from the nature of the service itself. Equally, carriers engaged even in interstate movement of unprocessed agricultural products can offer their services to the public without being classified and regulated as common carriers.

Under Interstate Commerce Commission regulations, common carriers are further classified by revenues. There are approximately 15,396 common carriers. Of these, 1,389 are Class I carriers with annual gross operating revenues in excess of $1 million, 2,769 are Class II carriers with revenues in excess of $200,000 and less than $1 million. The remaining 11,238 are Class III carriers with revenues below $200,000. The average motor common carrier obviously tends to be a much smaller operation than the average railroad counterpart. The trucking industry includes many very small operators, some ranging down to owners with single vehicle operations. The structuring of the transportation infrastructure, where rights-of-way and roadbed are government owned, enable small operations to flourish while the large capital investment required for railroad operation discourages the extremely small operator.

2–19. The Motor Carriers—Buses. Passenger carriers use the highways extensively. The form of operation differs greatly in type of service and in scale. *Intercity carriers* operate both on a regular service and special charter basis. Of the 338,000 buses registered in 1967, approximately 10 per cent only were engaged in regular intercity service. The remainder operated in intercity charter service, and, on a local level, in such categories as line-haul bus systems, local public and contract transit, and school buses. Intercity carriers which operate in interstate commerce are subject to regulation as common carriers by the Interstate Commerce Commission. The I.C.C. classifies bus carriers according to operating revenues. Class I carriers are those with gross operating revenues in excess of $200,000; Class II carriers are those within the range $50,000 to $200,000; Class III carriers are those with revenues less than $50,000. In 1970, there were approximately 70 Class I and 600 Class II and Class III carriers.

Intercity operations which do not cross state lines are subject to state regulatory bodies, as are those operations within most states which cross municipal boundaries. Within most municipal areas, passenger transportation is regulated by municipal franchise without recourse to federal and state level of control.

2–20. Railroad Carrier Classification. With the exception of a relatively insignificant amount of *private* carriage, mainly by small operations affiliated with individual industries, the railroad system of the United States is operated by *common* carriers. A common carrier is one who holds himself out to provide transport to all on a nondiscriminatory basis. Service is provided at reasonable demand according to published schedules, rates, and charges.

Most commonly, railroad carriers are classified by revenues, as previously discussed in Section 2–3.

2–21. The Air Carriers. The common carriers are normally classified on geographical basis as *domestic* and *international/territorial* operators, depending on the routes over which they operate, and are further subdivided by function, as shown in Table 2–4.

TABLE 2-4

Certificated Route Air Carriers (1967)

Domestic Operations:	
Trunk carriers	11
Local service carriers	12
Helicopter services	4
Intra-Alaska	7
Intra-Hawaii	2
All cargo carriers	3
Total passenger/cargo domestic	39
International/territorial passenger/cargo	14
Total Certificated Air Route Carriers	53*

Source: FAA Statistical Handbook of Aviation, 1967 ed. Federal Aviation Administration.

*Number of carriers includes 8 domestic and 1 all cargo carrier engaged in both domestic and international operations.

Trunk carriers operate between the major population areas of the United States over the most heavily traveled routes. The eleven trunk carriers have permanent operating rights over the routes shown in Fig. 2–8. In 1967, these trunk carriers accounted for 93 per cent of all domestic passenger mileage.

Local Service carriers, supplementing trunkline carriers, provide service for lower density movements. Since these lines connect into the trunk lines at major airports they are often referred to as *feeder* lines. Both trunk and local service carriers are primarily passenger modes, but do provide freight, mail, and express movements to a minor degree.

Intra-Alaska and *Intra-Hawaii* carriers are essentially local service carriers separately classified from other local service carriers because of their geographical isolation.

Helicopter services provide airport-to-downtown service in Chicago, Los Angeles, New York, and San Francisco. While primarily passenger operations, these lines also provide freight and express service.

All-Cargo carriers operate on certificated routes between major population areas of the United States. These lines are not permitted to operate passenger service.

International Passenger/Cargo carriers are those operations carrying the United States flag which engage in international carriage or operate over international waters. They do not include foreign lines which enter the United States. The operations of International Passenger/Cargo carriers are primarily passenger service.

Supplemental Air carriers and other *Commercial Operators:* In addition to the certificated carriers, *supplemental carriers* offer charter and non-scheduled service. Other commercial operators include such miscellaneous commercial operations as air taxis and sightseeing operations.

2–22. The Water Carriers. Domestic water carriers are classified by operating revenues into three divisions:

Class A: Annual gross operating revenues in excess of $500,000
Class B: Annual gross operating revenues between $500,000 and $100,000
Class C: Annual gross operating revenues less than $100,000

Water carriers are much smaller in scale than railroads, their chief competitors. Approximately 1700 companies are engaged in commercial carriage on the inland waterways system. Only 113 of these are certificated by the Interstate Commerce Commission as regular route common carriers, and 32 companies are permitted to act as contract carriers. Because of the nature of their operation and the type of goods carried, over 1100 companies are not under the jurisdiction of the I.C.C. and are, therefore, referred to as *exempt* carriers. The remaining carriers, numbering in the region of 400, are private carriers transporting their own products.

Considering both inland and coastal waterways, in 1966 there were 188 common carriers. Of these, 58 were Class A, 31 were Class B, and 99 were Class C. The Class A carriers alone accounted for 87 per cent of operating revenues, and carried over 94 per cent of all freight on a tonnage basis. The remainder of water-borne traffic was shared, with approximately 2 per cent being carried by Class B carriers, and the remainder by the Class C carriers. The small size of water carriers can be seen from statistics reported to the I.C.C. The largest reported operating revenue was approximately $28,800,000, yet 9 of the 58 carriers accounted for 54 per cent of all Class A revenues.

2–23. Pipeline Carriers. Oil-pipeline operators engage in both *private* and *common* carriage. Those engaged in common carriage are required to file schedules of rates and charges with the Interstate Commerce Commission in a manner similar to other common carriers.

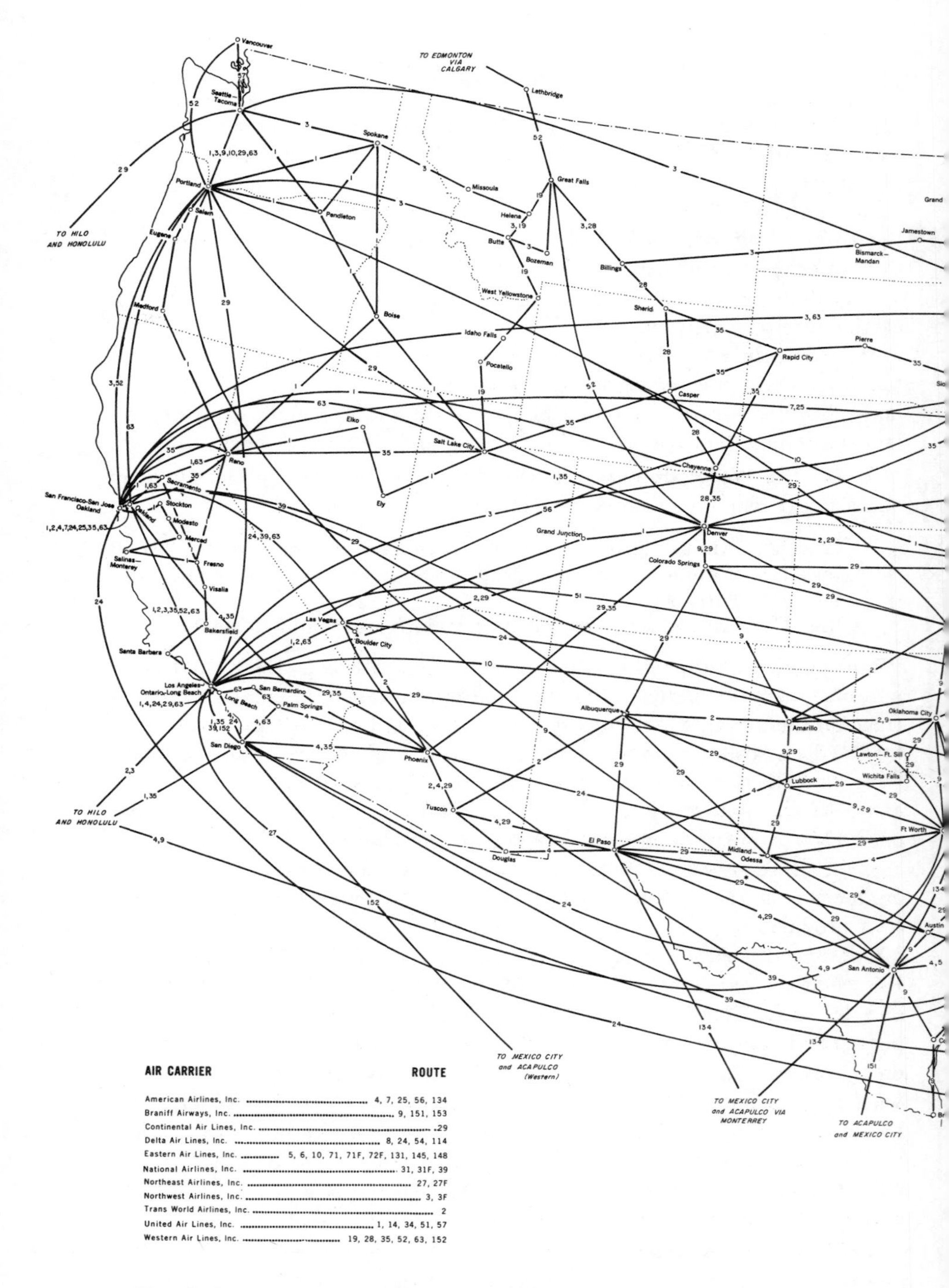

Fig. 2–8. United States air transportation system—routes certificated to Board.)

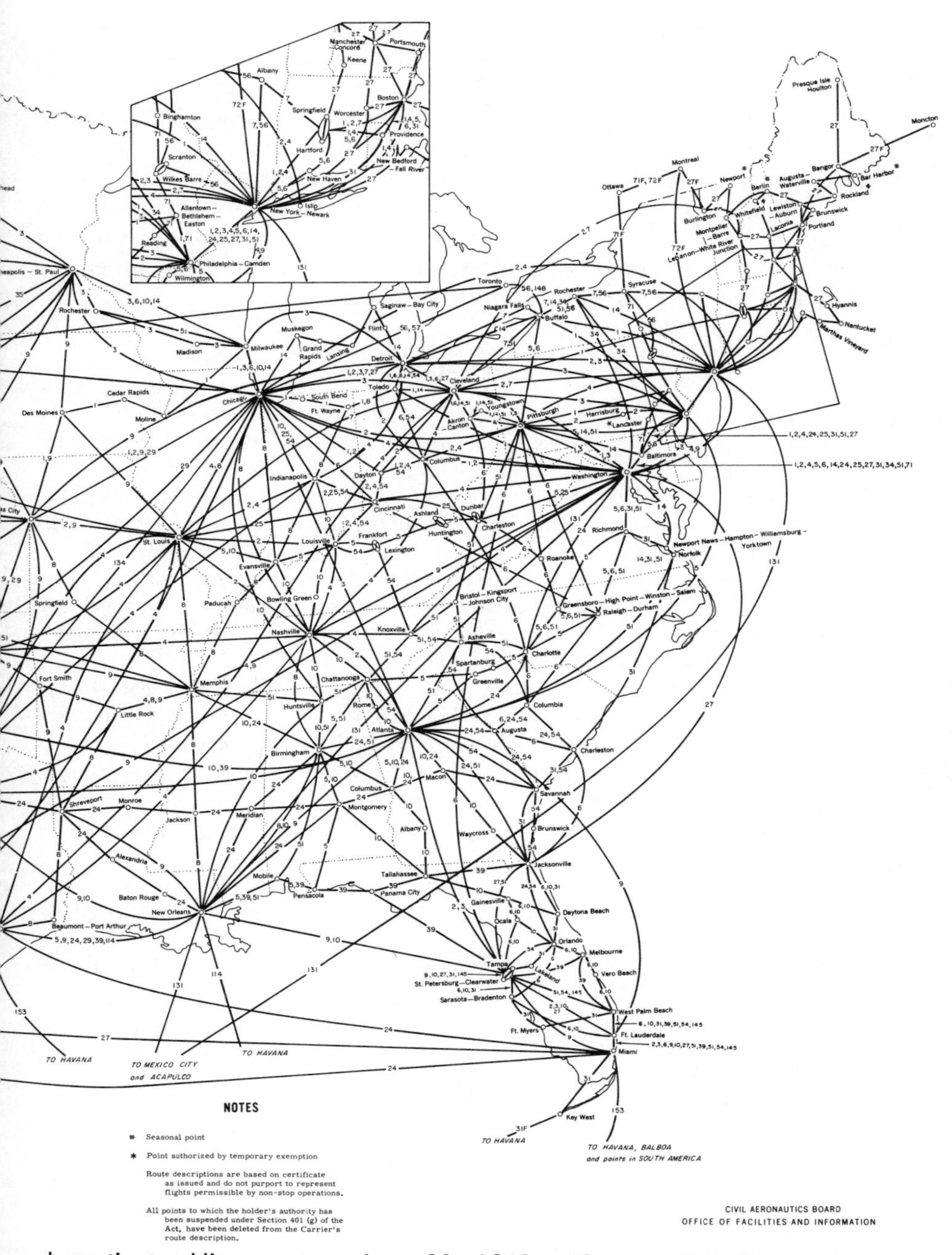

domestic trunkline carriers, June 30, 1969. (Source: Civil Aeronautics

2–24. Summary. The transportation system of the United States is seen to be a complex combination of several modes which stretches into the most remote sections of the nation. Few areas are dependent on only one principal mode; the shipper is, therefore, guaranteed transportation at various levels of service and price. In the interests of economic development, federal, state, and local governments have entered into the provision of transport systems by the provision of varying degrees of governmental aid. The result of this combination of private and public endeavor is an overall system which provides the American citizen and his goods with an unprecedented degree of mobility.

PROBLEMS

1. Describe the transportation system of some selected state. Show in the form of sketches the principal components of this system which account for the major intercity and interregional passenger and freight movements.
2. As the form of American society changes to an increasingly industrial society, what principal changes should be anticipated in the major modal networks on the national basis?
3. Outline the form of an idealized federal Department of Transportation. Discuss how this organizational structure differs principally from the existing structure and defend the structure that you propose. Relate your arguments to the existing physical systems and the carriers.
4. The National System of Interstate and Defense Highways was constructed partially on the assumption that such a system would strengthen the national defense. Discuss the arguments both for and against this premise.
5. Discuss the problems associated with the nationalization of the railway systems of the United States.

REFERENCES

1. Association of American Railroads, *Yearbook of Railroad Facts, 1970 Edition,* Washington, D.C., April, 1970.
2. *Highway Statistics,* Bureau of Public Roads, Department of Transportation, Washington, D.C., 1970.
3. *Aviation Forecasts,* F.A.A. Office of Policy Development, Washington, D.C., 1966.
4. *Transport Aircraft, Characteristics, Trends & Growth Projections,* Transport Aircraft Council, Aerospace Industries Association of America, Inc., Washington, D.C., March 1969.
5. *Statistical Handbook of Aviation,* 1968, Federal Aviation Administration, Washington, D.C., 1969.
6. *Transportation Facts and Figures—1969,* American Transportation Association, Washington, D.C., 1970.
7. *Transport Statistics in the United States for the year ending December 31, 1969, Part 6, Oil Pipe Lines,* Interstate Commerce Commission, Washington, D.C., 1970.
8. *F.A.A. Advisory Circular, AC-150-5090-2, National Airport Classification System,* Federal Aviation Administration, Washington, D.C., June, 1971.

3

Transportation and Development

Transport decisions are not made in isolation from the economic environment. The provision of accessibility can provide a base for economic development. Equally, growth itself creates activity patterns which pose a demand for transportation. The relationship between transportation and development patterns has long been of interest to planners and economists. Various theories of the nature of this relationship have been set forward.

3–1. Von Thunen's Isolated State [1]. One of the first attempts to rationalize the effect of distance (and therefore transportation) on regional development was the theoretical work published by Von Thunen in 1826. The author postulated an isolated state comprised of one central city located in the center of a large plain. The plain was equally fertile throughout its extent, had a homogeneous climate and similar terrain throughout. It was further assumed that the surrounding hinterland of the city supplied all required agricultural products, and that the prices set by the market in the city were not subject to farm control. One form of transportation was assumed, with radial movement directly to the city center. Freight rates were assumed to vary directly with the ton mileage of freight moved. Von Thunen investigated the effect of such assumptions upon the structure of agricultural economics. It can be seen that, under assumed conditions, whichever crop was most profitable at any particular point would also be most profitable at all other points equidistant from the central city or market point.

On analysis of the above structure, Von Thunen found that the first zone would be most profitable in the production of vegetables and dairy products. The second zone would support the lumber industry with products of high weight and low value. The third, fourth, and fifth zones would be used for the cultivation of grain, the intensity of cultivation declining with distance from the city. The decrease in intensity of cultivation is due to the relationship between profit and market price. Profit can be seen to be equal to market price less the sum of production and transport costs. Net price may be defined as market price less transport costs. The further the production area falls from the market the lower the net price. As the net price declines, there is less profit in the intensive use of capital and labor. Thus, profit can be attained only by increasingly extensive farming methods. The last zone would be suitable for the most extensive of all farming methods, pasture. The radial pattern of concentric zones of varying agricultural land use is shown in the upper half of Fig. 3–1. The lower half of this figure indicates the effect of the introduction of a navigable river into this symmetrical plain. With the lower cost of transport provided by water transportation along a lineal route, the zones of land use become elongated to follow the direction of the low cost transport route.

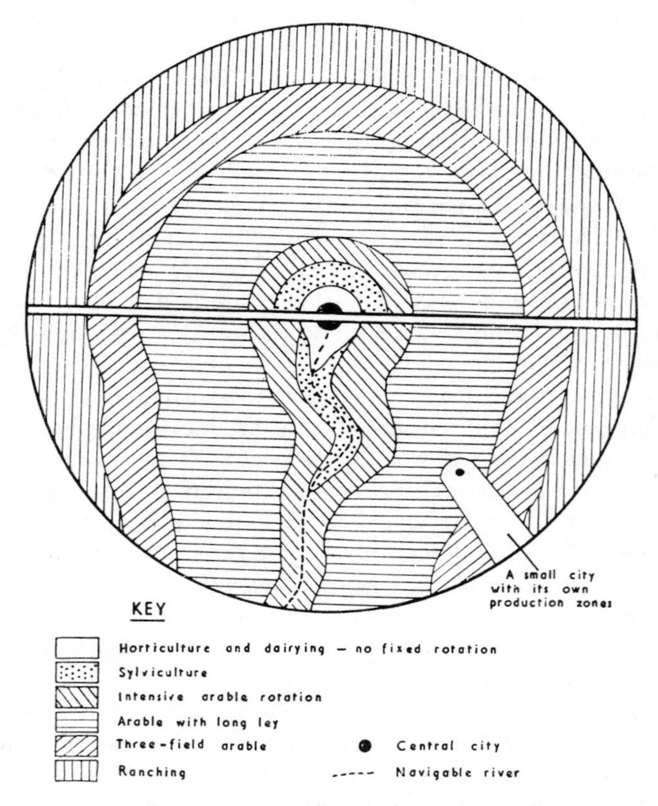

Fig. 3–1. Von Thunen's isolated state. (Source: *Rural Settlement and Land Use,* by Michael Chisholm, Hutchinson University Library, 1962.)

3–2. Theoretical Location of Centers Within a Regional Hierarchy. A theory of the development of a hierarchy of urban centers was developed by Christaller, assuming, like Von Thunen, a hypothetical agricultural economy [2]. The structure developed by this theory has been found applicable to modern automobile oriented cities [3].

Christaller found that each large central city was surrounded by smaller urban centers. These smaller centers themselves assumed central functions to a series of even smaller centers. Thus, a hierarchy of urban centers is built up. In an idealized region similar in topography, climate, and fertility, each urban center would develop a hexagonal area of influence (Fig. 3–2). The dominant central city contains all the activities required of the region such as wholesaling, specialized retailing, brokerage, medical facilities, and industrial banking. As com-

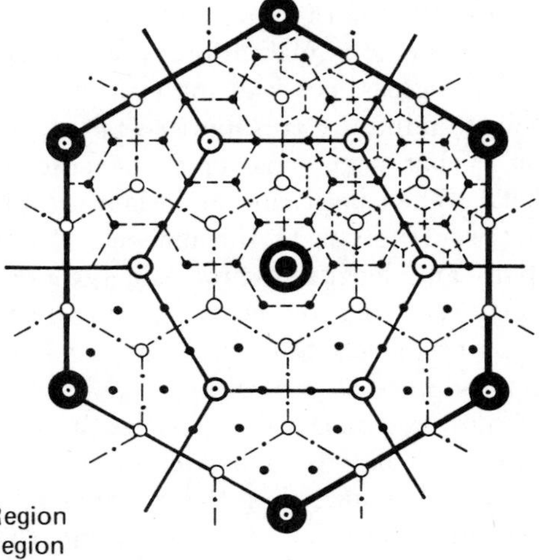

Fig. 3–2. Christaller's hierarchy of urban centers. (Adapted from Walter Christaller, *Central Places in South Germany,* 1966. Reprinted by permission of Prentice-Hall Inc., Englewood Cliffs, N.J.)

munities become smaller in the hierarchy the functions contained within these urban centers become less self-sufficient and simpler. The transportation routes for such a region show a hierarchy which corresponds closely to classifications observed in the real world. Large urban centers are linked by primary routes. Second order cities are linked to the primary routes by a secondary system, while third order and lesser centers require only minimal transport linkages.

3–3. Industrial Location—Von Weber. In 1909, Von Weber examined the problems of industrial location relative to markets and production areas in "Theory of the Location of Industries." This work clearly set out the strong influence of transportation costs on industrial location. According to Von Weber, raw materials are either *pure,* i.e., there is no weight loss in processing for market or *gross,* in which case weight loss does occur in the processing phase. Additionally, raw materials are either *ubiquitous* (available at all locations) or *localized* (available at certain locations only). Transport costs were assumed to be directly proportional to ton mileage only, and labor for processing was available at all locations.

Examining first the case of a single material which is ubiquitous, it can be seen that processing and production would occur at the market, avoiding needless transport costs. Pure material available at a point away from the market can be processed at the market point, the production point, or at any location between these two areas. The processing point for a gross localized material would be at the point of production in order to minimize transport costs.

Von Weber next considered the case of two raw materials and one market. When both raw materials are ubiquitous, production and processing would occur at the market. In the case where both materials are pure and localized, processing would take place at the market to minimize the transport costs of bringing the materials together. Where both materials are pure and only one is ubiquitous, processing would again take place at the market.

The most complicated case is that in which both materials are localized and gross. This case most obviously corresponds with real life situation. It was shown that the location of the optimum processing point fell within a location triangle joining the two production points with the market point as shown in Fig. 3–3. The exact location of the processing point can be determined by calculations which account for the percentages of weight loss for each raw material.

3–4. Urban Structure within the Region. The regional location theories discussed to this point attempt to explain the location of cities within their region. These theories postulate that the functional structure and interrelationships of cities within a region are related to distance and,

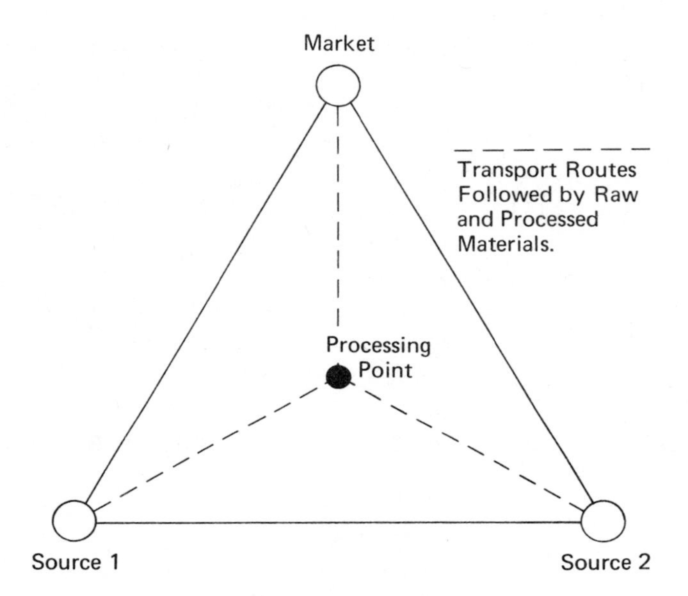

Fig. 3–3. Von Weber's locational triangle.

therefore, to transportation networks. A more apparent relationship between transportation and development patterns occurs at the metropolitan level since changes in metropolitan structure occur at a much more rapid rate than changes at the regional level. Empirical evidence to substantiate the theories of metropolitan structure is, therefore, more readily available.

3–5. Metropolitan Structure—Concentric Zone Theory. Based on observations of the development of cities, Burgess developed a theory which described metropolitan areas as they existed in the 1920's as areas of concentric zones [4]. Burgess found that in general terms socioeconomic status increased with increasing distance from the central city. Figure 3–4(a) shows the concentric city composed of five differentiated zones surrounded by a suburban commuters' zone.

The first zone is the Central Business District, composed of shops, offices, banks, and cultural and transportation activities. This is the zone of highest land values and highest accessibility. Next to the central zone comes the zone in transition. This area contains light warehousing and commercial activities which require more extensive land use, including market and wholesale functions. Some residences remain in this area, but they are in areas of rapid deterioration being held in speculation for redevelopment. Zone 3 is the area of low-income workers' residences. Normally this zone contains the residences of factory and

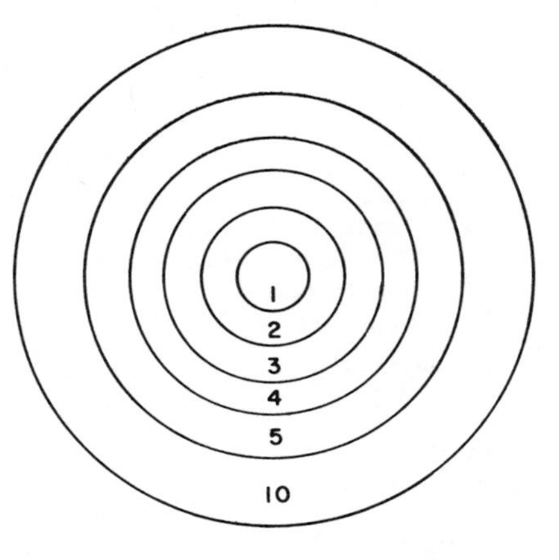

(a) CONCENTRIC ZONE THEORY

(b) SECTOR THEORY

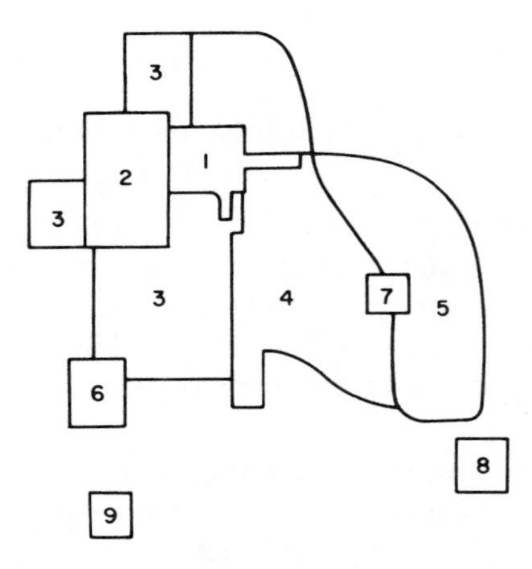

(c) MULTIPLE NUCLEI

1 CENTRAL BUSINESS DISTRICT
2 WHOLESALE LIGHT MANUFACTURING
3 HIGH DENSITY RESIDENTIAL
4 MEDIUM DENSITY RESIDENTIAL
5 LOW DENSITY RESIDENTIAL
6 HEAVY MANUFACTURING
7 OUTLYING BUSINESS DISTRICT
8 RESIDENTIAL SUBURB
9 INDUSTRAL SUBURB
10 COMMUTERS ZONE

Fig. 3–4. Land use patterns according to various theories of urban structure. (Source: *Future Highways and Urban Growth,* Wilbur Smith and Associates.)

wholesale workers, laborers, and other low-income workers who have been displaced by land use changes in the area of transition but still wish to remain close to the work opportunities of the central areas. The low-income zone merges into the fourth zone which contains the white collar or middle-income workers whose income permits a higher expenditure on transportation. These areas normally are at lower densities, returning to the residents in privacy and space what has been given up in accessibility to the urban center. The fifth and last urban zone is the low density zone of the upper income groups of the urban area. Finally, surrounding all other zones is the suburban commuters' zone, limited to the upper income groups.

3–6. Metropolitan Structure—Sector Theory. Another theory of urban development related to the presence of identifiable transportation corridors was advanced by Hoyt [5]. The sector theory, as indicated by Fig. 3–4(b), states that growth takes place in wedge-shaped sectors that follow the axes of transportation routes radiating from the Central Business District. Growth along any particular axis tends to be of a similar character of land use. Thus, high-rent areas tend to continue to grow to the periphery of the urbanized areas while medium-income areas and areas of light industry themselves expand in generalized sectors towards the edge of development. According to this theory, high-grade residential growth is responsive to several pressures. Among the strongest are progression to high ground free from floods, attraction to the fastest transportation routes, gravitation towards office buildings, banks and stores, and movement over long-term trends towards open country and away from areas that are limited by topographic barriers inhibiting expansion.

3–7. Metropolitan Structure—Multiple Nuclei Concept. Another theory of the structure of urban areas hypothesizes that the city does not develop from one single focus, but instead is the result of growth about several nuclei, as indicated by Fig. 3–4(c). This theory is helpful in explaining the form of urban development that has taken place in some of the more rapidly developing cities which have become metropolitan areas since the beginning of the twentieth century. Certain areas of the city apparently grow under stimuli for change which are independent of the patterns of development of the Central Business District. In some cities, the Central Business District itself is difficult to define, for example, in Los Angeles and Washington, D.C. Harris and Ullman cite the following factors acting either separately or in combination as stimuli of change [6]:

1. Certain activities require specialized locations. For example, retail areas need high accessibility, manufacturing areas require large

level sites close to rail and highway connections, and port activities must be located adjacent to the water.

2. Activities which benefit from proximity to one another tend to cluster. Warehousing is required close to port and manufacturing areas, doctors' and medical uses tend to cluster around hospitals, and legal offices group around court house facilities. Retail districts grow in areas where potential customers are located.

3. Just as some activities congregate together, others tend to polarize because of detrimental relationships. High-income residential areas and industrial uses are usually found to be incompatible neighbors.

4. While some uses could benefit from certain locations, land costs are too high for economic location within these areas. Bulk wholesaling and storage, for example, require too much land to be economic within the central areas of cities.

3–8. The Development of Metropolitan Form. Urban areas, although composed of apparently permanent physical structures, cannot be considered as static entities. Cities are physically dynamic in nature, and those elements which make up the physical structure of urban areas are in constant change. The casual visitor to a city receives a false image of the city form from the apparent durability and permanence of the individual structures and facilities which at any one time are the physical components of an urban area. The dynamism of cities is overlooked by individuals because the rate of change is slow compared to their own lifetime. The dynamic nature of urbanization is important to engineers and planners who intend to design and plan urban systems, for it is necessary that they view their work not simply as a physical entity with a given economic life, but rather as a link in a chain of urban change. The satisfactory development of urban areas is dependent upon how well the individual links fit the overall urban development patterns. The consequences of lack of adequate fit can be viewed in every urban area: slums, antiquated transportation modes, congested facilities, and undesirable arrangements and mixes of land uses.

Urban development takes place both in *spatial expansion* and by *overbuilding*. Spatial expansion, the most obvious form of urban development, takes place as the fringes of urbanization move into formerly rural areas. In the years since World War I, cities have rapidly increased their urbanized areas with the phenomenon of low-density suburban development. This form of development has continued at an accelerated pace since the 1950's.

Equally important, yet often overlooked in urban development, is the process of overbuilding. In conjunction with large programs funded by the federal government, this process in the United States has been called *urban renewal*. Urban renewal is not a new process peculiar to the twentieth century; it is inherent to the nature of cities. This can be more clearly seen by examining the central areas of the older European cities. In London, for example, within the very City itself, which has been inhabited for over two thousand years, with extensive development in mediaeval times, very few of the structures now to be seen date prior to the nineteenth century. In postwar years, many nineteenth-century structures themselves have been replaced by high-density modern office buildings. Only streets and street patterns remain relatively unchanged over time.

By intelligent planning, the dynamics of urban change can be used to move constantly towards a better urban environment. The degree of governmental activity required to promote the natural course of urban economics without the interim development of urban blight and slum conditions is open to question. It is hoped that eventually planners and engineers will be able to adopt policies which will permit the dynamics of urban change to continue without the cyclic development of slums which is now apparent.

3–9. Development of an Urban Area—Chicago [7]. To illustrate the degree of interrelationship between urban development and transportation facilities, Chicago can be used as a useful case study. Figure 3–5 shows the spatial form of urban development in the city from its founding until 1955. Initially the city started as a small settlement at the mouth of a river feeding into Lake Michigan. Up to 1860, the city grew in a reasonably concentric form around the initial settlement, with growth developing around the main roads which serviced horse drawn traffic. With the introduction of horse drawn cars and cable cars, development concentrated about the radial public transit routes on arterial streets. The finger structure reached a zenith with the construction and operation of suburban railroad lines. Isolated islands of development up to thirty miles from the city center grew up along these suburban railroad lines. Railroad suburbs were centered around the station stops with no intermediate development due to the limited access nature of rail lines. The strong radial form of the early twentieth-century city began to disintegrate with the general introduction of the motor car in the 1920's. The auto oriented suburban areas began to fill in the radial corridors with continuous development that had access at all points to the radial arterial streets, which were constructed to meet the needs of the growing number of auto commuters.

As the public increasingly chose to use automobiles as its means of personal transportation, arterial streets became seriously congested. The construction of freeways throughout the metropolitan region, together with widespread provision of high capacity arterial and feeder streets followed, making areas some distance from the radial axis still

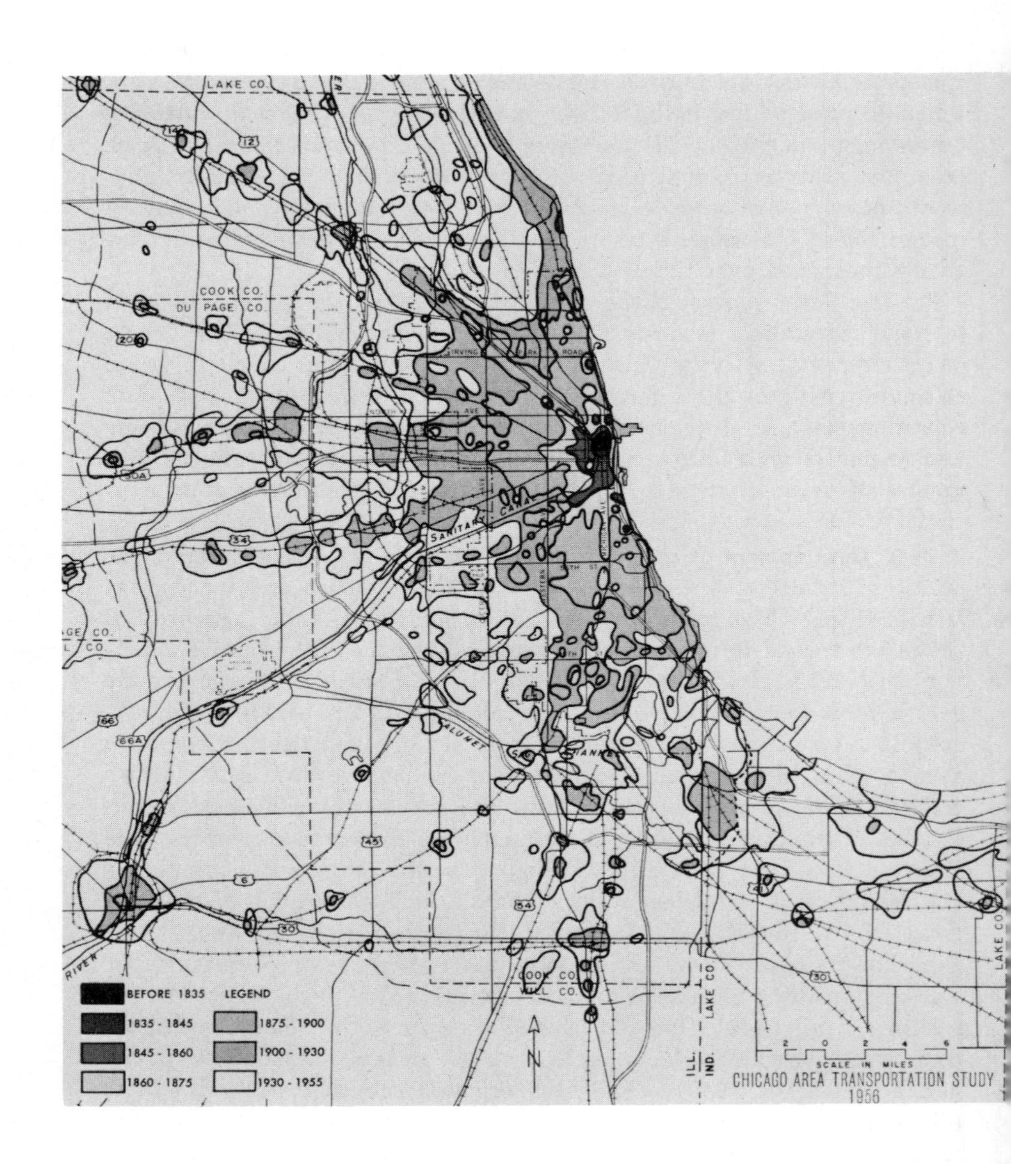

Fig. 3–5. Chicago growth patterns to 1955. (Source: Chicago Area Transportation Study.)

highly accessible. As a result the distinctive fingers of development became less well defined. By the year 1980, the many crosstown freeways and arterial streets will have virtually destroyed the radial patterns of growth brought about almost a century before by suburban railroads, as is indicated by the 1980 regional plan shown in Fig. 3–6. It is interesting to note that the introduction of a high-speed intraurban rapid transit system could radically change the form of development predicted

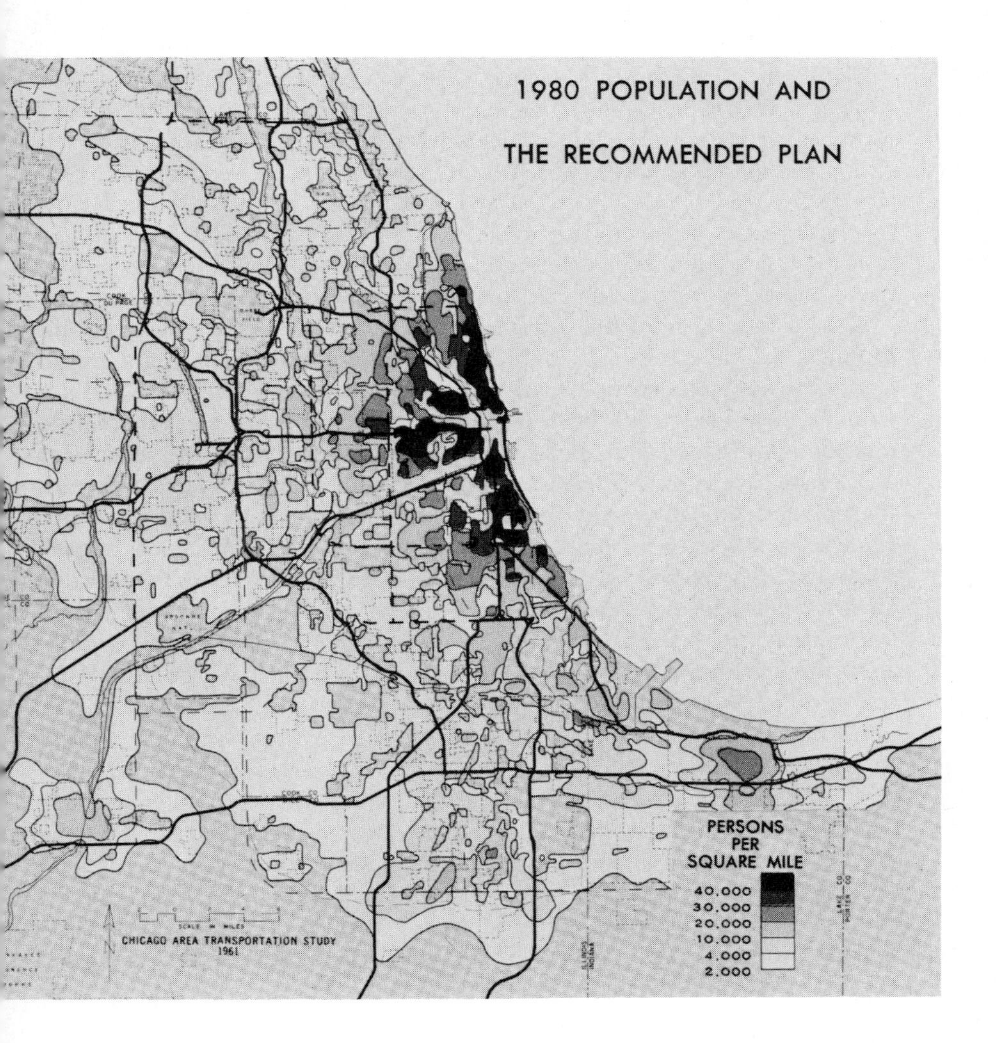

Fig. 3–6. Chicago recommended plan 1980. (Source: Chicago Area Transportation Study.)

in this plan. Such a new technology would tend to again bring about strong corridor developments.

3–10. Economic Shifts Within the Metropolitan Areas. The physical changes in the extent and density of urban development are accompanied by shifts in the various economic strata of the metropolitan populations [8]. In the infant stages of a metropolitan area the higher income groups lived in proximity to the Central Business District; at this stage transportation networks were rudimentary. This state of development is shown in Fig. 3–7(a).

With the development of high capacity arterial streets, bus and suburban railroad systems, the middle class moved to outlying central city locations while the more affluent rich leapfrogged to the newly developing commuter suburbs which were adequately, if somewhat expensively, serviced by rail lines. The lower income groups remained close to employment opportunities within the central areas as shown in Fig. 3–7(b). The poorer elements of the population continued to center most activities within the areas close to their place of residence.

The introduction of the automobile affected first the upper income groups. As shown in Fig. 3–7(c), the isolated suburban communities merged to become one suburban ring of wealthy development. The automobile was essential for this stage of development since access to rail commuter lines and to the Central Business District was dependent on personally owned transportation vehicles. Figure 3–7(d) indicates the changes brought about by increasing car ownership, which now included the majority of middle-income households. The increase of personal mobility permitted the middle class to move out to areas more distant from the city center. At this stage movement into and within the central areas tended to become more difficult. The outer middle-class ring generated a great deal of automobile traffic. Unlike the poor, the middle classes worked, shopped, and carried on other activities at some distance from their place of residence. As expressways and arterial streets were constructed to cater to this newly generated traffic, employment centers were generated at the newly created areas of high accessibility. A general decentralization of employment centers continued as the wealthy and middle classes continued to move further from the central area as shown in Fig. 3–7(e). This decentralization of employment, accompanied by declining levels of transit service as car ownership increased, provided serious accessibility problems to the urban poor with low car ownership. The problem of decreased accessibility and its consequent social ills have only lately been recognized.

3–11. Transportation Consequences of Various Regional Forms. The role of transportation in urban and regional development has been discussed at some length. When dealing with long term developmental planning of new towns, the reverse relationship is more important. Different forms of regional development create distinctive patterns of transportation demand. Research using simulation procedures has found that although the patterns of development are quite dissimilar, the overall demand for transportation services does not show the same variation. Variations in regional growth form does, however, require significantly different configurations of transport network.

Figure 3–8 shows the characteristics of six different proposals for the future development of Canberra, Australia [9]. Plan A concentrated expansion to one-half million around the existing city of 100,000. Plans B and C would create corridors of growth surrounding newly created *town centers* (secondary foci of commercial activity). Plan D created twin corridors around similar town centers. Plans A-1 and A-2 were variations on Plan A, the former with urban concentrations around planned town centers, the latter with a strong degree of centralization for the present city center. Using simulation procedures, the planners found that the transportation characteristics on assumed networks were remarkably similar (see Table 3–1). This contrasts with the diverse

TABLE 3–1

Transportation Characteristics on Assumed Networks

	Alternate Plan					
	A	B	C	D	A-1	A-2
Average trip length (min.)	11.3	11.5	11.5	11.7	10.8	11.6
Work	15.8	17.1	17.6	17.1	15.2	16.8
Shop	8.5	8.4	8.5	8.5	8.4	8.4
Other	11.2	11.1	10.7	11.2	10.7	11.3
Non-home based	10.9	10.8	10.5	10.8	10.7	10.7
Total vehicle miles (000)	9,750	9,934	9,766	9,841	ND*	9,966
Intertown Travel						
Percentage of all trips	43%	40%	37%	43%	ND	45%
Percentage of all vehicle miles	77%	72%	72%	77%	ND	78%
Airlines miles of travel (000)	6,690	6,397	6,098	6,735	ND	6,895
Ad						
Average daily travel costs (000)	$657	$693	$685	ND	ND	$675

Source: Table 4, *Canberra General Plan Concept*, A. M. Voorhees and Associates, Inc.
*ND = Not determined.

assumptions of the plans. Based on congestion, street and highway configuration, and public transportation potentials, findings indicated that:

1. Metropolitan growth should be directed into a limited number of corridors.

A SYNOPTIC VIEW OF METROPOLITAN GROWTH

a.

The rich live at high-access locations around the CBD and the poor cluster about job locations.

b.

The middle class split off to outlying central-city locations serviced by the internal transportation grid. The rich relocate to commuter suburbs serviced by the railroads.

c.

The rich fill the first suburban ring. The middle class and the poor take over most of the central city.

= upper-income groups

= middle-income groups

= lower-income groups

= employment locations

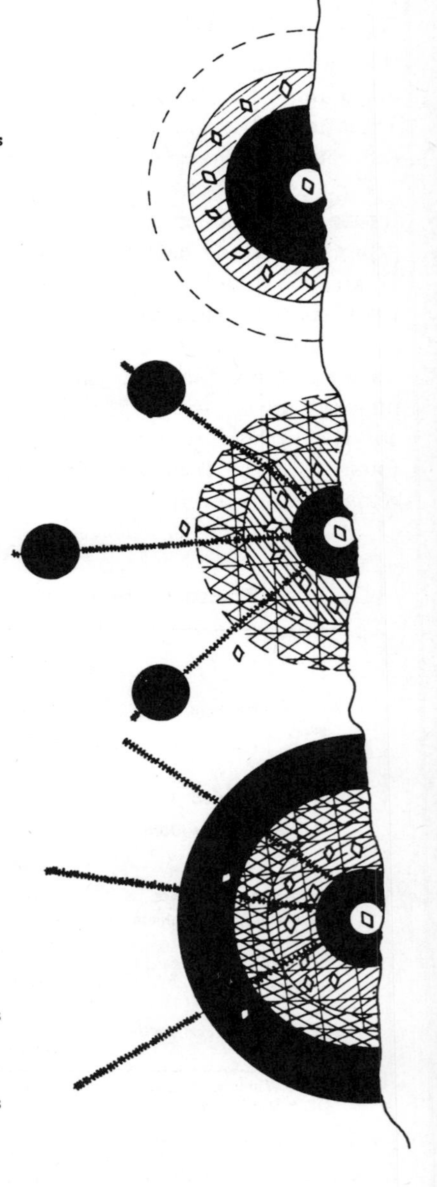

Fig. 3–7. Shift of economic classes within the growing metropolitan area. Praeger, Publishers.)

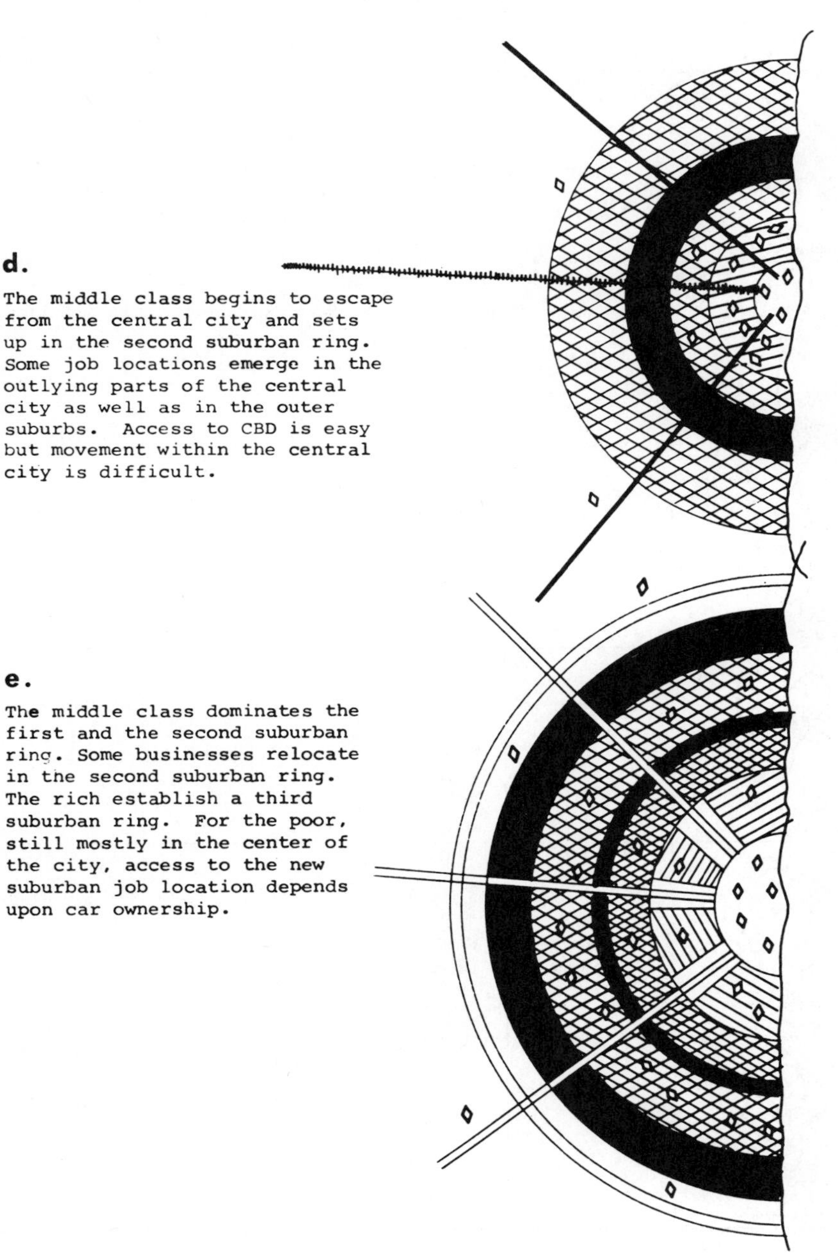

(Source: Oscar A. Ornati, *Transportation Needs for the Poor,* Frederick A.

Type of Urban Form	Residential Density in Towns	Retail Structure		Average Town Center Employment	Central Area Employment
		Town Centers	City Center		
 A Concentrate expansion around present city	Constant	In balance with Market Potential	14.3%	11 centers @ 11,400 Emp.	65,000 Emp.
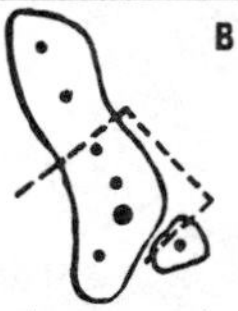 B Corridorize growth to N.W.	Constant	In balance with Market Potential	12.3%	8 centers @ 15,600 Emp.	65,000 Emp.
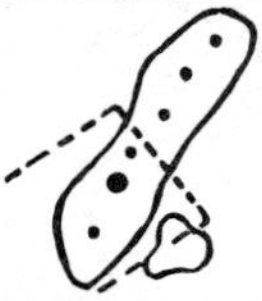 C Corridorize growth to N.E.	Constant	In balance with Market Potential	12.3%	11 centers @ 10,200 Emp.	65,000 Emp.
 D Growth to N.W., N.E., with minimum to South	Constant	In balance with Market Potential	12.3%	10 centers @ 11,800 Emp.	65,000 Emp.
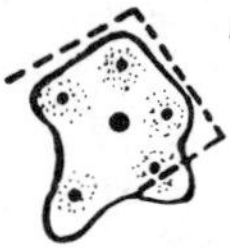 A-1 Expand around present city	Increasing around Town Centers	In balance with Market Potential	14.3%	11 centers @ 11,400 Emp.	65,000 Emp.
 A-2 Expand around present city and centralize	Constant	In balance with Market Potential	14.3%	11 centers @ 6,400 Emp.	120,000 Emp.

Fig. 3–8. Characteristics of various land development patterns for Canberra. (Source: Canberra General Plan Concept, Alan M. Voorhees and Associates.)

2. The optimum size for the Central Area is approximately 90,000 employees. In addition, it should contain approximately one million square feet of retail space.
3. Towns should contain populations of between 75,000 and 120,000 persons. Town Centers containing 750,000 square feet of retail space and providing employment for 10,000 to 15,000 persons should be established in the center of the towns on an arterial road public transportation spine.
4. The transportation system should consist of an articulated system composed of an arterial road public transportation spine and a peripheral freeway system.
5. Public transportation on separate rights-of-way offering high levels of transport service can be achieved if land development at the regional, town, and local level is organized to take advantage of the full potential of this mode of travel.

Limited studies were carried out in the Puget Sound Regional Transportation Study to examine the differences in transportation demand related to change in regional development policies [10]. Two basic plans were examined; Plan A was a trend pattern of unplanned spread of residential areas, similar to current conditions, while Plan B related to a planned pattern of cities and urban corridors. Although the underlying assumptions of the two development plans were vitally different, the major findings shown in Table 3–2 indicate that there were only minor differences in the transportation facilities required by these two plans. The locations of facilities, while somewhat responsive to forecasts, were found to be profoundly influenced by such topographic features as water barriers and by existing development.

3–12. National versus Regional Transport Planning. To this point discussion has centered on urban and regional development within a national context. While regions and smaller areas have a degree of choice of transportation modes, these areas act within the constraints of national policy. For example, an urban area retains the option of moving its population by freeways or by rapid transit; this decision is made subject to the knowledge that interstate routes are eligible for extensive federal subsidies which have not been available for the construction of rapid transit routes prior to 1970.

National transport planning is more open to real choice. Policies are determined chiefly within fiscal restraints. Decisions must be made concerning how much of the national budget is to be applied to the transport sphere, and how much of this share is to be appropriated to each mode. The former decision is determined within the context of national priorities. Modal selection depends partly on the level of total appropriation to the transport sphere. Frequently, the choice of

TABLE 3-2

Summary Findings of Puget Sound Analysis of Different Transport Plans

	Plan A with local, feeder, and express buses	Plan A with iocal, feeder, and express buses and rapid transit	Plan B with local, feed and express bu
Total Trips	6,795,000	6,795,000	6,792,000
Facility mileage			
Controlled access facilities	550	539	542
Surface arterials	1,748	1,748	1,746
Daily vehicle hours of travel x 1000			
Controlled access facilities	390	395	327
Surface arterials	270	276	257
Daily vehicle miles of travel x 1000			
Controlled access facilities	19,676	19,957	16,564
Surface arterials	8,133	8,293	7,802
Additional transit trips by rapid transit	–	24,000	–
Capital costs (000's)	1,587,000	1,662,000	1,511,000
Operating costs (000's)	1,427	1,454	1,293
Annual cost at 6% interest rate	1,551	1,583	1,412
Systemwide ratio of demand/capacity	0.67	0.70	0.6

Source: *Puget Sound Regional Transportation Study.*

modal investment is influenced by the amount of capital versus labor expenditures, the proportion of investment which will be spent on imported equipment and expertise, and a variety of other social and political factors.

3–13. Choice of Transport Technology. With limited public funds available for investment in the transport sphere, each nation must make a decision on how much investment is to be made in each transport mode. Freight movement can be made by road, rail, waterway, pipeline, or air. Air transport in all cases still carries an insignificant proportion of traffic due to the relatively high expense. Using United States prices as a basis, cost comparisons may be made on average ton mile costs, using pipeline costs as the basis [11]:

Mode	*Relative Unit Cost*
Pipeline	1.0
Ocean Tanker	0.5–0.8
Barge	1.0–3.0
Rail	4.0–8.0
Truck	6.0–8.0
Air	up to 128.0

Based on their various needs, countries have chosen to emphasize different modes in the national transportation systems. The bulk of freight transportation to any country falls on one of three modes—rail, road, or inland waterway as shown in Table 3–3. These statistics indicate significant shifts toward road transport for western nations in the eight-year period shown [12, 13].

Motor transport has the advantages of fast delivery, with low breakage and spoilage. There need be no inconvenience or damage involved in transshipment. In addition, the shipper has a wide flexibility of location and is not restricted to operation on land immediately adjacent to rail service. The mode is capable of adjustment to a wide variety of sizes of shipments and is of considerable advantage where small shipments are concerned. Because of speed and ease of shipment the mode is highly competitive in the short haul. Analysis of United States produce statistics indicates that 88 per cent of shipments less than 100 miles move by truck, compared to 7 per cent of shipments in excess of 2,000 miles [7, 14]. One significant characteristic of motor transport is the ability of both vehicles and roadway to be provided in staged levels. Light roads can be provided for low vehicle densities and construction standards can be upgraded as vehicle densities increase. This can be carried out with little loss of initial investment.

Motor transport has inherent disadvantages. Not the least is the capacity limitation which prevents the transport of large bulk shipments. Service is subject to interruption during periods of severe weather. Although service may be attainable with even a rudimentary road system, operating costs increase rapidly with inadequate roads. In developing countries, both the vehicles themselves and perhaps their fuel must be imported, placing a possible strain on international payment balances. A side effect of the provision of roads for road transport is the encouragement of private car ownership leading to a decline of rail utilization, which is a trend familiar to western nations. In an undeveloped nation the presence of roads may cause a demand for private cars which may seriously affect the viability of the rail system,

TABLE 3–3

Freight Transport in Six Selected Countries

United States

Type of Transport	Freight Traffic (in billions of ton-kms) 1940	1960	Percentage Distribution 1940	1960
Railways	606.7	926.6	61	44
Roads	99.2	479.0	10	23
Inland waterways	189.0	356.8	19	16
Pipelines	94.9	365.8	10	17
	989.8	2,128.2	100	100

Union of Soviet Socialist Republics

Type of Transport	Freight Traffic (in billions of ton-kms) 1940	1960	Percentage Distribution 1940	1960
Railways	415.0	1,504.4	89	86
Roads	8.9	98.5	2	6
Inland waterways	36.1	99.6	8	6
Pipelines	3.8	51.2	1	2
	463.8	1,753.7	100	100

France

Type of Transport	Freight Traffic (in billions of ton-kms) 1955	1963	Percentage Distribution 1955	1963
Railways	46.8	63.0	61	51
Roads	20.4	37.1	27	30
Inland waterways	8.9	11.4	12	10
Pipelines	—	11.0	–	9
	76.1	122.5	100	100

Source: Table 2, p. 20, *Introduction to Transport Planning*, United Nations.

TABLE 3-3

(Continued)

Netherlands

Type of Transport	Freight Traffic (in billions of ton-kms) 1955	1963	Percentage Distribution 1955	1963
Railways	3.6	4.2	17	14
Roads	2.1	4.9[a]	10	16
Inland waterways	15.2	20.2	73	66
Pipelines	–	1.2	–	4
	20.9	30.5	100	100

Poland

Type of Transport	Freight Traffic (in billions of ton-kms) 1955	1963	Percentage Distribution 1955	1963
Railways	52.0	74.4	96	95
Roads	1.5	3.0	3	4
Inland waterways	0.8	0.9	1	1
Pipelines	–	–	–	–
	54.3	78.3	100	100

United Kingdom

Type of Transport	Freight Traffic (in billions of ton-kms) 1955	1963	Percentage Distribution 1955	1963
Railways	34.9	27.0	48	32
Roads	37.6	57.0	52	67
Inland waterways	0.3	0.2	–	–
Pipelines	–	0.9	–	–
	72.8	85.1	100	100

[a]This figure expresses the total amount of internal transport for hire and reward, thus excluding transport on own account.

and can create patterns of urban and rural development which are difficult to service by a mode other than motor transport.

Rail transport, although twice as expensive as water transport, has the lowest fuel cost of any mode. The quality of service by rail is lower than by truck. Consequently, the rail mode most satisfactorily carries low value goods which are not subject to breakage and spoilage. The time in transit of low value goods is usually not important. From its operating characteristics, the mode is most suitable for high volume, large bulk, long haul traffic. Normally, the criterion for the construction of a new line is a traffic potential in excess of 1,000 tons/day, which is difficult to attain in a developing country except in conjunction with an extractive industry.

Water transport offers the lower level of service for general cargo shipment. The mode is slow, is subject to weather delays, and risk of spoilage and breakage is higher than by other modes. It is a prime transport mode in many developing countries where the principle economic commodities produced are low value staples and ores. Since the market value of these products is close to the value of production, profitable operation by the producers depends on cheap transportation. In the United States, revolutionary types of containerization (where barges are used as containers and lifted entirely for transportation by barge ships), promise to lower costs significantly while improving the quality of service of the mode by elimination of intermediate on- and off-loading. Costs are also declining with the extensive use of deeper draft barges for both inland waterways and transoceanic transport.

Air freight offers the highest quality of service of any mode. Costs (21 cents per ton mile), prevent the use of the mode except for very high value, low bulk items. In developing nations, costs of air transport can be reduced to approximately two-thirds of this figure by the use of vehicles which have been termed obsolete by western nations. With the introduction of the jumbo jet 747, the costs of air freight are likely to decline significantly. It would appear that above all other modes, the air transport share of the total market is most susceptible to changes in technology.

Pipeline transport requires high capital investment, and is limited in the number and type of commodities that can be carried. A significant advantage of the mode is the very low labor costs involved in operation, where few but highly skilled operators are required. High standards of maintenance are required to maintain flows. Under conditions of good maintenance deterioration of plant is minimal and replacement is more likely to be due to obsolescence. Both pipelines and air transport are similar in their inability to promote any form of development between terminal locations.

3–14. Transport Patterns Throughout the World. Transport patterns of the various nations on a worldwide basis have been divided into three main categories [11]. Transport types can be classified into one of the three following broad categories:

Type 1. Economically advanced countries such as Canada, Australia, the United States, and Western Europe.

Type 2. Densely populated countries such as India, Japan, Pakistan, Eastern Europe, and the Soviet Union.

Type 3. Less populated undeveloped countries.

Countries which fall in the first category show a tendency for increasing reliance on automobile and truck traffic for personal and freight movement. While the total volume of rail traffic increases, the rail share of the total transport market declines. Air transport on the other hand shows rapid increases both in passenger travel and freight movement. Containerization and multimodal freight traffic claims an increasing market share as new technology is developed to increase the efficiency of containerized movements. Countries that have reached this level of transport technology are relatively affluent [14]. However, within these countries, urban areas present the chief transport problems. Increased mobility continues to generate large numbers of personal movements within the urban areas. The new mobility has itself changed the form of modern industrialized cities, causing new environmental and social problems.

Transport systems of Type 2 show a high reliance on rail traffic even in urban areas. These systems can be subdivided into the *Soviet countries* and the *densely populated countries.* The level of rail reliance in the Soviet economies comes from both level of personal income and from political choice of level of public investment into the rail mode. Under current economic conditions in these economies, political planning excludes, for the moment, high levels of personal mobility and personal investment in motor transportation.

The heavy dependence on rail transport in the densely populated countries comes from a combination of both low personal income and economic feasibility. The very low per capita income of these nations precludes individual investment in autos and trucks to any significant degree. With low per capita car ownership, there is little political pressure for large public investment in roads. Road construction is not, therefore, a part of the local political process as in many western countries. Equally significant is the fact that in many densely populated nations, rail transport is unified under government ownership while motor transport under private hands presents a relatively disorganized front in the demand for public funds. High movements created by the dense populations permit the continued widespread use of rail for both passenger and freight traffic.

Countries of Type 3 rely principally on road transport, and have unsophisticated transport systems. Low densities of population and per capita incomes generate low traffic movements that are easily handled by a rudimentary road system which is usually provided for political and social reasons. Generally, funds for heavy capital investment in rail systems are unavailable and the light traffic volumes developed by underindustrialized nations cannot be economically handled by rail networks.

3–15. Transport Choice for Developing Regions. In dealing with the planning for underdeveloped and developing nations, the provision of transport is essential to economic development. Evaluation of systems for these regions requires special attention to determine whether the provision of transport alone will meet the goals. Very clearly transport is a *necessary* condition for development, but not a *sufficient* condition. For economic development to take place, the provision of transport must create a market for an otherwise unmarketable product. The provision of a road, for example, into a previously undeveloped area will not bring about desired development unless by the provision of transport some resource becomes economic in the general market. Nothing is gained, of course, if the cost of providing the transport is greater than the benefits accrued from the development it generates.

Underdeveloped economies offer significant differences from affluent nations in their transport needs. In areas with rudimentary economies, transport decisions relate chiefly to freight movements since personal car ownership is small. The area of choice, therefore, is related mainly to the selection of rail, road, or water transport linkages within the total system of demand [15]. Figure 3–9 indicates a model that has been suggested for the determination of transport needs in a developing region [16]. The economic model of the national economy is used to determine the supply and demand characteristics of each industry within the region. These total supply and demand figures are broken down into the seasonal supply and demand of goods by region. Each subcommodity is taken separately and the pattern of movements is established in terms of origins and destinations. Freight movements are next assigned to a mode (truck, rail or water, etc.) and a particular route on the assumed network. The summation of all movements over the network gives total system flows. Based on network flows and desired cost–performance relationships, it is possible to determine the level of investment in the individual modes.

Political and social conditions must also enter into the decision of choice of transport mode. Often minimum levels of road accessibility

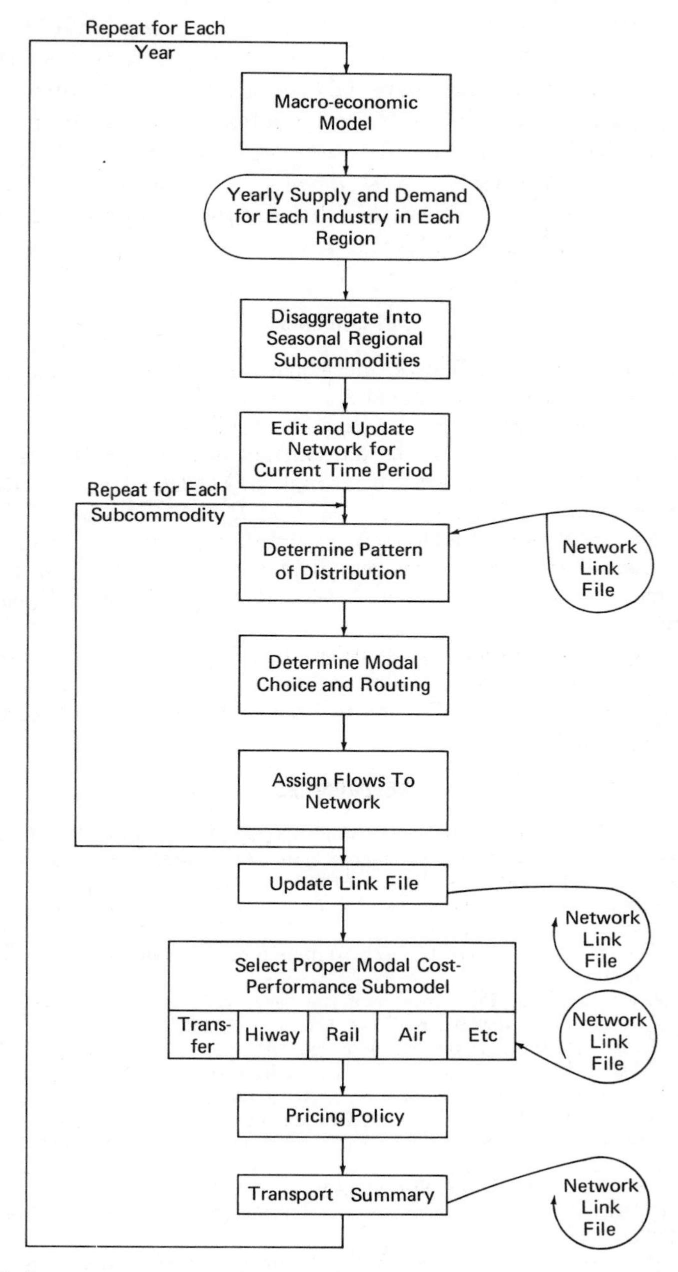

Fig. 3–9. A transport model for a developing region. (Source: *The Role of Transport in Developing Countries,* Paul O. Roberts, Transport Research Report, Discussion Paper No. 40, Harvard University.)

must be provided to all areas of a region. Even low grade roads may be able to handle the volumes of freight traffic moved in a developing region. Alternatively, the marginal costs involved in upgrading the minimally acceptable road system to a standard that can support anticipated traffic flows may be considerably less than the cost of constructing high capital cost railways. Most developing countries have opted for road oriented systems supplemented by waterways where natural, navigable waterways are already available.

PROBLEMS

1. Discuss the relationship between developments in transportation technology and the development of the United States.
2. Using the city in which you are living as an example, cite and describe the developmental impact of the introduction of a dual-mode system which will operate at average speeds of 30 mph in residential areas and 90 mph on the line-haul radial routes.
3. Some "anti-auto" advocates have suggested that automobiles should be completely banned from downtown central areas. Discuss the realism of such action and describe the impact that such action would have on developmental policies.
4. Discuss why most developing countries are choosing highway transportation as the basic method of providing access to backward areas, abandoning sections of rail systems. Why in the past did many colonial governments use rail systems?

REFERENCES

1. Chisholm, M., "Johann Heinrich Von Thunen," in Smith, R. T. H. *et al.* (editors), *Readings in Economic Geography: The Location of Economic Activity,* Rand McNally and Co., Chicago (1968).
2. Christaller, Walther, *Die Zentralen Orte Suddeutschelands,* Gustav Fischer, Jena (1933).
3. *Future Highways and Urban Growth,* Wilbur Smith and Associates, New Haven, Conn. (1961).
4. Burgess, Ernest W., "The Growth of the City," in Park, R. E. *et al.* (editors), *The City,* University of Chicago Press, Chicago (1925).
5. Hoyt, Homer, "City Growth and Mortgage Risk," *Insured Mortgage Portfolio,* Vol. I, Nos. 6–10, U.S. Federal Housing Administration, Government Printing Office, Washington, D.C., December 1936–April 1937.
6. Harris, Chauncey D., and Ullman, Edward L., "The Nature of Cities," *Annals of the American Academy of Political and Social Science,* CLXII, November, 1945.
7. *Chicago Area Transportation Study,* 1959.
8. Ornati, Oscar A., *Transportation Needs for the Poor,* Praeger Publishers, New York (1969).
9. *Canberra General Plan Concept,* A. M. Voorhees and Associates, Inc., January, 1967.
10. *Puget Sound Regional Transportation Study,* September, 1967.
11. Owen, Wilfred, *Strategy for Mobility,* The Brookings Institution, Washington, D.C., 1964.

12. *Introduction to Transport Planning,* Economic Commission for Asia and the Far East, United Nations, New York, 1967.
13. *Annual Bulletin of Transport Statistics for Europe,* United Nations, 1964.
14. GALBRAITH, JOHN KENNETH, *The New Industrial State,* Houghton Mifflin Company, Boston (1967).
15. SOBERMAN, R. L., *Transport Technology for Developing Regions,* M.I.T. Press, Cambridge, Mass., (1966).
16. ROBERTS, PAUL O., *The Role of Transport in Developing Countries: A Developmental Model;* Harvard Discussion Paper No. 40, Transport Research Program, Harvard University; Cambridge, Mass., 1966.

4

Operational Characteristics of Transport Modes

Although the civil engineer and planner have little professional involvement in the design of transport vehicles, the routeways and terminal facilities which must accommodate these vehicles and their traffic come within the domain of civil engineering design. The transport system is composed of a number of modes. Each mode has a variety of speeds and capacities depending on the type of vehicle being considered. From the viewpoint of understanding the impacts of vehicle demand on the geometric and structural design of the system a complete knowledge of operational characteristics and designs is desirable. This chapter attempts to provide a basic overview to some of the characteristics of the modes, while acknowledging that for actual design an in-depth study of system demands would be necessary.

HIGHWAY VEHICLES

4–1. Passenger Car Dimensions and Weights. The design of the passenger motor car reflects both secular changes in technology and changes in consumer tastes of styling. Since the automobile is personally owned, unlike most other forms of transportation, its design changes tend to be less related to technical improvements than other transportation vehicles. For passenger autos, after cost, styling is the most important factor in the consumer's decision to select any particular model. Passenger autos are, therefore, more luxurious and considerably less functional in design than mass transit vehicles or trucks.

Styling changes relate mostly, however, to somewhat superficial body and finish innovations. The operating characteristics of the vehicles change far less than the striking modifications in style might at first indicate. Over the long term, however, trends of changes in operating characteristics of the average highway vehicle become apparent. These are recognized by the transportation planner as placing changing demands upon the facilities which are to be provided. A natural transition is brought about whereby highways adapt over the long term to the secular trends in vehicular evolution. As highways continuously adapt to the vehicles using them, driver constraint is lessened and hazardous conditions are avoided. Certain physical limitations are placed upon vehicles using the highway system. These limitations, however, apply only to the largest vehicles on the system—trucks, while permitting a great deal of variation in individual passenger car models.

Figure 4–1 shows the variation of passenger auto vehicle length since 1925 [1]. Average vehicle lengths increased at a reasonably constant rate until World War II, after which, except for a brief period in the late fifties, they have remained relatively constant. Also shown are the average and range of values of vehicle height and width. The most significant long term trend of passenger automobiles has been a continuous decrease in vehicle height up to 1960. After this date vehicle height has remained relatively constant. Along with a decrease in overall vehicle height, the position of the driver's line of sight has also dropped over time. Lowering the driver's eye height has influenced changes in the geometric design requirements of the road designer. The trend of lower slung autos prompted the American Association of State Highway Officials in 1965 to change its passing sight distance criteria from distances based on a driver's eye height of 4.5 feet to 3.75 feet. Average vehicle weights can be seen to have remained fairly constant with time except for a moderate temporary increase during the latter part of the 1950's, as shown in Fig. 4–2.

4–2. Highway Acceleration Rates. The acceleration rates of the different vehicle types using roadways have a profound effect on both the design and operation of individual segments of highway. Two-lane highways can function well under low vehicular volumes where there is the ability to overtake slower moving vehicles. In multilane roads, the design of merging and weaving sections depends on the maneuverability and acceleration characteristics of the vehicular traffic. Figure 4–3 indicates the superior acceleration characteristics of passenger cars over commercial vehicles due to the much higher power to weight ratios of private automobiles. In the presence of even medium grades, the lower performance of commercial vehicles is accentuated. Trucks are found to decelerate to very low operating speeds over long uphill grades. Figure 4–4 shows the effect of grades of different length on trucks both from a standing start and from a previous operating speed of 47 mph. It can be seen that any grade appreciably greater than 2 per cent will result in low operating speeds even over a relatively short distance.

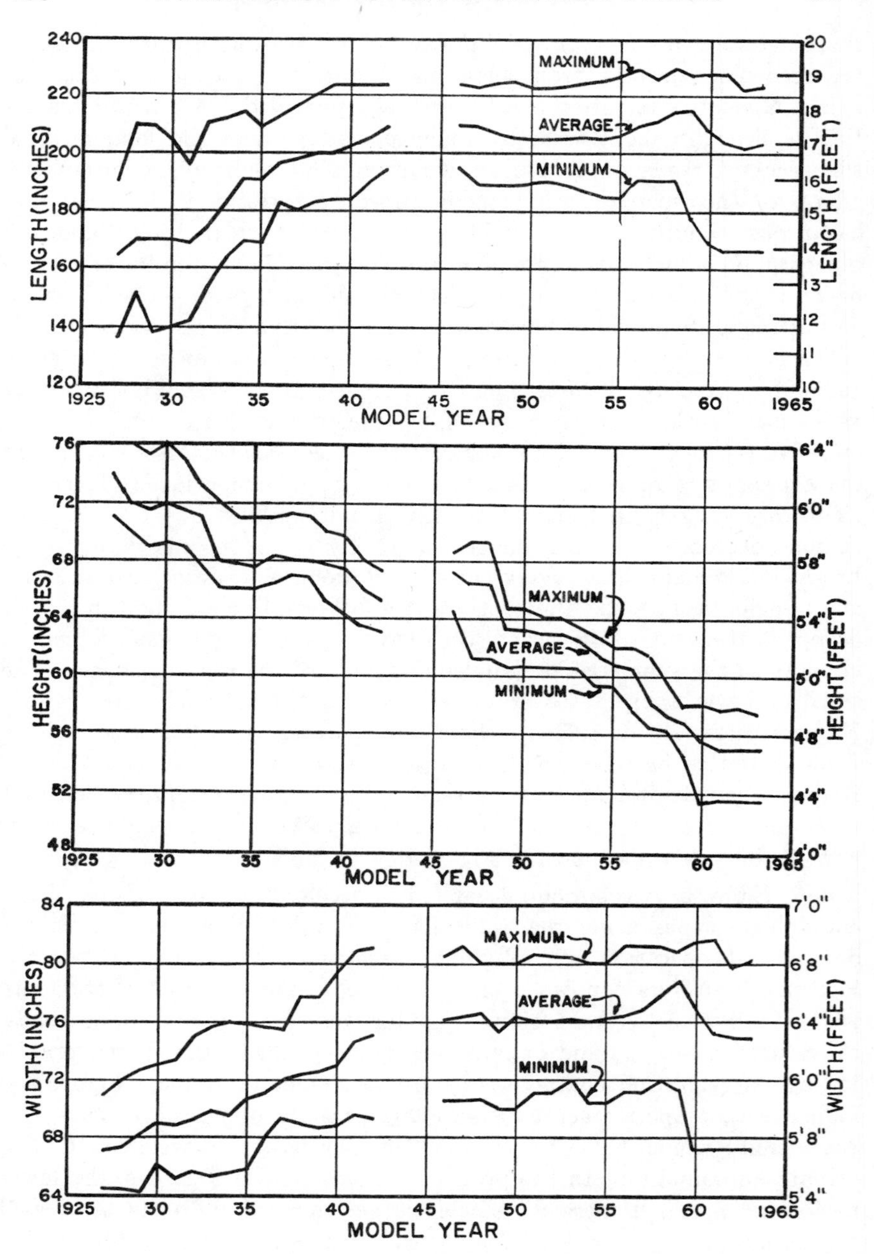

Fig. 4–1. Trends in passenger vehicle length, height, and width. (Source: *Traffic Engineering Handbook,* Institute of Traffic Engineers.)

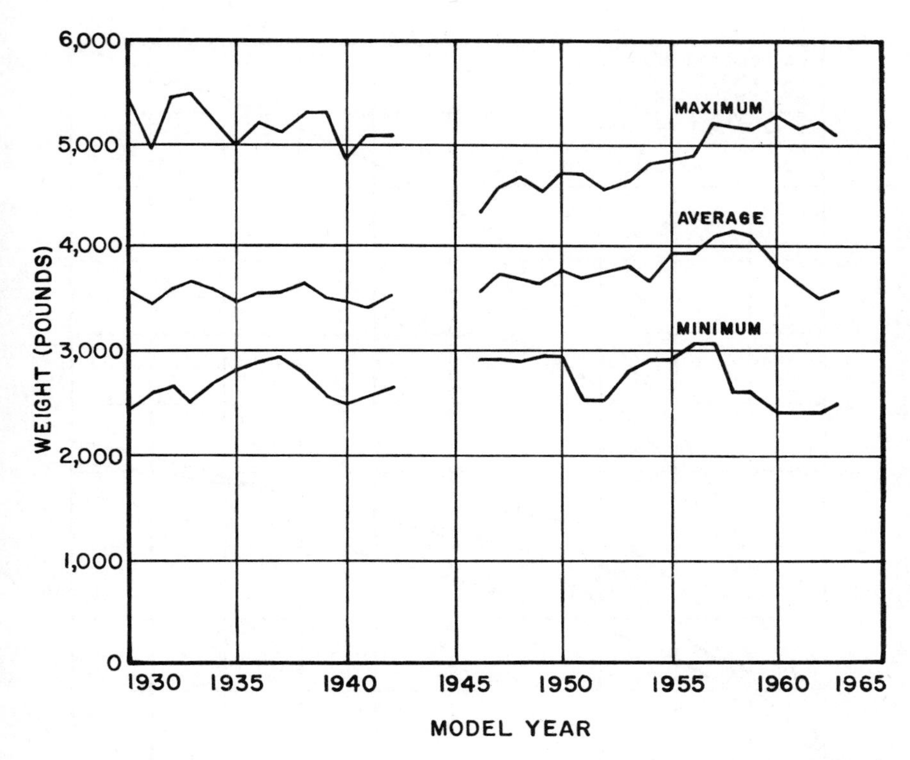

Fig. 4–2. Trends in passenger vehicle weight. (Source: *Traffic Engineering Handbook,* Institute of Traffic Engineers.)

Grades in excess of 4 per cent rapidly result in operation at crawl speeds and well below tolerable conditions. Where extensive lengths of uphill grades are unavoidable, the presence of low performance trucks requires additional climbing lanes to allow higher performance vehicles to overtake safely.

4–3. Highway Vehicle Deceleration Rates. The large range of highway vehicle types exhibits large variations in their abilities to decelerate. Empirical studies carried out indicate that, in general, as the size of a vehicle increases from the ordinary passenger car to vehicular combinations such as trucks with full trailers with multiple axles, the deceleration capability decreases significantly. Figure 4–5 shows that while 94 per cent of passenger vehicles can achieve a deceleration of 25 feet/second^2 or better, only 13 per cent of the class of trucks and full trailers were able to perform at the same standard. The large range of deceleration characteristics in a heterogeneous traffic stream leads to hazardous condi-

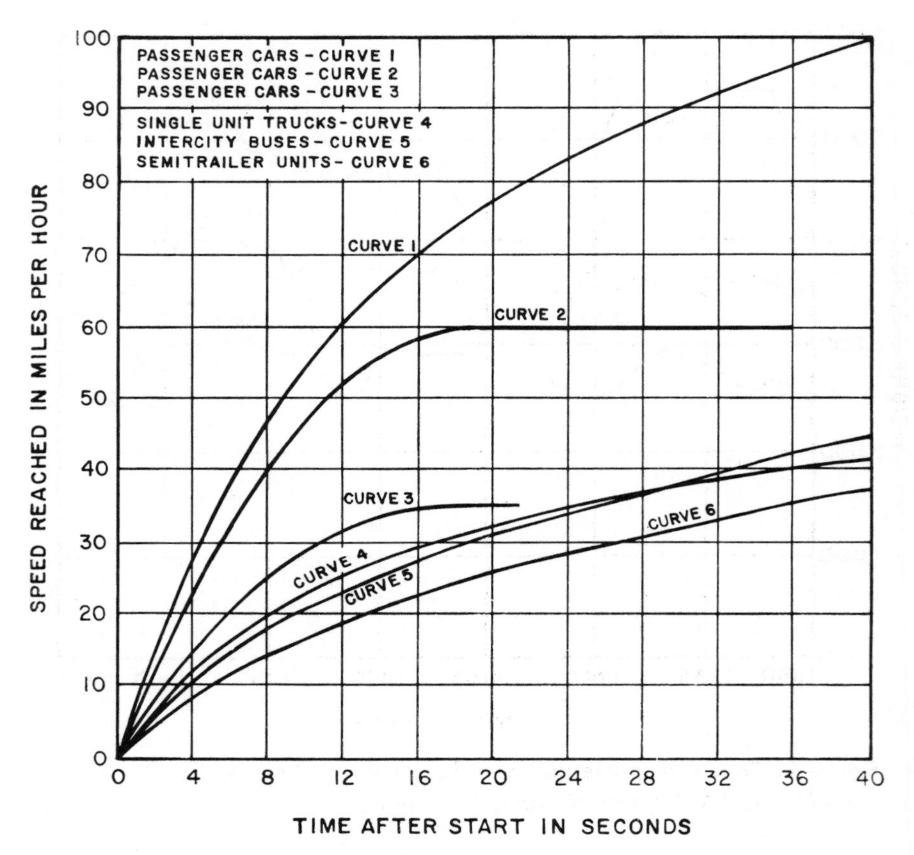

Fig. 4–3. Speed-time relationship during normal acceleration from a standing start. (Source: *Traffic Engineering Handbook,* Institute of Traffic Engineers.)

tions at high volumes and high speed. Under these conditions, it is not infrequent that the headway between vehicles falls below the amount required to compensate for the different deceleration characteristics, and rear-end collisions result.

4–4. Average and Maximum Automobile Speeds. The maximum speeds of automobiles increased continuously from their introduction up to the late 1950's when a peak of maximum speed capability was reached. Figure 4–6 shows the trends of maximum, average, and minimum top speeds of standard passenger vehicles since World War II. The design trends indicated show that the continued tendency toward more powerful higher speed cars has probably peaked out in the last few

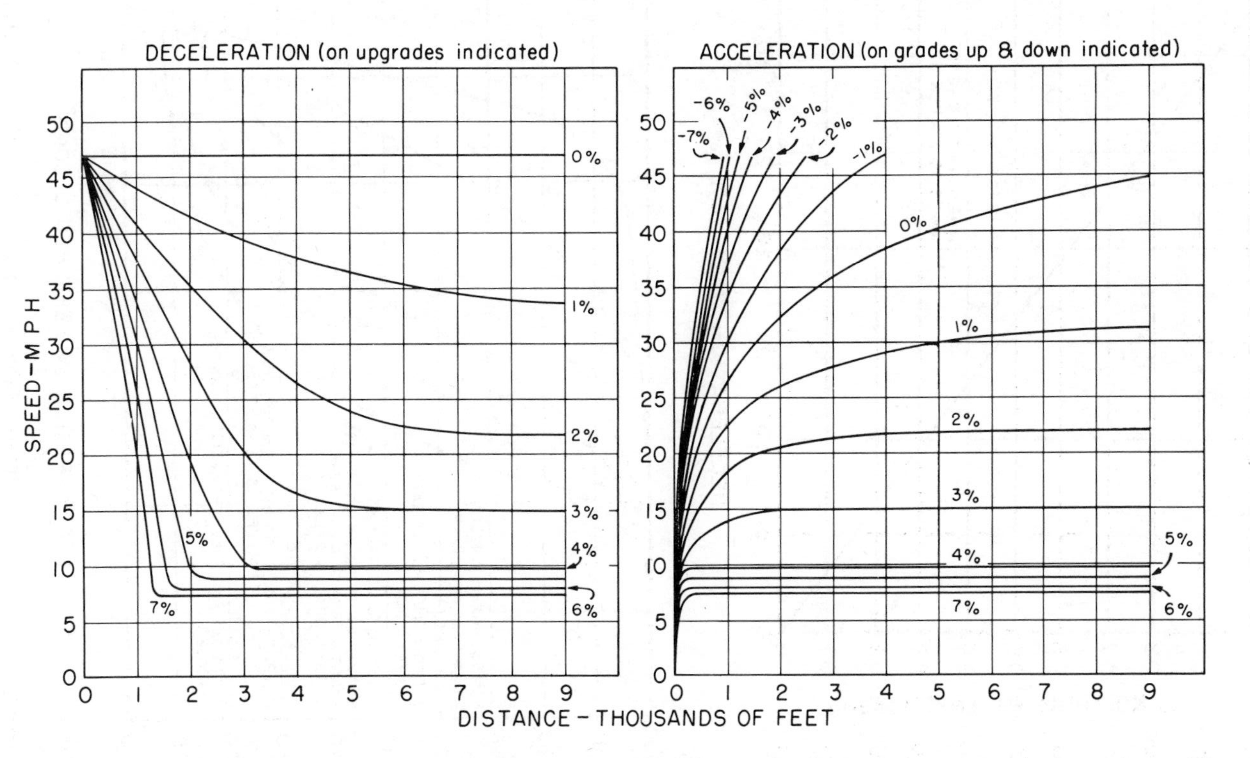

Fig. 4–4. Speed-distance curves from road test of a typical heavy truck operating on various grades. (Source: AASHO—*A Policy on Geometric Design of Rural Highways*, 1965.)

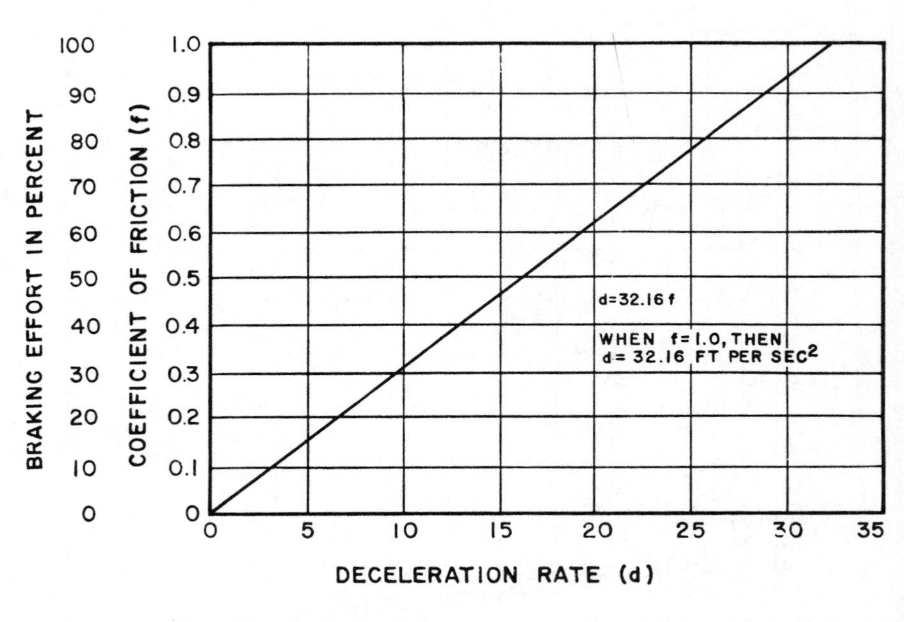

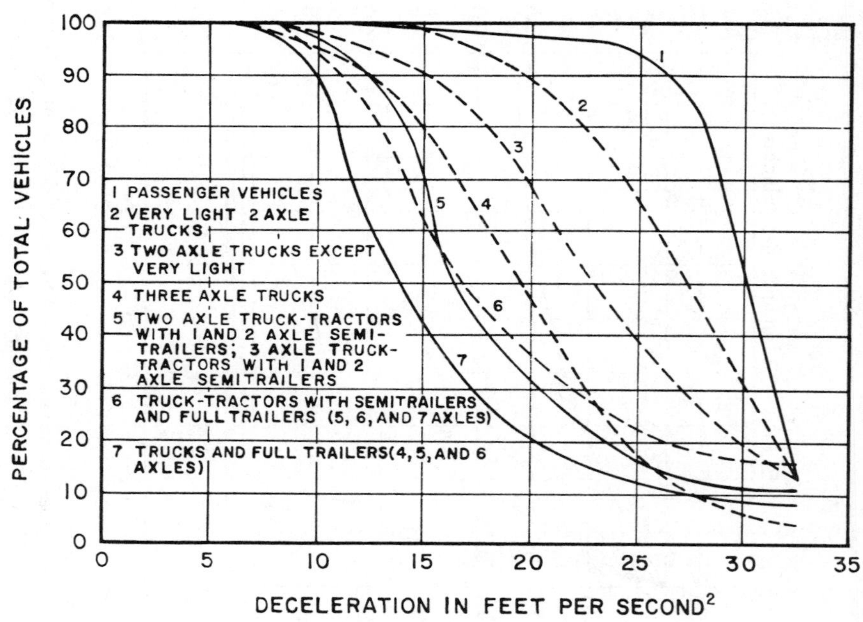

Fig. 4–5. Percentage of highway vehicles capable of a given or greater deceleration. (Source: *Traffic Engineering Handbook,* Institute of Traffic Engineers.)

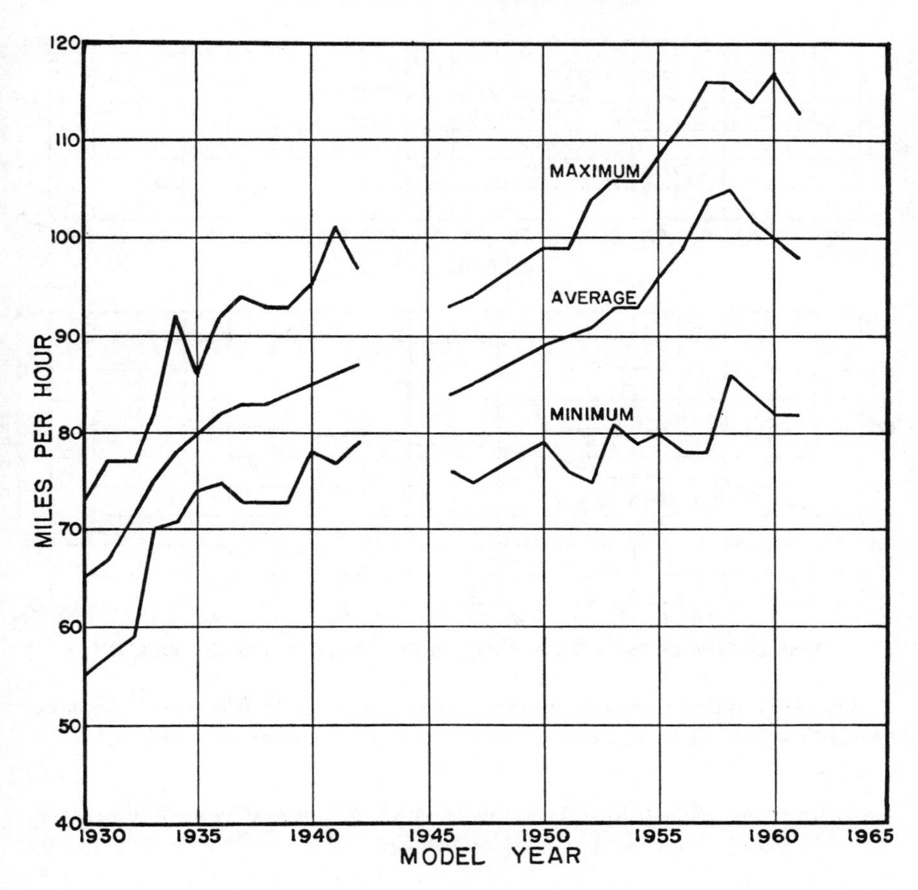

Fig. 4–6. Maximum speeds of standard passenger vehicles. (Source: *Traffic Engineering Handbook,* Institute of Traffic Engineers.)

years, and that the current top speed ranges may continue for some time. Obviously, the available maximum speed capabilities are well in excess of legal speed limits in the individual states. Figure 4–7 shows long term trends of average speed on rural highways since 1945. It is apparent that the combination of increasingly powerful vehicles and higher speed limits in rural areas has brought about long term constant increases in average speeds for all types of vehicles. This tendency of speed increase will undoubtedly continue while rural speed limits continue to rise but can be expected to level off in the near future.

4–5. Legal Maxima and AASHO Standards. The American Association of State Highway Officials has set standards for vehicle dimensions and

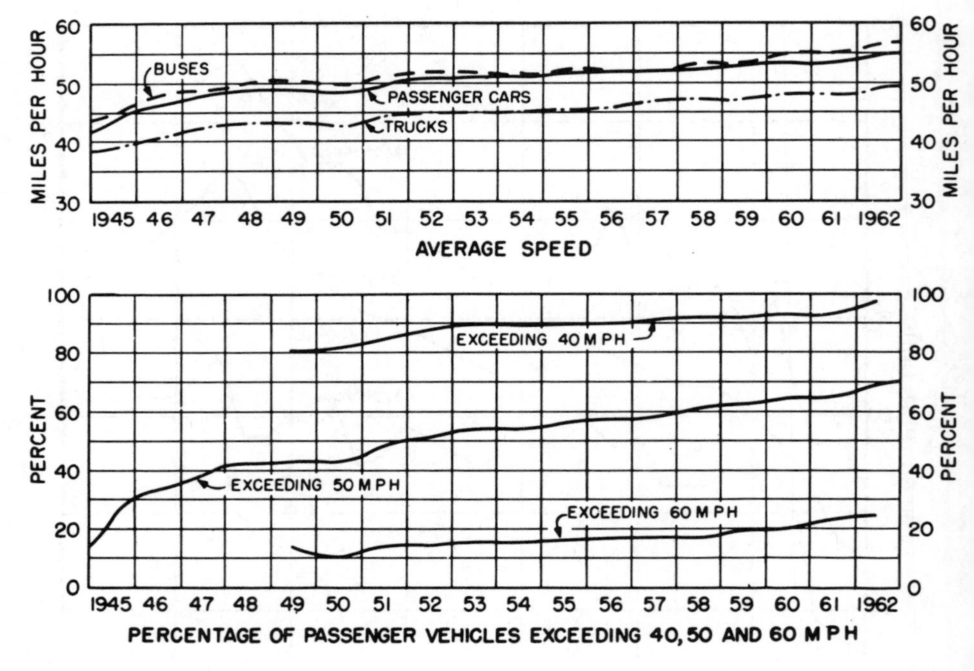

Fig. 4–7. Speed trends on main rural highways by vehicle type. (Source: AASHO—*A Policy on Geometric Design of Rural Highways,* 1965.)

weights to permit conformity between road design and vehicle characteristics. These proposed national standards encourage uniformity of design throughout the nation. Uniformity between states is essential to the unimpeded movement of interstate traffic. Geometric standards encountered in one state can, therefore, be expected in other states, and perhaps even more critically, the structural sufficiency of pavements and structures can be assured on a nationwide basis. In many cases, however, state standards vary considerably from AASHO standards, with the result that vehicles that can operate legally in some states are not permitted to use the highways of other states. Table 4–1 shows AASHO policy standards together with the number of states having higher or lower standards.

4–6. Design Vehicles. Two general classes of vehicle use the highways, namely passenger cars and trucks. Under the designation of passenger vehicles, it is normal to refer to all light vehicles including light panels, pickups, and delivery trucks. Trucks are taken to include the range of heavy vehicles including single unit trucks, trucks with semitrailer units, full trailer units and combinations, and also buses.

Highways are designed to accommodate the types of vehicles that will use them. Prior to undertaking the geometric design of any facility, the proportions of travel that can be anticipated by any particular vehicle class are estimated and projected from available road use data. As a result, the engineer can reasonably predict the type of usage any facility will receive prior to design. During the design phase, the facility is proportioned and arranged to meet the needs of the class of vehicle which it will seek to accommodate. To facilitate the design process, the American Association of State Highway Officials has determined the dimensions and geometric characteristics of certain "design vehicles."

A design vehicle is defined as a selected motor vehicle whose weight, dimensions, and operating characteristics are used to establish highway design controls to accommodate vehicles of a designated type [2]. To cover the large range of motor vehicles using the highways, AASHO has designated four design vehicles. These vehicles represent the passenger vehicle (P), the single unit truck and bus (SU), a vehicle representative of the majority of medium to large truck tractor-semitrailer combinations (WB-40), and a vehicle nearly inclusive of all truck tractor-semitrailers in use (WB-50). The dimensions of the various design vehicles are shown in Table 4–2.

The geometrics of the passenger design vehicle (P) and its turning characteristics were determined from an analysis of passenger vehicles in the 1962 production year. Only one model of car failed to fall within the 19 feet of overall length, wheel base of 11 feet, and rear overhang of 5 feet of the design vehicle. Three per cent of all vehicles had a front overhang in excess of 3 feet. The dimensions of the design vehicle, therefore, include the great majority of all vehicles within the passenger vehicle class. Figure 4–8 shows the turning characteristics of the largest and smallest design vehicles.

Based on findings of a nationwide survey in 1963, it was found that 98 per cent of all single unit trucks, excluding panels and pickups, fell within the 20 feet wheel base dimension of the SU design vehicle. This dimension was greater than the wheel base of 69 per cent of all buses manufactured during the period 1961–63. The overall length and turning radii were also found to cover all but an insignificant number of vehicles within this class. The width of 8.5 feet reflects the allowed legal widths rather than the actual width of vehicles in this class.

In a similar manner, both the WB-40 and WB-50 design vehicle dimensions and turning radii reflect the maximum dimensions which can be attributed to the overwhelming majority of vehicles within their respective class. In developing the dimensions of the design vehicles, trends of motor vehicle size were analyzed to give some indication of the utility of present designs to future traffic. Three significant trends

TABLE 4-1

AASHO Dimensions and Weight Standards for Motor Vehicles

	Width (*in.*)	Height (*ft*)-(*in.*)	Length (*ft*): Single-Unit: Truck	Length (*ft*): Single-Unit: Bus	Length (*ft*): Semi-Trailer or Trailer	Length (*ft*): Truck Tractor Semi-Trailer	Length (*ft*): Other Combination	Nu… Se Tra…
AASHO Policy	102	13-6	40	40	40	55	65	
Number of States:								
Higher	1	4	7	10	35	14	1	
Same	3	37	12	37	7	20	9	4
Lower	48	11	33	5	10	18	42	

Source: American Association of State Highway Officials.

in dimensions were found despite the fact that the P and SU design vehicles were essentially the same as the design vehicles of a decade earlier:

1. Passenger vehicles have become approximately six inches wider in the last twenty years.
2. There is a trend of increasing length of truck combinations requiring increased width of pavement on curves.
3. States are following an upward trend in setting height limitations.

AIR VEHICLES

4–7. Need for Aircraft Characteristics. Although the civil engineer and planner is not engaged in the design of aircraft, the characteristics of aircraft have very significant influences upon the design of airports. Satisfactory design of these facilities depends not only on a knowledge of carrier aircraft characteristics, but also on an understanding of the trends in these characteristics. The rapidly changing technology of aviation has made many airports technically obsolete long before the end of their economic lives. While aircraft weight, and specifically wheel loadings, largely determine whether an aircraft can use a facility, the size, capacity, and range also significantly affect the details of airport planning [3].

TABLE 4-1

(Continued)

ᶠ Towed Units		Axle Load—Pounds				
		Single		Tandem		Operating Tire Inflation Pressure (*psi*)
Full ᵀrailer	Semi-Trailer and Full Trailer	Statutory Limit	Including Statutory Enforcement Tolerance	Statutory Limit	Including Statutory Enforcement Tolerance	
1	2	20,000	...	32,000	...	95
7	6	16	...	26	...	52
42	18	2	...	25	...	0
3	28	34	...	1	...	0

Weight affects the design of pavement thicknesses for runways, taxiways, aprons, and hardstands.

Size, as demonstrated in fuselage length, wingspan, deck height, and height of empennage, affects the design of aprons and parking areas, runway and taxiway widths, turning radii, and hangars and maintenance sheds.

Capacity in terms of passenger and cargo capacity, together with fuel requirements, determines the size and capacity of ground services which must be supplied to minimize the turn around time of aircraft. Facilities affected by aircraft capacity are terminal size, baggage handling facilities, departure lounges and gate positions, off-loading facilities for cargo and freight, and fuel storage.

Range has an impact on frequency of operations, mix of type, and size of aircraft to be serviced by the airport with consequent effects on runway and gate capacities.

Tables 4–3 to 4–6 show selected characteristics of various classes of aircraft currently in use or in the stages of immediate development.

4–8. Weight Trends of Air Carrier Vehicles. The introduction of the jet engine permitted a continuing growth in gross weights of transport airplanes which otherwise would have been limited by a power ceiling on conventional internal combustion engines. Figure 4–9 indicates past and future possible trends in gross weight over the years 1947–87. Aircraft manufacturers anticipate that gross weights in the region of 1.5

TABLE 4–2

Design Vehicle Dimensions Recommended by AASHO Policy

DESIGN VEHICLE DIMENSIONS							
Design Vehicle		**Dimensions in Feet**					
Type	**Symbol**	**Wheel-base**	**Front Overhang**	**Rear Overhang**	**Overall Length**	**Overall Width**	**Heigh**
Passenger car	P	11	3	5	19	7	–
Single unit truck	SU	20	4	6	30	8.5	13.5
Semitrailer combination, intermediate	WB-40	13+27 =40	4	6	50	8.5	13.5
Semitrailer combination, large	WB-50	20+30 =50	3	2	55	8.5	13.5

Source: *A Policy on Geometric Design of Rural Highways 1965,* Table II-5, p. 86.

million pounds are possible by 1987 or even prior to this time if market demand is sufficient to warrant the production of such large vehicles. Although gross weights of the aircraft are continuing to rise, there is no similar increase in required thicknesses of airfield pavements. By means of elaborate landing gears, pavement thicknesses required to support the Boeing 747 are less than those for the 707, in spite of the fact that the jumbo jet is in excess of twice the weight of its predecessor.

4–9. Fuselage Lengths. Figure 4–10 shows trends in fuselage lengths since 1945. After the introduction of jet transports, aircraft fuselage lengths have grown at a steady rate. The most significant rate of increase has been associated with long range intercontinental transports with a quantum jump taking place in 1969 when the Boeing 747 was introduced. The first supersonic transport to go into service, the Concorde, will counter the tendency of increasing fuselage length, but if designs proposed by U.S. manufacturers reach development, the lengths will continue current trends. Subsonic planes will most likely soon reach a ceiling in body length as planes with multiple decks are introduced.

4–10. Wing Span. Wing span of transport aircraft has increased only slightly compared to increases in gross weight and carrying capacity. This can be attributable to more efficient design of wings, and greatly augmented power systems. Even with the relatively small increases, aprons have needed considerable expansion to handle the turning movements and parked aircraft. In addition, runway shoulders have experienced unexpected erosion due to jet blast from the outermost engines of the larger airplanes. Figure 4–11 shows wing span trends for long and medium range transports from 1945 to 1970.

4–11. Runway Length Requirements. Until the late 1950's, runways that had been designed for conventional piston aircraft were up to about 8,000 feet long. With the introduction of the large jet transports such as the Boeing 707 and the Douglas DC-8 it became necessary to design and provide runways of over 12,000 feet. Longer runways were required by the new jet aircraft because of low thrust characteristics at low speeds and the introduction of swept wing aircraft with high wing loadings. The trend of increasing runway lengths, however, was ended by the introduction of turbofan engines, as shown in Fig. 4–12. With the provision of a fan either before or after the turbojet, there is a significant increase in thrust. As a result, the aircraft has a much improved climb-out capability with a subsequent decrease in runway requirement. An added advantage of turbofan engines is the decrease of perceived noise due to take-off operation on account of the step climb out. While more airports can be expected to construct longer runways to handle existing aircraft, no continued trend of increased runway lengths is expected. Supersonic aircraft will be able to operate on existing runways at major air centers.

4–12. Passenger Capacity. Figure 4–13 shows the trends of capacity for long and medium range subsonic aircraft as well as estimated capacity trends for supersonic vehicles. Air vehicle technologists predict continued increases in the capacity of air transports at a rate of 10 to 12 per cent per year. With such rates of increase, an 800-passenger aircraft is possible by the year 1980 [4].

The steady increase in passenger capacity has led to significant problems in the air terminals. Although the use of high capacity aircraft tends to ease air traffic control problems, the air terminals on the land side have found themselves inadequately designed to handle the jumbo jets that first went into service early in 1970. Departure lounges, baggage handling, access and egress procedures have not kept pace with the rapidly changing capacities of the air vehicles. The economic life of airport terminals greatly exceeds that of the air vehicles. It is, therefore, likely that air terminals currently under construction will be required to handle very high capacity transports in the early 1980's. It is apparent that severe congestion problems are likely to occur in the air terminals of the future.

The drive to higher capacity aircraft is motivated by the ability of such vehicles to reduce operating costs and fares while increasing

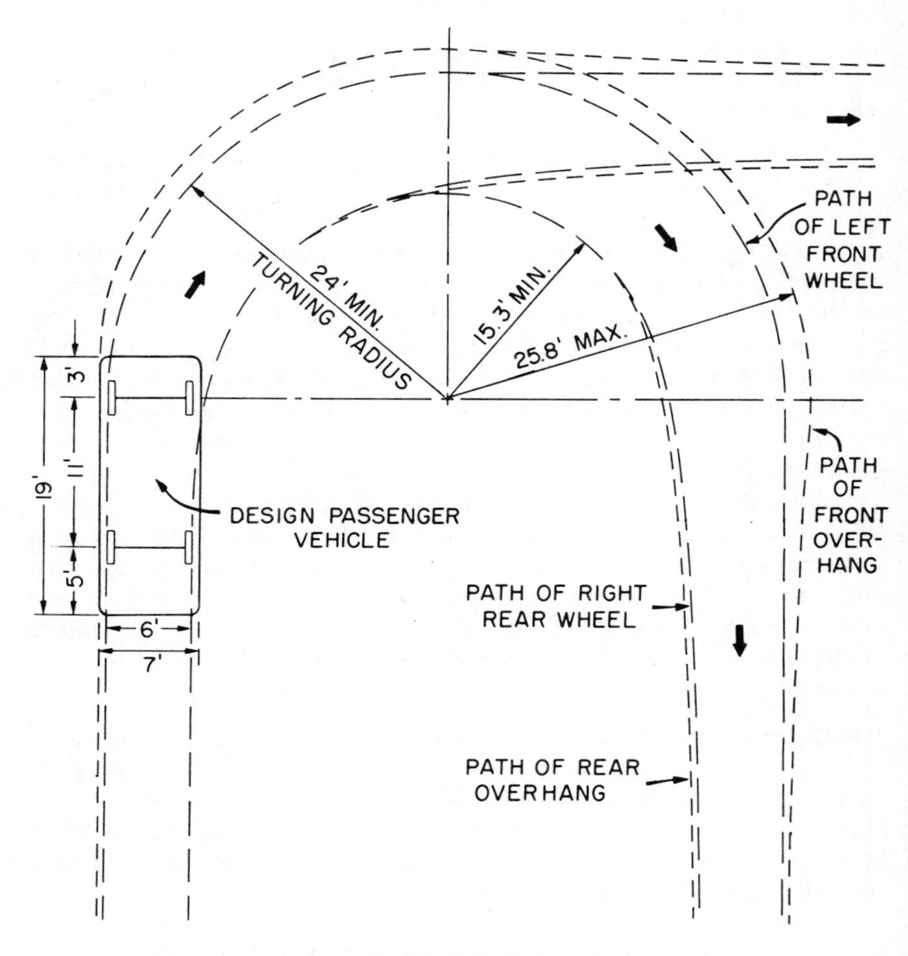

Fig. 4–8. Turning characteristics and dimensions of two design vehicles. 1965.)

capacity under constant load factors. This trend is expected with virtual certainty to continue.

4–13. Cruising Speed and Cruising Altitude. Since the beginning of commercial aviation there has been a long term trend toward higher cruising speeds for air transports, as is shown by Fig. 4–14. The first significant step in the general trend took place with the introduction of the large monoplanes in the early 1930's at the time of the DC-3. Speeds increased from an upper limit of approximately 190 miles per

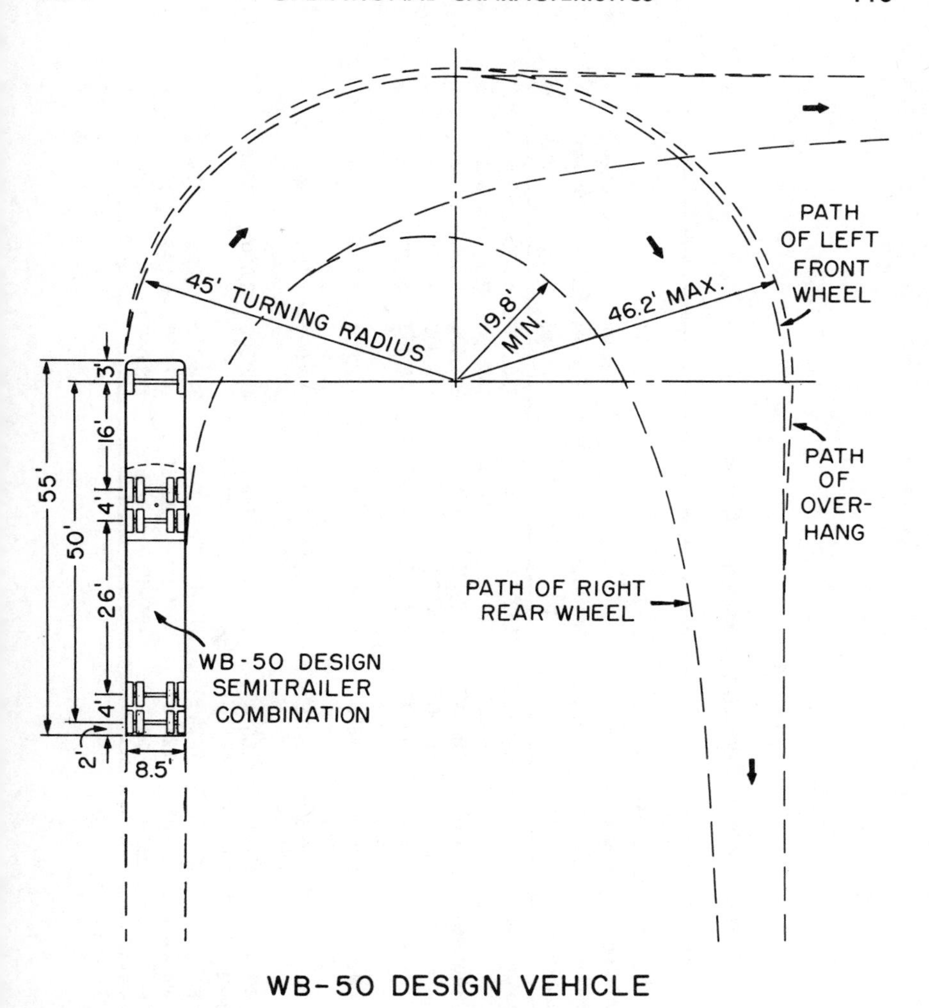

WB-50 DESIGN VEHICLE

(Source: AASHO—*A Policy on Geometric Design of Rural Highways,*

hour in 1934 to 350 miles per hour with increasingly advanced piston powered aircraft. With the introduction in 1952 of the Comet I, the first commercial jet transport, upper limits of speed increased to 500 miles per hour overnight. Improvements in jet transports have raised cruising speeds to a point just below the speed of sound. The Boeing 707, 727, and 747 all have cruising speeds between 600 and 630 miles per hour. With the introduction of the Concorde in 1972, the upper limit of speed will take another significant step to 1,450 miles per hour.

TABLE 4–3

Characteristics of Selected U.S. Transport Aircraft

Mfr.	Model	Name or Series Designation	No. of Passengers and/or (Cargo Capacity)	Wingspan (ft.)	Max. Lgth. (ft.)	Max. Ht. (ft.)	Gross or Take-off Wt. (lb.)	Range (mi.)	Cruising Speed (m.p.h.)
Boeing	707-320C	Intercontinental	202	145′-9″	152′-11″	42′-5″	336,000	4,000	600
Boeing	727-100		131	108′-0″	133′-2″	34′-0″	161,000	2,500	600
Boeing	2707-300	SST Prototype	350	180′-4″	306′-0″	46′-3″	675,000	4,000	1,800
Boeing	737-200		125	93′-0″	100′-0″	37′-0″	114,000	2,024	575
Boeing	747		490	195′-8″	231′-4″	63′-5″	710,000	5,750	625
McDonnell Douglas	DC-8	Super 62	189 (1615)	148′-5″	157′-5″	42′-5″	338,000	8,500	600
McDonnell Douglas	DC-9	Series 30	115 (895)	93′-5″	119′-3½″	27′-6″	98,000	1,725	565
McDonnell Douglas	DC-10	Series 10	270-345 (3045)	155′-4″	181′-5″	58′-1″	413,000	3,670	600+
McDonnell Douglas	DC-8	Jet Trader- (Cargo)	24-114 (8810)	142′-5″	150′-6″	42′-4″	315,000	7,090	579
Lockheed	749A	Constellation	46-54	123′-0″	97′-2″	22′-5″	107,000	5,100	328
Lockheed	188A	Electra	66-98	99′-0″	104′-6″	32′-9½″	113,000	3,400	406

TABLE 4–4

Characteristics of Selected North American Short Take-off and Landing (STOL) Aircraft

Mfr.	Model	Passenger Capacity	Max. Payload	Wingspan (ft.)	Length (ft.)	Height (ft.)	Max. T.O. Weight (lb.)	Cruise Speed (m.p.h.)	Landing Speed (Kt)	STOL Takeoff Distance to 50 ft. at Best Speed	STOL Landing Distance from 50 ft. at Best Speed	Range (mi.)
DeHavilland of Canada	DHC-5 Buffalo	53	18,000	96′-0″	77′-4″	28′-7″	49,000	290	70	1,265	1,170	1,900
DeHavilland of Canada	DHC-6 Twin Otter	20	5,300	65′-0″	51′-9″	18′-7″	12,500	204	63	1,200	1,050	118
DeHavilland of Canada	DHC-4 Caribou	30	8,750	95′-7½″	72′-7″	31′-9″	28,500	170	65	1,185	1,235	236
Fairchild Hiller	Heliporter	7	3,502	49′-10″	35′-9″	10′-6″	4,850	140	–	560	560	545
Helio Aircraft Div. of Gen. Aircraft Corp.	Helio Super Courier H295	5	–	39′-0″	31′-0″	8′-10″	3,400	150	–	635	515	615

TABLE 4–5

Characteristics of Selected U.S. General Aviation Aircraft

Mfr.	Designation	No. of Seats	Overall Span (ft.)	Overall Lgth. (ft.)	Max. Ht. (ft.)	Normal Gross Wt. (lb.)	Cruising Speed (m.p.h.)	Range (mi.)
Beech Aircraft Corp.	Musketeer Super R	4-6	32′-9″	25′-1″	8′-3″	2,750	162	880
Beech Aircraft Corp.	Bonanza	4-6	33′-5½″	26′-4½″	6′-6½″	3,400	203	1,111
Beech Aircraft Corp.	Queen Air 70	7-11	50′-4″	35′-6″	14′-2½″	8,200	214	1,660
Cessna Aircraft Co.	150 Commuter	2	32′-8½″	23′-9″	8′-7½″	1,600	93	725
Cessna Aircraft Co.	Skylane	6	35′-10″	28′-0½″	8′-10½″	2,950	160	1,160
Cessna Aircraft Co.	Centurion	4-6	36′-9″	28′-3″	9′-8″	3,800	187	1,250
Cessna Aircraft Co.	402B	9	39′-10¼″	35′-10″	11′-8″	6,300	218	1,186
Piper Aircraft Corp.	Super Cub	2	35′-3½″	22′-6″	6′-8½″	1,750	115	460
Piper Aircraft Corp.	Cherokee 180F	4	30′-0″	23′-6″	7′-3½″	2,400	143	725
Piper Aircraft Corp.	Twin Com. C.	4-6	36′-0″	28′-9½″	8′-2-7/8″	3,200	185	1,130
Piper Aircraft Corp.	Navajo	6-8	40′-8″	32′-7½″	13′-0″	6,500	247	1,150

TABLE 4-6

Characteristics of Selected Turbine Powered General Aviation Aircraft

Mfr.	Designation	Passengers	Wingspan (ft.)	Length (ft.)	Height 3-pt. (ft.)	Max. Gross Wt. (lb.)	Normal Cruise (m.p.h.)
Boeing Co.	737 Business Jet	25	93′-0″	94′-0″	37′-0″	97,000	515
Gates Lear Jet Corp.	24B	6	35′-7″	43′-3″	12′-3″	13,500	507
General Aviation Div. No. Amer. Rockwell	Sabreliner-60	4-10	44′-4″	47′-0″	16′-0″	20,000	500
Beech Aircraft Co.	Beechcraft 99 Executive	17	45′-10½″	44′-7″	14′-4¼″	10,400	254

GROSS WEIGHT GROWTH

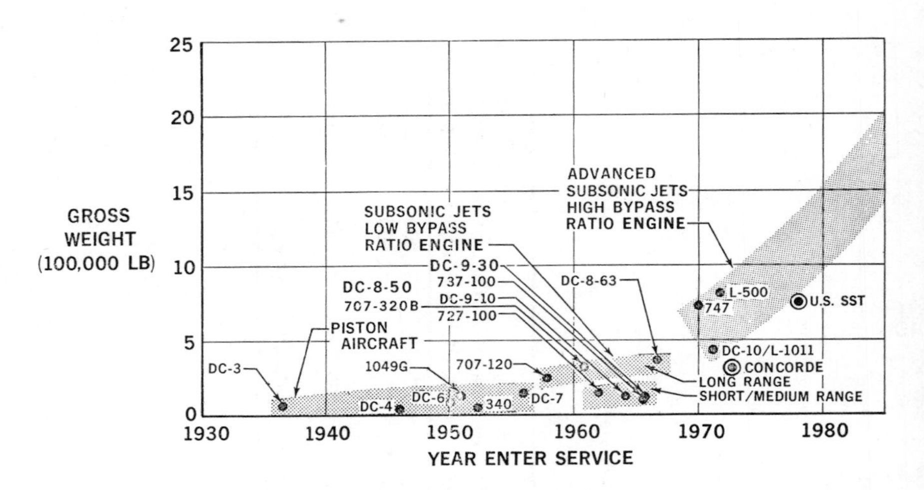

Fig. 4–9. Airplane weight trends. (Source: *Transport Aircraft, Characteristics, Trends, and Growth Projections,* Transport Aircraft Council, Aerospace Industries Association of America, Inc., March, 1969.)

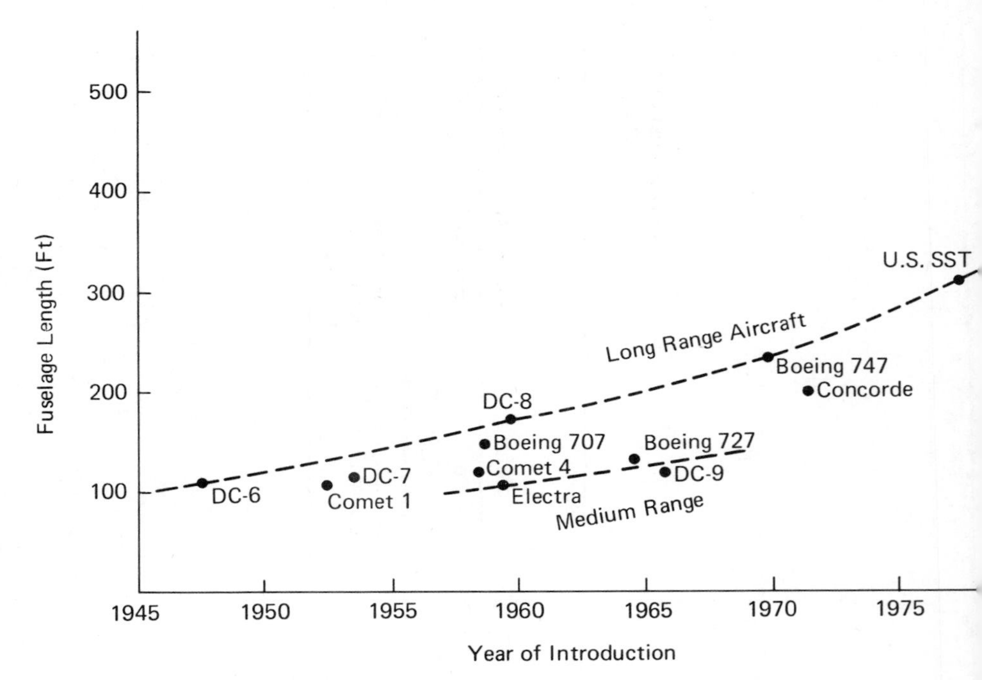

Fig. 4–10. Trends in fuselage length, 1945–1977.

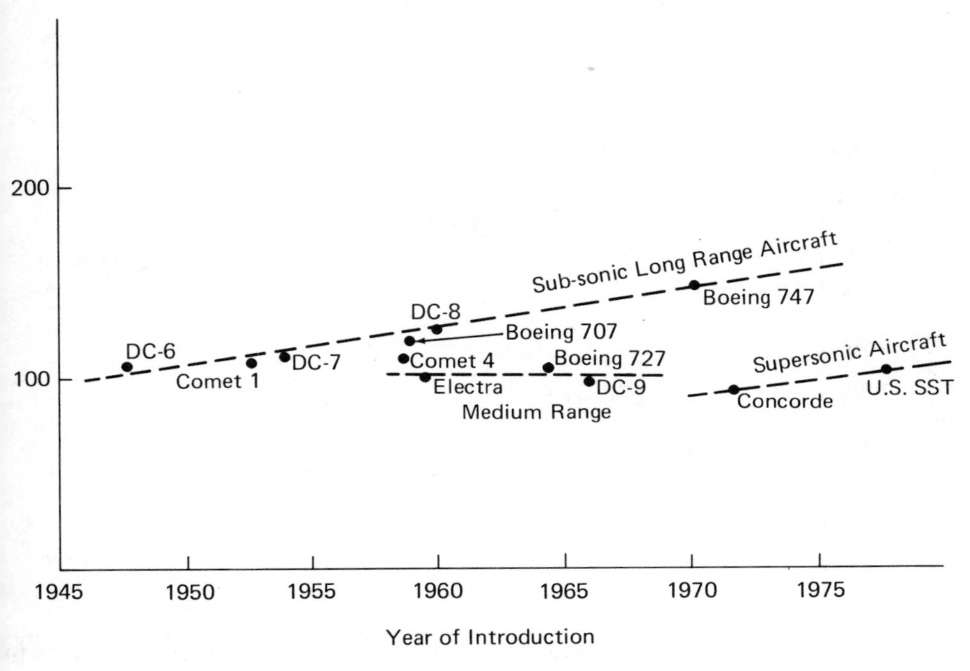

Fig. 4–11. Trends in wingspan, 1945–1977.

SST designs proposed by U.S. manufacturers indicate another stepped increase to a design cruising velocity of 1800 miles per hour. In the distant future, there is discussion of the hypersonic transport with speeds in excess of 3,000 miles per hour.

In spite of past long term trends the logic of further increasing cruising speed is open to serious question. Increased speed gives steadily decreasing marginal returns in savings of travel time. On domestic flights supersonic flight is uneconomic with the exception of transcontinental runs. Even on these very long trips the actual time saved would be relatively small. The use of supersonic transports over the continental United States is likely to generate considerable unfavorable reaction similar to that which has caused the banning of such flights in the United Kingdom and other Western European nations. Development of high speed aircraft hinges on the creation of an adequate intercontinental demand for very fast aircraft. It would appear that this scale of demand is unlikely to be generated in the near future.

Cruising altitudes have followed similar trends to increases in speeds (see Fig. 4–15). With the introduction of jet aircraft, cruising altitudes in excess of 30,000 feet became required for maximum efficiency of the aircraft. A similar quantum increase in cruising altitudes will be ob-

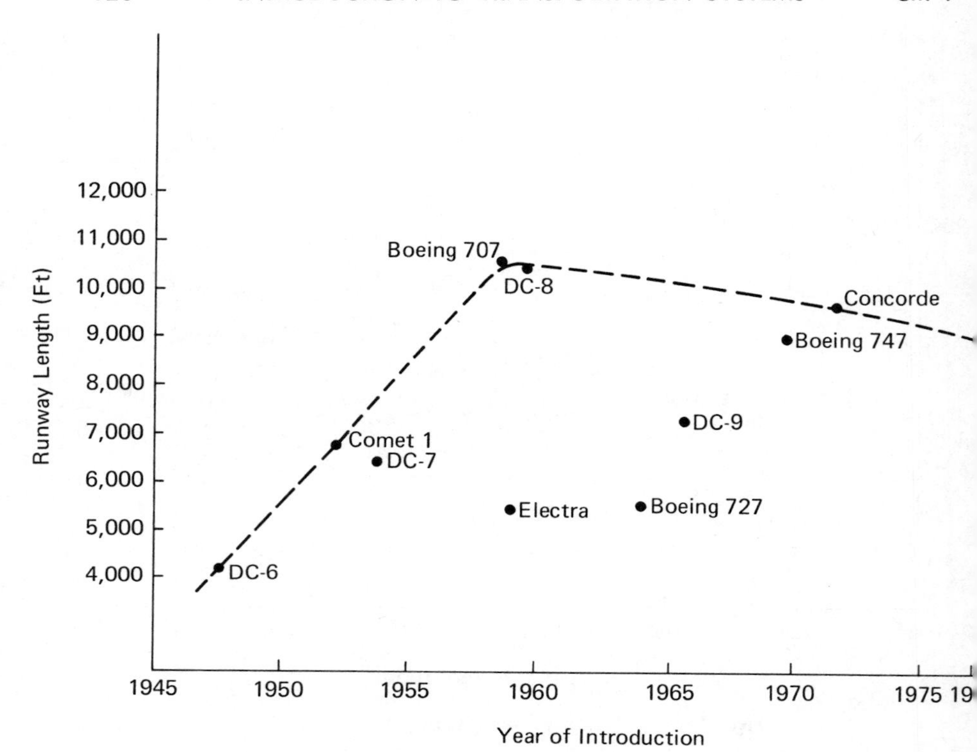

Fig. 4–12. Runway length requirements of various air carrier vehicles.

served with the introduction of supersonic craft in 1972. These aircraft will operate in the upper fringes of the atmosphere.

4–14. Future Trends in Air Transport Characteristics. The long term trends in the characteristics of air transports are not clear. Figure 4–16 shows the effect of increases in two dominant variables which currently are affecting the development of air transports—capacity and speed. The graph shows the relationship between passenger miles per hour and year of introduction for various selected aircraft. The upper limit of the relationship defines the trend for long distance transports; it would appear that this trend could continue. Both factors, however, seem to be reaching limiting values. Increased speeds appear to have marginal economies. Increased capacities offer the possibility of major air disasters with a shocking loss of life. Further changes in characteristics are likely to occur within the zone of change. Future improvements in technology are likely to concentrate more on decreased costs of operation and increased passenger comfort than on increased speeds or capacities in excess of 800 persons.

AIRCRAFT CAPACITY GROWTH TREND

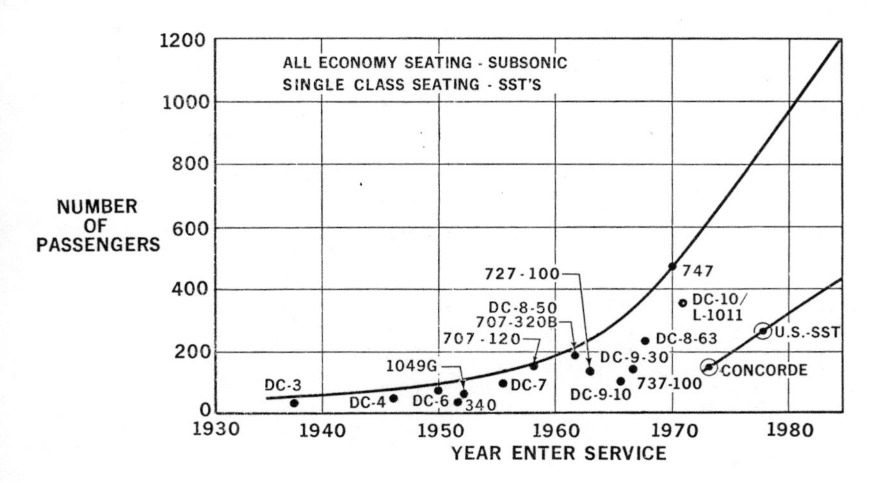

Fig. 4–13. Trends in passenger capacities of air carrier vehicles. (Source: *Transport Aircraft Characteristics, Trends, and Growth Projections,* Transport Aircraft Council, Aerospace Industries Association of America, Inc.)

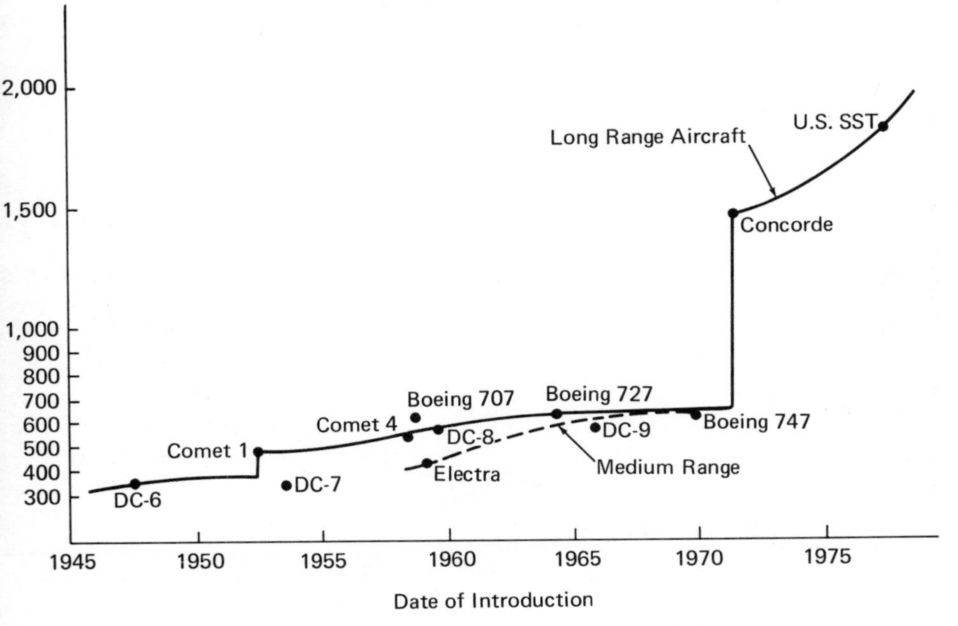

Fig. 4–14. Cruising speeds of air carrier aircraft, 1945–1977.

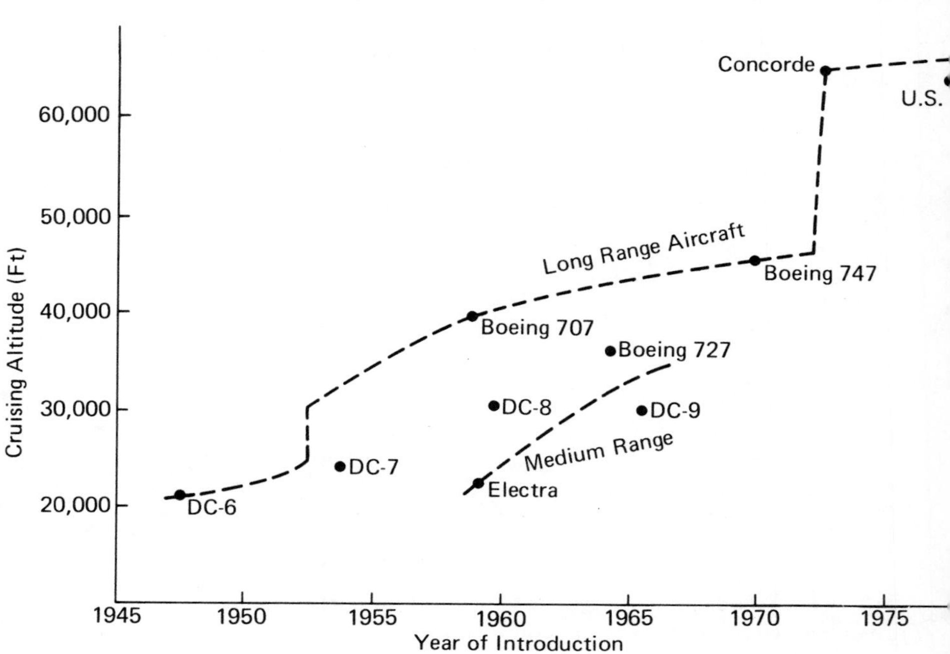

Fig. 4–15. Cruising altitudes of carrier aircraft, 1945–1977.

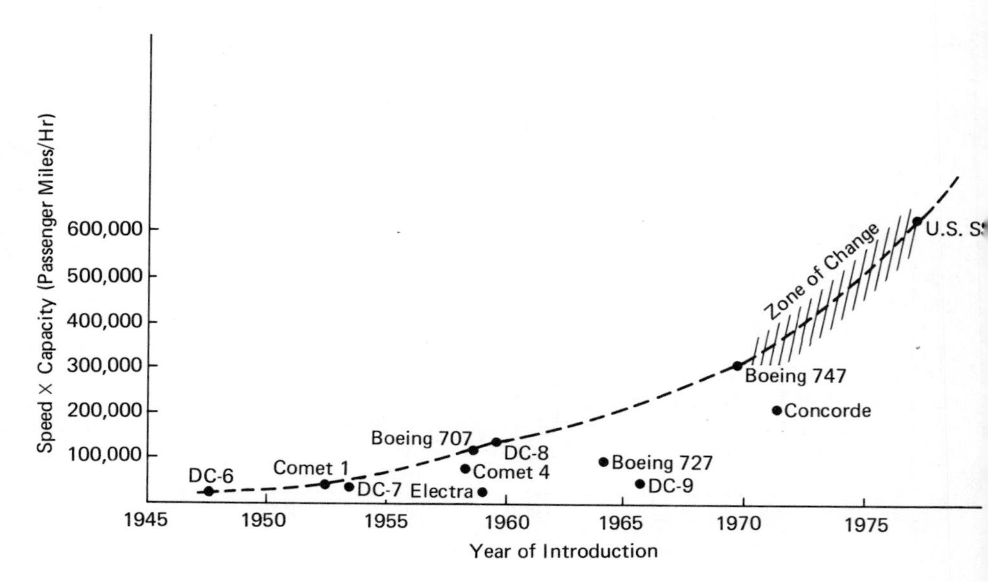

Fig. 4–16. Trends in passenger miles/hour of airline carriers.

WATERBORNE VESSELS

4–15. Components of Total Resistance. As a ship or other waterborne vessel moves through water it encounters resistance to movement which is overcome by the propulsive force driving the vessel. The total resistance to any vessel is made of four component parts [5]:

1. Frictional resistance
2. Wave resistance
3. Eddy resistance
4. Air resistance.

Frictional resistance is incurred both between the underwater surfaces of the ship and the water, and between the layers of water in the immediate vicinity of the vessel's hull which move over one another as the ship drives through the water.

Wave resistance is encountered as a vessel moves in a liquid medium. As the bow of a ship moves through the water, bow waves are produced which involve a loss of energy and a consequent drain on the propulsive force of the engine. It is advantageous for a ship to generate wave patterns that provide a bow wave crest at the stern when operating at design speed, since the additional pressure at the stern acts as a forward driving force.

Eddy resistance occurs whenever eddies are produced by a ship moving through the water. Typically, the stern post and the propeller struts produce eddies if not properly streamlined. It is possible that such unstreamlined vessels as square ended barges will produce stern eddies. Eddy resistance in a properly designed vessel is a small portion of total resistance.

Air resistance, while normally a small portion of total resistance, can be appreciable, and can be reduced by streamlining the vessel's superstructure.

4–16. Passenger Ships. Passenger liners are vessels whose primary function is the transport of persons. The U.S. Maritime Administration classifies any vessel that has accommodations for 100 or more passengers as a passenger vessel. Some cargo vessels carry less than 100 persons and, for certain legal purposes, are recognized as passenger vessels. Only when a vessel carries 12 or less passengers can it avoid the more stringent regulations pertaining to passenger ships. Many vessels, therefore, carry accommodations for exactly 12 passengers.

There is a large variation in the characteristics of passenger liners. Examples of the world's largest prestige liners are the *S.S. United States* and the *S.S. Queen Elizabeth 2* (shown in Fig. 4–17). The *S.S. United*

Fig. 4–17. *S. S. Queen Elizabeth 2.* (Source: Cunard Lines Ltd.)

States won the Blue Ribbon of the North Atlantic in 1952. Carrying at capacity 2,008 passengers and 1,093 crew, this vessel, with a gross tonnage of 53,329 tons, cruised at 30 knots. Despite the luxury accommodations and high speed of this vessel, however, it was withdrawn from service in 1969 because of a history of operating losses during the late 1960's. There appears to be considerable doubt that passenger liners will be able to compete against air passenger carriers. Improved technology, which permitted the introduction of jumbo jets early in 1970, has brought decreased fares through lowered operating costs on a seat mile basis. Passenger liners face stiff competition not only on the basis of speed but also now on a basis of cost. Under these conditions, it seems unlikely that large prestige ocean liners can continue to attract sufficient passenger traffic for continued economic operation. The *S.S. Queen Elizabeth 2,* introduced into service in 1969, is an attempt to reverse the trend to air transport. Whether passenger liner service of this type will continue in the seventies and eighties is open to doubt. The *Queen Elizabeth 2,* a strikingly modern ship, carries 2,025 passengers and a crew of 906. This vessel, whose gross tonnage is 65,000, cruises at 28½ knots. Many passenger liners are much smaller than the large prestige vessels already discussed. Some of their operating characteristics are shown in Table 4–7.

Traditionally, passenger liners have been converted in wartime to troop carriers. The *S.S. United States,* for example, is capable of carrying an army division. Whether such conversions will be carried out in the future is also in doubt. The slow moving ships are much more vulnerable to missile attack than the large fast moving air transports such as the Lockheed C-5A.

TABLE 4–7

Operating Characteristics of Passenger Liners

Ship	Engine	Gross Tonnage	Passengers	Cruising Speed in Knots
S.S. Flandre	Turbine	20,500	700	22
S.S. Independence	Turbine	23,719	1000	22 1/2
S.S. Santa Paula	Turbine	15,371	300	20
S.S. Ryndam	Turbine	15,024	915	16 1/2
S.S. Olympia	Steam Turbine	23,000	1300	21
M.V. Kungsholm	Diesel	22,000	1000	19

Source: Ladage, J. H., *Merchant Ships*, Cornell Maritime Press, 1968.

4–17. General Cargo Carriers. Figure 4–18 shows a modern high speed general cargo carrier operating in transoceanic trade. The "Constellation" class ship shown is one of the world's most modern designs, having operating speeds greater than 24 knots. Automation devices which have been incorporated into the design include direct bridge control of the engine, temperature registers in reefer spaces, automatic boiler control, constant tension winches, air conditioned crew and passenger accommodations, and hydraulic quick-acting hatch covers. Typically, general cargo carriers have become more sophisticated since World War II. In addition to a great deal of automated control and operation, ships now have such innovations as reinforced main and 'tween decks for forked lift truck operation, triple hatches, dehumidifying equipment, and

Fig. 4–18. Fast cargo liner, Mormacargo. (Source: Moore-McCormack Lines, Inc.)

modern ship's gear for rapid and efficient on-loading and off-loading the cargo compartments.

General trends in cargo ships indicate that modern freighters now exceed 500 feet in length compared with lengths of slightly under 400 feet in the 1930's. Operational speeds of modern carriers are much higher than 30 years ago. Most freighters at that time operated at speeds of 14 knots or less. Modern powerful cargo liners are capable of speeds in excess of 20 knots with the faster vessels operating at 24 knots. Increased speed and length have been accompanied by increased vessel size of typical general cargo vessels. The C1-A Cargo Vessel, one of the original designs in the U.S. Maritime Commission long range building program of the 1930's, had a dead-weight tonnage of 7,343 tons. The Liberty ships of World War II had a comparable tonnage of 10,920 tons. By the middle of the 1960's, dead-weight tonnages of typical modern general cargo vessels had grown to over 13,000 tons.

The container ship is an innovation in freighter design, brought about with the rapid growth of containerization. Figure 22–12 shows a modern container ship completed in 1968 which serves the North Atlantic run between Europe and the United States. These high speed 21,000 D.W.T. ships cruise at 21 knots, making round trips from Europe in three weeks. Fully loaded, such a fully automated ship can carry over 1,200 containers, each 20 feet long. The world's largest container ship fleet is operated by Sea Land Service, which in 1968 had 23 trailerships, all converted from older conventional ships. These converted vessels are typically smaller and slower vessels than the container liner pictured here. Cruising speeds of the 15,700 D.W.T. converted T3 ships are 16 knots, carrying 476 trailers. Converted C4's are somewhat longer and carry 609 standard 35 feet long Sea-Land containers [6].

4–18. Bulk Carriers. Bulk carriers are special purpose vessels that are designed to carry one cargo or one cargo type only. They include *ore carriers* and *tankers*. Tankers are vessels whose primary function is the carriage of liquid cargo, such as petroleum, oil, asphalt, bitumen, gasoline, molasses, and chemicals. Since World War II the change in characteristics of tankers has been astounding.

In 1945, the largest tanker could carry approximately 16,000 tons with drafts of 30 feet. By 1950, the largest tanker in the world was designated a "supertanker" and carried 30,155 tons with a draft of approximately 35 feet. In 1966, the scale of tankers had jumped to the order of "mammoth tankers," with the launching of the *Idemitsu Maru* having a carrying capacity of just over 200,000 tons, a length of 1,122 feet and a draft of 58 feet. Tankers continue upward in size as vessels with a capacity over 300,000 tons are being constructed in the late sixties and early seventies, and there is discussion of 1 million ton vessels in the near future. Obviously, few ports have the capability of handling mammoth tankers with their extensive draft requirements and large lengths. It is becoming more frequent for mammoth tankers to moor at special dolphins some distance off-shore to take on and discharge cargo via pipeline to shore.

4–19. Inland Waterway Vessels. A variety of vessels ply the inland waterways: towboats, tugboats, barges of various types, carfloats, and scows. Origin-destination speeds on a systemwide basis average close to six miles per hour. On the longer rivers and waterways giant "integrated tows" composed of powerful towboats and numerous barges make speeds of 15 miles per hour. Figure 4–19 shows a typical large tow

Fig. 4–19. An integrated tow. (Source: American Waterways Operators, Inc.)

composed of open hopper barges. Almost all towboats and tugboats are powered by diesel engines; few steamboats remain in service. With the introduction of powerful diesel units, large barges and integrated tows became possible. Many towboats now use the Kort nozzle, a funnel shaped structure enclosing the propellers which can improve propeller thrust up to 25 per cent in favorable conditions. One of the most important technological advances came during World War II with the introduction of reversing-reduction gears permitting the operation of the engines at the most favorable rpm. Prior to this time the high efficiency of diesel engines at high rpm was not matched to maximum propeller efficiency at low propeller rpm. Still unsolved is the problem of gaining maximum efficiency from diesel propulsion under large variations of load which can occur by different tow sizes. Techniques which are now being tested to overcome this problem are diesel electric power units, variable pitch propellers, and overload limiters [7].

Table 4–8 summarizes selected operating characteristics of inland waterway craft.

Tables 4–9 to 4–12 show summary statistics for the world's fleets by classification of Freighters, Bulk Carriers, Combination Passenger and Cargo Ships, and Tankers. It is interesting to note that though freighters, bulk carriers, and tankers show strong trends of increasing size with date of introduction into service, the largest passenger ships are relatively old [8].

RAILROAD LOCOMOTIVES[1]

The operational characteristics of railroad trains are dependent on the type of motive power supplied by the *power units* or *locomotives.* The most common types of power units used by the American railroads are:

1. Locomotives:
   - Electric
   - Diesel electric
   - Steam
   - Other types including gas turbine-electric, diesel-hydraulic
2. Railcars:
   - Electric
   - Diesel electric
   - Gas-electric

This section will limit its discussion to the operating characteristics of electric, diesel-electric, and steam locomotives which move almost entirely the total passenger and freight traffic on the North American continent.

In dealing with the different operating characteristics of locomotives several definitions are necessary to clarify the following discussion.

[1] This section is drawn extensively from Part 3, Vol. II, *Manual of Recommended Practice,* American Railway Engineering Association.

TABLE 4-8

Selected Operating Characteristics of Inland Waterway Craft

Towboats	Feet Length	Feet Breadth	Feet Draft	Horsepower	
	117	30	7.6	1000 to 2000	
	142	34	8	2000 to 4000	
	160	40	8.6	4000 to 6000	
Tugboats				**Horsepower**	
	65 to 80	21 to 23	8	350 to 650	
	90	24	10 to 11	800 to 1200	
	95 to 105	25 to 30	12 to 14	1200 to 3500	
	125 to 150	30 to 34	14 to 15	2000 to 4500	
Deck Barges				**Capacity Tons**	
	110	26	6	350	
	130	30	7	900	
	195	35	8	1200	
Carfloats				**Capacity Railroad Cars**	
	257	40	10	10	
	366	36	10	19	
Scows				**Capacity Tons**	
	90	30	9	350	
	120	38	11	1000	
	130	40	12	1350	
Open Hopper Barges				**Capacity Tons**	
	175	26	9	1000	
	195	35	9	1500	
	290	50	9	3000	
Covered Dry Cargo Barges				**Capacity Tons**	
	175	26	9	1000	
	195	35	9	1500	
Liquid Cargo (Tank) Barges				**Capacity Tons**	**Gallons***
	175	26	9	1000	302,000
	195	35	9	1500	454,000
	290	50	9	3000	907,200

Source: *Big Load Afloat,* American Waterways Operators, Inc.
*Based on an average of 7.2 barrels per ton and 42 gallons per barrel.

TABLE 4-9

Average Age, Speed, and Draft of the World's Merchant Type Freighters

Deadweight Tons	Average Age Years	Average Speed Knots	Average Draft Feet
Under 2,000	16	11	15
2,000–3,999	17	12	19
4,000–6,999	16	13	22
7,000–8,999	17	14	25
9,000–9,999	17	14	27
10,000–10,999	21	12	28
11,000–12,999	11	15	29
13,000–14,999	8	16	30
15,000 and over	9	16	31
Overall Average	16	13	24

Source: *A Statistical Analysis of the World's Merchant Fleets,* as of December 31, 1966, U.S. Department of Commerce, Maritime Administration, December, 1967.

TABLE 4-10

Average Age, Speed, and Draft of the World's Merchant Type Bulk Carriers

Deadweight Tons	Average Age Years	Average Speed Knots	Average Draft Feet
Under 10,000	15	11	20
10,000–19,999	11	14	29
20,000–29,999	8	15	32
30,000–39,999	4	15	35
40,000–49,999	4	16	38
50,000–59,999	3	15	39
60,000–79,999	2	16	42
80,000–99,999	2	15	43
100,000 and over	1	15	54
Overall Average	11	13	28

Source: *A Statistical Analysis of the World's Merchant Fleets*, as of December 31, 1966, U.S. Department of Commerce, Maritime Administration, December, 1967.

TABLE 4-11

Average Age, Speed, and Draft of the World's Combination Passenger and Cargo Ships

Gross Tons	Average Age Years	Average Speed Knots	Average Draft Feet
Under 4,000	23	13	16
4,000–6,999	21	16	20
7,000–9,999	21	16	25
10,000–14,999	19	17	27
15,000–19,999	20	19	27
20,000–29,999	15	21	29
30,000–49,999	11	24	31
50,000 and over	21	29	38
Overall Average	21	16	22

Source: *A Statistical Analysis of the World's Merchant Fleets,* as of December 31, 1966, U.S. Department of Commerce, Maritime Administration, December, 1967.

TABLE 4-12

Average Age, Speed, and Draft of the World's Merchant Type Tankers

Deadweight Tons	Average Age Years	Average Speed Knots	Average Draft Feet
Under 20,000	14	13	26
20,000–39,999	11	16	34
40,000–59,999	6	16	38
60,000–79,999	3	16	41
80,000–99,999	3	16	46
100,000–124,999	2	16	50
125,000–149,000	2	16	53
150,000–199,999	1	16	53
200,000 and over	1	16	57
Overall Average	11	14	31

Source: *A Statistical Analysis of the World's Merchant Fleets,* as of December 31, 1966, U.S. Department of Commerce, Maritime Administration, December, 1967.

Horsepower for the various types of locomotive is defined somewhat differently. The rated horsepower of electric locomotives is the power available at the rims of the driving wheels. For diesel or turbine electric locomotives, rated horsepower refers to the power available as input to the turbine or diesel engine. Tractive effort is more usually used in conjunction with steam locomotives.

Tractive Effort is the tangential force applied at the rims of the driving wheels by the locomotive.

Rail Horsepower is the power available at the rims of the driving wheels.

4–20. Electric Locomotives. Power is supplied to electric train systems by two types of distribution systems: direct current (d.c.) and alternating current (a.c.) systems. The method of current pick-up is either by collector shoes riding on a third rail or by means of a pantograph or trolley wheel passing under an overhead wire. Third rail systems are required where heavy currents are involved; these are normally suitable only where lower voltages are used. High-voltage systems require overhead wires from the viewpoint of safety. Four general types of electric motor traction are [9]:

1. d.c. power supply with d.c. traction motors
2. single phase a.c. power supply with single phase a.c. traction motors
3. single phase a.c. power supply, intermediate rectifiers and d.c. traction motors
4. single phase a.c. power supplying an intermediate d.c. generator driving a d.c. traction motor.

Electric locomotives can be used in single units or coupled as multiple units under one controller.

Electric motors rely on their power for an external source of supply. The capacity of the motor is, therefore, not limited internally but is governed either by the power that can be drawn from the supply system without slipping the driving wheels or by the temperature or commutation capacity. Because of the time delay involved, electric motors have a short-term overload capacity. This overload capacity, coupled with a high adhesion between rails and wheels with the nonpulsating form of torque supplied by an electric motor, makes this form of traction desirable under certain load conditions. Figures 4–20 to 4–22 show graphs relating tractive effort and horsepower to the various forms of electric motor traction already discussed.

Manufacturers supply rating curves which furnish the relationship of tractive effort to speed for the various locomotives under short term or continuous loading conditions. In the absence of these curves the

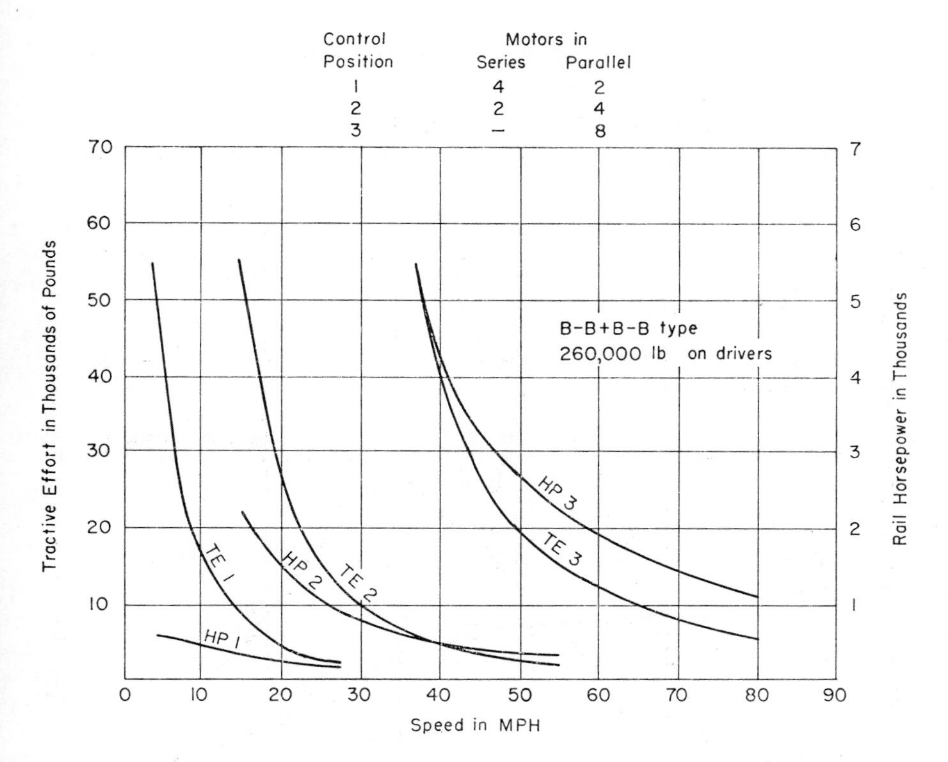

Fig. 4–20. **Tractive effort and horsepower curves for a 600 V. dc locomotive, with series-parallel control.** **(Source:** *A.R.E.A. Manual of Recommended Practice,* **1971 edition.)**

following equations can be used to describe the relationships:

$$V = \frac{d \times p \times S}{336 \times G}$$

$$TE = \frac{T \times G \times 24 \times e}{d \times p}$$

where:

V = locomotive speed in miles per hour
TE = locomotive tractive effort in pounds
S = number of armature revolutions per minute
T = motor torque in pound feet
d = diameter of driving wheels in inches
p = number of pinion teeth
G = number of gear teeth
e = gear efficiency (0.95 to 0.97)

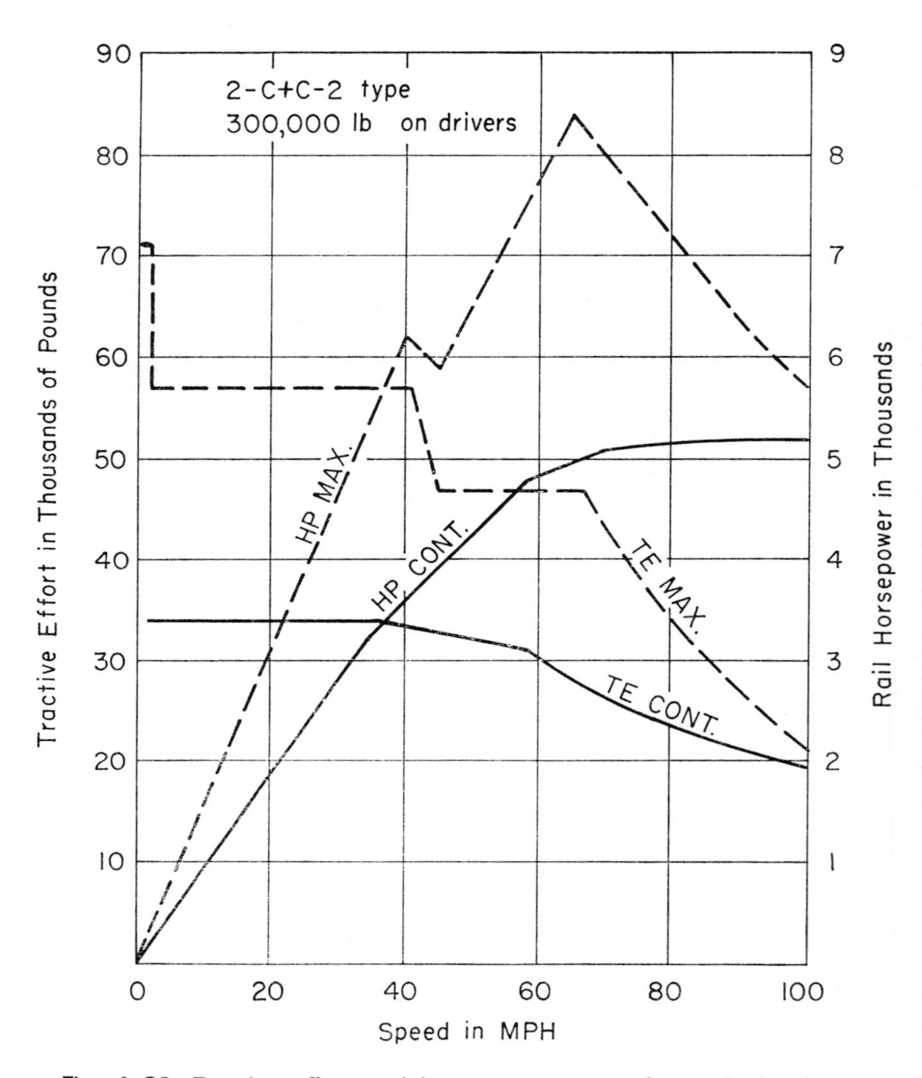

Fig. 4–21. Tractive effort and horsepower curves for a single-phase ac locomotive. (Source: *A.R.E.A. Manual of Recommended Practice,* 1971 edition.)

The overload capacity of electric locomotives combined with the high accelerations at low speed makes this form of traction most desirable for short runs with frequent stops and starts. Electric locomotives are, therefore, widely used for suburban and mass transit rail systems. They are of equal use on short steep grades where other locomotives would require the operation of multiple units to avoid overload.

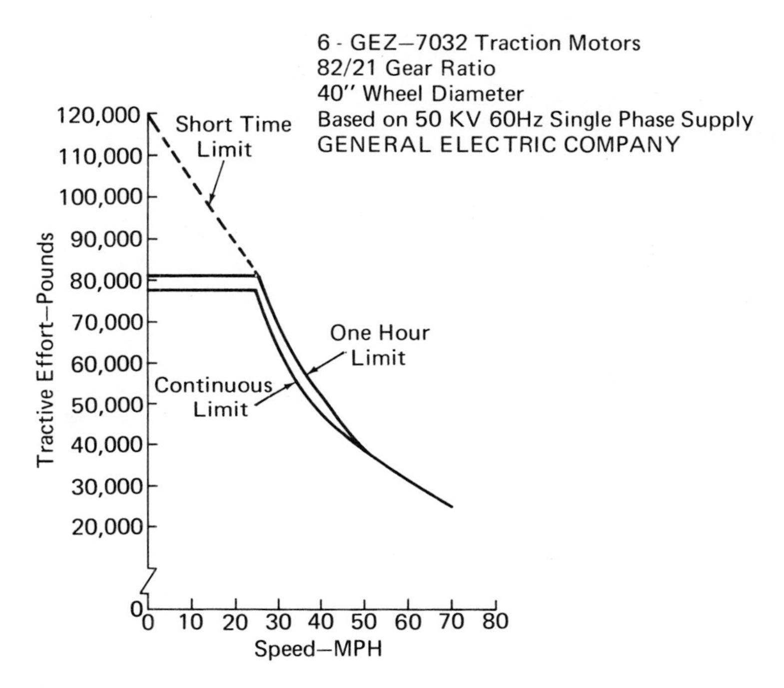

Fig. 4–22. Speed vs. tractive effort curve for rectifier type electric locomotive. (Source: General Electric Co.)

4–21. Diesel Electric Locomotives. The diesel electric locomotive is simply a complete power plant composed of a diesel engine prime mover connected directly to a d.c. generator. The d.c. generator itself feeds a series-wound d.c. traction motor. Each diesel electric locomotive unit is a self-contained system with both power plant and traction motor. For relatively low load conditions such as in yards and at switches, diesel electrics are usually operated singly; for line haul operation it is common to use multiple units under unit control from one cab.

In essence, diesel electric locomotives are constant maximum power units since their horsepower ratings are set according to the power delivered by the prime mover diesel to the main generator. This power is independent of the track speed of the locomotive. The capacity of the unit at low speeds is limited both by rail adhesion of the driving wheels and by overheating of the windings of the traction motor. At high speeds, there is again a decline in capacity due to overheating of the main generator in a similar manner. Because the actual losses are difficult to predict it is customary for manufacturers to supply curves which define the operating characteristics of their locomotives under short-term loading conditions. Figure 4–23 is a typical curve showing

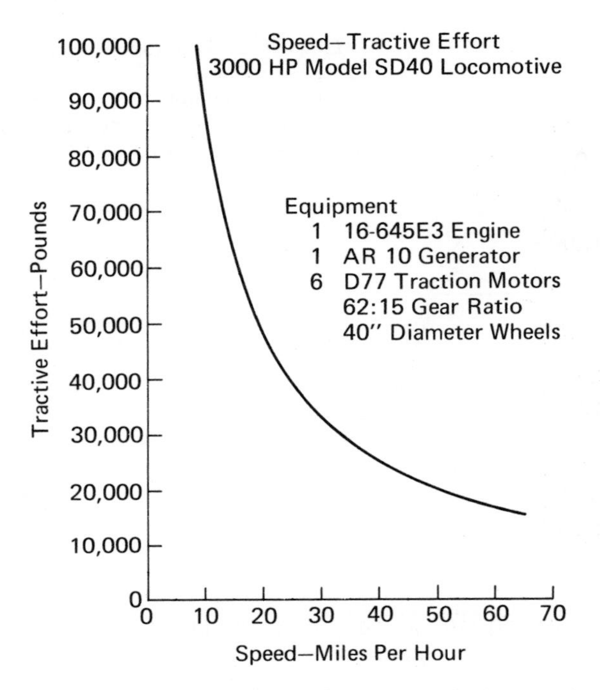

Fig. 4–23. Performance curve for a diesel-electric locomotive. (Source: Locomotive Division, General Motors.)

the short-term relationship between tractive effort and speed for a diesel electric unit, while Fig. 4–24 shows the relationship between locomotive horsepower and speed under short-term loading conditions. Where the rated shaft horsepower is known, the tractive effort can be calculated for predictive purposes from:

$$TE = \frac{375HPe}{V}$$

where:

TE = tractive effort in pounds
HP = rated horsepower of the diesel prime mover (0.93 of gross horsepower if not otherwise known)
V = track speed in miles per hour
e = efficiency of electro-mechanical drive system (0.82–0.83)

Because of their ability to take short-term overload, diesel electric locomotives are used in situations similar to those where electric motors are most useful. The reserve of power for diesel electrics is somewhat less than all electric systems. The principal advantage of the diesel electric locomotive is its ability to operate *without* the extensive power

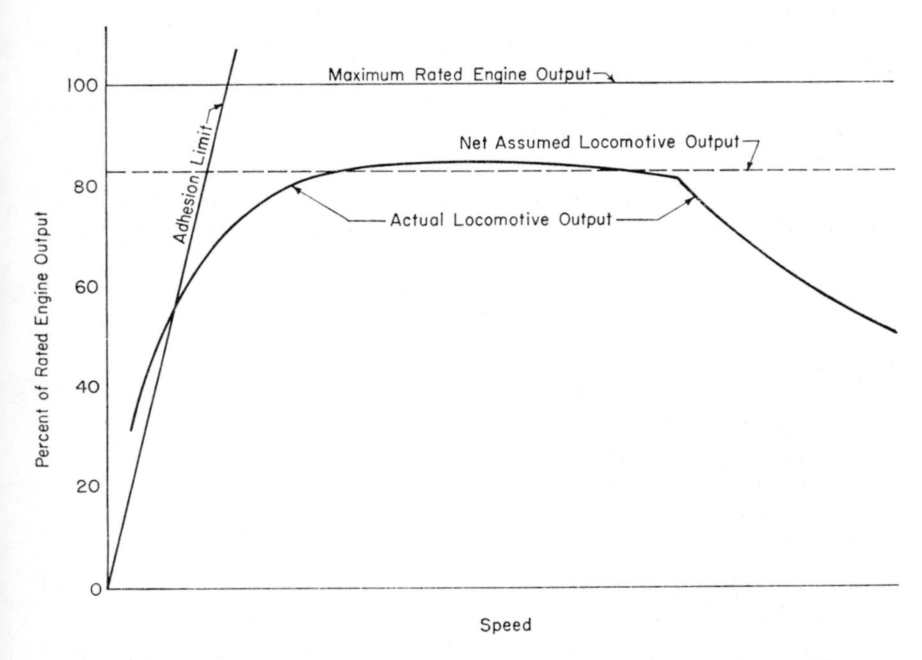

Fig. 4–24. Typical horsepower curve, diesel-electric locomotive. (Source: *A.R.E.A. Manual of Recommended Practice,* 1971 edition.)

distribution systems required by electric units. These distribution systems are both expensive and create safety hazards under certain conditions.

4–22. Steam Locomotives. Steam engines are no longer widely used in railroad locomotives in North America and the United Kingdom. Because of the inherently low efficiency of the reciprocating steam engine, they have been widely replaced by diesel electric units. Although the residual units represent an outmoded technology, the steam engine is worth discussion because it still forms the basic power unit for the rest of the world.

The advantages of the steam engine are relatively few but, in the case of capital-starved countries, may be decisive. The engine is a relatively simple unit and can be fabricated and maintained by a less advanced technology. However, the most compelling reason for continued use of the steam engine lies with low capital cost per unit of horsepower. In capital-scarce economies, the equipping of the railroad system with lower cost steam locomotives serves to free capital for other uses. The disadvantages of the steam locomotive lie in its low thermal efficiency, its inability to be easily adapted for different types of service by simple gear changes, the need for an adequate supply of water throughout its line of operation, and, finally, the sensitivity of engine efficiency to changes in the type of fuel used. Steam engines are customarily rated by tractive effort rather than by horsepower. There are two types of tractive effort, *cylinder tractive effort* and *driver tractive effort.*

Cylinder tractive effort relates to the force exerted by the steam pressure on the pistons. It may be calculated from the following formula [9]:

$$\text{Cylinder } TE = \frac{KPd^2S}{D}$$

where:

K = ratio of the mean effective pressure in the cylinders to that of the boiler (usually assumed at 0.85)
P = boiler pressure in p.s.i.
d = cylinder diameter in inches
S = length of stroke in inches
D = Diameter of the drivers in inches

Driver tractive effort is less than cylinder tractive effort by the amount of mechanical losses in the cylinders, valve gear, and driving rods.

The *capacity* of the steam locomotive is limited by the capacity of the cylinders at low speeds and by steaming capacity at high speeds. Cylinder capacity is related to both diameter of the cylinders and the length of available stroke. The steaming capacity of the boilers depends on the area of evaporating surface, the size of the firebox and grate, the fuel used, and the method of firing.

At low speeds, rail horsepower is limited by adhesion. The pulsating torque available from the reciprocating engine gives less favorable rail adhesion characteristics for steam engines than for other nonpulsating engines. Figure 4–25 shows the relation between track speed and both tractive effort and rail horsepower. For any particular locomotive it is usual for the manufacturer to supply such performance curves.

4–23. Train Resistance [10]. The total resistance to movement of trains is composed of *inherent* or *level tangent* resistance and incidental resistances due mostly to curvature, grades, and winds. Level tangent resistance is due to a combination of factors including speed, cross-sectional area, axle load, type of journal, winds, temperature, and condition of track. Based on analysis of tests, W. J. Davis [11] recommended the following formula for computing level tangent resistance for axle loadings in excess of 5 tons:

$$R = 1.3 + \frac{29}{w} + kV + \frac{KAV^2}{wn}$$

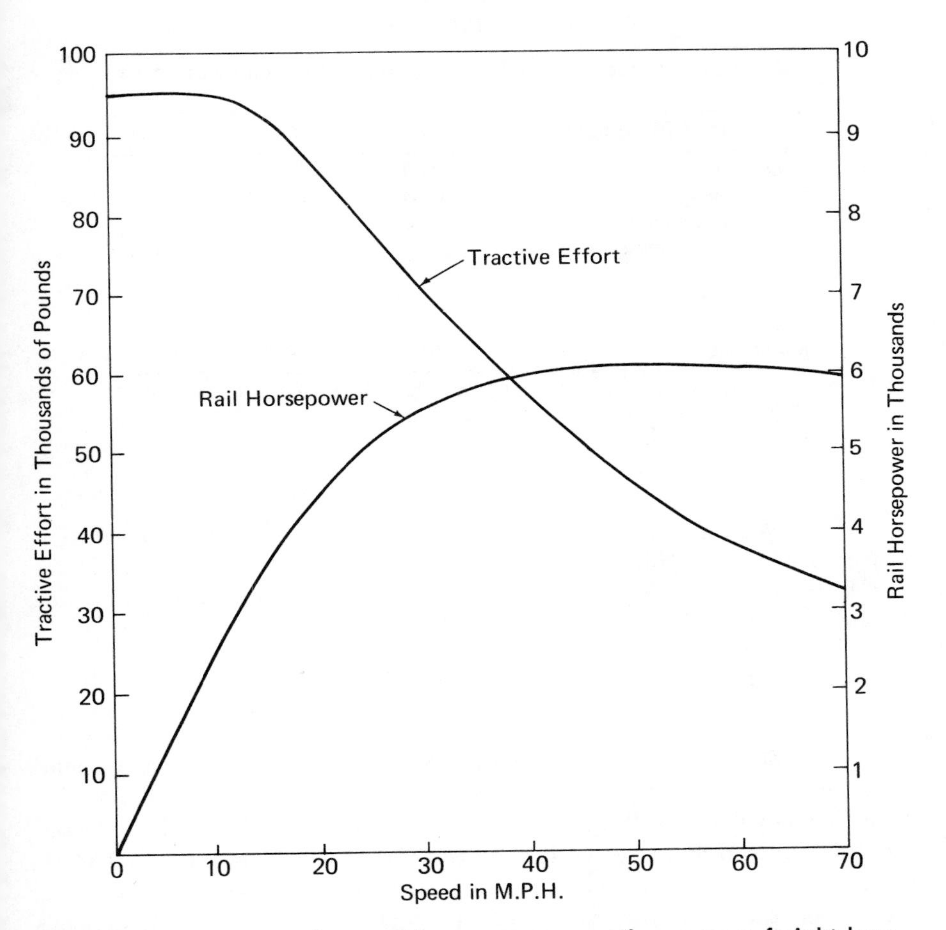

Fig. 4–25. Tractive effort and horsepower curve for a steam freight locomotive. (Source: *A.R.E.A. Manual of Recommended Practice;* illustration withdrawn from 1971 edition.)

where:

R = level tangent resistance in lb/ton
w = average load per axle in tons
n = number of axles
wn = average weight of locomotive or car in tons
A = gross section area in square feet
K, k = coefficients which vary according to units considered
V = track speed in miles per hour.

Values of K, k, and A are shown in Table 4–13.

TABLE 4-13

Coefficient Values for Davis Level Tangent Resistance Formula

Unit of Equipment	k	K^*	A
Locomotives	0.03	0.0024	120
Freight cars	0.045	0.0005	85–90
Passenger cars (vestibuled)	0.03	0.00034	120
Multiple unit trains, leading car (vestibuled)	0.045	0.0024	100–110
Multiple unit trains, trailing cars (vestibuled)	0.045	0.00034	100–110
Motor cars	0.09	0.0024	80–100

*For streamlined equipment 0.0016 may be substituted for 0.0024 and 0.00027 for 0.00034.

Grade resistance is usually taken at 20 lb per ton for each per cent of grade. Curve resistance is computed at 0.8 lb per ton for each degree of curvature. The computation of total train resistance is calculated from the relation:

$$\text{Total train resistance} = \text{Level tangent resistance} + \text{grade resistance} + \text{curve resistance}$$

4–24. Characteristics of Urban Rail Transit. Urban rail transit vehicles are significantly different in operational characteristics from intercity passenger vehicles. Where intercity passenger trains hold sustained speeds for considerable periods of time, the urban rail transit car is required to accelerate and decelerate constantly between station stops. It is not, therefore, surprising that the choice of transit vehicle depends on several variables which strongly reflect this type of operation. In selecting transit cars the engineer must account for:

1. Maximum acceptable acceleration rates
2. Desired station spacing
3. Limit of speed capability of train
4. Desired overall line-haul travel speeds including stops.

It has been found in practice that maximum acceleration rates should be limited to 8 ft/sec^2 where all passengers are seated and 5 ft/sec^2 where standees are anticipated. The selection of any two of the remaining criteria will determine the last. The procedure used to determine line haul characteristics of transit vehicles can best be described by an example.

EXAMPLE

A rapid transit vehicle has been selected which has a top speed capability of 60 mph, a capacity of 125 persons seated, and a length of 48 ft., 3 in. The maximum passenger volume to be carried is 40,000 persons in one direction. Manual control will be used, limiting headways to a minimum of 2 minutes. Station stops of ½ minutes are to be used. What is the minimum station spacing if the full top speed capability of the vehicle is to be used; and what station lengths must be designed?

Number of trains per hour = 30 at 2 minutes headway

Required train capacity = Hourly passenger volume/no. of passengers per hour
= 40,000/30
= 1,333 passengers

Required number of cars per train = Train capacity/car capacity
= 1,333/125
= 10.67—rounded off to 11 cars

Platform length required = Individual car length × no. of cars per train
= 48.25 ft. × 11
= 530 ft.

Assuming a maximum acceleration and deceleration rate of 5 ft./sec.^{2} was used

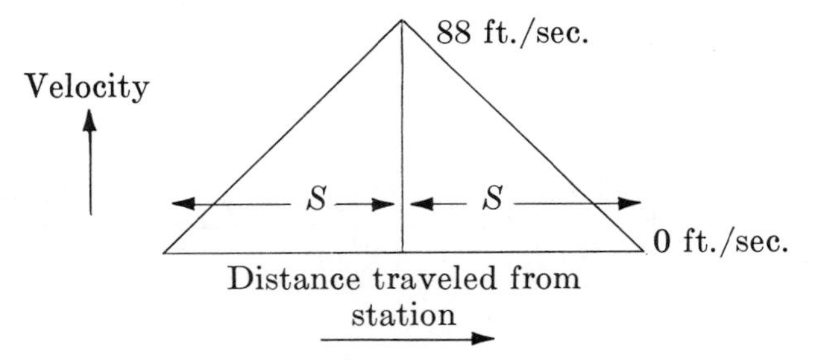

throughout the run, the distance–speed diagram would be as shown, attaining a maximum speed of 60 mph or 88 ft./sec. at the halfway point.

Time to reach midpoint = Velocity/acceleration
= 88/5
= 17.6 sec.
Total running time = 17.6 × 2 = 35.2 sec.
Distance to midpoint = $S = \frac{1}{2} \times \text{acceleration} \times \text{time}^2$
= $\frac{1}{2} \times 5 \times 17.6^2$
= 774 ft.

Minimum station spacing = 774 × 2
= 1548 ft.
= 0.30 miles
Average running speed = 30 mph

Time to travel between stations = Acceleration time + deceleration time
= 17.6 × 2
= 35.2 sec.

Total time per 1548 ft. of line = Running time + station stop time
= 35.2 + 30.0 sec.
= 65.2 sec.

Average overall travel speed = Station spacing/overall travel time
= 1548/65.2
= 16.2 mph

It is worth noting that with stations set at 1548 ft. spacing, there is no reason to use more powerful equipment, since overall travel speeds are limited by acceleration rates and station stop time. If high speed equipment with limiting speeds of 90 mph were used for this example the overall travel speeds would be identical. It is interesting to note the effect of increasing the station spacing of ½ mile. The additional 1,092 ft. are covered at top speed. The speed–distance diagram is now of the form:

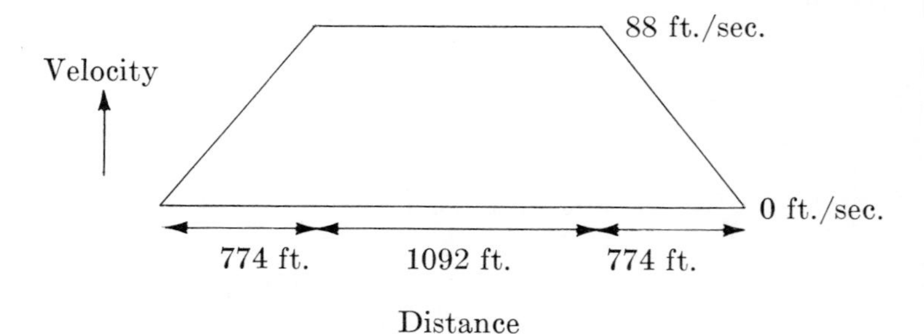

Additional distance traveled in 1092 ÷ 8.8 ft./sec. = 12.4 sec.
Total travel time = 65.2 + 12.4
= 77.6 sec.

Overall velocity = 2640/77.6
= 34.1 ft./sec. or 23.2 mph

An increase of approximately 0.2 miles to the station spacing results in a virtual doubling of overall line haul travel speeds. In practice, the calculations may become slightly more complicated by the nonuniform acceleration characteristics of the travel vehicles. Acceleration

TABLE 4-14

Selected Vehicle Capacities and Operating Speeds

	Freight (tons)	Passengers	Operating Speed
Highway Vehicles			
Autos		2–5	30–70
Minibus		30	25
U.S. urban transit bus		55–85	25
Streetcars, P.C.C.		125	20–30
Streetcars, articulated Europe		200	20–30
Trucks			20–30
Urban Rail Transit			
I.R.T. conventional duo-rail		180/car	20–30
Toronto duo-rail		220/car	20–30
San Francisco BART		150/car	
Transit expressway		54/car	23–39
Safege monorail			50–75
Alweg monorail			50
Interurban Rail			
Commuter railway—Budd Pioneer III		125	
U.S. passenger car		100+	
Japanese Tokaido car		1000/train	130
Open box car freight			10–40
Airplanes			
Transcontinental Boeing 707		185	623
Intercontinental Boeing 747		390	615
Concorde S.S.T.		148	1450
Beech Bonanza		5	195
Piper Cherokee		4	166
Transcontinental cargo, Boeing 707	111		600
Intercontinental cargo, Lockheed C-5A	100 for 2700 N.M.		460 KT
Water Transport			
Ocean liner, S.S. United States		2008	30 KT
Mammoth tanker	200,000		
Hovercraft, English Channel		700–800	80
Mississippi Barge	1500–3000		

TABLE 4-15

Route Capacities

Highways (uninterrupted flow)[a]	
Multilane	2000 passenger vehicles/hr./lane
Two lane—one way	2000 passenger vehicles/hr. total
Three lane—two way	4000 passenger vehicles/hr. total
Railroads (actual capacity passenger or freight trains)[b]	
A. Block signal system with train orders	
Single track	30 trains/day
Two or more tracks	60 trains/day
B. Centralized traffic control	
Single track	65 trains/day
Two tracks	120 trains/day
Airports and airways[c]	
Airway under instrumental flying rules, 10 minute separation	6/hr.
Airway under visual flying rules	Varies
Single runway airport	Up to 85 operations/hr.
Dual runways airport	Up to 160 operations/hr.
Pipelines[d]	
2 in.	4–150 bbl/hr.
4 in.	10–400 bbl/hr.
6 in.	50–2000 bbl/hr.
8 in.	100–4000 bbl/hr.
10 in.	100–4000 bbl/hr.
12 in.	400–5600 bbl/hr.

[a]*Highway Capacity Manual—1965, Special Report No. 87,* Highway Research Board, Washingto D.C., 1965.
[b]Conant, M. *Railroad Mergers and Abandonments*, University of California Press, Berkeley, 196
[c]*Airport Capacity Handbook, 2nd Edition*, Systems Research and Development Service, Federa Aviation Division, Washington, D.C., 1969.
[d]Hay, W. W., *An Introduction to Transportation Engineering,* John Wiley & Sons, New York, 1961.

and tractive effort decrease with speed, as has been previously indicated in the discussion of electric locomotives. Speed-distance relationships must be computed from the characteristic speed velocity relationships of the individual pieces of equipment. The overall approach is similar to the somewhat simplified example. It should be noted that the station spacings are normally greater in outlying areas than in the congested

TABLE 4–15

(Continued)

Belt conveyors[e]		
12 in. belt @ 300 ft./min.		129 tons/hr.
34 in. belt @ 450 ft./min.		843 tons/hr.
60 in. belt @ 600 ft./min.		8100 tons/hr.
Urban rapid rail transit[f]		
New York, IND Queens Line	(observed max. vol.)	61,400 persons/hr.
New York, 8th Ave. Express	(observed max. vol.)	62,030 persons/hr.
Toronto, Yonge St.	(observed max. vol.)	35,166 persons/hr.
Busways		
Individual vehicles, St. Louis design (estimated)[g]		7,200 persons/hr.
10-vehicle trains, 90–120 passengers/car[h]		36,000–48,000 passengers/hr.
Monorail and miniature rapid transit systems[i]		
Alweg Monorail (50 mph)		16,000–20,000 persons/hr.
Safege Monorail (50–75 mph)		12,000–48,000 persons/hr.
Transit Expressway (23–39 mph)		8,000–20,000 persons/hr.
Pedestrian conveyor[i]		
32" belt (1.5 to 2 mph)		3,000 persons/hr.
48" belt (1.5 to 2 mph)		10,000 persons/hr.

[e] *Plant Engineering Handbook*, William Stanier, ed., McGraw Hill Book Co., New York, 1950.
[f] *Urban Mass Transit Planning*, ed. Homburger, W. S., Institute of Transportation and Traffic Engineering, Berkeley, 1967.
[g] *St. Louis Metropolitan Area Transportation Survey Report*, W. C. Gilman and Company, September, 1959.
[h] Rainville, W. S. *et. al.*, "Preliminary Report of Transit Sub Committee on Highway Capacity," *Proceedings*, Highway Research Board, 1961.
[i] Richards, B., *New Movement in Cities*, Reinhold Publishing Corp., New York, 1966.

high-density central areas. Using constant acceleration rates throughout a system, overall travel speeds will vary significantly from CBD to suburban areas.

Figure 4–26 shows a rail rapid transit vehicle of conventional duo-rail design. The majority of vehicles in use in North America operates with one direct current motor to each axle. Current is normally supplied

Fig. 4–26. Rapid transit vehicle. (Source: Bay Area Rapid Transit District.)

to the d.c. motors from a collector shoe running over a "live" third rail. Alternatively, power can be supplied by a pantograph pick-up as described in Section 4–20. The direct current voltage used by most systems is 600-volt supply rectified and transformed from utility distribution voltage at special substations. For a fuller description of characteristics of the electric rail transit vehicles, the reader is referred to the section on electric traction locomotives.

4–25. Summary. To summarize some of the major characteristics of the different modes of transportation, Tables 4–14 and 4–15 show vehicle and route capacities of various types of vehicles and conveyances. These capacities can only be used as guideline figures since actual vehicle and route capacities in most cases will be limited by local conditions. The tables will serve, however, to show the immense variety of speeds and capacities that are currently available in the movement of passengers and freight.

PROBLEMS

1. The operating characteristics of European and American autos differ considerably. Why?
2. Discuss the practicality of the use of nuclear power plants for the various modes of transportation.
3. Discuss the problems involved in the fact that individual rapid transit systems have rolling stock peculiar to their own system, such as differences in guage

and power plant. Enumerate the advantages and disadvantages of standardization.

4. The capacities of systems are often quoted as measures of operational suitability. Discuss the pitfalls involved in using capacity as the only decision criteria for selection of a system. Relate to the problem of urban transportation corridors in your answer.

5. Currently there is considerable work on the development of a short haul air bus which have high capacity and low per seat mile costs. What impacts will a successful design have on intercity movements?

6. Research and development are currently underway on tracked air cushion vehicles for mass public transportation. What impact is this likely to have on:

a) long distance air travel?

b) intra-urban bus transportation?

7. Using the performance curve shown in Fig. 4–22, compute the rated horsepower curve from the relationship:

$$\text{RPH} = \frac{\text{Tractive Effort} \times \text{mph}}{375}$$

REFERENCES

1. *Traffic Engineering Handbook,* Institute of Traffic Engineers, Washington, D.C., 1965.
2. *A Policy on Geometric Design of Rural Highways 1965,* American Association of State Highway Officials, Washington, D.C., 1966.
3. Horonjeff, Robert, *The Planning and Design of Airports,* McGraw-Hill Book Co., New York (1962).
4. Schriever, Bernard A., and Seifert, William W. (eds.), *Air Transportation 1975 and Beyond,* The M.I.T. Press, Cambridge, Mass. (1968).
5. LaDage, J. H., *Modern Ships,* Cornell Maritime Press, Inc., Cambridge, Maryland, 1965.
6. LaDage, J. H., *Merchant Ships,* Cornell Maritime Press, Inc., Cambridge, Maryland, 1968.
7. *Big Load Afloat,* American Waterways Operators Inc., Washington, D.C., 1965.
8. *A Statistical Analysis of the World's Merchant Fleets,* U.S. Department of Commerce, Maritime Administration, December, 1967.
9. *Manual of Recommended Practice,* American Railway Engineering Association, Part 3, *Power,* 1961.
10. Urquhart, Leonard C. (ed.), *Civil Engineering Handbook,* Fourth Edition, McGraw-Hill Book Co., New York (1959).
11. Davis, W. J., Jr., *General Electric Review,* October, 1926, pp. 685–707.

5

Traffic Control Devices

Very light traffic, operating under ideal conditions, can move with ease and safety in any transportation network by following rudimentary rules of the road and strict adherence to schedules. As volumes and densities increase, conflicting movements bring both increased probability of collision and uneconomic operation through high average delays to vehicles in the system. The introduction of traffic control devices serves two principal purposes:

1. Increase of safety
2. Reduction of system delays and increase of capacity

The operating characteristics of the various modes and the variation in the degree of training required for the operators of each mode have resulted in a variety of control devices.

HIGHWAY TRAFFIC CONTROL DEVICES[1]

5–1. Pavement Markings. The chief purposes of pavement markings are to regulate and provide guidance to traffic and to channelize and separate movements with a minimum of physical equipment, relying on the psychological effect of the markings on behavior to bring about the desired driver reaction. The advantages of pavement markings lie in their lack of physical obstruction to the path of traffic, permitting their use on the traveled way itself, their low cost, and the ease of their removal when required. On the other hand, pavement markings are easily obliterated in snow conditions, are less effective under reduced visibility, can wear out rapidly under heavy traffic conditions, and provide no positive restraint against conflicting movements. Highway pavements in the United States are marked in a uniform manner to

[1] The material contained in this section relies heavily on Chapters 9 and 10, *Traffic Engineering Handbook* [1], and the *Manual on Uniform Traffic Control Devices for Streets and Highways* [2].

provide nationwide standards [2]. The conformity to national standards by the individual states and cities provides the interstate driver with uniform markings and signalization throughout the United States.

Center lines are broken or unbroken yellow lines used to separate opposing traffic flows. On multilane roads with undivided pavements of four or more lanes, center lines may be composed of two unbroken yellow lines.

Lane lines are broken white lines, with the recommended standard being 15 feet of line segment with 25 feet gap. At intersections where weaving is undesirable, lane lines are frequently marked as solid lines for a short distance as in Fig. 5–1.

No Passing Zone markings indicate road sections on two- or three-lane roads where limited sight distance or other dangerous situations require the prohibition of passing movements. The zone is designated by a solid yellow line on the driver's side (see Fig. 10–12). No passing in either direction is indicated by a double yellow center line.

Paved Shoulder markings are required to differentiate the shoulder from the traveled way. Typically, the shoulder is marked by a solid white line at the pavement edge and diagonal solid white lines at 100-foot intervals.

Pavement Edge lines are solid white lines that delineate edges to provide guidance to the driver under lowered visibility conditions especially in the rain and at night.

Pavement Width Transitions are solid yellow-line markings used in conjunction with edge and lane lines at sections where pavement width narrows. The length of the transition line is computed from the design speed of the road to relate the visibility of the line to operating conditions.

Approaches to Obstructions are marked by a solid yellow line flanking the usual broken center line. In the case of multilane roads a double solid yellow center line is used. In all cases the length of the approach marking is dependent on the design speed of the road.

Stop lines extend across the full width of intersection approaches. Normally running parallel to the *thin white crosswalk lines,* the broad white stop line provides a well defined stopping point for vehicles approaching intersections as shown in Fig. 5–1.

Other forms of pavement markings include *warnings at approaches to railroad crossings, route directions, parking space limits, lane use control markings,* and *channelization lines* to regulate traffic movements.

5–2. Curb Markings. Sections of the street where parking is prohibited at all times are often visually emphasized by the painting of the vertical face and top of the curb yellow. Unfortunately, lack of uniformity in practice, with the use of various colors to designate different types

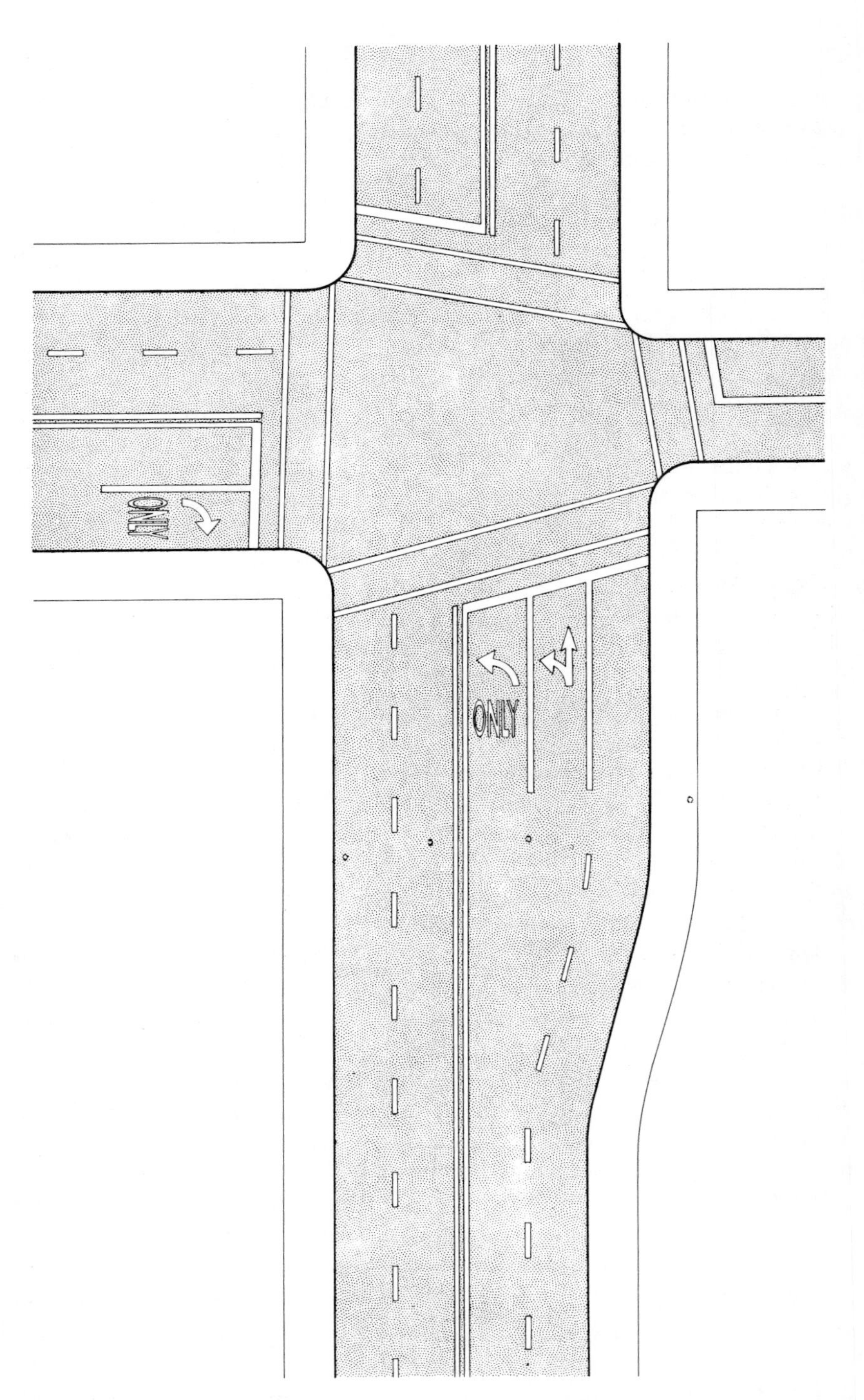

Fig. 5–1. Typical applications of lane use control markings. (Source: *Manual of Uniform Traffic Control Devices for Streets and Highways,* U.S. Department of Transportation.)

of prohibition, has proved confusing to the public. It is recommended that curve markings should be used only as previously described for areas of total prohibition [2].

5–3. Object Markings. Obstructions within or close to the traveled way need to be clearly marked to reduce the traffic hazard that is presented by their physical proximity to moving vehicles. Typically, these obstructions may be structures associated with the highway such as bridge abutments, low bridges, and underpass piers, or they may be elements of the traffic control system such as signal and sign supports, traffic islands, and traffic beacons. Obstructions are marked by reflectorized hazard markers and, where possible, are painted in diagonal black and white stripes facing the flow of traffic. Where obstructions occur close to the line of traffic, guardrails should be used to deflect direct collisions and additional warnings given by means of signing and pavement markings.

5–4. Reflectors. To aid the driver under the reduced visibility of nighttime driving conditions, traffic markings consisting of reflector units, either singly or in groups, are used to indicate hazardous conditions. Two types of reflectors are in principal use, *hazard markers* and *roadside delineators.*

Hazard markers are placed on obstructions within or close to the traveled way, in conjunction with other object markings which are less visible at night.

Delineators are placed on the right-hand side of two-way roads and on both sides of one-way roads where there are long sections with changes in grade and alignment. They are also used on short stretches where there is the possibility of confusion at night. The reflecting head is placed at a height of four feet above the roadway, between two and six feet from the roadway edge, always outside the shoulders.

5–5. Studs. While extensively used outside the United States, reflectorized pavement studs, sometimes called "cat's eyes," have had limited use as center, lane line, and pavement edge markings within this country. In areas with little snow clearing problems, the use of studs can greatly improve nighttime driving conditions, especially during poor weather conditions. The studs are, however, quite susceptible to damage by snow plows.

Other types of studs in some use in this country are a type which provides audible lane guidance by producing an unpleasant rumbling beneath the tire when driven on. These studs are used in urban areas to discourage weaving in high volume areas.

5–6. Signs. The most common and oldest method of traffic control is by means of the traffic sign. Signs function to control the movement of vehicles, to reduce the hazard of traffic operation, and to improve the quality of flow. These three functions are carried out by three different classifications of road sign, which are visually different to enable drivers to determine rapidly the category of any particular sign. The three classifications of traffic signs are:

1. Regulatory signs
2. Warning signs
3. Guide signs

5–7. Regulatory Signs. Regulatory signs inform the driver of the applicability of specific laws and regulations along indicated sections of road. In the absence of such regulatory signs, enforcement would not be possible, since the restrictions are not universally recognized. Thus, it is necessary to indicate a one-way street by posting a "Do Not Enter" sign. It is not, however, necessary to inform the driver of the universally accepted law[2] of driving on the right hand side of a street with two-way traffic. Regulatory signs are posted where they can be clearly seen, and the section of road over which the regulation applies should be clearly delineated. Regulatory signs can themselves be subclassified into:

1. Right-of-way signs
2. Speed signs
3. Movement signs
4. Parking signs
5. Pedestrian signs
6. Miscellaneous signs

The form of all signs is clearly defined by the Federal Highway Administration [2]. Figure 5–2 shows typical signs from each subclassification.

Right-of-way Signs—"Stop" and "Yield" signs are placed at the junctions of traffic flows to indicate which stream of traffic has right-of-way and the type of movement minor flows must observe before moving into the intersection.

Speed Signs indicate both daytime and nighttime speed limits and, in addition, delimit the beginning and end of speed zones.

Movement Signs indicate legal, mandatory, and prohibited movements.

Parking Signs state regulations governing the stopping and standing of vehicles. In general such signs may indicate times and days on which regulations are in force.

[2] In only a few nations of the world does the driving-on-the-left rule prevail. In recent years several countries have changed to the driving-on-the-right rule.

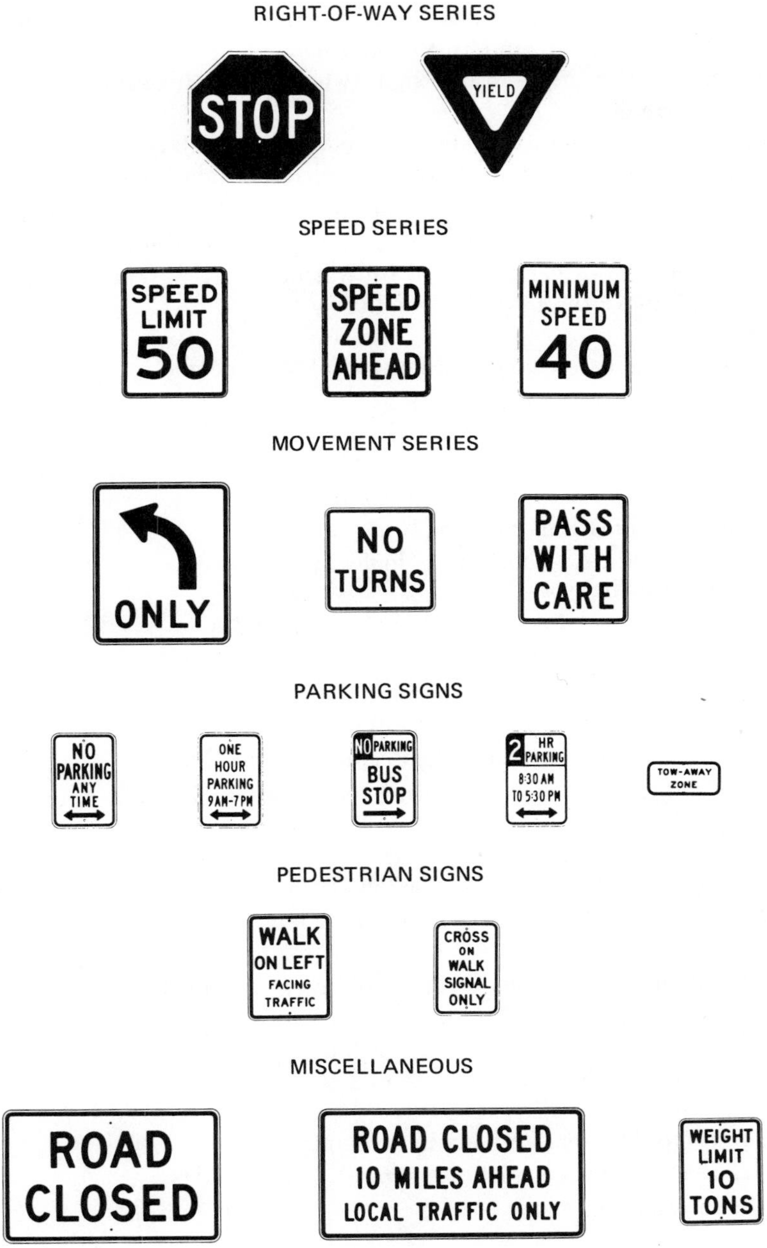

RIGHT-OF-WAY SERIES
STOP
YIELD
SPEED SERIES
SPEED LIMIT 50
SPEED ZONE AHEAD
MINIMUM SPEED 40
MOVEMENT SERIES
ONLY
NO TURNS
PASS WITH CARE
PARKING SIGNS
NO PARKING ANY TIME
ONE HOUR PARKING 9AM-7PM
NO PARKING BUS STOP
2 HR PARKING 8:30 AM TO 5:30 PM
TOW-AWAY ZONE
PEDESTRIAN SIGNS
WALK ON LEFT FACING TRAFFIC
CROSS ON WALK SIGNAL ONLY
MISCELLANEOUS
ROAD CLOSED
ROAD CLOSED 10 MILES AHEAD LOCAL TRAFFIC ONLY
WEIGHT LIMIT 10 TONS

Fig. 5–2. U.S. regulatory traffic signs. (Source: *Manual of Uniform Traffic Control Devices for Streets and Highways,* U.S. Department of Transportation.)

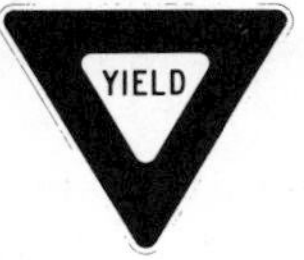

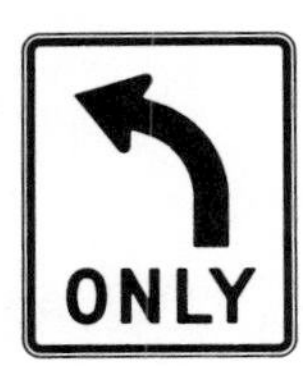

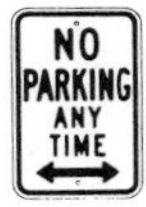

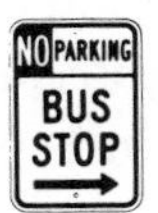

Pedestrian Signs are signs directed principally at pedestrian movements both in urban and rural movements. As such, their method of placement differs from other signs whose chief function is to inform vehicular traffic.

Miscellaneous Signs include a miscellany of regulations such as detour indications, information on road closings, and weight limits.

5–8. Warning Signs. Where caution is required, including in some cases the reduction of speed, a special alertness to conditions on or close to the road, the driver is alerted by warning signs. Warning signs may be erected under the following conditions [2]:

1. To indicate changes in horizontal alignment
2. To indicate an intersection
3. To give warning that the driver should expect traffic control devices
4. To warn of converging traffic lanes
5. To indicate narrow roadways
6. To indicate changes in highway geometry such as the end of a divided highway
7. To advise of unexpected or unusual grades
8. To indicate sudden changes in surface condition, or poor pavement condition
9. To advise of an at-grade rail crossing
10. To indicate unexpected entrances and crossings
11. Other signs such as advised speed, clearance limits, animal crossings, etc.

Warning signs are customarily yellow diamond-shaped with black edging and lettering. Exceptions to this are the advised speed signs which are rectangular in shape.

5–9. Guide Signs. Guide signs are erected along highways to enable the traveler to find and follow routes in rural and urban areas, and to identify and locate items of need and interest. This class of sign is normally considered to be composed of three categories:

1. Route markers and auxiliary markers
2. Destination and distance signs
3. Information signs

Examples of each type are shown in Fig. 5–3.

5–10. Channelization. One of the most effective and efficient methods of highway traffic control is the adoption of high standards of geometric

ROUTE MARKERS

AUXILIARY ROUTE MARKERS

DIRECTION AND DESTINATION MARKERS

INFORMATION

Fig. 5–3. Guide signs on U.S. highways. (Source: *Manual of Uniform Traffic Control Devices for Streets and Highways,* U.S. Department of Transportation.)

design at intersections. Islands and lane markings can be used to separate the intersecting, diverging, merging, weaving and turning movements that can occur within any individual intersection. In determining the design of channelized intersections, engineers observe all or some of the following principles [3]:

1. The areas of conflict must be reduced. By confining the driver's ability to make illogical maneuvers, much uncertainty of behavior can be eliminated.
2. Crossing streams at or near right angles. This reduces the size of the conflict area and crossing time, cuts down the impact energy

of any accidents that do occur, and puts drivers in the best position to judge the location and relative speed of the other car.

3. Where one traffic stream is to be bent to conform with channelization practice, the minor stream should be bent.
4. Merging streams shall converge at small angles, permitting the convergence at minimum speed differentials.
5. Turning and crossing vehicles need "shadowed areas" while waiting for an opportunity to clear the intersection.
6. Vehicles entering or leaving a high-speed roadway should have acceleration or deceleration lanes to reduce the relative speeds of vehicles or merging and diverging vehicles.
7. Prohibited movements can be blocked by channelizing islands.
8. For effective signal control at multiphase signal intersections, channelization may be required.

Figure 5–4 shows examples of the typical channelization techniques for various intersection geometries [4].

5–11. Traffic Signals. Intersections that carry large vehicular volumes cannot be safely and satisfactorily controlled without traffic signals. The installation of power-operated traffic signals at an intersection can effectively separate all or most conflicting flows bringing about a degree of orderliness and safety that would otherwise be impossible at higher traffic volumes. When designed and located properly, traffic signals have several advantages:

1. They provide for orderly movement of traffic. Where proper physical layouts and control measures are used, they can increase the traffic-handling capacity of the intersection.
2. They reduce the frequency of certain types of accidents (particularly right angle collisions).
3. Under conditions of favorable spacing, they can be coordinated to provide for continuous or nearly continuous movement of traffic at a definite speed along a given route.
4. They can be used to interrupt heavy traffic at intervals to permit other traffic, pedestrian or vehicular, to cross.
5. They represent a considerable economy, as compared with manual control, at intersections where the need for some definite means of assigning right-of-way first to one movement and then to another is indicated by the volumes of vehicular and pedestrian traffic, or by the occurrence of accidents [2].

Depending on their location and the type of traffic which they are designed to accommodate, individual traffic signals can be *pretimed, semi-actuated, fully actuated,* or *volume-density* signals, according to the type of controller which is used in their operation.

Pretimed signals, sometimes called fixed-time, operate by means of an electric motor that drives an adjustable dial timing mechanism capable of providing cycles for 30 seconds to 120 seconds in length. Signals with only one timing dial provide the same cycle length and split on a twenty-four-hour basis. More frequently, pretimed signal controllers operate with three dials, which are programmed to operate different cycle lengths and splits for the A.M. peak hour, the P.M. peak hour, and all off-peak times.

Semi-traffic-actuated signals operate on the basis of providing green time to the major artery until detectors on the minor roadway approaches actuate a demand for a signal. After providing green time for the minor roadway, the control returns to its normal position of permitting the main artery traffic to pass through the intersection. Traffic on the minor approaches receives variable lengths of green time up to some set maximum, dependent on the demand. Where all the traffic on the minor approach cannot clear the intersection during the maximum minor flow period, the green signal is returned to the major artery for some minimum predetermined time known as "minimum artery green," before returning again to the minor flow. In this manner the semi-traffic-actuated signal gives precedence to major arteries.

Full-traffic-actuated equipment can be useful where wide variations in traffic demand can occur on two or more conflicting approaches. Presence detectors are installed on all approaches. Maximum and minimum green phase times are selected. Arriving traffic triggers the detectors and the green phases adjust to reflect the arrival rates of the various approaches. Isolated intersections with large volume variations work well under full-traffic-actuated conditions.

Volume density equipment is a complex fully actuated system, permitting variation of phase and cycle lengths according to input data. Sensing devices record data that gives information on arrivals, waiting times, and headways at all intersection approaches. Green times and cycle lengths are constantly readjusted, enabling the signal to provide a maximum degree of response to traffic flows and fluctuations.

5–12. Signal Systems. Individually signalized intersections are frequently interconnected to form *signal systems.* The type of signal system adopted along the length of a route will influence the overall character of traffic flow through the area. In *progressive systems,* successive intersections along a street have identical cycle lengths. The cycle splits at individual intersections vary according to traffic demands of the cross streets. By off-setting the starting point of the cycles progressively

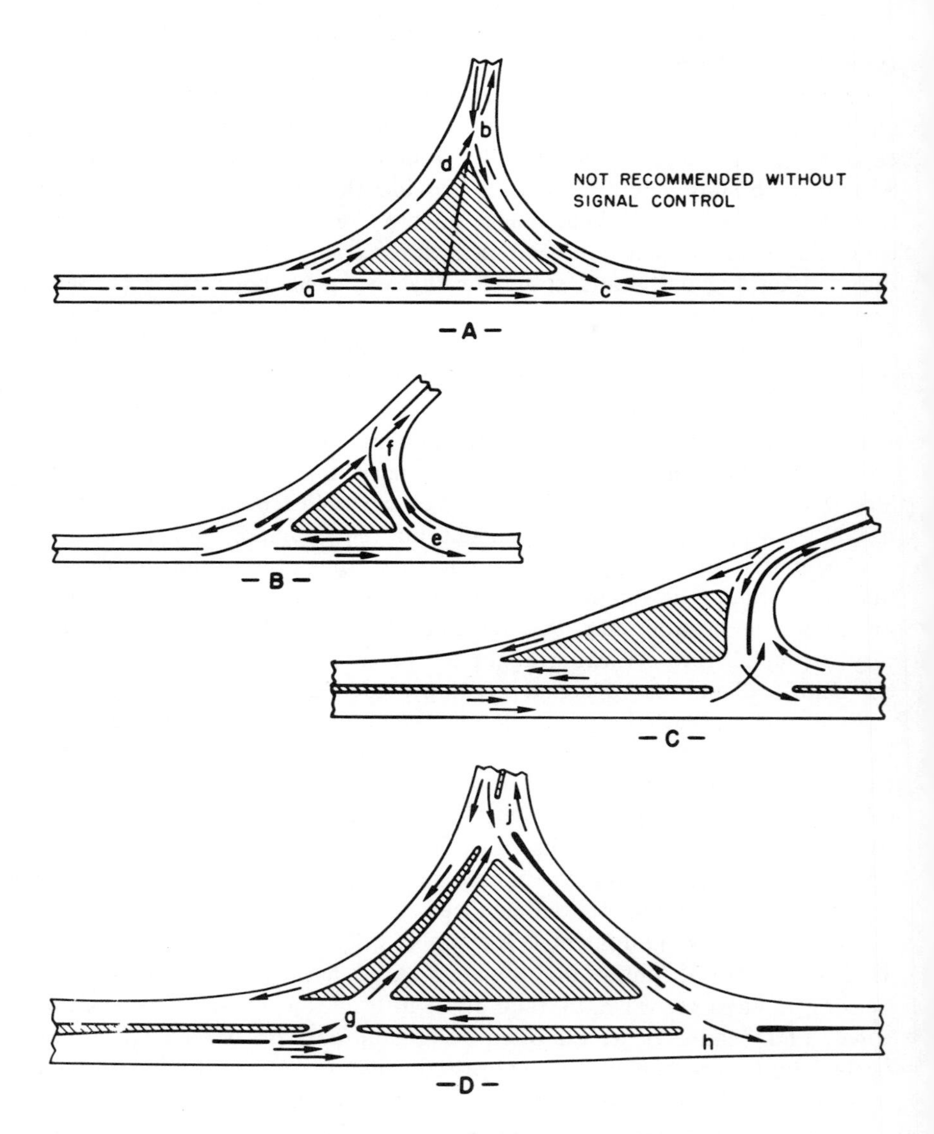

Fig. 5–4. Typical channelization patterns. (Source: AASHO—*A Policy on Geometric Design of Rural Highways,* 1965.)

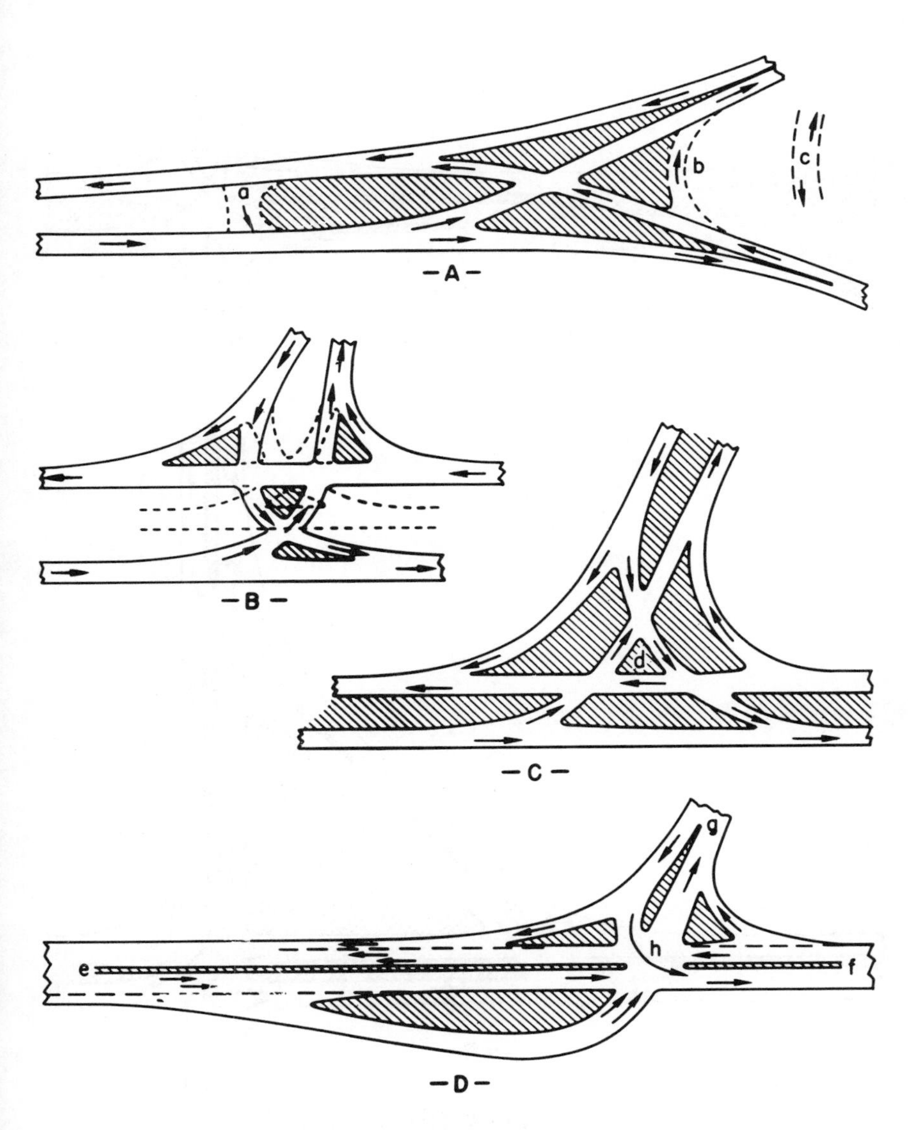

T OR Y INTERSECTIONS
CHANNELIZED - HIGH TYPE

Fig. 5–4. (*Continued.*)

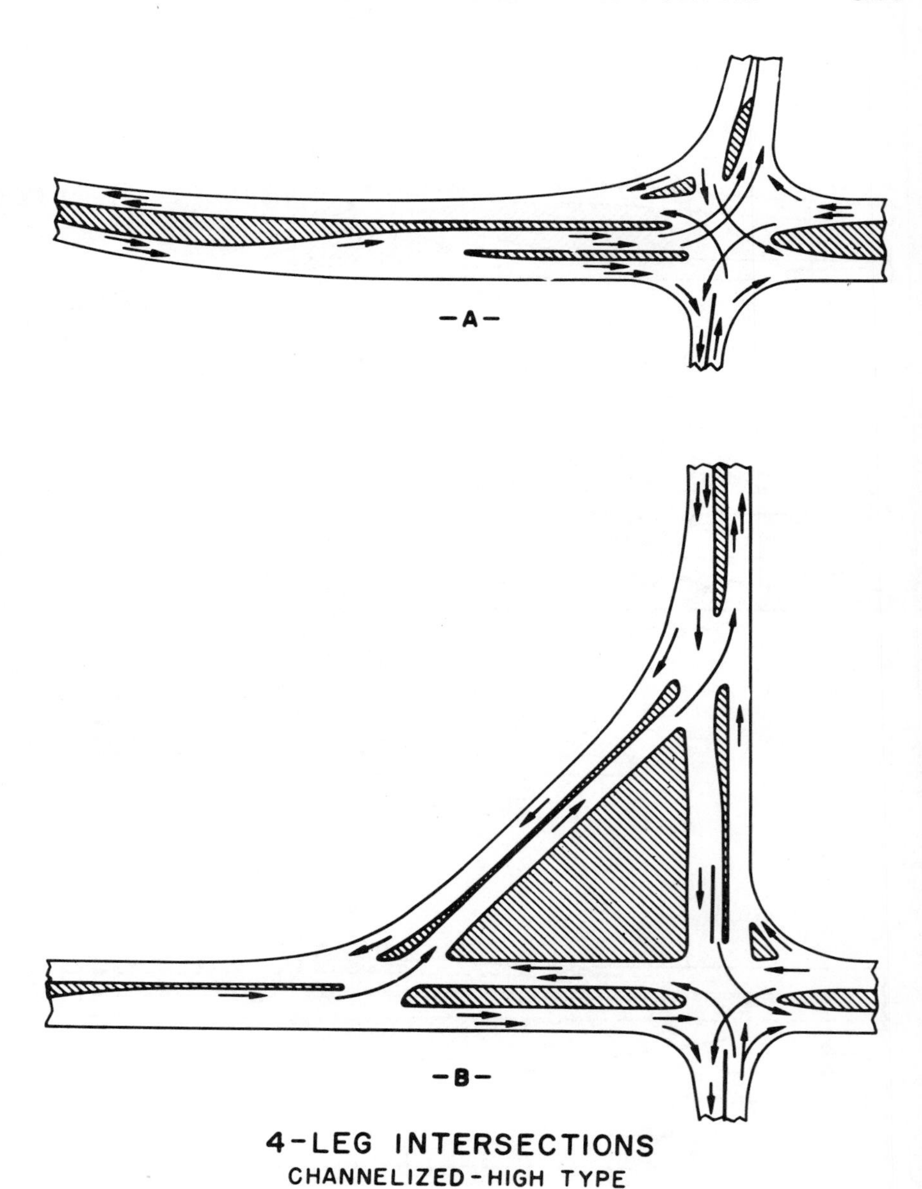

4-LEG INTERSECTIONS
CHANNELIZED-HIGH TYPE

Fig. 5–4. (*Continued.*)

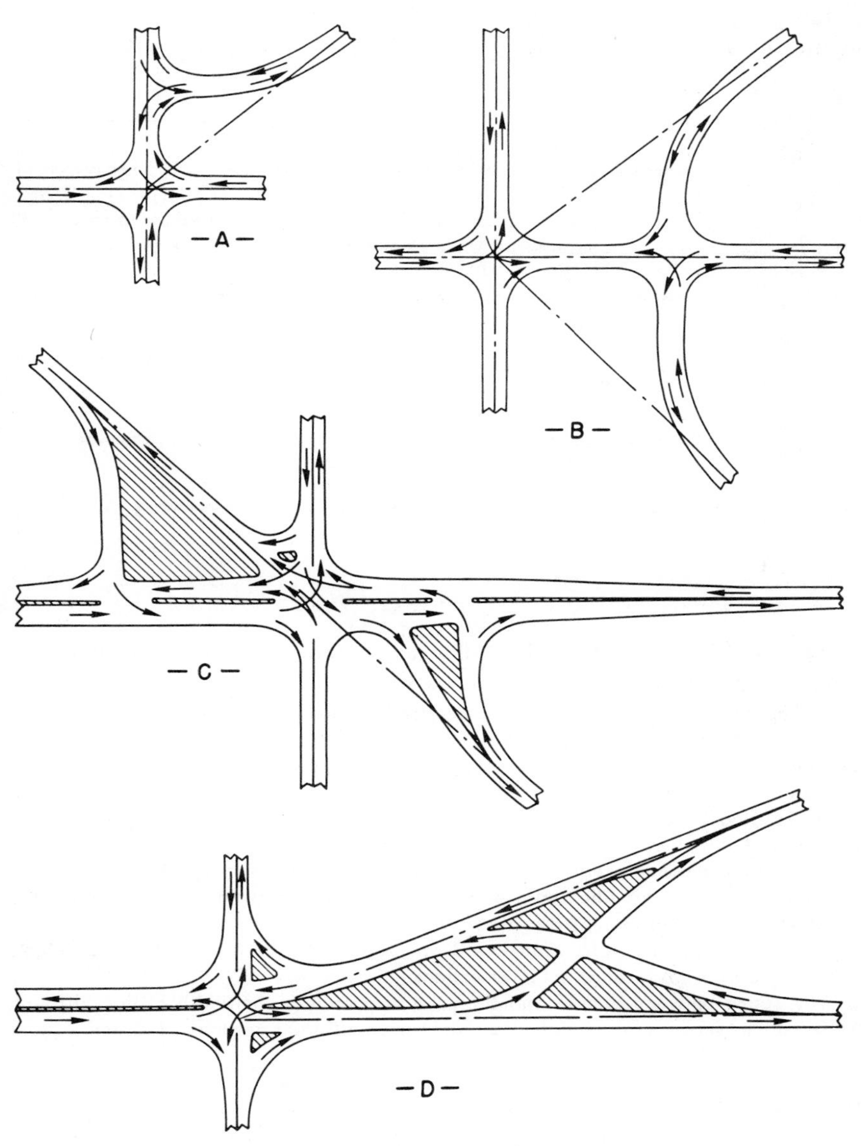

MULTILEG INTERSECTIONS

Fig. 5–4. (*Continued.*)

along the artery it is possible for a car maintaining the design speed of the progression system to pass through all intersections without encountering a red phase at any time. Figure 5–5 shows a time–space

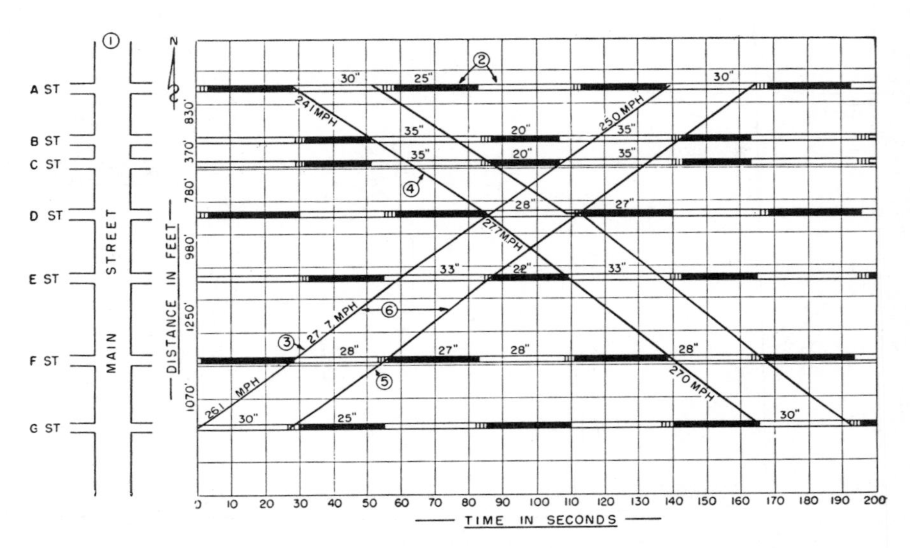

Fig. 5–5. Typical time space diagram. (Source: *Traffic Engineering Handbook*, Institute of Traffic Engineers.)

Note: The following items refer to corresponding numbers in Figure 5–5.

(1) Plan of street drawn to scale.
(2) "Go" intervals for Main Street designated by open space; "Stop" intervals for Main Street designated by solid line; "yellow" intervals for Main Street designated by crosshatching.
(3) Slope of this line indicates the speed and represents the first vehicle of a group or platoon moving progressively through the system from G Street to A Street.
(4) As in 3 above, this line represents the first vehicle moving from A Street to G Street in the southbound direction.
(5) This line is roughly parallel to line 3 and represents the last northbound vehicle in a group to go through all intersections.
(6) The space between lines 3 and 5 on the time scale is the width of the "through band" in seconds; this width may vary over different sections of street.

diagram for a progressive system on a two-way street. While progressive systems work reasonably well for light to medium-heavy traffic, they have been found to break down under very heavy loading.

Simultaneous systems, where all signals along one street are green or red at the same time, are often used under extreme loads, since all traffic tends to move at the same time along the street and there is little speed variation between blocks. Obviously, in the operation of a simultaneous system all signal cycles along a street must be identical, and their starting points synchronized.

Under heavy loading conditions, *alternate systems* have also been known to work more effectively than progressive systems. In the alternate system, adjacent signals or signal groups show opposite color phases. Under favorable conditions, the system can lead to continuous movements of platoons of vehicles along the street.

The most sophisticated traffic signal systems are those where entire networks are *computer controlled.* Detectors throughout the system provide data on current traffic conditions to a central computer. The computer can then calculate the best sequence of signal changes throughout the whole system. In such a system, the computer has complete and direct control of all signals in the system. Constant readjustment of signal timing takes place, providing immediate response to system demand. In one area the adoption of a computer controlled system resulted in a 25 per cent decrease in delay to vehicles and a 28 per cent decrease in overall congestion [5].

AIDS TO MARINE NAVIGATION

5–13. Purpose of Aids to Marine Navigation. Because of the low volumes of vessel traffic that use the waterways and harbors, aids to navigation function more in the area of guidance devices than controls. Maneuvering controls are provided chiefly by rules of seamanship. Navigation aids, therefore, serve two main purposes:

1. To provide information that will permit overseas vessels to determine their exact position at landfall, and to allow precise positioning within the waterways and close to coastal waters.
2. To mark channels clearly and warn of hazards such as wrecks, sandbars, and other sudden changes in bottom topography.

5–14. Buoys. Buoys are anchored floating markers which may be lighted or may give out some sound signal to indicate their position under conditions of limited visibility. A variety of buoy shapes is used, as indicated by Fig. 5–6, which shows standardized buoy shapes and markings [6].

Spar buoys are in the shape of large logs, floating with one end out of the water. *Can buoys* are cylindrical markers built up from steel plates, while *nun* buoys are distinguished by their truncated conical top. *Bell, gong,* or *whistle* buoys are steel floats supporting small skeleton towers housing the signal making equipment. Most buoys making sound signals are operated by the motion of the sea itself. Less com-

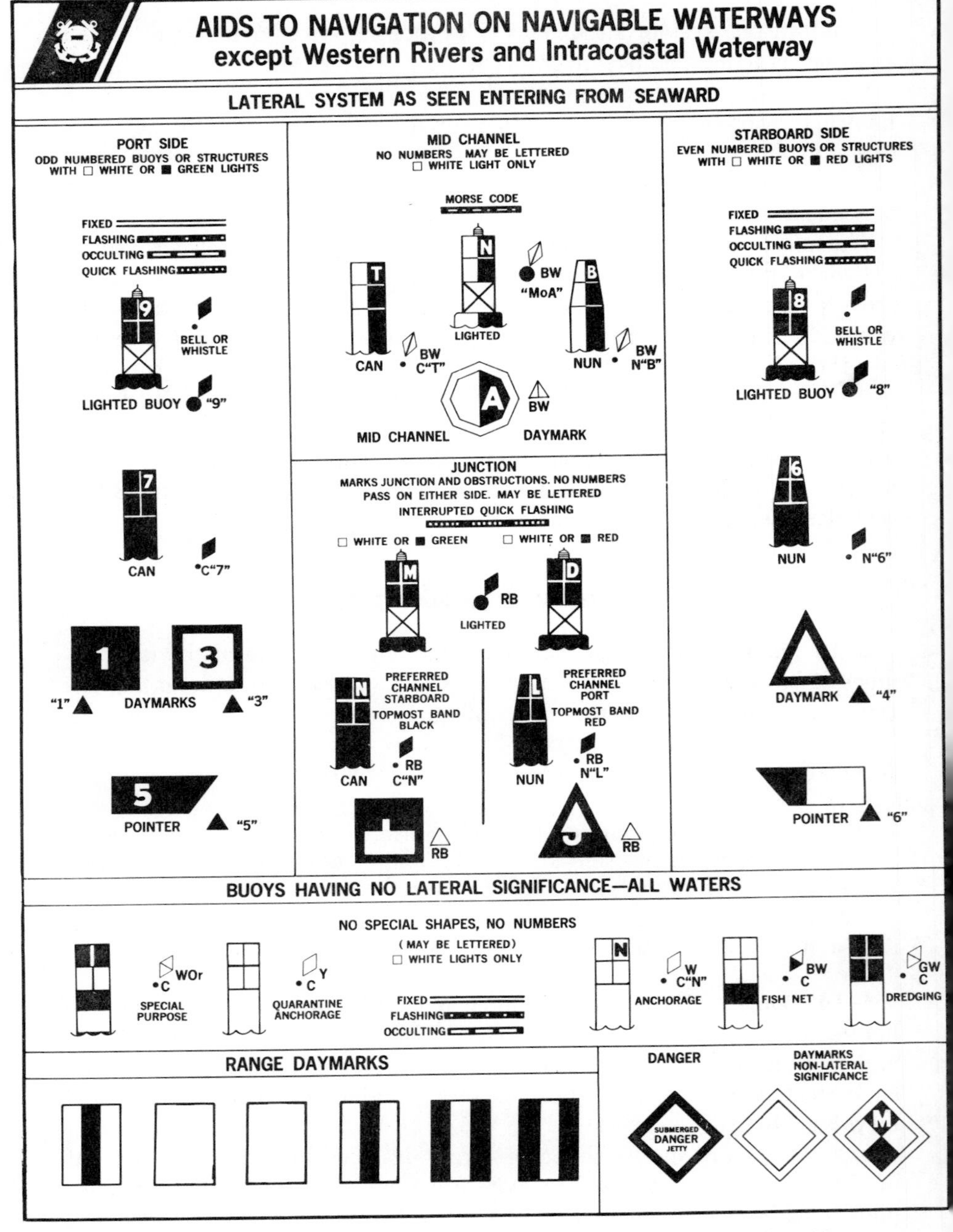

Fig. 5–6. The lateral system of buoy marking of the U.S. (Source: *Marine Aids to Navigation,* 1971. U.S. Department of Transportation, United States Coast Guard.)

monly compressed gas or electric power from batteries is used to produce the signals.

United States navigable waters are marked by buoys according to the *lateral system.* Buoys are numbered and marked in navigable channels from the point of entry to the head of navigation. Where channels do not lead from seaward, a direction is assumed for enumeration purposes. For example, directions of proceeding are southward on the Atlantic coast and the Atlantic Intracoastal Waterway, and northward and westward on the Gulf coast.

In addition to the numbering and lettering by which buoys can be referenced to charts, coloring systems are used to denote both purpose and side on which they are to be passed. Black buoys mark port sides of channels or obstructions requiring passage to the starboard. Red buoys mark the starboard sides of channels and passage to the port. Figure 5–6 shows marking preferred channels, quarantine and other anchorage in addition to other markings of special significance.

Often buoys are equipped with reflectors. *Radar reflectors* improve the radar response of the buoy for easier location in limited visibility [7]. *Optical reflectors* of different colors are used to enable buoys to be picked up by ships' searchlights.

Lighted buoys may have red or green lights depending on the color of the buoy on which they are mounted. In addition, white lights are mounted on buoys to convey signals of special significance. For example, flashing lights indicate black, red, or special purpose buoys and quick-flashing lights on black or red buoys indicate points at which special caution is required. Interrupted quick-flashing lights indicate junctions in channels or wrecks. Morse code flashes, groups of one short flash followed by one long flash, are placed only on buoys marking fairways and channel center lines which may be passed on either side.

5–15. Lighthouses. Lighthouses are used on prominent points and at areas of special hazard to aid ships in locating their position and to warn of possible danger. The function of a lighthouse is simply to support a light which is sufficiently powerful that it can be seen at great distances. The type of structure used varies greatly according to location. This variation depends on the height at which the light must be supported, the supporting foundation conditions for the structure, the availability of materials, and the physical loads that the structure must sustain. In addition, the type of design will depend on whether the light is to be manned at all times. Many lighthouses which were originally manned continuously are now operated by electrical remote control machinery. The structures of these lights reflect, in these cases, an obsolete technology, yet there is no reason to replace such existing structures until it becomes uneconomic or undesirable to continue their use. Many light structures or lighthouses emit sound signals under foggy conditions. Each station emits patterns of blasts at prescribed intervals which permit vessels to identify the sending station.

Just as lighthouse fog signals uniquely identify the sending station within the area, the lights themselves vary in their periods of lights and darknesses. Such "light characteristics" enable a vessel within any area to determine with certainty which signal is being observed. Light colors used are white, red, and green and can be either continuous in one color or alternating. Figure 5–7 shows the variations of phasing of light characteristics including fixed, flashing, occulting, and Morse code patterns.

5–16. Range Lights. Range lights are a simple yet effective navigation device consisting of two powerful lights some distance apart set along the axis of a channel, either within the channel or inland. Any vessel which aligns itself with the lights and steers a course that keeps the two lights aligned moves within a safe channel. Range lights are usually visible in one direction only and rapidly lose brilliance as the ship diverges from the range line. The lights can be any of the three standard colors and have characteristic phases which permit differentiation from other lights. When using range lights it is usually necessary to use, in addition, charts which designate the stretches of channel over which the range lights can safely be used.

5–17. Lightships. Lightships are functionally identical to lighthouses, except that they can be placed in positions where the construction of a lighthouse would be expensive and difficult. Lightships are manned vessels carrying lights, fog signals, and radio beacons. They are painted red with white superstructures, and carry the name of the station in white on both sides. When lightships must leave their station for servicing or some other reason, they are replaced by relief lightships which conform in coloring but carry the work "Relief" on either side. Any lightship under way or off station flies the international signal PC which signifies a lightship off station. Lightships ride on a single anchor and when on station exhibit only two lights, a brilliant masthead light, and a less brilliant forestay light. The position of the forestay light relative to the masthead light indicates the direction of current.

5–18. Fog Signals. In periods of limited visibility, lights are supplemented by sound signals. Each station emits a signal characteristic to itself to help the mariner determine his exact position. When synchronized with lights, the time interval between seeing the light and hearing the signal will enable a vessel to determine its distance from the station. Fog signals vary in type. Commonly diaphragm horns, reed horns, sirens, diaphones, whistles, and bells are used at prescribed

Illustration	Symbols and meaning		Phase description
	Lights which do not change color	Lights which show color variations	
	F. = Fixed	Alt. = Alternating.	A continuous steady light.
	F. Fl. = Fixed and flashing.	Alt. F. Fl. = Alternating fixed and flashing.	A fixed light varied at regular intervals by a flash of greater brilliance.
	F. Gp. Fl. = Fixed and group flashing.	Alt. F. Gp. Fl. = Alternating fixed and group flashing.	A fixed light varied at regular intervals by groups of 2 or more flashes of greater brilliance.
	Fl. = Flashing	Alt. Fl. = Alternating flashing.	Showing a single flash at regular intervals, the duration of light always being less than the duration of darkness. Shows not more than 30 flashes per minute.
	Gp. Fl. = Group flashing.	Alt. Gp. Fl. = Alternating group flashing.	Showing at regular intervals groups of 2 or more flashes.
	Qk. Fl. = Quick flashing.	----------	Shows not less than 60 flashes per minute.
	I. Qk. Fl. = Interrupted quick flashing.	----------	Shows quick flashes for about 4 seconds, followed by a dark period of about 4 seconds.
	S-L. Fl. = Short-long flashing.	----------	Shows a short flash of about 0.4 second, followed by a long flash of 4 times that duration.
	Occ. = Occulting.	Alt. Occ. = Alternating occulting.	A light totally eclipsed at regular intervals, the duration of light always equal to or greater than the duration of darkness.
	Gp. Occ. = Group occulting.	----------	A light with a group of 2 or more eclipses at regular intervals.

Fig. 5–7. Characteristic light phases for marine navigation lights. (Source: *Marine Aids to Navigation,* 1971. U.S. Department of Transportation, United States Coast Guard.)

intervals. On buoys, bells and whistles operated by the motion of the sea operate at irregular intervals.

5–19. Radio Beacons. Electronic devices provide all-weather navigation aid both to coastal and deep sea navigation. There are now a large number of installations of marine radio beacons operating in the 285 to 325 kilocycle range. In 1969, close to two hundred stations were operated by the U.S. Coast Guard on the Atlantic, Gulf, and Pacific Coasts and on the Great Lakes. The chief advantage of the radio beacon system is its essential simplicity. All that is required is a relatively inexpensive receiver with a directional antenna. Signals from radio beacons are received on a short-to-medium range up to 100 miles. Most beacons transmit one out of six minutes continually while other stations transmit continuously twenty-four hours a day. Each stations carries a characteristic signal permitting unique identification.

The LORAN system is a long range radio system made up of transmitting pairs of stations which permit vessels to position themselves at great distances from shore. Location is carried out by determining the difference between time of reception of signals from pairs of stations transmitting a simultaneous signal. The locus of this time difference is known as a loran line. By receiving signals from two pairs of stations, intersecting loci can be plotted giving a firm position fix. Two LORAN systems are currently available, LORAN A previously known as standard loran, operating at 1850 to 1950 kc, and LORAN C which operates at a lower frequency of 100 kc [6].

5–20. Radar Devices. Since World War II, the field of electronics has supplied several powerful additional tools to maritime navigation devices. Most important of these is undoubtedly radar. Radar, originally a shortening of Radio Direction and Ranging, makes use of the echo principle with electronic waves rather than the audible sound wave. Marine radar works on the basis of generating radio frequency oscillations of specific form and transmitting them into space in a narrow beam of radio waves. The beam is continuously rotated in azimuth about the sending aerial. Radio echoes from any suitable object return to a receiver and are displayed on a screen in such a way that the distance and direction of the surrounding objects can be directly determined. The essential equipment for radar, therefore, is a transmitter for generating radio waves, a rotating aerial or scanner to send out direct waves and pick up echoes, a receiver and a display for converting the received waves into directly readable information in the form of a radar plot [8].

Radar is available both on board ship and at shore stations. Ships, therefore, are able to navigate close to shore under the worst visibility conditions and can determine the presence of other ships. On-shore radar enables close ship-to-shore cooperation at all times provided the ship has at least radio communication with land.

RAILROAD SIGNALS AND TRAFFIC CONTROL

5–21. Railroad Signal Systems. Railroads have the most extensive control system of any major transportation mode excluding pipelines. At all times, the locomotives remain under a high degree of control. This is made possible by the restricted freedom of movement of the vehicles and the operation of a self-contained system over an essentially exclusive right-of-way with a minimum of conflicts from other modes. Without the elaborate forms of traffic control currently used, the *modus operandi* of railroads would require extensive modifications from a safety viewpoint. With existing procedures, the railroads can operate high speed, high capacity freight, and passenger service, often with opposing movements on single track with a high degree of safety. Passenger trains can have capacities in the region of 1,000 persons and operate at running speeds of 80 miles per hour. The fatality rate per passenger mile of rail operations is approximately one twenty-fifth of that for highway movements by private automobile. This remarkable safety record is built upon an elaborate signal and control system.

In areas where only light densities of traffic occur, operations are possible without signals by adherence to operating rules and time tables. (At low densities of air traffic, airports operate without radar by using time separations between aircraft operations.) Under heavier densities signal systems help provide train control. The two types of signal used almost exclusively are *semaphore* and *light signals* [9]. The Association of American Railroads has published the Standard Code of Operating Rules, Block Signal Rules, and Interlocking Rules which shows both the *aspect* and *indications* recommended for both semaphore and light signals.

The *aspect* of a signal is the position of the lights or semaphore blades and their relative arrangements. The *indication* of a signal is the message conveyed by the aspect of the signal according to the rules, indicating how the operator is to control the movement of the train. Figure 5–8 depicts the aspects of the AAR Standard codes showing the standard *indications* and *applications* (points where these aspects may be displayed). While the number of aspects is numerous, each signal light or semaphore has three basic positions.

Almost all semaphore signals currently in operation use the upper right-hand quadrant. The vertical position is used for *Clear,* the 45-degree position for *Approach* (requiring speed reduction), and the

RULE	ASPECT	NAME	INDICATION	APPLICATION
281	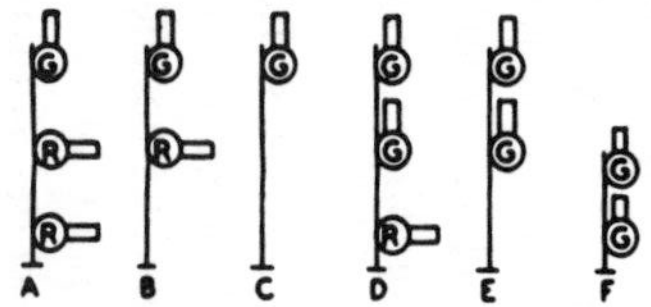 A B C D E F	Clear	Proceed	At entrance of normal speed route or block, to govern train movements at normal speed.
281A	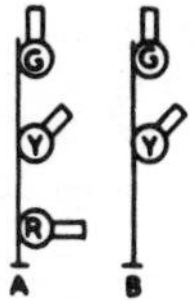 A B	Advance Approach Medium	Proceed approaching second signal at medium speed	At entrance of normal speed route or block, to govern the approach to approach-medium signal.
281B		Approach Limited	Proceed approaching next signal at limited speed	At entrance of normal speed route or block, to govern the approach to limited-clear signal.
281C		Limited-Clear	Proceed; limited speed within interlocking limits	At entrance of limited speed route or block, to govern train movements at not exceeding limited speed.

Fig. 5–8. Aspects of association of American railroads standard code of signals. (Source: *Railway Track and Structures Cyclopedia,* 8th ed., 1955, Simmons Boardman Publishing Corporation.)

282		Approach Medium	Proceed approaching next signal at medium speed	At entrance of normal speed route or block, to govern the approach to Medium-Clear, Approach, Medium-Approach, Advance-Approach or Approach Medium signals.
282A		Advance Approach	Proceed preparing to stop at second signal	At entrance of normal speed route or block, to govern the approach to approach signal.
283	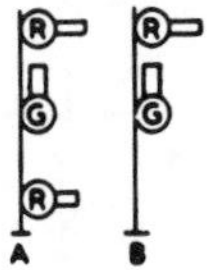	Medium-Clear	Proceed; medium speed within interlocking limits	At entrance of medium speed route or block, to govern train movements at not exceeding medium speed.
283A		Medium-Advance Approach	Proceed preparing to stop at second signal; medium speed within interlocking limits	At entrance of medium speed route or block, to govern approach to approach or medium-approach signal.
283B	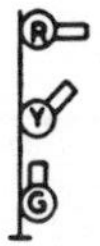	Medium-Approach Slow	Proceed at medium speed approaching next signal at slow speed	At entrance of medium speed route or block, to govern the approach to slow-clear or slow-approach signal.

Fig. 5–8. (*Continued.*)

RULE	ASPECT	NAME	INDICATION	APPLICATION
284	 Y R G	Approach Slow	Proceed approaching next signal at slow speed. Train exceeding medium speed must at once reduce to that speed	At entrance of normal speed route or block, to govern the approach to Slow-Clear, or Slow-Approach signals.
285	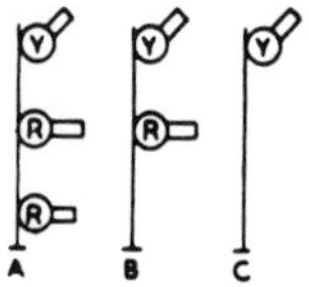 A: Y R R; B: Y R; C: Y	Approach	Proceed preparing to stop at next signal. Train exceeding medium speed must at once reduce to that speed	At entrance of normal speed route or block, to govern the approach to Slow-Clear, Slow-Approach, Permissive, Restricting, Stop and Proceed, or Stop signals; red switch lamp and end of signaled territory.
286	 R Y R	Medium-Approach	Proceed at medium speed preparing to stop at next signal	At entrance of medium speed route or block, to govern the approach to Slow-Clear, Slow-Approach, Permissive, Restricting, Stop and Proceed, or Stop signals and end of signaled territory.
287	A: R R G; B: G R; C: G	Slow-Clear	Proceed; slow speed within interlocking limits	At entrance of slow speed route or block, to govern train movements at slow speed.

Fig. 5–8. (Continued.)

288	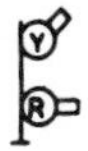	Slow-Approach	Proceed preparing to stop at next signal; slow speed within interlocking limits	At entrance of slow speed route or block, to govern the approach to Permissive, Restricting, Stop and Proceed, or Stop signals.
289		Permissive	Block occupied, proceed prepared to stop short of train ahead Designate by 1. Letter plate, or 2. Marker light, or 3. Shape of arm, or 4. Combination of these distinguishing features	At entrance of a manual block to govern trains entering and using that block.
290	 A B C D	Restricting	Proceed at restricted speed	At entrance of normal speed, medium speed, or slow speed route or block, to permit trains to proceed prepared to stop short of train, obstruction, or anything that may require the speed of the train to be reduced.
291	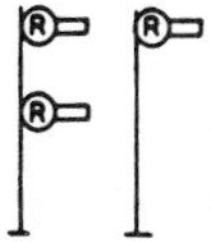 A B	Stop and Proceed	Stop; then proceed at restricted speed Designate by: 1. Number plate, or 2. Marker light, or 3. Pointed blade, or 4. Combination of these distinguishing features Note: Railroads desiring to avoid stopping trains may arrange accordingly	At entrance of a route or an automatic block requiring trains to stop and after stopping, permitting them on two or more tracks to proceed at restricted speed and on single track in accordance with the rules.
292	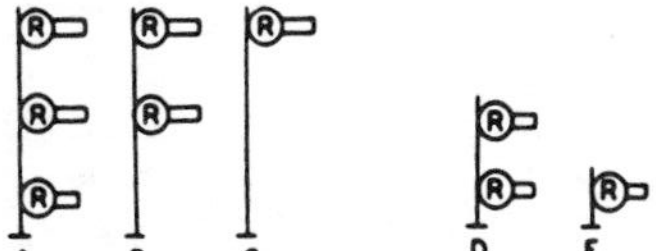 A B C D E	Stop	Stop	At entrance of a route or block requiring trains to stop until authorized to proceed by train order, clearance card, more favorable indication than Stop, or in accordance with the rules.

Fig. 5–8. (Continued.)

horizontal position for the most restrictive aspect, *Stop* or *Stop and Proceed.*

Light signals are arrangements of electric lights and optical lenses which are capable of producing bright colored lights visible even in bright sunlight. Three colors of lamp are used. Green is used for *Clear,* yellow for *Approach,* and red for the most restrictive aspect.

5–22. Automatic Block Signaling. The basic method of preventing rear-end collisions on double track sections of railroad, and both head-on and rear-end collisions on single track sections is the system of automatic block signaling. Through the use of track circuits a definite space interval can be provided between following or opposing trains. The term *automatic* is applicable because the control of the space interval is provided by the presence of the trains themselves. Under this system, sections of track are divided longitudinally into smaller sections called *blocks.* By providing a signal system in which no two trains are permitted to occupy the same block at the same time, collision situations are avoided. Initially, block signal systems were controlled manually. The introduction of automatic block signals reduced the possibility of human error, affording a higher level of protection. In addition to the increased safety, additional benefits accrued from the increased capacity which resulted from reduced delays and lower time intervals between meets and passes.

Figure 5–9 shows in simplified form the track circuitry required for a two-position one-block signal system suitable for operation for one direction of movement. The circuit consists of both track rails, insulated from rails of adjacent blocks by insulated joints at the ends of the block. At one end of the block the rails are joined to a battery. At the other end the rails are connected to the coil of a high-voltage resistance relay. With no train in the block the current runs through the relay, keeping the contact in the signal circuit closed. With the contact closed, the signal at the beginning of the block shows the Clear aspect. The presence of a train within the block provides a low-resistance shunt across the rails. The contact within the signal circuit is broken and the signal changes to Stop.

Figure 5–10 indicates an arrangement which allows a higher capacity for single direction track. Trains are informed under this system of the presence of trains two blocks ahead. Three-aspect two-block signaling permits a train to receive a Stop and Proceed signal if another train is in the block immediately ahead. An Approach aspect is displayed if the nearest preceeding train is two blocks ahead. If no trains occupy the next two blocks, then the Clear aspect is displayed. This type of signaling permits the second train, B, to slow when approaching the first train, A, without coming to a complete stop unless the distance

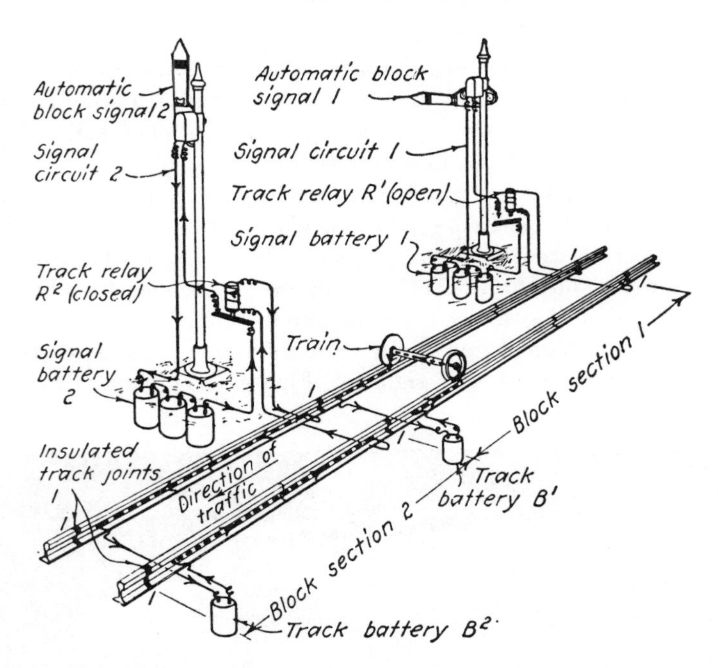

Fig. 5–9. A simplified typical circuit of an automatic block signal system. (Source: *Railway Track and Structures Cyclopedia,* 8th ed., 1955, Simmons Boardman Publishing Corporation.)

between them falls below a minimum stopping distance. With this type of block system, blocks can be shorter and unnecessary stopping minimized.

5–23. Centralized Traffic Control. By the use of centralized traffic control, a dispatcher at a central location controls the operation of

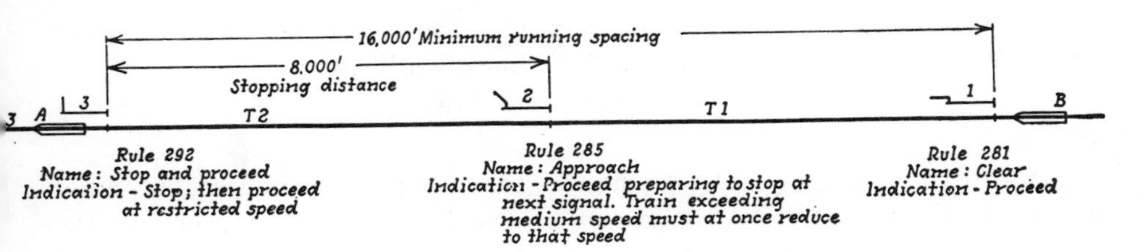

Fig. 5–10. Three-aspect two-block signaling. (Source: *Railway Track and Structures Cyclopedia,* 8th ed., 1955, Simmons Boardman Publishing Corporation.)

switches and signals at key points such as the ends of sidings and junctions. The dispatcher can track the location of trains on each section by visual displays of lights on a track diagram. Using constantly changing displays, the operator can remain informed concerning the position of progress of each train in each territory.

Older forms of CTC operated with single wire connections from the control panel to the switches and signals. With the introduction of coded impulses to code-following relays, the amount of wiring could be significantly reduced and CTC systems were extended to control hundreds of miles of rail systems. The dispatcher supervises a large control board having direct operation of all switches and signals which remotely control the line switches and signals. Figure 5–11 shows the display

Fig. 5–11. Centralized traffic control. (Source: Kansas City Terminal Railway and WABCO Signal and Communication Division.)

board of a large central traffic control system for hundreds of miles of track. By judicious operation of the buttons and switches at the central board the dispatcher can produce the most desirable meets and passes. CTC is obviously applicable to multiple-track operation, but has found its prime application to be the increase of the capacity of single-track sections and lines.

Increase of capacity became important with the introduction of diesel electric locomotives. With these more powerful engines, fewer trains could be run at higher speeds. By operating longer, faster trains at capacity, the number of tracks could be reduced. This has produced extensive saving in the area of maintenance in the face of mounting costs.

AIR TRAFFIC CONTROL

5–24. Purpose of Air Traffic Control. Two basic factors in Air Traffic Control (ATC) are safety and efficiency. Sufficient airspace must be provided to avoid collision, yet in order to maintain efficiency airspace cannot be wasted. ATC must give consideration to all users of airspace: the military, the commercial carriers, and general aviation.

Air traffic control measures have been instituted only after the need was demonstrated by human failures. As air traffic activity grows, increasing traffic problems will probably require further air traffic regulation and system development to provide for the expeditious movement of all aircraft. Greater attention will need to be given to the allotment of airspace and the compatibility of equipment between air carriers and general aviation as well as types of aircraft such as STOL/VTOL, SST, and new types of conventional aircraft. Safety, technology, regulation, and financing are four important factors that affect the development of a responsive air traffic control system.

Formal Federal government involvement in air traffic came into existence with the Air Commerce Act of 1926 which provided for the establishment, maintenance, and operation of lighted civil airways using beacon lights. Today the Federal Aviation Administration (FAA) provides control and navigation assistance for the movement of air traffic. The Federal governing authority in air traffic control exists by virtue of the fact that implications of air travel are nationwide and do not respect state boundaries.

5–25. Visual Flying Rules and Instrument Flying Rules. The Federal government prescribes two basic types of flight rules for air traffic which depend on weather conditions as well as on the location and altitudes of flight paths. These are known as visual flight rules (VFR) and instrument flight rules (IFR). In general, VFR means that the weather conditions are good enough for the aircraft to be operated by visual reference to the ground. On the other hand, IFR conditions prevail when the visibility or the ceiling (height of clouds above ground level) falls below those prescribed for VFR flight. In VFR conditions, there is essentially no enroute air traffic control except where prescribed; aircraft fly according to "rules of the road" using designated altitudes for certain headings and are responsible for maintaining their own separation. Positive traffic control is exercised always in IFR conditions and designated control areas. Essentially, these rules require the controlled assignment of specific altitudes and routes and minimum separation of aircraft flying in the same direction at common altitudes.

5–26. Aeronautical Use of Airspace. Airspace is both controlled and uncontrolled, the difference being that in controlled airspace, flight is conducted in accordance with promulgated altitude and heading combinations as shown in Fig. 5–12. Controlled airspace, extending upward from 700 feet AGL (above ground level) and, in a few areas from 1200 feet AGL, exists in almost all areas of the contiguous United States due to the designation of Control Areas and Transition Areas. In addition, controlled airspace extends upward from the ground in areas immediately surrounding an airport in Control Zones. The demand for airspace exhibits the need for shared airspace. Increased communications have aided and will become more important as they link computers and automatic air traffic control systems. In order to achieve greater airspace utilization and safety, an area above 14,500 feet altitude MSL (mean sea level) has been designated as a Continental Control Area. Aircraft flying above this altitude are higher performance aircraft, usually jet powered. In Positive Control Areas, above 18,000 feet altitude MSL, all aircraft are controlled by continuous surveillance and are required to have certain equipment to permit the higher aircraft densities of the higher performance aircraft.

Terminal Control Areas, such as the one shown in Fig. 5–13, are being designated around major aviation hub areas to impose special operating requirements on all flights in this airspace. In addition, special purpose areas are delineated either to prohibit or to caution flight operations. These are respectively designated as Restricted Areas, such as weapons ranges, and Warning Areas, such as areas of intense air traffic or student pilot training. Public interest in safety and efficiency with the designation and use of Positive Air Traffic Control in accordance with IFR flight rules in the Continental Control Area and transition areas at air terminals as well as other designated areas, is the cause for regulation in the aeronautical use of airspace.

5–27. Non-aeronautical Use of Airspace. Non-aeronautical use of airspace includes physical obstructions such as towers, buildings, and bridges as well as military bombing and gunnery ranges. Weather blocks (hazardous areas due to severe weather, i.e., turbulence, icing, thunderstorms, etc.) reduce the amount of airspace available. Accurate weather forecasts can aid in more effective use of airspace. Noise limitations have also curtailed the use of airspace creating congestion and

VFR ALTITUDES/FLIGHT LEVELS—CONTROLLED AND UNCONTROLLED AIRSPACE

Under Visual Flight Rules (VFR)
At 3,000 feet or more above the surface.

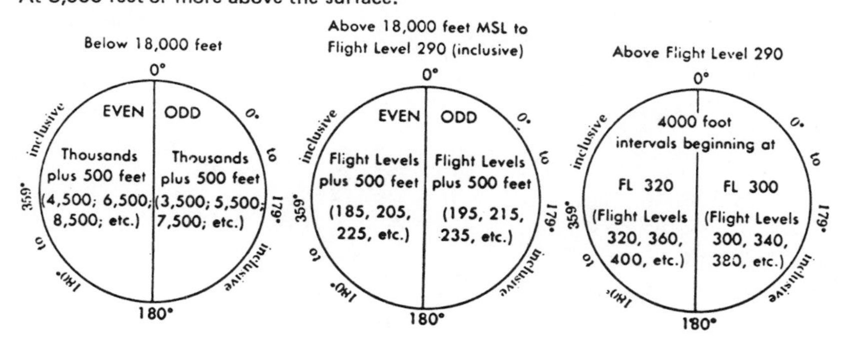

IFR ALTITUDES/FLIGHT LEVELS—UNCONTROLLED AIRSPACE

Outside Controlled Airspace.

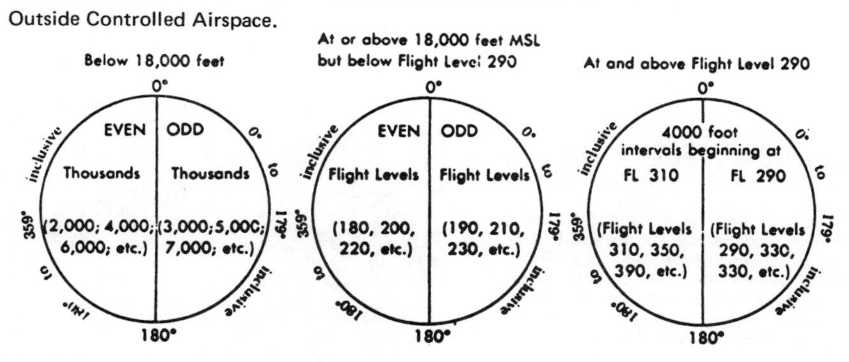

Fig. 5–12. IFR and VFR altitudes/flight levels controlled and uncontrolled airspace. (Source: *Airman's Information Manual,* Part I, May 1970, Federal Aviation Administration.)

inefficiencies. These factors must all be accommodated and accounted for in the aeronautical use of airspace.

5–28. Navigational Aids. Navigation has not been of major concern in air traffic control in previous years because of the sufficient margin for navigational error in vertical or horizontal separation. However, rapid growth in air traffic and increasing congestion is causing a continual

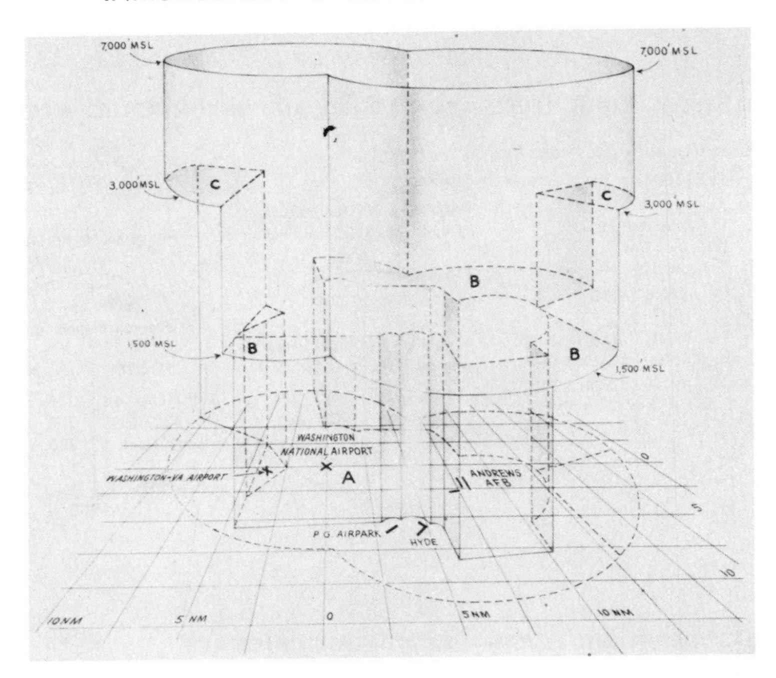

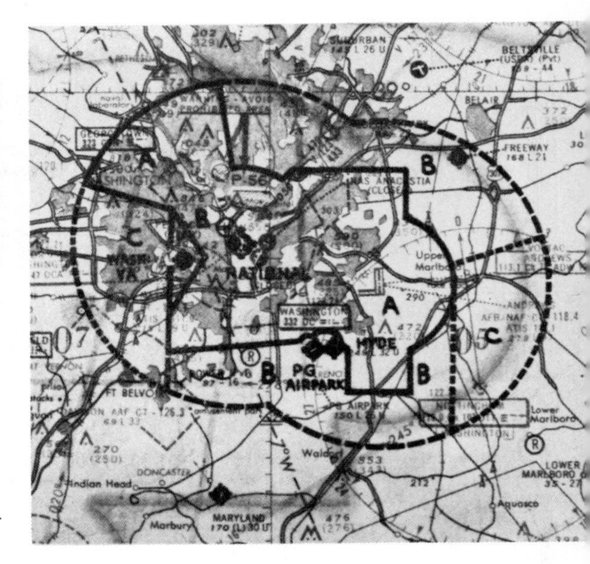

Isometric view (above) of diagram on chart (right) shows how new terminal control area would apply to the airspace around Washington National Airport and nearby Andrews AFB. The solid line shows how free access would be provided for use of the three smaller airports close to downtown Washington — Washington / Virginia, Prince Georges Airpark, and Hyde Field.

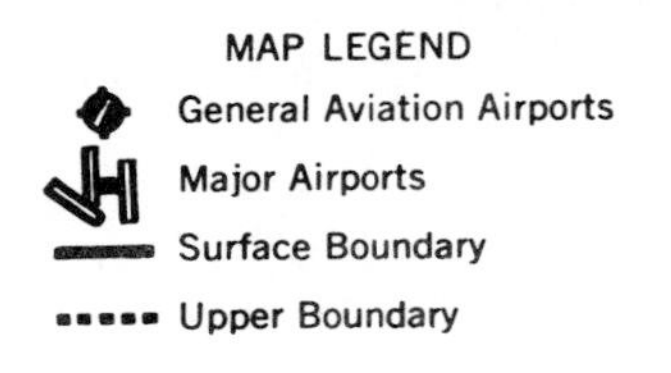

Fig. 5–13. Example of a terminal control area at a major hub. (Source: FAA Aviation News, November, 1969.)

re-examination of standards and procedures in air traffic control. Navigation technology is considered a major possible contributor to the al-be classified as that concerned with enroute navigation and terminal area navigation. They are described as follows:

Enroute Navigation Aids

Automatic Direction Finding (ADF)—a general purpose, low-frequency beacon or radio station on which aircraft home or aim and which is rather unsophisticated and inexpensive. It is subject to atmospheric noise and communications interference, although it is useful for long ranges (200 miles).

Very-high-frequency Omnidirectional Range (VOR)—a very high frequency, day-night, all weather, static free radio station which determines magnetic bearing to the VOR facility. The unit, which is pictured in Fig. 5–14, is limited to line-of-sight reception. High altitude aircraft

Fig. 5–14. VOR Station. (Source: FAA Aviation News, January, 1970.)

will suffer mutual VOR interference (multiple reception of facilities with like frequency assignments) due to their greater radio horizons. Figure 5–15 shows the VOR receiver display which is utilized by the pilot. The VORs form the basis for the present system of airways, called "Victor Airways," a section of which is shown in Fig. 5–16. The numbering system for Victor Airways is even numbers for East and West, odd numbers for North and South.

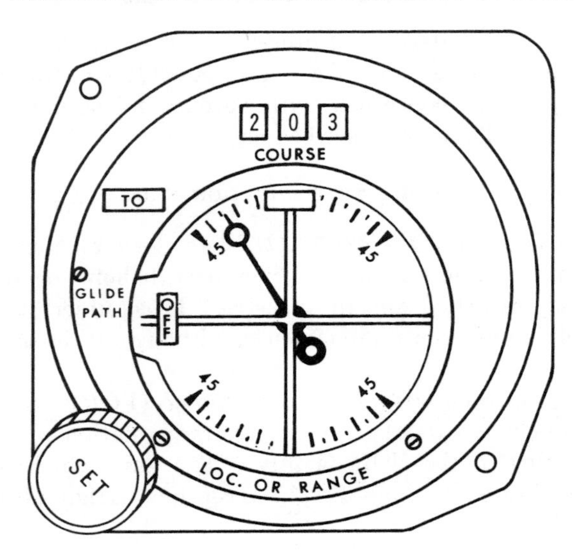

Fig. 5–15. Typical VOR receiver display. (Source: R. Horonjeff, *Planning and Design of Airports,* 1962, McGraw-Hill Book Co.)

Distance Measuring Equipment (DME)—measures the slant range (maximum of 200 miles) to the DME facility located at the VOR site and is subject to the same performance criteria as the VOR.

Tactical Air Navigation (TACAN) and VORTAC—incorporates VOR and DME functions into a single channelized system utilizing frequencies in the ultra-high-frequency range. TACAN is subject to line of sight ranges but designed for high-altitude flying with radio stations located at longer or greater separations to preclude station interference. If a station has full TACAN equipment and also VOR, it is designated as a VORTAC. High altitude airways, called "High Airways," similar to those described for use with the VOR have been created using the highly separated VORTACs.

Marker Beacons—fan markers, small radio transmitters which send coded identification signals, are aids to determine exact location on a given course. The markers are primarily used in instrument approaches or departure procedures as holding fixes or position reporting points.

Communications—are accomplished by radio receivers and transmitters located both in the aircraft and on the ground. Civilian aircraft primarily use VHF (very high frequency) radio ranges, and military aircraft, especially jet aircraft, utilize UHF (ultra high frequency) radio ranges. Air-ground communications are necessary to enable a pilot to

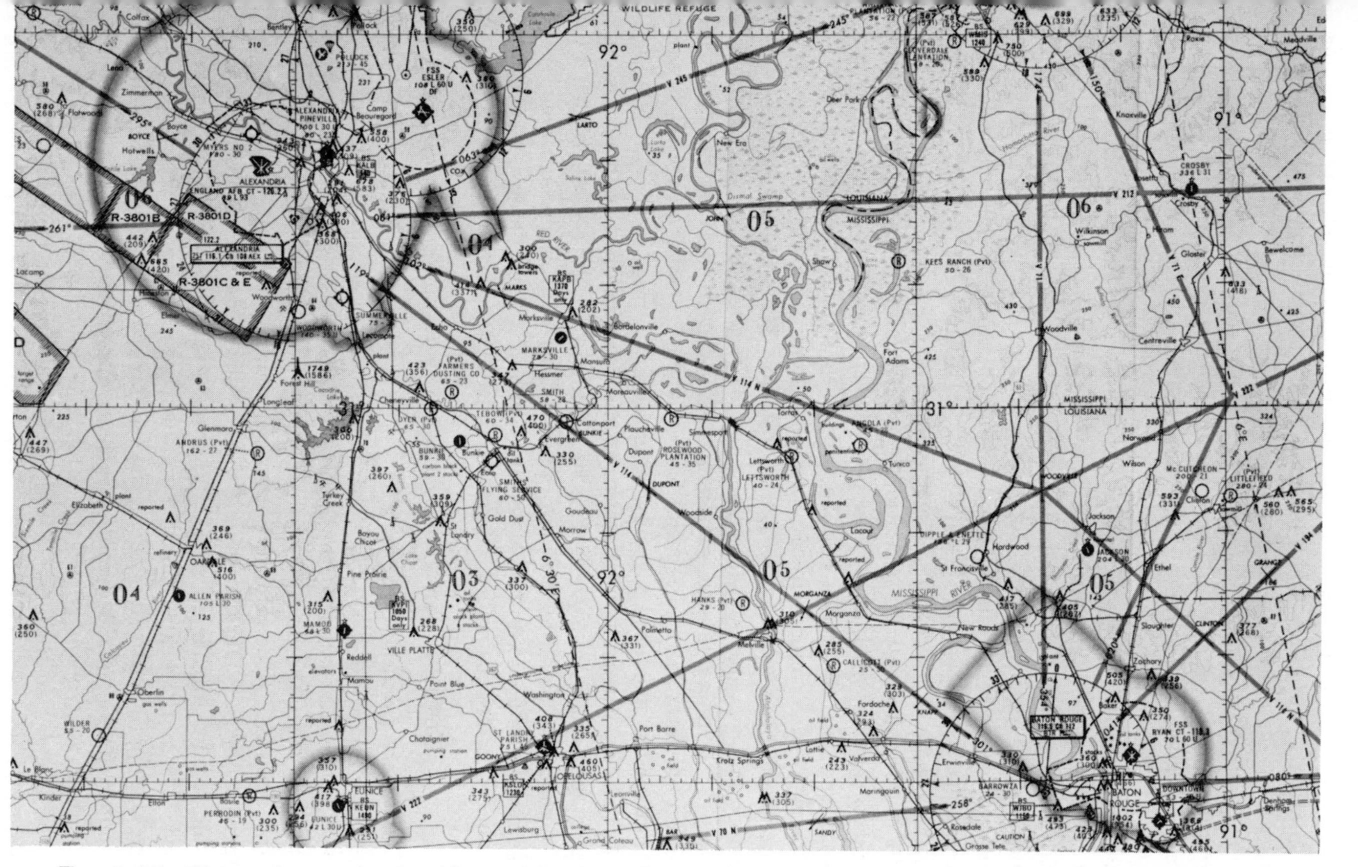

Fig. 5–16. Victor airways in the Alexandria-Baton Rouge Area. (Source: Federal Aviation Administration.)

receive flight instructions as he progresses along the airway or to his destination if not on a flight plan, to obtain reports of weather ahead, and to alter his flight planning as required.

Air Route Surveillance Radar—a long range radar for tracking enroute aircraft along an airway. These radars with a range of 200 miles are being installed nationwide. They will provide the controller in the air traffic control center with accurate information on the azimuth and distance position of each aircraft along the airway and will thus reduce the frequency of communication between the pilot and the controller. Prior to radar, the controller had to rely upon the pilot's accuracy in reporting his position. The long-range radars will permit the separation between aircraft flying at the same altitude to be reduced, thus increasing the capacity of the airways.

Air Traffic Control Transponder, Identification: Friend or Foe (IFF)—provides coded information when illuminated by air surveillance radar to enable tracking of aircraft. Often codes are assigned by air controllers, but the codes can be linked to altitude assignment.

Loran C—determines aircraft positions using a velocity input as well as magnetic bearing from the Loran station. The system uses a hyperbolic ground wave of low frequency for long range utilization (approximately 2,000 miles) and is extremely useful for over-water flights with position checking being accomplished by using cross bearings on Loran stations. It is also used for ship navigation.

Omega—an experimental system, similar to Loran C with extremely long range (6,000 miles). The system will probably be superseded by the developing satellite network with its better communications.

Doppler Navigation—an expensive system which determines vector distance traveled using an on-board radar and auxiliary equipment.

Terminal Area Navigation and Landing Aids

Instrument Landing System (ILS)—is an adaptation of the VOR and it is the most widely used system for instrument landings. It consists of two radio transmitters: the localizer transmitter situated at the end of the runway, and the glide slope transmitter situated near the side of the runway. The information display in the cockpit can be incorporated with the VOR instrument such as shown in Fig. 5–15. In order to help a pilot further on an ILS approach, two low-power fan markers called ILS markers are installed so that the pilot may know how far along the approach to the runway he has progressed. The first is called the outer marker (LOM) and is located approximately 4.5 miles from the end of the runway, and the other, the middle (LMM) marker, is located about two-thirds of a mile from the end of the runway. The ILS system is diagramed in Fig. 5–17. Advanced landing systems are

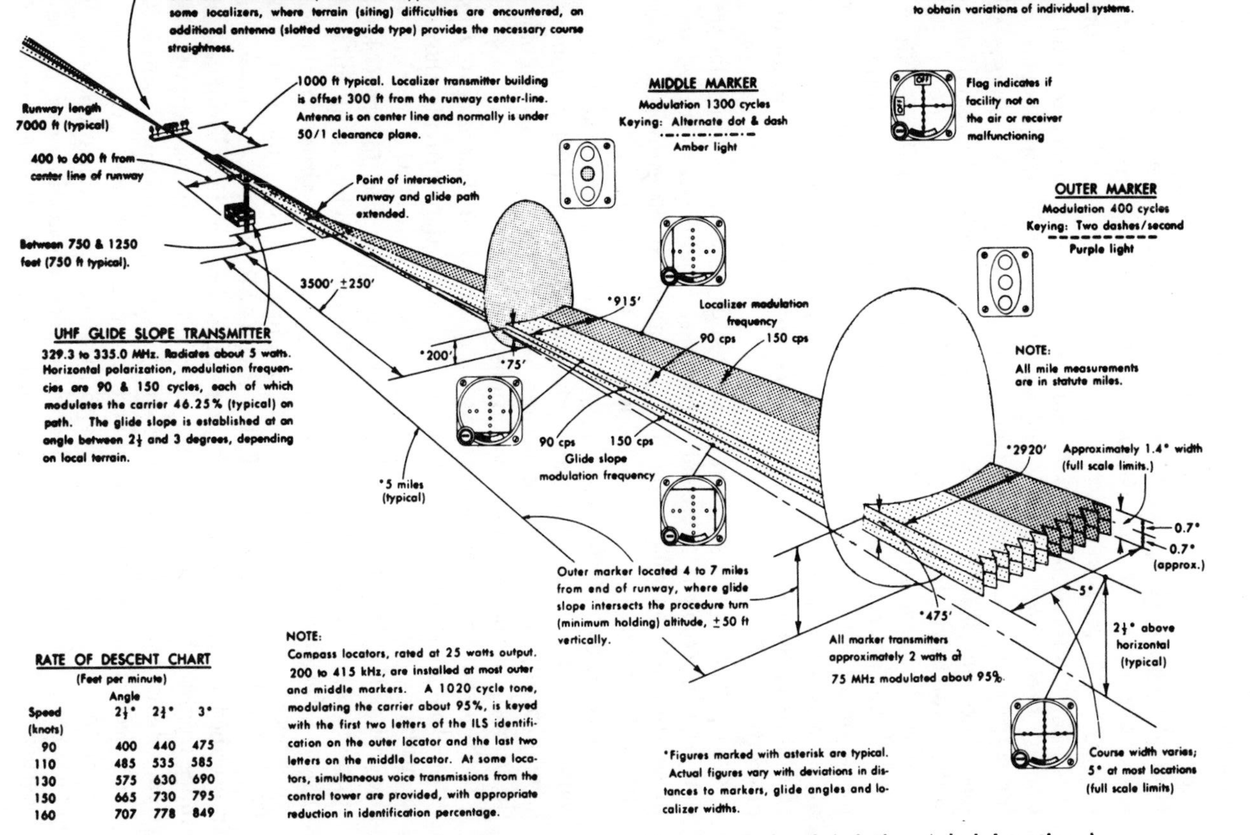

Speed (knots)	Angle 2½°	2¾°	3°
90	400	440	475
110	485	535	585
130	575	630	690
150	665	730	795
160	707	778	849

Fig. 5–17. FAA instrument landing system. (Source: Federal Aviation Administration.)

being developed which provide directional and distance information for all weather landings. They are experimental units using microwaves with later models utilizing laser beams.

Precision Approach Radar (PAR)—is more familiarly called the Ground Controlled Approach (GCA). It is not dependent on airborne navigation equipment as the radar is located in a van adjacent to the runway. The PAR radarscope gives the controller a picture of the descending aircraft in azimuth, distance, and elevation to enable determination of the aircraft's correct alignment and glide slope. Instructions for corrections are given by voice communications. At airports where both ILS and PAR are available, they are both used simultaneously with one being the back-up for the other.

Air Surveillance Radar (ASR)—provides for terminal area air traffic control and aircraft location information to airport tower operators. It, for example, will provide information for an aircraft transiting from an airway to a holding pattern to an ILS approach. The range of the ASR is 30 to 60 miles. It does not indicate the altitude of the aircraft.

Approach Lighting Systems (ALS)—aid pilots in making the transition from instrument to visual conditions. Lights are installed on and in the runways as well as on the approaches to the runways. Various approach lighting systems are described in Chapter 18.

Airport Surface Detection Equipment (ASDE)—a specially designed radar system for use at large, high density airports where controllers have difficulty in regulating taxiing aircraft because they cannot see the aircraft in poor visibility conditions.

Instrument Approach Procedures and Standard Instrument Departures—though these are not navigational aids, they provide the means for utilizing the enroute navigational aids as well as the terminal area navigational aids. They are not only indispensable for IFR landing approaches but are also helpful to the VFR pilot landing at an unfamiliar airport. Instrument approach charts diagram every airport in the country where some kind of instrument landing aid is installed, i.e., ADF, VOR, DME, TACAN, VORTAC, PAR/ASR, ILS, etc. They depict prescribed instrument approach procedures from a distance of about 25 miles from the airport and present all related data such as airport elevation, obstructions, navigational aid locations, procedural turns, etc. Each recommended procedure—and every airport has several—is designed for use with a specific type of electronic navigation aid. The pilot makes his choice depending upon instrumentation and prevailing weather conditions. To aid pilots on take-off, Standard Instrument Departures (SID) have been developed to facilitate the transition between takeoff and enroute operation, thus alleviating the need for a great amount of oral communication between controllers and pilots.

5–29. Air Traffic Control Facilities. Air Traffic Control Facilities provide the basis for communication with aircraft and the relay and clearance of flight plans for air traffic. There are three basic types of manned facilities: the air route traffic control center, the airport traffic control tower, and the flight service station.

Air Route Traffic Control Centers (ARTCC)

There are about 28 air route traffic control centers that control the movement of aircraft along the airways. Each center has control of a definite geographical area and is concerned primarily with the control of aircraft operating under instrument flight rules (IFR). At the boundary points marking the limits of the control area of the center, the aircraft is released either to an adjacent center or to an airport control tower. At present much of the aircraft separation is maintained by radar. With radar, off airways vectors can be utilized maintaining positive control of the aircraft, and thus the ARTCCs can accommodate more aircraft.

Each ARTCC is broken down into sectors in order to increase the efficiency of personnel in the center. Sectors are smaller geographic areas and air traffic is monitored in each sector by remote radar units at the geographic location. It can be observed that an aircraft flight plan is transferred between sectors within an air route traffic control center and between air route traffic control centers when crossing the ARTCC boundaries.

Airport Traffic Control Tower

Airport traffic control towers, approximately 327 in number, are the facilities which supervise, direct, and monitor the traffic within the airport area. The control tower provides a traffic control function for aircraft arriving at or departing from an airport for a 15-mile radius.

Some control towers have approach control facilities and associated air surveillance radar (ASR) which guide aircraft to the airport from a number of specific positions called fixes within approximately 25 miles of the airport. The aircraft are brought to these positions by the air route traffic control centers (ARTCCs). It is often at these fixes that aircraft are held or "stacked" for landing during periods of heavy air traffic. The control towers without approach control facilities differ in that, under IFR conditions, the clearing of waiting aircraft for landing is done by the ARTCC and they are turned over to the control tower after they have started their landing approach.

Flight Service Stations (FSS)

The flight service stations (FSSs), approximately 333 in number, are located along the airways and at airports. Their function may be described as follows:

1. Relay traffic control messages between enroute aircraft and the air route traffic control centers.
2. Brief pilots, before flight and in flight, on weather, navigational aids, airports that are out of commission, and changes in procedures and new facilities.
3. Disseminate weather information.
4. Monitor navigational aids.

PROBLEMS

1. Two urban four-lane undivided streets are to be designed to intersect at a rotary intersection. Draw a schematic intersection layout showing lane markings, channelization, and necessary signs and signals. Write a short report justifying the design.
2. The United States does not use international traffic sign designs. Present arguments both for and against U.S. adoption of these standards.
3. Under what conditions of traffic would you use:
   a) Traffic Signals
   b) Stop Signs
   c) Yield Signs

   (Refer to the *Traffic Engineering Handbook* for the preparation of this answer.)
4. What are the implications of the development of a "hands-off" take-off and landing system for commercial aviation?
5. Discuss the ramifications of the introduction of totally automatic railway locomotive control. In your answer deal with social problems (such as accident rates, reliability, and labor), economics, and technological feasibility.
6. What developments in marine navigation are foreseeable in the next fifty years?

REFERENCES

1. *Traffic Engineering Handbook,* Institute of Traffic Engineers, Washington, D.C., 1965.
2. *Manual on Uniform Traffic Control Devices for Streets and Highways,* U.S. Department of Transportation, Federal Highway Administration, Washington, D.C., 1971.
3. *Channelization, Special Report No. 2,* Highway Research Board, Washington, D.C., 1952.
4. *A Policy on Geometric Design of Rural Highways,* American Association of State Highway Officials, Washington, D.C., 1965.
5. Casciato, L., and Cass, S., "Pilot Study of the Automatic Control of Traffic Signals by a General Purpose Electronic Computer," *Bulletin* No. 338, Highway Research Board, Washington, D.C., 1962.

6. *Aids to Marine Navigation of the United States,* CG-193 United States Coast Guard, Washington, D.C., 1965.
7. QUINN, A. DE F., *Design and Construction of Ports and Marine Structures,* McGraw-Hill Book Co., New York (1961).
8. WYLIE, F. J. (ed.), *The Use of Radar at Sea,* American Elsevier Publishing Co., New York (1968).
9. *Railway Track and Structures Cyclopedia,* Eighth Edition, Simmons Boardman Publishing Corp., New York (1955).

OTHER REFERENCES

HORONJEFF, ROBERT, *Planning and Design of Airports,* McGraw-Hill Book Co., New York (1962).

SCHRIEVER, BERNARD A., and SEIFERT, WILLIAM W. (ed.), *Air Transportation 1975 and Beyond: A Systems Approach,* The M.I.T. Press, Cambridge, Mass. (1968).

U.S. Department of Transportation, Federal Aviation Administration, "Airport Design Requirements for Terminal Navigational Aids," Advisory Circular AC 150/5300-2, March 30, 1964.

———. "TCA: A New Concept" in *FAA Aviation News,* April, 1970, p. 3.

———. "Sky Maps," in *FAA Aviation News,* October, 1969, pp. 8–9.

———. "Terminal Control Areas," in *FAA Aviation News,* November, 1969, pp. 4–5.

———. "On Top of Old Snow Bird," in *FAA Aviation News,* January, 1970, pp. 4–6.

———. "Airman's Information Manual—Basic Flight Manual and ATC Procedures," Part 1, May, 1970.

II

URBAN TRANSPORTATION PLANNING

6

The Urban Transportation Problem

The "urban problem," which has been discussed extensively by many urban planners, touches upon the problems that must be solved by the urban transportation specialist. In the face of increasing local governmental costs due to such demands as social welfare, urban renewal, crime prevention, education, and a host of environmental problems, cities face a steadily eroding tax base due to a flight of both business and the affluent residents to surburban areas. The problem of the cities and urban transportation are tightly interwoven. Indeed, many planners feel that the changes in urban mobility that have been brought about in the last fifty years have, in part, caused the difficulties that face the cities in the latter part of the twentieth century, and state that until the cities can provide a minimum universal standard of mobility there will continue to be economic and racial segregation, urban decay, and a continuance of the pathological difficulties which have beset American cities. Conversely, while these conditions exist, it is said there cannot be a real solution to the transportation needs of the metropolis. This chapter will attempt to outline the basis of the urban transportation problem as it presents itself today.

6–1. The City Center. Much has been written about the decay of the urban centers of America. Some visualize the decay of the central cities as a natural phenomenon providing a transition phase from the highly centralized nineteenth-century city to a decentralized urban form more in tune with twentieth-century technology and mobility. Others view the decay as a result of improvident, if well meaning, governmental policies and programs which have subjected the central areas to overwhelming economic pressures at a time when urbanization of the poor rural population was taking place. All, however, agree that the central areas of the major American cities are suffering from a deepening urban blight. There is a continued replacement of middle-income groups in these areas by the poor rural migrants.

The question posed to the planner is not simple although it can be relatively easily stated. If urban blight is a symptom of an advanced technology where the need for the central business district will vanish, should the planner hasten this process by providing a transportation system which encourages decentralization, or is there justification for designing systems which will strengthen the central areas? In the coming age of videophones, computer linkages, and a general electronic revolution, will the need for face-to-face contact decline to the point that central business districts will become obsolete and the whole form of the city will change to specialized nodes outlying the central areas [1]? The answer is not clear, yet analysis of the changes that are currently taking place would indicate that what is happening to the central business districts of the cities is a change in specialization. Economic development of the last fifty years has brought about an even greater specialization in man's work. The work, recreation, living and shopping areas are more firmly segregated than ever before. There is strong evidence that while downtown areas are declining as general retail shopping areas and living areas, they gain in strength as commercial office centers, specialized culutral and entertainment centers, and specialty shopping areas. Alonso points out that, as economic development takes place, the worker becomes increasingly white collar in function. There is little evidence that these jobs are moving from downtown areas. As automation increases, the managerial workers are relieved of the need for physical presence on production lines, and are available in the central areas for the face-to-face contacts that enhance decision making and management. Alonso further states:

> And as our population becomes richer and better educated, it seeks the luxurious, the sophisticated and the specialized, which are the major attractions of downtown . . . there are very powerful forces which in the long run will mean the resurgence of the downtown area. [2]

6–2. The Area and Attractiveness of the Central Business District. Central business districts (CBD) in small urban areas assume a more dominant position than in larger areas. There is a general trend, shown in Fig. 6–1, that as the population of the urban area increases the size of the central business district increases at a decreasing rate, according to a logarithmic relationship. Only in the very largest American cities does the area of this CBD exceed one square mile. Despite its small size, however, the central area is vitally important to the metropolitan

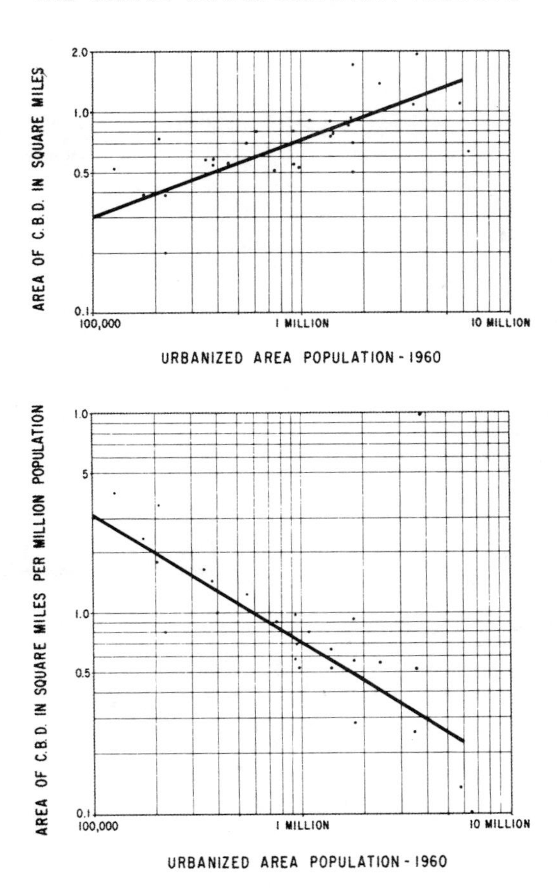

Fig. 6–1. Central business district area in relation to population. (Source: *Transportation and Parking for Tomorrow's Cities,* Wilbur Smith & Associates, 1966.)

area. As the metropolitan population increases in size, central areas become increasingly important regional financial and business centers attracting large volumes of high-income white collar workers and professionals. Redeveloping urban centers such as Toronto, Pittsburgh, and Atlanta generate large volumes of construction as extremely high-density office development is built over expensive central land.

The decreasing rate of increase of the area of the CBD is reflected by the maximum daily accumulation of persons within the area as shown in Fig. 6–2. This accumulation also shows a negative rate of increase with increasing urbanized populations. It is worthwhile noting that while Wilbur Smith and Associates found a significant difference in the

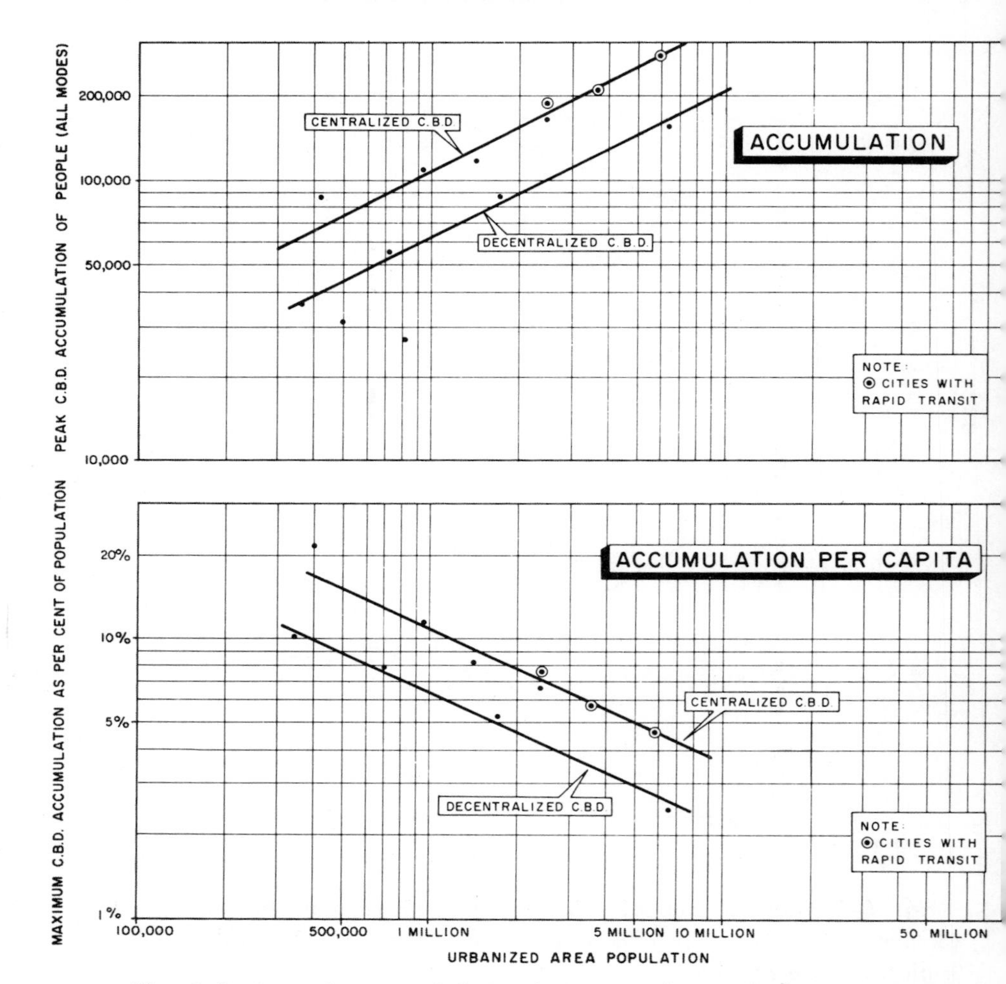

Fig. 6–2. Attractiveness of CBD relative to urbanized area populations. (Source: *Transportation and Parking for Tomorrow's Cities,* Wilbur Smith & Associates, 1966.)

total accumulation and accumulation per capita for centralized and decentralized central area, the rates of change in these relationships were virtually parallel.

6–3. The Relation of Urban Form to Transportation. The physical patterns of growth of a city strongly reflect the physical layout of the transportation system which is created to serve the urban area. Since facilities are designed to serve not simply existing demand, but rather demand at some point in future development, transportation systems not only serve the urban area, but also tend to shape the urban structure in their own likeness. The strong relation between urban growth and transportation is documented in the section on land development models in Chapter 9.

The strong correlation between urban growth and transportation has been documented elsewhere. Section 3–9 describes the relationships between historical changes in urban transportation technology and the development of a typical urban area. Sections 9–14 through 9–17 develop an empirical basis for the dependence of urban structure on transportation.

6–4. The Growing Importance of Subregional Centers. In 1967, there were 80 million privately and commercially owned automobiles in the United States. This amounted on the average to one automobile for every two and one-half persons. The high per capita car ownership has meant a dramatic increase in the accessibility of all parts of the urban area to individual urban residents. The average urbanite has found that his travel patterns are now no longer constrained by the direction and frequency of public transit; travel has become a function of the driving time between origin and destination.

Probably the most striking effect of automobile mobility is the decentralization of shopping habits. Its counterpart is the widespread construction of regional and community shopping complexes. Since the 1930's there has been a great change in the shopping habits of the average American family. The neighborhood store, which relied on pedestrian customers, has been replaced by more efficient supermarkets, oriented to a motorized clientele. Neighborhood or community shopping centers now are designed to service the convenience shopping needs of the urban residents within five to eight minutes driving time. More significant effects on urban form have been brought about by the decentralization of major shopping (furniture, appliances, etc.) at regional shopping centers. The position of the downtown business district as the retail center for all major shopping goods has continuously eroded as major department stores and specialty shopping have grown in satellite regional shopping centers. These centers cater to populations of up to 300,000 in their trade area, usually calculated on 15 to 20 minutes driving time. A typical regional center occupies a gross area of at least 30 acres including parking. The leasable area has been found to vary from 200,000 square feet to 1,300,000 square feet [3]. An example of a large regional shopping center is shown in Fig. 6–3. Because of their peripheral location, usually close to the urban freeway system, and the large number of free parking spaces, these centers have changed the national shopping patterns. As a result, the former dominance of the downtown area for

Fig. 6–3. A typical modern regional shopping center. (Source: Lockwood Survey Corporation Limited, Toronto, Canada.)

major shopping purchases has been broken and the travel for major shopping purchases no longer is oriented chiefly to the downtown area.

6–5. The Decline of the Central City. The changes in urban mobility brought about by almost universal car ownership have been accompanied by other social and technological trends which, acting in concert, have modified the function of downtown central areas. In addition to the erosion of the dominance of central areas in retail sales, central areas are no longer prime locations for wholesaling and industry. These two functions, which were extensive uses in nineteenth-century central cities, now tend to be located more economically in outlying areas. Shifts of land use reflecting this economy have been taking place over the last fifty years.

Wholesaling, warehousing, and industry, essentially very similar in their land use requirements, were located in central areas of most nineteenth-century industrial towns. Goods were moved mostly by rail, in intercity movement. Since most rail freight yards were located close to the downtown passenger stations, the transfer from rail to factory or warehouse was made by horsedrawn wagon. Ideally, these uses were located at some minimum distance from the railhead. The expense of purchasing downtown land could be economized by the use of multistory facilities so that land cost per unit of floor space could be brought to a reasonable figure. Mechanization of the twentieth century brought two significant trends which were to change radically the economics of downtown location.

The development of trucking reduced the costs of transfer of goods from rail to plant. Consequently, industrial and warehousing sites could be selected in outlying areas since very little additional cost was involved for the additional truck mileage. These incremental costs would be more than offset by decrease in land costs for purchases in outlying areas. The rise of the intercity trucking industry itself further influenced peripheral locations. There was little value in siting truck terminals inside the downtown areas, since intercity freight would have to move twice through congested streets.

Equally important to the decentralization of industry was the development of mechanization in materials handling. With the introduction of horizontal assembly lines and mechanized moving and lifting equipment, the old multistory factories and warehouses rapidly became uneconomic. Firms seeking to remain competitive were forced to mechanize their operations. Often, this involved horizontal plant expansion which would have been totally uneconomic if carried out in downtown areas. Industrial and wholesale uses, therefore, moved out to the peripheral fringes of the urban area, usually close to expressway or arterial streets which could easily accommodate the trucking which connected them with their markets.

Even as mechanization and mobility exerted direct influences on the spatial arrangements of land uses, the form of urban America has changed through the indirect influences of sociological and demographic pressures throughout the last thirty years. The two prime causes of sociological change in urban areas have been:

1. the increased area of developable land brought about by highway accessibility;
2. migration of the rural poor to urban areas.

The first of these factors is probably the more important. With the coming of the motor car, the average speed of travel in the urban area more than doubled. As a result, lower cost outlying land distant from fixed rail facilities became amenable to development. The direct result of increasing car ownership is the condition familiarly known as urban sprawl. Middle and upper income groups have found that lower land prices permit the low-density single-family housing of suburbia where half- to one-acre lots provide a pleasant and private environment around the individual's living space. Federal policy and programs have encouraged this form of development by the provision of low cost mortgages, tax relief, and federal subsidies to such facilities as roads and utilities to service the areas. The result has been a mass movement to the suburban areas by middle and upper income groups, leaving the central high-density areas to those of lower income.

The flight of the middle class from the cities has taken place at the same time that the United States has undergone rapid urbanization. The percentage of the population living in the urban areas was 40 per cent in 1900; this figure has risen to 70 per cent in the late 1960's and will be 75 per cent by the year 1980 [4]. Figure 6–4 shows past

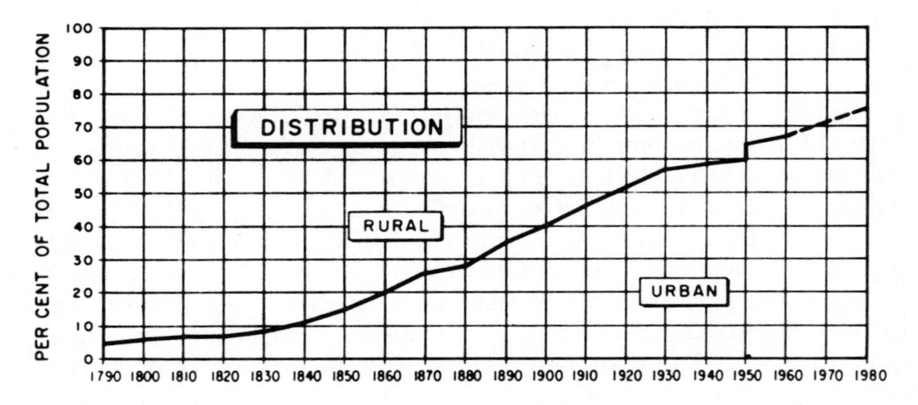

Fig. 6–4. The rural-urban balance of the U.S. (Source: *Future Highways and Urban Growth,* Wilbur Smith & Associates.)

and future trends of urbanization. Mechanization of farming has decimated the employment opportunities in rural areas. The result has been a rapid urbanization through migration of the rural poor to the cities. As the middle class has left the central city, it has been replaced by in-migrants, usually poor, often with low skills and predominantly black. By a combination of circumstances, the central American city has become a ghetto. To all planners this presents a problem, but to the transportation planner there are very special problems. Because of these peculiar spatial arrangements of different economic groups the need for travel is almost maximized. Low-income blue-collar workers are moving into the central areas precisely at the time that job opportunities in industry and commerce are moving out. White-collar middle- and high-income groups move out to suburbia and almost invariably work in the offices which have taken over the downtown core. The problem is further complicated by political fragmentation of metropolitan areas. Many suburban communities incorporate to provide themselves with political and fiscal separation from the financial problems of the central city where municipal costs tend to overrun the limited tax base. The planning of transportation facilities in multijurisdictional metropolitan areas has proved to be more difficult politically than technologically.

6–6. Inability of Transit to Satisfy Dispersed Travel Demands. The decentralization of the central business district and its decline in relative dominance of the urban scene leads to decentralized travel patterns. The scale of the problem can be seen by reference to Fig. 6–5, which

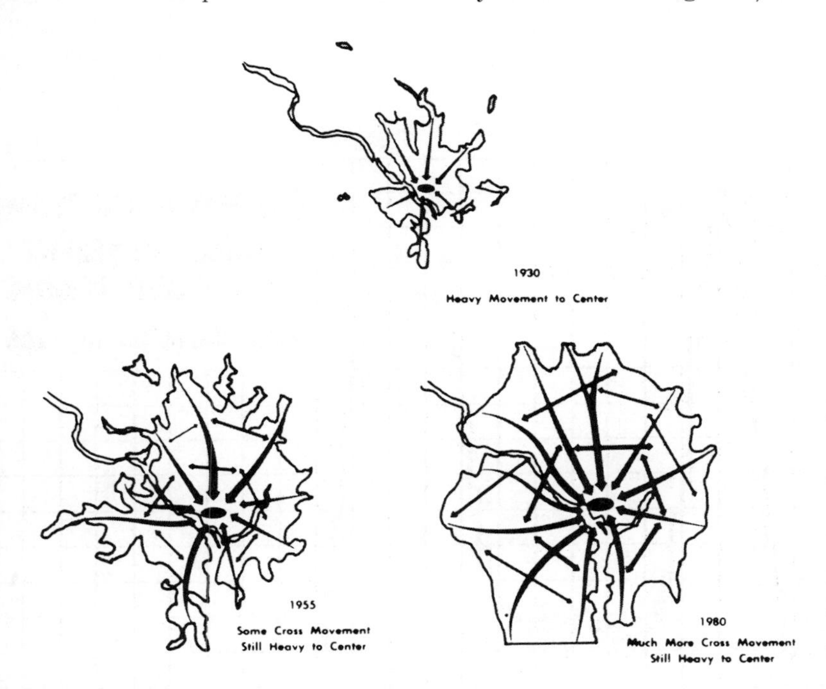

Fig. 6–5. Changing patterns of urban traffic flow. (Source: *Transportation and Parking for Tomorrow's Cities,* Wilbur Smith & Associates, 1966.)

denotes the changing patterns of urban movement in Washington, D.C. When car ownership was low, urban travel patterns were essentially radial from the central area. The urban form which was an outgrowth of the available transport system produced movements which were relatively easy to satisfy by transit. The dominant central area attracted trips throughout the whole day both during peak hours, when primarily work trips were made, and during off-peak periods, when shopping and other trips were made. As car ownership increased, decentralization brought peripheral movements in addition to radial movements. The central areas continue to attract more trips, but these trips tend to be more and more of one type, the downtown office worker. The result is a problem which is increasingly difficult to solve with transit. The remaining radial movements are high-peak movements occurring almost totally during the morning and evening "rush" hours. For the remainder of the day, trips to the downtown area are at a minimal level. To satisfy this type of demand, the transit operator is faced with the need to provide high capacity installations which are underused for the major portion of the day. Figure 6–6 shows the effect of peak demand on

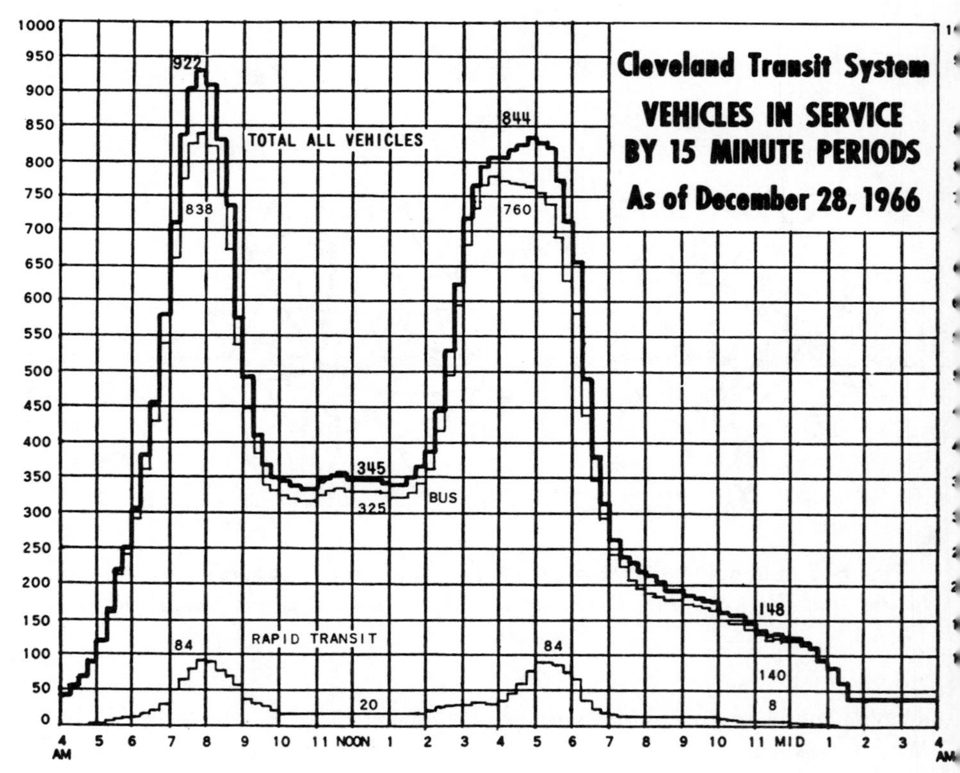

Fig. 6–6. Vehicles in service by 15-minute periods, Cleveland Transit Authority. (Source: *Highway Research Circular #91*, Highway Research Board.)

transit vehicles in service in the Cleveland area. The peaking is seen to be more severe for radial rapid transit service than for general transit throughout the whole urban area. Peripheral movements are extremely difficult to satisfy with transit, since the individual movement corridors are not dense, and the origins and destinations of these movements are highly dispersed.

6–7. The Decline of Mass Transit. Faced with the problem of attempting to cope with travel patterns and urban spatial separations caused mainly by the universal ownership of cars, transit has encountered secular decline in the United States. While the mileage of urban travel has increased at a rate approximating the growth of the Gross National Product, urban transit usage has suffered a steady decline since the late 1920's, with the exception of a short term period during the gas rationing period of World War II. Mass transit today is composed principally of rapid rail systems and surface bus systems. Less flexible surface systems such as trolley coaches and street cars have virtually disappeared. Alone of all modes of mass transportation, rapid transit has managed to hold ground against the long term decline which has affected the industry. Ridership of rapid transit lines in the mid-1960's was approximately the same as in the mid-1950's whereas all other forms had shown significant declines. Figure 6–7 indicates the trends in ridership of mass transportation modes from 1905 to 1965. Associated trends in the financial position are shown in Fig. 6–8. It can be seen that on a nationwide basis mass transit is an economically marginal operation. The demonstrated ability of relatively old rapid transit systems to hold ridership, and of new systems to attract riders in Toronto, Montreal, and Cleveland has spurred many major cities such as Baltimore, Los Angeles, San Francisco, Washington, Pittsburgh, Seattle, Miami, and Atlanta to plan and in some cases start the construction of rapid transit systems. The degree of success of these systems will depend on how well they can modify the urban form in which they are set. Successful operation of rapid transit lines depends on high densities of movements along the lines. In central Toronto, the subway is attributed with spurring $15 billion of development along the line itself [5]. This has been in the form of high-density residential areas and high-density office and retail commercial facilities. By changing the form of urban development, rapid transit facilities are able to reverse the trends of auto-dependence that suburban sprawl must inevitably bring. On a worldwide basis, the future of rapid transit appears strong. Over forty systems are now in operation, and more than twice that number are in the planning stages. Six cities are currently constructing new systems [6].

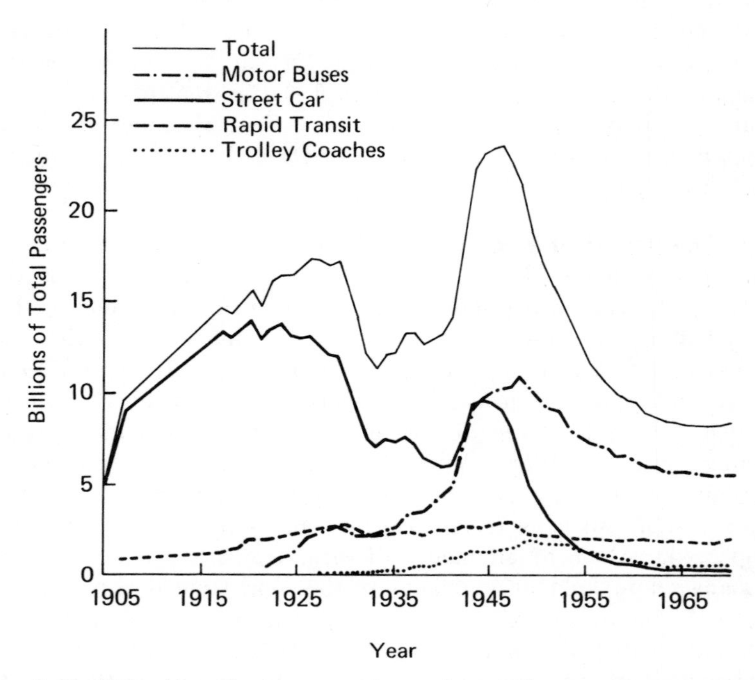

Fig. 6–7. Ridership of mass transit modes. (Source: American Transit Association.)

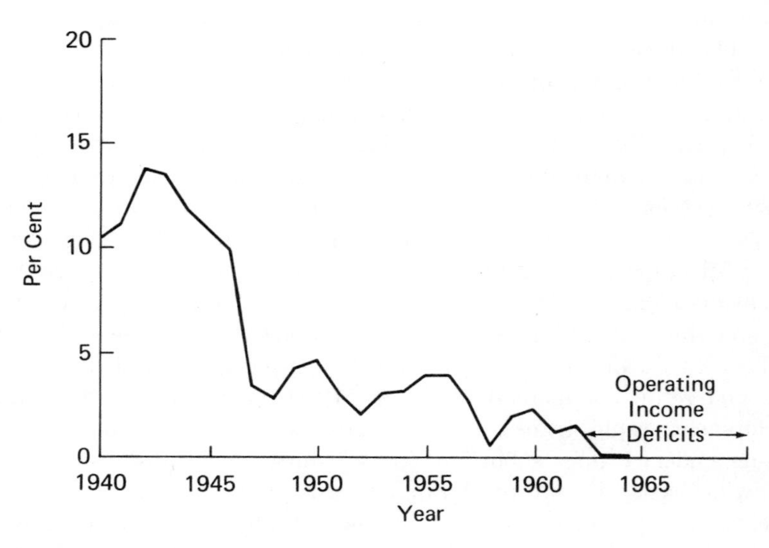

Fig. 6–8. Revenues for mass transit operations. (Source: American Transit Association.)

6–8. The Effect of Auto-oriented Planning for Urban Travel. The urban transportation system can be treated as four separate subsystems. In examining the effect of system design upon the urban structure the designer and planner must take note of the different forms of impact that these systems inevitably exert. The four subsystems are [7]:

1. The feeder system
2. The line haul system
3. The distribution system
4. The terminals

In a typical commuter's transit trip in a large metropolitan area such as New York, Chicago, or London the *feeder system* is that portion of the trip made up of walking and bus transit to the rapid transit system. The *line haul* section of the trip is carried out over a long distance by higher speed rapid transit from the rapid transit station nearest the commuter's home to the station nearest his workplace. The *distribution system* accommodates either a pedestrian or combined bus and pedestrian trip from this transit station to the work destination. In the case of this hypothetical trip the *terminal system* is the system providing interfaces between the other three subsystems, being the rapid transit stations at the beginning and end of the line haul system. In most existing rapid transit systems the feeder and distribution systems are of relatively simple design, being accommodated within the existing urban street systems. Consequently, existing rapid transit lines which have no facilities for long term parking at line haul stations, have given rise to high-density urban corridors within walking distance of the line itself.

The use of freeways as the line haul element of a transportation system has brought about significant changes in urban structure, changes which are a function of the altered mobility patterns. In the auto society, the motor car is the vehicle which provides the feeder system, the line haul system, and the distribution system. As a result, where an auto solution to the transportation problem is sought, the changes in urban form are wrought in the suburbs, the central city, and the downtown core. Urban transportation systems are designed to accommodate peak conditions, which are caused primarily by the morning and evening work trips. In most cities, the major problems are caused by the radial suburban to central core movements. When the auto is used totally for this trip, it provides its own feeder system over suburban streets. Since it can operate at fairly high overall speeds relative to feeder buses, it provides an efficient feeder service. Its impact on urban form at this end of the trip is significant but not obvious. High mobility permits low density urban development and the suburban areas are built in a manner which promotes the use of the auto. It is interesting to note at this point, that this type of suburban development which is so well adapted to the auto is most unsuitable for feeder buses which, in order to operate economically at high speeds, must have higher concentrations of potential riders. While the auto provides a highly desirable feeder system with minimal negative interaction with land use, for the other three subsystems the mode provides a less adequate solution to the mobility problem.

The chief problem with line haul freeways appears to be the relatively inefficient use of space for the volume of persons moved. One thirteen-foot lane can carry 2,000 vehicles per hour. At normal car occupancy this space permits movement of 3,200 persons per hour. Rapid transit lines, on the other hand, theoretically can move 72,000 persons per hour in the same space. One obvious result of this inefficiency is the need for large numbers of freeway lanes to provide adequate capacity where there are heavy movements into the downtown core area. A by-product of such large space requirements is severe social disruption of established communities closer to the downtown area. Community disruption has caused public antagonism to freeway plans in almost every large urban area. It would appear that joint development solutions (the multiple use of transportation lands), are only a partial answer to this problem [8]. As a major city continues to grow, if the downtown core is to maintain its dominance of the metropolitan area, additional freeways are required to provide additional access if the auto solution is chosen. This access can be provided only by cutting through the mature areas ringing the central city.

Auto usage presents equal problems for the distribution and terminal subsystems. Freeways in the past have discharged heavy traffic volumes into the downtown core, often into streets which were laid out for horse-drawn transport. Modification of the street pattern at the distribution end of the commuter's trip involves expensive rebuilding and remodeling of the high-value downtown core. Parking provides an even more troublesome problem. Gruen has pointed out that an automobile requires 1,400 square feet of space: 300 square feet at its origin, 600 square feet en route, 300 square feet at its destination, and a further 200 square feet for servicing, etc. [9]. Consequently, parking and streets become a major land use of the downtown area. In the central business district of Los Angeles, 24 per cent of all land is devoted to parking, and 35 per cent to streets [10]. Only the remaining 41 per cent is devoted to the commercial and living space necessary for human interaction. By attempting to provide accessibility by auto to the central areas, planners can destroy its essential cohesion.

6–9 Decision Point for Provision of Transit. It is self-evident that small urban areas have little need of mass transit. In such cities, transit facilities are usually marginal, often to the point that private companies are unable to make a reasonable return on investment, and operations must be carried out by the municipality. Almost universal car ownership has lowered the rate of transit usage in small cities, where lack of any real congestion problem promotes the use of the auto. Equally self-evident is the fact that without extensive mass transportation facilities such cities as New York, Chicago, and London would strangle in the congestion caused by their own auto traffic and could not hope to provide the parking spaces required. In 1956, for example, the number of persons entering downtown New York City on a typical weekday was in excess of 3.3 million while in Chicago the number was over 860,000 in 1960 [10]. The question often raised is: "At what point should a metropolitan area consider major investment in mass transit?" since apparently many cities are likely to grow to the point that their transportation needs can no longer be solved by autos alone. There is no simple answer to this question. However, as already stated, there are over forty rapid transit systems now in operation and over eighty in planning. It would appear that most cities whose populations are over the half-million mark are in the process of planning mass transit systems provided that there are reasonable expectations of continued growth and available capital for financing [6]. Surface mass transit systems are necessary in all size cities, but are likely to be unprofitable in small urban areas.

6–10. Summary. It has become apparent in the last thirty years that the forms of mass transportation which reasonably satisfied the essentially radial transportation patterns of the pre-automobile city are no longer competitive forms of transport where the individual has the alternative of the auto. Planners face the dilemma that while mass transport has proved unattractive to the urban dweller, the alternative of a heavily auto-oriented transportation system has caused major disruptions in some mature cities and an undesirable decentralized urban form in rapidly growing areas. Planners and engineers have responded to this challenge by devising and testing new forms of urban transit through programs of research and demonstrations under the auspices of the Urban Mass Transportation Act of 1964 and its subsequent amendments. Under the terms of this act, the Federal government underwrites the major part of the cost of testing new forms of transit, new levels of fare structure, differing levels of transit service, and other inducements to transit usage that cannot be reasonably tested by the currently marginal transit operations of the urban areas. The results of many of the research and demonstration programs have been both enlightening and encouraging [11, 12]. As this governmental program continues, it can be expected that the fruitful findings will be incorporated in existing and new operations on a nationwide basis.

PROBLEMS

1. There has been much discussion on the "latent demand" for transportation which indicates that in urban areas the young, the elderly, the poor, and the handicapped are inadequately served by auto-oriented transport systems. Discuss the problem and its possible solutions.
2. In order to reverse the rural to urban migration trend the Federal government has initiated demonstrations for *economic growth center development highways.* Discuss what types of facilities might be involved and the program's chances of success. (Refer to Section 1–16.)
3. Discuss the problems involved in construction of urban expressways while considering possible adverse effects on the environment. What are the legal requirements of State Highway Departments as imposed by Federal legislation?

REFERENCES

1. WHEATON, W. L. C., "Form and Structure of the Metropolitan Area," *Environment for Man,* Indiana University Press, 1966.
2. ALONSO, W., "Cities and City Planners," *Daedalus,* American Academy of Arts and Sciences, Fall, 1963.
3. *The Community Builders Handbook,* The Urban Land Institute, Washington, D.C., 1968.
4. *Future Highways and Urban Growth,* Wilbur Smith and Associates, New Haven, Conn., 1961.
5. HEENAN, G. W., *The Influence of Rapid Transit on Real Estate Values in Metropolitan Toronto,* The Institute for Rapid Transit, Chicago, 1966.
6. LISKAMM, WILLIAM H., "The Transit Planning Process," *Ekistics,* Volume 27, Number 159, February, 1969.
7. MEYER, KAIN, and WOHL, *The Urban Transportation Problem,* Harvard University Press, Cambridge, Mass. (1966).
8. "Joint Development and Multiple Use of Rights of Way," *Highway Research Special Report 104,* Highway Research Board, Washington, D.C., 1969.
9. GRUEN, V., "No More Offstreet Parking in Congested Areas," *Readings in Urban Transportation,* ed. G. M. SMERK, Indiana University Press, Bloomington (1968).
10. *Transportation and Parking for Tomorrow's Cities,* Wilbur Smith and Associates, New Haven, Conn., 1966.
11. PIGNATARO, L. J., "Summary of Results from HHFA/HUD Mass Transit Demonstration Studies," *Urban Mass Transit Planning,* ed. W. S. HOMBURGER, Institute of Transportation and Traffic Engineering, University of California, 1967.
12. *Tomorrow's Transportation, New Systems for the Urban Future,* United States Department of Housing and Urban Development, Washington, D.C., 1968.

7

Planning Studies and Economic Analysis

Prior to the actual construction of transportation facilities, considerable engineering and planning skill must be harnessed in the planning and design phases. In the past, lack of proper planning has often led to inadequate service from facilities which have in other respects been well designed and constructed. Satisfactory planning, however, is a goal not as easily achieved as adequate design or adequate construction. In these latter activities, the engineer is able to define fairly easily the limits of his considerations. Once such limits have been defined, he works to achieve the best solution within these confines. The limitations which must be set to planning considerations are not easily demarcated. Adequate planning must take account of physical, social, economic, and political considerations in the form of constraints on the solutions and impacts of the implemented plan.

Ideally, transportation planning must be comprehensive. Comprehensiveness requires an analysis of the effect of the environment on the planned facility and vice versa. Such an analysis is, in essence, the basis of *systems analysis;* all transportation planning should be carried out on such a basis. In practice, planning is carried out in a piecemeal manner because of the realities of political fragmentation which bear little relationship to the demands of personal and freight movements. The demands placed on a transportation system are on both a temporal and spatial continuum. The failure of planning to deal with the transportation problem on a continuum is a necessary result of the numerous governmental jurisdictions which have authority over the individual area.

A typical transportation plan within a metropolitan area is a good example of this problem. The principal funding for the study is available from the Federal Highway Administration. Understandably, the plan's major emphasis is placed on the most apparent urban transportation problem, the movement of autos. Surface transit is considered, to varying degrees, within the plan. Because of different sources of funding within the *Department of Transportation* and the *Department of Housing and Urban Development,* mass transit planning has usually been conducted separately from highway planning. The degree of coordination between these two plans has in the past varied from almost total coordination to total independence. In addition, airport, port, and pipeline planning have typically been conducted almost totally independently from the metropolitan transportation study. Not surprisingly, the final metropolitan transportation plan, which has been the synthesis of these relatively independent parts, often exhibited a "lack-of-fit" with resultant problems such as airports whose capacities are hampered by inadequate ground access, and ports whose truck movements congest the downtown areas.

It is not only modal separation that is artificial and, therefore, conducive to fragmentary planning. Traditionally, urban transportation planning has been treated separately from rural highway planning. To a great degree, this division is artificial, since there is competition between urban and rural areas for limited funds. In providing funds for one area, the decision maker is either consciously or unconsciously trading off benefits which could have been attained in other areas. Where planning for rural and urban areas is artificially separated, the values of these trade-offs are hidden.

The techniques of transportation planning outlined in this chapter are purposely not delimited to area. The somewhat artificial separation of rural and urban planning is avoided. Where techniques are applicable in any particular situation, they are used. In determining the possible use of a technique, the planner must consider its validity of application, rather than relying on an historical association with either urban or rural transportation.

7–1. The Design of a Comprehensive Transportation Study. The basic design for an urban transportation study was set out by the National Committee on Urban Transportation in the form of a guide with accompanying procedure manuals [1, 2]. These guidelines have been found generally workable for studies conducted in small- to medium-sized cities where auto transportation predominates. With some minor modifications in overall procedure they have provided the main rationale behind many studies carried out in even the major urban areas. Few transportation studies, however, are identical in procedure even in the smaller sized urban areas. In large metropolitan cities and urbanized regions, special study procedures must be designed to cope with the individual effects of large scale commercial and industrial developments. In addition, there are significant influences from other modes of transport in metropolitan areas of this size. While every transportation study differs in specific details, the program can be generalized into the following procedures:

1. Administrative organization
2. Data collection
3. Analysis of present and future status of transportation
4. Development of transportation plan and financial program
5. Implementation
6. Updating procedures

ADMINISTRATIVE ORGANIZATION

7–2. Study Staff. In the conduct of a study for a large metropolitan area it is customary to assemble a professional staff for this specific purpose. The study director, whose background should be sufficiently broad that he is sensitive to the broad spectrum of problems involved in such a study, is normally assisted by a number of professionals trained in engineering, economics, computer technology, and planning. In addition, there will be a need for clerical aid for coding, mapping, and data tabulation. Field survey staff can be hired on a temporary basis. However, more satisfactory field work is attained when staff fluctuations are kept to a minimum. This can be arranged by scheduling personnel requirements with Critical Path techniques [49]. In smaller urban areas, the assembly of a large staff is quite uneconomic. Overall direction is normally given by the local city planner with the major portion of the study conducted either by the state road departments and state transportation departments or by transportation planning consultants who can better handle the fluctuating staff requirements.

With the increasing number of state departments of transportation which are multimodal in their areas of concern, it is anticipated that a large number of statewide and regional multimodal studies will be carried out. These studies almost certainly will be conducted by the departments of transportation on an internal basis.

7–3. Auxiliary Committees. The work of the study staff is supplemented principally by two advisory committees:

1. Technical Coordinating Committee
2. Citizens Advisory Committee

The Technical Coordinating Committee of a multimodal study will contain representatives of all local government departments affected by the study and plan implementation. Typically, in a metropolitan area the committee would be comprised of the city engineer, city planner, city legal officer, the chief financial officer, traffic engineer, local transit representatives, airport, and port authority representatives and railroad officials. Where representatives of other bodies are required to guide the technical feasibility of the plans, they should be added.

Citizens Advisory Committees are formed to gain public support for plans and to serve as community inputs, communicating community goals and values to the study staff. Typically, such a committee would contain representatives of a wide spectrum of the community, e.g., business groups, service clubs, the League of Women Voters, local unions, mass media representatives, and community organizations. It is important that this committee be properly used to help in the formulation of the plan.

DATA COLLECTION

7–4. Highway and Mass Transit Data. By means of field studies and the assembly of current data used by various agencies within the area, the planner assembles a complete inventory of existing highway and transit facilities and a cross-section of current travel behavior in the urban area. Together with the Origin-Destination Survey and the Land Use Studies, described in Sections 7–17 and 7–19 respectively, data are collected to give past economic and demographic trends.

Street Classification and Street Use Map. The objective of this study is the identification of all streets and highways with significant travel and their classification according to their present use.

All streets within the urban area are tentatively classified as expressways and freeways, other arterial streets, collectors, or local streets. Rural highways are classified by their place in the hierarchy of state and Federal systems by functional class and by their design category as two-lane, three-lane, multilane, or freeway. Based on these tentative classifications, a street and highway use map is prepared.

Existing Traffic Service Studies. Those involved in traffic engineering in the area are required to provide an inventory of the traffic service provided on the existing facilities under current traffic demand. To permit an evaluation of current traffic service, the following studies are normally carried out:

1. *Traffic Volume Counts* [6]. Control counts are made at carefully selected stations, and sample counts on an areawide basis permit a reasonable estimation of traffic volumes throughout the urban area. A cordon-line count around the central business district can be coordinated with a parking study also carried out in this phase.

The CBD parking study determines the mode of entry to the central area and hourly variations. At least one screen line count should be taken with the dual purpose of providing data on the annual trends of vehicular trips and providing verification of ground movements as predicted from the origin-destination survey samples.

2. *Travel Time Studies* [7]. Measurement of travel time on the major street system enables comparison of the levels of service provided by the various road sections comprising the existing network. Travel time studies are conducted at different periods of the day to permit this comparison under peak and non-peak loading. This study is usually done by a driver and an observer using the "average speed" or "floating car" method.

3. *Street Capacity*. Based on the geometry of the roadway, the type of traffic control, and the vehicular composition of the traffic stream, capacity calculations can be made for all sections of major streets and highways. These calculations are normally based on the techniques explained in the *Highway Capacity Manual* [8]. Under unusual conditions, special techniques may have to be developed by the traffic engineer in the field. In urbanized areas, capacity analysis is carried out on all intersections of two major streets. This normally amounts to approximately 5 per cent of all intersections.

4. *Accident Study*. Information on automobile and truck accidents is collected and assembled on a comprehensive basis from existing data sources, such as the files of local authorities and police records. Accident information gives a measure of the safety of the street and highway system. Safety is another criterion used to determine level of service which is generally considered in addition to congestion and convenience.

5. *Parking Study*. The provision of adequate terminal parking facilities is an inherent part of any transportation plan which relies on auto and truck transportation for the movement of goods and people. A parking study is, therefore, an important element in the inventory studies. In the past, two principal techniques have been used, the *Comprehensive Parking Study* and the *Limited Parking Study*.

A Comprehensive Parking Study is carried out where a complete analysis is required of the auto terminal problem in the Central Business District and other critical areas. The general scope of the comprehensive parking study includes [9]:

a. An inventory of existing parking facilities.
b. An examination of the adequacy of existing laws and ordinances.
c. An analysis of the limitations of administrative responsibility.
d. Usage patterns for existing parking facilities including duration of parking and walking distances.
e. Current demand patterns.
f. Possible methods of financing.

g. Traffic flow.
h. Parking characteristics.
i. Effect of large traffic generators.

A Limited Parking Study is considerably narrower in scope and concerns itself chiefly with the examination of four elements only [10]:

a. Parking supply
b. Parking usage
c. Parking Duration
d. Parking Meter Revenues.

Conventionally, parking studies have been carried out using a combination of interview and tag techniques. Recently, significant savings have been achieved by the use of colored aerial photography to determine parking use and duration [11].

6. *Traffic Control Device Study.* Because of the significant effect that control devices exert on the capacity of a street network, the planner carries out a comprehensive study which determines:

a. Location, type, and functional characteristics of all major traffic control devices.
b. Areawide parking regulations on a block-face basis.
c. Transit routes and transit loading zones.

Existing Transit Studies [12]. In order to determine how well transit meets and stimulates passenger demand for service, it is essential that the planner have a comprehensive understanding of the existing level of transit service and demand. This can be achieved by carrying out the following studies:

1. *Routes and Coverage Study.* The existing route structure is inventoried to determine the relation of populated areas to areas within reasonable walking distance of service. Routes are further examined to determine whether service in general follows the desire lines of travel, and whether transit accommodates growing community needs.

2. *Transit Route Inventory.* A survey is made of the physical characteristics of each transit route.

3. *Service Frequency and Regularity, and Transit Running Time Studies.* From an evaluation of adherence of transit to schedule and the adequacy of the schedule itself to provide reasonable service, the analyst is able to compare existing service with recommended standards.

4. *Passenger Load Data.* Passenger load data can determine whether the service frequency is adequate to satisfy the existing demand and whether adequate standards of comfort are maintained. The feasibility of altered headways can be evaluated from this data.

5. *Transit Speed and Delay Studies.* By means of speed and delay studies conducted on actual transit runs, identifiable causes and areas

of delay can be delineated. Internal and external causes of delay are determined to assist future remedial measures.

6. *General Operating Data.* Six general yardsticks are used for comparative purposes:
   a. *Quantity of service* in vehicle miles per capita.
   b. *Quality of service* in vehicle miles per revenue passenger.
   c. *Efficiency of service* in terminal-to-terminal time per vehicle mile.
   d. *Use of service* in revenue passengers per capita.
   e. *Route coverage* in route miles per capita.
   f. *Time convenience* as indicated by operating system speed.
7. *Passenger Riding Habits.* By sample survey methods, data are gathered concerning the riding habits of the passengers, their origins, destinations, their social and economic status. Where necessary, attitude surveys can be incorporated into this study.

Physical Street System Inventory [13]. In order to evaluate properly the present and future traffic-carrying capacity of the street and highway system, it is necessary to conduct a physical inventory of the component parts including:

1. Street widths
2. Block lengths
3. Pavement conditions
4. Geometric design
5. Storm sewers and surface drainage.

7–5. Air Transport Data. The current status of air transportation can be analyzed from a series of studies which relate to existing physical plant, and existing demand and usage.

Airport Classification. An overall picture of air transport in an area can be gained from an examination of the existing FAA classification of all airport facilities within the area including reference to their spatial relationships among themselves and to the urbanized areas. Classification should also include the function and development plans of individual airports as indicated in the National Airport Plan [14].

Travel Patterns. Passenger travel patterns are determined by an analysis of carrier records indicating scheduled passenger service. In-depth analysis of travel patterns can be made by an origin-destination study of air travelers carried out on a sample basis. Figure 7–1 shows the form that the questionnaire might take. The form shown was for use at the airport terminal. Other techniques that have been used include the distribution of questionnaires in-flight [15].

UPPER GULF COAST
AIR TRANSPORTATION
ORIGIN SURVEY

conducted by

SOUTH ALABAMA REGIONAL PLANNING COMMISSION

ESCAMBIA-SANTA ROSA REGIONAL PLANNING
COUNCIL OF FLORIDA

GULF REGIONAL PLANNING COUNCIL OF
MISSISSIPPI

The information you provide in this survey is important in planning air transportation facilities in the coastal areas of Alabama, Mississippi, and Northwest Florida. In particular, this questionnaire requests information concerning:

1. The address of the place from which you began your trip to the airport,
2. The nature of your trip.

The information requested will take only a few minutes of your time. Your name or identification is not required.

When you have answered the questions, please deposit the questionnaire in box provided.

THANK YOU FOR YOUR COOPERATION.

R. DIXON SPEAS ASSOCIATES
Aviation Consultants

ABOUT YOUR FLIGHT TODAY

13. What is the **PRIMARY** nature of the trip you are taking today?
    (1) ☐ Business
    (2) ☐ Brief Pleasure (less than one week)
    (3) ☐ Vacation (more than one week)
    (4) ☐ Military orders or leave
    (5) ☐ Personal Matters
    (6) ☐ Other ____________
14. What is the total duration of this trip?
    (1) ☐ 1 day (5) ☐ 5 days
    (2) ☐ 2 days (6) ☐ 6 days
    (3) ☐ 3 days (7) ☐ 7 days
    (4) ☐ 4 days (8) ☐ More than 7 days
15. Other than yourself, how many persons accompanied you in the terminal? ____________
16. How many people who accompanied you will depart with you on this flight? ____________
17. At what city will you end your air travel today?

____

ABOUT YOURSELF

18. Do you reside within the area shown on the above map?
    1. ☐ Yes 2. ☐ No
19. About how many flights, including this one, have you taken from this Airport in the past 12 months?
    (1) ☐ 1 or 2 (4) ☐ 7 or 8
    (2) ☐ 3 or 4 (5) ☐ 9 or 10
    (3) ☐ 5 or 6 (6) ☐ More than 10

THANK YOU — PLEASE DEPOSIT THIS FORM IN THE BOX PROVIDED.

1. On which airline are you about to travel?

   (1) ☐ Eastern (3) ☐ Southern

   (2) ☐ National (4) ☐ United

2. Flight Number?____________

3. Date?____________

FOR PASSENGERS NOT STARTING THEIR AIR TRAVEL AT THIS AIRPORT TODAY

If you transferred to this flight from a different flight earlier today, or if you were on this flight when it arrived at this airport earlier today, please answer the questions in this box.

4. I was on this flight when it landed here today.

   (1) ☐ Yes (2) ☐ No

5. I was transferred to this flight at this airport from a connecting flight of:

   (1) ☐ This Airline

   (2) ☐ Another Airline____________ Please Specify

6. If you are a transferring passenger, did you leave the airport between flights?

   (1) ☐ Yes (2) ☐ No

If you answered Question 6 **YES,** please complete all of the remaining Questions.

If you answered Question 6 **NO,** do not answer the remaining Questions, and please deposit this form in one of the boxes provided.

ABOUT YOUR TRIP TO THIS AIRPORT TODAY

7. From what location in this area did you leave for the airport?

   ____
   No. and St. Address, Building Name, and/or nearest St. Intersection

   ____
   City County State Zip

8. This location was:

   (1) ☐ Private Residence

   (2) ☐ Hotel/Motel

   (3) ☐ Your Place of Employment

   (4) ☐ Business you were visiting

   (5) ☐ Other____________ Please Specify

9. What time did you leave for the Airport?

   ____________ (1) ☐ A.M. (2) ☐ P.M.

10. What time did you arrive at the Airport?

   ____________ (1) ☐ A.M. (2) ☐ P.M.

11. How did you travel to the Airport today?

   (1) ☐ Private Car (5) ☐ Taxicab

   (2) ☐ Rent-A-Car (6) ☐ Private Plane

   (3) ☐ Airport Limousine (7) ☐ Air Taxi

   (4) ☐ Bus (8) ☐ Hotel/Motel Courtesy Car

   (9) ☐ Other____________

12. If you arrived in a private car which was parked at the airport, about how long do you expect it will be parked there?

   (1) ☐ 0 to 4 hours (3) ☐ 10 to 24 hours

   (2) ☐ 4 to 9 hours (4) ☐ Over 24 hours

Fig. 7–1. Air passenger terminal origin-destination questionnaire. (Source: R. Dickson Speas Associates.)

General aviation patterns are best sampled by personal interview at the airport. Air freight and air cargo patterns can be established by a sampling of waybills with the cooperation of the carriers [16].

Surveys of Existing Traffic. Existing traffic patterns and their relation and impact on existing airport facilities are determined by a variety of studies.

1. *Volume.* Air traffic volumes are available from FAA approach control facility records and individual airport records. In addition, FAA data on individual carriers can be surveyed to permit a breakdown of emplanements and deplanements.
2. *Travel Time.* Airport-to-airport travel times for scheduled carriers can be determined from an examination of published schedules.
3. *Airport Capacity.* The hourly and annual capacities of the airport are the respective hourly movement rates which can be handled by the facility. Annual capacity is now recognized as a better measure of an airport's efficiency than hourly capacity, and is found to depend on:
   a. Hourly capacity
   b. Weather
   c. Hourly demand
   d. Variation of daily demand on an annual basis.

   Capacity calculations can be carried out following procedures developed in the *Airport Capacity Handbook* [17].
4. *Peak Loading.* The monthly, daily, and hourly variations of traffic demand and passenger movements are assembled from available airport data and from sample surveys where records are inadequate.

Physical Inventory of Existing Airports. A complete inventory of existing airports would record the physical characteristics of:

Runways, taxiways, turn-offs, lighting, drainage, aprons
Aircraft parking areas, hangars, and maintenance shops
Air traffic control facilities and devices
Terminal facilities for passengers, baggage, and freight
General aviation support facilities
Ground transportation facilities
Vehicular parking
Suitability of surrounding land use for expansion.

7–6. Port and Harbor Data. Analysis of the adequacy and functional performance of existing port and harbor facilities requires that the planner have an in-depth knowledge of:

the current level of traffic service;
the nature of waterborne commerce;
the physical characteristics of the port facilities.

Traffic Service. The evaluation of traffic service depends on the ability of the analyst to determine both how well the existing facilities serve current demand and how well they can be expected to fill future needs. It is suggested that studies be carried out to permit the collection of adequate data in the following areas:

1. *Volume.* Passenger and freight volumes should be estimated from the port operators' records. Freight volumes should be classified by categories because of the different requirements for cargo loading, transfer, storage, and ground transport facilities.
2. *Seasonal Variations.* Because of the seasonal nature of commerce, significant variations of traffic volumes occur throughout the year. Estimates of the seasonal variations of previous years can be gathered from the port operators' records.
3. *Berthing Capacity.* Berthing capacity data require classification by the type of marine terminal; i.e., public facilities operating on a "first-come, first-served" basis or private facilities under lease or private operation.
4. *Loading and Unloading Capacity.* The availability and general maneuverability of cargo handling equipment causes differentials of loading and unloading capacities from berth to berth. The presence of fixed cargo handling facilities limits certain unloading capacities to individual berths. These capacities must be determined.
5. *Storage Capacity.* Storage capacity can be classified into short-term transit shed storage, long-term storage, rail storage, open storage, and container storage. In addition, differentiation should be made between refrigerated and unrefrigerated space.
6. *Ground transportation interface.* The adequacy of the ground transportation interface should be examined. In the case of many older marine terminals, the facilities were designed for vertical freight movements rather than for the extensive horizontal movements which are more easily handled by modern materials handling devices. Many older terminals also suffer congestion problems in the transfer interface with trucks since they were designed at a time when virtually all land transportation was by rail. In such terminals there may be excess rail loading capacity along with a woeful shortage of trucking docks.
7. *Containerization.* Analysis should be made of the current containerization capacity. This rapidly expanding area of waterborne commerce poses conversion problems in some older terminals.
8. *Specialized Port Services.* Around any port, special services grow up which cater to the peculiar needs of the port itself. Since those services themselves serve as an attraction to waterborne commerce, it is essential that the planner include data on their availability in his analysis. Included within this category are such services as export packing, customs house brokerage, and freight forwarding.

Waterborne Commerce Analysis. The patterns of both freight and passenger origins and destinations should be determined to permit an understanding of the underlying factors causing port traffic volumes. The declining general use of water transportation for passenger traffic makes this mode relatively unimportant in most areas. Where data are required, an origin-destination interview at the terminal will give sufficient information. Data for freight origin and destination are available from waybills. A representative sampling of waybills with the cooperation of the port operators can satisfactorily identify the origins and destinations of freight traffic in the port [16].

Physical Inventory. An inventory of the port's facilities should include the physical characteristics of:

The harbor, channels, and turning basins
Slips, wharves, and dolphins
Navigational aids
Aprons
Transit and storage sheds
Warehousing
Rail and open storage
Ground transportation interfaces
Surrounding land and land use.

7–7. Railroad Data. Under normal circumstances, the transportation planner within the United States will find himself only obliquely concerned with planning for railroads. By reason of the private nature of railroad ownership and operation, the degree of influence which can be exerted on this mode is slight. Where planning is required, however, the following studies should be carried out [18]:

Classification. The rail facilities of the area can be classified in two ways: by I.C.C. classification and by functional classification (see Chapter 2).

Existing Traffic Service. In establishing the current level of traffic service supplied, analysis must be made of several elements:

1. Inventory of freight and passenger routings available
2. Service frequencies on individual routes
3. Route capacities
4. Capacities and service characteristics of classification yards
5. Terminal facilities
6. Passenger riding habits
7. General operating data

8. Route and classification yard volumes
9. Traffic control devices.

Physical Inventory. In order to establish the physical condition of existing railroad plant, an inventory should be made of such items as rolling stock, power plant, bridges, structures, roadbed, traffic control devices, geometric design, right-of-way widths, and terminal facilities. Most of this information is available from the records of individual railroad companies requiring only a small amount of field work.

Other Physical and Operational Patterns. After the completion of inventories, the intermodal and multimodal relationships should be established. This would include an analysis of joint facility operations, reciprocal switching agreements, interchange routes, and overall handling of traffic volumes.

ANALYSIS OF PRESENT AND FUTURE STATUS OF TRANSPORTATION

7–8. Formulation of Current Status of Transportation. For the most part, much of the data assembled in the data collection phase of the study must be synthesized into numerous displays which more easily portray the state of the area and its transportation infrastructure. This, however, is not the sole use of the large amount of data collected. Data will later be used for modeling the development and transportation demand of the area and also in designing, evaluating, and programming a new system. At this stage, in order to indicate clearly the scope of the study, the analyst prepares a variety of displays and tabular summaries which enable a more rapid comprehension of existing conditions. Although from study to study the form of these summaries varies, a typical study would include at least some of the following:

1. *Population.* Several studies have shown current population along with the growth patterns over the historic period of the area. Figure 3–5 shows the growth patterns of the Chicago area over its whole history.
2. *Intensity of Land Use.* Intensity of land use is as important as location of land use in an urban area. Land-use intensity is frequently displayed in terms of population density. Figure 7–2 shows a population density map used in the Pittsburgh study. For a transportation study, a useful surrogate for intensity is the trip-end density, which is indicative of all land-use intensity, rather than residential land use only.
3. *Land Use.* The spatial relationships of the various types of land use are normally most easily displayed on a land use map compiled from the dominant characteristics of areas determined in the land use survey.

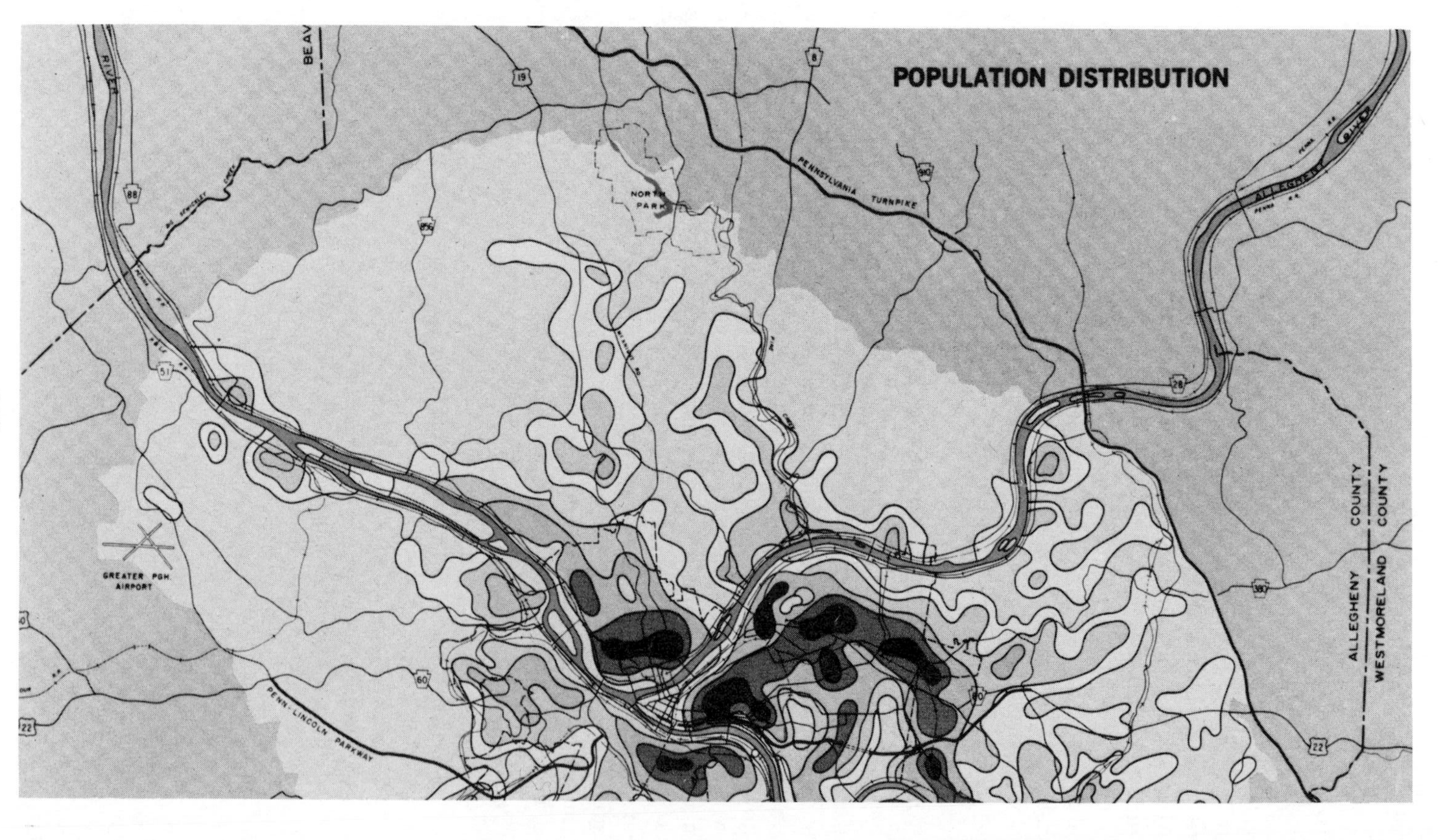
POPULATION DISTRIBUTION
PENNSYLVANIA TURNPIKE
NORTH PARK
ALLEGHENY RIVER
RIVER
BEAV
ALLEGHENY COUNTY
WESTMORELAND COUNTY
GREATER PGH. AIRPORT
PENN-LINCOLN PARKWAY

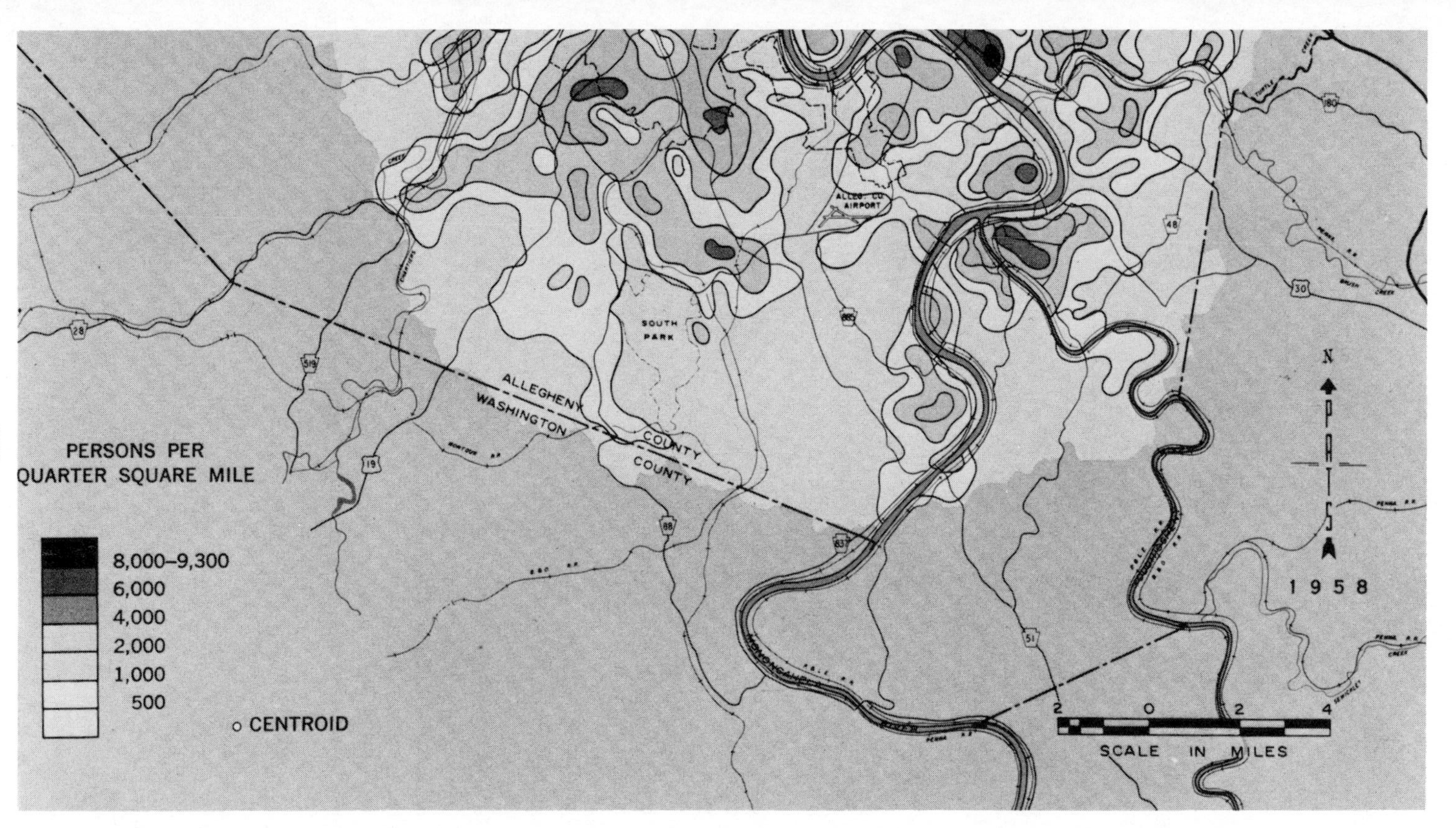

Fig. 7–2. Population distribution, Pittsburgh, 1958. (Source: *Pittsburgh Area Transportation Study*, Vol. 1.)

4. *Socio-economic Trends.* Under this category, it is customary to display those trends which are associated with increased demand for transportation facilities. Typically, in a comprehensive study some of the following would be displayed:

a. Population growth over time
b. Average income over time
c. Growth of car ownership
d. Transit usage and transit revenues over time
e. Vehicle travel over time
f. Composition of employment
g. Growth of emplanements and deplanements in aviation
h. Growth of air traffic activity, both general aviation and carrier
i. Air cargo trends over time
j. Air carrier, air cargo, and general aviation vehicle sizes and requirements over time
k. Rail passenger volume and revenues within the area
l. Major commodities originating and terminating by rail
m. Major commodities originating and terminating by water
n. Revenue trends of ports and harbors
o. Trends in modal split[1] of person miles travelled in the area
p. Trends in modal split of freight ton mileage in the area
q. Pertinent trends in other modes such as pipelines, etc.

5. *Origin-Destination Information.* As explained in Section 7–15, origin-destination information can be clearly displayed by superimposing desire lines on areawide maps. While customarily used for the display of auto and transit person trips, their use can be legitimately expanded to all modes of transportation for both person and freight movements. Figure 7–3 shows how this technique is suitably used to display national air travel patterns in the United States.

6. *Systemwide Facilities and Functional Characteristics.* The spatial arrangement of various systems over the study area are displayed by the use of systemwide maps. A concise exhibition of the various transportation systems permits the analyst to show clearly the relationships between land use, population, and transportation facilities.

There is also usually a need to show a variety of system characteristics which can help to identify problem areas. Three typical maps of system characteristics are shown in Fig. 7–4, which displays:

a. Functional classification of a highway system
b. Base year traffic volumes
c. Travel time contours or isochrones.

7–9. Land Use, Population, and Socio-economic Projections. Figure 7–5 outlines the transportation planning process as carried out in one large regional study. After the completion of the analysis of the data col-

[1] Modal split is defined in Chapter 9.

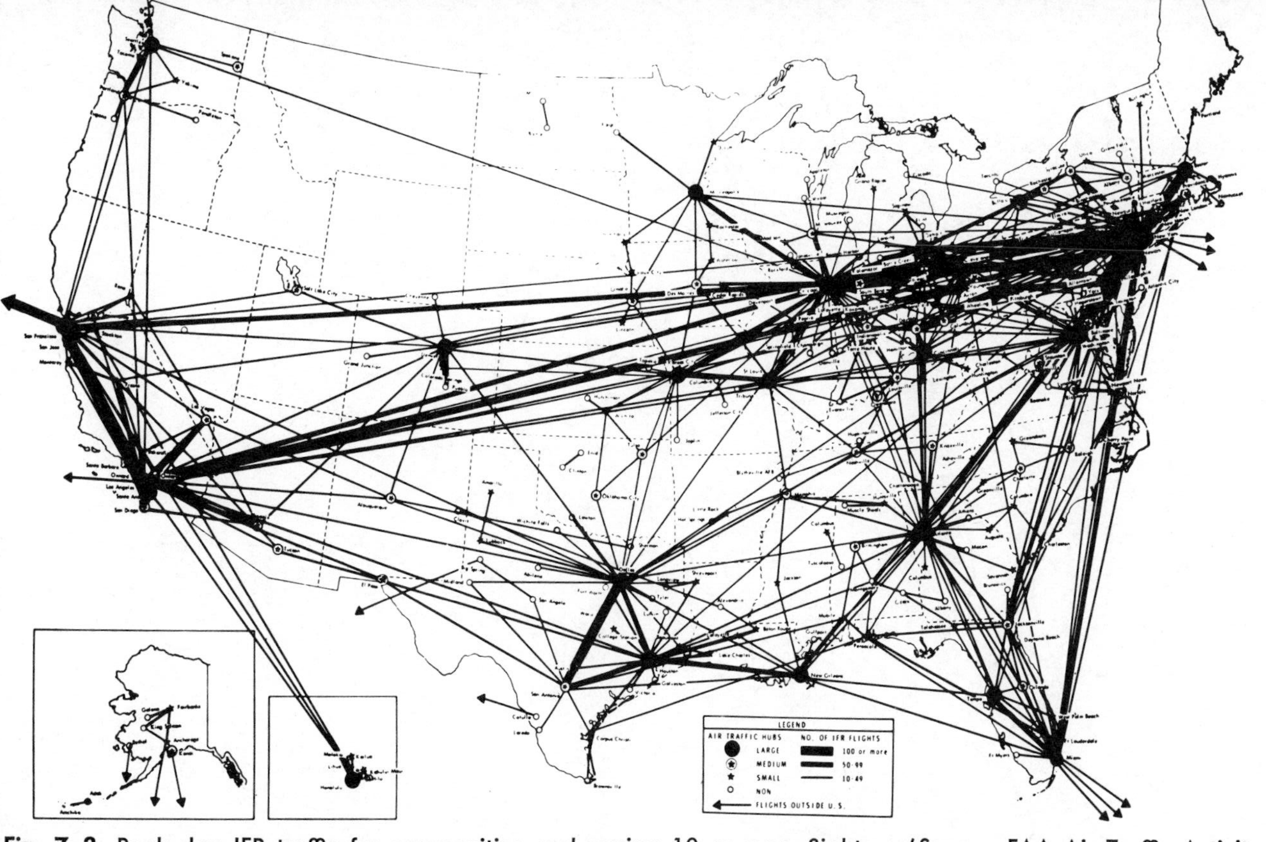

Fig. 7–3. Peak day IFR traffic for communities exchanging 10 or more flights. (Source: FAA Air Traffic Activity, 1966, Federal Aviation Administration.)

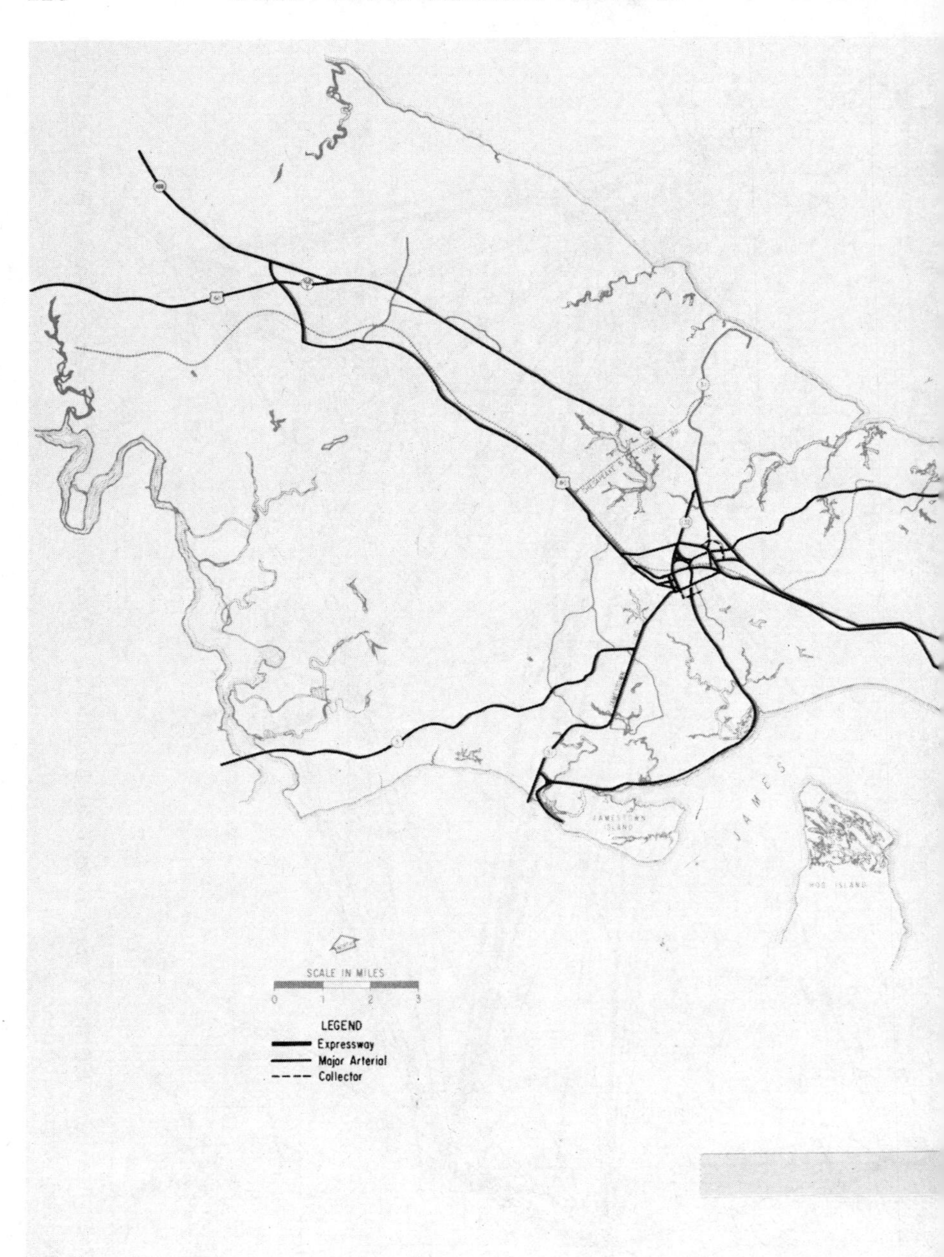

Fig. 7–4. Systemwide facilities and functional characteristics. (Source:

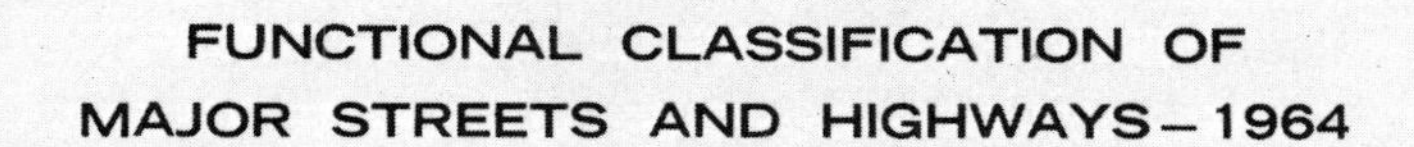

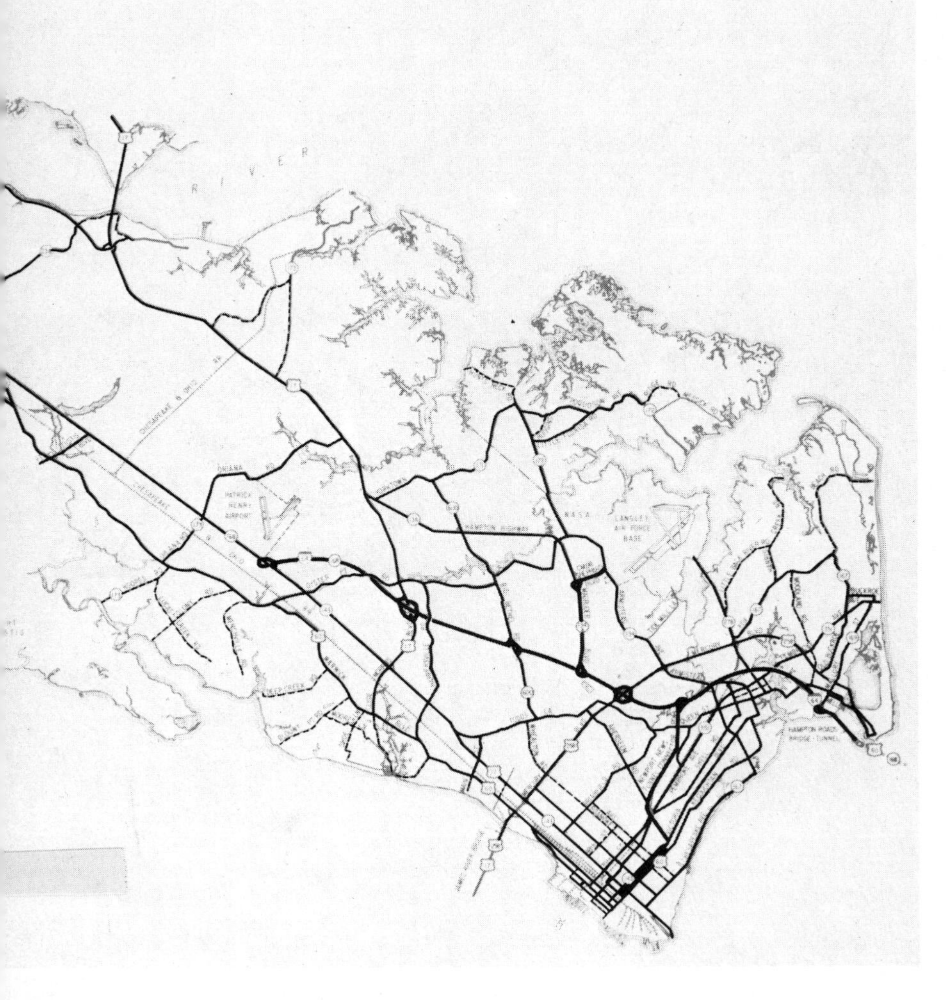

Peninsula Area Transportation Study, DeLeuw, Cather and Co.)

Fig. 7–4. (Continued.)

**1964 AVERAGE WEEKDAY TRAFFIC
MAJOR STREETS AND HIGHWAYS**

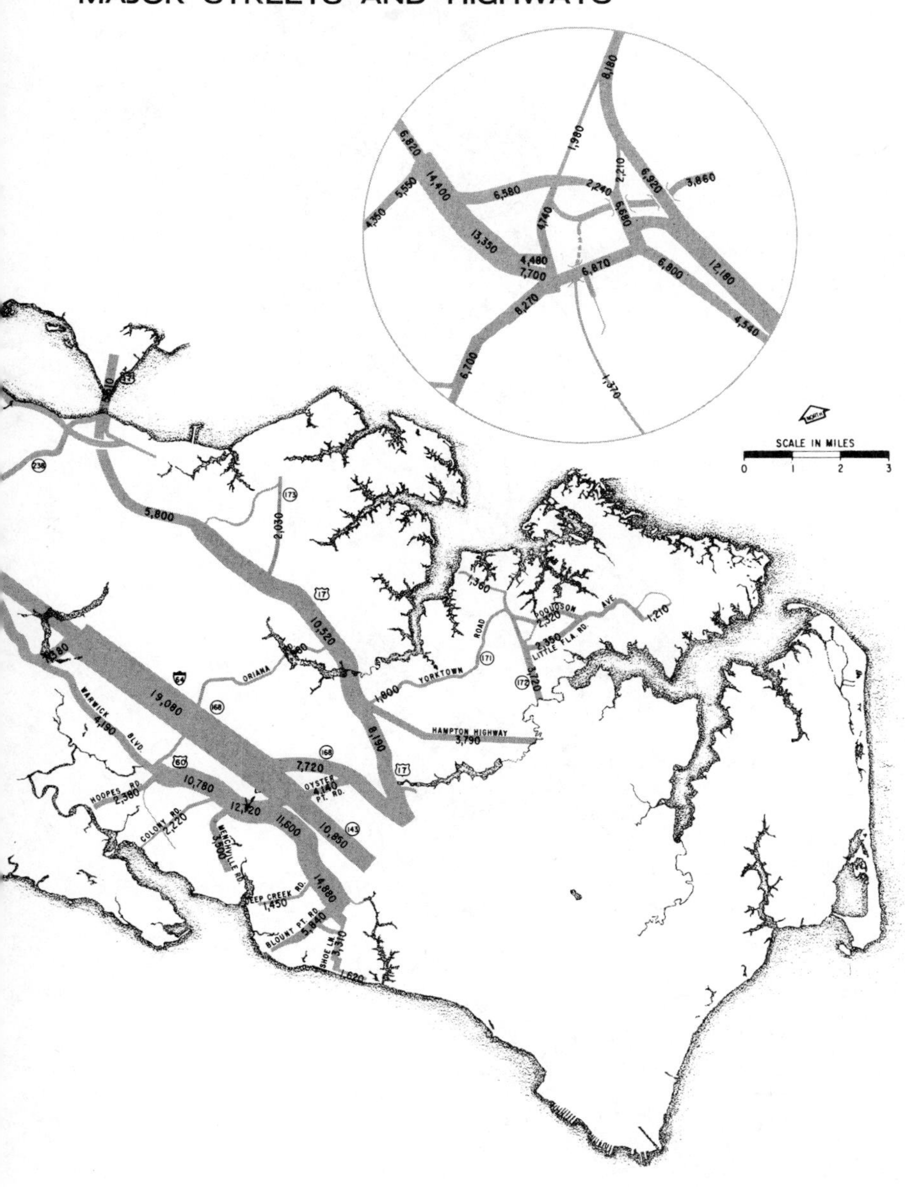

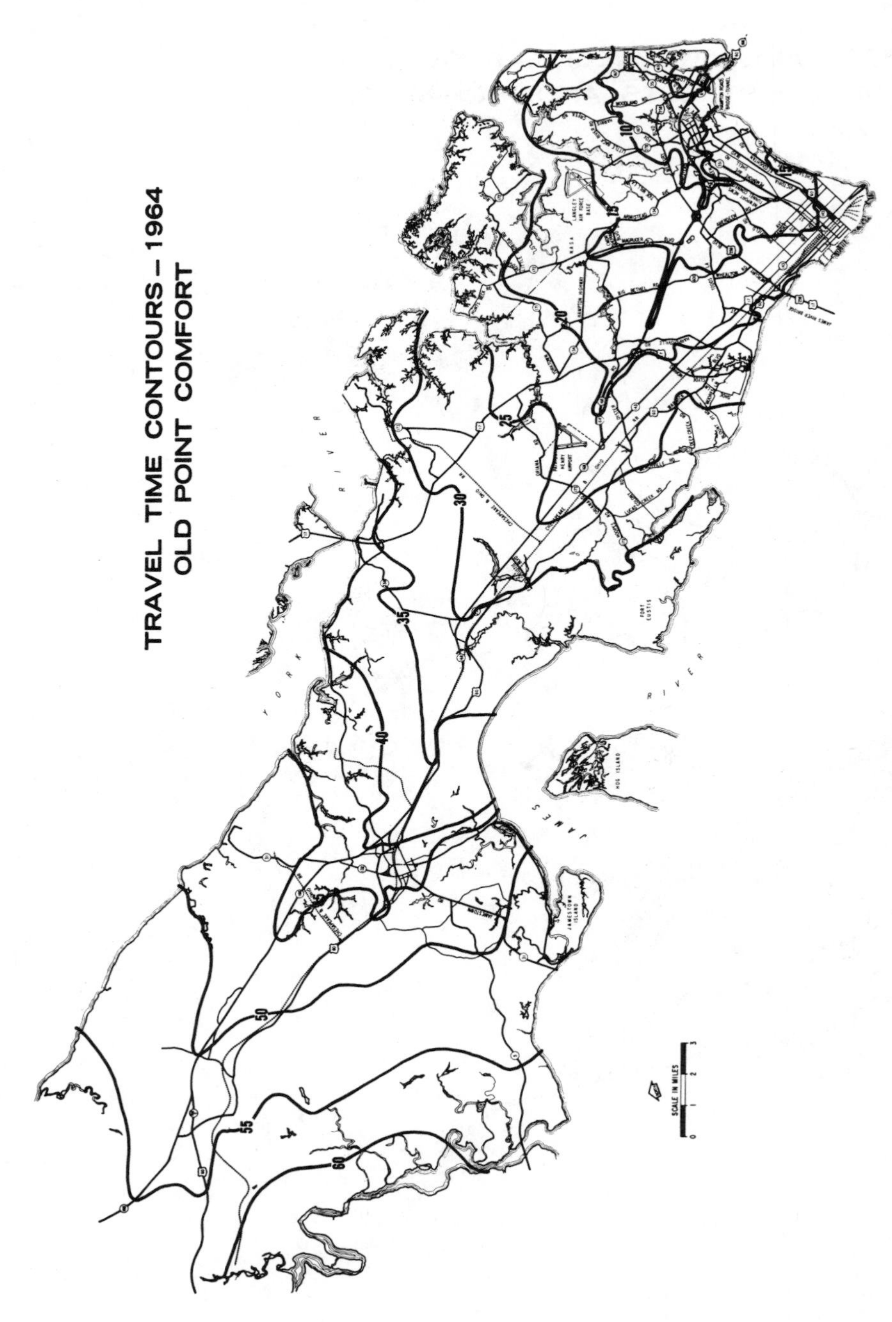

Fig. 7–4. (Continued.)

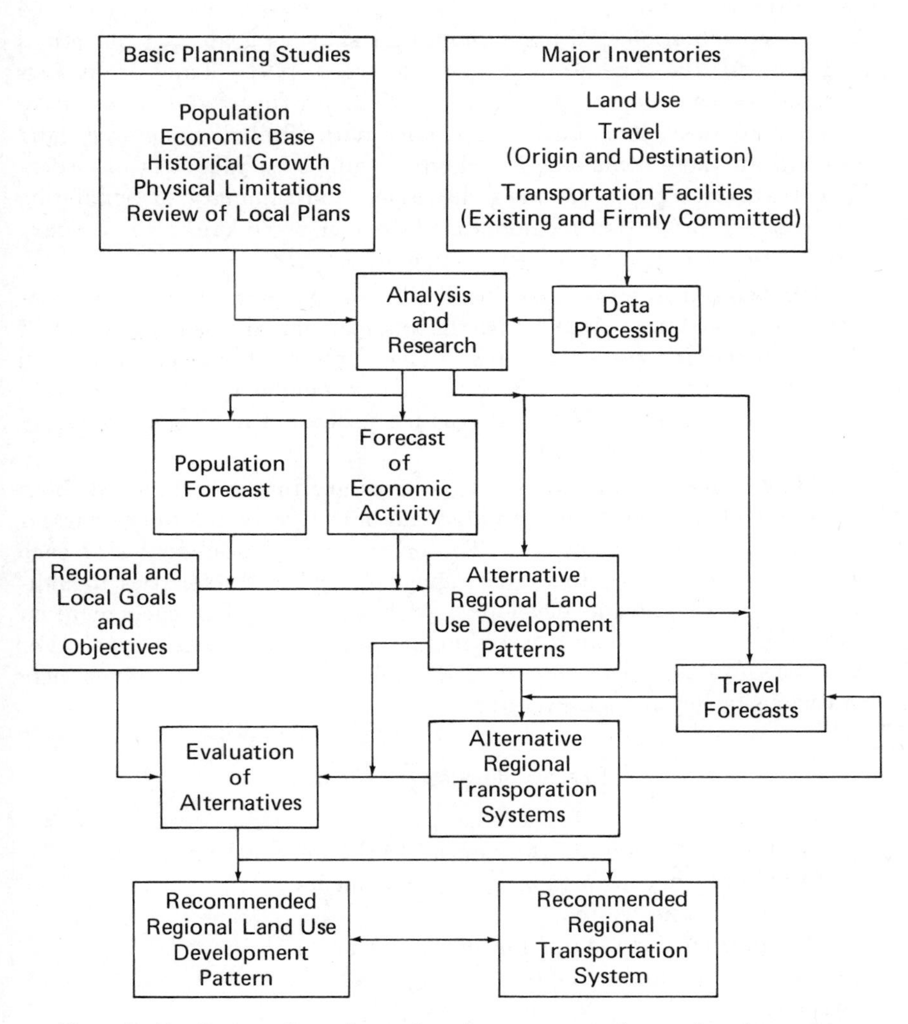

Fig. 7–5. Puget Sound regional transportation planning process. (Source: *Puget Sound Regional Transportation Study.*)

lected and generated from the basic planning studies and major inventories, the planner has reached the stage of projecting future land use, population, and socio-economic conditions in the study area. Population and economic projections are first made in conjunction with the goals and objectives of the region as articulated by local planning bodies. These forecasts are used to generate at least one future land use plan.

Generally, in a large study, several land use configurations are generated and, after evaluation, the plan which best satisfies goals and objectives is selected for the generation of travel forecasts [19]. Land use models as described in Chapter 9 may be used to produce valid alternatives in larger studies [20]. In small studies with limited manpower, land use plans are more commonly based on the subjective judgment of professional planners. For a complete discussion and summary of population and economic projection techniques, the reader should refer to standard works in the field of planning and economics [21, 22].

7–10. Travel Forecasts. On completion of the population, socio-economic, and land use projections, the planner can develop forecasts of travel demand for the study area. These interrelationships are shown in the flow chart, Fig. 7–5. Because of the complexity of the modeling process, the description of the forecasting procedure has been separately described in Chapters 8 and 9.

7–11. Adoption of Standards. It is necessary to adopt standards both of design and levels of service to allow an objective evaluation of current and future system deficiencies. The operational characteristics of each mode must be matched to the changing demand characteristics to indicate clearly existing and developing problem areas. For each mode of travel there will be numerous minimum standards of each type. For purposes of illustration only, a few typical standards are given here for various modes and submodes:

TYPICAL DESIGN STANDARDS

Highways
- Minimum Average Highway Speed (AHS) on rural sections
- Minimum wheel load capacities of pavements
- Minimum lane widths
- Recommended standards for bus lay-by areas
- Parking terminal standards

Airports
- Minimum runway lengths
- Minimum apron sizes to accommodate anticipated aircraft dimensions
- Terminal area space per 1,000 emplaned passengers

Waterways
- Recommended channel depths
- Loading apron widths on wharves
- Recommended areas of transit sheds per available berth

Railways
- Maximum rail curvature
- Recommended minimum overall operating speeds
- Maximum acceleration and deceleration

TYPICAL LEVEL OF SERVICE STANDARDS

Highways
- Overall travel speeds
- Level of service (as defined in *Highway Capacity Manual*)
- Accident level per 100 million vehicle miles
- Headway between transit vehicles
- Percentage of urban population within walking distance of transit
- Adequate parking supply/demand ratio

Airports
- Maximum delay to an individual aircraft operation
- Number of hours per year airport is closed due to atmospheric conditions
- Load factors on aircraft

Waterways
- Average delay to a ship arriving and waiting for a berth
- Rate at which a container ship can be unloaded
- Percentage of time that no storage will be available in port

Railways
- Service intervals
- Passenger demand to seat ratio
- Percentage of cancellations of service
- Age of rolling stock

7–12. Determination of Present and Future Deficiencies. Based on the information gathered in the basic planning studies and the major inventories, the planner is in a position to identify those areas of the transportation network with deficiencies. Deficiencies occur where the transportation facilities provided do not conform to the standards which the study has adopted as minimally acceptable. Where inadequate facilities are provided, either with respect to design or service levels, the results are manifest in operating delays, congestion, low overall travel times, large and inconvenient service headways, unsafe operating conditions, lack of comfort and convenience, and, in the extreme, complete lack of service. In practice, the transportation analyst examines the existing transportation network and evaluates this with reference to existing deficiencies. Current inadequacies are displayed by deficiency maps and a variety of tabulations. In evaluating existing highway facilities, for example, it would be necessary to tabulate existing operating speeds by facility type, accident rates by facility type, and parking deficiencies by location. It would also be necessary to display graphically areas with inadequate transit service or coverage, isochrones from the Central Business District for auto and transit trips, and the capacity-demand relationships for individual facilities or corridors.

In addition to determining current deficiencies it is usual practice to project future travel demand and to apply this demand to the "committed" system. The committed system can be defined in various ways, but normally it is regarded as that system of facilities that is so far into the process of implementation that there is little opportunity to introduce modifications. In some jurisdictions this is taken to be the point at which right-of-way purchase has started, in others it is taken as the beginning of the final contract design phase. By applying future traffic demand to the committed system, the planner becomes aware of the inadequacies which will arise if no further facilities are provided beyond those which are already in process. By including only the committed system, the options opened for planning the future system are at a maximum. Graphic displays of the areawide deficiencies are usually prepared to enable the planner to identify easily areas and corridors of insufficient facilities.

DEVELOPMENT OF TRANSPORTATION PLAN AND FINANCIAL PROGRAM

7–13. Development of Plans for Evaluation. There are quite obviously an infinite number of network and modal arrangements which will carry, with varying degrees of efficiency, the future transportation demands of a study area. These networks will themselves modify that demand since traffic generation, land use, and transportation facilities are highly interrelated. In attempting to develop an "optimal" system, however, the analyst is obviously unable to generate but a few of the infinite variations which could be used. The question that naturally arises, therefore, centers on how to generate plans for evaluation.

Fortunately, perhaps, from the viewpoint of study system generation, the transportation planner is not working "in a vacuum." The existing and fully committed network provides a firm starting basis from which the planner is obliged to work. It is in general unrealistic to abandon large amounts of existing capital investment and to abrogate plans to which substantial energy and capital have been devoted. The committed system, therefore, is generally used as a starting point for all future plans.

Even using the committed system as a basis, it is obviously quite possible to generate poorly conceived plans. With proper evaluation techniques these solutions would be rejected. In the interest of efficiency, however, it is necessary to develop a methodical and systematic way to develop study systems with reasonable likelihood of satisfying the evaluation process. Since the evaluation techniques are predicated on comprehensive planning goals, it is essential to take note of these goals in system development. Typically, one recent comprehensive plan generated study systems which would meet the following objectives:

1. A "best" plan for a given regional land use plan.
2. An "optimum" transportation system to meet regional transportation needs.
3. An encouragement and a service to desirable patterns of land use, promoting desirable development. [23]

Under normal circumstances, plan development typically would follow the following pattern:

1. Future traffic demand is applied by traffic assignments (see Section 9–6) to the committed system, and deficiencies noted.
2. A second round of solutions is tried which are based on the committed system with facilities added to overcome deficiencies. These second-round solutions are then tested for deficiencies. The types of solution often vary in their degree of dependency on different modes, e.g., in a metropolitan study second-round solutions might vary according to the extent of a rapid transit system.
3. Deficiencies in the second-round solutions are corrected by minor changes in system design and the third round of solutions is tested.
4. This process is repeated until satisfactory systems are available for comparative evaluation.

In the Atlanta Metropolitan Regional Transportation Study, system development was further limited by criteria which tested [24]:

Phase 1. The committed system
Phase 2. The committed system plus maximum traffic engineering
Phase 3. The committed system, maximum traffic engineering cost, plus improvements to existing alignments
Phase 4. Phase 3 plus new facilities on new alignments

7–14. System Analysis and Evaluation of Alternate Plans. It has been traditional in the area of public works to rely chiefly on cost-benefit analysis to determine the relative merits of various plans. Cost-benefit analysis has a long history of use in the selection of highway alternatives, and until recently the validity of the technique was not seriously questioned by practitioners within the field.

Benefit-cost analysis is a technique which determines the ratio of monetary benefits accruing from some action (or project) to the costs involved. The analysis is carried out for the entire economic life of the project, with both costs and benefits being discounted over time to reflect interest rates. In the case of a highway project, this type of engineering economic analysis usually considers at least the following costs and benefits:

Costs:
- Motor vehicle operating costs—fuel, tires, oil, maintenance, repairs and depreciation, stopping, idleing, and acceleration cost
- Construction costs including bonding, right-of-way costs, etc.
- Maintenance costs of highways
- Accident costs
- Travel time costs
- Highway operation and administration costs

Benefits:
- Net reduction in annual maintenance costs
- Net reduction in operating costs
- Net reduction in accident costs
- Net reduction in travel time costs

The advent of the interstate program in the late 1950's brought about extensive freeway construction within the urban areas. At the same time benefit-cost analysis was subject to severe criticism. Even relatively shallow reflection on the method indicated that some of the major problems caused by large freeway projects were in no way considered in conventional benefit-cost analysis. Further examination indicated that in some cases it was beyond the urban economists' current state of expertise to adapt the concept to account for some of these problems. To use this technique effectively, it is necessary to determine the results of all actions in the urban area in monetary terms. However, transportation plans were being attacked on the grounds that freeways disrupted communities and neighborhoods, polluted the air, reinforced *de facto* segregation, contributed to the decline of public transit, deteriorated the aesthetic environment, ennervated the vitality of the central areas and caused blight to their immediate environs. Obviously, such social costs existed as a by-product of some plans. However, it still remains beyond the ability of the planner to determine even a good estimate of the monetary costs involved. Cost-benefit analysis requires reasonably accurate costs to be of any use as a comparative tool. As a result cost-benefit analysis is now used mainly as a method of determining project priorities within the accepted plan, or as one element of consideration within the system evaluation process.

Several methods of system evaluation have been suggested [50, 51, 52]. Most methods are derived from traditional systems analysis techniques. The following technique has been suggested for metropolitan systems [53]:

1. Determine regional and metropolitan goals.
2. Determine objectives within the scope of the transportation plan which are related to the goal statements.
3. Determine by such means as attitude surveys the relative importance of the objectives so that some weighting can be put on each objective.
4. Select quantitative criteria which can be used to evaluate how well each objective is satisfied.
5. Evaluate each scheme by determining how well each criterion satisfies the overall objective. Total system effectiveness is determined in reference to all objectives by an effectiveness measure which is the sum of each criterion evaluation weighted by the importance of the objective to which it relates.
6. Select the system with maximum effectiveness.

Example: Assuming that regional and metropolitan goals have been determined, the following objectives have been determined [53]:

a. System should provide economy
b. Minimum disruption of individuals by relocation
c. Transit should provide a high level of comfort and convenience
d. The central area should be made highly accessible
e. Transit should be available to low-income areas.

The following criteria are selected to provide a measure of each objective:

a. Cost-benefit ratio
b. Number of persons relocated
c. Load factor on transit vehicles in peak hour
d. Accessibility index of core areas
e. Transit accessibility index to low-income traffic zones

From attitude surveys and rating panels objectives a through e have the following ratios of relative importance: 40 per cent, 20 per cent, 20 per cent, 10 per cent, 10 per cent.

Evaluation Matrix:

	Possible Effectiveness Score	*Score for Plan A*	*Score for Plan B*	*Score for Plan C*
1. Cost-benefit ratio	40	35	25	30
2. Persons relocated	20	10	20	5
3. Transit load factor	20	10	15	3
4. Core accessibility	10	3	5	10
5. Low-income transit availability	10	2	10	8
	100			
Total Effectiveness Score		60	75	56

Plan B is selected as being most responsive to the regional goal statements which the transportation plan is meant to strengthen.

Various planners have argued that some variables which require consideration can never be adequately quantified, and that the above type of analysis can be misleading. Other non-quantitative systems analysis techniques have been suggested. The reader is referred to References 51 and 52 for further discussion of this topic.

IMPLEMENTATION

7–15. Capital Programming and Implementation [25]. The work carried out in the evaluation phase of the comprehensive study indicates the particular system which best meets the goals and objectives of the study area. This is sometimes rather loosely referred to as the "optimal" system. Further work, however, is necessary to break this system down into individual projects and to schedule those projects and their inherent costs into the capital expenditures estimates of the area. Only when year by year estimates of capital expenditures are calculated does the ease or difficulty of implementation really become apparent. It is suggested that the planner and engineer should carry out capital programming in the following manner.

1. *Project Breakdown.* The selected system should be broken down into individual contract-size projects. Project size and boundaries should reflect the scale of construction projects normally carried out in the area.
2. *Project Evaluation and Priority Scheduling.* Each project is evaluated and based on its judged priority, and is programmed in sequence subject to the constraints of available funds. Programming can be carried out using the following procedures:
   a. Each project is estimated with reference to its time and cost requirements.
   b. Administrative constraints such as available funding and coordination required with the programming of transportation and other community facilities are clearly set out.
   c. Service considerations of existing facilities are evaluated to determine their maximum continued useful life, e.g., pavement conditions for roads and airports, useful life of rail roadbed, and structural condition of bridges.
   d. Evaluation of preliminary priority rankings based on benefit-cost ratios and community benefits is made.
   e. Establishment of the time period over which the capital programming is feasible. If a curtailed period is used, the design system is available at an earlier date at relatively high annual capital budgeting costs. If the catch-up period for construction is over the full period to the design year, annual costs are minimized but the interim level of service of the system is also minimized.

f. Final programming of individual projects based on preliminary priority rankings and other considerations which include:

Coordination with other public improvements
Equable geographic distribution of improvements
Continuity of minimum levels of service
Engineering considerations.

Minimum levels of service are verified by assigning interim year traffic demands on interim facilities. Where severe problems become apparent in the interim years, it is necessary to shuffle projects to maintain adequate service.
g. Calculation of final annual costs based on the accepted capital program.

UPDATING THE TRANSPORTATION PLAN

7–16. Updating Procedures. Modern transportation plans are conceived by their authors as dynamic entities. Faced with evolving technology and constantly changing social and economic conditions, a static plan has no chance of implementation. Urban planners, therefore, speak of their work in the context of a continuing process in which the details are constantly modified subject to the overall directions laid down by regional and community objectives. The design of any transportation study should incorporate procedures for continuous plan updating [5].

Updating procedures call for a day-to-day maintenance of data which affects the elements of the plan. These procedures require the establishment of an areal information system which can accommodate a continuous data flow. The information system should be either partially or totally computerized, and should be keyed to the lowest possible data collection unit, which is usually the individual land parcel. A general surveillance system would record changes in:

Land use
Transportation facilities
Social economic data
Population

Over the period of implementation of the plan, major updating studies such as origin-destination studies become necessary to verify that modeling procedures and travel forecasts hold valid for the study area over time. The detailed techniques of updating are beyond the scope of this chapter. For further information in this area, the reader is referred to Reference 54.

SELECTED PLANNING STUDY TECHNIQUES

7–17. Origin-Destination Studies. Perhaps the most important and certainly the most time-consuming and expensive planning study is the origin-destination study. Properly designed, this study will identify where and when traffic begins and ends, the socio-economic characteristics of the trip maker, the purpose of the trip, the mode of travel, the land use type at the beginning and end, and, for freight, the type of material moved. The origin-destination study is a sample cross-section of all travel on the average day within the area. This cross-section is taken to be representative of the average travel demand on the system at the time the survey is carried out. After relating the demand to the characteristics of the area and to its inhabitants the planner can use the derived relationships to determine the future travel demand in conjunction with projections of future economic development and population growth. The origin-destination survey is, therefore, regarded as the prime data source of the transportation study. The most comprehensive origin-destination studies have been carried out in conjunction with highway planning studies for cities over 50,000 population. Guidelines for these studies have been set out by the Bureau of Public Roads [26].

As the size of a city increases, the form of tripmaking within and through the city changes in nature. Table 7–1 indicates the changes in the nature of traffic approaching the urban area with increasing city size. There is a dramatic decrease in the percentage of bypassable

TABLE 7–1

Destination of Traffic Approaching Cities

City Population Group, thousands	Percentage of Total Approaching Traffic Destined to:		
	Central Business District	Other Points Within City	Points Beyond City
Under 5	29	22	49
5 to 10	29	29	42
10 to 25	28	37	35
25 to 50	26	49	25
50 to 100	24	57	19
100 to 250	21	62	17
250 to 500	18	70	12
500 to 1,000	15	77	8

Source: *A Policy on Arterial Highways in Urban Areas,* American Association of State Highway Officials, Washington, D.C., 1957.

traffic, from almost 50 per cent in areas with a population of less than 5,000 to only 8 per cent in cities over 500,000. This difference in patterns is reflected in the type of origin-destination survey that can be carried out to establish travel patterns satisfactorily. The American Association of State Highway Officials [27] currently recommends the following types of O-D surveys:

1. *External Cordon.* Normally carried out for cities under 5,000 population where external traffic forms the major traffic patterns. Sample sizes range from 20 per cent on high-volume roads to 100 per cent on low-volume roads.

2. *Internal–External Cordon.* Applicable for cities with populations of 5,000 to 50,000. Direct interviews are carried out on two cordon lines. The external cordon is set at the limit of the urbanized area, and the internal cordon set around the fringe of the Central Business District. Despite limitations, this type of survey can supply reasonably complete travel patterns. Typically, a 20 per cent sample is used.

3. *External Cordon-Parking Survey.* Applicable for cities of 5,000 to 50,000 with little transit and major problems in the central areas. The external cordon survey is carried out in the standard manner. The parking survey consists of three basic elements:
   a. Inventory of existing parking spaces
   b. Interview of all parkers at curb and off-street facilities to determine origin, destination, trip purpose, and length of time parked.
   c. A cordon volume count of traffic entering and leaving the central areas.

4. *External Cordon-Home Interview Survey.* This comprehensive type of study is applicable in areas of all sizes. The External Cordon Survey is carried out in the standard manner with a sample size of 20 per cent of all vehicles passing through the cordon. In addition, within the cordon, household interviews are carried out on a sample basis. The sample size varies with the size of the population of the area, and to some degree with the density of population. Recommended sample sizes vary from 20 per cent of all households in areas less than 50,000 total population to 4 per cent in areas over one million. Each sample dwelling is visited by an interviewer, and questions are asked concerning how many persons live at the address, their age and occupation. The interviewer also determines the number of cars at the address and usually records information concerning the number of trips made by all individuals, their purpose, origin, destination, mode of travel, and the times at the beginning and end of trips for the previous day. Figure 7–6 shows the recommended forms for the home interview.

When conducting a Home Interview Study, it is also necessary to conduct *Truck and Taxi Surveys;* otherwise, there will be no

METROPOLITAN AREA TRAFFIC SURVEY

DWELLING UNIT SUMMARY

Preceding number ____ Card ____ 1
Interview address ____ Tract No. ____
Succeeding number ____ Block No. ____
Sample No. ____
Date of travel ____ Subzone No. ____

A. How many passenger cars are owned by persons living at this address? ____
B. How many persons live here? ____
C. How many are 5 years of age or older? ____
D. Household information: Race ____

Person No.	✓ if inter-viewed	Sex and Race	Person Identification	Code		Occupation and Industry	Trips Yes	Trips No
01								
02								
03								
04								
05								
06								
07								
08								
09								
10								

E. Total number of trips reported at this address ____
1. Number of persons 5 years of age or older making trips ____
2. Number of persons 5 years of age or older making no trips ____
3. Number of persons 5 years of age or older with trips unknown ____

F. Comments and reason if complete information was not obtainable ____

G. Factor ____

Administrative Record

Interviewer ____

CALLS

	DATE	TIME
(1)		
(2)		
(3)		
(4)		

REPORT SUBMITTED INCOMPLETE
Date ____
Reason ____

Supervisor's comment ____

Remarks ____

Report completed ____ (Date) (Initial)
Interviews checked ____ (Initial)
Coded by ____ (Initial)

Fig. 7–6. Standard home interview questionnaire. (Source: *Survey Procedure Manual 2B,* National Committee on Urban Transportation, Public Administration Service, Chicago, 1958.)

record of truck and taxi trips which do not cross the external cordon. It is customary to contact a proportion of truck and taxi owners for these surveys and to determine travel patterns from the daily reports. Sampling must be at a higher rate than for the home interview: the sample rate may be calculated from the total taxi population. It is common practice to sample truck owners at twice the rate of taxi owners.

5. *Transit Surveys.* There are two chief types of study procedure used to define the patterns of transit usage.

a. *Transit Terminal Passenger Survey.* Passengers waiting to board a bus, streetcar, or tram are handed questionnaires which contain questions relating to the trip they are about to take. They are requested to mail these completed questionnaires which are on prepaid postcards. A variant of this system, in which the driver handed out survey forms to boarding passengers, was used by consultants in a transit patronage study in Washington [28].

b. *Transit Route Passenger Survey.* Questionnaires are handed to passengers by two field interviewers who ride the bus. The questionnaires are collected as the passenger leaves the bus. As might be expected, the response rate for this type of survey is higher than for the mailed questionnaire. In most cases, however, the added expense may not be warranted.

6. *Other Origin-Destination Surveys.* For rail, air, and water passenger modes, origin-destination data are best gathered by *terminal passenger questionnaires.* Figure 7–1 is an example of an air terminal passenger questionnaire used to gather a diversity of information on the trip maker and his travel habits. The determination of origin and destination patterns of freight presents significant problems which have not been altogether solved. For air and water carriage a sampling of waybills at the terminal points which these modes must use can give the desired information. With the cooperation of the railroads operating in the area, the same type of information is available but somewhat more difficult to obtain. Road motor freight, however, presents its own special problem. Because of the lack of localized terminals, there is no straightforward method of obtaining this information. Recent work in conjunction with descriptive sampling plans issued by the Interstate Commerce Commission indicates that continuous traffic study (CTS) data may be useful in the establishment of reliable commodity flow statistics on a regional basis [16].

7–18. Highway Capacity Studies. The capacity of a given section of roadway, given either in one direction or in both directions for a two-lane or three-lane roadway, may be defined as the maximum number of vehicles which has a reasonable expectation of passing over a given section of roadway during a given time period under prevailing roadway and traffic conditions [8]. While capacity denotes the maximum number of vehicles that a roadway can handle, there are lower volumes which can be accommodated on a section at higher operating speeds and decreased congestion levels. The different levels of operating speed and congestion are accounted for by the various levels of service at which a facility can operate.

Level of service denotes any of an infinite number of differing combinations of operating conditions which can occur on a given lane or roadway when it is accommodating various traffic volumes [8]. The various levels of service at which a facility can operate vary from level A, the free-flow condition, to level F, the congestion condition. For each level of service on a given facility, a service volume can be designated. The *service volume* is the maximum volume that can pass over the section while operating conditions are maintained at the specified level of service. Levels of service are related to both operating speed and volume/capacity ratios according to the general concepts outlined in Fig. 9–17.

At high operating speeds and low volumes, the driver is free to maneuver. The facility is said to operate at the highest level of service—level A. As volume increases, operating speeds decrease. Level B is the zone of stable flow with some slight reduction in driver freedom. Rural highways are often designed to accommodate the service volume of level B. Level C is still within the zone of stable flow, but the driver is subject to more restrictions due to increasing volumes. As a result, he has less freedom to pass and select his own operating speed. Level D approaches the zone of unstable flow. At this level of operation the speeds are still acceptable, but the driver has little freedom to maneuver within the traffic stream. Level E denotes volumes approaching capacity. The facility operates at low speeds and freedom to maneuver is almost totally eliminated. Small variations in volume can cause large speed changes as the facility operates in the range of unstable flow. Congestion and the familiar traffic jam are represented in level F, where densities increase rapidly as volume decreases and speeds drop from a crawl to a standstill. The description of methods of capacity analysis and the determination of service volumes for given facilities is beyond the scope of this section. For a full treatment of these areas, the reader is referred to sources which cover these procedures in sufficient depth [8]. Depending on the type of facility, the elements which determine level of service differ, as is shown by Table 7–2. It will be seen that capacity and service volume analysis can be divided into two main areas of concern:

1. Uninterrupted flow (freeways, expressways, and highways with access control)
2. Interrupted flow (at grade intersections).

It has been determined from extensive field observations of the behavior of traffic on operating facilities that capacities and level of service provided by roadways under uninterrupted flow are dependent on two groups of factors:

1. *Roadway factors* which relate to the physical characteristics of the design of the facility. The following roadway factors are found to be pertinent to capacity analysis:
   a. Lane widths
   b. Lateral clearance from edge of pavement
   c. Width of shoulders
   d. Presence and nature of auxiliary lanes:
      i. Parking lanes
      ii. Speed change lanes
      iii. Towing and storage lanes
      iv. Auxiliary lanes in weaving section
      v. Truck climbing lanes and passing bays

TABLE 7-2

Elements Used to Evaluate Level of Service

Element	Freeways	Multi-Lane Highways	Two- and Three-Lane Highways	Urban Arterials	Downtown Streets
Basic elements					
Operating speed for section	X	X	X		
Average overall travel speed				X	X
Volume-to-capacity ratio:					
(a) Most critical point	X	X	X	X	
(b) Each subsection	X	X	X	X	
(c) Entire section	X	X	X	X	
Related elements					
(a) Average highway speed	X	X	X		
(b) Number of lanes	X				
(c) Sight distance			X		

Source: *Highway Capacity Manual*, Table 4.2.

e. Surface condition
f. Alignment
g. Grades.

2. *Traffic Factors* are those which relate to the composition of the traffic using the facility and the types of controls in use. The following have been found to exercise control over capacities and are, therefore, used in analyses:
 a. Percentage of traffic constituted by trucks
 b. Percentage of traffic constituted by buses
 c. Lane distribution
 d. Variations in traffic flow
 e. Traffic interruptions.

At grade intersections facilities operate under interrupted flow conditions. The following factors have been found to affect the capacities and service volumes of such intersections.

1. Approach width of intersection
2. Location within the metropolitan area
3. Traffic flow variations
4. Size of metropolitan area
5. Percentage of left and right turn movements
6. Percentage of buses entering intersection
7. Percentage of trucks entering intersection
8. Location of bus stops in relation to intersection
9. Parking status.

7–19. Land Use Studies. Early in the 1950's, it became clear from a variety of research studies that the characteristics of travel could be related to the intensity and spatial separations of land use [29, 30, 31]. For the purpose of developing traffic models and to understand the physical impact of transportation facilities clearly, the planner needs up-to-date land use data on an areawide basis. Typically, the planner will employ at least the following techniques [32]:

1. Land use survey
2. Land use classification
3. Vacant land use study
4. Presentation and storage of land use data

Land Use Survey. Field studies are carried out to identify the type of use of all land parcels in the area. Two principle techniques are employed in the land use survey; these are *inspection* and *inspection-interview.*

The *inspection* method can be carried out on foot or, more commonly, "windshield inspection" is accomplished by a car with a driver and an observer. The use of individual parcels is noted directly on maps, or is recorded on schedules that are referenced in some manner to prepared parcel maps. Schedule sheets are necessary where land use data are stored as computer records.

The *inspection-interview* method can be used in areas of more intense land use, where more accurate estimates are needed of actual floor areas by type of usage. The results are recorded on schedule sheets.

Land Use Classification. In the process of differentiating the spatial arrangements and activity patterns of the urban area, it is necessary to arrange land uses in some form of standard classification. There are several land use classification systems in current use which permit detailed differentiation of usage [34, 35, 36]. It is recommended that in any study area one of these standardized classification procedures be adopted. For the purpose of the transportation planner, only major categories of land use are classified. It will, therefore, be necessary in most cases to group the results of the land use survey for analysis and presentation. Table 7–3 indicates eight broad major categories of

TABLE 7–3

Illustrative Major Urban Land Use Categories

Residence	
Low density	yellow
Medium density	orange
High density	brown
Retail Business	red
Transportation, utilities, communications	ultramarine
Industry and related uses	indigo blue
Wholesale and related uses	purple
Public building and open spaces	green
Institutionalized buildings and areas	gray
Vacant and non-urban land	uncolored

Source: Chapin, F. S. *Urban Land Use Planning*, 2nd Ed., University of Illinois Press, Urbana, 1965. Table 21.

land use. Included in this table are the color codings suggested for standardized presentation.

Vacant Land Use Study. Prior to the design of alternate future land use plans, it is necessary to determine the development capability of existing vacant land in the area. This type of analysis is called *Vacant Land Study* or sometimes a *Development Capability Analysis.*

Vacant land is classified by two chief criteria:

1. Capability as related to topographic and drainage standards
2. Capability as related to utilities and improvements available.

Standard planning works suggest division into *prime* and *marginal* land [33]. Prime land is all land which can be adequately drained with topographic slopes considered economic for building in the area. Low lying, steep, and derelict land is considered marginal because only with large capital investment is such land capable of conversion to the prime category. Prime land is often subcategorized to differentiate between different classes of slope. For example, land at slopes still satisfactory for residential areas, 10 to 15 per cent, would be unsuitable for extensive one-story industrial plants. Thus, the basic categories are expanded to Class I Prime, Class II Prime, and Marginal.

Further subcategorization is undertaken with respect to the factor of available improvements. Subclassification A could indicate all improvements available, B water and sewer only, C power only, etc. Based on the classification scheme of the planner it is possible to identify and summarize the amount of available land in each class and category by district over the whole area, e.g., Class II Prime A.

Presentation and Storage of Land Use Data. The assembled general land use patterns are stored and presented using the following techniques:

1. Land use maps. Figure 7–7 shows an existing land use plan for the city of San Juan, Puerto Rico.
2. Tabular summaries of land use studies which are readily made from data assembled in the various studies. Such summaries give the planner a clear idea of the nature of the area.
3. It is becoming widespread practice for cities to be in the process of setting up computerized data banks with information on:
   a. Persons
   b. Parcels
   c. Street facilities [40]

   Computerized data banks where information is coded and adapted on a street address basis can be used in conjunction with computer mapping techniques [41]. It is possible to produce by mechanical means graphic displays of the different characteristics of the urban area down to as fine a grain as the block face [42].

7–20. Traffic Volume Studies. The automobile and truck mode is unique in the fact that while public agencies provide the roadbed they have little or no control over the amount of usage this roadbed receives, and unless special studies are carried out, the volume of traffic using facilities is unknown. Motor vehicle volume studies have now been carried out for many years. The procedures used are well documented and, therefore, will be outlined only in the following section [46, 48].

Areawide Counts. The comprehensive traffic volume study is carried out largely on a sample basis [47]. In order to develop factors to permit adjustment of sample counts to reflect the average daily traffic, it is necessary to conduct *control counts* throughout the system. There are three categories of control counts:

1. *Key count stations.* Selected control stations are used to determine hourly, daily, and annual traffic variations. It is recommended that one non-directional 7-day count be carried out annually in conjunction with one non-directional 24-hour count monthly. There will be at least one key station on each category of street classification.
2. *Major control stations.* Each major street will have a major control station which serves to estimate on a sample basis the traffic flows on the major street system. Therefore, each freeway, expressway, major arterial, and collector street is sample counted with a 24-hour non-directional count every two years.
3. *Minor control stations.* These are set up to sample typical minor street traffic. In smaller cities it is recommended that three counts be taken on each of the three classes of minor street: industrial, commercial, and residential. A 24-hour non-directional count is carried out every two years.

Coverage Counts are the samples used to estimate average daily traffic (ADT) throughout the street system. On major streets, which would include freeways, expressways, arterials, and collectors, one non-directional 24-hour count is made on each control section at a maximum interval of four years. Minor streets are sampled by coverage counts at the rate of one count for every mile of local street. In this case, the count frequency should be every four years.

Rural Counts. Highways are generally classified by state, federal, and local system and also by function. Control stations are set up in a similar manner to urban major control stations, in such a way that the seasonal and long-term variations can be determined. Each class of highway would have several control stations. Sample counts are set up on a one- or two-day basis to provide annual average counts on all highway sections in the system.

The sample counts are adjusted to give ADT for all segments of the system. Results of the volume survey are normally summarized by section and plotted for graphical presentation in a form of a traffic flow map. Figure 7–8 shows a portion of statewide traffic volume map which is updated annually.

Cordon Counts. Cordon counts can be used to measure the traffic activity of specific areas. They are often used in conjunction with traffic

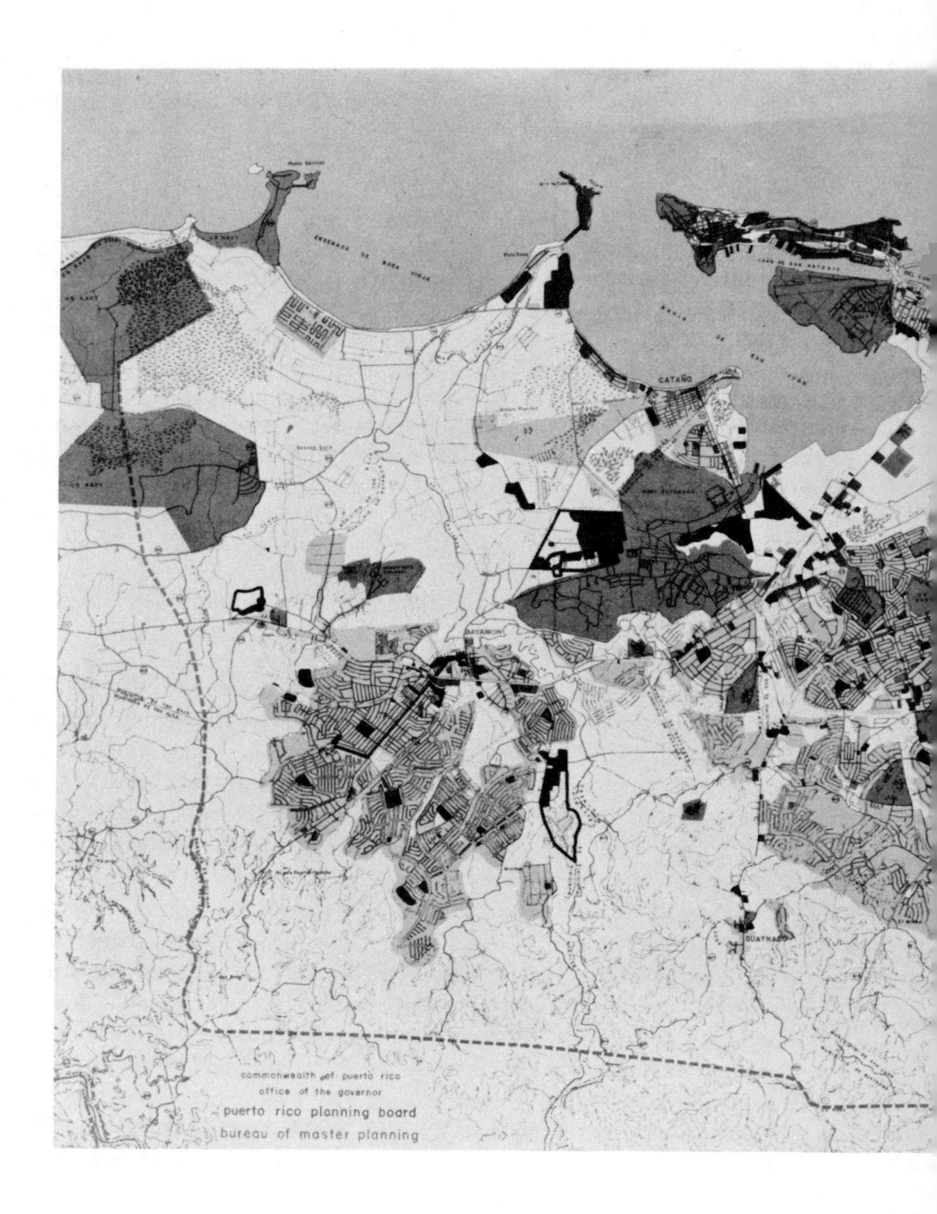

Fig. 7–7. Existing land use plan, 1965, San Juan, Puerto Rico. (Source: Associates.)

San Juan Metropolitan Area Transportation Study, Wilbur Smith and

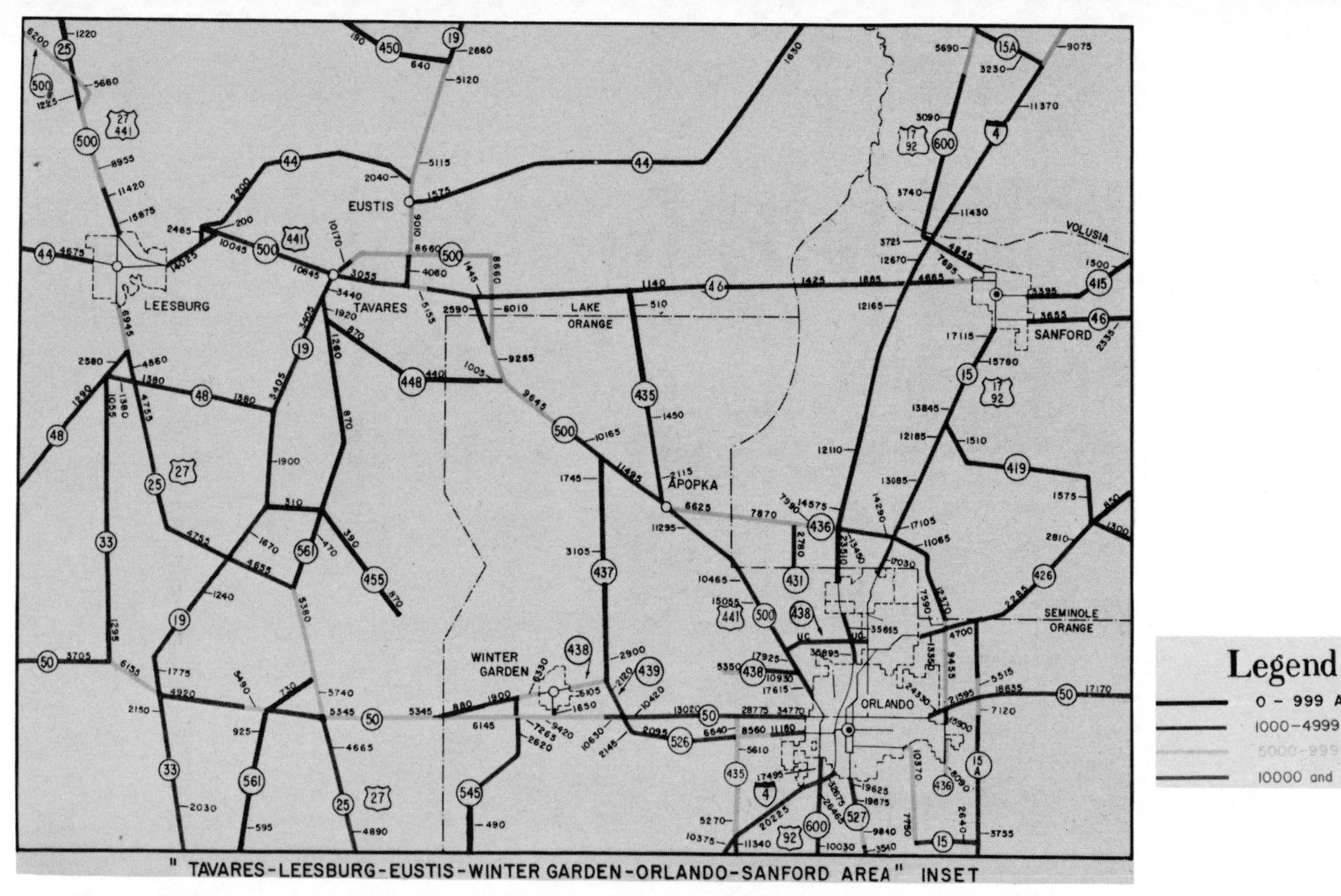

Fig. 7–8. Portion of statewide traffic volume map. (Source: State of Florida Department of Transportation, Division of Transportation Planning, in Cooperation with U.S. Department of Transportation, Federal Highway Administration.)

studies of Central Business Districts of cities, large institutions such as universities, and in the case of external cordons. Information from such counts can indicate:

1. Number of persons entering and leaving the area
2. Travel modes
3. Hourly variations
4. Accumulations within the cordon area of persons and vehicles.

Depending on the purpose of the cordon, it may be run on a 24-hour basis or for a shorter time, omitting the night hours during which there is little traffic movement. In general a cordon line is set which will minimize, as far as possible, the number of cordon crossing points, without loss of information. Figure 7–9 shows the graphical summaries from a central business district cordon.

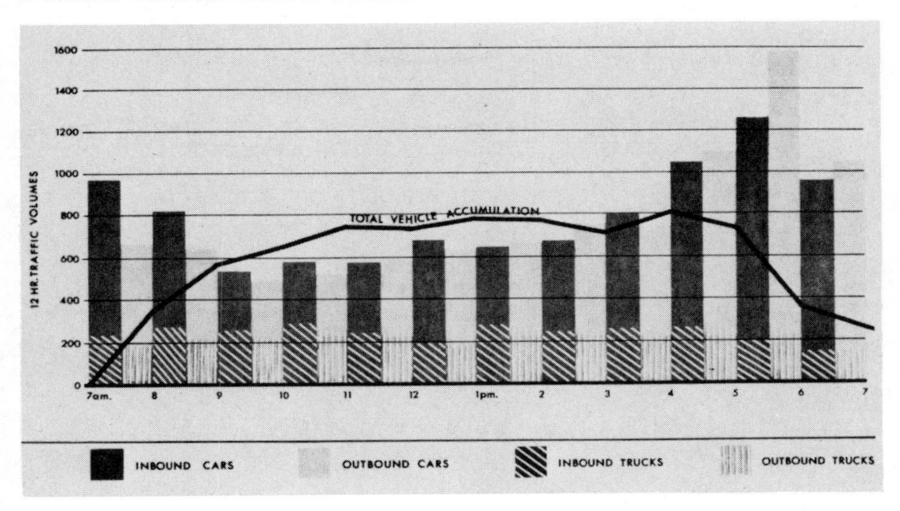

Fig. 7–9. Graphical analysis of vehicle movements through a CBD cordon. (Source: *City of Galt, Traffic Planning Report,* Damas and Smith, Ltd.)

Screen Line Counts. These are most commonly used to verify the volume of movements predicted by origin-destination sample surveys. Ground counts are compared with expanded sample data. The screen line which coincides with internal traffic zone boundaries can be used to determine the number of trips passing from one group of zones to another. Normally, in order to minimize the number of crossing points, the line incorporates such geographical and topographic barriers as rivers, lakes, railroad lines, and freeways. Figure 7–10 shows a graphical comparison of screen line ground counts and expanded survey data.

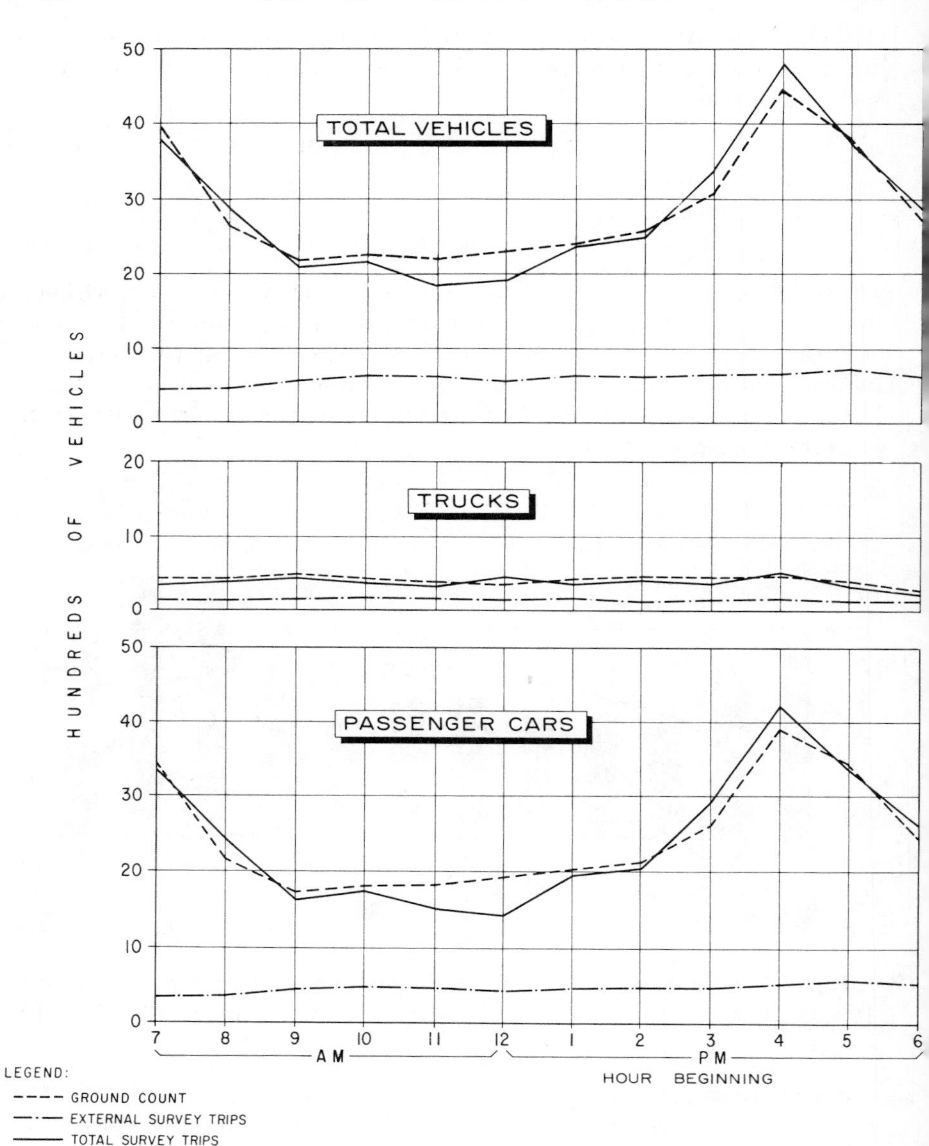

Fig. 7–10. Screen line comparisons. (Source: *Southeastern Virginia Regional Transportation Study,* 1965, Vol. 1, Wilbur Smith and Associates.)

7–21. Attitude Surveys. One of the most difficult areas in which the transportation planner necessarily finds himself is the evaluation of community values prior to preparing a plan. It is not uncommon for a facility where assiduous attention has been paid to alignment, design speeds, and capacity to come under furious attack because the proposed facility violates what the community feels is essential to the preservation of a proper environment. While the 1962 Federal Highway Act requires that the transportation plan take account of community values, planners find that the techniques for evaluation of attitudes are perhaps the least well defined of all planning techniques.

In determining community values it is necessary to determine the attitudes of those who live in the community. An attitude is a more basic feeling than an opinion, which can change over time when influenced by social pressure. Attitudes relate to basic value judgments. "An attitude is a learned predisposition to behave in a consistent manner in a given situation; as such, attitudes are much more enduring than opinions. Hence, attitude assessment is a more reliable basis for prediction of terminal action than opinion study" [43].

Various techniques have been used by transportation planners to evaluate basic attitudes. Using home interviews and mail return responses, planners have carried out surveys to determine attitudes towards housing, the town and state lived in, and leisure and recreation. The responses were analyzed to determine attitude patterns by persons of various income groups and living at different densities [44]. Other techniques which have attempted to determine basic community attitudes include tools which are more commonly used in the social and psychological sciences [43]:

1. Word association
2. Sentence completion technique
3. Semantic differential technique.

Other methods recommended for use in determining community values involved the use of groups and panels from specifically selected groups rather than random samples [45].

Working Review Committees, composed of elected officials, downtown leaders, business community spokesmen, and other community leaders, can evaluate preliminary proposals raising objections which can be surmounted at an early date.

Focus Groups are made up of people with common backgrounds and common interests. Under the direction of a discussion leader they articulate the viewpoints which the planner must consider in plans.

Rating Panels are commonly constituted of professionals in the field of planning. They have been used to evaluate the merit of alternative development patterns with respect to criteria related to areawide goals and objectives.

7–22. Highway Needs Studies. Immediately after World War II, the accelerating traffic demand rapidly outstripped the rate of provision of new road facilities. It became readily apparent that it was necessary to extend the scope of the original planning surveys carried out in the 1930's. Long-range planning programs for statewide highway networks were an obvious necessity to permit long-term planning and budgeting of new facilities and for the correction of current and future deficiencies of existing roads.

Out of this demand grew the *Highway Needs Study*. Usually carried out by committees of public officials and road user groups or other bodies closely associated with the provision or use of highways, the studies made an in-depth examination of the statewide highway program. A typical state needs study examined the following areas of concern [38]:

1. The economics of highway transport in the state
2. Highway usage—existing and future demand
3. A classification system of highways
4. Design standards
5. Existing and future deficiencies and needs
   a. Interstate and primary systems
   b. Secondary
   c. Local and county
   d. Urban streets
6. Requirements and extent of federal aid
7. Maintenance costs—present and future
8. Administration
9. Financial needs and highway revenue
10. Sources of additional income
11. Recommendations

Typically, the studies considered a number of catch-up periods in which funds and projects for the current backlog plus the developing deficiencies would be programmed. These catch-up periods varied from 5 to 20 years. At the end of the ten-year catch-up program, therefore, all current deficiencies plus those developing in that period with anticipated traffic expansion, would be accommodated by the plan, including required maintenance operations.

It was felt that the needs study, conducted as it often was by bodies outside state government, provided a reasonably objective estimate of highway needs for guiding legislative appropriation. Often these studies recommended that sweeping changes in highway administrative procedure were necessary before an efficient highway program could be achieved. In spite of their past usefulness, it is unlikely that many highway needs studies will be undertaken in the future for the following reasons:

1. Highway needs studies were the product of a period when a great deal of financing needed to be poured into a highway program which suffered from a post-depression and war-time backlog of deficiencies. The immense investment in the Interstate system and other federal road construction has basically altered this situation.
2. Rural highway needs are likely to be fairly well satisfied for many years ahead. The pressing areas of highway construction are in the urban areas.
3. There is increasing resistance to the solution of transportation problems by highways alone. Needs studies were unimodal in approach. Many states are currently embarking on statewide multimodal transportation studies [39]. It is likely that this type of study eventually will totally replace the needs study which will remain only as a portion of the total study.[2]

7–23. Highway Sufficiency Studies. In the past, and still in a few areas, the provision and improvement of highways is subject to political maneuvering by groups and individuals. The rapid growth of demand for highways after World War II and the increasing importance of highway user taxes in overall state revenues led many states to adopt an objective system for priority rating. These systems are called *highway sufficiency ratings.* More than half the states have such systems.

The objective of the sufficiency rating is to provide a standardized measure by which all segments of a highway system are classified in a non-subjective, uniform, and unbiased manner. While the actual method of calculating the rating differs from state to state, the basic concept is similar. All segments of the system are rated according to the following factors [37]:

1. Adequacy of structural condition
2. Safety
3. Service level for the traffic demand.

Each factor is scored according to adequately defined criteria. By combinations of the score for each factor, a sufficiency rating can be calcu-

[2] Under the terms of a Joint House–Senate Resolution in 1965, a summary of National Transportation Needs is prepared from a compilation of individual State Needs Studies determined on a biennial basis by State Departments of Transportation and State Highway Departments. Beginning in 1970, these studies were required to be multimodal in character, coordinated through the offices of the governors of the individual states.

lated which can have any value from 0 to 100. Those route segments attaining perfect scores in all three areas can be seen to be completely adequate and will produce a perfect score of 100. Depending on the degree of inadequacy, the combined score will fall short of this perfect rating. The systems of evaluating sufficiency ratings fall into two main categories. Either the rating is compared with a minimum acceptable rating which is established as a standard, or a particular section is compared to all other sections by establishing a priority ranking based on the sufficiency rating. Under the latter system, a section of highway with a rating of 80 is considered to be less adequate than one with a rating of 85.

There are serious flaws in the conceptual framework which went into most rating systems. The systems were constructed without a great concern for the interrelationships of the utilities of the various factors. For example, under many rating systems it is possible for a section of highway carrying no traffic to attain a higher priority ranking than another carrying a significant traffic volume. However, despite the inability to describe properly the true trade-off relationships between level of service, safety, and structural condition at all levels of adequacy, the sufficiency rating still does serve a function. By prescribing a set of elements to be considered and by defining how each element and factor is to be judged and scored, the bias of personal judgment is eliminated from the evaluation process when the adequacy of existing systems is under consideration. While the transportation planner may find it difficult to place total reliance on the absolute value of the sufficiency rating, he is perfectly justified in having great confidence in the individual scorings which make up the rating. It would appear, therefore, that sufficiency rating systems are likely to continue for many years.

7–24. Engineering Economic Analysis.[3] In order to determine the economic effects of transport investment, planners and engineers frequently conduct economic analysis. From the results of such analysis the relative economic viability of schemes can be demonstrated. Any economic analysis of an engineering project must deal with the problem of discounting cash flows over a period of time to reflect the effect of interest on both costs and benefits. Only when costs and benefits are compared on an equivalent basis is it possible to determine the economic impact or feasibility of a project.

Four different methods of economic analysis are discussed:

1. Net present worth
2. Equivalent uniform net annual return

[3] For underlying theory of interest and definition of interest factors the reader is referred to Reference 55.

3. Benefit-cost ratio
4. Internal rate of return

Although upon preliminary examination, the various methods described appear to be quite different, in fact the underlying rationale is similar. The different forms of the economic analysis reflect the type of information required by different decision makers.

Benefit-cost ratio has been extensively used since its adoption by the U.S. Corps of Engineers as an evaluation procedure under a Congressional mandate to show that the benefits accruing from public works projects exceed the costs of construction, maintenance, and operation. To some decision makers the relation of annual costs to annual benefits in the form of net annual benefits is more meaningful, since it can be related to annual budgeting costs. The calculations which support net present value more clearly indicate the size of total costs and benefits over the life of the project. Rate of return analysis is more meaningful to some in that it makes direct comparisons between the earning power of invested capital. All methods will, however, lead to the selection of the same project when used for comparative purposes, and are, therefore, all appropriate criteria for selection.

Net Present Worth (NPW). The net present worth method of analysis permits comparison of costs and benefits throughout the life of a project. The periodic costs and benefits are treated as cash flows which are brought to equivalent worths at the zero time point. The discount rate used in the interest formulae is the minimum attractive rate of return over the period of project analysis. The net present worth is defined as the difference between net present benefits and net present costs. When the net present worth of a project is positive, the project is economically viable since it earns more than the minimum attractive rate of return. In the comparison of mutually exclusive alternatives, the greatest net present worth represents the most desired choice.

In the simplest case, a project could be shown as a single cash flow, C, at time zero bringing a benefit, B, n years later, as shown below:

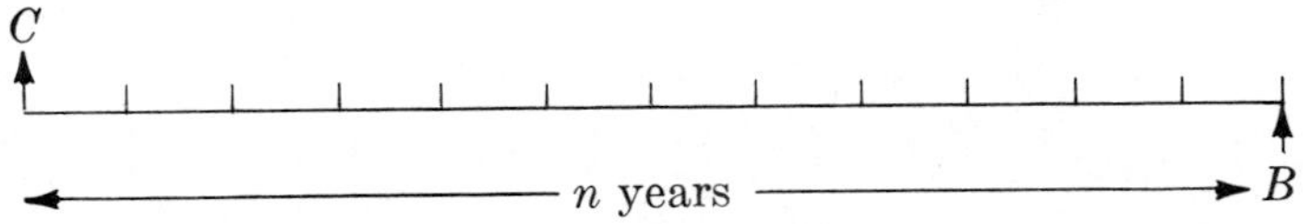

C

n years B

$$\text{Net Present Worth (NPW)} = B\left[\frac{1}{(1+i)^n}\right] - C$$

$$\text{NPW} = B(pwf' - i - n) - C$$

where:

C = Cost
B = Benefit
i = minimum attractive rate of return
n = period of project analysis
$(pwf' - i - n)$ = present worth factor for a single benefit at discount rate i for a period of n years

Where a single project cost produces an uniform flow of annual benefits, the cash flow diagram is of the following form:

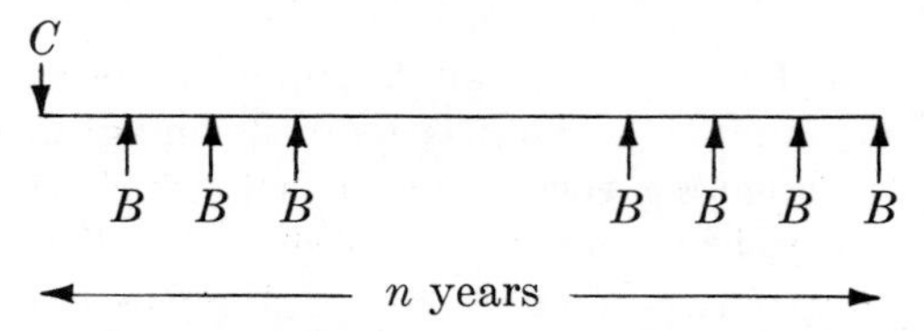

In this case the calculation of net present worth is made from the following equation:

$$\text{NPW} = B\left[\frac{(1+i)^n - 1}{i(1+i)^n}\right] - C$$

$$= B(pwf - i - n) - C$$

where:

B = Annual Benefit
C = Initial Cost
i = minimum attractive rate of return
n = period of project analysis
$(pwf - i - n)$ = present worth factor for a series of uniform benefits at discount rate i for a period of n years

In the general case, both costs and benefits will be irregular cash flows over the whole time period of analysis. The general equation can be written for an irregular cash flow diagram.

C_0 C_1 C_2 C_3 C_{n-2} C_{n-1} C_n

B_1 B_2 B_3 B_{n-2} B_{n-1} B_n

$$\text{NPW} = \sum_{k=0}^{n} B_k\left[\frac{1}{(1+i)^k}\right] - \sum_{k=0}^{n} C_k\left[\frac{1}{(1+i)^k}\right]$$

or:

$$\text{NPW} = \sum_{k=0}^{n} B_k(pwf' - i - k) - \sum_{k=0}^{n} C_k(pwf' - i - k)$$

Equivalent Uniform Net Annual Return (EUNAR). The equivalent uniform net annual return can be directly calculated from the net present worth by multiplying the NPW by the capital recovery factor. This factor is the compound interest relationship which converts a cash flow at time zero into a series of uniform flows over a defined time period, in this case the analysis period. The relationship is shown by the following equations and cash flow diagram.

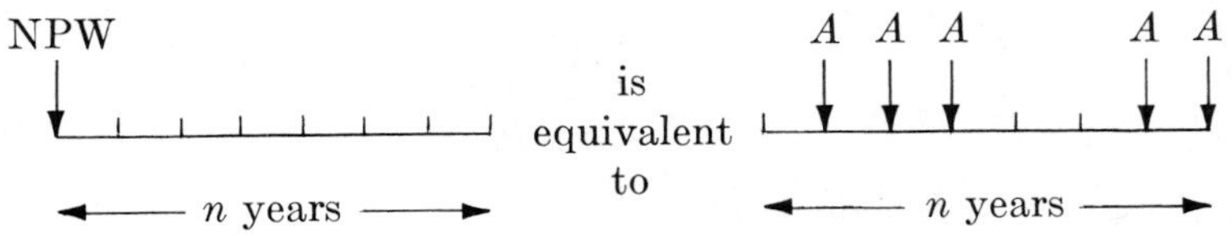

NPW — *n* years — is equivalent to — $A\ A\ A \ldots A\ A$ — *n* years

$$A = \text{NPW}\left[\frac{i(1+i)^n}{(1+i)^n - 1}\right]$$

or:

$$A = \text{NPW}\ (crf - i - n)$$

where:

A = Equivalent Uniform Net Annual Return
NPW = Net Present Worth
i = minimum attractive rate of return
n = period of project analysis
$(crf - i - n)$ = capital recovery factor for discount rate i for a period of n years

It can be seen that the equivalent uniform net annual return is the amount by which the uniform annual benefits exceed the uniform annual costs when discounted according to a minimum attractive rate of return. Provided that this difference is positive, the project is economically advisable since economic benefits outweigh costs at an attractive rate of return. Mutually exclusive alternate projects can be evaluated by comparison of the uniform annual net returns. The project with the higher net return is the more desirable.

Benefit-Cost Ratio (B/C). One of the most widely used forms of economic analysis is the benefit-cost ratio. As the name implies, the method determines the ratio of benefits to costs after each has been comparably discounted with respect to time at the minimum attractive rate of return. Those projects with benefit-cost (b/c) ratios greater than 1.0 are economically viable, while those with ratios below 1.0 are not feasible. The ratio is computed from comparably discounted cash flows, and can be applied either to equivalent uniform annual costs and benefits or to present worth of costs and benefits. This equivalency is illustrated by the following cash flow diagrams and equations, where the redundancy of the capital recovery factor in the second form of the ratio is obvious.

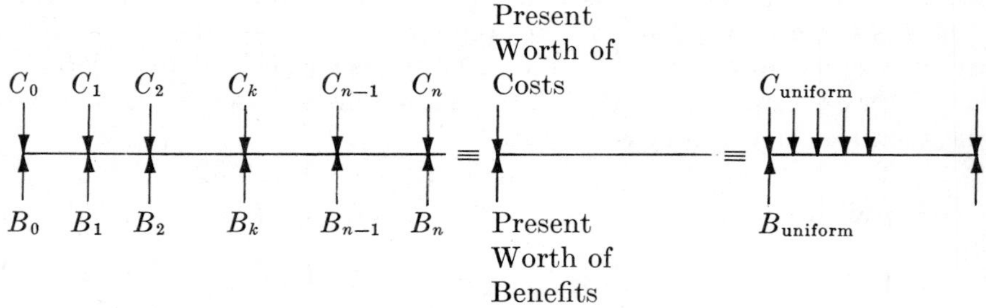

C_0 C_1 C_2 C_k C_{n-1} C_n ≡ Present Worth of Costs ≡ C_{uniform}

B_0 B_1 B_2 B_k B_{n-1} B_n Present Worth of Benefits B_{uniform}

Using Present Worth Values:

$$B/C = \text{Benefit-Cost Ratio} = \frac{\sum_{k=0}^{n} B_k \left[\frac{1}{(1+i)^k}\right]}{\sum_{k=0}^{n} C_k \left[\frac{1}{(1+i)^k}\right]} = \frac{\sum_{k=0}^{n} B_k(pwf' - i - k)}{\sum_{k=0}^{n} C_k(pwf' - i - k)}$$

or using Equivalent Uniform Annual Benefits and Costs:

$$B/C = \frac{(crf - i - n) \sum_{k=0}^{n} B_k(pwf' - i - k)}{(crf - i - n) \sum_{k=0}^{n} C_k(pwf' - i - k)}$$

Direct benefit-cost analysis can be used to determine the economic feasibility of a project, since the ratio specifies the relationship between the benefits and costs associated with undertaking the project as compared with no action at all. When attempting to evaluate the desirability of alternative projects with different associated costs and benefits, it is necessary to undertake incremental cost-benefit analysis. In this form of analysis the projects are examined in order of cost. The incremental cost-benefit ratio is the ratio between the incremental discounted benefits of the next least expensive alternative and its associated discounted costs. Where the incremental cost carries a cost-benefit ratio in excess of 1.0 the additional costs provided a return greater than the minimum attractive rate of return which has been assumed. Incremental cost-benefit ratios of less than 1.0 indicate that the additional investment is economically unwarranted.

Internal Rate of Return. One method of economic evaluation which makes no internal assumption concerning the minimum attractive rate of return on investment is called the internal rate of return method. In evaluating the economic feasibility of a scheme, the analyst computes, by a trial and error method, that discounting rate which exactly equalizes the discounted benefits and the discounted costs. While it is more usual to equalize the present worths of costs and benefits, the method can also be applied to equivalent uniform annual cash flows since these are directly calculable from present worths. It can be seen that the internal rate of return is that interest rate r which satisfies the following equation of cash flows:

$$\begin{array}{cccccc} B_0 & B_1 & B_2 & B_k & B_{n-1} & B_n \\ \downarrow & \downarrow & \downarrow & \downarrow & \downarrow & \downarrow \\ \uparrow & \uparrow & \uparrow & \uparrow & \uparrow & \uparrow \\ C_0 & C_1 & C_2 & C_k & C_{n-1} & C_n \end{array}$$

$$0 = \sum_{k=0}^{n} B_k \frac{1}{(1+r)^k} - \sum_{k=0}^{n} C_k \frac{1}{(1+r)^k}$$

Provided that the internal rate of return exceeds the minimum attractive rate of return, the scheme is judged to be economically feasible. The relative attractiveness of alternative schemes can be related directly to the size of their internal rates of return. The planner can select the most economically advantageous scheme by choosing that scheme which has the highest internal rate of return.

Internal Rate of Return Analysis requires an incremental procedure similar to Cost-Benefit Ratio Analysis to determine the marginal rate of return on incremental investments. For a more extensive discussion of all methods of economic analysis, the reader is referred to Reference 57.

The A.A.S.H.O. Method of Benefit-Cost Analysis. The American Association of State Highway Officials has set out a procedure for the evaluation of highway improvements by means of benefit-cost analysis [56]. As defined by the A.A.S.H.O.:

$$\begin{aligned} \text{Benefit Cost Ratio} &= \frac{\text{Annual Benefits}}{\text{Annual Costs}} \\ &= \frac{\text{Difference in Annual Road User Costs}}{\text{Difference in Annual Highway Costs}} \\ &= \frac{\Delta R}{\Delta H} \end{aligned}$$

Annual Road User Costs (U). Road users' costs are calculated for homogeneous sections of road. The individual costs are computed from eight factors:

1. The type of vehicle using the highway—passenger car, truck, or bus.
2. The type of area—rural or urban.
3. Facility type—two-lane or divided highway.
4. Type of operation—free, normal, or restricted flow as related to capacity.
5. Running speeds.
6. Gradients.
7. Type of surface—paved, gravel, or unsurfaced.
8. Alignment—tangent, flat curve, or curved.

Annual Highway Costs (H). Annual highway costs are composed of (1) annual capital costs (C), due to right-of-way, drainage, pavement, and major structures, and (2) annual maintenance and operating costs (M). This may be expressed as:

Annual Highway Costs = Annual Capital Cost + Annual Maintenance and Operating Costs

$$H = C + M$$

Example of the Conversion of Capital Costs to Equivalent Uniform Annual Capital Costs. A section of urban rapid transit line is to be constructed at a total cost of $11,050,000. The breakdown of capital costs, the service lives of the individual portions of capital investment, and the assumed percentage of initial costs recoverable in salvage value is given below. The minimum attractive rate of return over the life of the project is assumed to be 5 per cent.

Capital Investment Item	*Initial Cost*	*Service Life*	*Percentage of Initial Value Recoupable in Salvage at End of Service Life*
Land	$5,000,000	100 years	100%
Earthwork	$3,000,000	50 years	0%
Drainage	$ 750,000	25 years	10%
Signalization and structures	$3,000,000	40 years	10%
Roadbed	$ 100,000	25 years	0%
Track	$ 200,000	10 years	20%

The total equivalent uniform annual capital cost can be calculated by summing the annual capital costs derived from the six separate items:

Land:

$$\text{EUAC}_l = 5{,}000{,}000\ (crf - 5\% - 100) - 5{,}000{,}000\ (sff - 5\% - 100)$$

$$\text{EUAC}_l = 5{,}000{,}000\ (0.05038) - 5{,}000{,}000\ (0.00038)$$

$$\text{EUAC}_l = \$250{,}000$$

Earthwork: $\text{EUAC}_e = \$3{,}000{,}000\ (crf - 5\% - 50)$
$\text{EUAC}_e = \$3{,}000{,}000\ (0.05478)$
$\text{EUAC}_e = \$164{,}340$

Drainage: $\text{EUAC}_d = \$750{,}000\ (crf - 5\% - 25) - \$75{,}000\ (sff - 5\% - 25)$
$\text{EUAC}_d = \$750{,}000\ (.07095) - 75{,}000\ (.02095)$
$\text{EUAC}_d = \$17{,}484 - \$1{,}571$
$\text{EUAC}_d = \$15{,}913$

Signalization and Structures: $\text{EUAC}_{s-s} = \$3{,}000{,}000\ (crf - 5\% - 40) - 300{,}000\ (sff - 5\% - 40)$
$\text{EUAC}_{s-s} = \$3{,}000{,}000\ (.05828) - \$300{,}000\ (0.00828)$
$\text{EUAC}_{s-s} = \$174{,}840 - \$2484$
$\text{EUAC}_{s-s} = \$172{,}356$

Roadbed: $\text{EUAC}_r = \$100{,}000\ (crf - 5\% - 25)$
$\text{EUAC}_r = 100{,}000\ (.07095)$
$\text{EUAC}_r = \$7{,}095$

Track: $\text{EUAC}_t = \$200{,}000\ (crf - 5\% - 10) - \$20{,}000\ (sff - 5\% - 10)$
$\text{EUAC}_t = \$200{,}000\ (0.1295) - \$20{,}000\ (0.0795)$
$\text{EUAC}_t = \$25{,}900 - \$1590$
$\text{EUAC}_t = \$24{,}310$

$$\text{EUAC}_{\text{total}} = \text{EUAC}_l + \text{EUAC}_e + \text{EUAC}_d + \text{EUAC}_{s-s} + \text{EUAC}_r + \text{EUAC}_t$$
$$= \$250{,}000 + \$164{,}340 + \$15{,}913 + \$172{,}356 + \$7095 + \$24{,}310$$
$$= \$634{,}014$$

Example of an Economic Analysis using Benefit-Cost Ratio and Internal Rate of Return Methods. An engineer wishes to make an economic analysis of four mutually exclusive projects involving investments of \$2 million, \$4 million, \$5 million, and \$11 million. The following table indicates the capital, maintenance, and user costs for the proposed

		Annual Items (X \$1000) $i = 7\%$ $n = 30$ years				
Alternative	Investment (X \$1000) I	Capital Cost C	Maintenance Cost M	User Cost R	B/C Ratio $\frac{R_{basic}-R_i}{H_i-H_{basic}}$	Internal Rate of Return
(Col. 1)	(Col. 2)	(Col. 3)	(Col. 4)	(Col. 5)	(Col. 6)	(Col. 7)
Basic	–	–	60	2200	–	–
1	2000	161	30	1860	2.60	18.4%
2	4000	322	20	1690	1.80	13.5%
3	5000	403	30	1580	1.66	12.6%
4	11,000	886	50	1340	0.98	6.8%

projects and the no-investment as-is situation. Each has been discounted to equivalent uniform annual costs based on a project life of 30 years and minimum attractive rate of return of 7 per cent.

Benefit Cost Analysis:

$$\mathrm{B/C}_{1-\text{basic}} = \frac{R_{\text{basic}} - R_1}{H_{\text{basic}} - H_1} = \frac{R_{\text{basic}} - R_1}{(C_1 + M_1) - M_{\text{basic}}} = \frac{2200 - 1860}{(161 + 30) - 60} = \frac{340}{131} = 2.60$$

$$\mathrm{B/C}_{2-\text{basic}} = \frac{2200 - 1690}{(322 + 20) - 60} = \frac{510}{282} = 1.80$$

$$\mathrm{B/C}_{3-\text{basic}} = \frac{2200 - 1580}{(403 + 30) - 60} = \frac{620}{373} = 1.66$$

It can be seen from Col. 7 that Alternatives 1, 2, and 3 are economically advantageous at the assumed minimum attractive rate of return. The B/C ratio of 0.98 of Alternative 4 would indicate that this alternative falls short of an attractive investment and can be disregarded. At this point, incremental cost-benefit analysis must be used to determine which of Alternatives 1, 2, and 3 should be constructed. It would be incorrect to assume that the alternative with the highest benefit-cost ratio relative to the basic case is the most advantageous.

Alternative 1 is now regarded as the "defender," and the incremental costs and benefits are compared for the increasingly expensive "challengers."

$$\mathrm{B/C}_{2-1} = \frac{\text{Incremental Benefits}}{\text{Incremental Costs}} = \frac{R_1 - R_2}{H_2 - H_1} = \frac{1860 - 1690}{(322 + 20) - (161 + 30)} = \frac{170}{151} = 1.13$$

Since the incremental costs associated with Alternative 2 produce a rate of return that is greater than the minimum attractive rate of return, there is economic benefit in this incremental investment. Alternative 2, therefore, displaces Alternative 1 and becomes the *defender* against the next most expensive challenger, Alternative 3.

$$\mathrm{B/C}_{3-2} = \frac{R_2 - R_3}{H_3 - H_2} = \frac{1690 - 1580}{(403 + 30) - (322 + 20)} = \frac{110}{91} = 1.21$$

Alternative 3 displaces Alternative 2 as the defender and Alternative 4 is the new challenger.

$$B/C_{4-3} = \frac{R_3 - R_4}{H_4 - H_3} = \frac{1580 - 1340}{(886 + 50) - (403 + 30)} = \frac{240}{503} = 0.477$$

Clearly, Alternative 4 does not provide an economic investment of the incremental costs involved beyond the level of investment for Alternative 3. Alternative 3 is, therefore, accepted as the most appropriate level of investment.

Rate of Return Analysis:

Similar investment decisions would be reached if, instead of benefit-cost analysis, the analyst were to use internal rate of return. This method also requires the use of incremental analysis. The engineer compares the internal rate of return on incremental investments with the minimum attractive rate of return.

$$(crf - r - 30)_{1\text{-basic}} = \frac{\text{Savings due to Investment}}{\text{Investment}}$$

$$(crf - r - 30)_{1\text{-basic}} = \frac{(M + R)_{\text{basic}} - (M + R)_1}{I_1 - I_{\text{basic}}}$$

$$(crf - r - 30)_{1\text{-basic}} = \frac{2260 - 1890}{2000}$$

$$(crf - r - 30)_{1\text{-basic}} = \frac{370}{2000}$$

$$(crf - r - 30)_{1\text{-basic}} = 0.185$$

By interpolation $(cf - r - 30) = 0.185$ gives $r = 18.4\%$.

Using Alternative 1 as the "defender," incremental analysis is carried out for Alternative 2.

$$(crf - r - 30)_{2-1} = \frac{\text{Savings due to incremental investment}}{\text{Incremental Investment}}$$

$$(crf - r - 30)_{2-1} = \frac{1890 - 1710}{2000}$$

$$(crf - r - 30)_{2-1} = \frac{180}{2000}$$

$$(crf - r - 30)_{2-1} = 0.090$$

By interpolation $(crf - r - 30) = 0.90$ gives $r = 8.5\%$.

Assuming a 7% minimum attractive rate of return, Alternative 2 is economically advisable. Alternative 3 is next evaluated against the new defender, Alternative 2.

$$(crf - r - 30)_{3-2} = \frac{1710 - 1610}{1000}$$

$$(crf - r - 30)_{3-2} = 0.10$$

By interpolation $(crf - r - 30) = 0.10$ gives $r = 9.8\%$, making Alternative 3 economically attractive.

Finally, Alternative 4 is evaluated against the current defender, Alternative 3.

$$(crf - r - 30)_{4-3} = \frac{1610 - 1390}{6000}$$

$$(crf - r - 30)_{4-3} = \frac{420}{6000}$$

$$(crf - r - 30)_{4-3} = 0.07$$

By interpolation, $(crf - r - 30) = 0.07$ gives $r = 5.7\%$. Since the incremental rate of return is less than 7%, the designated minimum attractive rate of return, Alternative 4 is not economically advisable, and Alternative 3 is the selected alternative.

Multimodal Urban Transportation Analysis. When making economic analyses of urban transportation systems, meaningful computations require consideration of both direct and indirect costs and benefits. Many questionable analyses have been carried out in the past which have included only direct user benefits and direct project costs. Since the location of major transportation facilities has massive direct and indirect effects on both users and non-users, this type of analysis is relatively meaningless and often has drawn misleading conclusions. Haney points out that meaningful highway vs. transit cost benefit studies should consider a range of possible effects such as the following [58]:

1. Transit user effects;
2. Highway user effects, including savings in (a) travel time, (b) operating costs, (c) ownership costs, (d) accident costs, and (e) parking costs;
3. Unemployment effects;
4. Educational opportunity effects;
5. Business productivity effects;
6. Government productivity effects;
7. Real estate effects;

8. Life-style effects;
9. Environmental pollution effects;
10. Tax effects;
11. Disruptive effects, including those that are (a) temporary during construction, and (b) permanent in neighborhood division;
12. Construction labor effects;
13. Highway construction effects;
14. Aesthetic effects;
15. Property losses and relocation effects;
16. Regional and neighborhood growth effects;
17. Crime effects;
18. Civil defense effects;
19. Achievement of desired urban form;
20. Detailed nodal studies and projections;
21. Implementation evaluation;
22. Financing effects; and
23. Tourism effects.

Few studies currently attempt such sophisticated analysis. More typical is the case of the preliminary engineering analysis of the Puget Sound study which included the following [59]:

Costs:

1. Motor vehicle operating costs for the cost of fuel, tires, oil, maintenance and repairs, and depreciation. The additional operating and time costs for stopping, idling, and resuming speed at intersections (or as a result of traffic congestion delays) over uniform speed operation should be included. Motor vehicle operating costs in engineering economy analyses should exclude fuel taxes, since they are transfer payments used for highway construction.
2. Maintenance costs of all highway facilities. In the case of the cross-Sound bridge this included maintenance of the bridge, tollbooth operation, and insurance premiums.
3. Motor vehicle accident costs, including fatalities, injuries, and property damage on highways with various levels of access control.
4. Operating costs of bus transit facilities, including items for payrolls, maintenance, insurance, and overhead.
5. Operating costs of ferries, including items for payrolls, maintenance, insurance, and overhead.
6. Operating costs of parking facilities, including maintenance, insurance, and overhead.
7. Operating costs of rapid rail transit facilities, including payrolls, maintenance, insurance, and overhead.

Benefits:
Benefits were taken as net reduction in:
Total annual maintenance on all systems;
Total operating costs of all systems;
Total accident costs on all systems; and
Total travel time costs on all systems.

Engineering analysis based on the above parameters is highly user oriented and is hard to relate to overall development goals of the community [60].

Economic Evaluation in Transport Planning at the National and Regional Levels. At the regional or national level of transportation planning, economic analysis of transport facilities and networks centers around comparison of all accruing benefits and project investment costs [61].

Benefits are due to both reductions in the cost of transportation and increases in regional output due to transport investment. Reductions in transport costs that need consideration should include, for example:

1. Decreases in vehicular operating costs such as depreciation, repairs, fuel, maintenance, and operators' wages.
2. Decreases in facility costs due to depreciation, maintenance, and operation.

Benefits accruing from increased regional output are directly related to anticipated levels of production. It is essential that such benefits are not "double counted" in several analyses. Increased national output can be due to added investment in parallel facilities, such as water and power distribution systems. It is possible that several schemes could be erroneously justified independently, using the same benefits. In underdeveloped and developing countries the lack of reliable underlying statistical data makes the evaluation of benefits from decreased vehicle and facility costs difficult.

At the national level of planning, some authorities indicate that benefits should reflect true growth in national income [62]. This may be difficult to estimate without firm data support. More easily projected are specific benefits such as reduced user expenses, lower maintenance costs, fewer accidents, time savings, increased levels of comfort and convenience, and stimulation of economic development. On the cost side of the comparison, Adler points out that accounting at the national level should reflect "shadow" prices which differ from apparent costs by:

1. The exclusion of taxes, fees, and import duties, etc.
2. The reduction of the level of wages which does not reflect the true cost of labor to the country.
3. The difference between the "financial" costs of money and the "opportunity" costs available in alternate investments.

Other errors in national accounting can be introduced by the introduction of inflationary trends which cause no real costs, and by failure to include the costs of all necessary support systems not directly financed in the project costs, e.g., access roads, power distribution, and water supply.

7–25. Summary. It can be seen from the previous discussion that the subject of economic analysis is vast. While in theory the method would appear simple, the complexities involved in putting monetary values on costs and benefits have in practice proved extremely difficult. Too often analyses have neglected these complexities resulting in an oversimplification of the problem. The results of such economic analyses are of dubious value. Prior to conducting any analysis, the planner should be familiar with the extensive surrounding literature, which has been partially referenced in this discussion.

PROBLEMS

1. Use the benefit-cost ratio method to select the most economic scheme from the following three proposals. Use a minimum attractive rate of return of 5 per cent and an analysis period of 25 years.

Scheme	*Initial Capital Cost × $1000*	*Maintenance Costs × $1000*	*User Costs × $1000*
Basic	—	150	5,000
A	4,329	100	4,010
B	7,852	88	3,500
C	12,520	80	3,010

2. Using the figures of Example 1, conduct internal rate of return analysis for the selection of the most economic scheme.
3. Compute the net present value of the following three projects on the basis of 8 per cent minimum attractive rate of return.

Project	*Initial Investment × $ million*	*Salvage Value × $ million*	*Service Life*	*Annual Benefits Accruing Through Decreased Travel Costs × $ million*
A	$400	$50	25 years	$60
B	$480	$250	40 years	$65
C	$500	$500	50 years	$80

4. Compute the equivalent uniform net annual return of the following rapid transit project. Use a discount rate of 8 per cent.

Costs

Item	*Initial Cost* (× $ *million*)	*Salvage Value* (× $ *million*)	*Service Life*
Right of Way	200	—	infinite
Main Structures	250	25	50 years
Drainage	70	10	40 years
Earthwork	50	nil	50 years
Signalization	100	25	20 years
Roadbed	20	nil	25 years
Track	40	8	15 years

Annual Benefits (assume over an infinite period)

Decreased Direct User Costs	$40 million
Indirect Community Benefits	$55 million

What is the benefit/cost ratio of this scheme?

5. Discuss the main strengths and weaknesses of the evaluation method shown in the example in Section 7–12 with the methodology outlined in Reference 50.

REFERENCES

1. *Better Transportation for Your City,* National Committee on Urban Transportation, Public Administration Service, Chicago, 1958.
2. *Procedure Manuals,* National Committee on Urban Transportation, Public Administration Service, Chicago, 1958.
3. *Determining Street Use,* Procedure Manual No. 1A, National Committee on Urban Transportation, Public Administration Service, Chicago, 1958.
4. *Origin-Destination and Land Use,* Procedure Manual No. 2A, National Committee on Urban Transportation, Public Administration Service, Chicago, 1958.
5. "The Continued Cooperative Planning Process—Coordination with Day-to-Day Traffic Operations," *Informational Reports on Transportation Planning,* Institute of Traffic Engineers, Washington, D.C., April, 1967.
6. *Measuring Traffic Volumes,* Procedure Manual No. 3A, National Committee on Urban Transportation, Public Administration Service, Chicago, 1958.
7. *Determining Travel Time,* Procedure Manual No. 3A, National Committee on Urban Transportation, Public Administration Service, Chicago, 1958.
8. *Highway Capacity Manual,* Special Report 87, Highway Research Board, Washington, D.C., 1965.
9. *Conducting a Comprehensive Parking Study,* Procedure Manual No. 3D, National Committee on Urban Transportation, Public Administration Service, Chicago, 1958.
10. *Conducting a Limited Parking Study,* Procedure Manual No. 3C, National Committee on Urban Transportation, Public Administration Service, Chicago, 1958.
11. Syrakis, T. A., and Platt, J. R., "Aerial Photographic Parking Study Techniques," *Highway Research Record 267,* Highway Research Board, Washington, D.C., 1969.
12. *Measuring Transit Service,* Procedure Manual 4A, National Committee on Urban Transportation, Public Administration Service, Chicago, 1958.

13. *Inventory of the Physical Street System,* Procedure Manual 5A, National Committee on Urban Transportation, Public Administration Service, Chicago, 1958.
14. *National Airport Plan,* Department of Transportation, Federal Aviation Administration (published annually).
15. CORRADINO, J. C., and FERRERI, M. G., "In-Flight Origin-Destination Study at Philadelphia International Airport," *Highway Research Record 274,* Highway Research Board, Washington, D.C., 1969.
16. RICHARDS, H. A., and JONES, J. D., "Application of Motor Carriers Continuous Study Techniques to the Assembly of Intercity Freight Traffic Data," *Highway Research Record No. 175,* Highway Research Board, Washington, D.C., 1967.
17. *Airport Capacity Handbook,* Federal Aviation Administration, SRDS Report No. RD–68–14, Washington, D.C., 1969.
18. FOLK, E. H., *Railroad Problems in Urban Planning,* Unpublished Master's Thesis, Georgia Institute of Technology, Atlanta, 1960.
19. RAJANIKANT, N. JOSHI and UTEVSKY, FRED, *Alternative Patterns of Development—Puget Sound Region,* Staff Report No. 5, Seattle, Puget Sound Regional Transportation Study, 1964.
20. *Urban Development Models,* Special Report No. 97, Highway Research Board, Washington, D.C., 1968.
21. CHAPIN, F. S., *Urban Land Use Planning,* Second Edition, University of Illinois Press, Urbana (1965).
22. PERLOFF, H., and WINGO, L. (editors), *Issues in Urban Economics,* Johns Hopkins Press, Baltimore (1968).
23. Puget Sound Regional Transportation Study, *Summary Report,* Seattle, 1967.
24. *Atlanta Metropolitan Regional Transportation Study,* Unpublished Staff Manual on System Evaluation, Atlanta, 1967.
25. *Developing Project Priorities for Transportation Improvements,* Procedure Manual 10A, National Committee on Urban Transportation, Public Administration Service, Chicago, 1959.
26. *Conducting a Home Interview Origin-Destination Survey,* Procedure Manual 2B, National Committee on Urban Transportation, Public Administration Service, Chicago, 1958.
27. *A Policy on Arterial Highways in Urban Areas,* The American Association of State Highway Officials, Washington, D.C., 1957.
28. A. M. Voorhees and Assoc., *Washington, D.C. 1980 Rapid Rail Transit Patronage Forecasts,* Maclean, Va., 1967.
29. VOORHEES, A. M., "A General Theory of Traffic Movement," *Proceedings of the Institute of Traffic Engineers,* 1955.
30. WYNN, F. H., "Intra-city Traffic Movements," *Bulletin 119, Factors Influencing Travel Patterns,* Highway Research Board, Washington, D.C., 1955.
31. MITCHELL, R. B., and RAPKIN, C., *Urban Traffic, A Function of Land Use,* Columbia University Press, New York (1954).
32. International City Managers Association, *Principles and Practice of Urban Planning,* WILLIAM I. GOODMAN (ed.), Washington, 1968.
33. CHAPIN, F. S., *Urban Land Use Planning,* Second Edition, University of Illinois Press, Urbana (1965).
34. Land Classification Advisory Committee, Detroit Metropolitan Area Planning Commission, *Land Use Classification Manual,* Chicago, Public Administration Service, 1962.
35. Bureau of the Budget, *Standard Industrial Classification Manual,* U.S. Government Printing Office, Washington, D.C., 1967.
36. U.S. Urban Renewal Administration and U.S. Bureau of Public Roads, *Standard Land Use Coding Manual,* U.S. Government Printing Office, Washington, D.C., 1965.
37. Highway Sufficiency Ratings, *Bulletin No. 53,* Highway Research Board, Washington, 1953.

38. *A Highway Program for Kentucky,* Automotive Safety Foundation, Washington, D.C., 1955.
39. State-wide Transportation Planning, *Highway Research Record No. 264,* Highway Research Board, Washington, D.C., 1969.
40. Hearle, E. F. R., and Mason, R. J., *A Data Processing System for State and Local Governments,* Prentice-Hall, New York (1963).
41. U.S. Bureau of the Census, *Census Use Study: Computer Mapping,* Report No. 2, Washington, D.C., 1969.
42. U.S. Bureau of the Census, *Census Use Study: The DIME Geocoding System,* Report No. 4, Washington, D.C., 1969.
43. Shaffer, M.T., "Attitudes, Community Values and Highway Planning," *Highway Research Record No. 187,* Highway Research Board, Washington, D.C., 1967.
44. Barnes, C. F., "Living Patterns and Attitude Survey," *Highway Research Record No. 187,* Highway Research Board, Washington, D.C., 1965.
45. Voorhees, A. M., "Techniques for Determining Community Values," *Highway Research Record No. 102,* Highway Research Board, Washington, D.C., 1965.
46. *Manual of Traffic Engineering Studies,* Institute of Traffic Engineers, Washington, D.C., 1953.
47. *Measuring Traffic Volumes,* Procedure Manual 3A, National Committee on Urban Transportation, Public Administration Service, Chicago, 1958.
48. *Traffic Engineering Handbook,* Institute of Traffic Engineers, Washington, D.C., 1965.
49. Moder, J. J., and Phillips, C. R., *Project Management with CPM and PERT,* Reinhold Publishing Corporation, New York (1964).
50. Oglesby, Clarkson H. (*et al.*), "A Method for Decisions Among Freeway Location Alternatives Based on User and Community Consequences," *Highway Research Record No. 305,* Highway Research Board, Washington, D.C., 1970.
51. Irwin, Neal A., "Criteria for Evaluating Alternative Transportation Systems," *Highway Research Record No. 148,* Highway Research Board, Washington, D.C., 1966.
52. Manheim, Marvin L., "Principles of Transport Systems Analysis," *Highway Research Record No. 180,* Highway Research Board, Washington, D.C., 1967.
53. Jessiman, William (*et al.*), "A Rational Decision-Making Technique for Transportation Planning," *Highway Research Record No. 180,* Highway Research Board, Washington, D.C., 1967.
54. *Instructional Memorandum on Urban Transportation Planning,* PPM–50–9, U.S. Bureau of Public Roads, Department of Transportation, Washington, D.C., 1968.
55. Grant, Eugene L., and Ireson, W. Grant, *Principles of Engineering Economy,* Fourth Edition, The Ronald Press Company, New York (1961).
56. *A Report on Road User Benefit Analyses for Highway Improvements,* American Association of State Highway Officials, Washington, D.C., 1960.
57. Winfrey, Robley, *Economic Analysis for Highways,* International Textbook Company, Scranton, Pa., 1969.
58. Haney, Dan G., "Problems, Misconceptions and Errors in Benefit-Cost Analyses of Transit Systems," *Record No. 314,* Highway Research Board, Washington, D.C., 1970.
59. Niebur, Howard Duke, "Preliminary Engineering Economy Analysis of Puget Sound Regional Transportation Systems," *Record No. 180,* Highway Research Board, Washington, D.C., 1967.
60. Goldberg, Michael A., and Heaver, Trevor D., "A Cost Benefit Evaluation of Transportation Corridors," *Record No. 305,* Highway Research Board, Washington, D.C., 1970.
61. *Introduction to Transport Planning,* United Nations, New York, 1967.
62. Adler, Hans A., "Economic Evaluation of Transport Projects," *Transport Investment and Economic Development,* Gary Fromm (editor), The Brookings Institution, Washington, D.C., 1965.

8

Planning Models: Trip Generation and Distribution

8–1. Rationale. The complexity of the urban transportation problem can be seen to defy the simple design approach inherent in many engineering problems. The usual engineering rationale for design is the recognition of the demand to be placed upon a system, and the provision of sufficient capacity to satisfy anticipated levels of demand. The system under design is isolated and independent, enabling the engineer to compute the optimum solution in a prescribed sequence of steps. The formulation of a transportation plan is not as easily approached. First, the problem is not isolated and independent. The urban transportation problem is an aggregation of many smaller areas of difficulty which we could denote as transportation and traffic engineering problems. Furthermore, urban transportation systems are themselves a smaller part of the regional and national transportation infrastructure. Proper planning requires an examination of the problem at various levels, since policy decisions at any one of these levels may have severe effects on proposed plans.

The most striking difficulty in plan design is not, however, the multiplicity of levels on which the solution must be considered. The chief problem comes from the fact that unlike most engineering problems, the solution which is selected will affect the environment of the problem itself. This change of environment will change the demand upon the system, possibly invalidating the criteria and input used in the initial design. The cyclic interaction of transportation facilities and land use is shown in Fig. 8–1. Land use has been found to be the prime determinant of trip generation activity. The level of trip generation activity

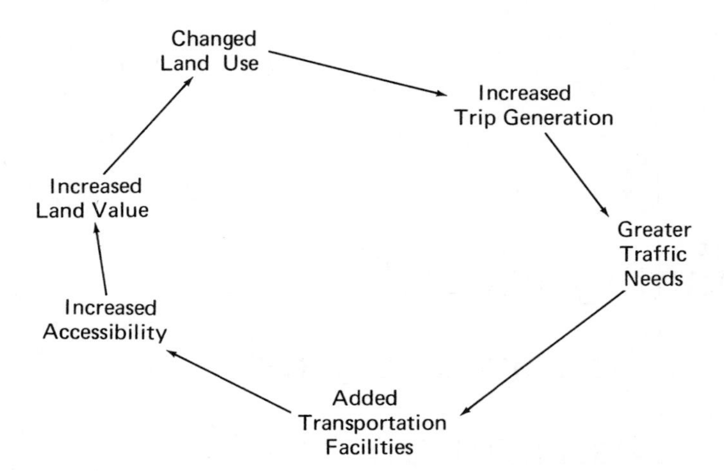

Fig. 8–1. The land use transportation cycle.

and the orientation of trips within the study area will determine the needs for facilities. Provision of facilities alters the accessibility of the land which in turn helps determine the value of land. Land value being the major determinant of land use, the planner is faced with a cycle in which alteration in any one element causes change both to all other elements and to itself.

The complex interaction of the problem and its own solution leads the planner to model formulation. The types of models are schematic and mathematical. *Schematic models* are flow diagrams characterizing relationships and decision processes; they are pictorial in form, composed of lines and symbols, providing a representational abstraction of the real world. A *mathematic model* is a set of equations, the solution of which explains and predicts the effect of changes in system. Among the advantages of the building and use of models, the following can be cited for transportation planning:

1. Models enable inexpensive simulation of the effect of planning and policy decisions. In this way costly mistakes can be foreseen and eliminated.
2. The planner, in determining the interactions involved between the various elements of the urban fabric, obtains a clear insight into the workings of the city.
3. A full examination of the effects of changes in solutions is possible. The planner can work towards an *optimal* solution by examining the effect of variations of various actions.

4. Subjectivity of decision can be eliminated to any desired degree by the use of objective criteria in evaluating decisions.

The transportation modeling process is shown in Fig. 8–2. The breakdown of the procedure into six separate blocks follows the groupings of models discussed in subsequent sections. The models discussed in detail in this chapter and the following chapters are:

1. Land use
2. Trip generation
3. Trip distribution
4. Modal split
5. Traffic assignment

Population and economic activity predictions are generally derived by demographic and economic projections outside the scope of the trans-

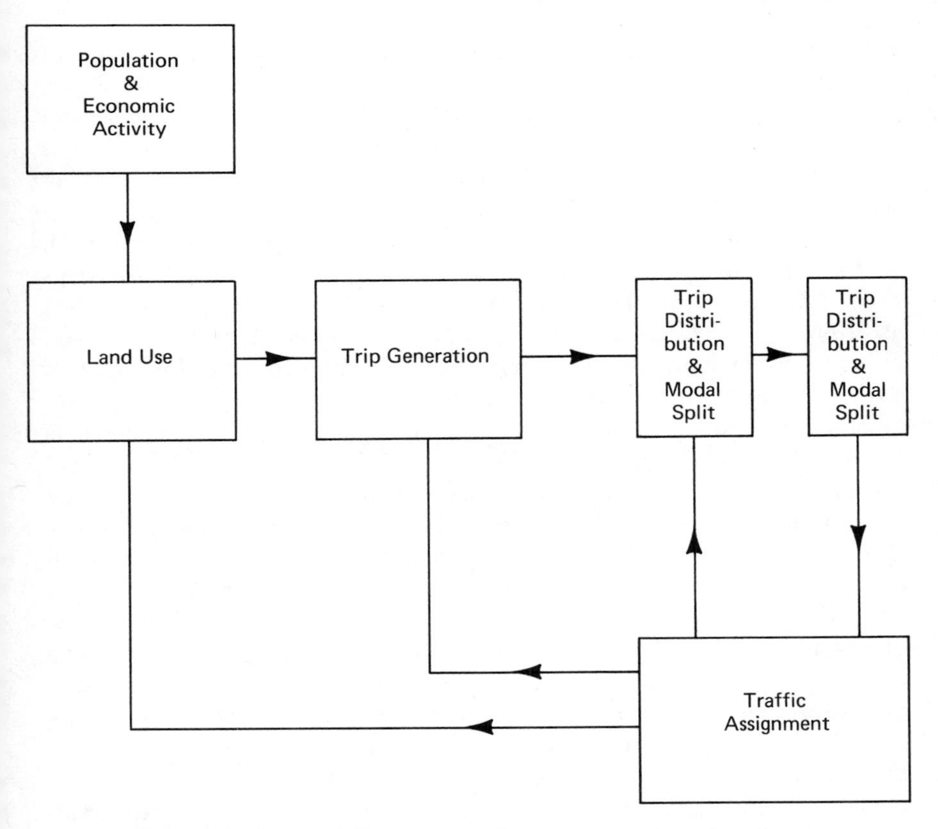

Fig. 8–2. The modeling process in transportation planning.

portation planner. Since these areas would not usually require detailed examination by the transportation specialist in the preparation of the typical urban or regional transportation plan, they are omitted from this discussion.

TRIP GENERATION MODELS

With the completed forecasts of socio-economic conditions and land use within a study area, the transportation planner is in a position to begin to predict the transportation demand that will be placed upon any future transportation system. For an adequately designed system it is necessary to know this demand both at the aggregate level of the total study area and at the level of the traffic zones which in combination form this total area. Trip generation is the analytical process by which the planner projects the number of trips which will originate or terminate within each zone. These origins and destinations are known as *trip ends.*

8–2. Origins, Destinations, Productions, and Attractions. With some of the distribution models to be discussed later, these trip ends are called *productions* and *attractions* which are not synonymous with *origins* and *destinations.* This requires some clarification. Origins and destinations need little definition, since the origin of the trip is the point at which the trip begins, and the destination is the point of termination. When the terms *production* and *attraction* are used, there is some implication of the land use at either the beginning or end of the trip. All trips that either begin or end at the home of the trip maker are classified as *Home Based Trips,* while those that have neither beginning nor end at the trip maker's place of residence are termed *Non-Home Based Trips.* In a typical study approximately 80 per cent of all trips are home based trips. In the case of the home based trips, the home is termed the production end of the trip, while the other end is termed the attraction end. Consider the example of a man who makes only two journeys in a day, to work and back home. If origins and destinations are considered, the home is both an origin and destination, and so is the place of work. In terms of productions and attractions, however, the place of residence generates two productions, and the place of work generates two attractions. Only in the case of the non-home based trip is the origin of the trip the production end and the destination the attraction end in all cases. Origins and destinations have customarily been used in conjunction with the opportunity-type of distribution models, while productions and attractions are used with gravity models. Productions, attractions, origins, and destinations all fall under the general term *trip end.*

8–3. Trip Purpose. In developing models to determine the intensity of generation of trip ends, experience has shown that more reliable models can be derived if separate estimates are made for different trip purposes. All trips are not homogeneous in character. In the same study area, for example, the length of work trips has a different frequency distribution from shopping trips, reflecting the willingness of the individual to spend more time in a work commuting trip than on a trip to buy cigarettes at the local drugstore. Figure 8–3 shows various distribution rate curves for different trips purposes in the Southeastern Virginia Regional Area Transportation Study. The stratification of trip purpose at the trip generation stage depends on the distribution model to be used. Typical purposes encountered in a large study using the gravity model are those used in the Baltimore study [1]:

1. Work
2. Personal business
3. School
4. Social
5. Recreational
6. Change travel mode
7. Shopping (convenience)
8. Shopping (general merchandise, apparel, furniture, appliances, furnishings)
9. Serve passenger
10. Non-home based

Many studies do not stratify trip purposes to the degree shown above. A small city may be amenable to modeling with only three purposes: home based work, home based other than work, and non-home based trips. In those cities where opportunity models have been used in the trip distribution phase, a different type of stratification has proved successful for model calibration. For example, in the CATS study [2], for purposes of distribution model calibration, trips were stratified into:

1. Intrazonal trips (short trips)
2. Other trips (long trips)

8–4. Trip Generation Analysis. The rate of tripmaking within an area depends primarily on the land use of the area. This land use, in conjunction with socio-economic information concerning the resident and working populations, has been found to be closely related to the demand that an area places on the transportation system. Ultimately, the function of trip generation analysis is to establish meaningful relationships between land use and trip making activity so that changes in land use

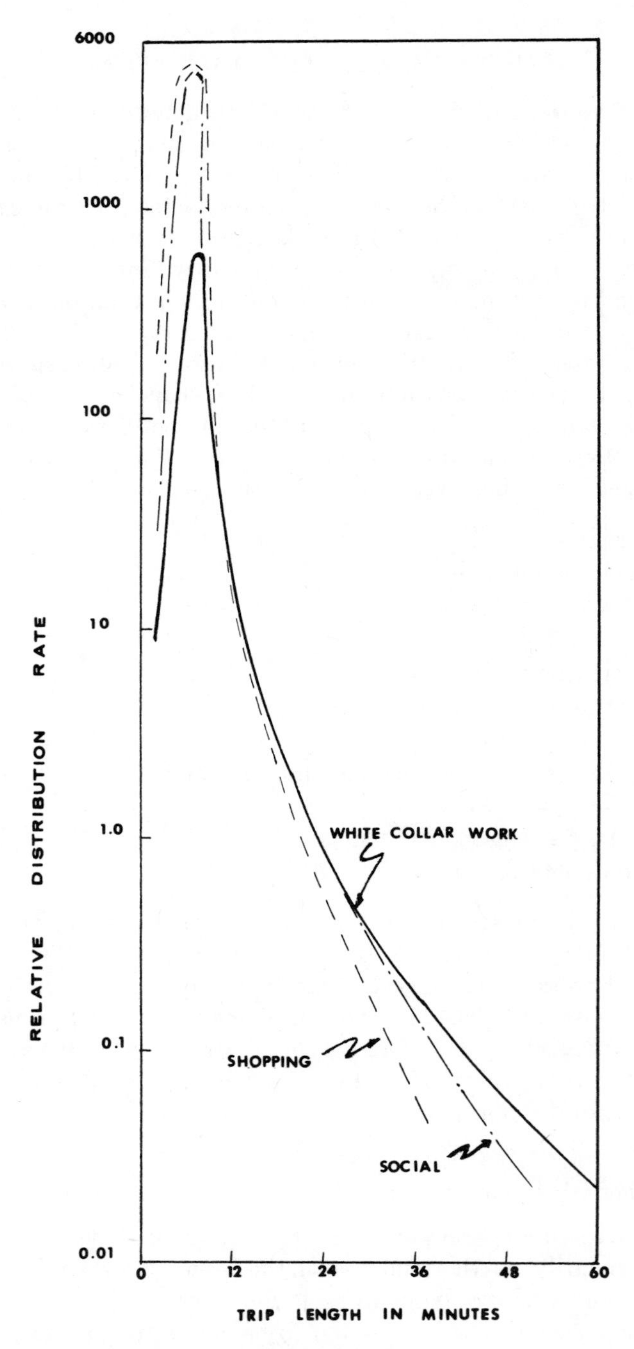

Fig. 8–3. Relative distribution rates for different trip purposes. (Source: *Southeastern Virginia Transportation Study*. Wilbur Smith and Associates.)

can be used to predict subsequent changes in transportation demand. The three characteristics of land use which have been found to relate closely to trip generation are [3]:

1. Intensity of land use
2. Character
3. Location of land use activity

Intensity of land use is usually expressed in such terms as dwelling units per acre, employees per acre, employees per 1000 sq. ft. of retail floor space. This factor has been found to have a profound bearing on trip making activity as shown in Fig. 8–4 [4, 5]. The Pittsburgh Study found that at typical row house density of 40 d.u./acre, each household produces only three person trips per day, while at the residen-

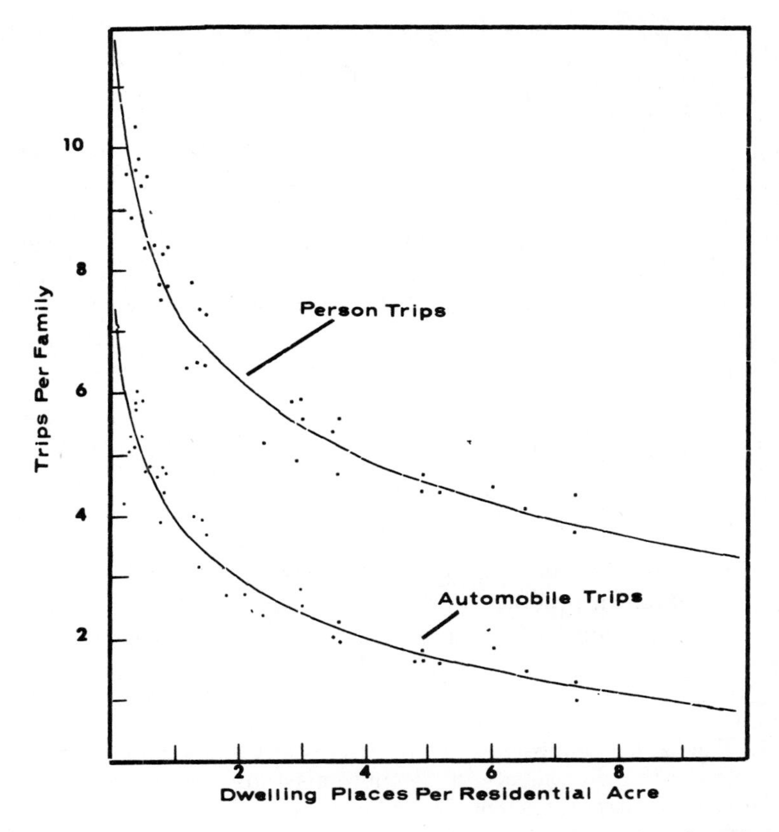

Fig. 8–4. Person and automobile trips made per family, related to density. (Source: *Chicago Area Transportation Study,* Vol. 1.)

tial densities of single family residences, 6 or less d.u./acre, productions per household were in the area of 7 person trips per day or more.

Character of land use may be quite different although intensity and location are similar. For example, in areas close to the central business districts of the larger cities, high rise residential development often replaces older delapidated housing and commercial space. In some cities, such as Atlanta and Toronto, this development has been for low-income families. In other areas of the same cities similar redevelopment has provided luxury apartments for high-income families. Not surprisingly, the trip making activity of the high-income households is found to be greater than that of the low-income households. The reason is to be found in different life patterns and different amounts of spendable surplus income. For residential land use, the two variables which have proved to have closest association to character are income and car ownership per capita.

Location of Land Use Activity can alter trip making activity. Often expressed in terms of distance from the Central Business District, location within an urban area has been found to be a variable which can express the combined effect of such variables as family size, stage in family life cycle, availability of parking, and index of street congestion [6].

Prediction of trip making activity is possible by a variety of available methods. All should take account of land use and the socio-economic conditions of the resident and working population which have been recognized to affect travel patterns. Three methods, somewhat similar in overall general approach but differing in detail, will be discussed at length in the following sections. These methods have come to be known as:

1. Land Area Trip Rate Analysis
2. Cross Classification Analysis
3. Regression Analysis

8–5. Land Area Trip Rate Analysis. Examples of the land area trip rate analysis method in practice can be found in the Chicago and Pittsburgh transportation studies. The methodology followed in each case was essentially similar, and can be covered by reference to the Chicago Study alone.

1. The traffic zones, into which the study area had been divided for sampling in the origin destination study, were aggregated into eight concentric rings, centered about the CBD.
2. Base year trip generation rates were calculated for residential and various non-residential land uses on a per acre basis for each

ring from the trip reports of the origin-destination study, see Table 8–1.

3. The base year trip rates were applied to the developed acreage in the design year. The total number of trips for the design year could then be calculated by aggregating the zonal totals. This *land-use* based total was compared with a *population* based total for the design year, which was calculated from an aggregation of zonal estimates based on car ownership and net residential density in dwelling units per acre. The population based estimate was found to be approximately 13 per cent higher than the land use estimate. The latter estimate was felt to be low since it took no account of increasing car ownership and the growing preference for low density residential development. The population based estimate was, therefore, accepted as the control figure for total trip generation in the design year.

4. The next step consisted of an allocation of the total trips for the forecast year to the individual traffic zones. This was done in a different manner for residential destinations and non-residential destinations.

a. *Residential Destinations.* Reference to the origin-destination information showed that the percentage of destinations occurring within a certain land use category was dependent on the total number of trips made in a household in a day. These patterns were assumed to carry over to the forecast year. For each traffic zone the average number of trips per day per household was calculated from projected car ownership and the net residential densities determined in the land use model. Table 8–2 shows the percentage trip rates which were applied to the residential land use areas.

b. *Non-Residential Destinations.* The existing land area trip rates were applied for non-residential land with adjustments to permit balancing the aggregate totals and to allow the trips generated to reflect the particular areas of development envisaged in the land use model. To cite two examples, half the increase in trips to transportation lands was assigned to the airports in the study area, with a significant assignment of trips to the expansion of O'Hare Field facilities, and to another airport located in the southeastern part of the region.

Industrial trips were calculated from employment figures for the industrial land use area of the land use model. Since manufacturing workers made 91.5 per cent of all trips to industrial lands in the base year, using the same trip rate per worker the industrial employment was a good predictive variable for the industrial trip category.

8–6. Cross Classification Analysis. The technique of cross classification analysis is a method of indicating the value of the response variable

TABLE 8-1

Person Trips Generated by Land Use Type and Ring In Person Trips Per Acre

Ring	Average Distance from CBD in Miles	Residential	Manufacturing	Transportation	Commercial	Public Buildings	Public Open Space
0	0.0	2228.5	3544.7	273.1	2132.2	2013.8	98.5
1	1.5	224.2	243.2	36.9	188.7	255.5	28.8
2	3.5	127.3	80.0	15.9	122.1	123.5	26.5
3	5.5	106.2	86.9	10.8	143.3	100.7	27.8
4	8.5	68.3	50.9	12.8	212.4	77.7	13.5
5	12.5	43.0	26.8	5.8	178.7	58.1	6.1
6	16.0	31.2	15.7	2.6	132.5	46.6	2.5
7	24.0	21.1	18.2	6.4	131.9	14.4	1.5
Study Area Average		48.5	49.4	8.6	181.4	52.8	4.2

Source: Chicago Area Transportation Study, Vol. 1, 1960.

TABLE 8-2

Percentage Distribution of Trips by Land Use Related To Trip Making Per Family

Trips per Family	Residential	Manufacturing	Transportation	Public Buildings	Public Open Space	Commercial
2	54.8	14.6	4.1	5.2	1.1	20.2
4	55.1	10.9	2.8	6.6	1.6	23.0
6	56.2	8.0	2.6	8.0	2.2	23.0
8	56.3	6.9	2.5	7.5	3.0	23.8
10	54.7	4.5	2.2	10.1	2.9	25.6
12	52.9	4.6	2.2	9.8	3.6	26.9
14	53.4	4.0	2.6	9.2	4.7	26.1
16	51.8	3.4	3.4	9.3	5.6	26.5

Source: Chicago Area Transportation Study, Vol. 1, 1960.

(in this case person trips), at various levels of one or more independent predictive variables. In its use of independent predictive variables the method resembles, in some ways, multiple regression techniques. Cross classification is essentially discreet in approach, however, with the number of values of the response variable being limited to the number of cells used. Regression techniques are continuous in character. A very simple example of the cross classification technique involving one variable is shown in Table 8–3. The technique is essentially non-parametric,

TABLE 8–3

One-Way Cross Classification Table for Six Trip Purposes

Home-Based Trip Generation and Income Level					
	Relative Income Level[1]				
Trip Purpose	**I**	**II**	**III**	**IV**	**V**
Work	0.38	0.44	0.48	0.48	0.52
Personal Business	0.15	0.17	0.21	0.22	0.23
Shopping	0.23	0.28	0.38	0.46	0.54
Social-Recreation	0.28	0.36	0.38	0.42	0.47
School	0.07	0.13	0.24	0.32	0.25
ALL PURPOSES	1.45	1.95	2.42	2.88	3.14

Source: Southeastern Virginia Regional Transportation Study.

since no account is taken of the distribution of the individual values which compose the cells.

A rather sophisticated approach to cross classification was used in the Puget Sound Regional Transportation Study. Two types of variables were selected as independent variables for prediction of the residential trip productions. These types were variables relating to household characteristics and variables relating to environmental characteristics:

Household characteristics
1. Average household size
2. Average automobiles per household
3. Median income of head of household

Environmental characteristics
1. Population per acre
2. Population per net residential acre.

Each of the zones was then assigned a rank on the basis of each of the individual measures of both household characteristics and environmental characteristics. The ranking was based on its order from the lowest observed value (rank 1) to the highest observed value (rank equaling the number of internal zones). The zonal ranks for both household and environmental characteristics were determined from a weighted average of the rankings of the individual variables making up the set. The frequency distribution of the calculated household characteristic was plotted, and was divided into 12 frequency intervals. These frequency intervals were themselves grouped into 6 levels of categorization. Similarly, the environmental characteristics were grouped into 10 frequency intervals, which were consolidated into three levels of categorization—low, medium, and high. Table 8–4 shows the matrix into which

TABLE 8–4

Classification Matrix for Trip Production Rates

Matrix into Which Analysis Zones were Classified for Trip Production Rate Purposes

Household Characteristics		Environmental Characteristics Grouped Category and Frequency Interval		
Grouped Category	Frequency Interval	(Low) 1, 2, 3	(Medium) 4, 5, 6, 7	(High) 8, 9, 10
Low	1, 2	(1)	(2)	(3)
	3, 4	(4)	(5)	(6)
Medium	5, 6	(7)	(8)	(9)
	7, 8	(10)	(11)	(12)
High	9, 10	(13)	(14)	(15)
	11, 12	(16)	(17)	(18)

Note: Numbers in parentheses are the cell numbers assigned to the matrix.
Source: Staff Report No. 16, Puget Sound Regional Transportation Study, 1964.

the analysis zones were classified for trip rate production purposes. The matrix has 18 cells, with six levels of household characteristics and three levels of environmental characteristics. Table 8–5 shows the trip production rates for each of the seven home based trip purposes which was used in conjunction with Table 8–4. The Puget Sound method extends the more more basic cross classification approach by the effective weighting of different variables in the calculation of the indices used in determining cell classification.

TABLE 8–5

Matrix of Trip Production Rates

1961 Home Based and Non-Home Based Trip Production Rates (Person Trips Per Household Per Average Weekday)

Trip Production Grouping Cell No.'s	TRIP PURPOSE							
	Total Home Based	Home Based Work	Home Based Shop	Home Based Soc. Rec.	Home Based Misc.	Home Based School	Non-Home Based	Total Person Trips
01	–	–	–	–	–	–	–	–
02	2.29	0.62	0.39	0.59	0.60	0.09	0.59	2.88
03	2.54	0.85	0.41	0.62	0.55	0.10	0.65	3.19
04	4.38	1.15	0.87	1.03	0.90	0.43	1.13	5.51
05	4.80	1.40	0.89	1.09	1.05	0.37	1.23	6.03
06	4.21	1.50	0.68	0.95	0.86	0.22	1.08	5.29
07	6.02	1.55	1.14	1.17	1.24	0.93	1.55	7.57
08	5.52	1.58	1.04	1.26	1.17	0.47	1.42	6.94
09	4.98	1.68	0.93	1.07	0.99	0.31	1.28	6.26
10	6.33	1.56	1.07	1.43	1.27	1.00	1.63	7.96
11	5.99	1.78	1.09	1.43	1.19	0.51	1.54	7.53
12	5.44	1.73	0.96	1.26	1.15	0.33	1.40	6.84
13	6.73	1.63	1.09	1.60	1.38	1.04	1.73	8.47
14	6.66	1.85	1.25	1.53	1.34	0.69	1.71	8.38
15	6.19	1.76	0.95	1.67	1.30	0.50	1.59	7.79
16	7.22	1.79	1.37	1.62	1.31	1.14	1.86	9.08
17	7.59	1.85	1.33	1.85	1.49	1.06	1.95	9.54
18	–	–	–	–	–	–	–	–

Source: Staff Report No. 16, Puget Sound Regional Transportation Study, 1964.

As a method for trip generation, the cross classification method has inherent advantages found in neither of the other two methods discussed here. "Families of curves" can be plotted, showing the effect of change of one independent variable at constant levels of the other independent variables. These plots are most useful to the planner enabling him to get a good feeling for the importance of the independent variables. This is sometimes difficult with multiple regression techniques, where the strength of the independent variable is described by significance levels and partial correlation coefficients (see Section 8–7). Another advantage of the technique comes from the fact that there is no assumption of linearity between the dependent and independent variables; the technique is, therefore, suitable for application where the effect of the independent variable is non-linear and its actual form not certain. Knowledge of the form of dependence is needed for transformation when multiple regression is used.

There are, however, disadvantages to the method which must not be overlooked. Perhaps the greatest disadvantage is the fact that the amount of the total variance explained by the independent variables is unknown with this non-parametric approach. Equally, there is no examination of the underlying distribution of the individual values which make up the mean value entered in each cell. Where these distributions are highly skewed, the sample size to assure a meaningful cell entry may have to be large. One of the most serious weaknesses of the method, however, is the possibility that the "independent" variables that the analyst selects are not truly independent. The resultant relationships and predictions may well be spurious.

This method of trip generation is most useful where proper caution is exercised, and promises widespread use as the art progresses.

8–7. Multiple Regression Analysis. The most widespread method of trip generation analysis is the use of multiple regression techniques. Regression is a mathematically based procedure which has been programmed for most high-speed electronic computers. The technique is, therefore, readily available to the analyst. With reasoning similar to the cross classification method, the trip generation rate is treated as a *dependent* variable which is a function of one or more *independent* variables. The relationship assumed is linear, of the form:

$$Y = A_0 + A_1X_1 + A_2X_2 + \cdots + A_NX_N \qquad (8\text{–}1)$$

where:

Y = the trip rate

$X_1 \cdots X_N$ = independent variables

$A_0 \cdots A_N$ = constants

For example, in the Rapid City Transportation Study [7], the following equation was used to predict home based personal business trips:

$$\begin{aligned} Y = {} & 0.45 \times \text{Population} + 1.16 \times \text{Cars} - 1.38 \times \text{Labor Force} \\ & - 0.39 \times \text{Public and Semi-Public Acres} - 0.03 \times \text{Undeveloped} \\ & \text{Acres} + 38.65 \end{aligned} \qquad (8\text{–}2)$$

The approach is entirely mathematical; therefore, statistical tests of reliability of the derived relationships can be applied with ease. Two basic assumptions are made concerning the form of the independent variables. First, all variables are assumed to be random variables with a normal distribution. Second, the predictive variables are assumed to be independent of each other. The effect of violation of either of these assumptions is discussed later.

Possibly the greatest advantage of the regression approach is the ease with which the analyst can determine the degree of relationship between the dependent and independent variables, and can define the accuracy of the predictive equation itself. Some of the most common measures which are used in analysis need explanation:

1. *The Coefficient of Multiple Determination* (r^2) is a measure of the amount of variance described by the model expressed as a decimal ratio of the total variance observed in the dependent variable. The value of this coefficient has an upper limit of 1.0, which would be the value for a perfect model.
2. *The Standard Error Estimate* (S_y) is a measure of the deviation of observed trip values from values predicted by the model. Where the model values agree exactly with observed values, the standard error reaches its lower limit, zero, the value for a perfect model.
3. *The Partial Correlation Coefficient* (r_j) of an independent variable describes the relation between the dependent variable and the particular independent variable under consideration. Descriptively, it is a measure of this association, with the effect of all other independent variables eliminated.

The method of multiple regression can be well illustrated by a simple example relating shopping destinations to office and retail sales floor space, with only two independent variables. The form of the recommended model equation was in this case [8]:

$$Y = 4.53X_1 + 5.57X_2 - 71 \tag{8–3}$$

where:

Y = 24-hour person destinations to CBD zone for shopping
X_1 = Area of floor space within zone used for retail sales
X_2 = Area of floor space within zone used for offices.

The ability of the equation to model the observed shopping destinations can be seen in a qualitative manner by reference to Table 8–6, and quantitatively by examination of the statistics of the equation.

Standard error, (S_y) = 186
Coefficient of multiple determination (r^2) = 0.860
Partial correlation coefficient of X_1, (r_1) = 0.882
Partial correlation coefficient of X_2, (r_2) = 0.764

An interpretation of the calculated statistics gives the analyst a clear indication of the suitability of the model. The coefficient of multiple determination indicates that the model explains 86 per cent of the total

TABLE 8-6

CBD Shopping Trips Related to Sales, Office Floor Space Use—Gainesville, Georgia

Zone of Origin	Computed Shopping Destinations	Observed Shopping Destinations
006	312	368
010	803	426
003	1037	876
001	1068	1407
011	19	58
013	116	52
004	0	3
009	62	26
007	26	199
009	94	17
005	0	40
008	60	17

Source: Relationship of Traffic Attracted to a City's C.B.D. to Intrazonal Floor Space, P. H. Wright, 1964.

variation, leaving only 14 per cent variation due to the deviation of the computed values from those observed.

The standard error signifies that approximately 67 per cent of all observed values can be expected to fall within ±186 trips of the computed values. The two partial correlation coefficients denote that each independent variable is closely associated with the dependent variable, total person shopping trips. The partial correlation coefficients are important indicators. It is entirely possible for a regression equation to model observed values apparently with suitable accuracy, with a low standard error and a high value of r^2, yet to contain an independent variable which does not have a close relationship to the dependent variable. The equation could be improved with the deletion of the spurious variable. For a full discussion of multiple regression techniques the reader is referred to a standard text on correlation and regression analysis [9], where the extensive treatment that the subject deserves is fully set out before the student.

The application of regression analysis to modeling of trip ends stratified by purpose can be exemplified by reference to Tables 8–7 and 8–8, which indicate the variables used for production and attraction equations, respectively. The person trip relationships derived were for all modes of transportation, including transit and auto. Table 8–9 shows some examples of the types of equation built from the relationships

TABLE 8-7

Regression Analysis Input Variables
Productions—All Modes

Trip Purpose	INDEPENDENT VARIABLES[1]										
	Labor Force				School Enrollment				Persons Per Sq. Mi.	Population	Cars Per D. U.
	White-Collar	Blue-Collar	Military	Total	Grade School	High School	College	Total			
White-Collar Work	X										
Blue-Collar Work		X									
Military Work			X								
Personal Business Incomes I-IV										X	
Income V				X						X	
Convenience-Shopping Incomes I-IV										X	
Income V									X	X	
G.A.F. Shopping										X	
Social										X	
Recreation								X		X	
Grade School					X						
High School						X					
College							X		X	X	X
Other School								X			
Miscellaneous				X					X	X	X

[1] "X" indicates variable(s) were used in "best fit" regression equation.

Source: Southeastern Virginia Regional Transportation Study, Wilbur Smith and Assoc., 1965.

TABLE 8-8

Regression Analysis Input Variables
Attractions—All Modes

Trip Purpose	INDEPENDENT VARIABLES[1]															
	Employment				School Attendance				Retail Sales							
	White-Collar	Blue-Collar	Military	Total	Grade School	High School	College	Total	Conv. Sales	G.A.F. Sales	Total	Total School Enroll.	Total Labor Force	Persons Per Sq. Mi.	Popu-lation	Cars Per D. U.
White-Collar Work	X															
Blue-Collar Work		X														
Military Work			X													
Personal Business											X			X	X	
Convenience-Shopping				X					X		X					
G.A.F. Shopping				X						X	X					
Social														X	X	X
Recreation								X				X		X	X	X
Grade School					X											
High School						X										
College							X									
Other School								X								
Miscellaneous				X				X			X	X	X		X	

[1] "X" indicates variable(s) were used in "best fit" regression equation
Source: Southeastern Virginia Regional Transportation Study, Wilbur Smith and Assoc., 1965.

TABLE 8-9

Example Equations for Trip Productions and Attractions

		Equation	r^2 Value
Productions			
Home based white-collar work trips	=	18.12 + 1.46 X white-collar labor force	0.95
Home based convenience shopping trips (income level V)	=	21.40 + 0.40 X persons per square mile + 0.02 X population	0.66
Home based college trips	=	-23.62 - 0.70 X college enrollment -0.01 X persons per square mile + 0.05 X population + 29.10 X cars per dwelling unit	0.82
Attractions			
Home based white-collar work trips	=	12.29 + 1.31 X white-collar employment	0.97
Home based convenience shopping (income level III)	=	139.68 - 0.02 X total employment + 1080.90 X convenience - goods retail sales + 78.12 X total retail sales	0.54
Home based social trips	=	76.94 + 0.02 X persons per square mile + 0.11 X population - 20.93 X cars per dwelling unit	0.68

Source: Southeastern Virginia Regional Transportation Study, 1965.

of the previous two tables. The equations are obtained by regression on the independent variables, with one observation in each zone.

Multiple regression techniques have been attractively packaged in computer program routines to the point that very little effort is required of the analyst to produce equations which will adequately fit the zonal data. From this viewpoint, multiple regression is more convenient than the two other techniques covered in previous sections. The method has several pitfalls awaiting the unwary user that can produce adequate looking but erroneous relationships. By re-examining the assumptions of regression, the user will gain an insight into errors to be avoided. Restating these, the method assumes a linear relationship between the dependent variable and one or more normally distributed independent variables.

When the independent variables are not independent of each other, they are said to be *co-linear*. Equations containing independent variables that demonstrate a high degree of co-linearity must be avoided if spurious relationships are to be eliminated. The result of using two such variables is in effect to count the same factor twice. This can render the equation useless for projection purposes. Co-linearity can be avoided by examining the correlation coefficients of the independent variables among themselves. If two independent variables in an equation have a high correlation coefficient, they may be presumed to be co-linear; one variable should be eliminated. A distinct possibility, in the case where co-linear variables are used, is the occurrence of an incorrect sign in the derived equation. A variable which apparently should contribute to trip generation and should, therefore, be associated with a positive coefficient will be found to have a negative coefficient. This frequently happens if care is not taken to assure independence of predictive variables. The need for wariness in variable selection should indicate to the analyst that hidden interactions may come into play if too many variables are used in the regression equation. Usually, three variables are a limiting condition for most predictive equations. Introducing further variables, while marginally increasing the r^2 value, can result in a significant decrease in partial correlation coefficients. Such "overstuffed" equations should be avoided.

Non-normality of data is a lesser problem. The analyst must examine the distribution of the variables used. Where a choice exists between a variable with a skewed distribution, and one normal in character, the latter is preferable. The result of using non-normal data is to render the evaluation statistics inaccurate, but the form of relationship itself is not badly affected in most cases. In cases where the underlying distribution of an independent variable is very highly skewed, the planner is advised to avoid the use of such a variable in regression analysis.

Multiple linear regression is a process of fitting a regression plane to observed results with a minimization of the squares of the residuals. The process will force the results into the chosen linear model, whether the relationship is truly linear or not. Where the relationship is not linear, but its form is known, the method can still be applied by a transformation of the involved variable. For example, if the form is felt to be:

$$Y = A_0 + A_1X_1 + A_2X_2^2 \tag{8-4}$$

the variable Z_2 can be introduced, where

$$Z_2 = X_2^2 \tag{8-5}$$

and regression on the form:

$$Y = A_0 + A_1X_1 + A_2Z_2 \tag{8-6}$$

can be performed in a standard manner. Transformation of variables is useful in increasing the flexibility of the method. The difficulty lies, however, in the recognition that a non-linear relationship exists. Its presence may be difficult to determine when mixed with the effects of other variables. In the case where the non-linear effect is ignored, multiple regression will force the data into an incorrect linear relationship.

TRIP DISTRIBUTION MODELS

At the end of the trip generation phase of the modeling process, the planner has knowledge of how many trips will be generated in each zone of the study area in the forecast year. In addition, it is known for what purpose these trips are to be made. At this stage of projection, however, the answers to the questions, "to what zones are these trips going and by what route?" are unknown. Trip distribution models solve the first of these two questions by calculating trip interchanges according to criteria which are found to govern interchanges in the base year. Most of the models discussed in the subsequent sections depend greatly upon the origin-destination study for calibration purposes, just as future trip generation was predicted from behavior patterns observed in base year generation. The importance of the origin-destination study is underlined by the strong dependence of the modeling process on the data collected during the survey.

8–8. Growth Factor Methods. The earliest and simplest type of trip interchange models were known as growth factor methods. Their basic assumptions were similar. They related increases in zonal interchanges to the growth of zones at either end of the interchange.

1. Uniform Growth Factor Method. Using this method, an areawide average growth factor is calculated. This average can be a simple mean value of the growth factor for all zones or may be a weighted average calculated from the zonal factors. Expressing the model in mathematical form:

$$T_{ij} = t_{ij} \times F_{(\text{average})} \tag{8–7}$$

where:

T_{ij} = the future trips between zone i and zone j
t_{ij} = base year trips between zone i and zone j
$F_{(\text{average})}$ = the average areawide growth factor

Example 8–1. The trip interchange between zone 8 and zone 100 of a study area is 100 trips. The growth factor for the city over the 20-year design period has been computed to be 1.6. What is the trip interchange between the two zones in the forecast year?

$$\begin{aligned} T_{8,100} &= t_{8,100} \times F_{\text{average}} \\ &= 100 \times 1.6 \\ &= 160 \text{ trips} \end{aligned}$$

In summary, the areawide growth factor method is obviously a limited approach to the determination of trip interchange. The factor is usually based on past trends and may be a poor predictor of increase of activity. Large errors will be encountered where there is significant change in zonal characteristics, such as rapidly developing areas. Equally, there will be overestimation when both zones are already intensively developed. It should be obvious to the student that there is no value in attempting sophisticated trip generation techniques when such a rudimentary distribution method is used, since the zonal trip ends will all be increased by the same growth factor.

2. Average Factor Method. Where growth factors for individual zones could be calculated, an average of the two zonal growth factors could be applied to the base year interchange to arrive at a forecast for the design year. Expressing this model in equation form:

$$T_{ij} = t_{ij} \times \frac{F_i + F_j}{2} \tag{8–8}$$

where:

T_{ij} = the future trip interchange between i and j
t_{ij} = the base year trip interchange between i and j
F_i = growth factor at i
F_j = growth factor at j

Example 8–2. The trip interchange between zone 10 and zone 33 in the base year is 100 trips. In the 20-year forecast period, zone 10 is expected to have a growth rate of 1.6, and zone 33 is expected to grow at a rate of 1.4. Compute the future trip interchange between 10 and 33.

$$T_{10,33} = t_{10,33} \times \frac{F_{10} + F_{33}}{2}$$

$$= 100 \times \frac{1.6 + 1.4}{2}$$

$$= 150 \text{ trips}$$

The average growth factor method suffers from shortcomings similar to those discussed for the uniform factor method. Growth factors can be notoriously unreliable, since they normally are simply projections of past trends. With this method, there is the added difficulty of trying to determine accurately individual factors for each zone. All growth factor methods usually fail to estimate adequately growth in developing areas, with the result that in areas that are almost completely undeveloped in the base year there may be a severe underestimation of trips.

3. Detroit Growth Factor Method. The most extensive study in the last decade to use a growth factor method of trip distribution was the Detroit Area Transportation Study. The form of the Detroit model was expressed as:

$$T_{ij} = t_{ij} \times \frac{F_i \times F_j}{F_{\text{average}}} \tag{8–9}$$

where:

T_{ij} = trip interchange between zones i and j in forecast year
t_{ij} = trip interchange between zones i and j in base year
F_i = growth factor at i
F_j = growth factor at j
F_{average} = average areawide growth factor

Example 8–3. The trip interchange between zone 10 and zone 33 in the base year is 100 trips. In the 20 year forecast period zone 10 is expected to have a growth rate of 1.6 and zone 33 is expected to grow at a rate of 1.4. The average rate of growth areawide is expected to be 1.7. Compute the future trip interchange between 10 and 33.

$$T_{10,33} = t_{10,33} \times \frac{F_{10} \times F_{33}}{F_{\text{average}}}$$

$$= 100 \times \frac{1.6 \times 1.4}{1.7}$$

$$= 132 \text{ trips}$$

The Detroit method introduced into its computations a method whereby the trips emanating from a zone would sum to the base year trips multiplied by the zone's growth factor. It can be seen that if, for example, the Uniform Growth Factor method is used, the forecast year trips will be increased to the degree of areawide growth, even though little growth would be expected in some zones. In the DATS study, an iterative procedure ensured that at the completion of computations, the relationship $\sum_{\text{all } j} T_{ij} = F_i \sum_{\text{all } j} t_{ij}$ was obtained at all zones.

4. Fratar Method. The most widely used method of trip distribution to utilize growth factors is the Fratar Method, introduced in 1954 [10].

While the method is now seldom used as the studywide distribution model, it is still in extensive use for predicting interchanges between external stations of a study area. These are known as the external-external trips. The method is applied iteratively with the interchange being computed according to the relative attractiveness of each interzonal movement from the point under consideration. Calculation of the future interchange is made considering separately the growth factors at each end of the trip. The average of these two estimates is accepted as the future interchange for each successive approximation. Based on the newly calculated trip ends, and the projected trip ends for the design year, the calculation can be reiterated using new growth factors. The method may be formulated in the following way:

t_{AB} = trip interchange between A and B in base year
T_{AB} = future trip interchange

$$\frac{\sum_{\text{all } x} T_{AX}}{\sum_{\text{all } X} t_{AX}} = \text{Initial growth factor at } A = F_A$$

$$T_{AB(A)} = \frac{t_{AB} \times F_B}{\sum_{\text{all } X} t_{AX} \times F_X} \times \sum_{\text{all } X} t_{AX} \times F_A \text{ considering growth at } A \text{ only} \quad (8\text{–}10)$$

$$T_{AB(B)} = \frac{t_{AB} \times F_A}{\sum_{\text{all } X} t_{BX} \times F_X} \times \sum_{\text{all } X} t_{BX} \times F_B \text{ considering growth at } B \text{ only} \quad (8\text{–}11)$$

$$T'_{AB} = \frac{T_{AB(A)} + T_{AB(B)}}{2} \tag{8–12}$$

$$\frac{\sum_{\text{all } X} T_{AX}}{\sum_{\text{all } X} T'_{AX}} = F'_A, \text{ the new growth factor for the next iteration.}$$

The method converges when $\dfrac{F_X}{F'_X}$ is acceptably close to unity throughout the network.

Example 8–4. Three zones A, B, and C are shown with interchanges $AB = 10$, $BC = 15$, and $CA = 20$. These are non-directional interchanges. Growth factors of 2, 3, and 1 are forecast for zones A, B, and C, respectively. Using Fratar's iterative method, compute the zonal interchanges in the forecast year.

Solution:

$\sum_{\text{all } x} t_{ax} = 30$

$F_a = 2$

$\sum_{\text{all } x} T_{ax} = 60$

Zone A — $t_{ab} = 10$ — Zone B

$t_{ac} = 20$

$t_{bc} = 15$

Zone C

$\sum_{\text{all } x} t_{bx} = 25$

$F_b = 3$

$\sum_{\text{all } x} T_{bx} = 75$

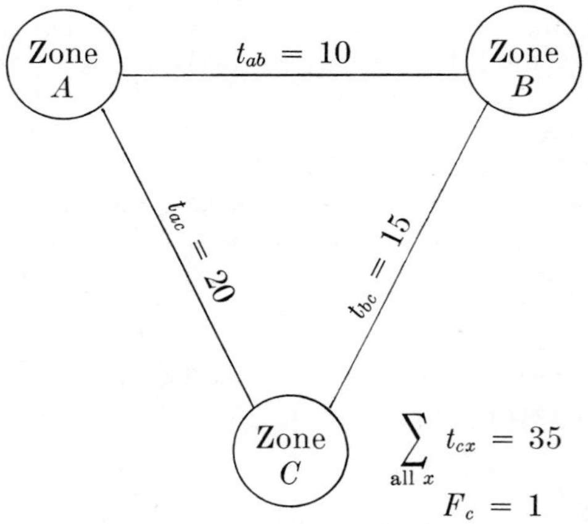

$\sum_{\text{all } x} t_{cx} = 35$

$F_c = 1$

$\sum_{\text{all } x} T_{cx} = 35$

$$\begin{aligned} T_{ab(a)} &= \frac{(t_{ab} \times F_b)\left(\sum_{\text{all } x} (t_{ax} \times F_a)\right)}{\sum_{\text{all } x} (t_{ax} \times F_x)} \\ &= \frac{(10 \times 3)[(10 \times 2) + (20 \times 2)]}{(10 \times 3) + (20 \times 1)} \\ &= 35 \text{ trips} \end{aligned}$$

Considers growth at A only

$$T_{ab(b)} = \frac{(t_{ab} \times F_a)\left(\sum_{\text{all } x} (t_{bx} \times F_b)\right)}{\sum_{\text{all } x} (t_{bx} \times F_x)}$$

$$= \frac{(10 \times 2) \times [(10 \times 3) + (15 \times 3)]}{(10 \times 2) + (15 \times 1)}$$

$$= 43 \text{ trips}$$

Considers growth at B only

$$T'_{ab} = \frac{T_{ab(a)} + T_{ab(b)}}{2} = \frac{35 + 43}{2}$$

$$= 39$$

Similarly:

$$T_{bc(c)} = 18.6$$
$$T_{bc(b)} = 32.2$$
$$T'_{bc} = 25.4$$
$$T_{ac(a)} = 24$$
$$T_{ac(c)} = 16.5$$
$$T'_{ac} = 20.25$$

At the end of the first iteration:

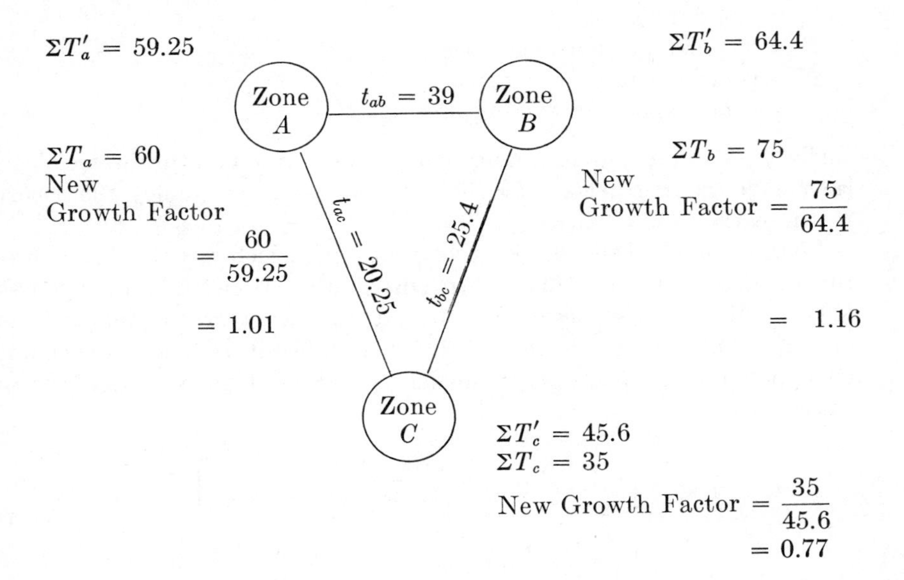

It can be seen that, after one iteration, the new growth factors for A, B, and C are 1.01, 1.16, and 0.77. The rapidity of convergence is indicated below:

Iteration No.	*Computed New Growth Factors*		
	A	*B*	*C*
1	1.01	1.16	0.77
2	0.96	1.07	0.93
3	0.97	1.04	0.98
4	0.98	1.03	0.99
5	0.98	1.02	0.99
6	0.99	1.01	0.99

The method is reiterated until the degree of convergence required is obtained.

8–9. The Gravity Model. Because of a similarity in behavior to Newton's theory of gravity, human interaction has for some time been described by a model formulation which states that interaction between two populations is a function of the populations involved and the distance between them. In mathematical terms this may be stated:

$$I_{ij} = \frac{K \times P_i \times P_j}{D^n} \tag{8–13}$$

where:
I_{ij} = the interactions between i and j
P_i = the population at i
P_j = the population at j
D = the distance between i and j
K = some constant
n = some exponent

Early work in the area of gravity models was carried out in the sociological and marketing fields. Traffic engineers began to employ the theory in the early 1950's.

Voorhees applied the gravity theory in work which was the direct forerunner of the Gravity Model Trip Distribution Model now widely used [11]. This work suggested a constant exponent for the influence of distance, but recognized the need for a different exponent depending on trip purpose. The gravity model as suggested by Voorhees was of the form:

$$T_{1-a} = \left[\frac{\dfrac{S_1}{(D_{1a})^n}}{\dfrac{S_1}{(D_{1a})^n} + \dfrac{S_2}{(D_{2a})^n} + \cdots + \dfrac{S_m}{(D_{ma})^n}}\right] \times T_A \tag{8–14}$$

where:

T_A = Total trips produced at A for a purpose
T_{1-a} = Trips for this purpose between 1 and a
S_i = Measure of attraction to ith zone for trips of this purpose
D_{ia} = Distance from A to ith zone
n = Some exponent which varies with purpose

The effect of trip purpose, both on the variable used to describe the measure of attraction and on the exponent of distance, is shown in Table 8–10. There is a significant change in the distance exponent, with the

TABLE 8–10

Effect of Trip Purpose upon Attraction Variable and Distance Exponent

Purpose of Trip	Unit to Express Size of Attractor	Exponent of Distance
Work	Number of workers employed	0.5
Social	Dwelling units	3
Shopping		
Shopping goods	Commercial floor area	2
Convenience goods	Commercial floor area	3
Business	Floor area	2
Recreation	Floor area	2
Other	Floor area	2

Source: Institute of Traffic Engineers. A. M. Voorhees, "A General Theory of Traffic Movement," Proceedings of I.T.E., 1955.

smallest exponent for work trips and the greatest exponent occurring with social and convenience shopping trips. Interpreting this effect, it can be said that distance has a greater effect upon the less important trip type than upon important and necessary trips such as work trips. The principal findings of Voorhees have been substantiated in numerous transportation studies carried out across the United States and in many other countries where the gravity model has been successfully used as trip distribution model.

The gravity model now in current use, as recommended by the Bureau of Public Roads, is somewhat different in form to the relationship previously indicated.[1] The differences have evolved from experience gained

[1] The Bureau of Public Roads, which formulated standard procedures for the use of the gravity and other transportation models, was abolished after the creation of the Federal Department of Transportation. Much of the literature detailing these procedures will be found in titles published by this now defunct agency.

in numerous studies which have been completed to date. Currently, the recommended formulation can be stated, for each purpose considered:

$$T_{ij} = P_i \times \frac{A_j F_{t_{i-j}} K_{ij}}{\sum_{\text{all } r} (A_r F_{t_{i-r}} K_{ir})} \tag{8-15}$$

where:

T_{ij} = trip interchange between zone i and zone j
P_i = total productions at zone i
A_j = total attractions at zone j
K_{ij} = social economic adjustment factor between zones i and j
$F_{t_{ij}}$ = travel time factor for travel time between zones i and j

The difference in form of these two models need clarification.

1. Experience has shown that a constant exponent of distance is not observed in practice. Referring to Fig. 8–5, where the logarithm of travel time is plotted against the logarithm of the travel time factor, it can be seen that if the observed relationships were of the form:

$$F_t = t^{-n}$$

the travel time factor curve would be linear. Observed curves are nonlinear in logarithmic form, showing an increase of negative exponent with increasing time.

2. Travel time, rather than distance, has been determined to be more closely related to decay in travel demand than distance.

3. Rather than measures of attraction, such as floor area or number of persons employed, trip attractions themselves are used in the equation. These trip ends for each zone are predicted in the trip generation phase of modeling. Early gravity model studies used the formulation shown in Equation 8–14.

4. Zone-to-zone social economic adjustment factors are included for each zone to allow the analyst to adjust the model to fit observed travel patterns. An example of the need for such a factor would be the situation of a poor neighborhood close to the downtown area of a state capital. Assuming the capital complex was an entire traffic zone, it would provide many work attractions, the majority of which, however, would mostly be filled by medium- to high-income white-collar workers. Without zone-to-zone adjustment factors, the gravity model would assign the nearest workers to these positions, in this case the low-income workers.

Unlike the growth factor methods previously described, the gravity model and the opportunity models discussed in Sections 8–11 and 8–12 must be "calibrated." By calibration we mean the establishment of a distribution parameter based on observed travel patterns. This parameter

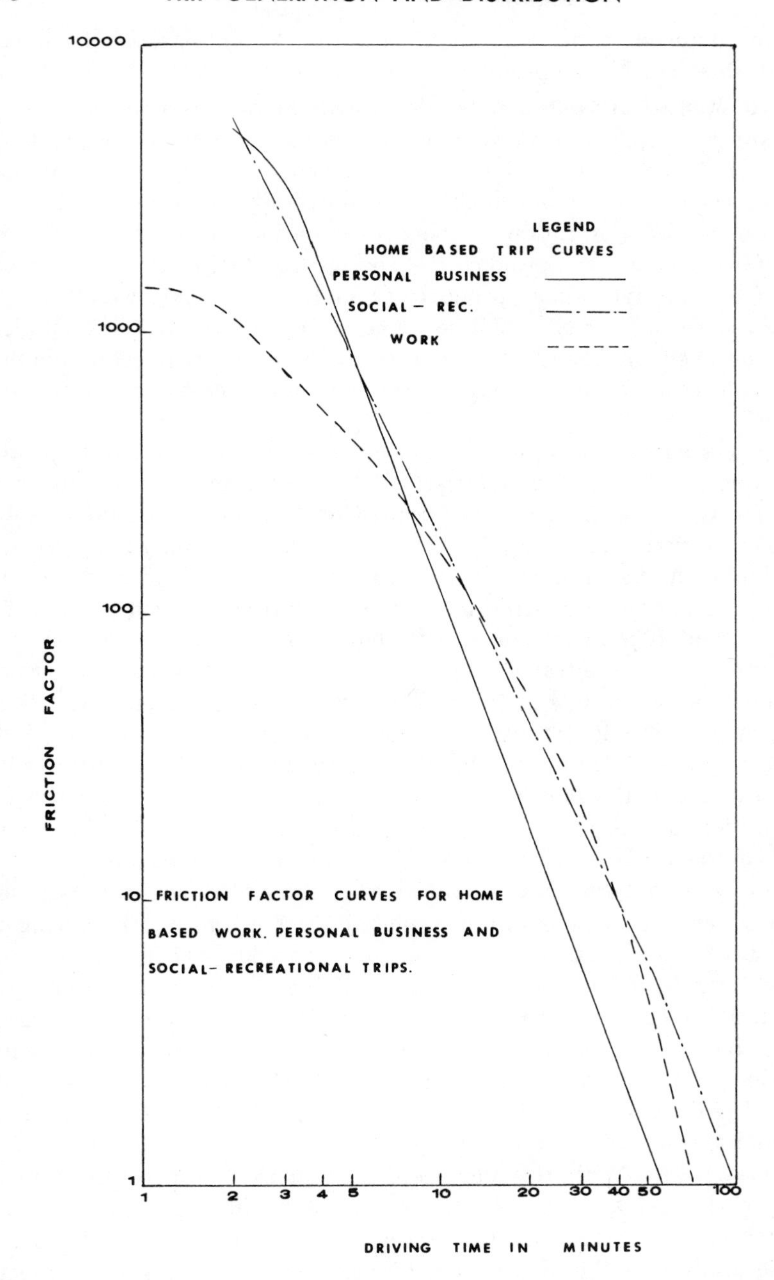

Fig. 8–5. Travel time factor curves for three purposes, New Orleans. (**Source:** *New Orleans Metropolitan Area Transportation Study,* Louisiana State Department of Highways.)

will be assumed to be valid in the future. In the case of the gravity model, this distribution parameter is the travel time factor.

8–10. Method of Calibration of the Gravity Model. The origin-destination study is used as a data base for the calculation of the particular travel time factors for a study area. Each trip purpose has an individual travel time factor which is calculated by an iterative procedure. The process of calibration is the determination of the form of travel time factor which will produce the zone-to-zone trip tables of the base year from the trip ends (productions and attractions), which are observed in the base year. These travel time factors are then assumed constant over time, and by applying them to the trip ends computed for the forecast year, the planner can compute zone-to-zone future trip tables.

A suggested method of calibration has been formulated by the Bureau of Public Roads. The analyst starts the calibration phase with an assumed travel time factor curve, from which travel time factors at one-minute intervals can be read. Using these travel time factors, in conjunction with the productions and attractions at each zone determined in the O-D survey, a distribution of trips is carried out using the gravity model. The trip interchanges will not agree with those observed in the survey unless the travel time factor curve assumed is correct. Figure 8–6 indicates the results of the first calibration run for Non-Home Based Trips in the Atlanta Study. The graphs shown are the observed trip length frequency distribution for the trips sampled in the origin-destination study, and the distribution of trip lengths obtained from the first distribution of the O-D trip ends by the gravity model. The general form of the curve is correct, yet it would appear reasonable that by the choice of a better travel time curve, agreement could be arranged to be closer than that obtained from this first estimate. Following the Bureau of Public Roads recommended procedure [12], the assumed travel time factor curve is modified. Using travel time factors from the modified curve, a second iteration of the gravity model is carried out. The travel time factor curve is then again readjusted and a third iteration of the gravity model is run with the O-D trip ends. After this third iteration the difference had shrunk to 0.34 per cent for the person hours of travel and 0.46 per cent for the average trip length (see Fig. 8–7). With this degree of agreement the travel time factor curve of the last iteration is accepted as the "calibrated" curve. These factors are assumed constant over the planning period and will be used by the analyst as the distribution criteria for distributing the projected zonal trip ends of the design year.

Before the gravity model can be assumed to be totally calibrated, the analyst must determine whether socio-economic adjustment factors

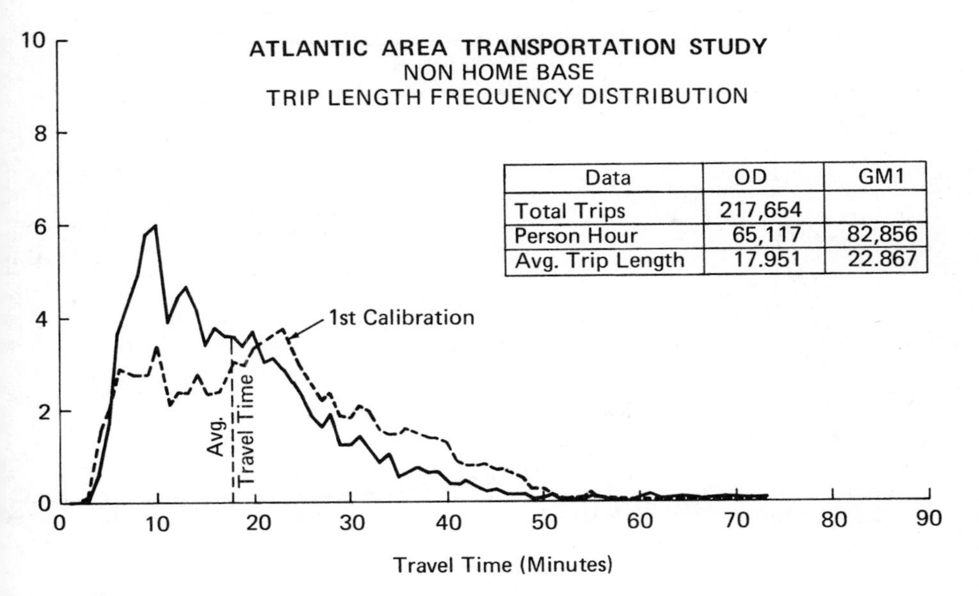

Fig. 8–6. Comparison of O-D and gravity model trip length frequency distributions; first calibration. (Source: *Atlanta Area Transportation Study*, Georgia State Highway Department.)

are required to bring individual gravity model zonal interchanges into agreement with the zonal interchanges. These are calculated where necessary and inserted into the model. Projections of what K-factors will be for individual zones in the future are obtained from regression equations. These equations are developed from zonal socio-economic data in the base year and base year K-factors.

One other adjustment may be necessary for complete calibration. The presence of mountains, rivers, large open spaces, toll bridges, and heavily congested areas may introduce resistance to travel which is not accountable in observed travel times. These are known as topographical barriers. Such phenomena are handled by the introduction of time penalties into the transportation network. Washington (Potomac River crossings) [13], and Hartford (toll bridges) [14] are studies which may be cited as models that have required topographical barriers. Figure 8–8 indicates the sequencing of operations for gravity model calibration as carried out in a typical study.

Example 8–5 (Gravity Model Calculation). Calculate the interzonal interchanges due to 100 productions at zone 1, with 250 attrac-

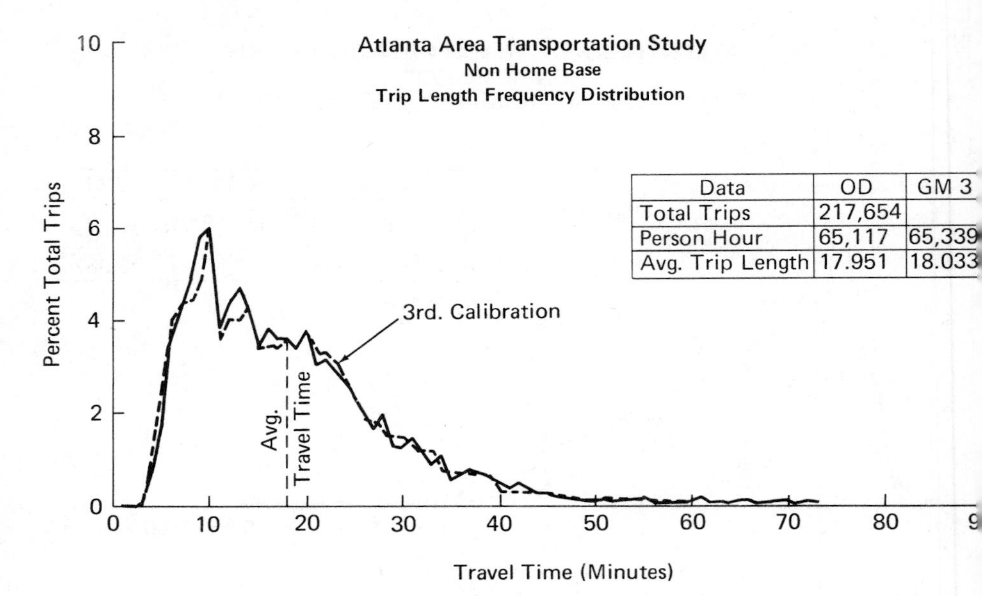

Fig. 8–7. Comparison of O-D and gravity model trip length frequency distributions; third calibration. (Source: *Atlanta Area Transportation Study*, Georgia State Highway Department.)

tions at zone 2, 100 attractions at zone 3, and 600 attractions at zone 4. Assume that 1–2 is 5 minutes, 1–3 is 10 minutes, and 1–4 is 15 minutes. Assume all K_{ij} factors are unity, and that F_{ij} factors are as shown below.

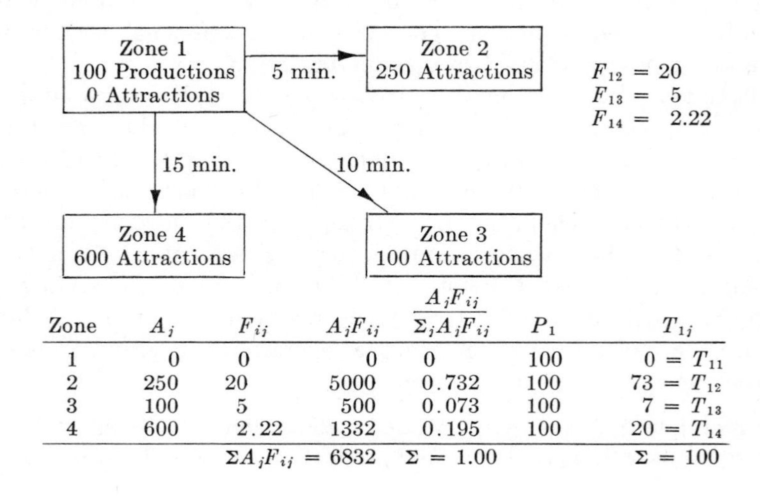

$F_{12} = 20$
$F_{13} = 5$
$F_{14} = 2.22$

Zone	A_j	F_{ij}	A_jF_{ij}	$\frac{A_jF_{ij}}{\Sigma_j A_jF_{ij}}$	P_1	T_{1j}
1	0	0	0	0	100	$0 = T_{11}$
2	250	20	5000	0.732	100	$73 = T_{12}$
3	100	5	500	0.073	100	$7 = T_{13}$
4	600	2.22	1332	0.195	100	$20 = T_{14}$
		$\Sigma A_jF_{ij} =$	6832	$\Sigma = 1.00$		$\Sigma = 100$

The reader is advised to verify that the exponent of time for the travel time factor is 2.

8–11. The Intervening Opportunities Model. The gravity model reflects the confining effect of distance upon travel demand. The two probability models which will be considered in this chapter approach the trip distribution problem from a different viewpoint, applying probabilities that a trip originating at one designated zone will find a destination at another designated zone. The first probability model to be considered is the *intervening opportunities model,* which was used successfully in two large urban areas, Chicago and Pittsburgh. The model is based on the assumption that total travel time from a point is minimized given that each destination has a specified probability of being suitable for the origin in question. A trip considers all destinations. Failing to be satisfied by the nearest, it will consider the next nearest and so on until a suitable destination is found. From this premise, a mathematical model is derived [15].

Consider the trips within the trip volumes V and $V + dV$ as shown in Fig. 8–9.

The probability that a trip will terminate within some volume of destination points is equal to the probability that this volume contains an acceptable destination multiplied by the probability that an acceptable destination closer to the origin of the trip has not been found.

Let:

l = probability of acceptability of a destination;

P = probability that a trip has terminated prior to the zone under current consideration. This is a function, not a constant.

Then, with respect to elemental volume dV, this probability of termination may be stated as an elemental change in P:

Change in $P = Pr$ (acceptable destination is found in volume dV) $\times$ Pr (acceptable destination has not been found prior to the incremental volume)

$$dP = (l \times dV)(1 - P)$$

In order that the mathematics remain tractable, even for high-speed computers, it is necessary to make the assumption that l is a constant in the above equation. How well this agrees with reality will be discussed later.

Therefore:

$$\frac{dP}{1 - P} = l \times dV$$

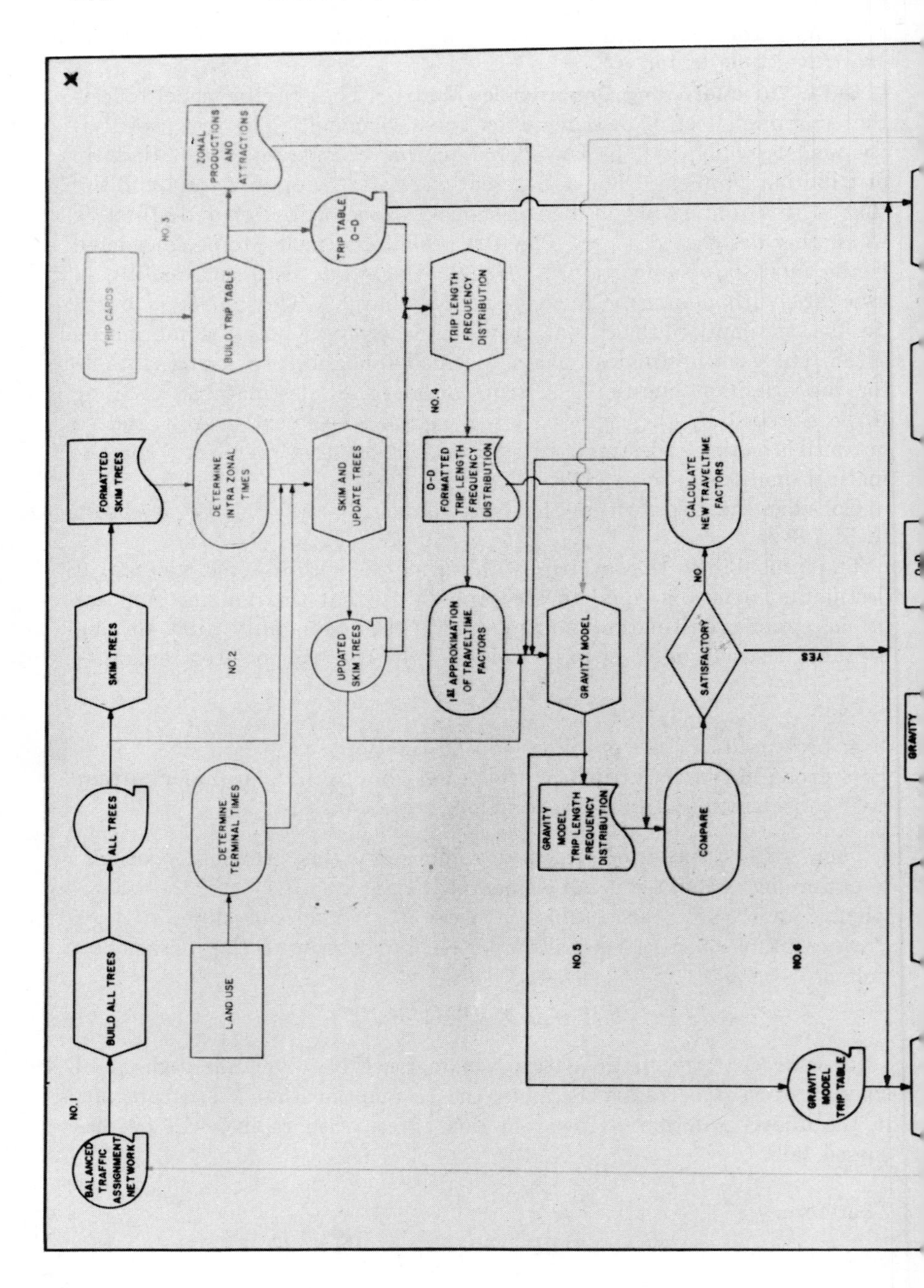
NO.1
BALANCED TRAFFIC ASSIGNMENT NETWORK
BUILD ALL TREES
ALL TREES
SKIM TREES
FORMATTED SKIM TREES
NO.2
LAND USE
DETERMINE TERMINAL TIMES
DETERMINE INTRA ZONAL TIMES
NO.3
TRIP CARDS
BUILD TRIP TABLE
ZONAL PRODUCTIONS AND ATTRACTIONS
TRIP TABLE O-D
SKIM AND UPDATE TREES
UPDATED SKIM TREES
NO.4
TRIP LENGTH FREQUENCY DISTRIBUTION
O-D FORMATTED TRIP LENGTH FREQUENCY DISTRIBUTION
1ST APPROXIMATION OF TRAVELTIME FACTORS
NO.5
GRAVITY MODEL
GRAVITY MODEL TRIP LENGTH FREQUENCY DISTRIBUTION
NO.6
COMPARE
SATISFACTORY
NO
YES
CALCULATE NEW TRAVELTIME FACTORS
GRAVITY MODEL TRIP TABLE
GRAVITY

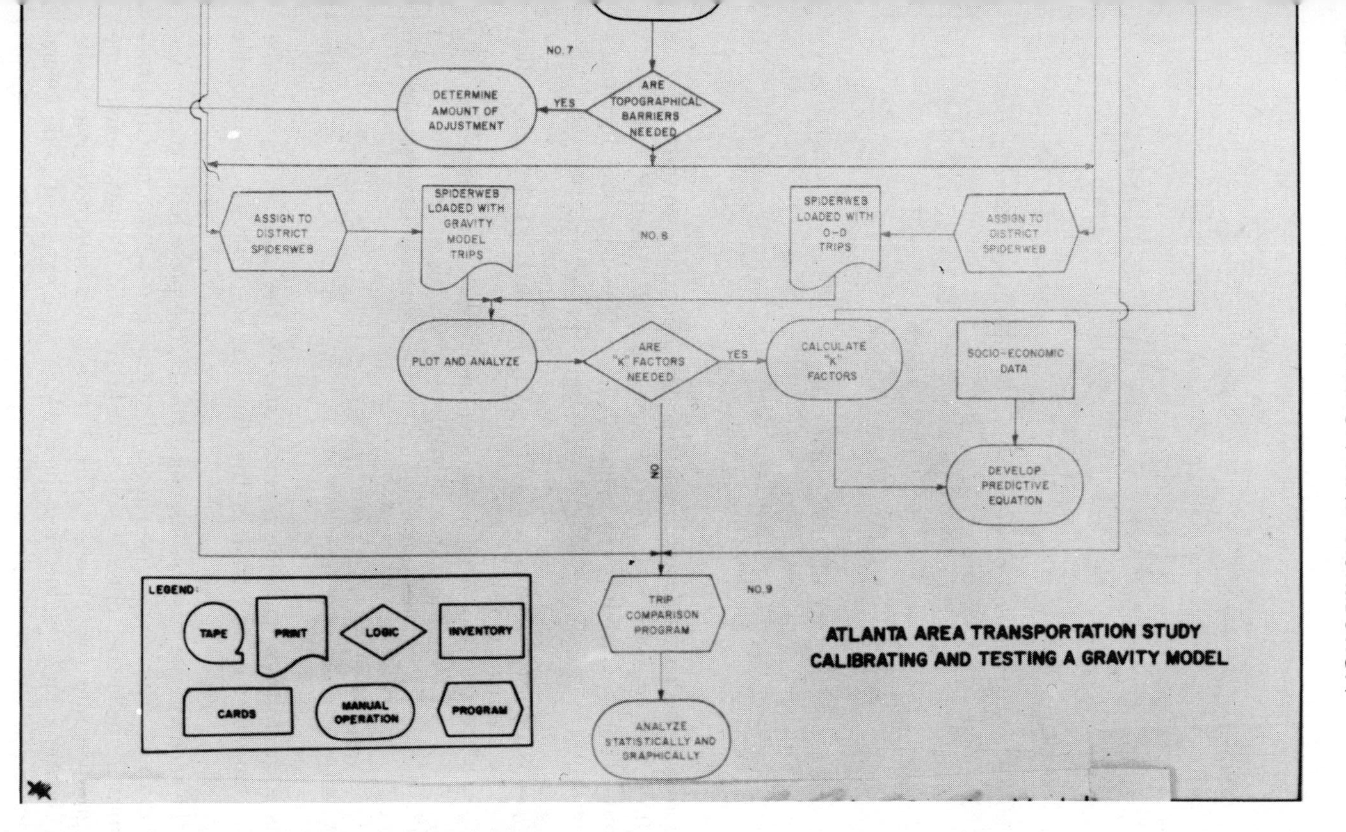

Fig. 8–8. Flow chart of the calibration process for the gravity model. (Source: *Atlanta Area Transportation Study*, Georgia State Highway Department.)

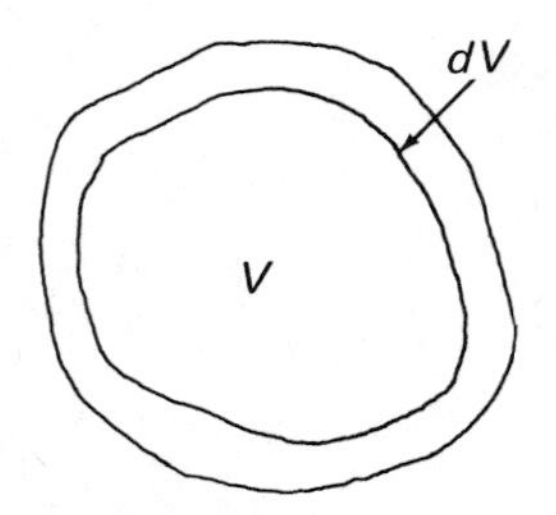

Fig. 8–9. Accumulation of destinations by time ring around origin.

By integration:

$$-\ ln(1 - P) = lV + k$$

where k is some constant of integration.
or:

$$1 - P = -ke^{-lV}$$
$$P = 1 + ke^{-lV}$$

The value of k can be determined from the boundary condition, when $V = 0$, $P = 0$, and $k = -1$. The relation may be stated $P = 1 - e^{-lV}$ The number of trips from zone i to zone j, T_{ij}, can be determined from the relationship:

$$T_{ij} = \text{Number of origins at } i \times Pr \text{ (a trip terminates in volume which contains zone } j)$$
$$T_{ij} = P_i \times Pr \text{ (trip terminates in volume which contains zone } j)$$
$$T_{ij} = P_i \times \begin{bmatrix} Pr \text{ (acceptance in volume including zone } j) - \\ Pr \text{ (acceptance in volume immediately prior to zone } j) \end{bmatrix}$$
$$T_{ij} = P_i[Pr(V + V_j) - Pr(V)]$$
$$T_{ij} = P_i[(1 - e^{-l(V+V_j)}) - (1 - e^{-lV})]$$
$$T_{ij} = P_i[e^{-lV} - e^{-l(V-V_j)}] \qquad (8\text{–}16)$$

This is the form of equation normally used in the model. An example of the computations involved in the distribution is shown in Example 8–6).

In the model theory it was assumed that l, the probability parameter was constant. Experience has shown that this can be assumed in practice only if the model operates on stratified trips. Trips at each zone were grouped into two classes. Those trips which were expected to be short, having a high l value, and those which were expected to be longer with more selective destinations, with a low l value. Further assumptions were made with respect to the longer trips. First, all longer trips could be modeled using a single value of l, and second that the longer trips of a specialized nature could not have both origin and destination of residential or non-residential use. Figure 8–10 shows the effect of using a single value of l. The value of l chosen was that which would equate the model and survey trips distributed at the 50 per cent level. The model does not distribute the shorter trips rapidly enough, yet the l value is obviously too high for longer trips. By stratifying in short and long trips, the actual prediction curve was obtained, an admirable fit.

In distributing trips for the future year, the l values of the base year are not used. The high l value, initially computed to give the correct proportioning of short or intrazonal trips, would give erroneous answers when applied to the future trips. With the increased densities and increased destinations of the design year, the same l value would result in shorter intrazonal trips, which is unlikely. Distribution of future trips was, therefore, carried out in two steps:

1. Determination of that l value which would keep the same proportion of trips within the zone. This l value was the l_{high} which was applied to intrazonal trips.
2. The total travel mileage was computed from social economic estimates, and that l value which satisfied this estimate in conjunction with the computed l value was accepted as the future value of l_{low}. The values of l used in the base and future years in the Chicago study is shown in Table 8–11.

TABLE 8–11

l Values for the Chicago Area

	1956	1980
high	21×10^{-6}	18×10^{-6}
low	2.3×10^{-6}	1.75×10^{-6}

Source: Chicago Area Transportation Study.

Example 8–6 (Intervening Opportunities Model Calculation). Calculate the interzonal interchanges due to 100 productions at zone 1, with 250 attractions at zone 2, 100 attractions at zone 3, and 600 attractions at zone 4. Assume that 1–2 is 5 minutes, 1–3 is 10 minutes, and 1–4 is 15 minutes. Use a value of l, the probability measure, of 0.006. Note the trip ends are the same as those used in the gravity model and competing opportunities models.

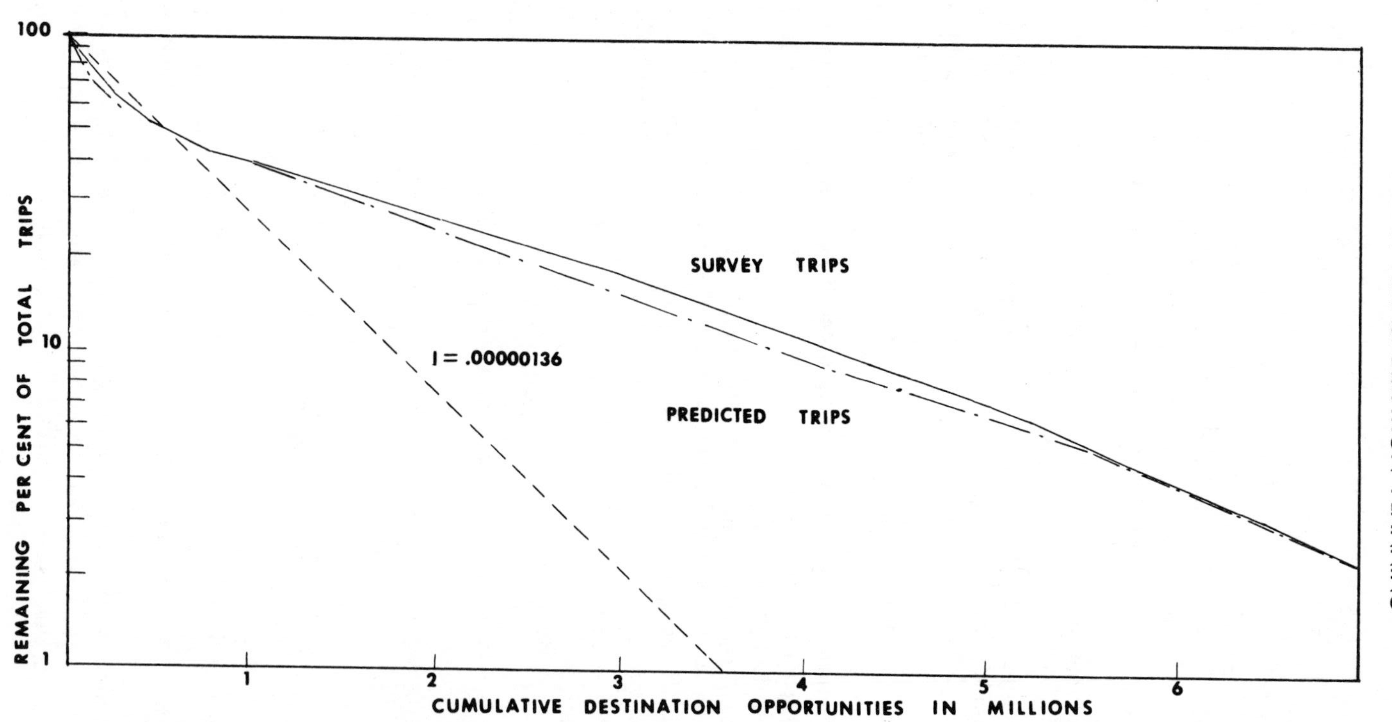

Fig. 8–10. Decrease of unsatisfied destinations with increase of distance from the origin. (Source: *Chicago Area Transportation Study*, Figures 39, 40 Vol. II.)

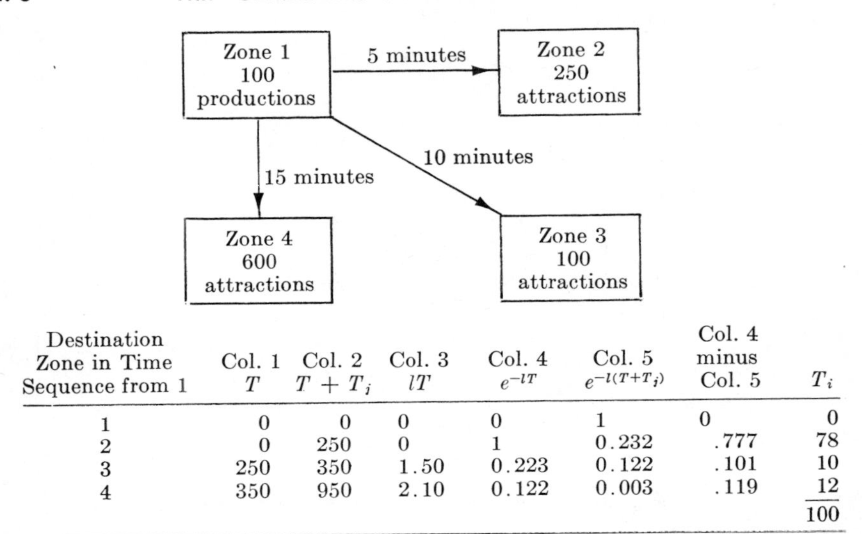

Destination Zone in Time Sequence from 1	Col. 1 T	Col. 2 $T + T_j$	Col. 3 lT	Col. 4 e^{-lT}	Col. 5 $e^{-l(T+T_j)}$	Col. 4 minus Col. 5	T_i
1	0	0	0	0	1	0	0
2	0	250	0	1	0.232	.777	78
3	250	350	1.50	0.223	0.122	.101	10
4	350	950	2.10	0.122	0.003	.119	12
							100

8–12. The Competing Opportunities Model. Another probability model which has been of interest in the last few years is the model used in distribution phase of the Penn Jersey Study [16]. This model has had limited application because of difficulties of calibration. It is, however, of more than passing interest to the student of distribution models, for it presents an approach to the use of probability in the distribution of trip ends that is quite different from the Chicago model. The basic assumption of the competing opportunities model is the probability that a trip originating in a given district and ending in another district is given by the ratio of destination opportunities within the destination district to all destination opportunities within the same time zone up to and including the destination district in question. This assumption can be illustrated by Fig. 8–11.

S_1 represents the destination opportunities within a district in time zone t_1. H_1 is the total of all opportunities within the time zone t_1. S_2 represents the destination opportunities within a district in time zone t_2, while H_2 is the sum of all opportunities within time zone t_2, including those within t_1. S_3 and H_3 are similarly defined. Conditional probabilities can be calculated from a knowledge of the destinations in the area:

$$Pr(S_1|H_1) = \frac{S_1}{H_1}$$

$$Pr(S_2|H_2) = \frac{S_2}{H_2}$$

$$Pr(S_3|H_3) = \frac{S_3}{H_2}$$

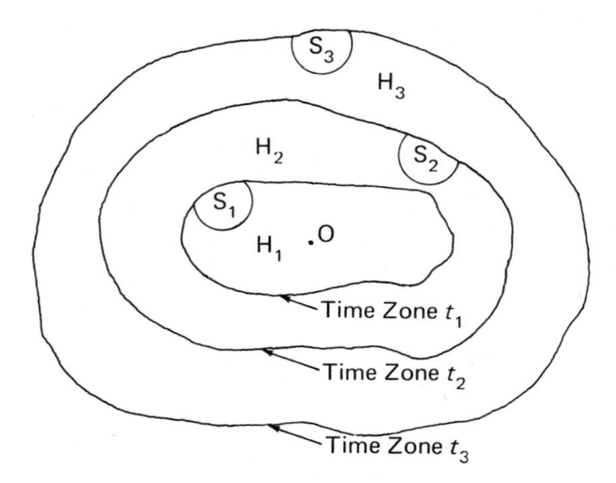

Fig. 8–11. Relation of districts to time zones.

These conditional probabilities can be converted to absolute probabilities by imposing the condition that the sum of the absolute probabilities over the region is unity, which can be attained by dividing the individual conditional probabilities by their sum over the region.

Converting terminology for the purposes of trip interchange calculation, the form of the model can be expressed:

$$T_{ij} = T_i \times \frac{A_j/A_x}{\Sigma(A_j/A_x)} \tag{8–17}$$

where:

T_{ij} = the trip interchange from zone i to zone j.
T_i = the total number of origins at zone i.
A_j = the number of destination opportunities in zone j.
A_x = the total number of destination from zone of origin i within the time band containing the zone of destination

The method of calculation is illustrated in Example 8–7. The suggested method of calibration for this model is adjustment of the time band intervals until agreement is obtained between the origin-destination trip length frequency distribution and that generated by the model. These time band intervals are then held constant from the base year to the forecast year. The distribution obtained using the calibrated time bands is considered applicable to the design year. In one noteworthy study which compared various distribution models applied to the Washington, D.C. data, fairly good agreement was obtained between the Fratar, Gravity, and Intervening Opportunity Models. It was found that calibration of the Competing Opportunity model was impossible and the attempt was abandoned [17].

EXAMPLE 8–7 (Competing Opportunities Model Calculation). Calculate the interzonal interchanges due to 100 productions at zone 1, with 250 attractions at zone 2, 100 attractions at zone 3, and 600 attractions at zone 4. Assume that 1–2 is 5 minutes, 1–3 is 10 minutes, and 1–4 is 15 minutes.

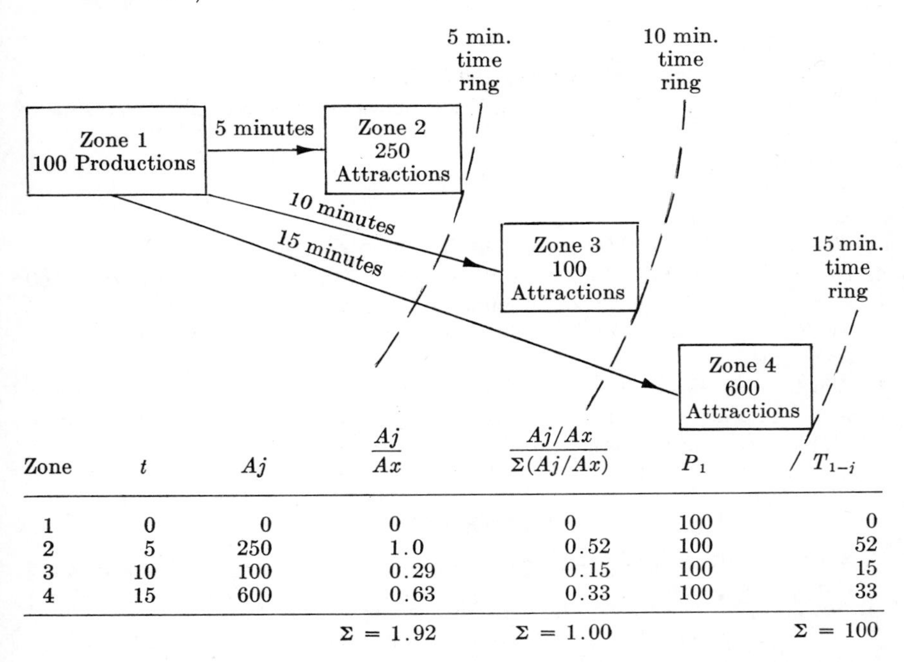

Zone	t	Aj	$\frac{Aj}{Ax}$	$\frac{Aj/Ax}{\Sigma(Aj/Ax)}$	P_1	T_{1-j}
1	0	0	0	0	100	0
2	5	250	1.0	0.52	100	52
3	10	100	0.29	0.15	100	15
4	15	600	0.63	0.33	100	33
			$\Sigma = 1.92$	$\Sigma = 1.00$		$\Sigma = 100$

PROBLEMS

1. In an urban area the average growth factor over a 20-year period is 1.8. For four selected zones the following data were collected or projected.

Zone	*Growth Factor*		*Base Year Trip Interchange*
1	2.5	(1–2)	1000
		(1–3)	2000
		(1–4)	0
2	1.0	(2–1)	1000
		(2–3)	2000
		(2–4)	0
3	1.5	(3–1)	2000
		(3–2)	2000
		(3–4)	0
4	no growth	(4–1)	0
		(4–2)	0
		(4–3)	0

Compute all zonal interchanges by:
a. Average Factor Method
b. Uniform Factor Method
c. Detroit Method
d. Fratar Method (use 2 iterations)

2. Using Example 8–6, determine the effect on the trip interchanges of using an l value of:
a. 0.06
b. 0.6
c. 0.006
Discuss the effects of changing the l value.

3. Rework Example 8–5 assuming the following socio-economic adjustment factors:
$K_{12} = 0.85$
$K_{13} = 2.00$
$K_{14} = 1.00$
Discuss the effect of these factors on the trip interchanges.

4. If, in Example 8–7, Zone 5 were added at 20-minutes distance with 300 attractions, and Zone 6 at 25 minutes distance with 100 attractions, what would be the effect on the interchanges 1–2, 1–3, 1–4, assuming 5-minute time rings?

5. Which equations in Table 8–7 would require the closest scrutiny to ensure that co-linearity did not exist between the independent variables? State your reasons for suspecting possible co-linearity.

REFERENCES

1. *Baltimore Metropolitan Area Transportation Study,* Wilbur Smith and Associates, 1964, p. 100.
2. *Chicago Area Transportation Study,* Vol. II, 1960, p. 84.
3. *Guidelines for Trip Generation Analysis,* Department of Transportation, Bureau of Public Roads, June 1967, p. 7.
4. *Pittsburgh Area Transportation Study,* Vol. I, 1961, Fig. 40.
5. *Chicago Area Transportation Study,* Vol. I, 1960, Fig. 37.
6. *Guidelines for Trip Generation Analysis, Department of Transportation, Bureau of Public Roads,* June, 1967, p. 12.
7. *Rapid City Metropolitan Area Transportation Study,* Wilbur Smith and Associates, 1965.
8. WRIGHT, P. H., *Relationship of Traffic Attracted to Zones in a City's Central Business District to Interzonal Floor Space,* Ph.D. Dissertation, Georgia Institute of Technology, 1964, p. 115.
9. EZEKIEL, M., and FOX, K. A., *Methods of Correlation and Regression Analysis,* Third Edition, John Wiley and Sons, New York (1959).
10. FRATAR, T. J., "Vehicular Trip Distribution by Successive Approximations," *Traffic Quarterly,* January, 1954.
11. VOORHEES, A. M., *A General Theory of Traffic Movement,* Proceedings of the Institute of Traffic Engineers, 1955.
12. *Calibrating and Testing a Gravity Model for Any Size Urban Area,* Department of Commerce, Bureau of Public Roads, October, 1965.
13. HANSEN, W. G., *Evaluation of Gravity Model Trip Distribution,* H.R.B. Bulletin 147, 1962.

14. Barnes, C. F., *Integrating Land Use and Traffic Forecasting,* H.R.B. Bulletin 297, 1961.
15. *Chicago Area Transportation Study,* Vol. II, p. 111.
16. Tomazinis, A., "A New Method of Trip Distribution in an Urban Area," *Highway Research Board Bulletin 347,* 1962.
17. Heanue, K., and Pyers, C., "A Comparative Evaluation of Trip Distribution Procedures," *Record No. 114,* Highway Research Board, 1965.

9

Planning Models: Modal Split, Assignment, and Land Use Models—The Systems Approach

The models described up to this point have concentrated on the planner's approach to answering questions concerning how many trips, by purpose, will be generated in a particular part of the urban area and where these trips are destined. In the generation phase, person trips are modeled and later vehicles are assigned to routes by the traffic assignment models. An interim model is needed to convert person trips to vehicle trips. *Modal split* supplies answers to the questions: "By what mode of transport will trips from this zone be made?" and "What is the vehicle occupancy that can be expected?" Knowledge of the variables affecting modal choice will enable the planner to convert person trips into vehicle trips for the purposes of traffic assignment. Unlike the other models discussed here, the analyst has a choice of position of the modal split model.

9–1. Types of Modal Split Analysis. In the case of a *pre-distribution split,* sometimes called a *trip end* model, the first step is generation of person trips on a zonal basis. These trips are then split into auto vehicle trips and transit person trips based on the criteria developed in the analysis of modal choice. Each mode is separately distributed using one of the trip distribution models previously discussed. It is worth noting at this point that since the transit system and the highway system may well have significantly different overall travel times between similar zones, there may be a great difference in the relative distributions of zone-to-zone interchanges. The auto vehicle trips and the transit person trips are then separately assigned to their respective networks. A generalized model flow diagram is shown in Fig. 9–1. Such an approach has been found useful only in cities which have appreciable transit usage. Where there is minimal transit usage, highly unstable distribution models can be derived due to the small number of trip ends involved in their calibration. Under these circumstances, another type of modal split analysis is likely to be more reliable.

The second form of modal analysis is known as a *post-distribution split,* or sometimes is referred to as a *trip interchange* model. This model is applied after the distribution phase of modeling and, therefore, acts upon trip interchanges, rather than on the trip ends themselves, immediately after the generation phase. Figure 9–2 indicates the general form of the post-distribution split model. This flow diagram shows that immediately after the generation phase a distribution analysis is carried out to compute interzonal trip interchanges. These interchanges are then split into interzonal transit trips and, after an analysis of auto occupancy analysis, the remaining trips are converted into interzonal vehicle trips. Both transit and auto trips are finally assigned to the system network.

9–2. Factors Affecting Modal Split. Just as the other phenomena of travel examined in this chapter have been amenable to analysis, modal choice is predictable. The decision of a tripmaker to take transit rather than go by automobile is not a random event. This decision is influenced by several factors. Our experience would indicate that a social trip by a middle to higher income white-collar worker living in a low-density subdivision on the outer suburban areas of a city is less likely to be made by transit than a CBD-oriented work trip of a low-income blue-collar worker, who lives in the fringe area of the CBD in a high-density housing development. In fact, some of the differences contained in this trip comparison are found to be variables which are a measure of the principal factors which affect the modal split.

The factors chiefly influencing modal choice are:

1. The type of trip
2. The characteristics of the trip maker
3. The relative levels of service of the transportation system

Trip type has been considered in studies in several ways. By a combination of various variables associated with trip type, this factor can be taken into account. Among the variables used in various studies, the following are noted:

1. Trip purpose
2. Length of trip

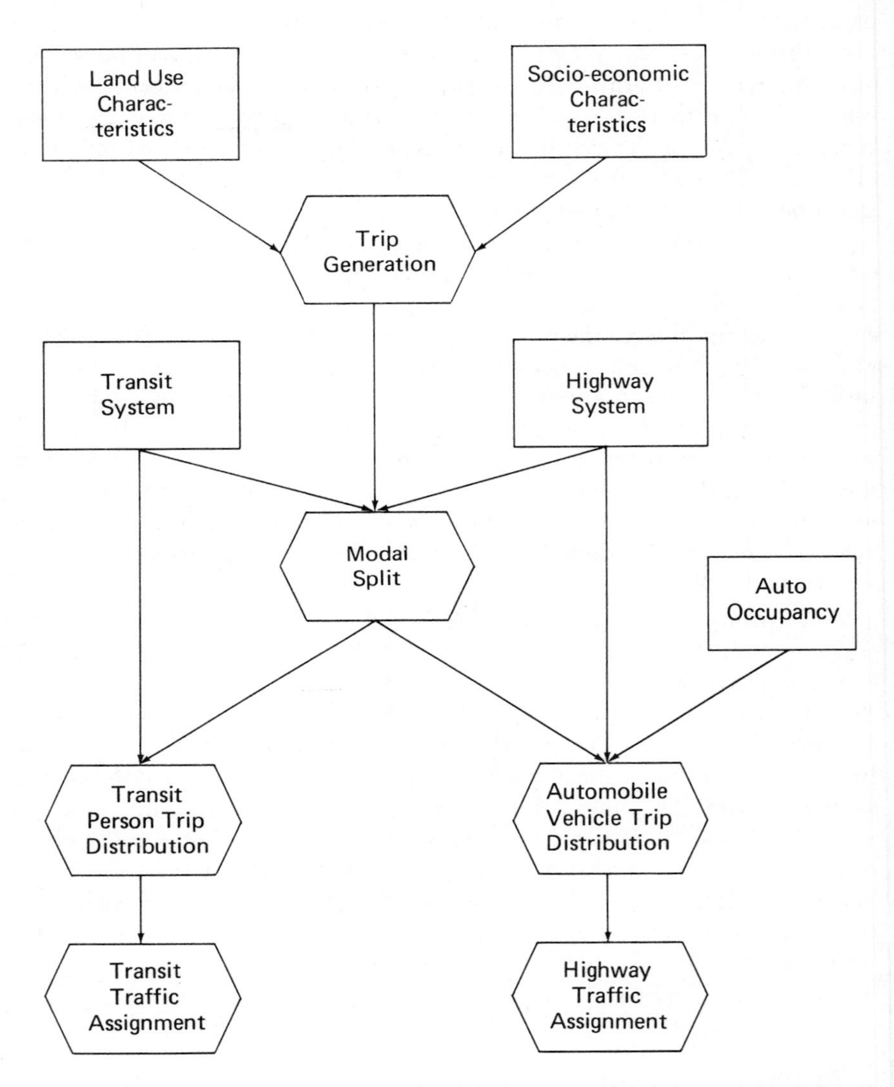

Fig. 9–1. Flow chart of the modeling process for a trip end modal split model. (Source: *Modal Split,* U.S. Department of Commerce, Bureau of Public Roads.)

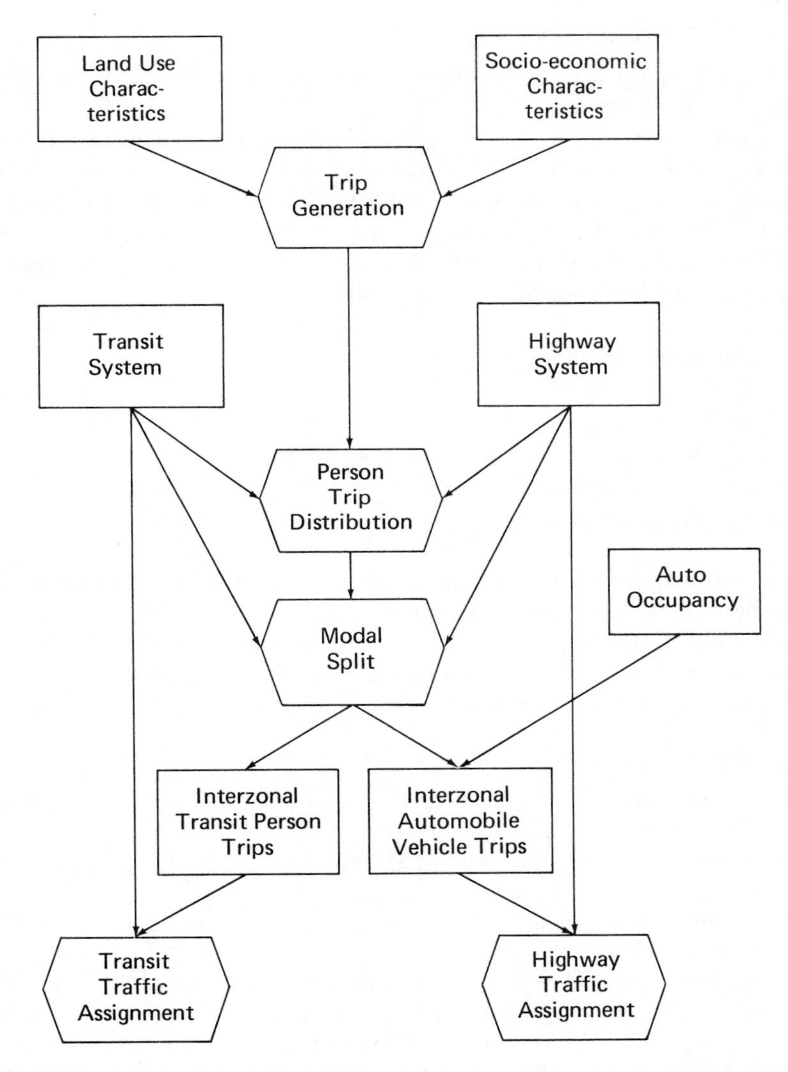

Fig. 9–2. Flow chart of the modeling process for a trip interchange modal split. (Source: *Modal Split,* U.S. Department of Commerce, Bureau of Public Roads.)

3. Time of day
4. Orientation to the CBD

These variables are usually taken into consideration by stratification, where this is necessary.

The characteristics of the trip maker have been determined to be of primary importance in determining transit usage. The variables which have been used to express this factor are both personal (e.g. income), environmental (residential density), and locational (distance to the CBD). The Bureau of Public Roads reports the following variables used in nine assorted modal split models [1]:

1. Auto ownership
2. Residential density
3. Income
4. Workers per household
5. Distance to CBD
6. Employment density

The relative levels of service of the transportation system ideally must be reflected in any meaningful model. Regardless of the type of trip and tripmaker, there will be a great variation in transit usage between areas of high and low transit service. Variables that have been used for measures of relative service include:

1. Travel time
2. Travel cost
3. Parking cost
4. Extra travel time in terms of time spent outside the vehicle during the total trip. This would include walking, waiting, and transfer time for transit trips while for auto trips it includes time walking to and from car.
5. Accessibility—The measure of accessibility frequently used is the "accessibility index," which is the denominator of the gravity model for each zone. The reader should check for himself that this index will be relatively high for zones with minimum travel times and vice versa.

9–3. Example of a Trip End Model [2]. Traditionally, the trip end model has concentrated more on the first two factors, trip type and trip maker characteristics, for modal analysis. An early predistribution split in the Chicago Study ignored system characteristics. Transit usage was predicted from auto ownership with stratification by trip purpose and orientation to the CBD. Later trip end models still tend to put their major emphasis on the characteristics of the trip maker, yet include to some degree the characteristics of the transportation system itself. A suitable example of a fairly recent trip end model is the modal analysis used in the Erie Transportation Study in 1964. Five major steps were involved in this analysis which is outlined in the flow diagram Fig. 9–3.

1. The decision was made to predict only those transit trips made for work purposes. It was reasoned that since the current modal split accounted for only 4 per cent of all trips, and 58 per cent of these trips were work trips, it would not be worthwhile to develop models for other purposes. Only the work trips placed a peak demand on the network.
2. Accessibility indices were calculated, computing the accessibility of each zone to employment centers. This index was computed for both auto and transit service.
3. A curve was developed relating the percentage of transit usage to the ratio of transit to highway accessibility (see Fig. 9–4).
4. Car occupancy was related to car ownership (see Fig. 9–5). This occupancy ratio was used in converting auto person trip ends to auto vehicle trip ends.
5. Future zonal modal split was computed from the relations developed in steps 3 and 4. For the purposes of modal split prediction, accessibility ratios were calculated assuming:
   a. Future transit service would be at present transit levels.
   b. The future highway system would comprise the existing system plus all committed facilities.

Step 5 reflects the problem of the analyst who wishes to take system characteristics into account in trip end analysis. The final proposed system will reflect the demands put on facilities as calculated from the distribution and assignment analyses. However, the final system has to be known prior to the distribution phase in the calculation of a modal split model which takes into account system characteristics. The assumptions made in the Erie model show how this problem has been tackled in the past.

9–4. Example of a Trip Interchange Model. Post-distribution modal split models were initially very simple diversion-type curves that assigned a percentage of traffic to transit depending upon the ratio of travel times by each mode. This simple approach, even if accounting for different trip purposes by stratification, could not account for the third factor which is known to affect modal split—the characteristics of the trip maker. As with the trip end models, more sophisticated methods of analysis have been attempted in recent studies. An example

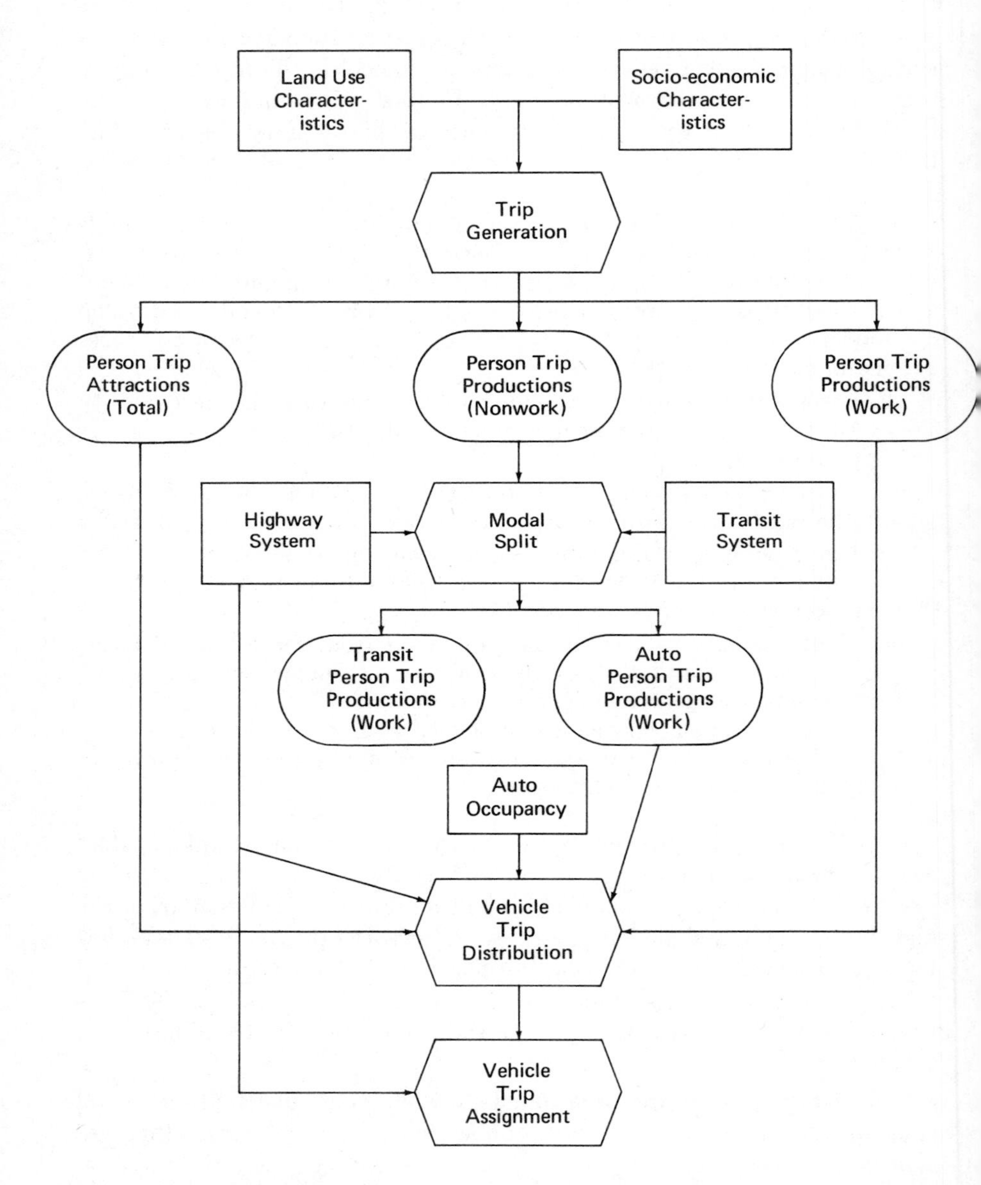

Fig. 9–3. Generalized model split diagram, Erie, Pennsylvania. (Source: *Modal Split,* U.S. Department of Commerce, Bureau of Public Roads.)

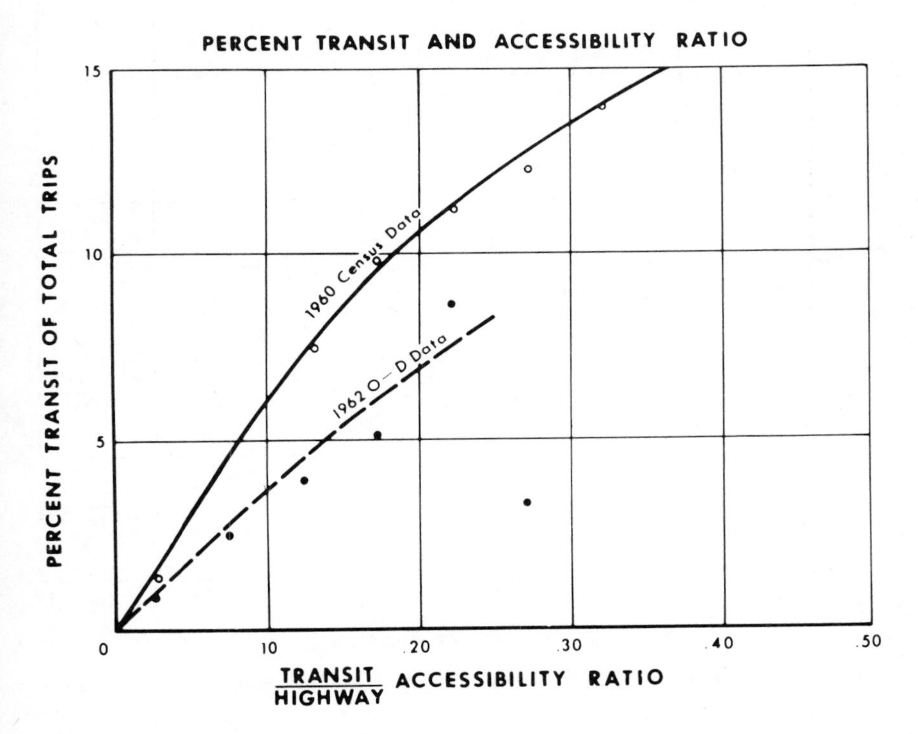

Fig. 9–4. Relationship between transit usage and ratio of accessibility by transit and by auto. (Source: *Modal Split,* U.S. Department of Commerce, Bureau of Public Roads.)

of a model which attempted to recognize the three major factors affecting modal split is the form of analysis that was used in Washington, D.C. in 1962.

The Washington model was essentially a stratified diversion model. The degree of stratification used was so large that, in all, 160 diversion curves were calculated. The future modal split was determined by finding which of the 160 curves applied to the zone and trip purpose concerned. The transit split was read from the chart by entering the graph at the relative travel time ratio of transit to auto travel (see Fig. 9–6).

The variables considered and their degree of stratification was as follows:

1. Trip purpose—2 levels, work and non work
2. Relative travel time—continuous variable

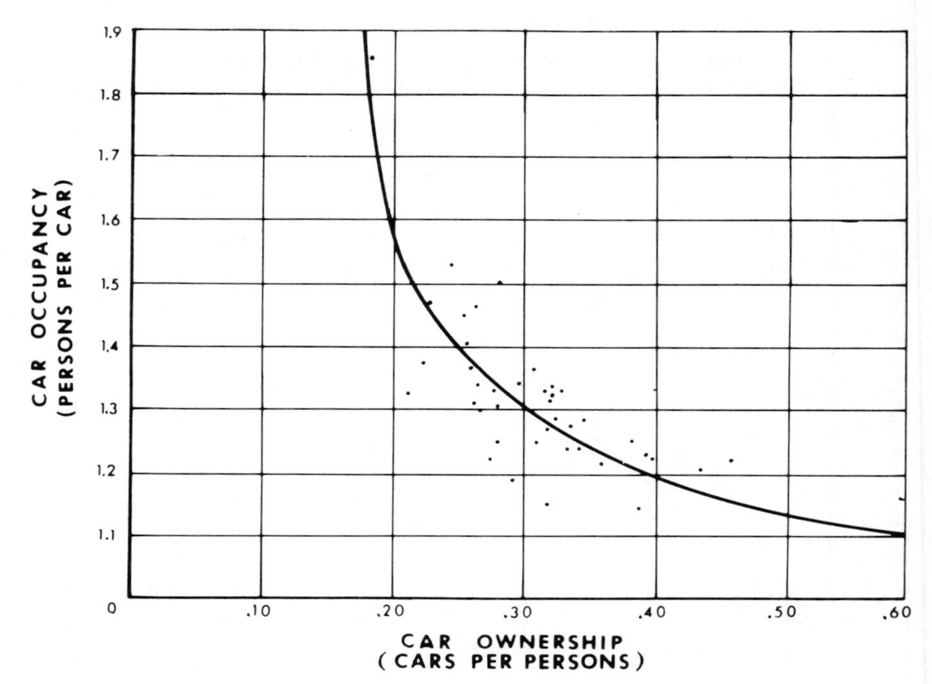

Fig. 9–5. Relationship between car occupancy and car ownership. (Source: *Modal Split,* U.S. Department of Commerce, Bureau of Public Roads.)

3. Relative travel cost—4 levels:
   0.0 to 0.5 = CR_1
   0.5 to 1.0 = CR_2
   1.0 to 1.5 = CR_3
   1.5 and over = CR_4

This variable represents the ratio of transit fare to out-of-pocket auto travel costs.

4. Travel service—4 levels, relation of time spent outside transit vehicle to parking and walking time for auto trip:
   L_1 = 0 to 1.5
   L_2 = 1.5 to 3.5
   L_3 = 3.5 to 5.5
   L_4 = 5.5 and over
5. Economic Status of Trip maker—5 levels:
   $0–$3100 per annum
   $3100–$4700 per annum
   $4700–$6200 per annum
   $6200–$7500 per annum
   $7500 per annum and over

It can be seen that the first variable reflects trip type. The second, third, and fourth variables are measures of the level of service of transportation system, while the characteristics of the trip maker are represented by the last variable.

It is interesting to note that, in an examination of this model, several possible defects became apparent. The model appeared to be oversensitive to walk and parking times for auto trips. An increase of two minutes to walking time caused a 33 per cent decrease in auto trips. Lack of sensitivity appeared in the cost ratio where a 35 per cent fare increase caused only a 5 per cent reduction in riding. Another weakness of the model appeared to be an inability to model non-CBD oriented trips accurately.

9–5. Summary. Modal split analysis has been conducted in a variety of forms. Diversion type curves, cross classification matrices, and regression analysis have all been used to develop curves to model the observed transit usage. Unfortunately, there is evidence that those models are not as effective in prediction as the planner could wish. The number of variables involved combined with the sample size of the origin-destination study would indicate that some degree of model instability is likely to remain until the transportation planning field has more experience and maturity.

TRAFFIC ASSIGNMENT

The final step in the simulation of travel patterns in the planning area is the assignment of zonal trip interchanges to the individual transportation facilities. The basic resulting output is in the form of traffic volumes on each portion of the transportation system. As a result, the planner can determine for the base year how well the assignment process has simulated observed traffic volumes. When the traffic assignment is used with future trip interchanges, the model indicates how well the proposed facilities will serve the anticipated travel demand. As it has now evolved, the assignment process converts trip interchanges into traffic volumes by the application of specified criteria in an objective manner. By elimination of subjective decisions in route choice, the transportation planner can be more confident that the traffic volumes he predicts are reasonably likely to occur.

9–6. Development of the Assignment Process. The current methods of objective assignment are relatively new. As late as 1952, traffic assignment was described as "more of an art than a science" [3]. At this time, the process consisted chiefly of the assignment of an interchange volume to a selected route between the two points. The selection of the route was based on the "experienced judgment" of the traffic

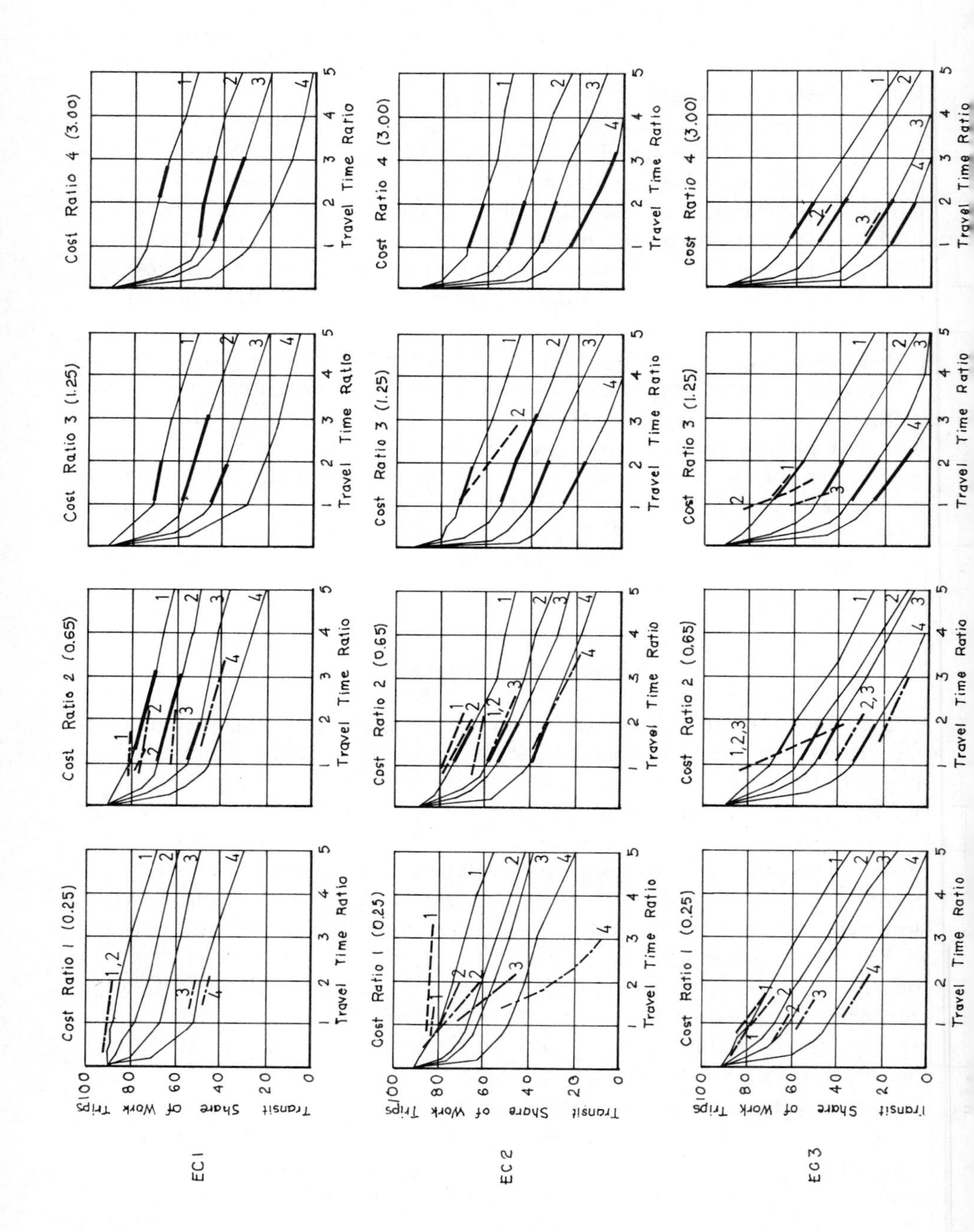

Cost Ratio 1 (0.25)
Cost Ratio 2 (0.65)
Cost Ratio 3 (1.25)
Cost Ratio 4 (3.00)
Travel Time Ratio
Transit Share of Work Trips
EC1
EC2
EC3

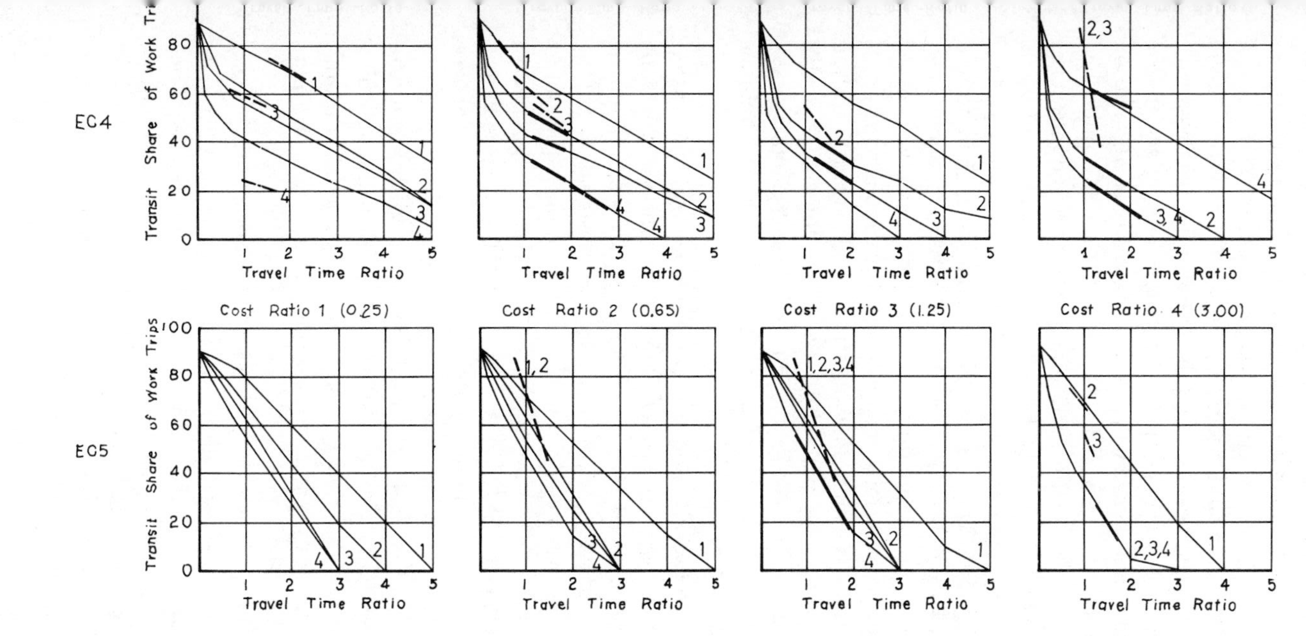

Fig. 9–6. Work trip modal split relationships with adjusted extrapolation (Washington). (Source: Peat, Marwick, Mitchell and Co.)

planner. Not surprisingly, route choice was somewhat subjective, and two different planners often arrived at different assignments under similar conditions. With the development of the need for major urban transportation studies after World War II, it became apparent that a new approach was needed which would provide a substitute for the "hand assignment" methods. Machine-based methods became feasible with the widespread introduction of high-speed electronic computers in the late 1950's. At this time, another development of immediate application to the assignment process occurred. E. F. Moore, in 1957, published an algorithm for the calculation of minimum paths through networks [4]. "Moore's Algorithm," as it has become known, was seen to be well-suited to computer use, and was rapidly adopted by transportation planners for determining minimum time, minimum distance, or minimum cost routings between two points in an urban area. With the continued refinement and development of computer technology, there are other methods of minimum path selection, and programs available which will enable the planner to take account of congestion on the transportation system (*capacity restraint*), and will also permit assignment partially to the freeway system and partially to the surface street system (*diversion*). These two important facets of current assignment techniques are discussed in Sections 9–9 and 9–10.

9–7. The Assignment Network. For the purposes of trip generation and distribution, the study area is broken down into a number of supposedly homogeneous traffic zones (see Fig. 9–7) [5]. In the assignment process, the centers of activity of these zones are designated as the *zone centroids* which are connected into a network representation of all functional elements of the transportation system considered. Figure 9–8 shows a sample network for the city shown in the previous figure. The network shown is a highway network for the assignment of highway trips. All important elements of the street system are incorporated into such a network. Each road is made of a series of *links* which represents a section of road of essentially similar characteristics throughout. The ends of a link are denoted by *nodes,* which must appear at all link intersections. In this way, any particular section of road can be referenced by the nodal points at both ends. For example, the most northerly stretch of road in the study shown is the section of route 312–34. Also indicated on the network are the centers of zonal traffic activity, the zone centroids, which are connected into the network by dashed lines called centroid connectors. These centroid connectors need not necessarily correspond with streets, since they are merely representative of directions of movement within the traffic zone. All links other than centroid connectors correspond to elements of the physical street system. The class of street designated by a link can be segregated by judicious

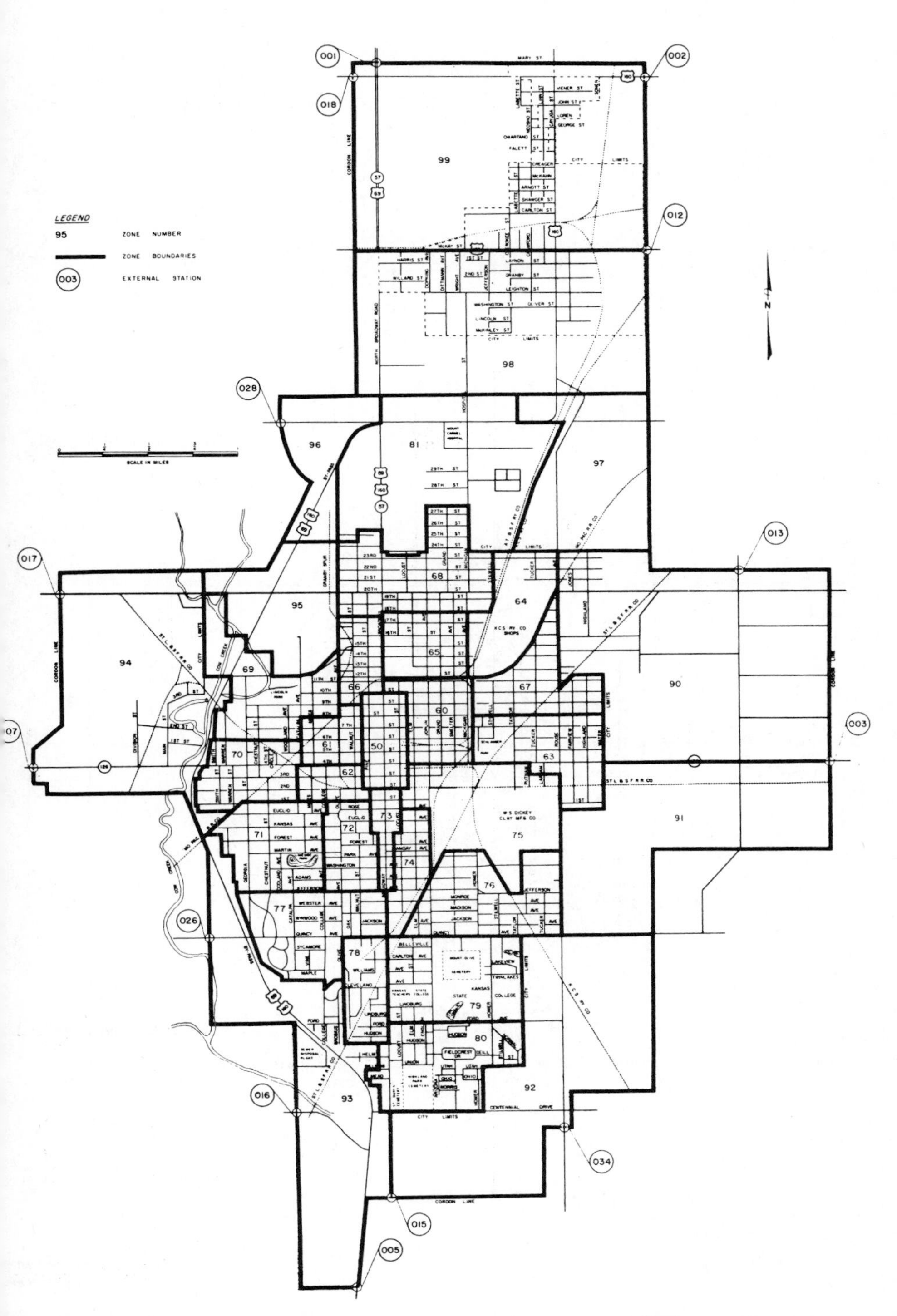

Fig. 9–7. Typical arrangement of traffic zones in a small city. (Source: *Traffic Assignment Manual 1964,* U.S. Department of Commerce, Bureau of Public Roads.)

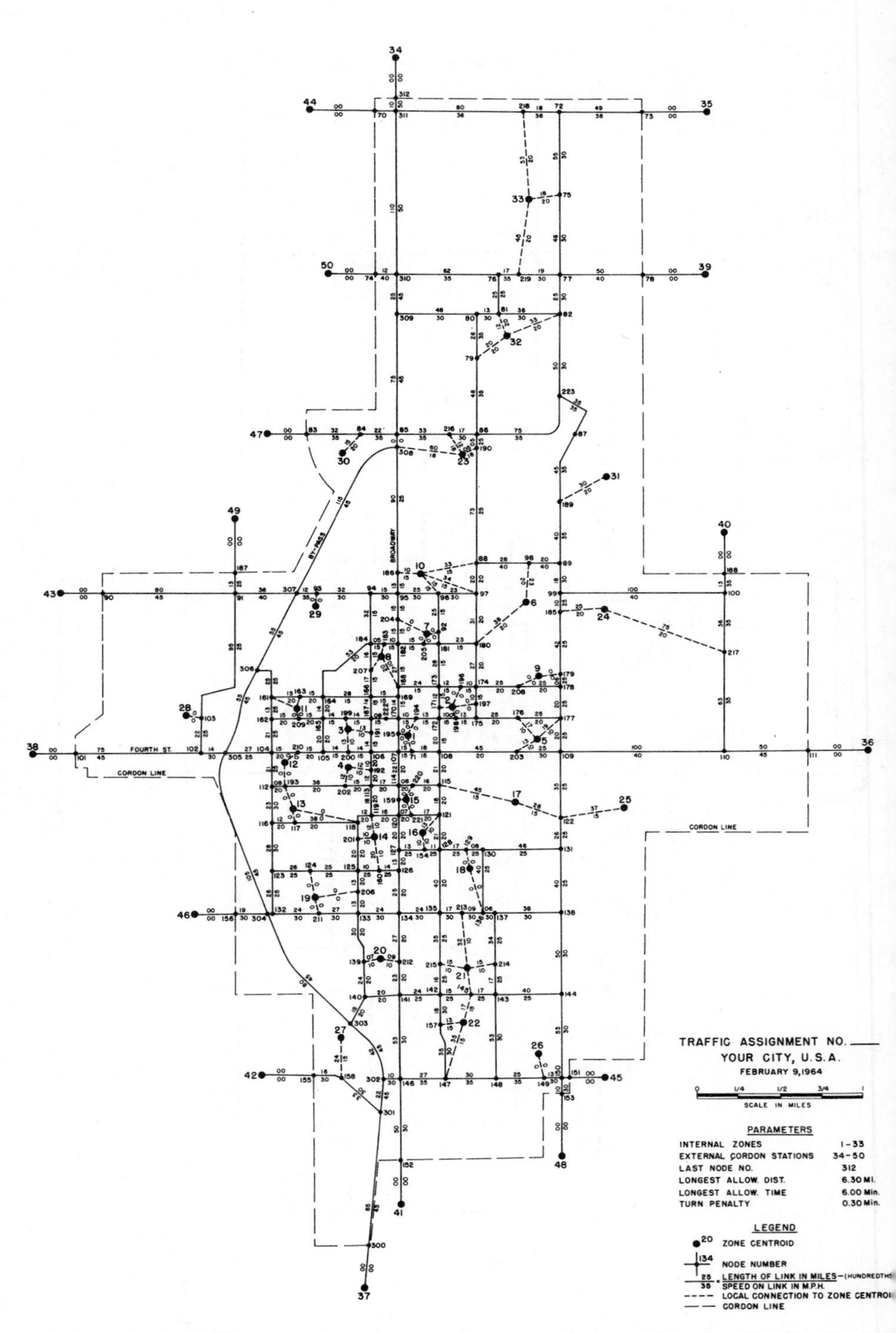

Fig. 9–8. Typical network layout for a small city. (Source: *Traffic Assignment Manual 1964,* U.S. Department of Commerce, Bureau of Public Roads.)

use of the numbering system. For example, if all freeway nodes carry a number about 800, then the planner would know that a freeway link is any link with both nodes in excess of 800. A freeway ramp, on the other hand, would have one node above 800 and one node below 800. An arterial or a collector street would be designated by two nodes below 800. In drawing out and specifying a network, the planner must consider which elements of the street system should be included. Usually, all streets with signalization or carrying a significant amount of traffic must be included. If insufficient roads are included, the process will result in assigned volumes on facilities being higher than those which are realized. Such a network is sometimes referred to as "too gross" in detail. When a network is "too fine" the assigned volumes are less than those observed in reality. Experience in the analyst is required to strike a proper balance.

9–8. The Assignment Process, "All-or-Nothing" Assignment. Using Moore's Algorithm, the minimum paths from each zone centroid to all other zone centroids are calculated by the computer and are stored within memory. These records of minimum time paths are known as "trees" due to their graphical similarity to the plan view of a tree. Figure 9–10 shows a tree for zone 1 in the simplified problem of the four-zone area, shown in Fig. 9–9. The zone-to-zone minimum times, without

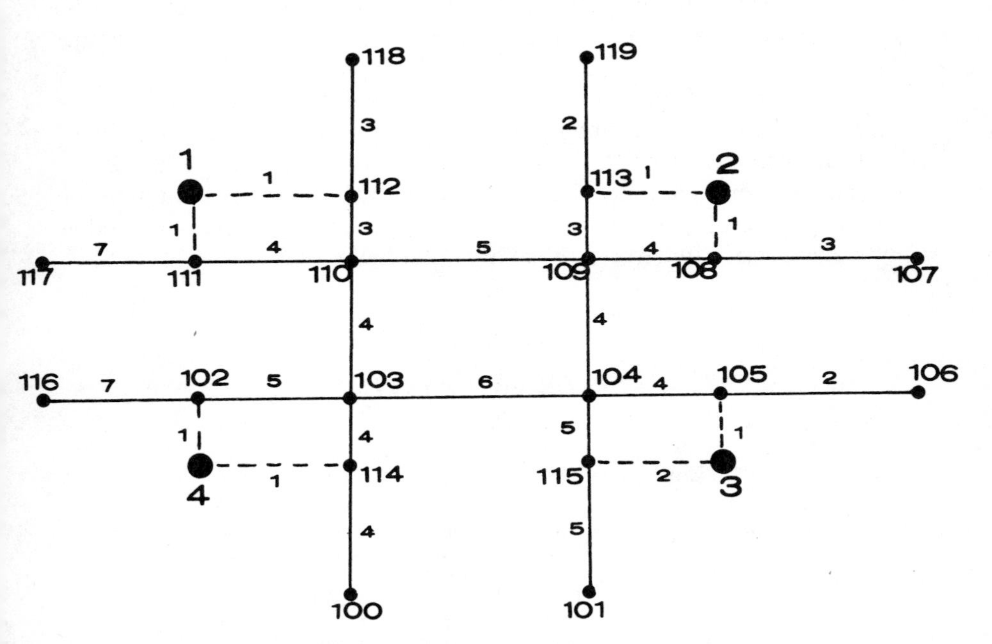

Fig. 9–9. Network for a four-zone area.

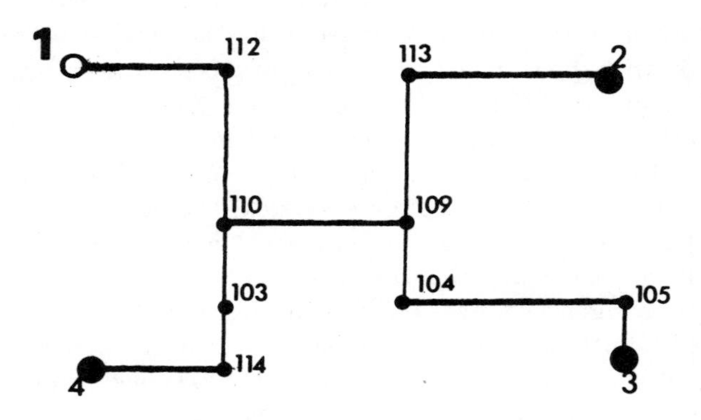

Fig. 9–10. Tree for zone one.

reference to the intermediate nodes, are called "skim trees." For example, the skim tree for zone 1 in the problem can be easily calculated:

1 to 2	13 minutes
1 to 3	18 minutes
1 to 4	13 minutes

Skim trees are required input to the gravity model for the computation of interzonal travel times which are needed for the travel time factors (F_{ij}'s).

Assignment using the "all-or-nothing" assignment method is a simple process. The minimum time path between two zone centroids is assigned the total volume of the trip interchange. The assigned volumes are accumulated for each link, and the total line volume is the accumulation of all individual link volumes for each trip interchange. The methodology is most easily illustrated by a simple example.

TABLE 9-1

Trip Interchanges

From Zone \ To Zone	1	2	3	4
1	–	1000	5000	3000
2	4000	–	3000	4000
3	5000	6000	–	1000
4	2000	1000	3000	–

EXAMPLE 9–1. Assign the trip interchanges shown in Table 9–1 to the network of the four-zone area shown in Fig. 9–9. The times shown in this figure are the travel times along these links, expressed in minutes.

Step 1. Compute minimum path trees (from Moore's Algorithm):

Tree No. 1	*Time*	*Skim Tree*
1–112–110–109–113–2	13 min.	1–2 13 min.
1–112–110–109–104–105–3	18 min.	1–3 18 min.
1–112–110–103–114–4	13 min.	1–4 13 min.
Tree No. 2		
2–113–109–110–112–1	13 min.	2–1 13 min.
2–113–109–104–105–3	13 min.	2–3 13 min.
2–113–109–110–103–114–4	18 min.	2–4 18 min.
Tree No. 3		
3–105–104–109–110–112–1	18 min.	3–1 18 min.
3–105–104–109–113–2	13 min.	3–2 13 min.
3–105–104–103–114–4	16 min.	3–4 16 min.
Tree No. 4		
4–114–103–110–112–1	13 min.	4–1 13 min.
4–114–103–110–109–113–2	18 min.	4–2 18 min.
4–114–103–104–105–3	16 min.	4–3 16 min.

Step 2. A sample tree can be plotted using tree no. 1.

Step 3. Assign zonal interchanges to each link comprising the minimum path for that interchange. The total volume for each link for the assignment is the summation of the individual zonal interchange link volumes.

Table 9–2 shows a breakdown of the individual assignments. The total assigned link volumes for the loaded network are displayed in Fig. 9–11.

All-or-nothing assignment, if used as the sole means of assigning traffic, will arrive at unrealistic traffic volumes. The sole reflection of the physical street system in the network is travel time. There are no constraints placed upon the volume which any individual link can handle, nor is the operating speed related in any way to the ratio of the volume carried to the capacity. Another point overlooked in the assignment method is the preference which some drivers show for arterial streets when travel time is close to that of an alternative freeway routing. In order to obtain reasonable simulations of the traffic volumes caused by trip interchanges, one or both of these points must be considered. They are termed, respectively, *capacity restraint* and *diversion.*

Discussion to this point has centered on the use of minimum time paths as the criteria of route selection. The use of the travel time criterion is now fairly traditional in transportation and is not often questioned. Distance has been found to be a poor criterion, because it discounts the higher service level of operation of the freeway and high type arterial street. Much thought has been given to the use of

TABLE 9-2

Assignment of Interchange Volumes by All-or-Nothing Assignment

Link / Interchange	1 -112	113- 2	117-111	111-110	110-109	109-108	108-107	116-102	102-103	103-104	104-105	105-106	4 -114
1-2	1000	1000			1000								
1-3	5000				5000						5000		
1-4	3000											3000	
2-1	4000	4000			4000								
2-3		3000									3000		
2-4		4000			4000							4000	
3-1	5000				5000						5000		
3-2		6000									6000		
3-4										1000	1000	1000	
4-1	2000											2000	
4-2		1000			1000							1000	
4-3										3000	3000	3000	
	20000	19000			20000					4000	23000	14000	

minimum cost paths where costs include vehicle running costs, accident costs, and the opportunity cost of the driver's time. Such a total cost is probably a better criterion of selection than time; it has not received general acceptance due to the difficulty of accurately measuring these costs. This is especially true of the value of time saved, which can swamp the effect of some of the other costs involved, depending on the assumptions made. It is likely that minimum time will continue to be used for traffic assignment for the forseeable future. Existing computer programs can, however, make assignments by any criterion the planner chooses.

9–9. Diversion. In the context of transportation models, diversion refers to the allocation of a trip interchange to two possible routes in

TABLE 9–2

(Continued)

115– 3	4 –102	111– 1	100–114	114–103	103–110	110–112	112–118	101–115	115–104	104–109	109–113	113–119	3 –105	108– 2	
						1000					1000				1000
						5000				5000			5000		5000
				3000	3000	3000									3000
						4000					4000				4000
										3000	3000		3000		3000
				4000	4000						4000				4000
						5000				5000			5000		5000
										6000	6000		6000		6000
				1000									1000		1000
				2000	2000	2000									2000
				1000	1000						1000				1000
				3000									3000		3000
				14000	10000	20000				19000			23000		

a designated proportion which depends on some specified criterion. Usually the two possible routes are the fastest all-surface route and the fastest route comprised partially or totally of expressway links. Experience has shown that there are those who will travel a little longer both in time and distance in order to travel in the less congested atmosphere of a freeway. Equally, even though a freeway route offers time and distance savings, there is a proportion of drivers, often the elderly, who prefer to use surface routes. Recognition of this phenomenon has produced several methods of diversion which have been used or are in current usage.

1. California Diversion Curves. The percentage of freeway usage is related to savings in both distance and time in this method diver-

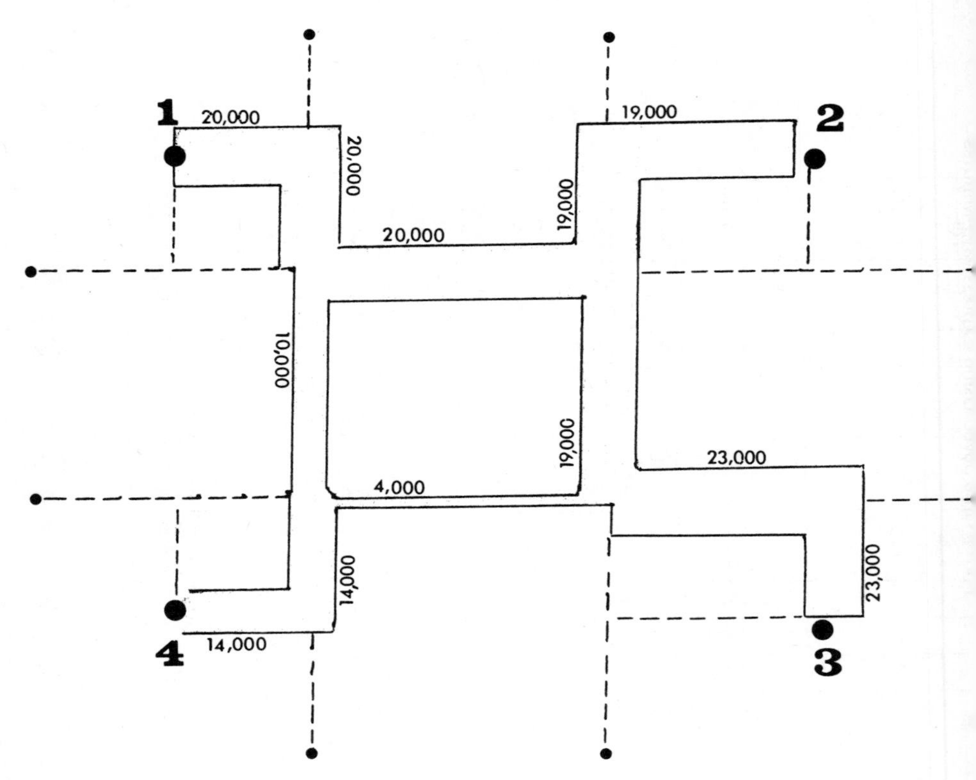

Fig. 9–11. Network flows resulting from an all-or-nothing assignment.

sion. The use of two parameters, distance and time, as criteria for freeway use is represented by the equation:

$$P = 50 + 50(d + \tfrac{1}{2}t)[(d - \tfrac{1}{2}t)^2 + 4.5]^{-1/2} \tag{9–1}$$

where:
- P = per cent of trips via freeway
- d = distance saved in miles via the freeway
- t = time saved in minutes via the freeway

This equation is the basis of the family of hyperbolic curves shown in Fig. 9–12.

2. Detroit Diversion Curves. The Detroit Area Transportation Study estimated diversion from a somewhat different viewpoint, still using a two-parameter approach (see Fig. 9–13). In this case, the parameters found to be related to freeway usage were the ratio of

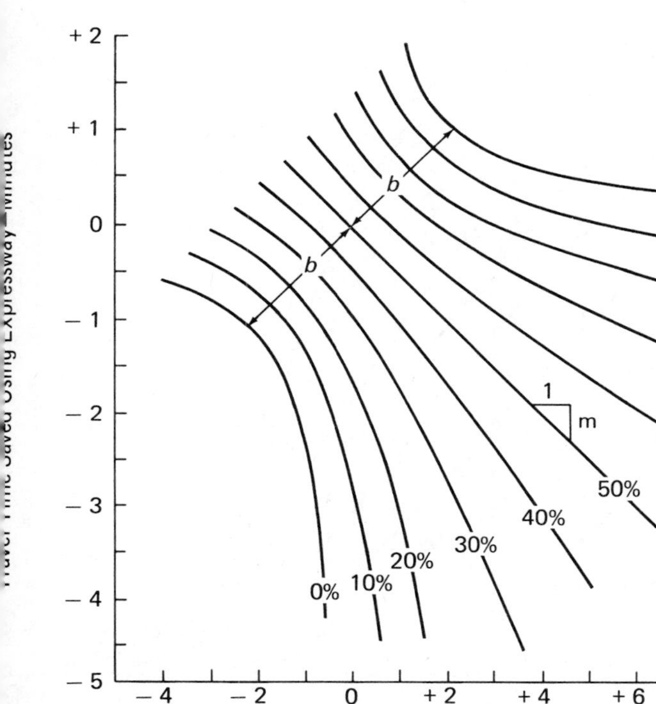

Fig. 9–12. California diversion curves. (Source: Martin, Memmott, and Bone, *Principles and Techniques of Predicting Future Demand for Urban Area Transportation,* M.I.T. Press.)

expressway speed to arterial speed, and the ratio of expressway distance to arterial distance. In each case the minimum applicable path was used for the ratio computations.

It should be stated here that the two methods do not give similar answers. Examination of the curves displayed in Fig. 9–14 indicates that for equal distances, the California method diverts less for longer trips and tends to divert more traffic in the case of short trips.

3. Bureau of Public Roads Diversion Curve. Undoubtedly, the most widely used method of diversion is that which is available in the Bureau of Public Road's series of traffic planning computer programs. This form of diversion is dependent on one parameter only, the ratio of travel times by the quickest combined arterial-freeway route to the quickest arterial-only route. With a one-parameter relationship, one single diversion curve defines the relationship; this

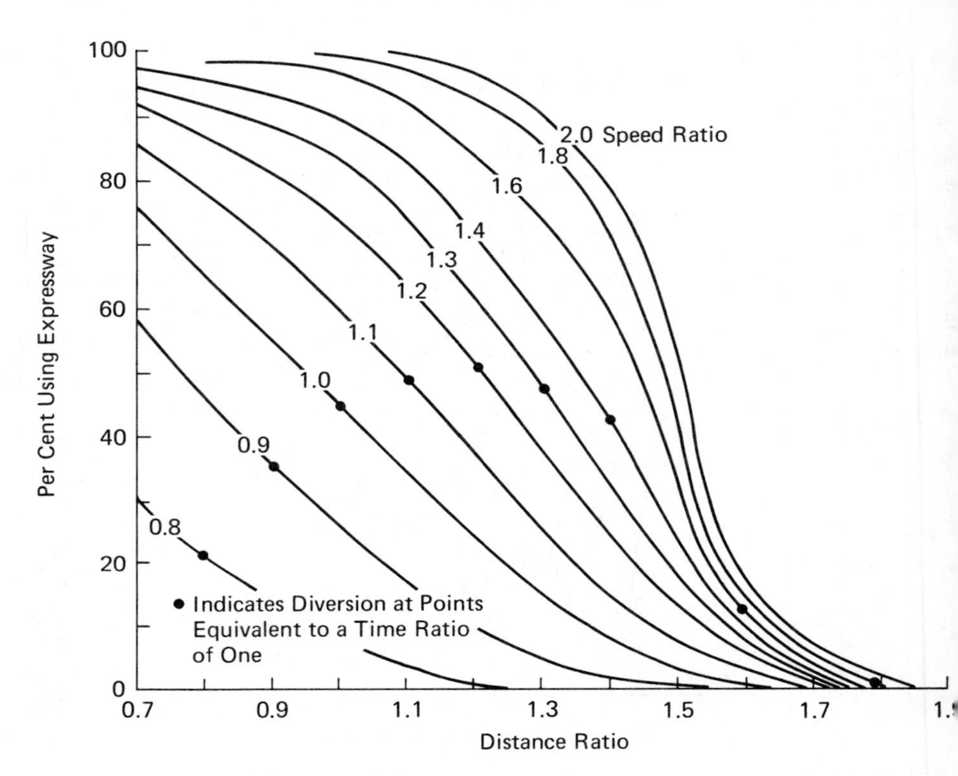

Fig. 9–13. Detroit area transportation study diversion curves. (Source: Martin, Memmott, and Bone, *Principles and Techniques of Predicting Future Demands for Urban Area Transportation,* M.I.T. Press.)

is displayed in Fig. 9–15. The form of the S-shaped diversion curve is similar to those used in the Detroit Study for higher speed ratios. Total freeway usage occurs when the travel time ratios fall below 48 per cent while no freeway usage can be anticipated when the travel time ratio exceeds 150 per cent of the quickest surface route.

9–10. Capacity Restraint. The failure of all-or-nothing assignment to reflect the traffic carrying capability of the network elements results in unrealistic assignments. The assigned volumes on some links are likely to be well in excess of their capacities while parallel links carry no traffic, due to an initial assumption of some slight speed differential. More realistic assignments are obtained by relating the travel time on a facility to the ratio of the volume on that facility to its capacity. At low volume/capacity ratios, traffic travels at a "free" speed on any facil-

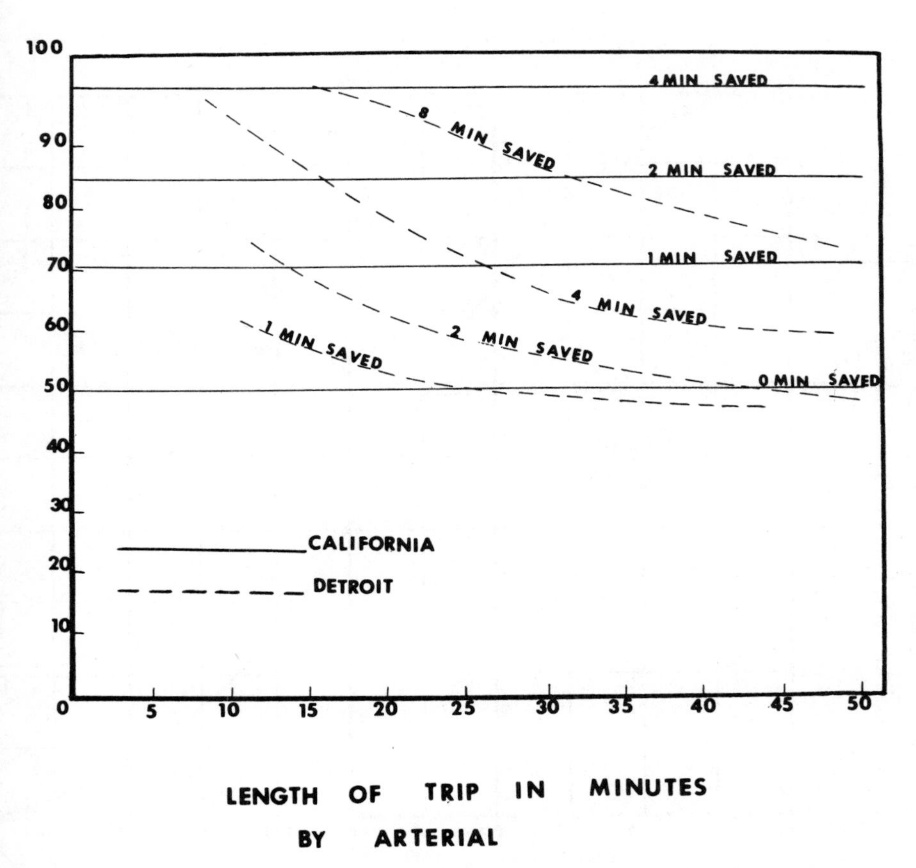

Fig. 9–14. Comparison of the California and Detroit diversion curves.

ity. As volume increases, so does the volume/capacity ratio and an increasing restraint is placed on the driver's freedom of action with a resulting decrease in speed (see Fig. 9–16). At a volume/capacity ratio of 1.0, the flow is in an unstable condition. Speed may drop to zero and volume declines rapidly, as shown in the level of service *F* in Fig. 9–17. This condition is the familiar "traffic jam." The following sections will discuss three methods of capacity restraint; which take into account observed variations of operating speed with differing volume/capacity ratios.

9–11. Bureau of Public Roads Capacity Restraint Methods. The S-shaped curve shown in Fig. 9–18 [5] shows the relationship between

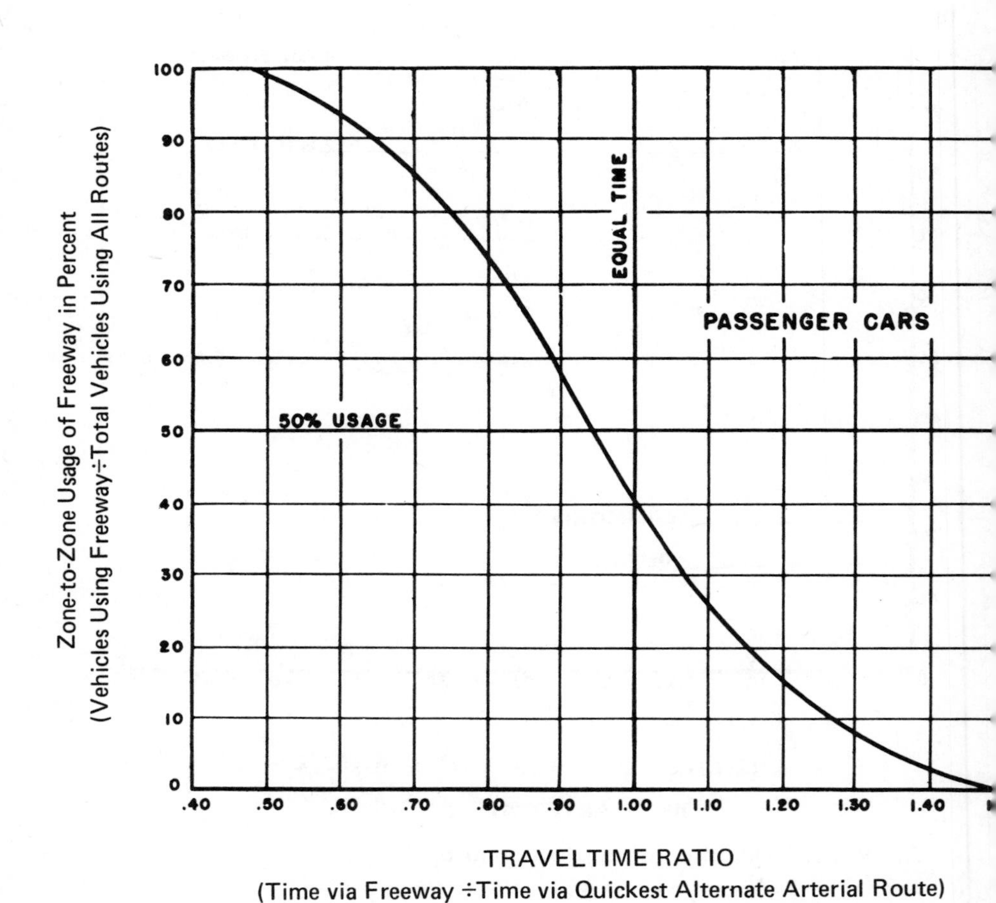

Fig. 9–15. Bureau of public roads diversion curve. (Source: *Traffic Assignment Manual 1964,* U.S. Department of Commerce, Bureau of Public Roads.)

the speed ratio and volume ratio that is used in the Bureau method for computing the effect of volume level upon operating speeds. When converted to an equation expressed in terms of travel time, the relationship is of the form:

$$T = T_0\left[1 + 0.15\left(\frac{\text{Assigned Volume}}{\text{Practical Capacity}}\right)^4\right] \qquad (9\text{–}2)$$

where:

T = Traveltime at which assigned volume can travel on the subject link

T_0 = Base traveltime at zero volume

It can be seen that even at very low volumes, the equation assumes that speeds will be no greater than 15 per cent above speed at capacity. At high volume/capacity ratios, the speed falls asymptotically to zero.

Assignment using capacity restraint is an iterative process when using the Bureau method. A typical process flow diagram is shown in Fig. 9–19. A first assignment is made on assumed or ideal travel times. This is known as an "unrestrained" assignment. Following an assignment, new travel times are computed for each link based on the assigned volume/capacity ratio for that link. These recomputed travel times are used for the calculation of new minimum time trees and the traffic is again assigned. The process is reiterated as many times as the analyst desires. The accepted assigned volume on a link is the average value of all assignments. In practice, four iterations are usually sufficient for the system to settle down to the point that average volumes on each link are virtually constant. Large oscillations in link volumes

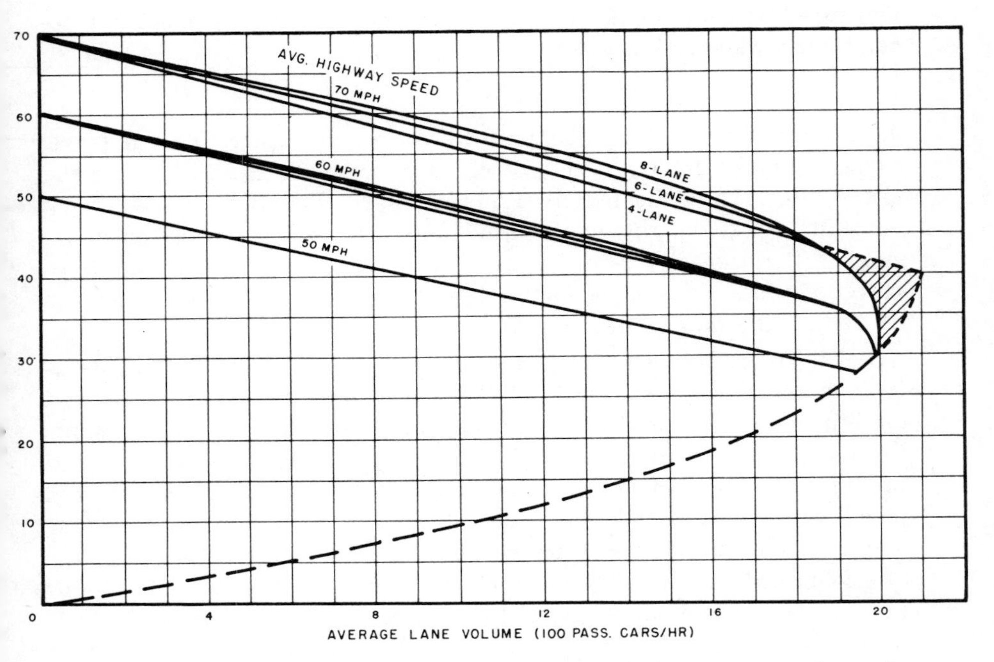

Fig. 9–16. Relationship of operating speed to volume to capacity ratio. (Source: *Highway Capacity Manual*, Highway Research Board.)

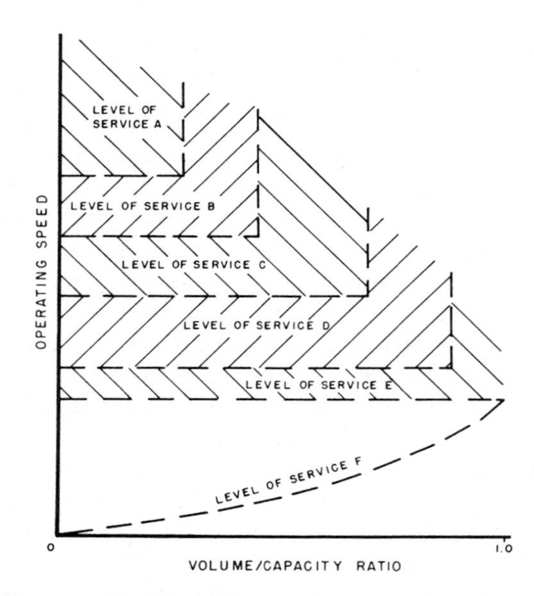

Fig. 9–17. General concept of relationships of levels of service to operating speed and volume capacity ratio. (Source: *Highway Capacity Manual,* Highway Research Board.)

have been eliminated in the past by decreasing the change in travel times computed for each link, after an assignment, to one quarter the change in value indicated by Equation 9–2. The Bureau assignment program can be operated in conjunction with a diversion option if desired. Since capacity restraint tends to reproduce realistically observed travel patterns, this is seldom necessary, but is used where diversion of traffic is known to take place.

9–12. Traffic Research Corporation Method. A different type of iterative approach was developed by the Traffic Research Corporation and used in its study for Toronto, Canada. The TRC method attempts to overcome one obvious theoretical failing of the methodology outlined in Fig. 9–12. It can be seen that the trip generation and distribution blocks are independent of the traffic assignment process. However, true travel times between zones are not known until congested travel times are computed in the assignment block. The distribution of trip ends must be based on assumed travel time where the distribution and assignment blocks are independent. The validity of such a distribution is obviously questionable. In an attempt to overcome this illogical independence, the TRC method followed an eleven-step sequence where out-

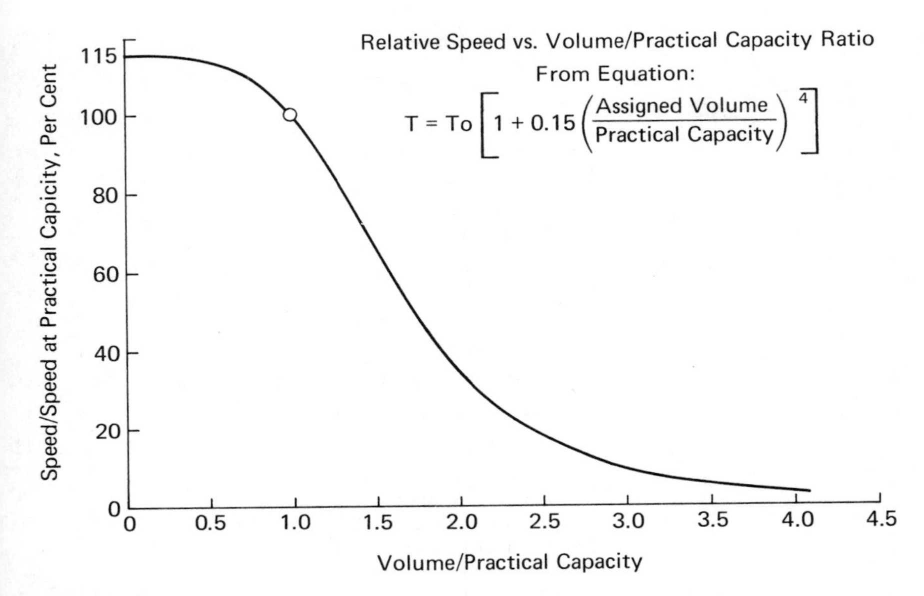

Fig. 9–18. Capacity restraint relationship. (Source: *Traffic Assignment Manual, 1964,* U.S. Department of Commerce, Bureau of Public Roads.)

put of congested travel times was recycled back as input to the distribution block.

Figure 9–20 is a process flow diagram of this method of capacity restraint, described in the following steps:

1. Trip generation phase is carried out using land use and socio-economic data. This step is never repeated.
2. Ideal or unrestrained routes are calculated by tree-building programs based on ideal travel times. These travel times assume uncongested conditions.
3. An initial trip distribution is carried out based on uncongested travel times.
4. The trip interchanges computed in Step 3 are assigned to minimum paths computed from ideal travel times.
5. Based on assigned volume/capacity ratios, new congested travel times are computed.
6. A new set of minimum time paths is computed from congested travel times.
7. Another assignment is carried out, with vehicles divided between available routes inversely proportionate to travel times.

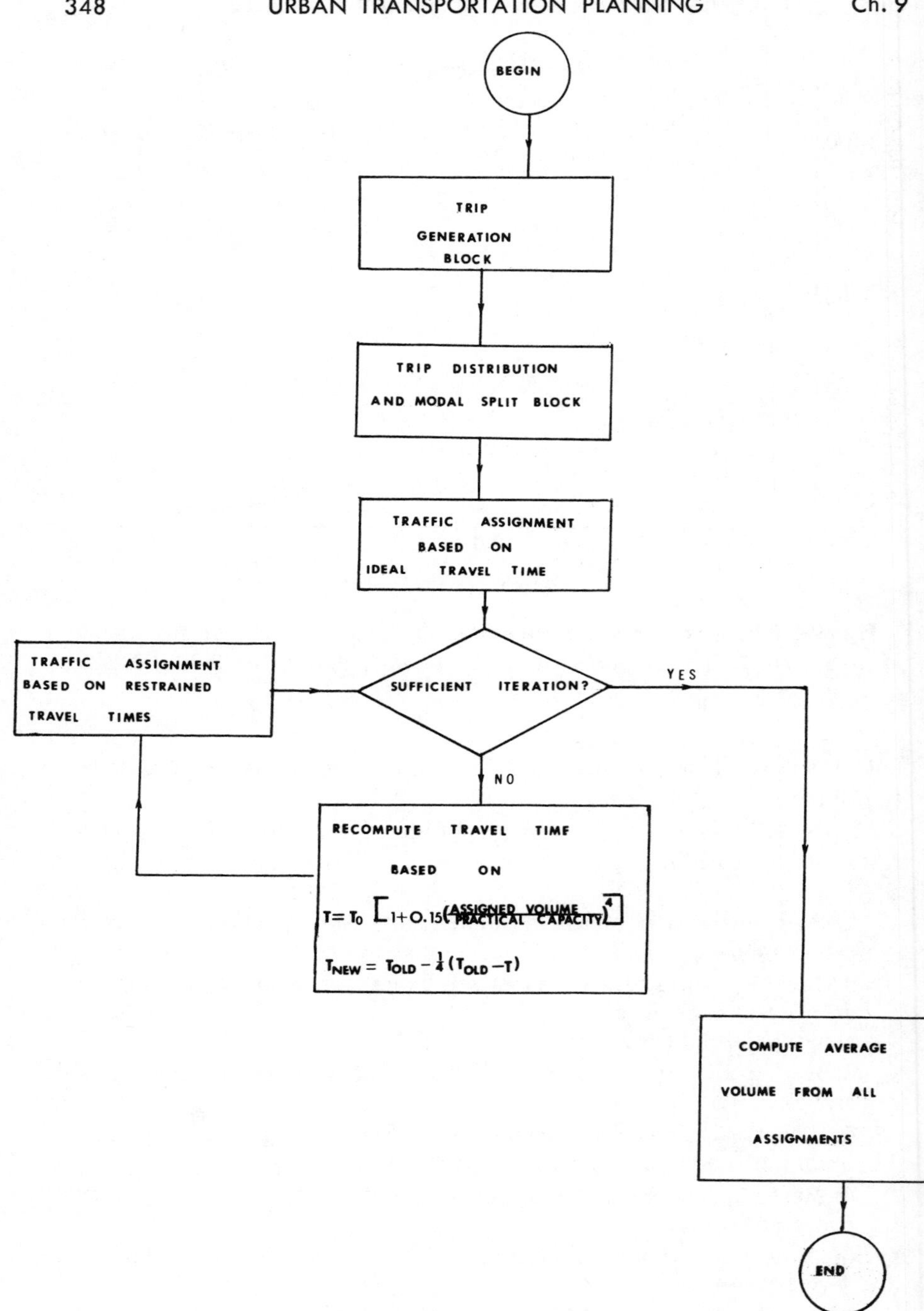

Fig. 9–19. Bureau of Public Roads capacity restraint method.

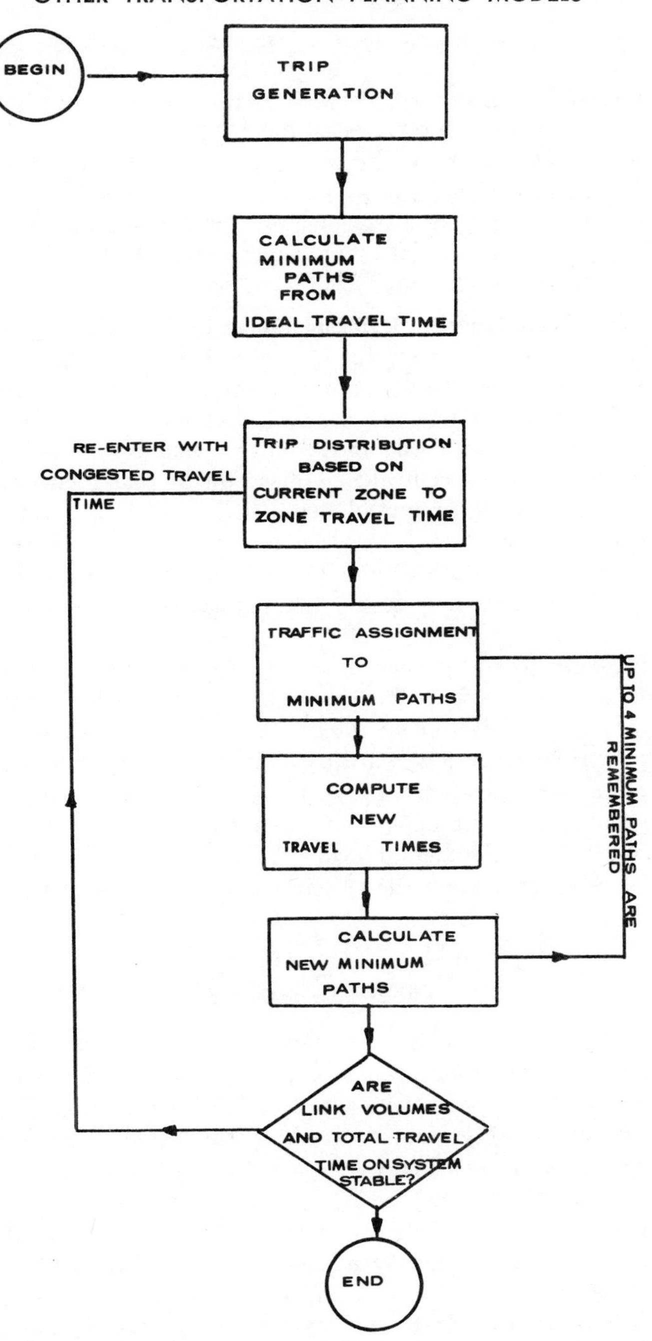

Fig. 9–20. Traffic Research Corporation capacity restraint method.

8. New travel time calculations are made on the basis of the last assignment.

9. Steps 6, 7, 8 are iterated twice more to produce up to four minimum paths which are retained in memory.

10. A new distribution is carried out using congested travel times.

11. Steps 7 and 8 are reiterated until the system settles down.

Criteria for convergence are constant link loading and constant total system vehicle-minutes of travel time. Steps 10 and 11 may require iteration to obtain convergence.

9–13. The CATS Capacity Restraint Method. The Chicago Study used an approach to capacity restraint basically different from the two iterative procedures discussed in preceding sections. In the assignment process, the first trip interchange was selected at random. Using initial uncongested travel times, the interchange volume was assigned to the minimum path between centroids. Based on the assigned volumes, new link travel times were computed for each loaded link according to the volume/capacity ratio. Another trip interchange was selected at random, and assigned to its minimum path using updated travel times. After this assignment, new travel times were computed for the loaded links.

The procedure was continued until all trip interchanges were assigned, with adjustment of network-link travel times after each assignment. By the use of random selection of interchanges, the network was gradually loaded with traffic with downward adjustment of travel times throughout as congestion in the system increased. Since the constraints of the system were considered throughout one assignment of all interchanges, no reiteration of the assignment block was necessary. This one-step procedure differs sharply from the two preceding methods which have been outlined.

LAND USE MODELS

In its correct sequence in the modeling process, land use modeling would follow directly upon projection of population and economic activity. However, the types of models that have been developed for land use activity are closely related to those that have been described in some depth in the areas of trip generation and distribution. Therefore, the following sections have been placed out of sequence so that the student will already be familiar with the underlying theories of the model forms used in land use activity simulation. The models discussed in this section have been used in various transportation studies. Often they are referred to as first-generation land use models, since some of their shortcomings have been recognized and more sophisticated models are now under development and testing. These later or second-generation models generally tend to be theoretical refinements of the four types of models presented in this section:

1. Multiple Regression Model
2. Accessibility Model
3. Schneider Opportunity Model
4. Density Saturation Gradient Model

9–14. Multiple Regression Land Use Model. Using standard multiple regression techniques, the analyst can form equations which relate growth in a study area zone to the characteristics of that zone at the beginning of the growth period. The characteristics of the zone are used as independent variables for forecasting growth, the dependent variable. Model development is easily carried out using available computer programs such as the UCLA Bio-Medical Center program, BJMD 34.

The method takes advantage of the ability of the multiple regression techniques to take advantage of the influence of a large number of independent variables upon a single dependent variable. It would appear that such a complex activity as the development of land is likely to involve a large number of influential variables. Regression is one of the few modeling techniques that is able to account for the influence of many variables. Typically, the form of the multiple regression model can be stated:

$$Y = A_0 + A_1X_1 + A_2X_2 + A_3X_3 + \cdots + A_nX_n \qquad (9\text{–}3)$$

As already discussed in Section 8–7, it is important for model stability that the independent variables should be independent of each other, avoiding the error of co-linearity. Variables used in the past for the prediction of land use growth include:

Accessibility to employment
Percentage of available vacant land
Land value
Intensity of land use
Measures of zone size
Amount of land in different uses
Net density of development in the base year
Land employment
Time and distance to highest valued land of study area
Degree of zoning protection expressed on a quantitative scale
Transit accessibility
Quality of water and sewer service

Some of the dependent variables which have been predicted include:

Increase in dwelling units
Increase in dwelling units per unit of available land

The independent variables are used to form linear equations. Where linear relationships are unsuitable, the analyst from an *a priori* knowledge of the form of relationship would transform the variables to enable linear analysis. Typically, the equation would have a form similar to that found suitable for Greensboro, N.C. [7]:

$$Y = -2.3 + 0.061X_1 + 0.00066X_2 + 1.1X_3 - 0.11X_4 - 0.0073X_5 \tag{9-4}$$

where:

Y = logarithm of growth in dwelling units over a planning period
X_1 = degree of zoning protection in base year
X_2 = per cent of total land in residential use in base year
X_3 = logarithm of accessibility of employment
X_4 = dwelling unit density in the base year
X_5 = per cent of land in use devoted to industrial use in base year

This equation was found to be significant at the 5 per cent level, and the correlation coefficient was 0.61. It can be seen that for the development of a model of this form the analyst must have dwelling unit data at two periods in time in order to compute variables as growth rate.

The fairly low correlation coefficient is typical of most land use models of this type, to date. The rather low value of the correlation coefficient indicates that only 37 per cent of the total variation is explained by the model. More sophisticated techniques such as the Empiric Model, a simultaneous multiple regression model, have been found to have correlation coefficients in excess of 0.9.

9–15. Accessibility Land Use Model. In generalized form the accessibility model bears a close resemblance to the gravity model previously discussed under trip generation. It may be formulated as follows:

$$G_i = \sum_{\text{all } i} G_i \cdot \frac{A_i^n V_i}{\sum_{\text{all } i} A_i^n V_i} \tag{9-5}$$

where:

G_i = the forecast growth for zone i
A_i = accessibility index for zone i
V_i = vacant, available land in zone i
n = some constant to be determined

The *accessibility index* referred to in the above equation is a derivative of the gravity model in the trip distribution phase, with slight modification. Using employment per zone as the measure of activity in a zone, the *accessibility index* (in this case with reference to employment), is defined as:

$$A_i = \sum_{\text{all } j} E_j F_{ij}$$

where:

A_i = accessibility index of zone i
E_j = total employment within zone j
F_{ij} = friction factor for travel time between zone i and zone j as computed from gravity model calibration

The model is calibrated by the use of data at two periods of time to provide value of G_i for each zone. The value of the exponent n is determined by a "best fit" to the relationship in Equation 9–5. In work for Washington, D.C. [8], it was determined that the value of the exponent should be 2.7.

9–16. Schneider Opportunity Land Use Model. The Schneider formulation of the intervening opportunity model discussed in an earlier section on trip distribution has also been used in distributing growth of land use activity throughout a study area. As applied, for example, to residential development, the formulation would be of the form of the differential equation:

$$d(G_j) = g_t[e^{-l0} - e^{-l(0+0_j)}] \qquad (9\text{–}6)$$

where:

G_j = total number of units located from central point to district under consideration j
g_t = total growth to be allocated
l = probability that an opportunity will be accepted for location
0 = total number of opportunities aggregated from central point up to but not including the district under consideration
0_j = number of opportunities within district under consideration

By integration, Equation 9–6 becomes:

$$G_j = g_t[1 - e^{-l0}]$$

By subtracting g_t from both sides and expressing in logarithm form, the following equation is obtained:

$$ln\,(g_t - G_j) = ln\, g_t - l0 \qquad (9\text{–}7)$$

The left-hand side of the equation is seen to be the logarithm of the dwellings to be located. The right-hand side has a slope of $(-l)$ and there is an intercept of $(ln\ g_t)$. This relationship provides an indication of a method of calibrating the model. If a graph on semilogarithmic scale is plotted with dwelling units remaining to be located by the time district j has been reached in the time sequence versus accumulated opportunities for location up to district j, a straight line relationship is obtained. The slope of the line of best fit to the plotted points is taken as the l value for the opportunity model. For calibration, the analyst must have data from two points in time. Figure 9–21 shows the calibration of a Schneider model for Greensboro, N.C. carried out by Swerdloff and Stowers. It is notable that the land use model requires two values of l for calibration, similar to the need for two l values in the trip distribution mode. Values found to fit the Greensboro data

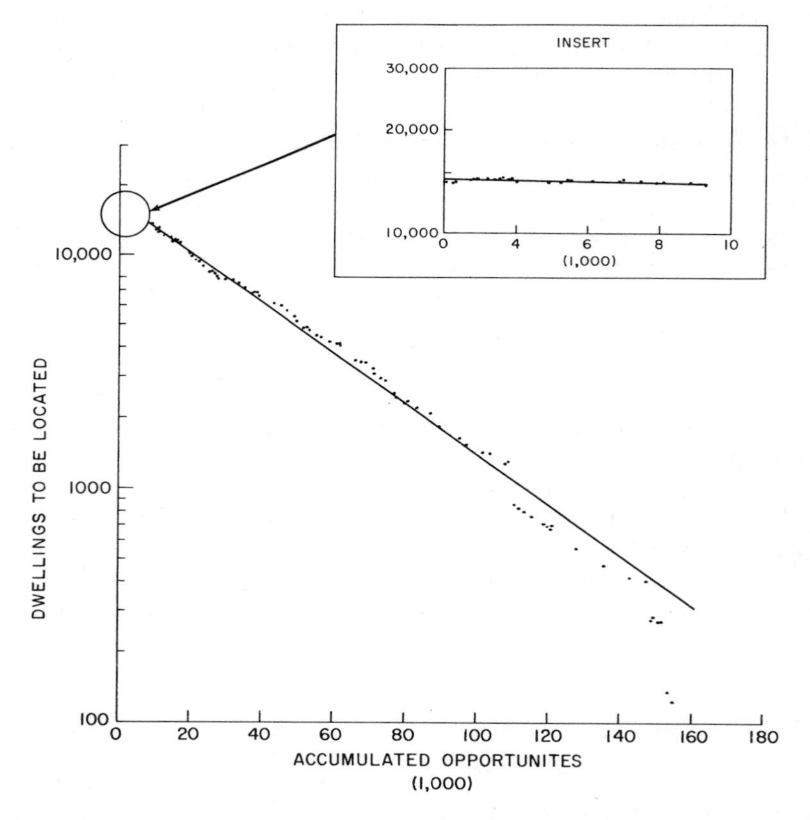

Fig. 9–21. Calibration curve for the Schneider Model. (Source: *Highway Research Record 126,* Highway Research Board.)

were 1.707×10^{-6} for the central city area and 10.9×10^{-6} for the outer area of the city.

9–7. Density Saturation Gradient Land Use Model. The Chicago Area Transportation Study made use of a more subjective method of predicting future land use. For comparison with the previously described residential land use models, the same land use type will be discussed in this section. In a step-by-step procedure, the method can be described as follows:

1. Determine the relationship between net residential density and distance from the CBD. The relationship for Chicago data is shown for two points in time in Fig. 9–22.

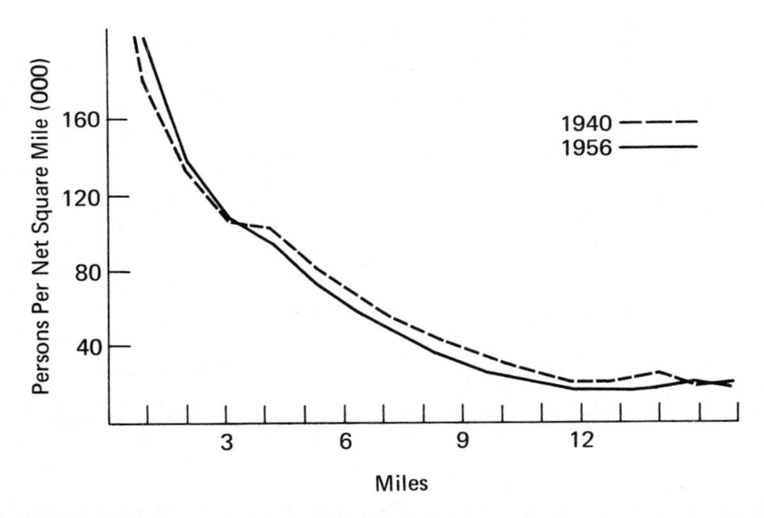

Fig. 9–22. 1940 and 1956 net residential density in Chicago by distance from the Loop. (Source: *Chicago Area Transportation Study,* Vol. II.)

2. Determine the percentage of developable land that has been developed for each district. This percentage, known as the percentage residential saturation, is an indicator of the degree of possible development that has taken place. Figure 9–16 shows the percentage residential saturation for the Chicago area plotted against distance from the CBD.
3. Project the relationship between net residential density and distance from the CBD in the forecast year. Unlike the previously discussed land use models, there is no "calibration" phase. Both net residential density and residential saturation must be computed for the design year with subjective decisions on the part of the

analyst. In the case of the Chicago Study, it was felt that essentially similar densities to those obtained in 1956 would fit the conditions of the design year. It is unlikely that net residential densities will change substantially within any developed portion of an urban area unless widespread redevelopment is envisaged.

4. Next, the planner must project the percentage residential saturation. Figure 9–23 shows the assumed relationship between residen-

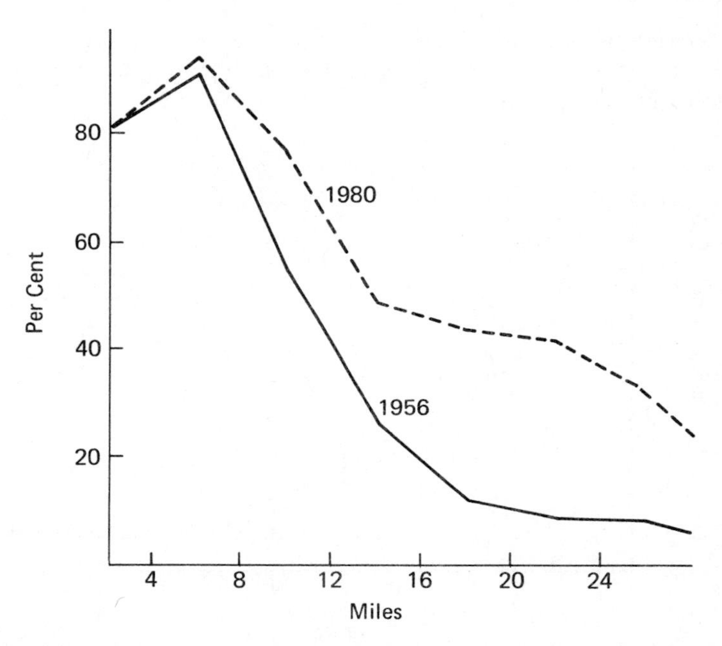

Fig. 9–23. 1956 and estimated 1980 usage of designated manufacturing land by distance from the Loop. (Source: *Chicago Area Transportation Study*, Vol. II.)

tial saturation and distance from the CBD. The projection of the form of this curve is the most subjective element in this type of model. The only mathematical constraint put upon the planner is that the study area multiplied by the ordinates of both the residential saturation curve and the net residential density curve must sum to the total projected dwelling units within the study area. However, a rational assumption of the form of residential saturation curve is maximum increase within the median distances, lower increase in the peripheral areas, and virtually no increase, perhaps even a decrease, in the central areas. Figure 9–24 indicates an assumed curve for Greensboro by Swerdloff and Stowers.

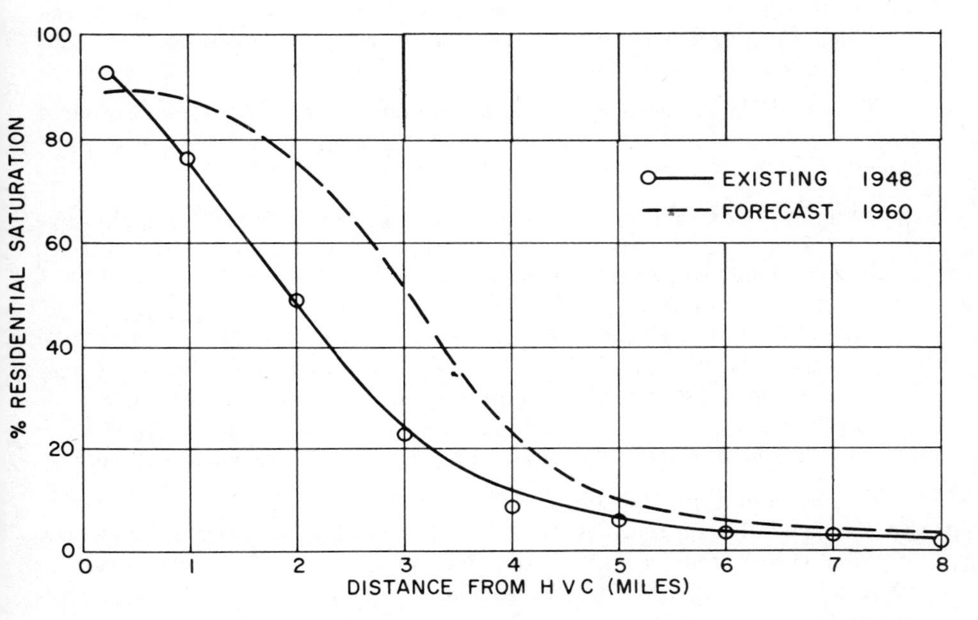

Fig. 9–24. Residential saturation by distance from the highest value corner of the C.B.D. (Source: *Highway Research Record No. 126,* Highway Research Board.)

5. Having determined the added residential units by distance or ring from the CBD, the increments must be assigned to individual districts. These assignments can be made by a subjective weighting of the zone's attractiveness according to such factors as:

Distance to convenience shopping
Available residential capacity
Distance to the major street system
Percentage of industrial development in the zone
Percentage of residential development in the zone

9–18. Modeling in Perspective, the Systems Approach [9]. The urban transportation problem is complex. In the last decade it has become apparent that only a systems approach can effectively tackle a problem of such involvement. The systems approach is a decision-making process for complex problem solving composed of:

Systems analysis—a clear evaluation of the combination of all elements which structures the problem, and those forces and strategies needed for the achievement of an objective.

Systems engineering—organizing and scheduling the complex strategies for problem solution, and the development of procedures for effecting alternate solutions.

The modeling process is an important portion of a systems approach to the solution of the urban problem. In general, we may say that systems analysis has three major characteristics:

1. The team tackling the problem is interdisciplinary in make-up. All facets of the problem are considered, not simply those facets which conveniently fall within one discipline.
2. Throughout the analysis, the team uses scientific method. This requires that a theory must be formulated to account for a set of observed facts. This theory is checked to determine whether it actually explains known facts and is also checked to determine its predictive validity.
3. The work is carried out according to a predetermined sequence.

The modeling process discussed at length in this chapter conforms to the general requirements of theory formulation in systems analysis. System theory requires that the effect on a system, E, is the result of interaction of both the independent system variable X_i and the dependent variables Y. In equation form, this is denoted:

$$E = f(X_i, Y_j) \tag{9–8}$$

Optimization of the system is achieved by adjustment of the variables, within the systems model, representing alternative approaches for arriving at a final design which provides optimum performance with respect to stated criteria. System optimization is brought about by continual feedback of information from later models to produce refined input to the earlier models. At the subsystem level, the requirement for feedback is clearly brought out in the interdependence of the trip distribution and traffic assignment models as recognized by the Traffic Research Corporation Capacity Restraint Model. A macroscopic view of the system with feedback and re-evaluation of the inputs at the various levels of analysis is shown in Fig. 9–25. The system is optimized by continual adjustment from feedback information.

The modeling process set out here for study by the reader is not without faults, as some eminent authorities have properly pointed out [10]. The prediction of traffic volumes on alternate systems is effected by a string of individual models as shown in Fig. 9–26. Each model has an associated degree of error, the residual error, amounting to the difference between the variance of the observed data and the variance explained by the model. For linear models the amount of variation explained is described by the *coefficient of multiple determination* (R^2).

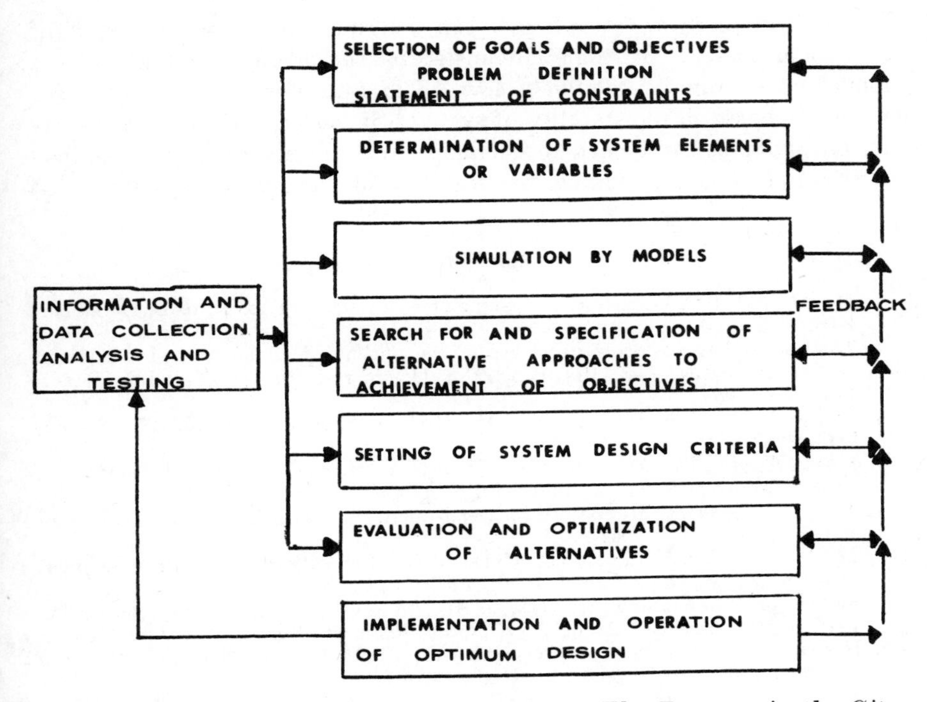

Fig. 9–25. The systems approach. (Source: *The Freeway in the City*, Federal Highway Administration.)

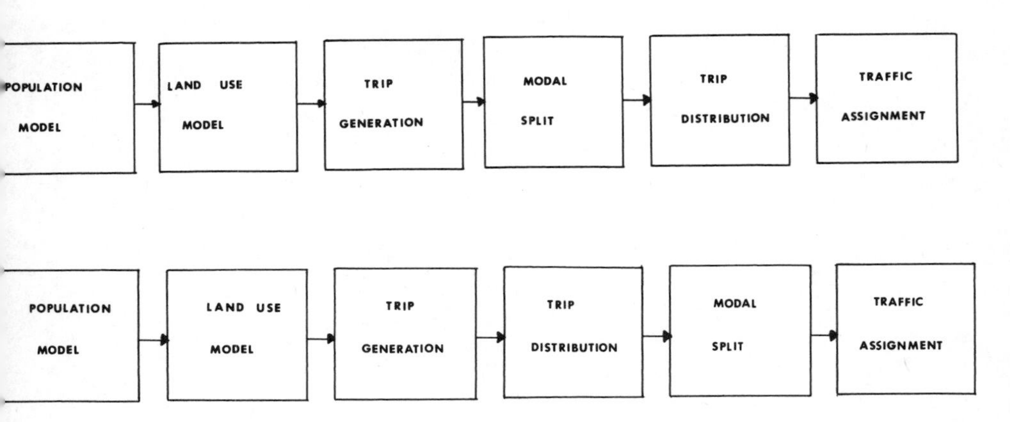

Fig. 9–26. Model chains in transportation planning.

Using error analysis, the planner can determine the effect of chains of regression models. In some circumstances, the overall reliability of the model chain can be relatively low. The future will undoubtedly see more emphasis on the stability of systems of models. Current transportation planning has tended to overemphasize the importance of the inddividual model, overlooking the stability of the overall system.

REFERENCES

1. *Modal Split,* U.S. Department of Commerce, Bureau of Public Roads, December, 1966, p. 3.
2. The author has made extensive use of the outline of modal analysis contained in "Modal Split," U.S. Department of Commerce, Bureau of Public Roads, December 1966.
3. Highway Research Bulletin No. 16, Highway Research Board, Washington, D.C., 1952.
4. MOORE, E. F., *The Shortest Path Through a Maze,* International Symposium on the Theory of Switching Proceedings, Harvard University, 1957.
5. *Traffic Assignment Manual,* U.S. Department of Commerce, U.S. Bureau of Public Roads, June, 1964.
6. IRWIN, N. A., DODD, N., and VON CUBE, H. G., "Capacity Restraint in Assignment Programs," *Highway Research Board Bulletin 297,* 1961.
7. SWERDLOFF, C. N., and STOWERS, J. R., "A Test of Some First Generation Residential Land Use Models," *Highway Research Record No. 126,* 1966.
8. HANSEN, W. G., "Land Use Forecasting for Transportation Planning," *Highway Research Bulletin 253,* 1960.
9. *The Freeway in the City,* U.S. Department of Transportation, 1968.
10. ALONSO, W., "The Quality of Data and the Choice and Design of Predictive Models," *Highway Research Board Special Report 97,* 1968.

III

DESIGN OF LAND TRANSPORTATION FACILITIES

10

Design of Streets, Highways, and Railways: Surveys and Route Layout

In relocating old land transportation facilities or establishing new ones, the field work and the layout of routes are basically the same for all systems. Surveys are needed to be able to prepare plans for the project and to estimate the cost. These surveys for a given facility may include (1) a reconnaissance, (2) a preliminary survey, and (3) a final location survey. These are discussed in the following sections.

10–1. Available Techniques. The determination of the alignment may be done either by conventional ground survey methods, or by aerial photogrammetry. While the older techniques are still used on projects of medium and small size, the use of photogrammetric surveys on large works is almost universal and is becoming more frequent on smaller projects. The adaptability of photogrammetry to computer operations has enabled substantial savings over conventional methods in terms of time and money.

10–2. Reconnaissance. The term "reconnaissance" may be defined as an exhaustive preliminary study of an area in which the improvement is to be made. The first step in any reconnaissance is to procure all available pertinent data. These data may be in the form of maps, aerial photographs, charts, or graphs and so on, and may require the application of a large variety of engineering and economic knowledge. All reconnaissance work is not of the same magnitude. In the planning of an expressway or other high type of improvement, it can readily be seen that many factors must be considered. On minor streets or highways only a few of the engineering and economic factors may warrant consideration.

10–3. Engineering and Economic Data. Engineering data help determine the feasibility of a project, engineering problems that may be encountered, natural or man-made physical features, and the probable cost of items which may be included. Much information may be readily available in the form of topographical and geological maps of the particular area to be investigated. Stream and drainage basin maps, along with climatic records, may prove to be valuable in the study of drainage problems. Preliminary survey maps of previous projects in the area may also be a source of information which can be applied to the contemplated improvement. For street and highway projects, engineering data from planning and traffic surveys should also be incorporated in the reconnaissance. Thse should include maps, charts, and tables showing the type, density, and volume of traffic to be expected. Information as to the capacities of highways and streets is also usually available. To estimate the probable cost, cost data may be readily obtained from existing tabulations of unit costs for roadway and bridge construction. Economic data should provide a means of evaluating the benefits derived from the contemplated improvement. Data relative to the possible reduction in vehicle operating costs because of time saved, distance shortened, or improvement in alignment or reduction in grade will aid in determining the dollar saving which would result from the proposed improvement. Records showing the density, distribution, and volume of population are a source of information relative to benefits to be received by establishing a new or improved facility. Records of planning by other local governmental agencies should be studied so that economies may be effected by cooperative planning. Records of the various transportation media serving the community, such as buses, freight-truck lines should also be considered. The density, distribution, and dollar volume of agriculture, manufacturing, and retail sales will in part determine the benefits that may result from the proposed improvement.

10–4. Analysis of Data. When all the available engineering and economic data have been assembled, a detailed analysis should reveal much information pertinent to the proposed project. For example, analysis of the available information may allow the engineer to determine the advisability of selecting an entirely new location or improving the existing one. After an exhausting study of aerial photographs, topographical maps, drainage maps, soil maps, and other data is made in the office, a series of proposed locations may then be made from photogrammetric maps or from a field investigation. This may be considered as a preliminary survey.

10–5. Preliminary Survey. The purpose of a preliminary survey is to gather information about all the physical features which affect the tentatively accepted route, and to prepare a preliminary map upon which the final location survey can be made. This can be accomplished by setting a line down on photogrammetric maps or by having a field survey party do the work. The most likely route selected is that route which best satisfies the directness of the route, the suitability of the terrain encountered, and the adequacy and economy of crossings at water courses and other transportation routes. Upon the selection of the preliminary line the final location survey is made.

10–6. Final Location Survey. The final location survey serves the dual purpose of permanently establishing the centerline and collecting the information necessary for the preparation of plans for construction. Centerline stakes are placed at 100-foot intervals and curves are staked out by deflection angles. Bench marks are established, cross sections are taken at 100-foot intervals, and at any intermediate points with abrupt slope changes. This information is necessary for accurate determination of existing grade line, design of drainage structures, computation of earthwork quantities, and in general, estimation of work to be done. Here again the use of photogrammetric methods can effect economies in the final location.

10–7. Geometric Design. Geometric design is the design or proportioning of the visible elements of a highway, street, or railway. It relates to horizontal alignment and profile, intersections, clearances, and the dimensions of the roadway cross-section. These features of design are directly affected by traffic volume and speeds, as well as by vehicle weights and dimensions. Many of the geometric design standards and procedures for railroads were developed several decades ago to meet needs of an expanding railroad system. Most of these criteria, which were based on fundamental concepts of force and motion have been refined and improved by empirical studies and experience. Studies in the dynamics of highway movements by planning surveys, along with other factual data pertaining to traffic, form the basis of the newer geometric standards in highway design. Research and experience are also reflected in the highway design standards currently in use.

10–8. Design Controls and Criteria for Streets and Highways. Topography and physical features, the character and extent of man-made developments, and traffic to be served from the major controls for determining the general type of highway design. Specific controls will also include the details of traffic, the design vehicle, the design speed, capacity, safety, and economic and social considerations.

For purposes of design, the American Association of State Highway Officials has suggested a classification of highways on the basis of (1) traffic volume, (2) character of traffic, (3) assumed design speed, and

(4) weight of traffic. The latter, which chiefly affects the design of pavements for weight-carrying capacity, is not considered in this text. For those interested in this topic, the reference given at the end of this chapter should provide the necessary information.

Traffic Volume. Traffic volume, one of the factors which influence the design of a highway, is indicated by a numerical figure representing the number of vehicles expected to use the highway in a fixed period of time (one hour).

Traffic volume varies greatly from hour to hour and from day to day. It is for peak hours that highway standards should be developed. It has been found that the thirtieth highest hourly traffic in one year generally is the most suitable value to use for design purposes. Studies made in the United States show that for any particular location the relationship between the thirtieth hour and the annual average daily traffic remains substantially unchanged from year to year.

The value most often used for traffic volume for design purposes, then, is the thirtieth highest hourly volume for the year for which the road is designed. On minor low-volume roads, design may be based on the average daily traffic. On highly seasonal (resort) highways, the eightieth to one-hundredth highest hourly volume may be used. The selected value is also dependent upon the probable life of the roadway. Table 10–1 gives a summary of the traffic elements pertinent to design, and applies generally to rural and urban areas. As indicated the ADT (average daily traffic) is the traffic volume normally given to a route and the DHV (design hourly volume) is expressed as a percentage (K) of the ADT. The K value in urban areas ranges from 7 to 18 per cent while in rural areas the range is from 11 to 20 per cent.

Directional distribution of traffic is also necessary for geometric design. While the total traffic in each direction may be equal for the day, this is not always true for peak hour traffic conditions. Two-way design hourly volumes may be expressed as a percentage (D) of traffic in the heavier direction. As indicated in Table 10–1 this may average 67 per cent in rural areas and 50 per cent in urban areas.

Character of Traffic. The character of traffic affects the choice of the type of highway and also many details of design. Highways may be divided into three general groups with respect to the character of traffic using them.

The symbol *"P"* denotes highways on which the traffic is predominantly composed of passenger vehicles, the percentage of trucks being such that they have little or no effect on the movement of passenger vehicles. The symbol *"T"* denotes highways on which the traffic consists of a relatively high percentage of trucks and buses. The character of this traffic is such as to hinder the speed of traffic to the extent that

TABLE 10–1

Traffic Elements and Their Relation—Rural and Urban Highways

Traffic Element	Explanation and Nationwide Percentage or Factor
Average daily traffic: ADT	Average 24-hour volume for a given year; total for both directions of travel, unless otherwise specified. Directional or one-way ADT is an average 24-hour volume in one direction of travel only.
Current traffic	ADT composed of existing trips, including attracted traffic, that would use the improvement if opened to traffic today (current year specified).
Future traffic	ADT that would use a highway in the future (future year specified). Future traffic may be obtained by adding generated traffic, normal traffic growth, and development traffic to current traffic, or by multiplying current traffic by the traffic projection factor.
Traffic projection factor	Future traffic divided by current traffic. General range, 1.5 to 2.5 for 20-year period.
Design hour volume: DHV	Future hourly volume for use in design (two-way unless otherwise specified), usually the 30th highest hourly volume of the design year (30HV) or equivalent, the approximate value of which can be obtained by the application of the following percentages to future traffic (ADT). The design hour volume, when expressed in terms of all types of vehicles, should be accompanied by factor T, the percentage of trucks during peak hours. Or, the design hour volume may be broken down to the number of passenger vehicles and the number of trucks.
Relation between DHV and ADT: K	DHV expressed as a percentage of ADT, both two-way; normal range 11 to 20, 7–18% in urban areas. Or, DHV, expressed as a percentage of ADT, both one-way; normal range, 16 to 24.
Directional distribution: D	One-way volume in predominant direction of travel expressed as a percentage of two-way DHV. General range, 50 to 80. Average, 67. In urban areas this approaches a 50–50 ratio.
Composition of traffic: T	Trucks (exclusive of light delivery trucks) expressed as a percentage of DHV. Average 7 to 9 on rural highways. Where weekend peaks govern, average may be 5 to 8.

Source: A Policy on Geometric Design of Rural Highways, American Association of State Highway Officials, 1965, page 70.

other vehicles are slowed appreciably, or a condition is created where there is continual passing of the slower moving vehicles with accompanying danger to the faster moving vehicles. "*M*" designates mixed traffic and includes such a proportionate amount of *P* and *T* vehicles that the slow-moving vehicles impede traffic only occasionally. It is evident that highways of a *T* classification will require a design treatment in regard to width of traffic lanes, shoulder widths, grades, and pavement thickness different from those of a *P* classification.

The percentage of trucks required to change a highway into a *T* classification cannot be fixed, but it is a variable that depends upon other factors such as alignment and grade.

Assumed Design Speed. The assumed design speed for a highway, according to the American Association of State Highway Officials, may be considered as "the maximum safe speed that can be maintained over a specified section of highway when conditions are so favorable that the design features of the highway govern."

The main factor that affects the choice of a design speed is the character of the terrain through which the highway or street is to pass. Thus, a 40 mile per hour design speed for a state two-lane arterial might be tolerated in mountainous terrain, while a 70 mile per hour design speed might be specified in flat terrain. Choice of design speed should also be logical with respect to the type of highway being designed. In contrast to high-speed arterials, local residential streets should be designed so as to discourage excessive speeds. A design speed as low as 30 miles per hour is, therefore, recommended for local streets in ordinary terrain, while 20 miles per hour is specified for hilly terrain. Other considerations determining the selection of an assumed design speed are economic factors based on traffic volume, traffic characteristics, costs of rights of way, and other factors which may be of an aesthetic nature. Consideration must also be given to the speed capacity of the motor vehicle. Approved design speeds, as adopted by the American Association of State Highway Officials, are 30, 40, 50, 60, 65, 70, 75, and 80 miles per hour.

The design speed chosen is not necessarily the speed attained when the facility is constructed. Speed of operation will be dependent upon the group characteristics of the drivers under prevailing traffic conditions. Running speeds on a given highway will vary during the day depending upon the traffic volume. The relationship between the design speed and the running speed will also vary. This is illustrated in Table 10–2.

Design Designation. The design designation indicates the major controls for which a highway is designed. As discussed above, these include in a broad sense traffic volume, character or composition of traffic, and

TABLE 10-2

Assumed Relation Between Design Speed and Average Running Speed—Main Highways

Design Speed, mph	Average Running Speed, mph—Main Highways: Low Volume	Intermediate Volume	Approaching Possible Capacity
30	28	26	25
40	36	34	31
50	44	41	35
60	52	47	37
65	55	50	
70	58	54	
75	61	56	
80	64	59	

Source: A Policy on Geometric Design of Rural Highways, AASHO, 1965, page 97.

design speed. Traffic volume should be designated in terms of annual daily traffic (ADT) for the current year and for the future design year. The design hourly volume (DHV), a two-way value, should be given along with the directional distribution of traffic expressed as a percentage (D) of the DHV. The character or composition of traffic is indicated by the percentage of trucks (T) of the DHV. Design-speed designation is basic to overall standards, and with the DHV and T it gives the necessary information to be incorporated in the final plans.

The ratio (K) of the design hour volume (DHV) to the annual daily traffic (ADT), the directional distribution (D) of the design hour volume (DHV) expressed as a percentage, and the percentage of trucks (T) should be shown on the plans.

Where access is fully or partially controlled, this should be shown in the design designation; otherwise, no control of access is assumed.

An example of such designations is given below; the tabulation on the left is for a two-lane highway; the tabulation on the right is for a multilane highway:

Two-lane highway	Multilane highway
	Control of access = full
ADT (1966) = 2500	ADT (1966) = 10,200
ADT (1986) = 5200	ADT (1986) = 22,000
DHV = 720	DHV = 2950
D = 65%	D = 60%
T = 12%	T = 8%
V = 60 mph	V = 70 mph

Vehicle Design. The dimensions of the motor vehicle also influence design practice. The width of the vehicle naturally affects the width of the traffic lane; length has a bearing on roadway capacity and affects the turning radius; the height of the vehicle affects the clearance of the various structures. Weight, as mentioned previously, affects the structural design of the roadway. Vehicle weights, dimensions, and operating characteristics are given in Chapter 4.

10–9. The Nature of Railroad Track Design. In the United States, the major system of railways has, for all practical purposes, been completed. With the rare exception of new passenger rail lines in urban areas, the construction of extensive mileage of rail trackage is no longer needed. The emphasis has shifted to such activities as the construction of spur tracks to new industrial sites and newly developed mines and power plants. As railroad companies merge, short connecting lines are being constructed to improve movements over the new system. Considerable emphasis is also being placed on improving geometric design features and the riding quality of existing tracks.

As with highways, the design of rail lines is strongly influenced by such factors as topography and man-made developments. The nature of the traffic, whether it be freight or passenger, is of signal importance. The factors of train operating speeds, freight tonnages, and type of rolling stock also significantly control the elements of geometric design for new railroad lines.

10–10. Alignment. An ideal and most interesting roadway is one that generally follows the existing natural topography of the country. This is the most economical to construct, but there are certain aspects of design that must be adhered to which may prevent the designer from following this undulating surface without making certain adjustments in a vertical and horizontal direction.

The designer must produce an alignment in which conditions are consistent. Sudden changes in alignment should be avoided as much as possible. For example, long tangents should be connected with long sweeping curves, and short sharp curves should not be interspersed with long curves of small curvature. The ideal location is one with consistent alignment where both grade and curvature receive consideration and satisfy limiting criteria. The final alignment will be that in which the best balance between grade and curvature is achieved.

Terrain has considerable influence on the final choice of alignment. Generally, the topography of an area is fitted into one of the following three classifications: level, rolling, or mountainous.

In level country, the alignment, in general, is limited by considerations other than grade, i.e., cost of right-of-way, land use, waterways requiring expensive bridging, existing roads, railroads, canals and power lines, and subgrade conditions or the availability of suitable borrow.

In rolling country, grade and curvature must be considered carefully. Depths of cut and heights of fill, drainage structures, and number of bridges will depend on whether the route follows the ridges, the valleys, or a cross-drainage alignment.

In mountainous country, grades provide the greatest problem, and in general, the horizontal alignment (curvature) is conditioned by maximum grade criteria.

10–11. Circular Curves. Circular curves may be described by giving either the radius or the "degree of curve." In highway design, the degree of curve is defined as the central angle subtended by an arc of 100 feet. This is known as the "arc definition." Books on railroad location define the degree of curve as the central angle subtended by a chord of 100 feet, and some highway departments follow this procedure. According to the arc definition, it can be shown that the curve radius, R, and the degree of curve, D, are related by the following equation:

$$D = \frac{5729.58}{R} \tag{10–1}$$

use

The chord definition of degree of curve leads to a different equation:

$$\sin \tfrac{1}{2}D = \frac{50}{R} \tag{10–2}$$

Arc or chord measurements can be considered alike for all curves less than 4 degrees without appreciable error. An examination of tables will show that the following chords may be assumed to be equal to the arcs without appreciable error:

100-foot chords up to	4 degrees
50-foot chords up to	10 degrees
25-foot chords up to	25 degrees
10-foot chords up to	100 degrees

Figure 10–1 shows a simple circular curve and its component parts, with the necessary formulas for finding the values of the various elements. A combination of simple curves can be arranged to produce compound or reverse curves.

10–12. Horizontal Alignment Design Criteria for Streets and Highways. In general, the maximum desirable degree of curvature for highways is from 5 to 7 degrees in open country and not over 10 degrees in moun-

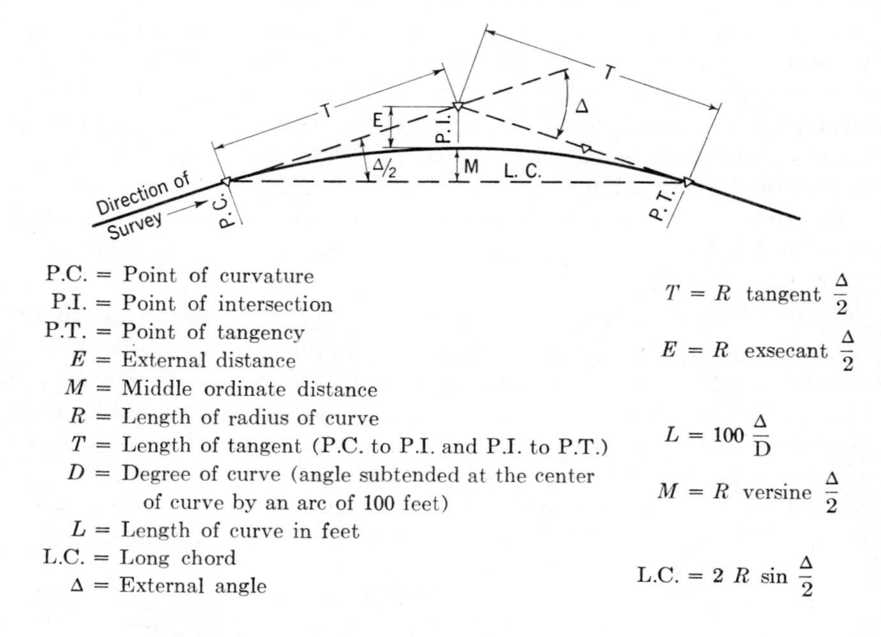

Fig. 10–1. A simple highway curve. (Source: Leo Ritter and Radnor Paquette, *Highway Engineering,* 3rd ed., The Ronald Press Co., 1967.)

tainous areas. Many states limit curvature to 3 degrees on principal highways. In contrast, a minimum centerline radius of 350 feet is permitted for collector streets in ordinary terrain, while local residential streets in hilly terrain may have a centerline radius as short as 100 feet.

Two curves in the same direction connected with a short tangent known as "broken back" curves, should be combined into one continuous curve. This may be accomplished by compounding the curve, provided that the difference between the two branches of the curve does not exceed 5 degrees.

Where reversed curves are necessary, they should be separated by a tangent distance of at least 200 feet in order to allow proper easement from one curve to another.

For small deflections a curve should be at least 500 feet long for a 5-degree central angle, with an increase of 100 feet for each decrease of one degree in the central angle.

Widening of Curves. Extra width of pavement may be necessary on curves. As a vehicle turns, the rear wheels follow the front wheels on a shorter radius, and this has the effect of increasing the width of the vehicle in relation to the lane width of the roadway. In the study of traffic behavior of a large number of drivers, it has also been found that they do not drive as close to the edge of curved sections as when on a tangent. This fact is hard to evaluate, and justification for widening on this basis is questionable.

Present practice requires no widening when the degree of curvature is less than 10 degrees on a two-lane pavement 24 feet wide. For greater curvatures and/or narrower pavements, widening normally is from 2 to 4 feet, depending on design speed and pavement width. It is suggested that no pavement be widened less than 2 feet.

10–13. Horizontal Alignment Design Criteria for Railways. Sudden changes in horizontal alignment cannot be tolerated in railroad design. Horizontal curvature limits the speed of trains and increases the probability of derailments and overturning accidents. Tracks on curves also require greater horizontal clearances and may create difficulties in coupling. These undesirable effects of curvature stem from the necessity to design for a vehicle unit (the passenger or freight car) which is typically 85 feet long and measures 60 feet between truck centers.

The operation of trains around curves causes wear on the rails and increases train resistance. Curve resistance increases in direct proportion to the degree of curve and amounts to about 0.8 pounds per degree of curve. It has been estimated that a 12-degree curve approximately doubles the train resistance likely to be experienced on a straight level track [1].

When horizontal curvature is imposed on a section which has steep gradient, it may be necessary to "compensate" for the curvature by lessening the grade. Since a 1-per cent grade results in a resistance of 20 pounds per ton, a decrease in grade of 0.8/20 or 0.04 per cent will compensate for the resistance of a 1-degree horizontal curve.

Generally speaking, 1-degree to 3-degree railroad curves are considered relatively flat curves while 8-degree to 10-degree curves are considered relatively sharp. Curves greater than about 10 degrees are seldom used for main railroad lines, although curves on this sharpness have sometimes been utilized in mountainous areas. Even in mountainous topography, seldom will a horizontal curve greater than about 24 degrees be used. Curves as sharp as 40 degrees, however, have been used in railroad yards [1].

To facilitate the movement of long wheel-based engines, some railroads widen the gage through curves sharper than about 6 degrees.

10–14. Superelevation of Highway Curves. On rural highways, most drivers adopt a more or less uniform speed when traffic conditions will permit them to do so. When making a transition from a tangent section to a curved section, if the sections are not properly designed, the vehicle must be driven at reduced speed for safety as well as for the comfort of the occupants. This is due to the fact that a force is acting on the vehicle which tends to cause an outward skidding away from the center of the curve. Most highways have a slight crowned surface to take care of drainage. It can be readily seen that when these crowns are carried along the curve, the tendency to slip is retarded on the inside of the curve because of the banking effect of the crown. The hazard of slipping is increased on the outside of the curve, however, due to the outward sloping of the crown. In order to overcome this tendency to slip and to maintain average speeds, it is necessary to "superelevate" the roadway sections; i.e., raise the outside edge or "bank" the curve.

Analysis of the forces acting on the vehicle as it moves around a curve of constant radius indicates that the theoretical superelevation is equal to:

$$e + f = \frac{V^2}{15R} = \frac{0.067V^2}{R} \tag{10-3}$$

where:

e = rate of superelevation in feet per foot
f = side-friction factor
V velocity in miles per hour
R = radius of curve in feet

Research and experience have established limiting values for e and f. Use of the maximum e with the safe f value in the formula permits determination of minimum curve radii for various design speeds. Safe side-friction factors to be used in design range from 0.16 at 30 miles per hour to 0.11 at 80 miles per hour. Present practice suggests at a maximum superelevation rate of 0.12 feet per foot. Where snow and ice conditions prevail, the maximum superelevation should not exceed 0.08 foot per foot. Some states have adopted a maximum superelevation rate of 0.10; however, other rates may have applications on some types of highways in certain areas.

The maximum safe degree of curvature for a given design speed can be determined from the rate of superelevation and the side-friction factor. The minimum safe radius R can be calculated from the formula given above:

$$R = \frac{V^2}{15(e + f)} \tag{10-4}$$

Using D as the degree of curve (arc definition):

$$D = \frac{5729.6}{R} \quad \text{or} \quad D = \frac{85{,}900(e + f)}{V^2} \tag{10–5}$$

The relationship between superelevation and degree of curve is illustrated in Fig. 10–2 and Table 10–3. Figure 10–3 is used to determine superelevation rates in Michigan.

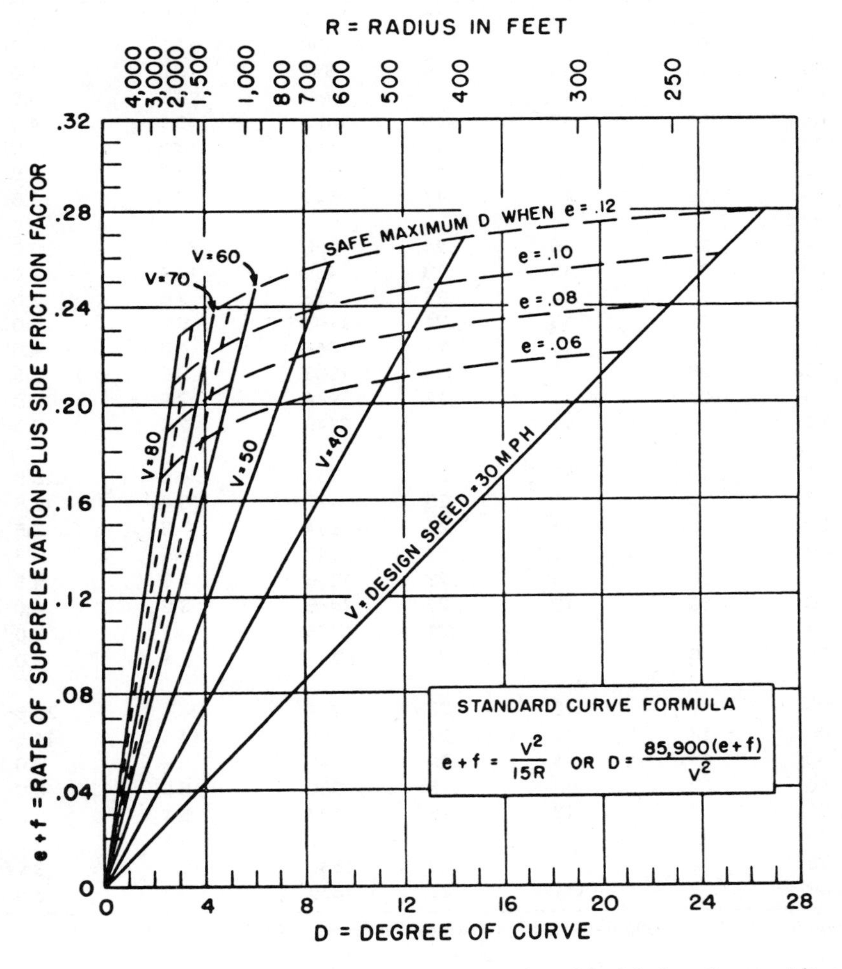

Fig. 10–2. Relation of superelevation (e) plus side-friction factor (f) to degree of curve for different design speeds. (Source: A Policy on Geometric Design of Rural Highways, 1965.)

TABLE 10-3

Maximum Degree of Curve and Minimum Radius Determined for Limiting Values of *e* and *f*

Design Speed	Maximum *e*	Maximum *f*	Total (*e* + *f*)	Minimum Radius, ft.	Max. Degree of Curve	Max. Degree of Curve, Rounded
30	.06	.16	.22	273	21.0	21.0
40	.06	.15	.21	508	11.3	11.5
50	.06	.14	.20	833	6.9	7.0
60	.06	.13	.19	1263	4.5	4.5
65	.06	.13	.19	1483	3.9	4.0
70	.06	.12	.18	1815	3.2	3.0
75	.06	.11	.17	2206	2.6	2.5
80	.06	.11	.17	2510	2.3	2.5
30	.08	.16	.24	250	22.9	23.0
40	.08	.15	.23	464	12.4	12.5
50	.08	.14	.22	758	7.6	7.5
60	.08	.13	.21	1143	5.0	5.0
65	.08	.13	.21	1341	4.3	4.5
70	.08	.12	.20	1633	3.5	3.5
75	.08	.11	.19	1974	2.9	3.0
80	.08	.11	.19	2246	2.5	2.5
30	.10	.16	.26	231	24.8	25.0
40	.10	.15	.25	427	13.4	13.5
50	.10	.14	.24	694	8.3	8.5
60	.10	.13	.23	1043	5.5	5.5
65	.10	.13	.23	1225	4.7	4.5
70	.10	.12	.22	1485	3.9	4.0
75	.10	.11	.21	1786	3.2	3.0
80	.10	.11	.21	2032	2.8	3.0
30	.12	.16	.28	214	26.7	26.5
40	.12	.15	.27	395	14.5	14.5
50	.12	.14	.26	641	8.9	9.0
60	.12	.13	.25	960	6.0	6.0
65	.12	.13	.25	1127	5.1	5.0
70	.12	.12	.24	1361	4.2	4.0
75	.12	.11	.23	1630	3.5	3.5
80	.12	.11	.23	1855	3.1	3.0

Source: A Policy on Geometric Design of Rural Highways, AASHO, 1965, page 158.

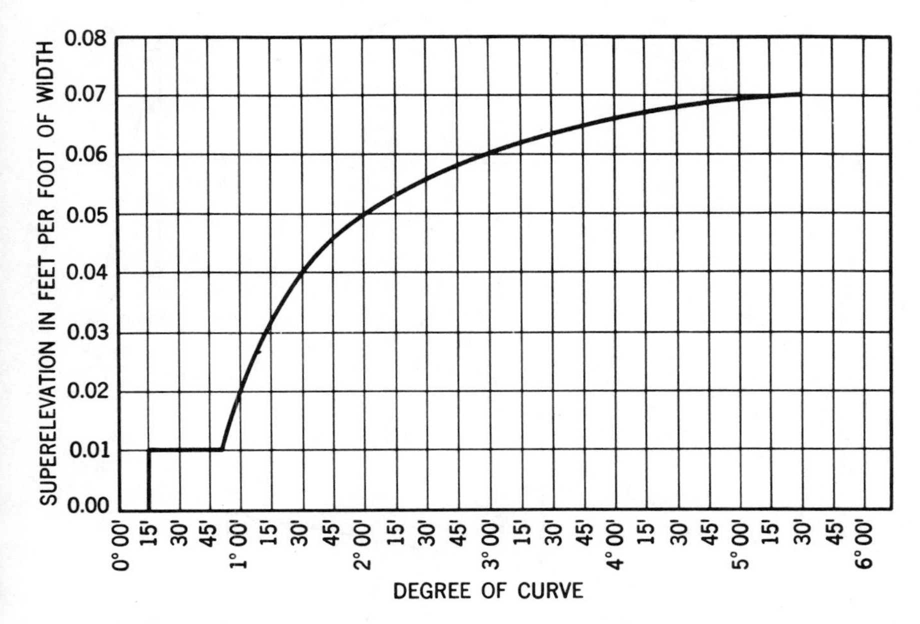

Fig. 10–3. Relationship between superelevation and degree of curve used for design in one state.

10–15. Superelevation of Railway Curves. As shown by Fig. 10–4, a train car traversing a curve is subjected to forces of its weight, the resisting forces exerted by the rails and the centrifugal force:

$$F = \frac{Wv^2}{gR} \tag{10-6}$$

where:

W = weight of car, pounds
v = velocity in feet per second
R = radius of curve, feet
g = acceleration due to gravity, ft./sec.^{2}

A state of equilibrium is said to exist when both wheels bear equally on the rails. Under these conditions, E, the equilibrium elevation is just sufficient to cause the resultant force, R, to be perpendicular to the plane of the top of the rails. Using an effective gage (center to center of rails), G, of 60 inches, by simple proportion:

$$\frac{E}{60} = \frac{F}{W} = \frac{v^2}{gR}$$

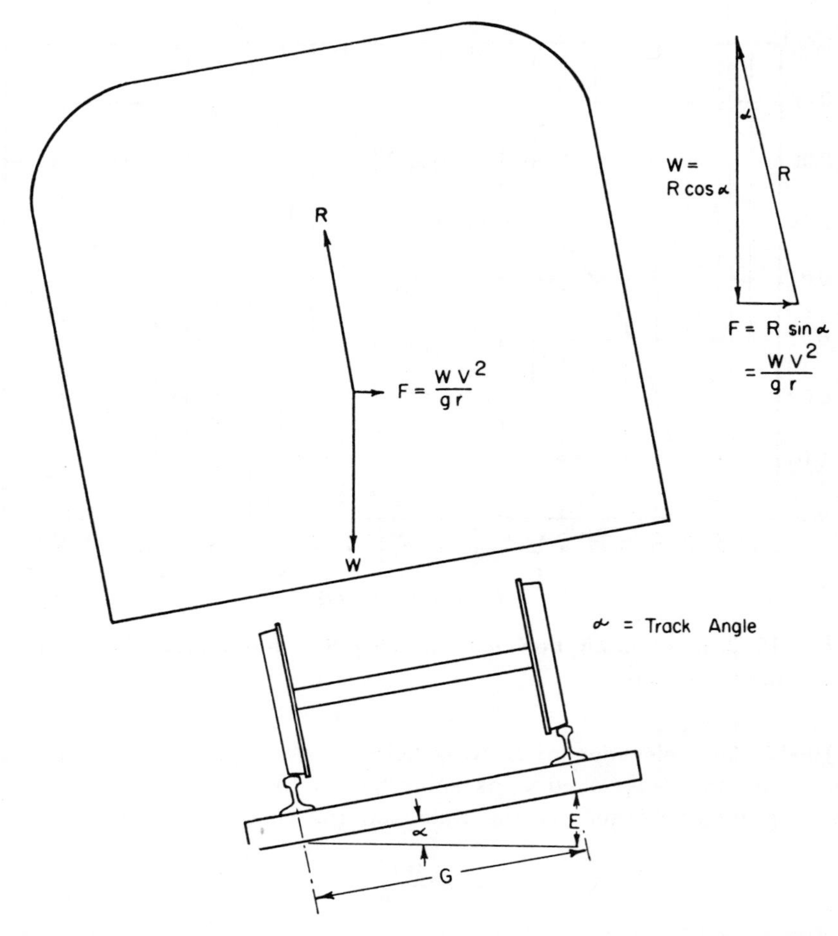

Fig. 10–4. Forces on a Car Body Traversing a Curve at Equilibrium Speed.

Recalling that $R = \frac{5730}{D}$, it can be shown that:

$$E = 0.0007V^2D \tag{10–7}$$

where:

E = equilibrium elevation of the outer rail, inches
V = speed, miles per hour
D = degree of curve.

Values of equilibrium elevations for various speeds and degrees of curvature are given in Table 10–4.

TABLE 10-4

Equilibrium Elevation for Various Speeds on Curves

	E = Equilibrium Elevation for Various Speeds on Curves																
D = Degree	V = Speed in Miles per Hour																
of Curve	10	20	30	35	40	45	50	55	60	65	70	75	80	85	90	95	100
0° 30′	0.04	0.14	0.32	0.43	0.56	0.71	0.88	1.06	1.26	1.48	1.72	1.97	2.24	2.53	2.84	3.16	3.50
1° 00′	0.07	0.28	0.63	0.87	1.12	1.42	1.75	2.12	2.52	2.96	3.43	3.94	4.48	5.06	5.67	6.32	7.00
1° 30′	0.11	0.42	0.95	1.29	1.68	2.13	2.63	3.18	3.78	4.44	5.15	5.91	6.72	7.59	8.51	9.48	10.50
2° 00′	0.14	0.56	1.26	1.72	2.24	2.84	3.50	4.24	5.04	5.92	6.86	7.88	8.96	10.12	11.34	12.64	
2° 30′	0.18	0.70	1.58	2.14	2.80	3.54	4.38	5.29	6.30	7.39	8.58	9.84	11.20				
3° 00′	0.21	0.84	1.89	2.57	3.36	4.25	5.25	6.35	7.56	8.87	10.29	11.81					
3° 30′	0.25	0.98	2.21	3.00	3.92	4.96	6.13	7.41	8.82	10.35							
4° 00′	0.28	1.12	2.52	3.43	4.48	5.67	7.00	8.47	10.08								
5° 00′	0.35	1.40	3.15	4.29	5.60	7.09	8.75	10.59									
6° 00′	0.42	1.68	3.78	5.15	6.72	8.51	10.50										
7° 00′	0.49	1.96	4.41	6.00	7.84	9.92											
8° 00′	0.56	2.24	5.04	6.86	8.96	11.34											
9° 00′	0.63	2.52	5.67	7.72	10.08												
10° 00′	0.70	2.80	6.30	8.58	11.20												
11° 00′	0.77	3.08	6.93	9.43													
12° 00′	0.84	3.36	7.56	10.29													

E in inches = $0.0007\ V^2 D$

Source: American Railway Engineering Association, *Manual of Recommended Practice,* 1965.

It has been found that a rail car will ride comfortably and safely around a curve at a speed which requires an elevation about three inches[1] higher for equilibrium. It will be noted from Table 10–4, for example, that for a 60 mile per hour speed and a 2-degree curve, the outer rail should be elevated 5.04 inches. Experience indicates that a train will ride safely and comfortably around the curve at about 76 miles per hour, the value corresponding to an elevation of 8.04 inches.

Not all trains, of course, travel at the same speed. Even high-speed trains must occasionally slow down or stop on horizontal curves because of traffic interferences or other reasons. Because passengers experience discomfort when the train stops or moves slowly around superelevated curves, a maximum elevation of the outer rail of 8 inches is recommended and a maximum elevation value of 6 or 7 inches is desired.

Trains traveling at speeds less than the equilibrium speed tend to cause excessive wear on the inside rail. Similarly, trains which exceed the equilibrium speed cause excessive wear on the outer rail, and in extreme cases, can cause overturning. The velocity which would cause overturning can be computed by Equation 10–8 which is based on the assumption that the center of gravity of the car is 7.0 feet[2] above the top of the outer rail.

$$V = \frac{170}{\sqrt{D}} \tag{10–8}$$

10–16. Spirals or Transition Curves for Highways. Transition curves serve the purpose of providing a gradual change from the tangent section to the circular curve, and vice versa. A vehicle which enters a circular curve with transitions travels smoothly and naturally along a curve which gradually changes from zero to curvature of some finite value, D, which is maintained throughout the length of the circular curve. As the vehicle emerges from the circular curve, the curvature is gradually diminished from D to zero. The most commonly used transition curve is the spiral, the radius of which at any point is inversely proportional to its length.

Being unrestrained laterally, automobiles and trucks are free to shift across the traffic lane allowing the driver to effect artificially a transition from a tangent section (of infinite radius) to a curve section (of finite radius). For this reason, transition curves are not universally used for highway design. Although there is no consistent practice on the use of transition curves by the various highway agencies, spiral transition

[1] Because of the trend toward the usage of freight cars with higher centers of gravity, some railroads now use a maximum of 1.5 to 2.0 inches of unbalanced elevation.

[2] A combined center of gravity (of car and its contents) of 98 inches is now considered satisfactory for new freight car design specifications.

curves are normally used only on high volume highways where the degree of curvature exceeds about 3 degrees.

When used in combination with superelevated sections, the superelevation should be attained within the limits of the transition.

The minimum length of the transition curve is given as:

$$L_s = 1.6 \frac{V^3}{R} \tag{10–9}$$

where:

L_s = length of the transition in feet
V = speed in miles per hour
R = radius in feet

Barnett [3] has published tables from which required highway transition curves can be chosen and located without extensive calculations.

10–17. Spirals or Transition Curves for Railways. A railroad car is strictly restrained between the two rails and is, therefore, not free to shift laterally and artificially effect a transition in horizontal curvature. For this reason, spiral transition curves are extensively used in railroad design. The use of spiral curves is recommended by the American Railway Engineering Association [4] on all mainline tracks between tangent and curve and between different degrees of curvature where compound curves are used.

According to the A.R.E.A. [4]:

> The desirable length of the spiral for main tracks where the alignment is being entirely reconstructed or where the cost of the realignment of the existing track will not be excessive should be such that when passenger cars of average roll tendency are to be operated the rate of change of the unbalanced lateral acceleration acting on a passenger will not exceed 0.03 g per sec. Also, the desirable length in this case needed to limit the possible racking and torsional forces produced should be such that the longitudinal slope of the outer rail with respect to the inner rail will not exceed 1/744, which is based on an 85-ft-long car.

Two formulas are given by A.R.E.A. to achieve these results:

$$L = 1.63 E_u V \tag{10–10}$$

where:

L = desirable minimum length of spiral, feet
E_u = unbalanced elevation, inches
V = maximum train speed, miles per hour

and

$$L = 62 E_a \tag{10–11}$$

where:

E_a = actual elevation, inches

The maximum length of spiral computed by the two formulas should be used.

The A.R.E.A. Manual of Recommended Practice [4] lists the various formulas for the calculation of spiral curve data.

10–18. Attainment of Superelevation in Highway Design. The transition from the tangent section to a curved superelevated section must be accomplished without any appreciable reduction in speed and in such a manner as to insure safety and comfort of the vehicle and occupants.

In order to effect this change, it will be readily seen that the normal road cross section will have to be tilted to the superelevated cross section. This tilting usually is accomplished by rotating the section about the centerline axis. The effect of this rotation is to lower the inside edge of the pavement and to raise the outside edge without changing the centerline grade. Another method is to rotate about the inner edge of the pavement as an axis so that the inner edge retains its normal grade but the centerline grade is varied or rotation may be about the outside edge. Rotation about the centerline is used by a majority of the states, but for flat grades too much sag is created in the ditch grades by this method. On grades below 2 per cent, rotation about the inside edge is preferred. Regardless of which method is used, care should be exercised to provide for drainage in the ditch sections of the superelevated areas.

The roadway on full superelevated sections should be a straight inclined section. When a crowned surface is rotated to the desired superelevation, the change from a crowned section to a straight inclined section should be accomplished gradually. This may be done by first changing the section from the centerline to the outside edge to a level section. Second, the outside edge should be raised to an amount equal to one-half the desired superelevation, and at the same time the inner section changed to a straight section and the whole section put in one inclined plane. Third, rotation about the centerline should be continued until full superelevation is reached.

The distance required for accomplishing the transition from a normal to a superelevated section, sometimes called the "transition runoff," is a function of the design speed, degree of curvature, and the rate of superelevation. However, all roadways are not superelevated at the same rate. The American Association of State Highway Officials has recommended maximum relative slopes between the centerline and edge profiles of a two-lane highway. These recommended values, shown in Table 10–5, are dependent upon the design speed.

Superelevation is usually started on the tangent at some distance before the curve starts, and full superelevation is generally reached beyond the P.C. of the curve. In curves with transitions, the superelevation

TABLE 10-5

Superelevation Design Data

Design Speed	Maximum Relative Slope Between Profiles of Edge of Two-Lane Pavement and Centerline
30	0.66%
40	.58
50	.50
60	.45
65	.41
70	.40
75	.38
80	0.36%

Source: A Policy on Geometric Design of Rural Highways, AASHO, 1965, page 136

can be attained within the limits of the spiral. In curves of small degree where no transition is used, between 60 and 80 per cent of the superelevation runoff is put into the tangent.

On multilane undivided roads or roads with wide medians, increased runoff lengths are necessary. In order to obtain smooth profiles for the pavement edges, it is recommended that the breaks at cross sections be replaced by smooth curves (see Fig. 10–5). Different states adopt differing methods to attain smooth edge profiles. Figure 10–6 indicates the method adopted by one state.

10–19. Attainment of Superelevation in Railway Design. As in the case of highway curves with transitions, the superelevation of railway curves is attained or run out uniformly over the length of the spiral. This is accomplished by elevating the outer rail. The inner rail is normally maintained at grade.

10–20. Grades and Grade Control. The vertical alignment of the roadway and its effect on the safe and economical operation of the vehicle constitutes one of the most important features of highway and railway design. The vertical alignment, which consists of a series of straight lines connected by vertical parabolic or circular curves, is known as the "grade line." When the grade line is increasing from the horizontal, it is known as a "plus grade," and when it is decreasing from the horizontal it is known as a "minus grade." In analyzing grade and grade controls, the designer usually studies the effect of change in grade on the centerline profile.

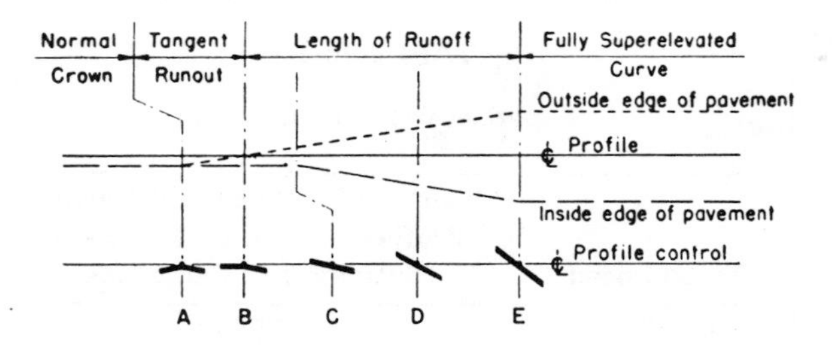

A-PAVEMENT REVOLVED ABOUT CENTER LINE

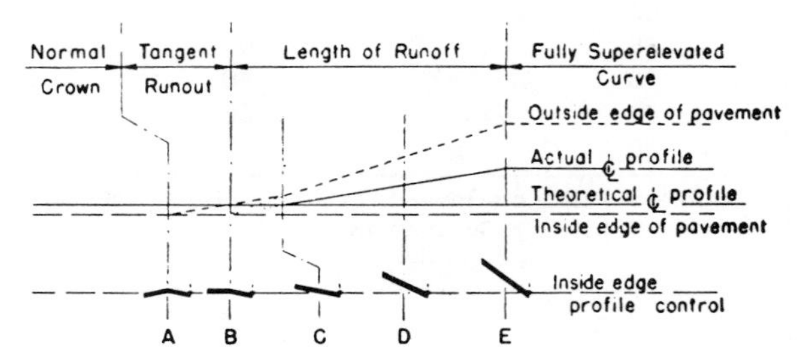

B-PAVEMENT REVOLVED ABOUT INSIDE EDGE

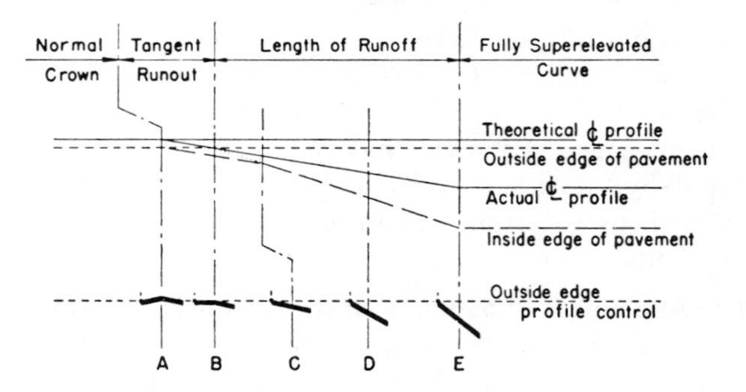

C-PAVEMENT REVOLVED ABOUT OUTSIDE EDGE

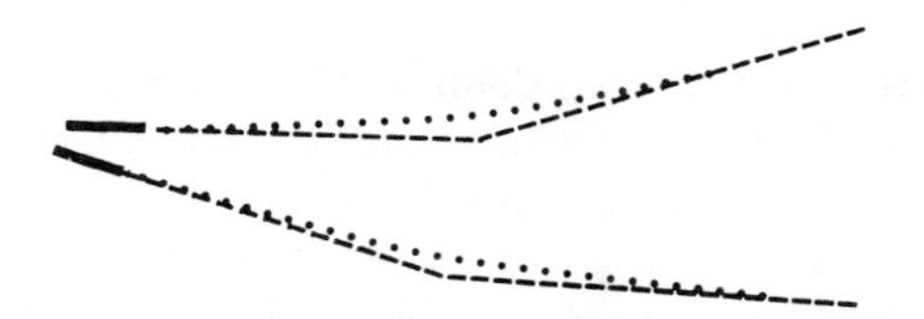

Note: Angular breaks to be appropriately rounded as shown by dotted line.

Fig. 10–5. Diagrammatic profiles showing methods of obtaining superelevation, (Courtesy AASHO). (Source: A Policy on Geometric Design of Rural Highways, 1965.)

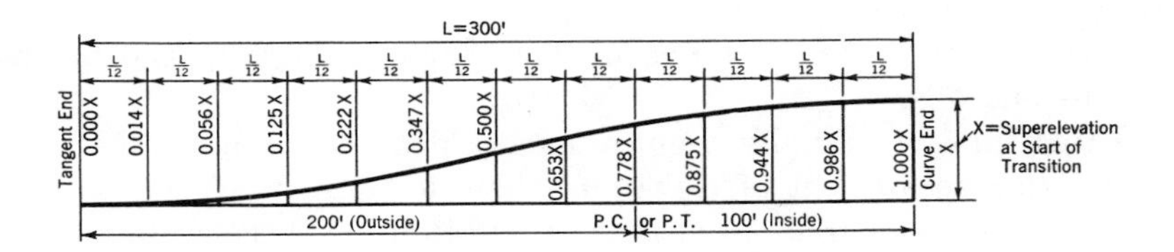

To be used on curves where the rate of superelevation is less than .04 ft. per ft. unless otherwise shown on plan.

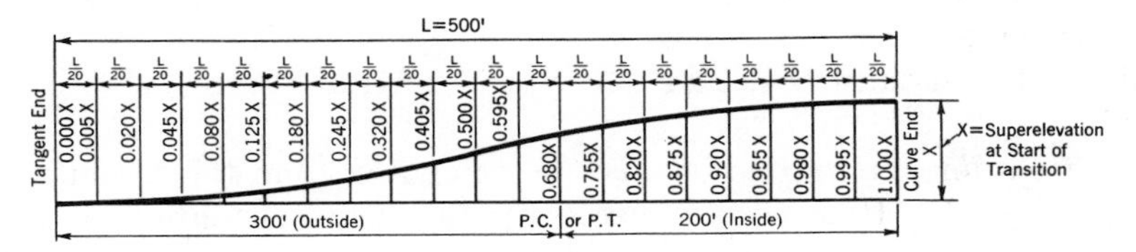

To be used on curves where the rate of superelevation is .04 foot per foot or greater unless otherwise shown on plans.

TRANSITION VERTICAL REVERSE CURVES

The transition from normal section of metal to the superelevated section shall be made by means of the proper reverse vertical curve shown above. This curve shall be applied to all grade lines which are to be depressed or raised in the superelevation transition.

For spiraled curves, the length of the superelevation transition equals the length of the inside spiral.

Fig. 10–6. Superelevation using reversed parabolic curves. (Courtesy Michigan Department of State Highways.) (Source: Leo Ritter and Radnor Paquette, *Highway Engineering*, 2nd ed., The Ronald Press Co., 1960.)

In the establishment of a grade, an ideal situation is one in which the cut is balanced against the fill without a great deal of borrow or an excess of cut to be wasted. All hauls should be downhill if possible, and not too long. Ideal grades have long distances between points of intersection, with long vertical curves between grade tangents to provide smooth riding qualities and good visibility. The grade should follow the general terrain and rise and fall in the direction of the existing drainage. In rock cuts and in flat swampy areas it is necessary to maintain higher grades. Further possible construction and the presence of grade separations and bridge structures also control grades.

Changes of grade from plus to minus should be placed in cuts, and changes from a minus grade to a plus grade should be placed in fills. This will generally give a good design, and many times it will avoid the appearance of building hills and producing depressions contrary to the general existing contours of the land. Other considerations for determining the grade line may be of more importance than the balancing of cuts and fills.

Urban projects usually require a more detailed study of the controls and a fine adjustment of elevations than do rural projects. It is often best to adjust the grade to meet existing conditions because of additional expense when doing otherwise.

10–21. Vertical Curves. The parabolic curve is used almost exclusively in connecting grade tangents because of the convenient manner in which the vertical offsets can be computed. This is true for both highway and railway design. A typical vertical curve is shown in Fig. 10–7.

Offsets for vertical curves may be computed from the formulas given in Fig. 10–7. It is usually necessary to make calculations at 50-foot stations, while some paving operations require elevations at 25-foot inter-

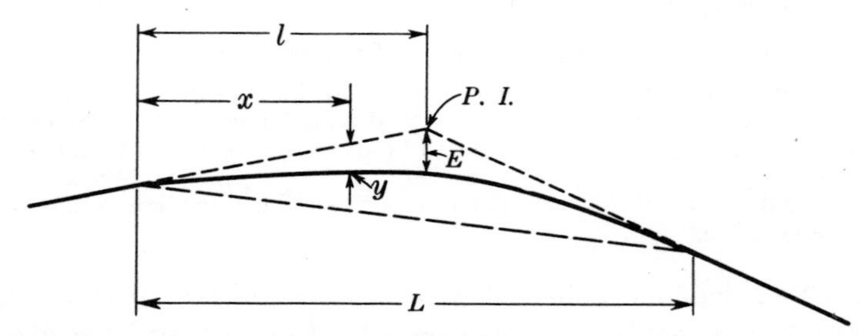

The distance from the P.I. to the middle of the parabolic curve may be found from the following formula:

$$E = \frac{gL}{8}$$

where:

E = the external distance in feet
g = the algebraic difference in grade in per cent of the intersecting grades
L = the length of the curve in stations

The vertical offset at any point on the curve may be computed by the following:

$$y = \left(\frac{x}{l}\right)^2 E$$

where:

y = offset in feet
x = any distance from the P.C. in feet
$l = \frac{L}{2} \times 100$ ft or half the length of the curve
E = external distance in feet

Fig. 10–7. A typical vertical curve. (Source: Leo Ritter and Radnor Paquette, *Highway Engineering,* 3rd ed., The Ronald Press Co., 1967.)

vals or less. It is often necessary to compute other critical points on the vertical curve in order to ensure proper drainage or clearances, such as at sags and crests. The low point or high point of a parabolic curve is not usually vertically above or below the vertex of the intersecting grade tangents, but is to the left or right of this point. The distance from the P.C. of the curve to the low or high point is given as:

$$x = \frac{Lg_1}{g_1 - g_2} \tag{10-12}$$

and the difference in elevation is given as:

$$Y = \frac{Lg_1^2}{2(g_1 - g_2)} \tag{10-13}$$

where:

x = distance from the P.C. to the turning point in stations
L = length of the curve in stations
g_1 = grade tangent from the P.C. in per cent
g_2 = grade tangent from the P.T. in per cent
Y = the difference in elevation between the P.C. and the turning point in feet

The above formulas apply only for the symmetrical curve, that is, one in which the tangents are of equal length. The unequal tangent or unsymmetrical vertical curve is a compound parabolic curve. Its use is generally warranted only where a symmetrical curve cannot meet imposed alignment conditions.

10–22. Vertical Alignment Design Criteria for Streets and Highways. In the analysis of grade and grade control, one of the most important considerations is the effect of grades upon the operating costs of the motor vehicle. An increase in the gasoline consumption and a reduction of speed are apparent when grades are increased. An economical approach would be to balance the added annual cost of grade reduction against the added annual cost of vehicle operation without grade reduction. An accurate solution to the problem depends on the knowledge of traffic volume and type, which can be obtained only by means of a traffic survey.

While maximum grades vary a great deal in various states, AASHO recommendations make maximum grades dependent on design speed and topography. Present practice limits grades in flat country to 6 per cent at 30 mph design speed and 3 per cent for 80 mph. In mountainous terrain 9 per cent at 30 mph and 5 per cent at 70 mph should be used.

Residential streets, which have the primary function of providing access to land areas and only the secondary function of traffic service, tend to conform to and blend with undulations in the existing terrain. Maximum grades as steep as 15 per cent may be permitted for local streets in hilly terrain.

Whenever long sustained grades are used, the designer should not substantially exceed the critical length of grade without the provision of climbing lanes for slow-moving vehicles. Critical grade lengths vary from 1700 feet for a 3 per cent grade to 500 feet for an 8 per cent grade.

Long-sustained grades should be less than the maximum grade used on any particular section of a highway. It is often preferred to break the long-sustained uniform grade by placing steeper grades at the bottom and lightening the grades near the top of the ascent. Dips in the profile grade in which vehicles may be hidden from view should also be avoided.

Minimum grades are governed by drainage conditions. Level grades may be used in fill sections in rural areas when crowned pavements and sloping shoulders can take care of the pavement surface drainage. It is preferred, however, to have a minimum grade of at least 0.3 per cent under most conditions in order to secure adequate drainage.

Sight Distance. Safe highways must be designed to give the driver a sufficient distance of clear vision ahead so that he can avoid hitting unexpected obstacles and can pass slower vehicles without danger.

Sight distance is the length of highway visible ahead to the driver of a vehicle. When this distance is not long enough to permit passing an overtaken vehicle, it is termed "stopping (or nonpassing) sight distance." The stopping distance is the minimum distance required for stopping a vehicle traveling at or near the design speed before reaching an object in its path. This stationary object may be a vehicle or some other object on the roadway. When the sight distance is long enough to enable a vehicle to overtake and pass another vehicle on a two-lane highway without interference from an oncoming vehicle, it is termed "passing sight distance."

Minimum Stopping Sight Distance. Sight distance at every point should be as long as possible, but never less than the minimum stopping sight distance. The minimum stopping sight distance is based upon the sum of two distances: one, the distance traveled from the time the object is sighted to the instant that the brakes are applied, and two, the distance required for stopping the vehicle after the brakes are applied. The first of these two distances is dependent upon the speed of the vehicle and the perception time and brake reaction time of the operator. The second distance depends upon the speed of the vehicle; condition of brakes, tires, and roadway surface; and the alignment and grade of the highway.

Braking Distance. There is a wide variation among vehicle operators as to the time that it takes to react and to apply the brakes after an obstruction is sighted. Investigations seem to indicate that the minimum value of perception time can be assumed to vary from 2 seconds at 30 miles per hour to 1 second at 70 miles per hour. Brake reaction time is less than perception time, and tests indicate that the average brake reaction time is about ½ second. To provide a factor of safety for operators whose brake reaction time is above average, a full second is assumed as the total brake reaction time. Some investigators feel that perception and brake reaction time should be combined and have assigned 2½ seconds for this value.

The approximate braking distance of a vehicle on a level highway is determined by:

$$d = \frac{v^2}{2fg} \tag{10–14}$$

where:

d = braking distance in feet
v = velocity of the vehicle in feet per second when the brakes are applied
g = acceleration due to gravity
f = coefficient of friction between tires and roadway

Changing v in feet per second to V in miles per hour, and substituting 32.2 for g, we have:

$$d = \frac{V^2}{30f} \tag{10–15}$$

It is assumed that the friction force is uniform throughout the braking period. This is not strictly true; it varies as some power of the velocity. Other physical factors affecting the coefficient of friction are the condition and pressure of tires, type and condition of the surface, and climatic conditions such as rain, snow, and ice. Friction factors for skidding are assumed to vary from 0.62 at 30 miles per hour to 0.55 at 70 miles per hour for dry pavements. For wet pavements these are lower, as shown in Table 10–6. Recommended minimum stopping sight distances are shown in Table 10–6. In this table, perception and brake reaction time are combined.

Effect of Grade on Stopping Distance. When a highway is on a grade, the formula for braking distance is modified as follows:

$$d = \frac{V^2}{30(f \pm g)} \tag{10–16}$$

in which g is the percentage of grade divided by 100. The safe stopping distances on upgrades are shorter and on downgrades longer than hori-

TABLE 10-6

Minimum Stopping Sight Distance

Design Speed (mph)	Assumed Speed for Condition (mph)	Perception and Brake Reaction		Coefficient of Friction (*f*)	Braking Distance on Level (feet)	Stopping Sight Distance	
		Time (sec.)	Distance (feet)			Computed (feet)	Rounded for Design (feet)
Design Criteria—Wet Pavements							
30	28	2.5	103	.36	73	176	200
40	36	2.5	132	.33	131	263	275
50	44	2.5	161	.31	208	369	350
60	52	2.5	191	.30	300	491	475
65	55	2.5	202	.30	336	538	550
70	58	2.5	213	.29	387	600	600
75	61	2.5	224	.28	443	667	675
80	64	2.5	235	.27	506	741	750
Comparative Values—Dry Pavements							
30	30	2.5	110	.62	48	158	
40	40	2.5	147	.60	89	236	
50	50	2.5	183	.58	144	327	
60	60	2.5	220	.56	214	434	
65	65	2.5	238	.56	251	489	
70	70	2.5	257	.55	297	554	
75	75	2.5	275	.54	347	622	
80	80	2.5	293	.53	403	696	

Source: A Policy on Geometric Design of Rural Highways, AASHO, 1965, page 136.

zontal stopping distances. Table 10–7 shows the effect of grade on stopping distance. Where an unusual combination of steep grades and high speed occurs the minimum stopping sight distance should be adjusted to provide for this factor.

Measuring Minimum Stopping Sight Distance. It is assumed that the height of eye of the average driver, except drivers of buses and trucks, is about 3.75 feet above the pavement. The height of the stationary object may vary, but a height of 6 inches is assumed in determining stopping sight distance.

TABLE 10-7

Effect of Grade on Stopping Sight Distance: Wet Conditions

Design Speed, mph	Assumed Speed for Condition mph	Correction in Stopping Distance, Feet					
		Decrease for Upgrades			Increase for Downgrades		
		3%	6%	9%	3%	6%	9%
30	28	—	10	20	10	20	30
40	36	10	20	30	10	30	50
50	44	20	30	—	20	50	—
60	52	30	50	—	30	80	—
65	55	30	60	—	40	90	—
70	58	40	70	—	50	100	—
75	61	50	80	—	60	120	—
80	64	60	90	—	70	150	—

Source: A Policy on Geometric Design of Rural Highways, AASHO, 1965, page 139.

Minimum Passing Sight Distance. The majority of our highways carry two lanes of traffic moving in opposite directions. In order to pass slower moving vehicles, it is necessary to use the lane of opposing traffic. If passing is to be accomplished safely, the vehicle driver must be able to see enough of the highway ahead in the opposing traffic lane to permit him to have sufficient time to pass and then return to the right traffic lane without cutting off the passed vehicle and before meeting the oncoming traffic. The total distance required for completing this maneuver is the passing sight distance.

When computing minimum passing sight distances, various assumptions must be made relative to traffic behavior. On two-lane highways it may be assumed that the vehicle being passed travels at a uniform speed, and that the passing vehicle is required to travel at this same speed when the sight distance is unsafe for passing. When a safe passing section is reached, a certain period of time elapses in which the driver decides whether it is safe to pass. When he decides to pass, it is assumed that he accelerates his speed during the entire passing operation. It is also assumed that the opposing traffic appears the instant the passing maneuver starts and arrives alongside the passing vehicle when the maneuver is completed.

Accepting these assumptions, it can be shown that the passing minimum sight distance for a two-lane highway is the sum of four distances: d_1 = distance traveled during perception and reaction time and during the initial acceleration to the point where the vehicle will turn into the opposite lane; d_2 = distance traveled while the passing vehicle occupies the left lane; d_3 = distance between the passing vehicle at the end of its maneuver and the opposing vehicle; d_4 = distance traveled by the oncoming vehicle for two-thirds of the time the passing vehicle occupies the left lane.

The preliminary delay distance d_1 is computed from the following formula:

$$d_1 = 1.47t_1\left(v - m + \frac{at_1}{2}\right) \tag{10–17}$$

where:

t_1 = time of preliminary delay, seconds
a = average acceleration rate, mphps
v = average speed of passing vehicle, mph
m = difference in speed of passed vehicle and passing vehicle, mph

The distance

$$d_2 = 1.47vt_2$$

where:

t_2 = time passing vehicle occupies the left lane, seconds
v = average speed of passing vehicle, mph

The clearance distance d_3 varies from 110 to 300 feet. These distances are adjusted as shown in Table 10–8 and Fig. 10–8. The distance $d_4 = 2d_2/3$.

TABLE 10–8

Elements of Safe Passing Sight Distance—Two-Lane Highways

Speed group, mph Average passing speed, mph	30–40 34.9	40–50 43.8	50–60 52.6	60–70 62.0
Initial maneuver:				
a = average acceleration, mphps*	1.40	1.43	1.47	1.50
t_1 = time, seconds*	3.6	4.0	4.3	4.5
d_1 = distance traveled, feet	145	215	290	370
Occupation of left lane:				
t_2 = time, seconds*	9.3	10.0	10.7	11.3
d_2 = distance traveled, feet	475	640	825	1030
Clearance length:				
d_3 = distance traveled, feet*	100	180	250	300
Opposing vehicle:				
d_4 = distance traveled, feet	315	425	550	680
Total distance, $d_1 + d_2 + d_3 + d_4$, feet	1035	1460	1915	2380

Source: A Policy on Geometric Design of Rural Highways, AASHO, 1965, page 144.
*For consistent speed relation, observed values adjusted slightly.

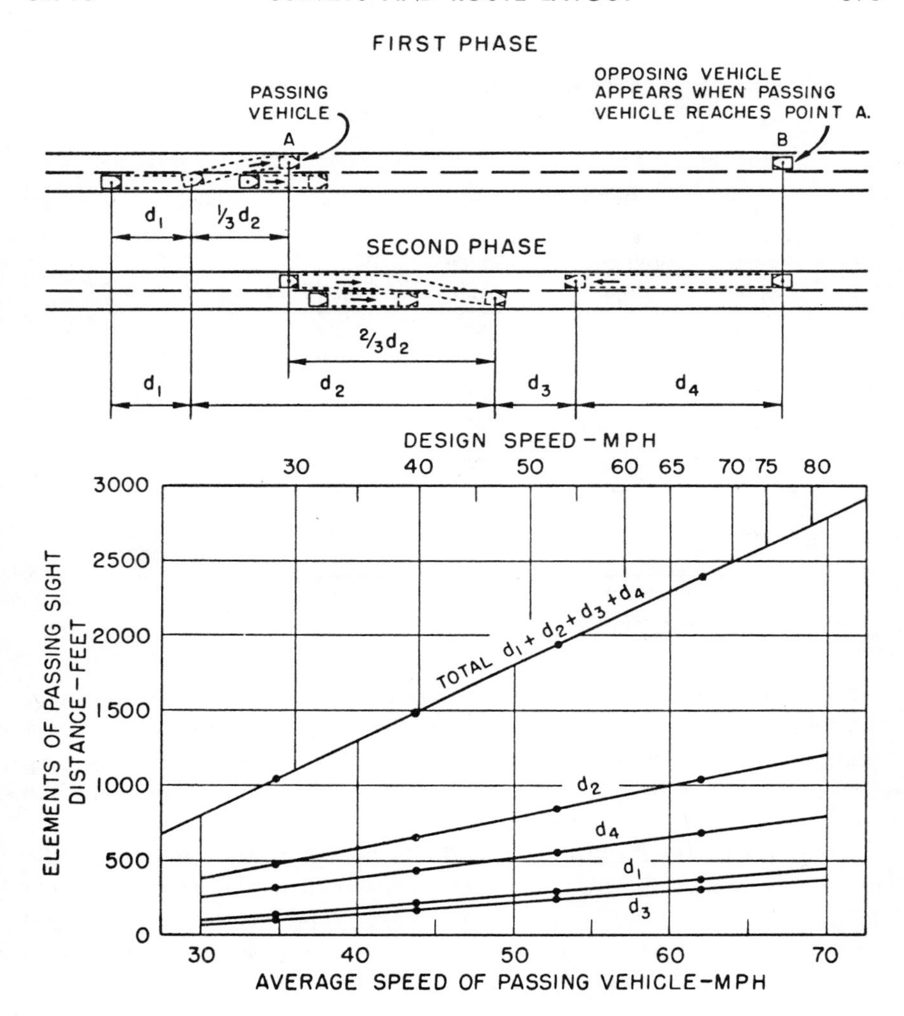

Fig. 10–8. Elements of and total passing sight distance, two-lane highways. (Source: A Policy on Geometric Design of Rural Highways, 1965, AASHO, page 143; Leo Ritter and Radnor Paquette, *Highway Engineering*, 3rd ed., The Ronald Press Co., 1967.)

The relationship of d_1, d_2, d_3, and d_4 is illustrated in Table 10–8 and Fig. 10–8.

Minimum passing sight distances for the design of two-lane highways are shown in Table 10–9. Here the passing speed is related to the design speed of the highway.

TABLE 10-9

Minimum Passing Sight Distance for Design of Two-Lane Highways

Design Speed, mph	Assumed Speeds		Minimum Passing Sight Distance, Feet	
	Passed Vehicle, mph	Passing Vehicle, mph	Fig. 10-11	Rounded
30	26	36	1090	1100
40	34	44	1480	1500
50	41	51	1840	1800
60	47	57	2140	2100
65	50	60	2310	2300
70	54	64	2490	2500
75	56	66	2600	2600
80	59	69	2740	2700

Source: A Policy on Geometric Design of Rural Highways, AASHO, 1965, page 145.

Sight Distances for Four-Lane Highways. A four-lane highway should be so designed that the sight distance at all points is greater than the stopping minimum. The crossing of vehicles into the opposing traffic lane on a four-lane highway should be prevented by a median barrier and prohibited by enforcement agencies.

Extra Lanes for Passing. Inability to pass indicates that the traffic density is greater than the capacity of the road at the assumed design speed and that a wider road is indicated. The capacity of the road may be increased by adding an extra lane at safe passing sections. The added lane should be at least as long as the minimum length of safe passing section to which it is added. Under such conditions it must be remembered that the primary purpose of the added lane is to provide a passing lane at the safe passing section. When additional lanes for passing are used on steep grades, the length of the additional lane should have as great a sight distance as possible.

Measuring Minimum Passing Sight Distance. As previously stated, the height of eye of the average driver is assumed to be about 3.75 feet above the pavement. As vehicles are the objects that must be seen when passing, it is assumed that the height of object for passing sight distance is 4.5 feet.

Horizontal Sight Distances. Horizontal sight distances may be scaled directly from the plans with a fair degree of accuracy. However, where the distance from the highway centerline to the obstruction is known, it is possible to determine the sight distance for a known radius or degree of curve from the following relations:

$$R = \frac{m}{1 - \cos \dfrac{SD}{200}} \tag{10–18}$$

$$S = \frac{200}{D} \cos^{-1} \frac{R - m}{R} \tag{10–19}$$

where:

m = distance from center of roadway to edge of obstruction in feet
S = stopping distance along the center of the road in feet
R = radius of curvature in feet
D = degree of curvature

The formulas do not apply when the length of the circular curve is less than the sight distance, or when the radius of curvature is not constant throughout.

Figure 10–9 shows a horizontal sight distance chart developed from the above formula.

Figures 10–10 and 10–11 show a chart used for the design of crest and sag vertical curves, respectively.

Marking and Signing Nonpassing Zones. The marking of pavements for nonpassing zones is intended to show the motorist that sight distance is restricted. This restriction may be caused by vertical or horizontal alignment or by a combination of both.

The American Association of State Highway Officials has adopted standards for the marking of pavements. A nonpassing zone for two- and three-lane pavements is one in which the sight distance is less than 500, 600, 800, 1000, and 1200 feet for assumed design speeds of 30, 40, 50, 60, and 70 miles per hour, respectively. An example of marking and signing nonpassing zones is shown in Fig. 10–12.

10–23. Vertical Alignment Design for Railroads. Grade design for railroads is similar in many respects to that for highways. As in highway design, there is a need to provide smooth and consistent vertical alignment and to consider controlling elevations of crossing and connecting railroad highways and bridges and drainage structures. Vertical parabolic curves are used to connect intersecting railroad grade lines, and the calculation of curve elevations is accomplished as described in Section 10–21.

Railroad vertical alignment design differs significantly in several respects from the profile grade design of highways. These differences arise from inherent vehicle differences and result in more stringent design

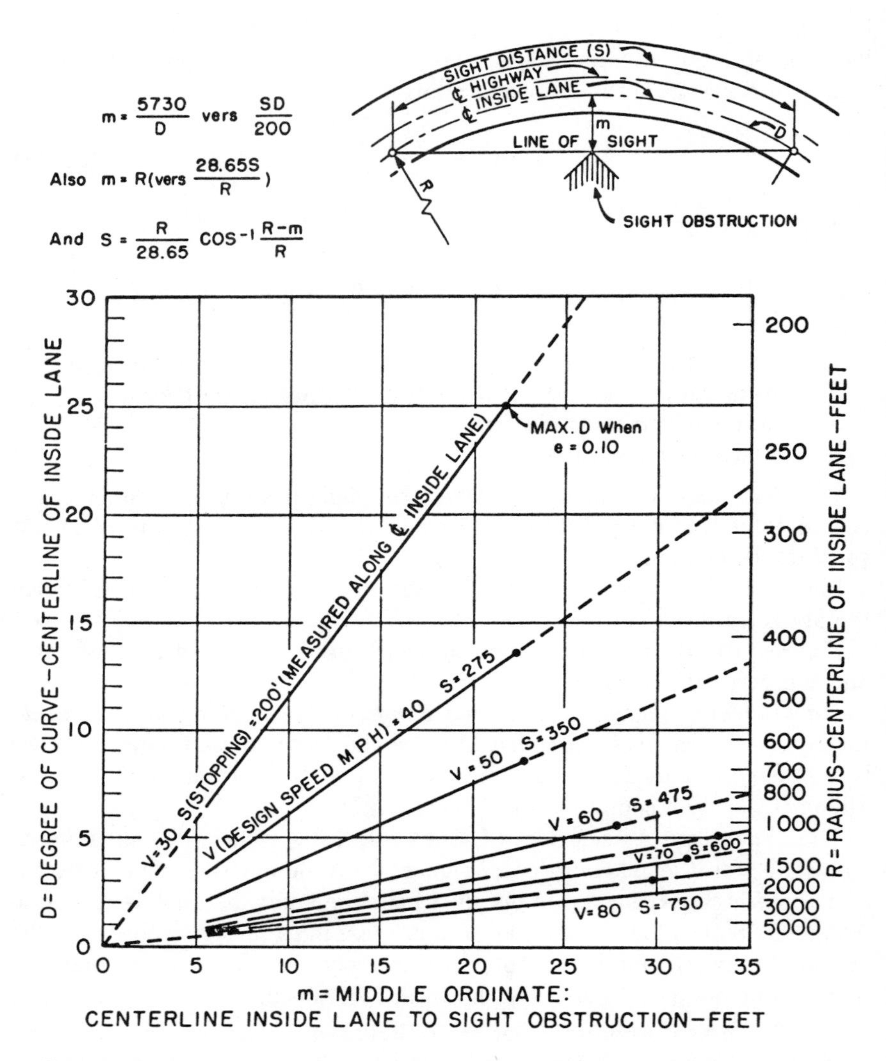

Fig. 10–9. Stopping sight distance on horizontal curves. (Source: A Policy on Geometric Design of Rural Highways, 1965, AASHO, page 188; Leo Ritter and Radnor Paquette, *Highway Engineering*, 3rd ed., The Ronald Press Co., 1967.)

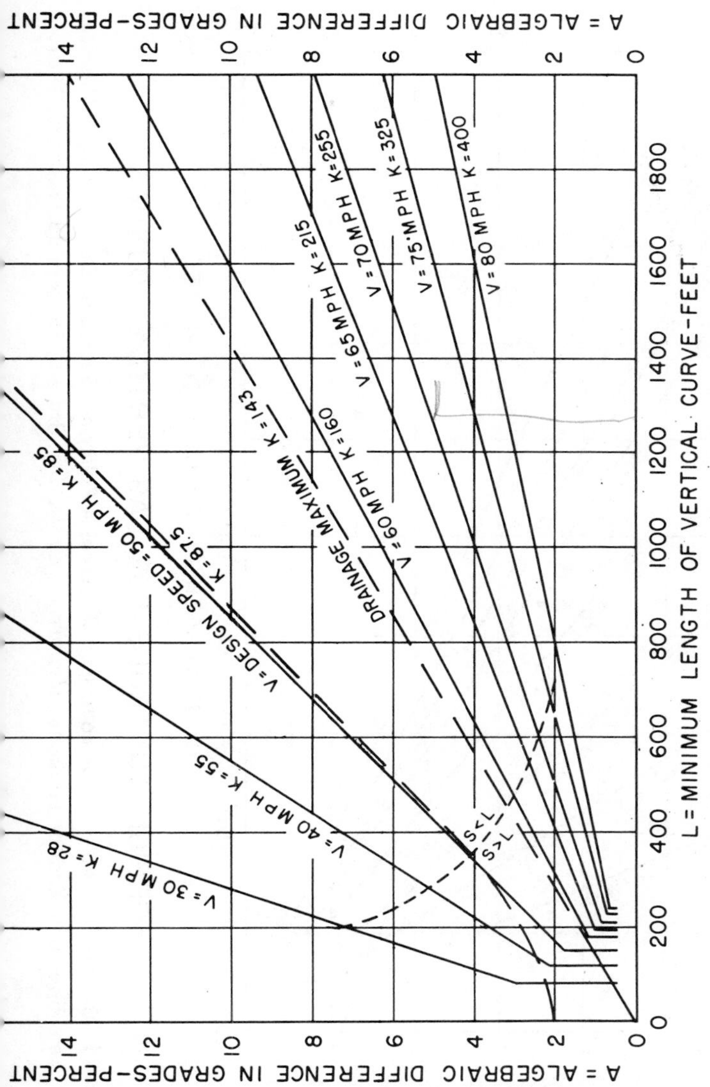

controls for crest vertical curves. (Source: A Policy on Geometric Design of Rural Highways, 1965, AASHO, page 205; Leo Ritter and Radnor Paquette, *Highway Engineering*, 3rd ed., The Ronald Press Co., 1967.)

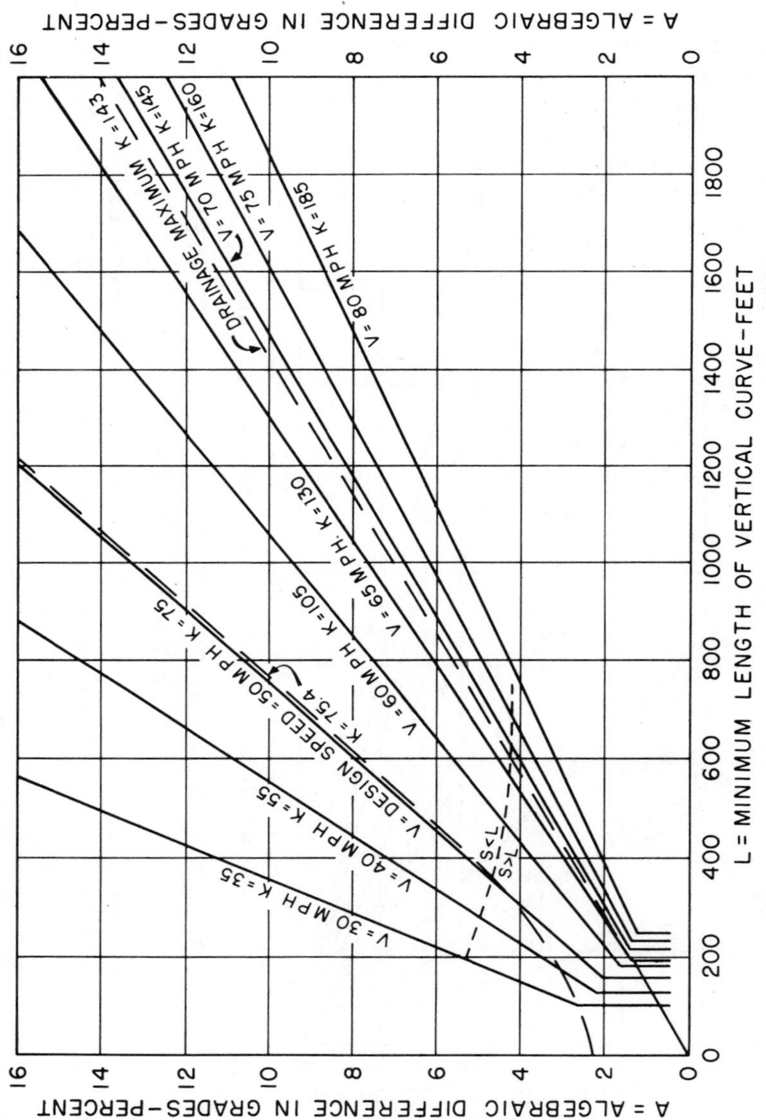

Fig. 10–11. Design for sag vertical controls. (Source: A Policy on Geometric Design of Rural Highways, AASHO, page 210; Leo Ritter and Radnor Paquette, *Highway Engineering*, 3rd ed., The Ronald Press Co., 1967.)

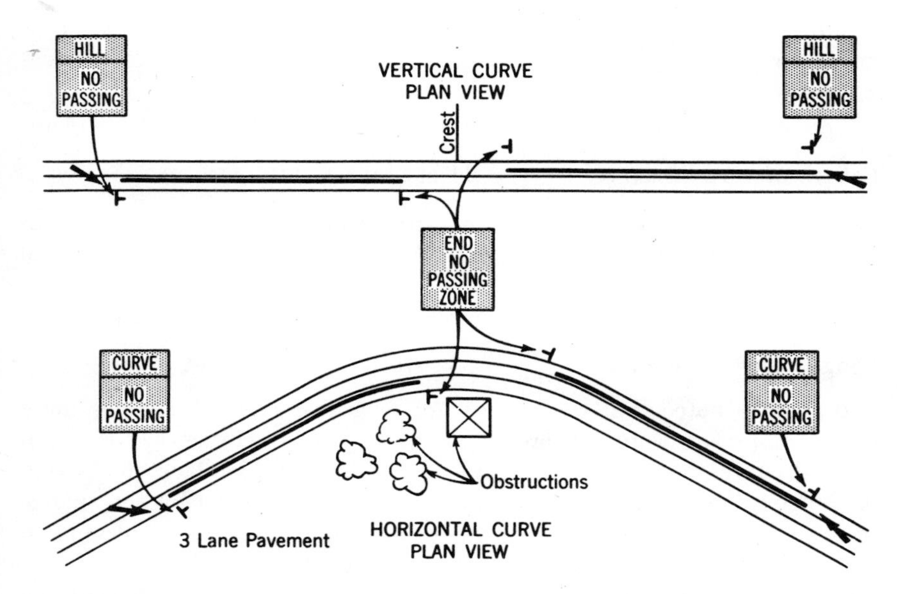

Fig. 10–12. Signs and pavement markings for non-passing zones. (Source: Leo Ritter and Radnor Paquette, *Highway Engineering,* 3rd ed., The Ronald Press Co., 1967.)

criteria for railroads. The need for stricter design criteria for railroads is principally attributed to two considerations:

1. the much longer and heavier railroad vehicle
2. the relatively low coefficient of friction between the driver wheels and the rails.

Railroad design is characterized by much smaller maximum grades and much longer vertical curves than highways. Generally, steep grades cannot be tolerated in railroad design. The maximum grade for most main lines is about 1.0 per cent, although grades as high as 2.5 per cent may be used in mountainous terrain. Even in mountain districts and on lines of secondary importance, the use of grades greater than 4.0 per cent is rare.

The A.R.E.A. [4] gives the following specifications for the calculation of the lengths of vertical curves:

	Maximum Rate of Change of Grade, Feet per Station	
	In Sags	*On Crests*
High Speed Main Tracks	0.05	0.10
Secondary Main Tracks	0.10	0.20

Example 10–1. A plus 0.8 per cent intersects a minus 0.3 per cent grade on a high-speed main track. What minimum length of vertical curve should be used?

This curve is on a crest. The total change of grade is 1.1 per cent.

$$\text{Length of vertical curve} = \frac{1.1}{0.1} = 11 \text{ stations or 1100 feet.}$$

Example 10–2. A minus 0.4 per cent grade intersects a plus 1.2 per cent grade on a high speed main track. What minimum length of vertical curve is required?

This curve occurs in a sag. The total change of grade is 1.6 per cent.

$$\text{Length of vertical curve} = \frac{1.6}{0.05} = 32 \text{ stations or 3200 feet.}$$

It will be noted that the criteria for length of vertical curve is more critical for sags than for crests. Longer vertical curves are required for sag curves because of the tendency of undesirable slack to develop in the couplings as the cars in the front are slowed down by the change in grade. The subsequent removal of the slack may cause jerking to occur which, in extreme cases, may cause a train to break in two.

Finally, the suitability of a railroad profile grade design depends on the ability of trains to operate over the line smoothly and economically and whether trains of a given size (tonnage) will stall on the maximum grade.

In studying performance characteristics of a locomotive and train of cars over a particular stretch of track, the construction of a *velocity profile* may be helpful. The velocity profile is based on the fact that the total energy of a moving train is the sum of the elevation head (potential energy) and the velocity head (kinetic energy). An example of a velocity profile is shown in Fig. 10–13. In this figure, the solid line represents the actual profile of the track, the elevation head. The dashed line represents the virtual or velocity profile, the elevation head plus the velocity head.

The velocity head, h, of a moving train is approximately equal to that of a freely falling body:

$$h = \frac{v^2}{2g} \qquad (10\text{–}20)$$

where:

v = train velocity in feet per second

According to the A.R.E.A., this value should be increased about 6 per cent to allow for energy stored in the rotating wheels. This results in the equation:

$$h = 0.035V^2$$

where:

V = train speed in miles per hour

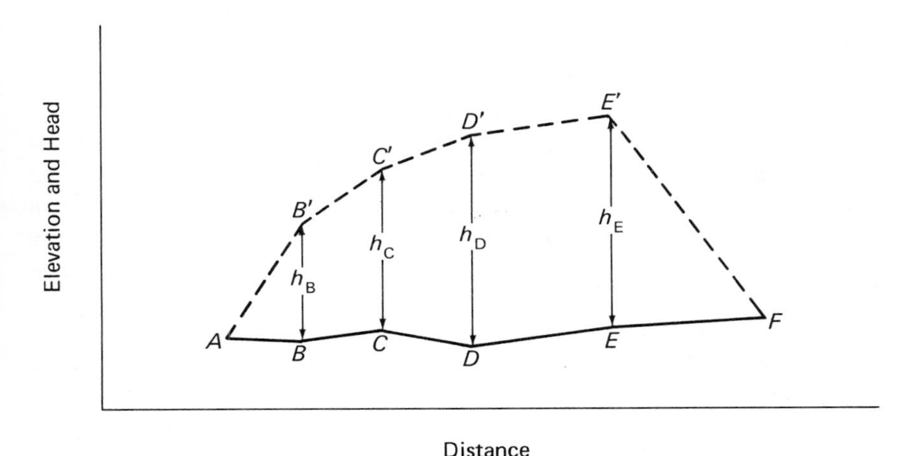

Fig. 10–13. A velocity profile.

Line AB' represents the maximum grade the particular locomotive and train of cars can negotiate at a speed corresponding to point B. This grade is called the *acceleration grade.* Acceleration grade is computed by the following steps:

1. Determine the net tractive effort available for acceleration[3] by subtracting the train resistance (on tangent level track) from the drawbar pull of the locomotive.
2. Express the net tractive effort available for acceleration in pounds per ton by dividing by the total weight of the train.
3. Divide the value obtained in step 2 by 20 pounds/ton/per cent grade, the resistance due to grade. The result is the acceleration grade.

It will be noted that acceleration grade decreases with increase in speed because of similar decreases in drawbar pull.

The general procedure for the construction of a velocity profile is to lay off on a graph of the actual profile the equivalent acceleration grades for speed increments of 0 to 5 mph, 5 to 10 mph, etc. For example, consider first the 0 to 5 mph increment. From the first point on the survey profile (point A), a slope equal to the acceleration grade is laid off until the distance between the two grade lines is equal to the velocity head corresponding to 5 mph, $h_B = 0.035(5)^2 = 0.88$. From point B, a slope equal to the acceleration grade until the distance between the velocity grade line and the actual grade line is equal to the velocity head corresponding to 10 mph, $h_c = 0.035(10)^2 = 3.5$. This

[3] This effort may also be used for climbing grades.

procedure is repeated until the maximum speed is reached. When this occurs, the velocity profile is parallel to the actual profile and continues to be parallel until a steeper grade is reached or the locomotive decelerates due to braking or a reduction of the throttle.

Given a freight train of a certain tonnage and the minimum desirable climbing speed, the velocity profile can be used to determine the *ruling grade*. The ruling grade is defined as the maximum gradient over which a given locomotive pulling maximum tonnage can be hauled at a given constant speed. On a ruling grade, the velocity profile is parallel to the track profile as line $D'E'$ in the figure.

On a grade greater than the ruling grade, termed a *momentum grade,* a part of the momentum of the locomotive and train is used in ascending the grade, resulting in a reduction in speed. The extent of this reduction can be determined by the velocity profile. On a momentum grade, the train decelerates, and the actual and velocity profiles converge.

An extensive discussion of the velocity profile and its application to railroad vertical alignment design has been given by Hay [5].

PROBLEMS

1. A plus 5.2 per cent grade intersects a minus 4.2 per cent grade at station 452 + 85 at an elevation of 1045.09. Calculate the centerline elevations for every 50 foot station for a 600 foot curve.
2. A vertical parabolic curve is to be used under a railroad grade separation structure. The curve is 1000 feet long. The minus grade from left to right is 4.7 per cent and the plus grade is 5.8 per cent. The intersection of the two grades is at station 450 + 25 at an elevation of 1586.50. Calculate the low point at which to put a catch basin for drainage purposes.
3. Given an intersection angle $I = 16°22'$ and a Degree of Curve $D = 1°40'$, determine the various components of a simple curve by the arc definition.
4. Given an I angle $= 20°34'$ and Degree of Curve $D = 2°05'$, determine the various components of a simple curve by the arc definition.
5. Discuss the differences in the superelevation of a highway with that of a railway.
6. Determine the equilibrium superelevation for a 2°45′ railroad curve, given a design speed of 65 miles per hour. What would be the overturning speed for these conditions?
7. It is the policy of a certain railroad to use a 6-inch maximum elevation of the outer rail. Given a 60 mile per hour design speed and a 3°00′ curve, determine the superelevation which should be used and the minimum length of spiral.
8. A plus 1.6 per cent grade intersects a plus 0.2 per cent grade on a secondary main railroad track. What minimum length of vertical curve should be used?

REFERENCES

1. Raymond, William G.; Riggs, Henry E. and Sadler, Walter C., *Elements of Railroad Engineering,* Sixth Edition, John Wiley and Sons, New York (1947).
2. *Proceedings,* 54th Annual Convention, American Railway Engineering Association, Vol. 56. pp. 127–214, 1955.
3. Barnett, Joseph, *Transition Curves For Highways,* U.S. Government Printing Office, 1940.
4. American Railway Engineering Association, *Manual of Recommended Practice,* 1965.
5. Hay, William W., *Railroad Engineering,* John Wiley and Sons, New York (1953).

OTHER REFERENCES

1. A Policy on Geometric Design of Rural Highways, AASHO, 1965.
2. A Policy on Arterial Highways in Urban Areas, AASHO, 1957.
3. Highway Capacity Manual, 1965, Special Report No. 87, 1966.
4. Manual of Uniform Traffic Control Devices for Streets and Highways, AASHO, 1961.
5. Highway Design and Operation Related to Highway Safety, AASHO Special Report, Feburary, 1967.
6. Yoder, E. J., *Principles of Pavement Design,* John Wiley and Sons, New York (1959).
7. "Recommended Practices for Subdivision Streets," *Traffic Engineering,* January, 1967.

11

Design of Streets, Highways, and Railways: Sections and Intersections

11–1. The Highway Cross-Section. The width of the surfaced road should be adequate to accommodate the type and volume of traffic anticipated. Roads currently in use consist of two-lane, three-lane, and multilane and may be undivided or divided.

Many three-lane roads have been built in previous years and are still in use. The great advantage seemed to stem from the large improvement in capacity over the two-lane road with only a moderate increase in construction and right-of-way costs. Accident statistics indicate, however, a high accident rate and the construction of these roads is no longer recommended.

While most design standards permit the lane width to be less than 12 feet, there is general agreement that the 12-foot width is more desirable. On high-speed and high-volume roads 13-foot and 14-foot widths have been used. Widths in excess of 14 feet are not recommended because some drivers will use the roadway as a multilane facility. On some multilane divided highways combinations of 12-foot and 13-foot lanes are used especially where large truck combinations are likely to occur. Table 11–1 gives the AASHO minimum widths of surfacing for two-lane highways.

In urban areas, recommended pavement width varies primarily with the classification of street or highway and the volume of traffic served. The latter factor depends primarily on the development density. Local residential streets vary in width from about 27 to 36 feet, providing safe movement of one lane of traffic in each direction, even where occasional curb parking can be expected to occur.

TABLE 11-1

Minimum Widths of Surfacing for Two-Lane Highways

Design Speed, mph	Minimum Widths of Surfacing in Feet for Design Volume of:				
	Current ADT 50-250	Current ADT 250-400	Current ADT 400-750 DHV 100-200	DHV 200-400	DHV 400 and over
30	20	20	20	22	24
40	20	20	22	22	24
50	20	20	22	24	24
60	20	22	22	24	24
65	20	22	24	24	24
70	20	22	24	24	24
75	24	24	24	24	24
80	24	24	24	24	24

Source: A Policy on Geometric Design of Rural Highways, AASHO, 1965, page 261.

A minimum pavement width of 36 feet is recommended for residential collector streets except in high density developments where at least 40 feet of pavement is specified. These widths provide two traffic lanes, one in each direction, plus space for curb parking on each side. By prohibiting curb parking in the vicinity of intersections, an additional lane may be provided to facilitate turning movements.

Although the four-lane highway is the basic multilane type, traffic volumes may warrant the use of highways having six or even eight lanes, particularly in urban areas.

11–2. Divided Highways. To provide protection against the conflict of opposing traffic, highways are frequently divided by a median strip. On such highways the lane width should be a minimum of 12 feet and, as mentioned previously, 13-foot lanes are desirable where large truck combinations are likely to occur. It is highly desirable that all multilane highways should be divided.

The width of these median strips varies from 4 to 60 feet, or more. A median strip less than 4 to 6 feet in width is considered to be little more than a centerline strip and its use, except for special conditions, should be discouraged. The narrower the median, the longer must be the opening in the median to give protection to vehicles making left turns at points other than intersections. Where narrow medians must be used, many agencies install median barriers to separate physically opposing streams of traffic to minimize the number of head-on collisions.

While medians of 14 to 16 feet are sufficient to provide most of the separation advantage of opposing traffic, medians of 16 to 60 feet are now recommended. The median should also be of sufficient width to maintain vegetation and to support low-growing shrubs that reduce headlight glare of opposing traffic. Median strips at intersections should receive careful consideration and should be designed to permit necessary turning movements.

Divided highways need not be of constant cross-section. The median strip may vary in width; the roads may be at different elevations; and the superelevation may be applied separately in each pavement. In rolling terrain, substantial savings may be affected in construction and maintenance costs by this variation in design. This type of design also tends to eliminate the monotony of a constant width and equal grade.

Medians may be classified as traversable, deterring, or barrier. A traversable median may be an area of pavement of contrasting color, painted stripes, or the use of buttons. A deterring median is one in which a minor physical barrier is used such as a mountable curb or corrugations in the pavement. A barrier median is one in which the use of guard rail, shrubbery, or some type of wall acts as a definite barrier to traffic.

11–3. Parking Lanes and Sidewalks. A parking lane is a lane separate and distinct from the traffic lane. Parking should be prohibited on rural highways but in some rural areas parking adjacent to the traffic lane cannot be avoided. Parallel parking in this case should be permitted and extra lanes provided for this purpose. Where it is desirable to provide parking facilities in parks, scenic outlooks, or other points of interest, off-the-road parking should be provided. In urban and suburban locations the parking area often includes the gutter section of the roadway and may vary from 6 to 8 feet.

The minimum width of parking lane for parallel parking is 8 feet, with 10 feet preferred. For angle parking, the width of the lane increases with the angle. When the angle of parking exceeds 45 degrees, it is necessary to use two moving traffic lanes for maneuvering the vehicle into position. Angle parking should be used only in low-speed urban areas where parking requirements take precedence over the smooth flow of traffic. Parking at the approaches to intersections should be prohibited.

The use of sidewalks in the highway cross-section is accepted as an integral part of city streets. In rural areas little consideration has been given to their construction since pedestrian traffic is very light. Serious consideration should be given to the construction of sidewalks in all areas where the number of pedestrians using the highway warrants it. The Institute of Traffic Engineers has recommended that sidewalks be provided along subdivision streets where the development density exceeds two dwellings per acre. Where the residential density warrants the use of sidewalks, a sidewalk width of 5 feet is recommended.

11–4. Special Cross-Sectional Elements. In the design of multilane highways and expressways, it is necessary to separate the through traffic from the adjacent service or function roads. The width and type of separator is controlled by the width of right-of-way, location and type of overpasses or underpasses, and many other factors. The border strip, which is that portion of the highway between the curbs of the through highway and the frontage or service road, is generally used for the location of utilities.

Special consideration for off-the-road parking facilities, mailbox turnouts, or additional lanes for truck traffic on long grades causes corresponding changes in the cross-sections. Highways involving these elements of the cross-section are usually treated as special cases and receive consideration at the time of their design.

11–5. Right-of-Way. Right-of-way requirements are based on the final design of the cross-sectional elements of the facility. On two-lane secondary highways with an average daily traffic volume of 400 to 1,000 vehicles, a minimum of 66 feet is required, with 80 feet desirable. On the Interstate system minimum widths will vary depending on conditions from 150 feet without frontage roads to 250 feet with frontage roads. An eight-lane divided highway without frontage roads will require a minimum of 200 feet, while the same highway with frontage roads will require a minimum of 300 feet. In rural areas of high-type two-lane highways, a minimum width of 100 feet with a desirable width of 120 feet is recommended. A minimum of 150 feet and a desirable width of 250 feet is recommended for divided highways.

A right-of-way width of 60 feet is generally recommended for local subdivision streets, while collector streets should have 70 feet.

Right-of-way should be purchased outright or placed under control by easement or other means. When this is done, sufficient right-of-way is available when needed. This eliminates the expense of purchasing developed property or the removal of other encroachments from the highway right-of-way. A wide section of right-of-way provides a safer highway, permits gentle rounded slopes, and in general lowers maintenance and snow removal costs.

11–6. Pavement Crown or Slope. Another element of the highway cross-section is the pavement crown or slope. This is necessary for the proper drainage of the surface to prevent ponding on the pavement. Pavement crowns have varied greatly through the years. On the early low-type roads, high crowns were ½ inch or more per foot. Present day high-type pavements with good control of drainage now have crowns as low as 1/8 inch per foot. This has been made possible with the improvement of construction materials, techniques, and equipment which permit closer control. Low crowns are satisfactory when little or no settlement is expected and when the drainage system is of sufficient capacity to remove the water quickly from the traffic lane. When four or more lanes are used, it is desirable to provide a higher rate of crown on the inner lanes in order to expedite the flow of water.

Crowns may be formed by intersecting tangent lines or by curved lines emanating from the road centerline. When curved lines are used, they may be circular arcs of long radii as well as parabolic arcs. It makes little difference which is employed; however, the parabolic arc lends itself better to making computations for the initial offsets or ordinates in constructing templates or setting grades.

11–7. Shoulders. Shoulders should be provided continuously along all highways. This is necessary to provide safe operation and to develop full traffic capacity. Well-maintained, smooth, firm shoulders increase the effective width of the traffic lane as much as 2 feet, as most vehicle operators drive closer to the edge of the pavement in the presence of adequate shoulder. Shoulders should be wide enough to permit and encourage vehicles to leave the pavement when stopping. The greater the traffic density the more likely will the shoulder be put to emergency use. A shoulder width of at least 10 feet and preferably 12 feet clear of all obstructions is desirable for all heavily traveled and high-speed highways. In mountainous areas where the extra cost of providing wide shoulders may be prohibitive, a minimum width of 4 feet may be used, but a width of 6 to 8 feet is preferable. Under these conditions, however, emergency parking strips should be provided at proper intervals. In terrain where guard rails or retaining walls are used, an additional 2 feet of shoulder should be provided.

The slope of the shoulder should be greater than that of the pavement. A shoulder with a high-type surfacing should have a slope at least 3/8 inch per foot. Sodded shoulder may have a slope as high as 1 inch per foot in order to carry water away from the pavement.

11–8. Guard Rails. Guard rails should be provided at points where fills are over 8 feet in height, when shoulder slopes are greater than 1 on 4, and at locations where great reduction in speed is necessary. Deep roadside ditches with steep banks, or right-of-way limitations, often make it necessary to steepen the side slopes and require the use of guard rails. Where guard rails are used, the width of the shoulders is increased approximately 2 feet to allow space for placing the posts.

Various types of guard rails are in use at the present time. The most important of these are the ordinary wooden guard rail, the cable guard rail, the wire mesh rail, the aluminum and steel plate railing, and masonry railing. A great deal of research has been done in recent years to develop safer and more effective guard rails and median barriers. This work is continuing.

11–9. Slopes. The slope from the edge of the shoulder toward the ditch is called the side slope. The slope from the edge of the ditch upward is called the back slope. Side slopes and back slopes may vary a great deal depending upon the type of material and the geographical location. Well-rounded flat slopes present a pleasing appearance and are most economical to build and maintain. Side slopes of 1 on 4 are used a great deal in both cut and fill sections up to about 10 feet in depth or height, but where the height of cut or fill does not exceed 6 feet a maximum side slope of 6:1 is recommended. Extremely flat slopes are sometimes used in swamp sections. Where guard rail is used, slopes may be as high as 1 on 1.5. Slopes as high as 1 on 1 are generally not satisfactory and present an expensive maintenance problem. In certain fill sections, special slopes may be built with riprap, dry masonry rubble, reinforced concrete curbing, and various types of retaining wall.

The back slope in cut areas may vary from a 6 on 1 to vertical in rock or loess formations to 1 on 1.5 in normal soil. It is advisable to have back slopes as flat as 1 on 4 when side borrow is needed. Slope transitions from cut to fill should be gradual and should extend over a considerable length of roadway.

11–10. Curbs, Curb and Gutter, and Drainage Ditches. The use of curbs is generally confined to urban and suburban roadways. The design of curbs varies from a low, flat, lip-type to a nearly barrier-type curb. Curbs adjacent to traffic lanes, where sidewalks are not used, should be low and very flat. The face of the curb should be no steeper than 45 degrees so that vehicles may mount them without difficulty. Curbs at parking areas and adjacent to sidewalks should be 6 to 8 inches in height, with faces nearly vertical. Clearances should be sufficient to clear fenders and bumpers and to permit the opening of doors. When a barrier or nonmountable curb is used, it should be offset a minimum of 10 feet from the traffic lane.

Drainage ditches, in their relation to the highway cross-section, should be located within or beyond the limits of the shoulder and under normal conditions be low enough to drain the water from under the pavement. The profile gradient of the ditch may vary greatly from that of the adjacent pavement. A rounded ditch section has been found to be safer than a V-type ditch, which also may be subject to severe washing action. Ditch maintenance is also less on the rounded ditch sections.

In summing up the various elements of the cross-section, it can readily be seen that the opportunity for wide variations in design presents itself in the selection of crowns, pavement and shoulder widths, slopes, and other features. Every feature, from the centerline to the extremity of the right-of-way, must be given careful consideration for a balanced design. Typical designs of cross-sectional elements are illustrated in Figs. 11–1 and 11–2. Table 11–2 gives geometric design standards for

TABLE 11–2

Geometric Design Standards for a Heavily Traveled Two-Lane Highway

GEOMETRIC DESIGN STANDARDS
Class II
Traffic: 2000 V.P.D. and over (2 lanes)

	Desirable				Minimum			
	F	R	H	M	F	R	H	M
Design Speed (mph)	70	70	60	60	70	60	50	40
Max. Curvature (degree)*	3	3	5	5	3	5	7	12
Max. Gradient (percent)	3	3	5	6	3	4	5	6
Min. Stopping Sight Distance	600′	600′	475′	475′	600′	475′	350′	275
Min. Passing Sight Distance	3200′	3200′	2300′	2300′	2900′	2100′	1800′	1500′
Clear Recovery Area	30′				10′			
Width of Lanes	12′	12′	12′	12′	12′	12′	12′	12
Width of Usable Shoulder	10′	10′	10′	10′	8′	8′	8′	8
Width of Graded Shoulder**	12′	12′	12′	12′	10′	10′	10′	10
Width of Ditch Invert (min.)	2′	2′	2′	2′	2′	2′	2′	2
Right-of-Way Width (min.)***	130′	130′	130′	130′	100′	100′	100′	100

Source: Georgia State Highway Department.
*Min. Length Curve = 1000 ft. for desirable.
**Increase width of shoulder 2′ where guard rail is required.
***60′ acceptable in municipalities on existing right-of-way—or 64′ minimum on acquired right-of-way.
Note: The letters F, R, H, and M refer to the classification of terrain:
F—Flat R—Rolling H—Hilly M—Mountainous

a heavily traveled two-lane highway as promulgated by the Georgia State Highway Department, while Table 11–3 gives recommended standards for subdivision streets with medium development density.

11–11. Limited-Access Highways. A limited-access highway may be defined as a highway or street, especially designed for through traffic, to which motorists and abutting property owners have only a restricted right of access. Limited or controlled access highways may consist of [1] freeways which are open to all types of traffic, and [2] parkways

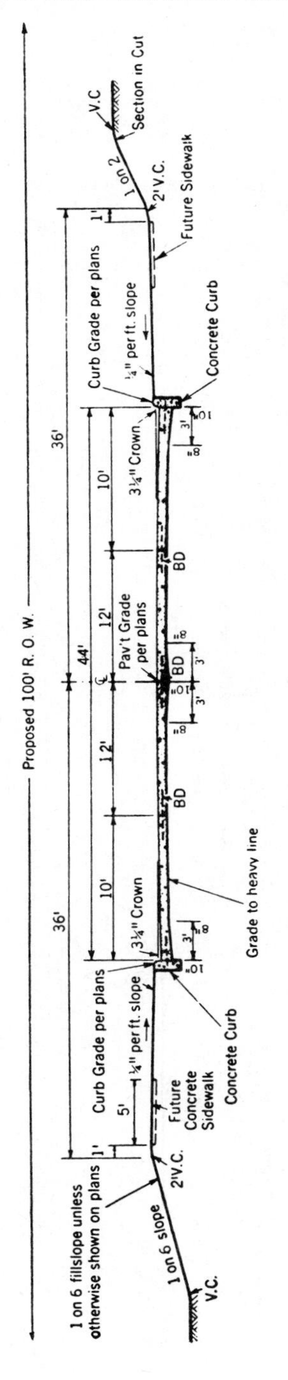

Fig. 11–1. Typical cross-section urban. (Courtesy Michigan Department of State Highways)

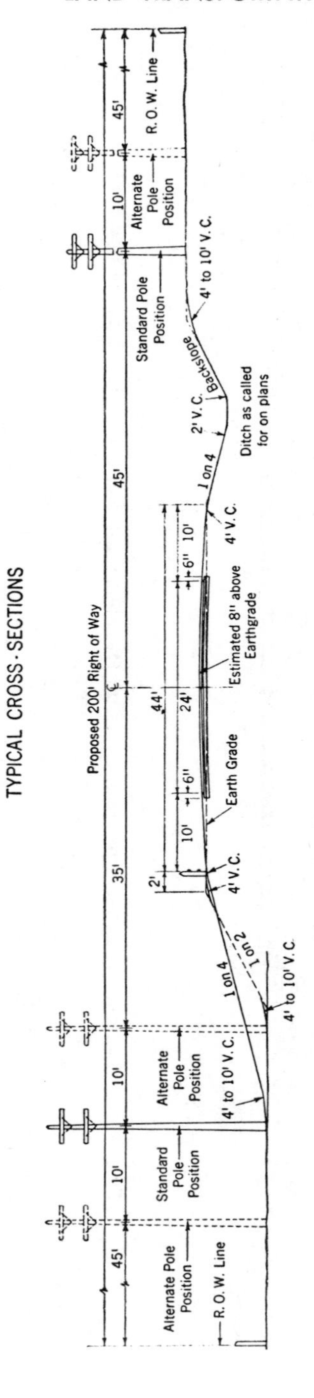

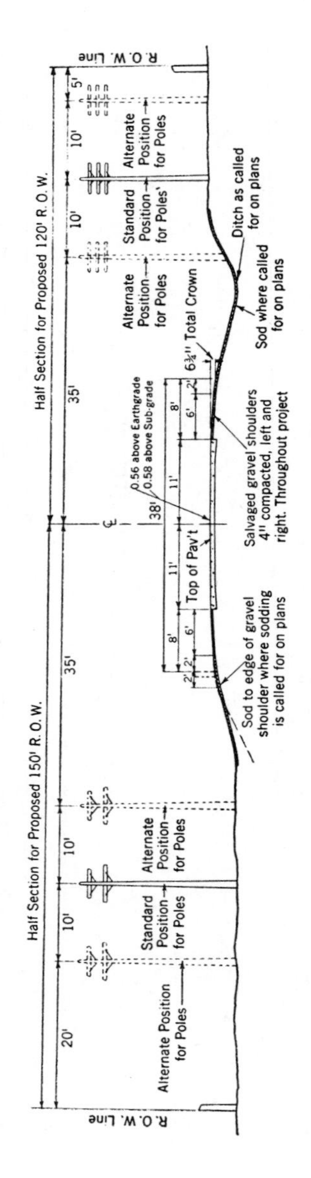

Fig. 11–2. Typical cross-section rural. (Courtesy Michigan Department of State Highways)

TABLE 11-3

Design Standards for Subdivision Streets for Medium Development Density, 2-6 Dwelling Units/Acre

	Local Streets			Collector Streets		
Terrain, Cross Slope, %	0–8	8–15	15	0–8	8–15	15
Right-of-way Width, Ft.	60	60	60	70	70	70
Pavement Width, Ft.	32	34	34	36	36	36
Sidewalk Distance from Curb Face, Ft.	6	6	6	10	10	10
Minimum Sight Distance, Ft.	200	150	110	250	200	150
Maximum Grade, %	4	8	15	4	8	12
Design Speed, MPH	30	25	20	35	30	25
Minimum Centerline Radius, Ft.	250	175	110	350	230	150
Maximum Cul-De-Sac Length, Ft.	500	500	500	–	–	–
Minimum Cul-De-Sac Radius, R.O.W., Ft.	50	50	50	–	–	–

Source: *Traffic Engineering*, January, 1967.

from which all commercial traffic is excluded. Most of our present expressway systems have been developed as freeways.

Limited-access highways may be elevated, depressed, or at grade. Many examples of the various types may be found in the United States in both rural and urban areas.

The control of access is attained by limiting the number of connections to and from the highway, facilitating the flow of traffic by separating cross-traffic with overpasses or underpasses, and by eliminating or restricting direct access by abutting property owners.

The need for limited access in urban and suburban areas is apparent in planning for present and future traffic needs. High traffic densities, congestion, delays, loss of life, and property damage are factors demanding better traffic facilities. Due to traffic congestion, property values in the heart of many municipalities are diminishing and blight areas are increasing. Control of access is necessary to maintain the design capacity of the facility, reduce accidents, and protect the public investment in highways and streets. Rural highways also need control of access but for reasons slightly different from those in urban and suburban areas. While it is desirable to facilitate the flow of traffic at moderate speeds on rural routes, consideration should also be given to their use. It is important to maintain esthetic and scenic values to control roadside encroachment. Very often, when a new highway is constructed without access control, many new businesses spring up along its route. The effect of this so-called "ribbon development" in rural areas tends to make a highway obsolete long before the physical structure wears out.

Many states have adopted legislation aiding the concept of limited access and the control of highway development. The adoption of such legislation was given a tremendous boost by the Federal-Aid Highway Act of 1956, with its basic requirement of control of access on all mileage of the Interestate System. One approach to the problem which is worthy of note has been made by the Ohio Department of Highways. By means of reservation agreements between the Ohio Department of Highways and the owner, the state acquires specific rights in designated areas for a nominal consideration. The owner is permitted the use of the reserved areas for normal purposes, and the state is expected to prevent the erection of buildings and public utility facilities that would increase the cost of acquiring the easement. Areas thus reserved serve to control developments along the highway and make possible the easy taking of this land as needed.

The design of limited-access routes should provide for adequate width of right-of-way, adequate landscaping, prohibition of outdoor advertising in the controlled access proper, and provisions for controlling abutting service facilities such as gas stations, parking areas, and other roadside appurtenances.

In urban areas, the design of a limited-access facility is usually accompanied by the design of frontage roads, parallel to the facility, which may serve local traffic and provide access to adjacent property. Such roads may be designed for either one-way or two-way operation. Reasonably convenient connections should be provided between through-traffic lanes and frontage roads.

11–12. Design Standards for the Interstate System. The National System of Interstate and Defense Highways carries more traffic per mile than any other comparable national system and includes the roads of greatest significance to the economic welfare and defense of the nation. The highways of this system must be designed with control of access to insure safety, permanence, utility, and flexibility to provide for possible future expansion. A minimum of four traffic lanes shall be provided.

All interstate highways are required to meet the minimum standards established by the *Policy on Design Standards* adopted July 12, 1956, by AASHO, and later revised. The following has been abstracted from this *Policy* as revised September 15, 1966.

Traffic Basis. Interstate highways shall be designed to serve safely and efficiently the volumes of passenger vehicles, buses, and trucks, including tractor-trailer and semitrailer combinations and corresponding military equipment, estimated to be that which will exist 20 years beyond that in which the plans, specifications, and estimates for actual construction of the section are approved, including attracted, generated, and development traffic on the basis that the entire system is completed.

The peak-hour traffic used as a basis for design shall be as high as the thirieth highest hourly volume of the design year, *DHV*. Unless otherwise specified, *DHV* is the total, two-direction volume of mixed traffic.

Control of Access. On all sections of the Interstate System, access shall be controlled by acquiring access rights outright prior to construction.

Gradients. For design speeds of 70, 60, and 50 miles per hour, gradients generally shall be not steeper than 3, 4, and 5 per cent, respectively. Gradients 2 per cent steeper may be provided in rugged terrain.

Medians. Medians in rural areas in flat and rolling topography shall be at least 36 feet wide. Medians in urban and mountainous areas shall be at least 16 feet wide. Narrower medians may be provided in urban areas of high right-of-way cost, on long, costly bridges, and in rugged mountainous terrain, but no median shall be less than 4 feet wide.

Curbs or other devices may be used where necessary to prevent traffic from crossing the median.

Where continuous barrier curbs are used on narrow medians, such curbs shall be offset at least 1 foot from the edge of the through-traffic lane. Where vertical elements more than 12 inches high, other than abutments, piers, or walls, are located in a median, there shall be a lateral clearance of at least 3.5 feet from the edge of a through-traffic lane to the face of such element.

Shoulders. Shoulders usable by all classes of vehicles in all weather shall be provided on the right of traffic. The usable width of shoulder shall be not less than 10 feet. In mountainous terrain involving high cost for additional width, the usable width of shoulder may be less, but it must be at least 6 feet. Usable width of shoulder is measured from the edge of a through-traffic lane to intersection of shoulder and fill or ditch slope except where such slope is steeper than 4:1, where it is measured to beginning of rounding.

Slopes. Side slopes should be 4:1 or flatter where feasible and not steeper than 2:1 except in rock excavation or other special conditions. (Note: Under new criteria from the special AASHO report of February, 1967, entitled *Highway Design and Operation Related to Highway Safety,* the slopes should be 6:1 for 18 feet past the shoulder point. No obstructions, including signs, should be not less than 18 feet from the shoulder point. The use of breakaway posts to be used where necessary.)

Control of Access. On all sections of the Interstate System, access shall be controlled by acquiring access rights outright prior to construction or by the construction of frontage roads, or both. Control of access is required for all sections of the Interstate System, including the full length of ramps and terminals on the crossroad. Control for connections to the crossroad should be effected beyond the ramp terminals by purchasing of access rights, providing frontage roads to control access, controlling added corner right-of-way areas, or denying driveway permits. Such control should extend along the crossroads beyond the ramp terminal about 100 feet or more in urban areas and about 300 feet or more in rural areas. Under all of the following conditions, intersections at grade may be permitted in sparsely settled rural areas which are a sufficient distance from municipalities or other traffic generating areas to be outside their influence, and where no appreciable hazard is created.

1. The interstate highway has a *DHV* of less than 500.
2. The intersection at grade is with a public road or private driveway with little potential for traffic increase and on which the current *ADT* does not exceed 50 vehicles.
3. Such intersections do not exceed two per mile of the Interstate highway.
4. At the time of initial construction additional right-of-way is obtained or controlled so as to permit the ultimate construction of an interchange in place of an intersection with a public road.
5. The right to eliminate, terminate, or reroute each private driveway that intersects at grade with an Interstate highway is vested in the appropriate public authority at the time of initial construction.

Railroad Crossings. Railroad grade crossing shall be eliminated for all through-traffic lanes.

Intersections. All at-grade intersections of public highways and private driveways shall be eliminated, or the connecting road terminated, rerouted, or intercepted by frontage roads, except as otherwise provided under Control of Access.

Design Speed. The design speed of all highways on the system shall be at least 70, 60, and 50 miles per hour for flat, rolling, and mountainous topography, respectively, and depending upon the nature of terrain and development. The design speed in urban areas should be at least 50 miles per hour.

Curvature, Superelevation, and Sight Distance. These elements and allied features, such as transition curves, should be correlated with the design speed in accordance with the *Policy on Geometric Design of Rural Highways* of the AASHO (1965).

Right-of-Way. Fixed minimum widths of right-of-way are not given because wide widths are desirable, conditions may make narrow widths necessary, and right-of-way need not be of constant width. The following minimum widths are given as guides.

In rural areas right-of-way widths should be not less than the following, plus additional widths needed for heavy cuts and fills:

	Minimum Width in Feet	
Type of Highway	*Without Frontage Roads*	*With Frontage Roads*
Four-lane divided	150	250
Six-lane divided	175	275
Eight-lane divided	200	300

In urban areas, right-of-way width shall be not less than that required for the necessary cross-section elements, including median, pavements, shoulders, outer separations, ramps, frontage roads, slopes, walls, border areas, and other requisite appurtenances.

Bridges and Other Structures. The following standards apply to interstate highway bridges, overpasses, and underpasses. Standards for crossroad overpasses and underpasses are to be those for the crossroad.

1. Bridges and overpasses, preferably of deck construction, should be located to fit the overall alignment and profile of the highway.
2. The clear height of structure shall be not less than 16 feet over the entire roadway width, including the usable width of shoulders. In urban areas, this clearance shall be applied to a single route (primarily for defense purposes). On other urban routes, the clear heights are to be not less than 14 feet. Allowance should be made for any contemplated resurfacing.
3. The widths of all bridges, including grade separation structures, between inside faces of curbs, when used, or between rails or parapets when curbs are not used, shall equal the full roadway width on the approaches, including usable width of shoulders. Curbs, if used, shall not exceed 18 inches in width. Major longspan bridges may have lesser width which shall be individually analyzed.
4. Barrier curbs on other bridges and curbs on approach highways, if used, shall be offset at least 2 feet. Offset to face of parapet or rail shall be at least 3.5 feet measured from edge of through-traffic lane and shall apply on right and left.
5. The lateral clearance from the edge of through-traffic lanes to the face of walls or abutments and piers at underpasses shall be the usable shoulder width but not less than 8 feet on the right and 4.5 feet on the left.
6. A safety walk shall be provided in tunnels.

11–13. The Railway Cross-section. New railroad construction in the United States has practically ended. Most work today consists of improving existing lines and relocating lines made necessary by government dams and irrigation projects. Minor work in the construction of spur line facilities to industrial sites continues but the main work in railroads consists of the maintenance-of-way and the operation of its rolling stock. The cross-section elements are given here because of the relattionship to other transportation modes necessary for an integrated transportation system. Figure 11–3 shows typical cross-sections for tangent track sections.

Standards for the width of the subgrade are determined for mainline, secondary line, and light traffic branch lines and spurs. Many railroads have adopted 20 feet as the width of single-lane main lines. This width will also vary according to the height of fill. A width of 22 feet has been used for fills under 20 feet; 24 feet for fills 20 to 50 feet; and 26 feet for fills over 50 feet. Standard widths in cuts including side ditches is 30 feet. In some locations 40 feet is used to permit the use of off-track equipment for maintenance purposes. Common widths of rights-of-way in open country are 50, 60, 80, 100, and sometimes 200 and 400 feet.

11–14. Ballast. Track ballast is a structural element of the railroad permanent way. Its prime function is to spread the wheel loadings from the base of the cross ties to the subgrade at pressures that will not cause subgrade failure. In addition, ballast serves to anchor the track, preventing longitudinal and transverse track movements under dynamic train loadings, to provide immediate drainage of the permanent way under the ties, and to provide a road material that inhibits vegetation growth and minimizes dust. The A.R.E.A. has set quality standards on ballast with reference to the following criteria:

1. *Wear resistance.* Under the Los Angeles abrasion test, percentage of wear of any ballast material is limited normally to 40 per cent.

2. *Cleanliness.* Deleterious substances are limited in prepared ballasts to the following amounts:

Soft and friable pieces	5 per cent
Material finer than No. 200 Sieve	1 per cent
Clay lumps	0.5 per cent

3. *Frost resistance.* Ballast must be capable of resisting freeze-thaw cycles. A.R.E.A. requires an average weight loss of not more than 10 per cent after 5 cycles of the sodium sulfate soundness test.

4. *Unit weight.* Specifications require compacted weights of not less than 70 and 100 lb. per cubic foot for blast furnace and open hearth slugs, respectively.

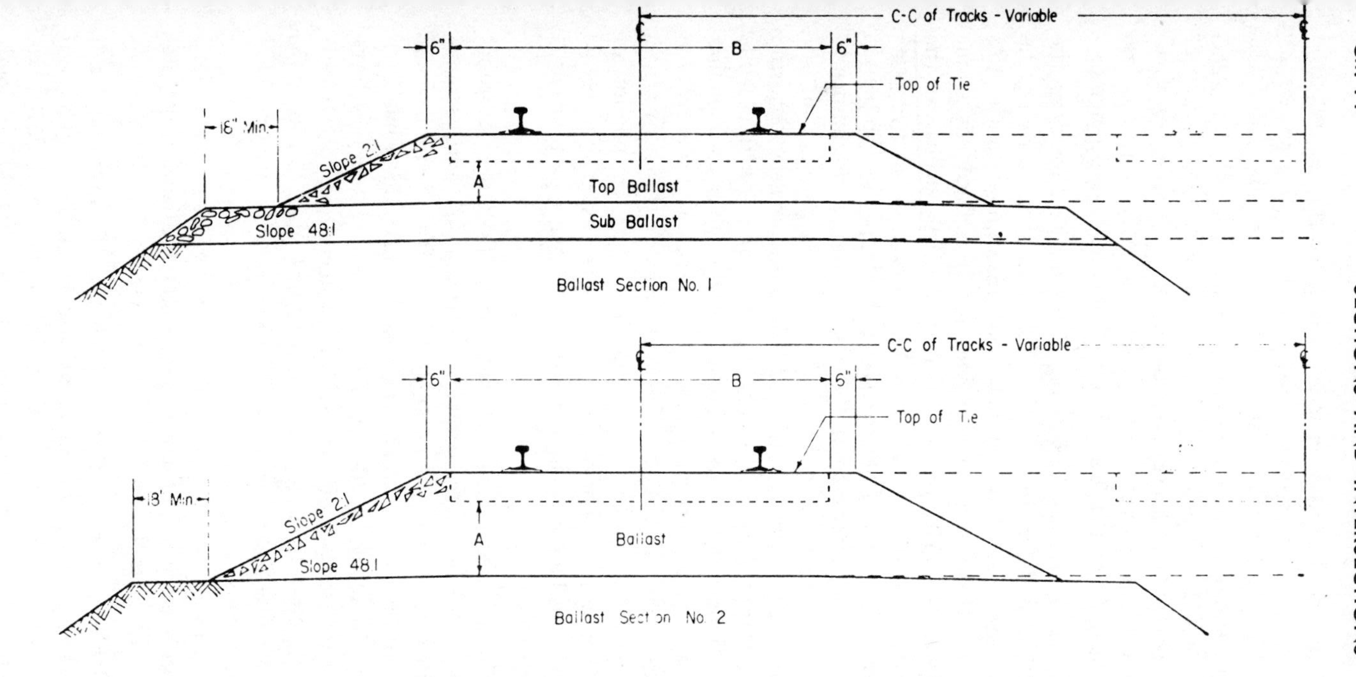

AREA ballast sections, single and multiple track, tangent

Notes:

Depth of ballast section to be used will depend on conditions peculiar to each railroad or location.
Sections apply to all types of ballast.
Sections for use with jointed or continuous rail.
Top of ballast determined by the use of various mechanized ballast distributing operations.

Fig. 11–3. Tangent track sections. (Source. *A.R.E.A. Manual,* Volume I, p. 1–2–7.)

Open graded materials that can satisfactorily perform the required functions of ballast are crushed stone, washed river or pit run gravel, and furnace slags. Typically, material varies in grain size from 1½ to 1¾ inches. Where ballast material is expensive or in short supply, or where subgrade strength is sufficiently low that excessive depths of ballast would be required, a layer of *sub-ballast* is frequently used. Material for the sub-ballast layer of the permanent way can be a less openly graded material meeting less stringent quality requirements.

11–15. Crossties. The crosstie serves the function of spreading the horizontal and vertical rail loadings to the ballast and maintaining the correct gauge between rails. Vertical loadings are applied to the ties by the train weight. Horizontal longitudinal loadings occur as trains accelerate and decelerate, while transverse loadings are applied as the vehicles traverse curved sections. Additional transverse loads are present due to the "barreling" effect of locomotives at high speed. To permit the horizontal transfer of forces from the tie, ballast is mechanically tamped between ties as shown in Fig. 11–3.

Typically, ties are made of wood, treated with both preservative and coating materials for protection against weathering and splitting while in service. Tie sections vary from 6 inches × 6 inches to 7 inches thick × 9 inches wide. Tie lengths also vary. Standard sizes are 8 feet, 8 feet 6 inches, and 9 feet long. Tie replacement, which averages about 4 per cent of all ties yearly, accounts for a large proportion of total track maintenance. Although initially more expensive, some European railroads have had satisfactory experience with prefabricated prestressed concrete cross ties.

11–16. Rails. The steel rail has a characteristic T-shape. It functions as a continuous steel beam, transmitting vertical loads and horizontal shears to the ties via the tieplates and spikes. Rail sections come in standard lengths of 39 feet. Lately, where new construction has been carried out with continuously welded rails, lengths of 1,440 feet have been used. The design section of the rail varies according to the railroad. Rails are designated by weight in lbs. per yard (e.g., 115 lb./yd.). Available section weights vary from 50 to 155 lbs./yd. Figure 11–4 shows a typical rail section.

Rail gauge, the distance between the inside rail heads, is a standard 4 feet, 8½ inches in the United States. Gauge standardization permits free interchange of rolling stock on a nationwide basis.

11–17. Spikes, Tieplates, and Joint Bar Assemblies. Traditionally, rail sections are connected by the use of joint bar assemblies. The rail is connected to the crosstie by the device of the tieplate and rail spike. The tieplate serves as a metal base plate to bear the vertical rail loads

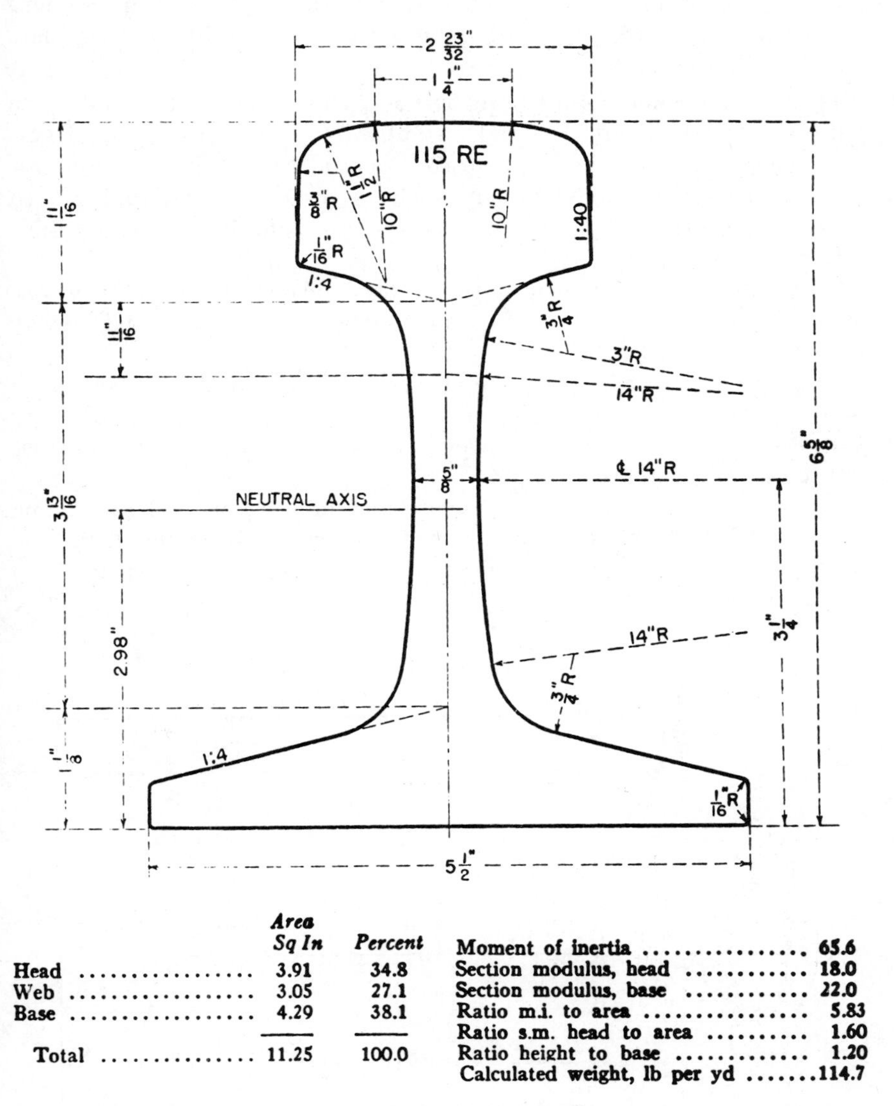

	Area Sq In	*Percent*
Head	3.91	34.8
Web	3.05	27.1
Base	4.29	38.1
Total	11.25	100.0

Moment of inertia	**65.6**
Section modulus, head	**18.0**
Section modulus, base	**22.0**
Ratio m.i. to area	**5.83**
Ratio s.m. head to area	**1.60**
Ratio height to base	**1.20**
Calculated weight, lb per yd	**114.7**

Fig. 11–4. 115 RE rail section. (Source: *A.R.E.A. Manual,* p. 4–1–3.)

while the spikes maintain the rail guage against possible movements due to horizontal shears. Although the spike and the plate system were developed with the introduction of the locomotive, it still functions satisfactorily. Where new sections of track have been laid with continuously welded rail, specialized track arrangements have generally been found necessary differing from the conventional track described in Sections 11–11 through 11–16. The rail spike, joint bar assembly and tie plate are shown in Fig. 11–5.

11–18. Urban Rail Transit Cross-Sections. The cross-sections of urban rail transit systems are essentially similar to conventional railroad systems where duorail systems are used. Special cross-sections are necessary for non-conventional designs such as suspended and rubber-tired systems. Figure 11–6 gives typical cross-sections for suspended and supported systems.

In designing the alignment of rail transit and limited access freeways, the engineer faces the choice of elevated, at-grade, depressed, and subway sections.

At-grade alignments being the cheapest to construct have large economic advantages. This type of facility causes the most community disruption from the viewpoint of severing existing street patterns. Noise and fume levels can also be high.

Depressed systems tend to cut noise and fume levels and, depending on the number of bridge structures provided, can decrease the disruption of existing traffic patterns. Depressed facilities are more expensive to construct than at-grade alignments.

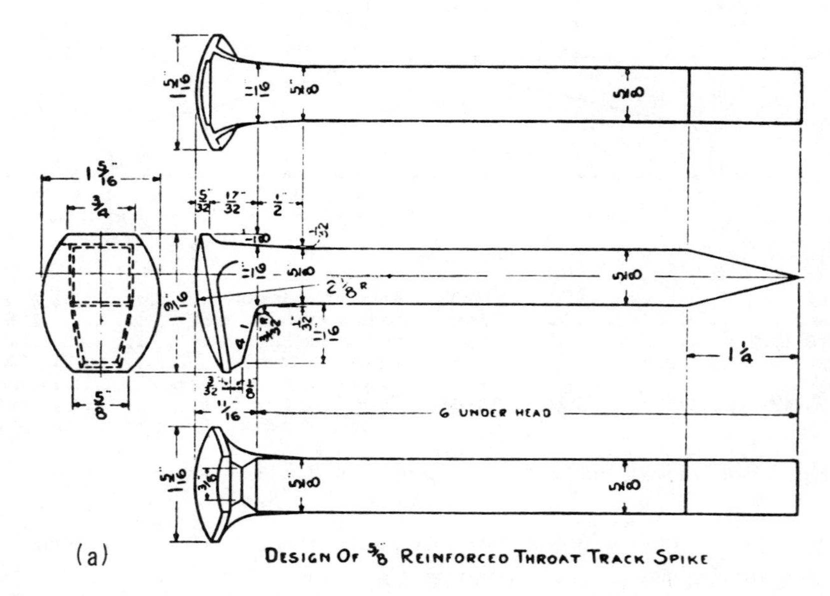

Fig. 11–5. Design of (a) 5/8″ reinforced throat track spike, (b) joint bar assembly, and (c) rail tie plate. (Source: *A.R.E.A. Manual,* pp. 5–2–6, 4–1–8, 5–1–11.)

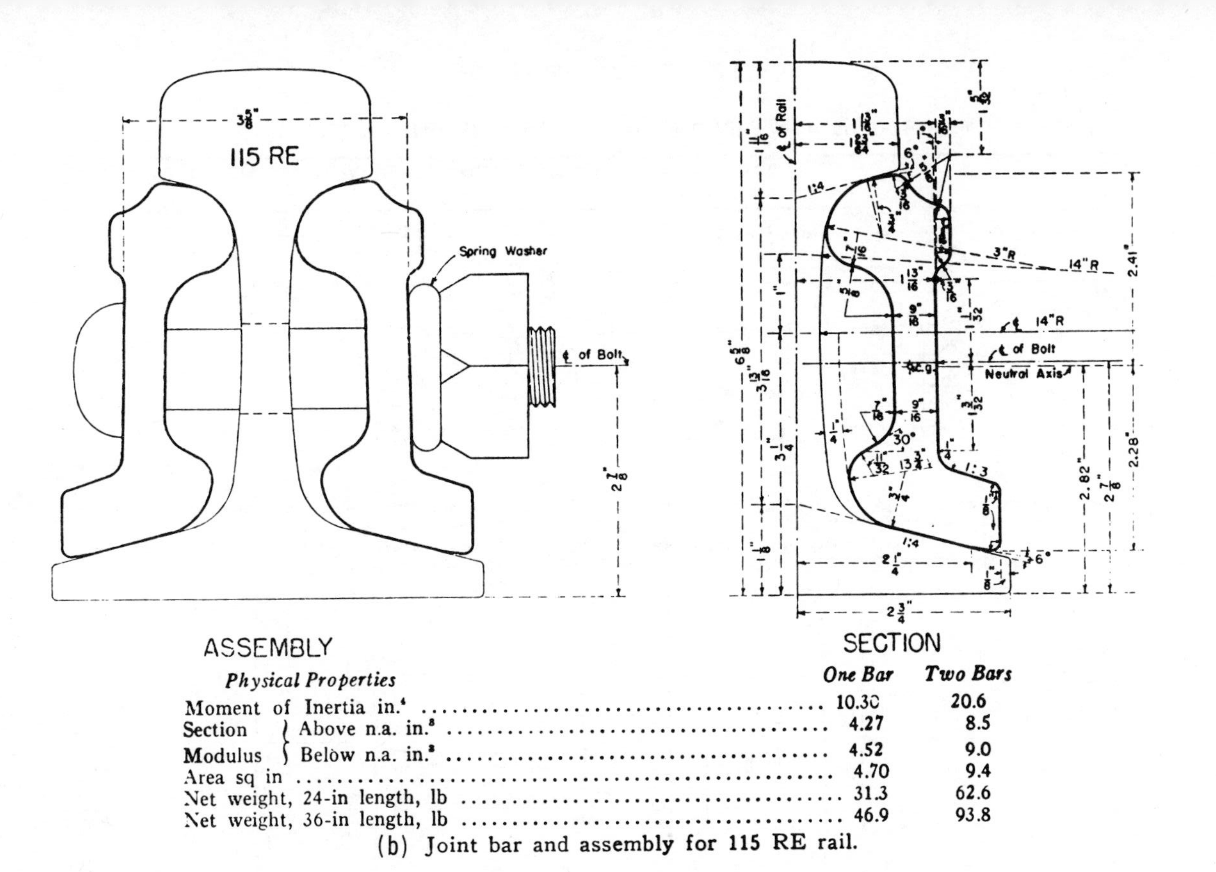

Physical Properties	*One Bar*	*Two Bars*
Moment of Inertia in.[4]	10.30	20.6
Section Modulus { Above n.a. in.[3]	4.27	8.5
Section Modulus { Below n.a. in.[3]	4.52	9.0
Area sq in	4.70	9.4
Net weight, 24-in length, lb	31.3	62.6
Net weight, 36-in length, lb	46.9	93.8

(b) Joint bar and assembly for 115 RE rail.

Fig. 11–5. (Continued.)

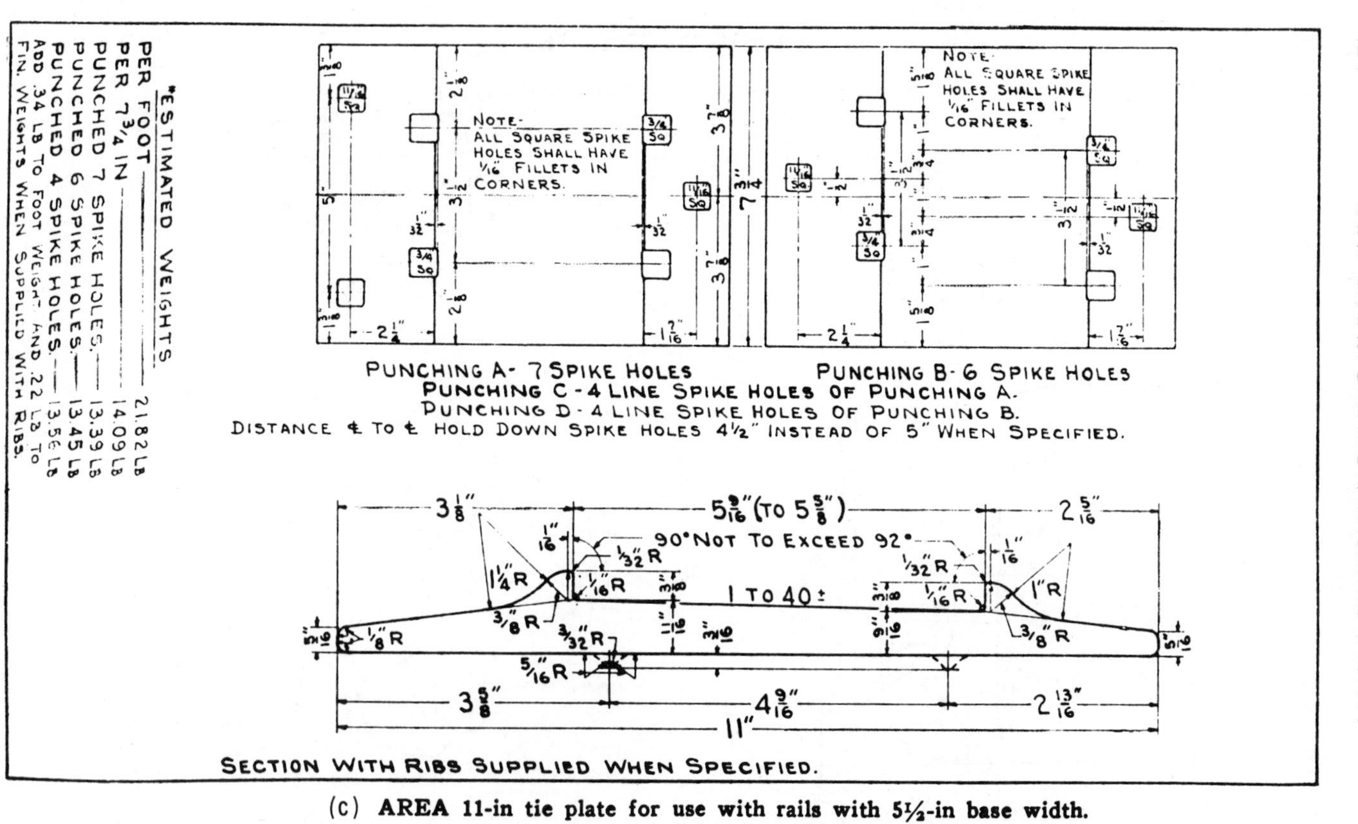

(c) **AREA 11-in tie plate for use with rails with 5½-in base width.**

Fig. 11–5. ***(Continued.)***

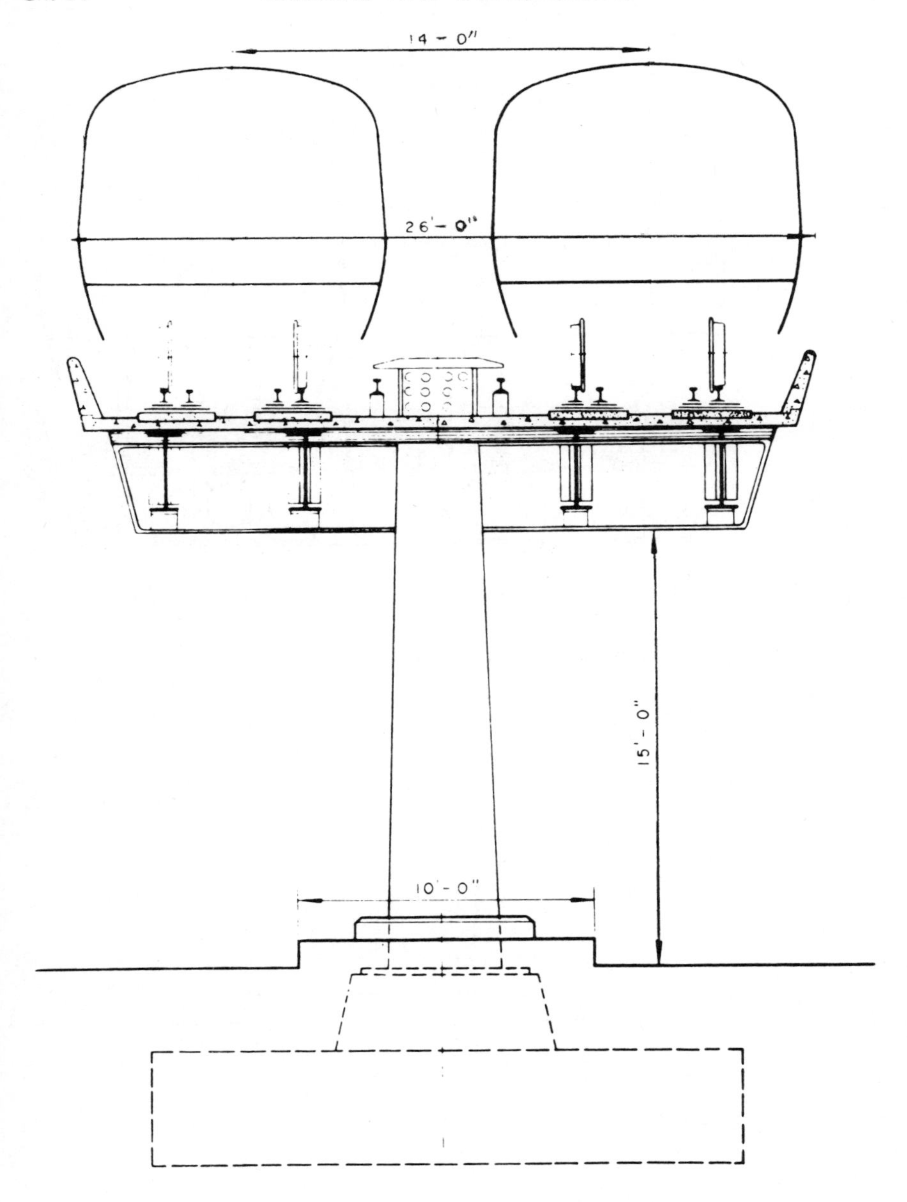

TYPICAL STEEL ELEVATED STRUCTURE FOR THE SUPPORTED SYSTEM

Showing concrete deck on steel beams.

Fig. 11–6. Typical urban rail transit roadbed cross-sections. (Source: *Regional Rapid Transit for Bay Area,* Parsons, Brinkerhoff, Quade & Douglas, pp. 84–88.)

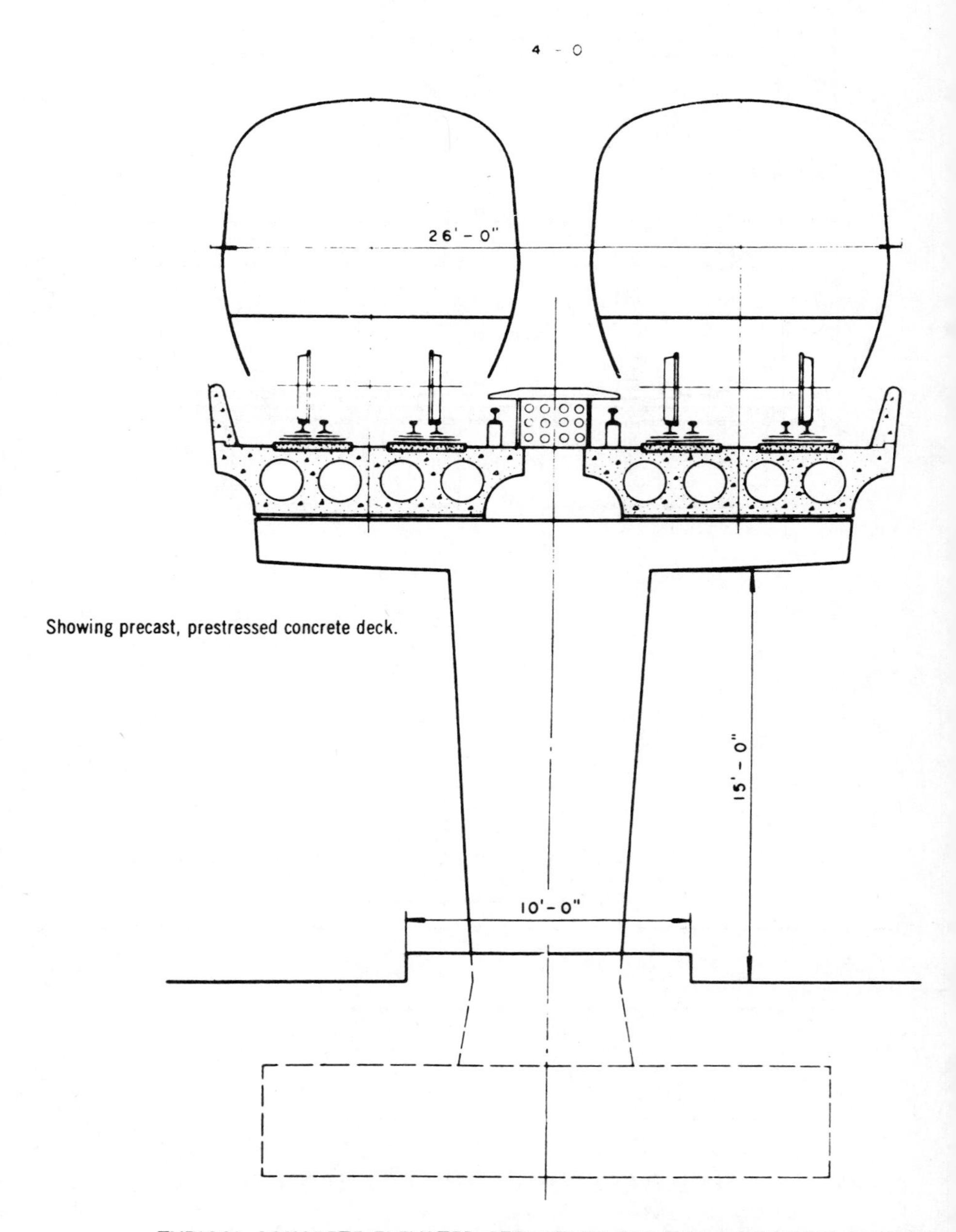

TYPICAL CONCRETE ELEVATED STRUCTURE FOR THE SUPPORTED SYSTEM

Fig. 11–6. (*Continued.*)

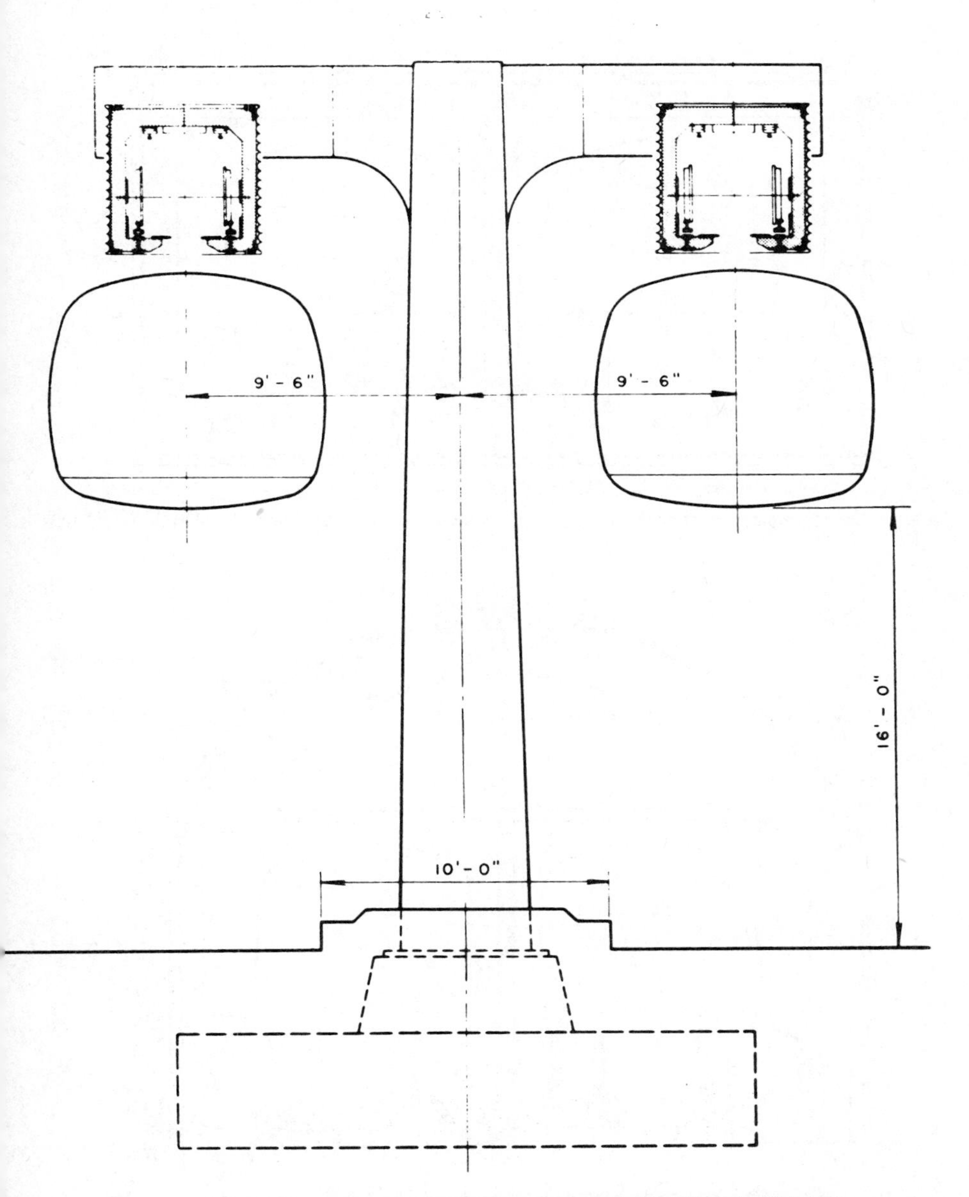

TYPICAL STEEL ELEVATED STRUCTURE FOR THE SUSPENDED SYSTEM

Fig. 11–6. *(Continued.)*

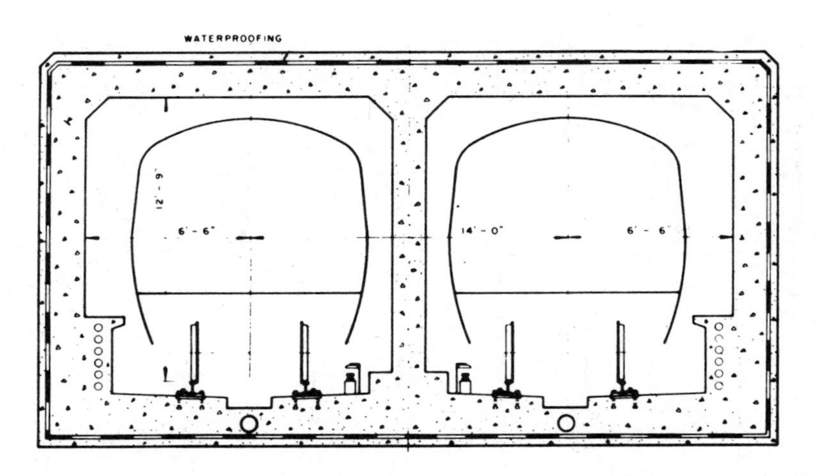

Showing application for supported system, using third-rail current collection. UNDERGROUND

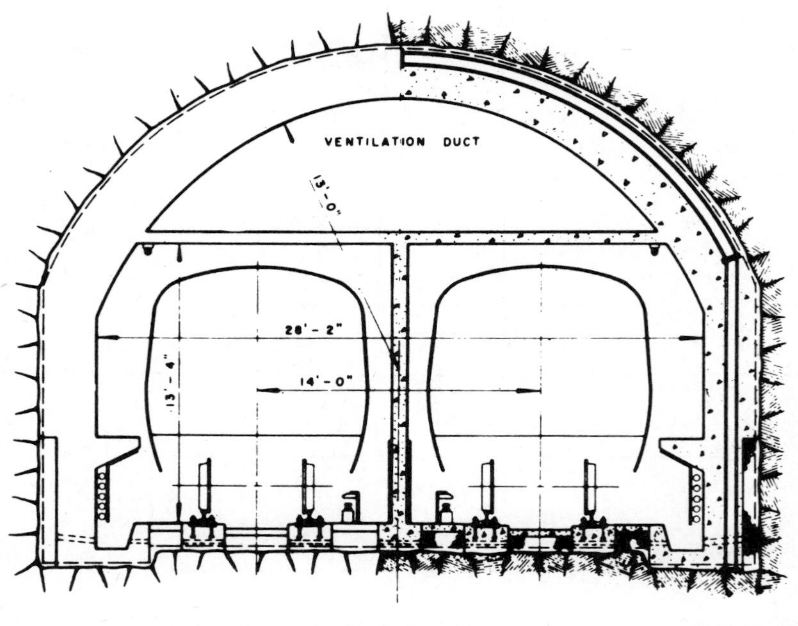

SUPPORTED SYSTEM CROSS SECTION

Fig. 11–6. ***(Continued.)***

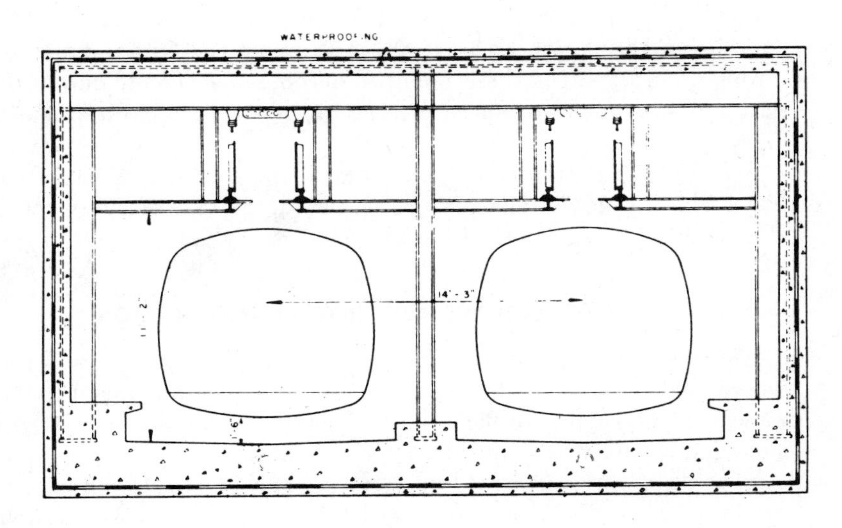

CROSS SECTIONS Showing application for suspended system, using overhead current collection.

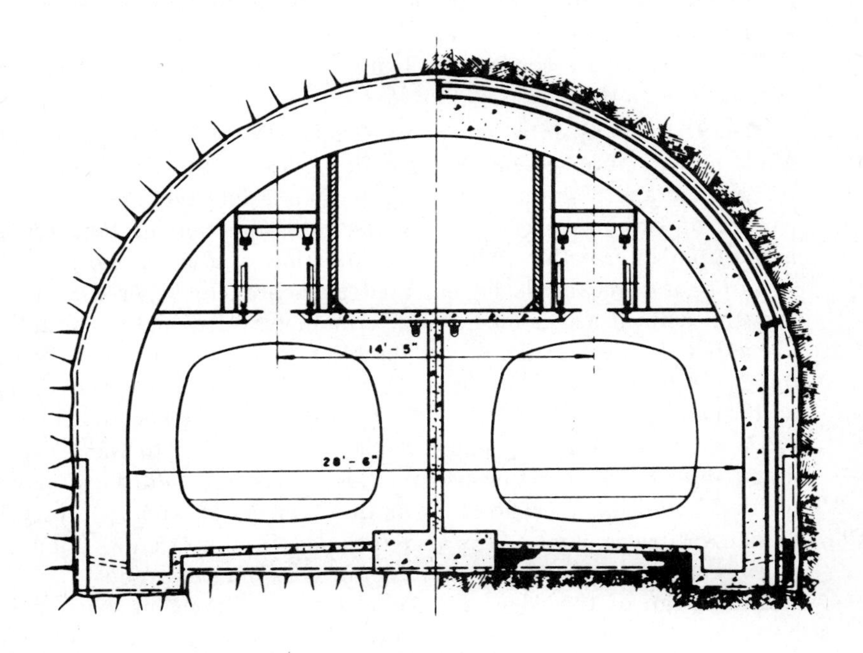

OF ROCK TUNNELS SUSPENDED SYSTEM

Fig. 11–6. (*Continued.*)

Elevated facilities can almost completely eliminate surface traffic disruption. Noise, fume, and aesthetic pollution levels tend to be high in spite of the designer's best efforts to make the structures acceptable. In high-density areas these facilities become uneconomic due to the difficulty of providing station space for rail systems and ramp spacing for freeways.

Subway alignments cause minimum disruption of existing circulation. Their chief disadvantages lie in the very large cost of construction, and the increased noise levels within the vehicles.

INTERSECTIONS AND INTERCHANGE DESIGN

An important part of a highway is the intersection. This is the place where two or more highways meet and provides an area for the cross movement of vehicle traffic. The efficiency, safety, speed, cost of operation, and capacity are dependent upon its design.

There are three general types of intersections: (1) intersections at grade, (2) grade separations without ramps, and (3) interchanges. Design standards for these groups have been developed by the American Association of State Highway Officials. With the large number of intersections and interchanges being constructed throughout the United States, these standards are continually being improved from the experience obtained through their use.

11–19. Highway Intersections at Grade. Most highways intersect at grade and the intersection area should be designed to provide adequately for turning and crossing movements. Simple intersections at grade may consist of two intersecting highways, or a junction of four roads and a multiple intersection, which is a junction of five or more roads. A junction of three roads is indicated as a branch, *T*, or *Y*. A branch may be defined as an offshoot of a main traveled highway, and it has a small deflection angle. A *T* intersection is one in which two roads intersect to form a continuous highway and the third road intersects at or nearly at right angles. A *Y* intersection is one in which three roads intersect at nearly equal angles. In addition to these types, the flared intersection may be used. This consists of additional pavement width of additional traffic lanes at the intersection area. Rotary intersection or traffic circles are also included in this group. Figure 11–7 gives examples of the general types of intersections at grade.

The design of the edge of pavement for a simple intersection should provide sufficient clearance between the vehicle and the other traffic lanes. It is frequently assumed that all turning movements at intersections are accomplished at speeds of less than 20 miles per hour and the design is based on the physical characteristics of the assumed design

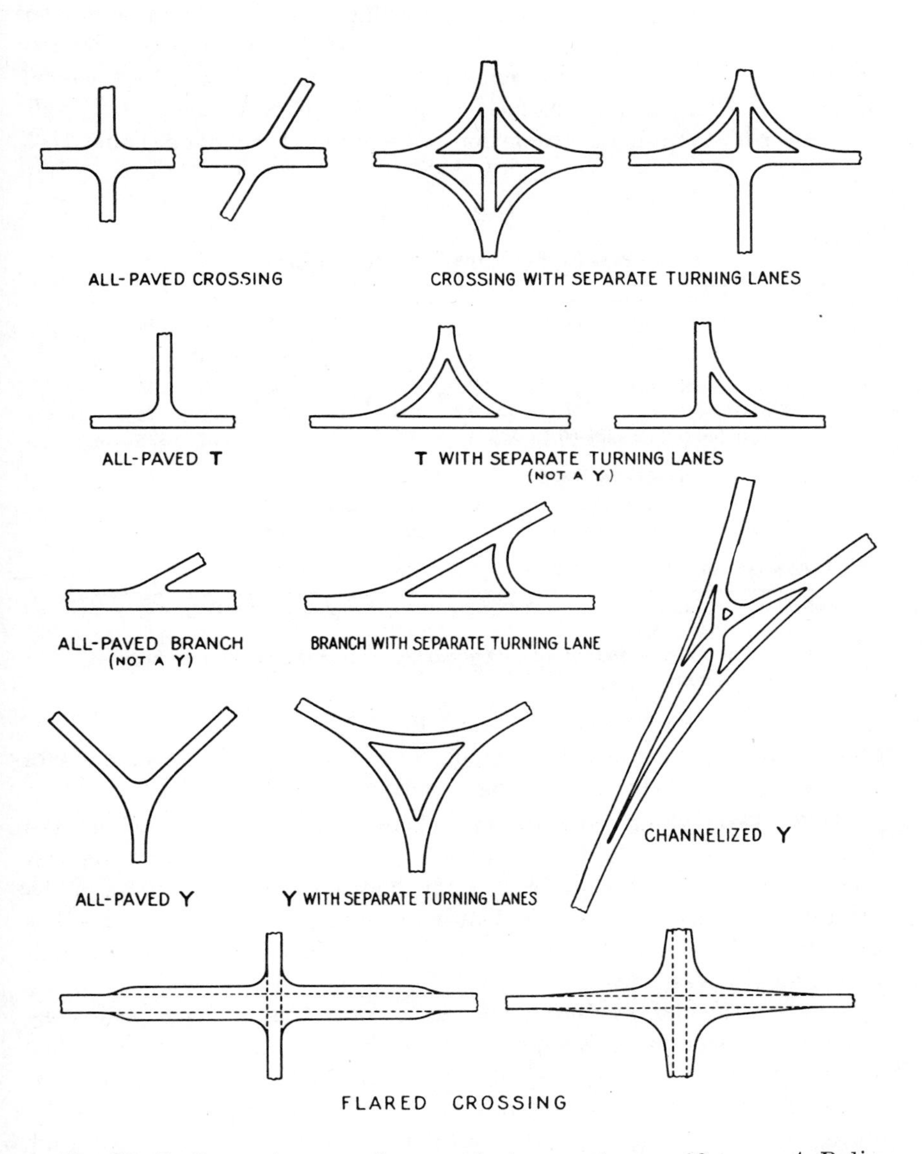

Fig. 11–7. General types of at-grade intersections. (Source: *A Policy on Geometric Design of Rural Highways,* 1965, AASHO, p. 388.)

vehicle. The radius needed for the turning movement of a passenger vehicle is less than that needed for trucks and greatest for semi-trailer-truck combinations. A minimum curb, radius in 90-degree turns of 30 feet for *P* traffic and 50 feet for *T* traffic is recommended. Table 11–4 gives the minimum radius for intersection curves, and Table 11–5

TABLE 11–4

Minimum Radii for Intersection Curves

Design (turning) speed (*V*), mph	15	20	25	30	35	40
Side friction factor (*f*)	0.32	0.27	0.23	0.20	0.18	0.16
Assumed min. superelev. (*e*)	.00	.02	.04	.06	.08	.09
Total $e + f$	.32	.29	.27	.26	.26	.25
Calculated min. radius (*R*), ft.	47	92	154	231	314	426
Suggested curvature for design:						
Radius—minimum, ft.	50	90	150	230	310	430
Degree of curve—maximum	—	64	38	25	18	13
Average running speed, mph	14	18	22	26	30	34

Source: A Policy on Geometric Design of Rural Highways, AASHO, 1965, page 325.
Note: For design speeds of 40 mph and over, values same as for open highway conditions.

gives the minimum design for the edge of pavement. Figure 11–8 shows the effect of curb radii and parking on turning paths.

11–20. Channelized Intersections. Channelization, as defined by the Highway Research Board Committee on Channelization, "is the separation or regulation of conflicting traffic movements into definite paths of travel by the use of pavement markings, raised islands, or other suitable means to facilitate the safe and orderly movements of both vehicles and pedestrians."

Islands in an intersection can separate conflicting movements or control the angle at which conflict may occur. They can regulate traffic, indicate the proper use of the intersection, and often reduce the amount of pavement required. They provide for the protection of motorists, protection and storage of turning and crossing vehicles, and space for traffic control devices.

Islands are generally grouped into three major classes: directional, divisional, and refuge. General types and shapes of islands are shown in Fig. 11–9.

Directional islands are designed primarily to guide the motorist through the intersection by indicating the intended route. The placing

TABLE 11-5

Minimum Edge of Pavement Designs for Turns at Intersections

Design Vehicle	Angle of Turn (degrees)	Simple Curve Radius (feet)	3-Centered Compound Curve, Symmetric		3-Centered Compound Curve, Asymmetric	
			Radii (feet)	Offset (feet)	Radii (feet)	Offset (feet)
P	30	60	–	–	–	–
SU		100	–	–	–	–
WB-40		150	–	–	–	–
WB-50		200	–	–	–	–
P	45	50	–	–	–	–
SU		75	–	–	–	–
WB-40		120	–	–	–	–
WB-50		170	200–100–200	3.0	–	–
P	60	40	–	–	–	–
SU		60	–	–	–	–
WB-40		90	–	–	–	–
WB-50		–	200-–75–200	5.5	200–75–275	2.0–6.0
P	75	35	100–25–100	2.0	–	–
SU		55	120–45–120	2.0	–	–
WB-40		85	120–45–120	5.0	120–45–200	2.0–6.5
WB-50		–	150–50–150	6.0	150–50–225	2.0–10.0
P	90	30	100–20–100	2.5	–	–
SU		50	120–40–120	2.0	–	–
WB-40		–	120–40–120	5.0	120–40–200	2.0–6.0
WB-50		–	180–60–180	6.0	120–40–200	2.0–10.0
P	105	–	100–20–100	2.5	–	–
SU		–	100–35–100	3.0	–	–
WB-40		–	100–35–100	5.0	100–35–200	2.0–8.0
WB-50		–	180–45–180	8.0	150–40–210	2.0–10.0
P	120	–	100–20–100	2.0	–	–
SU		–	100–30–100	3.0	–	–
WB-40		–	120–30–120	6.0	100–30–180	2.0–9.0
WB-50		–	180–40–180	8.5	150–35–220	2.0–12.0
P	135	–	100–20–100	1.5	–	–
SU		–	100–30–100	4.0	–	–
WB-40		–	120–30–120	6.5	100–25–180	3.0–13.0
WB-50		–	160–35–160	9.0	130–30–185	3.0–14.0
P	150	–	75–18–75	2.0	–	–
SU		–	100–30–100	4.0	–	–
WB-40		–	100–30–100	6.0	90–25–160	3.0–11.0
WB-50		–	160–35–160	7.0	120–30–180	3.0–14.0
P	180	–	50–15–50	0.5	–	–
SU	U-Turn	–	100–30–100	1.5	–	–
WB-40		–	100–20–100	9.5	85–20–150	6.0–13.0
WB-50		–	130–25–130	9.5	100–25–180	6.0–13.0

Source: A Policy on Geometric Design of Rural Highways, AASHO, 1965, page 318.

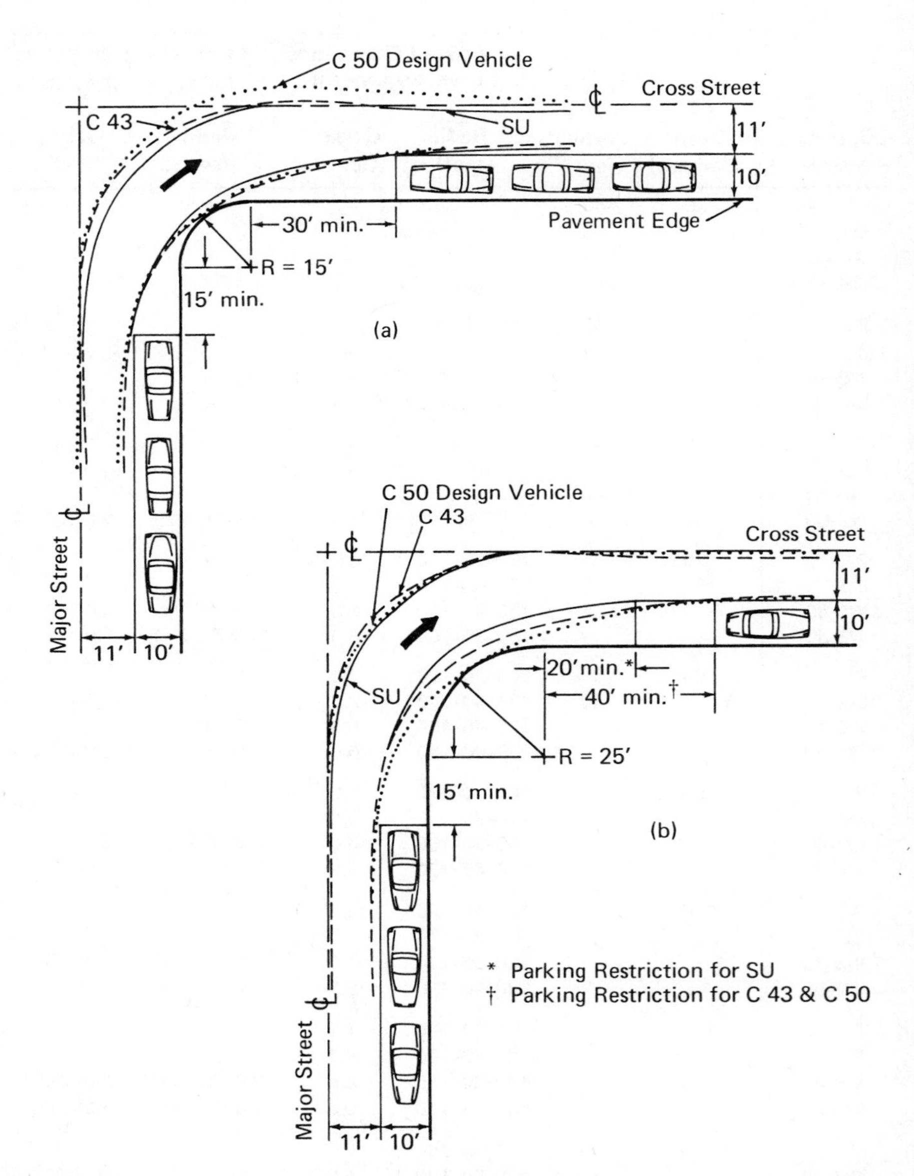

Fig. 11–8. Effect of curb radii and parking on turning movements. (Source: *A Policy on Arterial Highways in Urban Areas*, 1957, AASHO.)

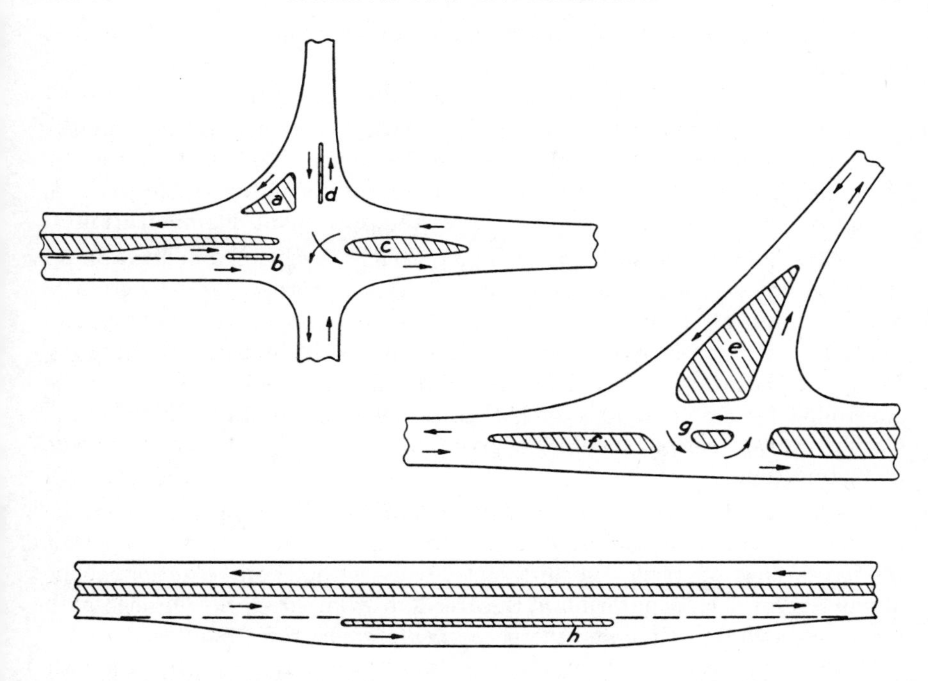

Fig. 11–9. General types and shapes of islands. (Source: *A Policy on Geometric Design of Rural Highways*, 1965, AASHO.)

of directional islands should be such that the proper course of travel is immediately evident and easy to follow. A complicated system of islands where the desired course of travel is not immediately evident may result in confusion and may be more hindrance than help in maintaining a steady traffic flow. Islands should be so placed that crossing streams of traffic will pass at approximately right angles and merging streams of traffic will converge at flat angles. By the use of such angles there will be less hindrance to traffic in the thoroughfare and the possibility of accidents in the intersection will be decreased.

Divisional islands are most frequently used in individual highways approaching intersections. They serve to alert the driver to the intersection and regulate the flow of traffic into and out of the intersection. Their use is particularly advantageous for controlling left-turning traffic at skewed intersections.

A refuge island is located at or near crosswalks to aid and protect the pedestrian. These islands are most generally used on wide streets in urban areas for loading and unloading transit riders. The design of refuge islands is the same as that of other types of islands, except that a higher barrier curb is necessary.

A good approach in the design of channelization is to make a comprehensive study of field conditions. Then, with the use of pavement markings, observation of the traffic patterns is done. This is followed by the placing of sand bags and another observation of the traffic pattern, which is then followed by the placing of the permanent channelization.

11–21. Rotary Intersections. One approach to the channelization of traffic is by the use of a rotary intersection. A rotary intersection is one in which all traffic merges into and emerges from a one-way road around a central island. Rotary intersections have their advantages. Many of those previously built as rotary intersections have had to be controlled by traffic signals and, thus, become a signalized intersection. Careful consideration should be given in the design of a rotary type of installation.

11–22. Grade Separations and Interchanges. Intersections at grade can be eliminated by the use of grade-separation structures which permit the cross-flow of traffic at different levels without interruption. The advantage of such separation is the freedom from cross-interference with resultant saving of time and increase in safety for traffic movements.

Grade separations and interchanges may be warranted (1) as a part of an express highway system designed to carry heavy volumes of traffic, (2) to eliminate bottlenecks, (3) to prevent accidents, (4) where the topography is such that other types of design are not feasible, and (5) where the volumes to be catered to would require the design of an intersection, at grade, of unreasonable size.

An interchange is a grade separation in which vehicles moving in one direction of flow may transfer direction by the use of connecting roadways. These connecting roadways at interchanges are called ramps.

Many types and forms of interchanges and ramp layouts are used in the United States. These general forms may be classified into four main types:

1. T- and Y-interchanges
2. Diamond interchanges
3. Partial and full cloverleafs
4. Directional interchanges

T- and Y-Interchanges. Figure 11–10 shows typical layouts of interchanges at various junctions. The geometry of the interchange can be altered to favor certain movements by the provision of large turning radii, and to suit the topography of the site. The trumpet interchange has been found suitable for orthogonal or skewed intersections. Figure

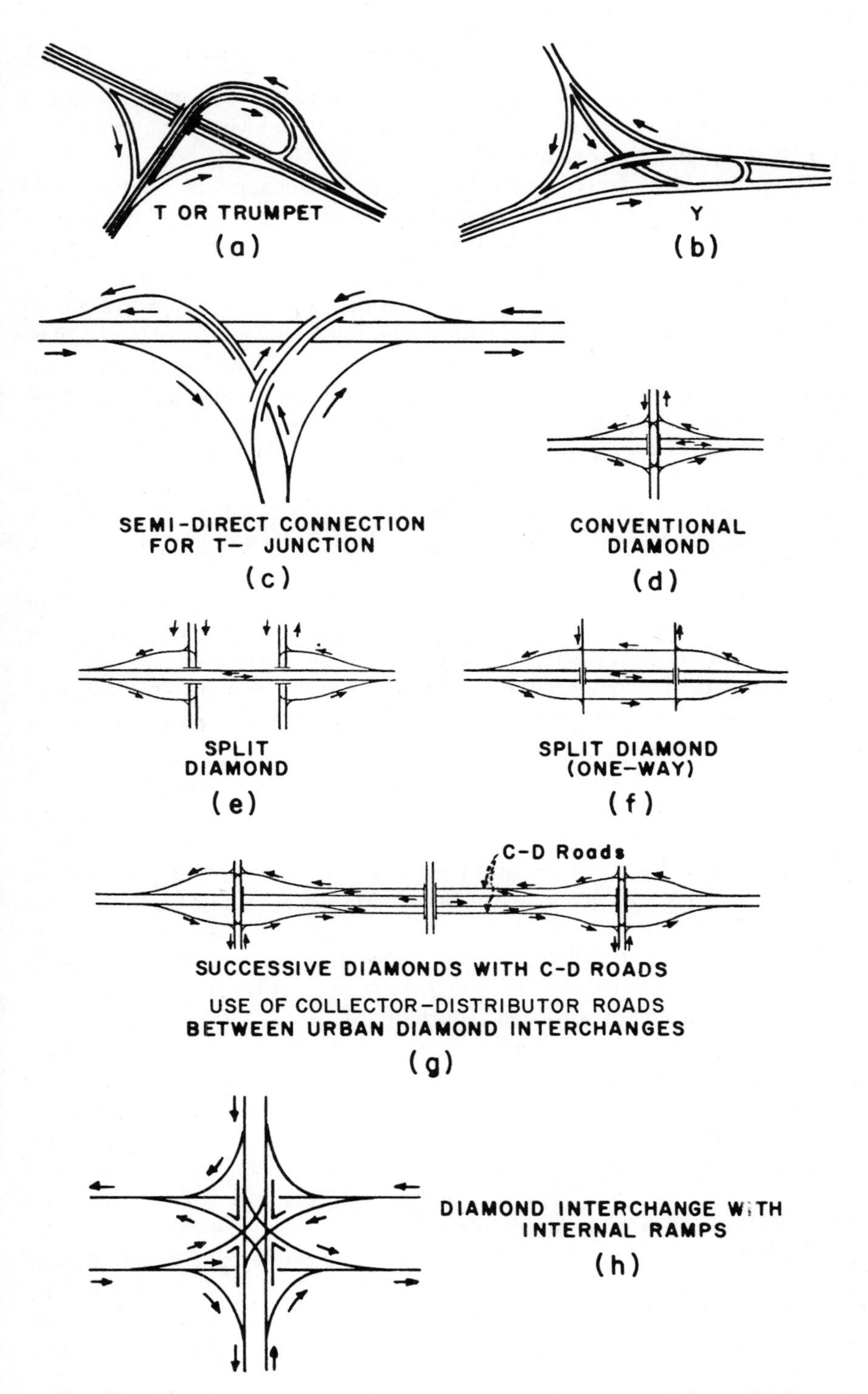

Fig. 11–10. Highway interchanges. (Partially adapted from "Adaptability of Interchanges to Interstate Highways," Volume 124, page 558, *Transactions*, American Society of Civil Engineers.)

11–10(a) favors the left turn on the freeway by the provision of a semi-direct connecting ramp. Figure 11–10(c) indicates an intersection where all turning movements are facilitated in this way.

Diamond Interchanges. The diamond interchange is adaptable to both urban and rural use. The major flow is grade-separated, with turning movements to and from the minor flow achieved by diverging and merging movements with through traffic on the minor flow. Only the minor flow directions have intersections at grade. In rural areas, this is generally acceptable, owing to the light traffic on the minor flow. In urban areas, the at-grade intersections generally will require signalized control to prevent serious interference of ramp traffic and the crossing arterial street. The design of the intersection should be such that the signalization required does not impair the capacity of the arterial street. To achieve this, widening of the arterial may be necessary in the area of the interchange. Care must also be taken in the design of the ramps, so that traffic waiting to leave the ramp will not back up into through lanes of the major flow.

One disadvantage of the diamond interchange is the possibility of illegal wrong-way turns, which can cause severe accidents. Where the geometry of the intersection may lead to these turns, the designer can use channelization devices and additional signing and pavement marking. Wrong-way movements are, in general, precluded by the use of cloverleaf designs.

Figure 11–10(d) shows the conventional diamond interchange. Increased capacity of the minor flows can be attained by means of the arrangement shown in Fig. 11–10(e) or Fig. 11–10(f). The arrangement shown in Fig. 11–10(g) is suitable where two diamond ramps are in proximity. Weaving movements, which would, in this case, inhibit the flows of the major route, are transferred to the parallel collector distributor roads. Figure 11–10(h) indicates how the diamond interchange can be adapted for internal entry and exit ramps. Figure 11–11 shows a typical depressed expressway in an urban area.

Partial and Full Cloverleafs. The partial cloverleafs shown in Fig. 11–12 are sometimes adopted in place of the diamond interchange. Traffic can leave the major flow either before or after the grade-separation structure, depending on the quadrant layout. The intersections at grade for the minor road are present as for the diamond interchange, but the probability of illegal turning movements can be reduced. By the provision of two on-ramps for each direction of the major route as in Fig. 11–12(c), left-turn traffic on the minor route can be eliminated. The more conventional arrangement of the full cloverleaf, which can be adapted to nonorthogonal layouts, eliminates at-grade crossings of all traffic streams for both major and minor roads. The ramps may

Fig. 11–11. Edsel Ford Freeway in Detroit. (Courtesy Michigan Department of Highways.)

be one-way, two-way separated, or two-way unseparated roads. Although all crossing movements are eliminated, the cloverleaf design has some disadvantages: (1) the layout requires large land areas, and (2) decelerating traffic wishing to leave the through lanes must weave with accelerating traffic entering the through lanes. Figure 11–12(e) is a layout using collector-distributor roads to overcome this second disadvantage.

Figure 11–13 shows a typical rural full cloverleaf at the intersection of two four-lane divided highways.

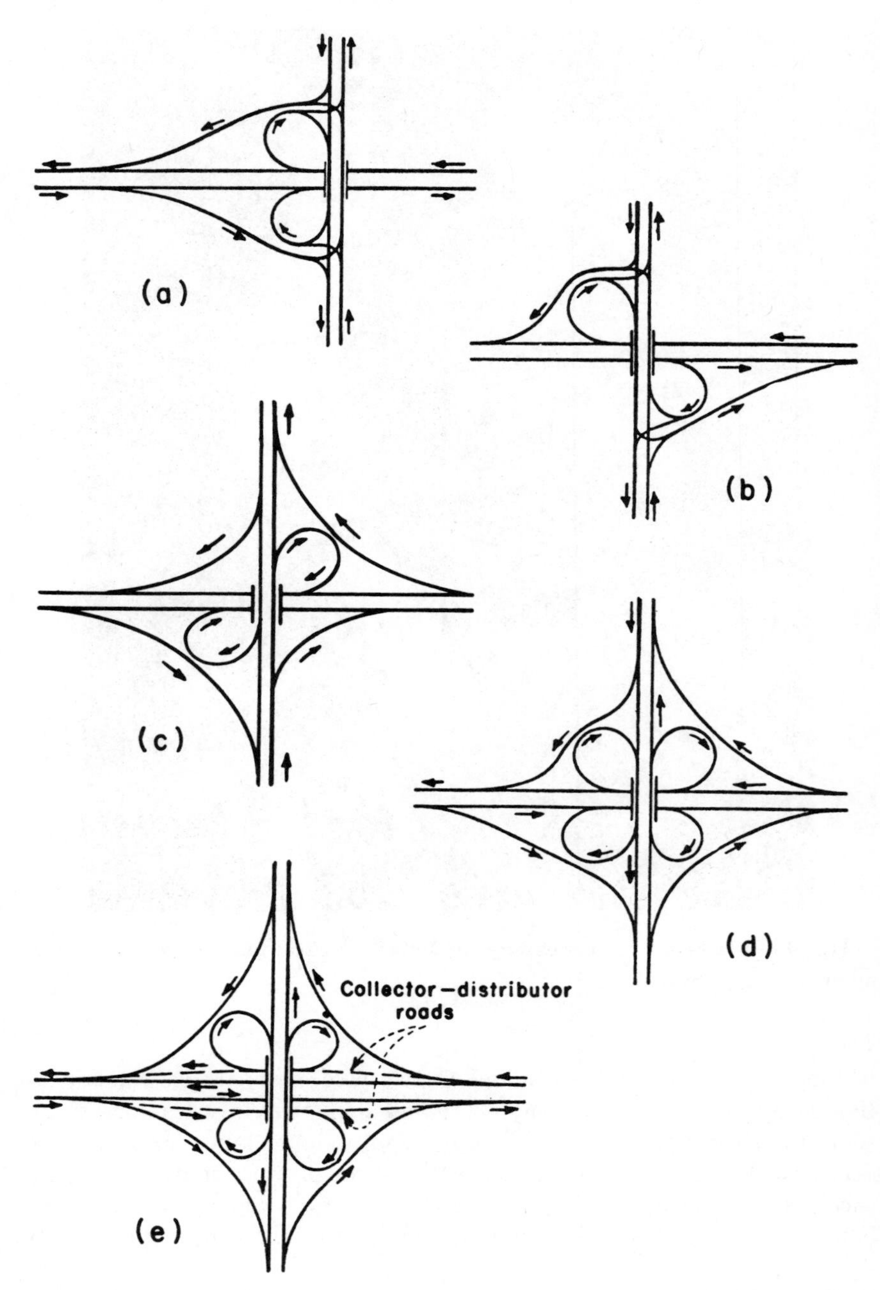

Fig. 11–12. Cloverleaf interchanges. (*A Policy on Geometric Design of Rural Highways*, 1965, AASHO.)

Fig. 11–13. A cloverleaf interchange 194 and M78 in Michigan. (Courtesy Michigan Department of Highways.)

Directional Interchanges. Directional interchanges are used whenever one freeway joins or intersects another freeway. The outstanding design characteristic of this type of interchange is the use of a high design speed throughout, with curved ramps and roadways of large radius. The land requirements for a directional interchange are, therefore, very large. In cases where volumes for certain turning movements are small, design speeds for these movements are reduced and the turnoff is effected within a loop. Figure 11–14 shows examples of interchanges with and without loops. In the highest type of design weaving sections are eliminated. Figure 11–15 shows a multilevel interchange.

Two directional interchanges are shown in Fig. 11–16, one at either end of a bridge over the East River in New York City to join the Cross Bronx Expressway to the East River Drive and the Major Deegan Expressway. In the background, the expressway passes beneath three apartment houses and a bus terminal, then to the George Washington Bridge over the Hudson River.

11–23. Railroad Intersections. Railroad tracks intersect at *turnouts, crossovers,* and *crossings.* Turnouts are curved sections of track that permit the diversion of rolling stock from one track to another. Where

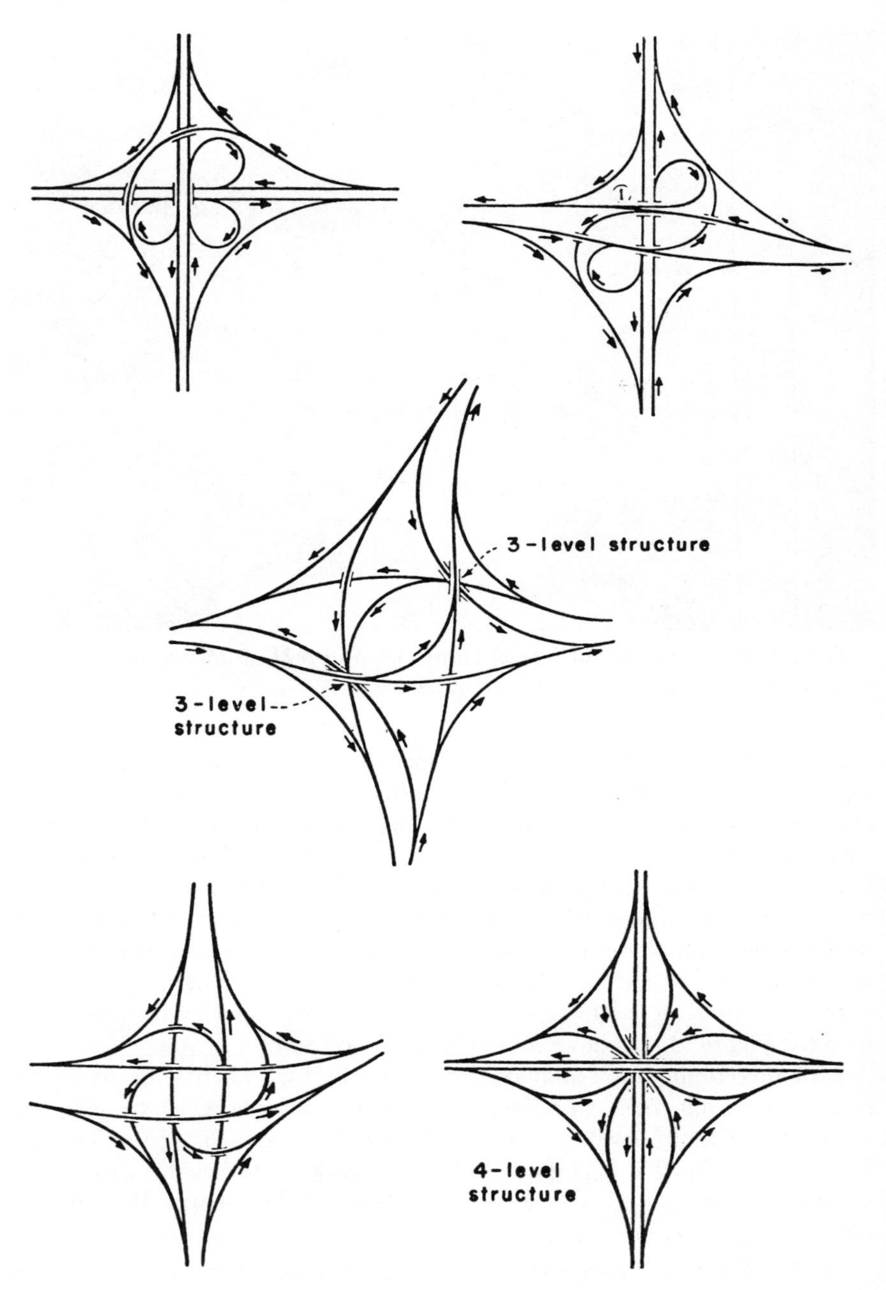

Fig. 11–14. Directional interchanges. (Source: *A Policy on Geometric Design of Rural Highways*, 1965, AASHO.)

Fig. 11–15. San Diego-Santa Monica Freeway Interchange, West Los Angeles. (Courtesy California Division of Highways.)

the turnout provides an intersection with another continuous parallel or nonparallel track it is called a crossover.

At crossings, tracks intersect permitting movement of the rolling stock on one track across the alignment of the other. Figure 11–17 shows schematic arrangements for the various types of railroad intersections discussed in the following sections.

11–24. Switches. The device which determines the diversion of rolling stock movement through a turnout is known as a switch. A switch is designated as a left-hand or right-hand switch depending on the direction of diversion of the rolling stock into the turnout. Switches are relatively simple devices principally composed of switch rails, rods which hold the points in their proper position and relationship, gage and switchplates which support the switch rails at their proper elevation, and heel blocks which effect a rigid joint at the head of the switch.

Various types of switches are available, only one of which is of standard railroad use in the United States.

Fig. 11–16. An urban highway complex. (Courtesy New York Department of Transportation.)

1. *Stub switches* are used to some extent in industrial tramways. Both switch rails are mainline rails. This type of switch is not considered safe for high-speed train movements. Its use is, therefore, extremely limited. It is found in a few switch yards and other relatively low-speed areas.

2. *Tongue switches* are used in paved street locations. Designed for slow-moving traffic, they consist of a movable tongue on one side of the track with either another movable tongue or a fixed mate on the other side.

3. *Spring switches* permit point movement to allow the passage of trailing wheels through the switch in the reverse direction of movements prevented by the points for facing movements. After the passage of the trailing wheels, the points are moved back into position by spring devices.

4. *Split switches* are the standard switch in use on American railroads. They have proven safe for very high-speed movements. Switching is carried out by the use of one mainline rail and one

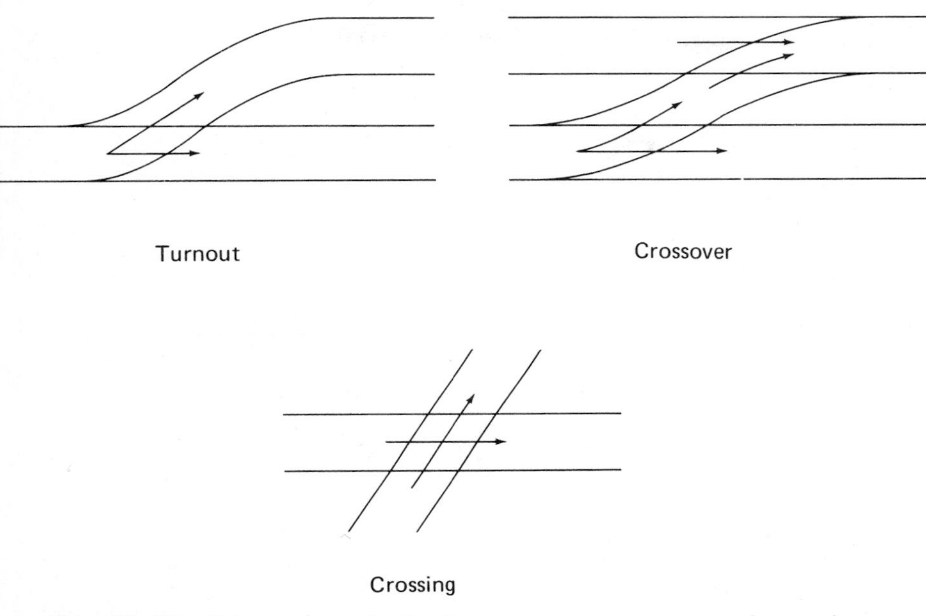

Fig. 11–17. Schematics of simple turnout, crossover and crossings. (Original drawings.)

turnout rail for switch rails as shown in the schematic diagram in Fig. 11–18.

The parts and arrangement of a typical left-hand turnout are shown in Fig. 11–19.

11–25. Frogs. The turnout frog is a device which permits rolling stock wheels on one rail to cross the rail of a diverging track. It performs two functions, supporting the wheel over the intersection of the flangeways and providing continuous channels for the wheel flanges. Two principal frog types are in common use, the rigid frog and the spring frog. Figure 11–20 shows typical arrangements of rigid and spring frogs. In the rigid frog, both flangeways are always open and both wing rails are bolted to the body of the frog. In the design of the conventional spring frog, the main flangeway only is always open. The turnout direction is opened when one wing moves, establishing a flangeway for the wheels.

Frogs are normally designated by a frog number which is defined as one-half the cotangent of one-half the frog angle or the ratio of the

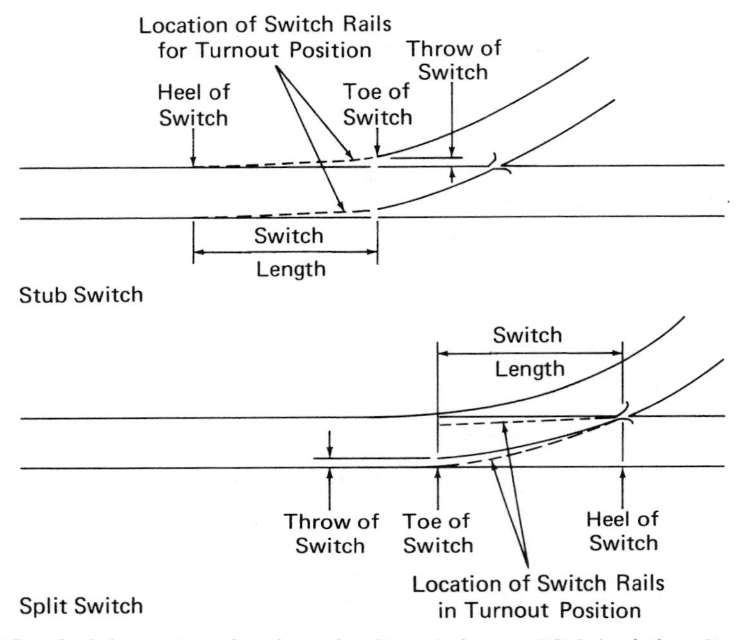

Fig. 11–18. Schematics of stub and split switches. (Original drawings.)

spread at any point to the length of a bisecting line between that point and any theoretical point of frog:

$$n = \frac{1}{2} \cot \frac{\phi}{2}$$

where:

n = frog number
ϕ = frog angle

Frog angles in common use in railwork are betwen 9 deg. 32 min. (No. 6 frog) and 2 deg. 52 min. (No. 20 frog). No. 18 and No. 20 frogs are in standard use on large railroad systems with mainline tracks. Slower movements can be accommodated with No. 10 and No. 12 frogs while No. 6 and No. 8 frogs are in use in sidings and industrial tramways.

11–26. Crossings. Where two tracks intersect and cross, specially designed and fabricated *crossings* are necessary. Since the angle of intersection is usually non-standard, specially designed frogs are necessary. In general crossings can be designated into four general classifications:

1. Bolted rail crossings. All members are heat treated or open hearth rails bolted together.

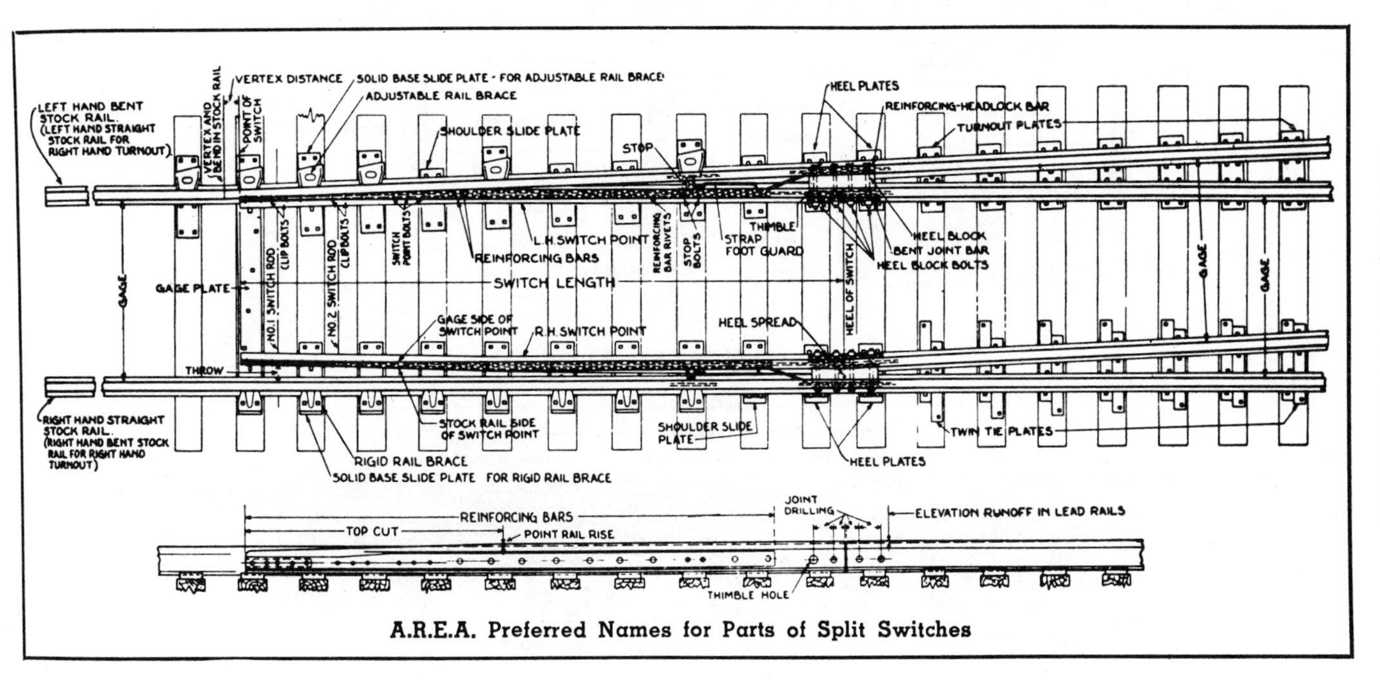

Fig. 11–19. A typical split switch left-hand turnout. (Source: *Railway and Track Structures Cyclopedia*, p. 361.)

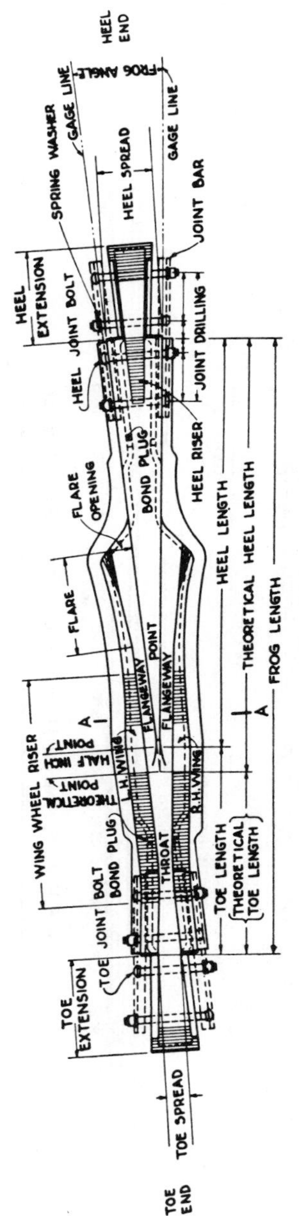

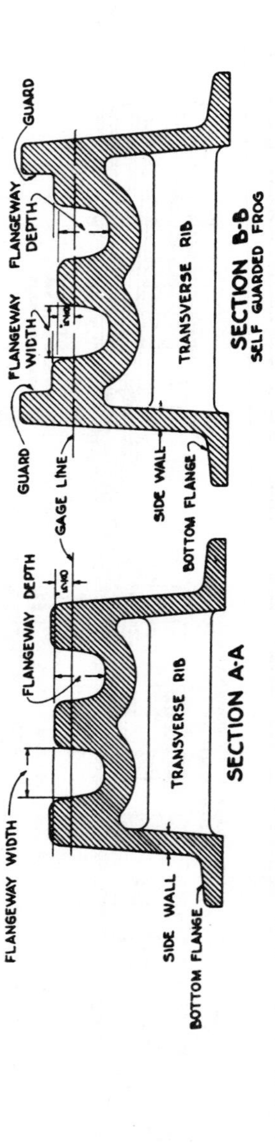

A.R.E.A. Preferred Names of Parts for Rigid Bolted, Railbound Manganese; and Solid Manganese Frogs

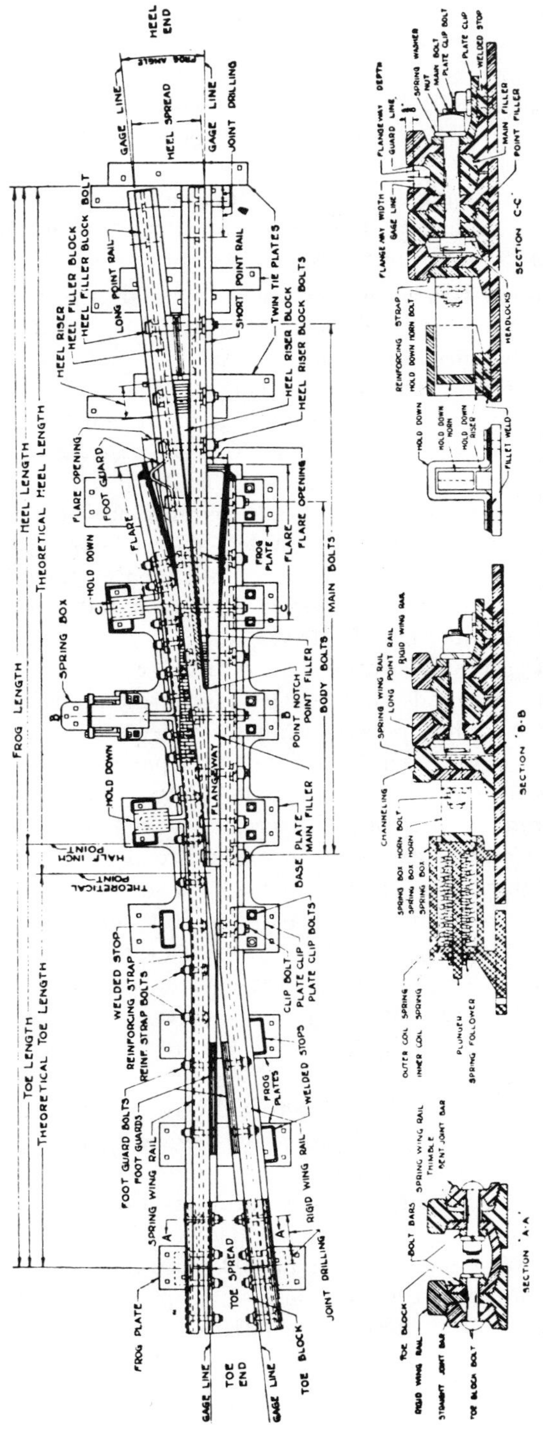

A.R.E.A. Preferred Names of Parts for Spring Rail Frogs

Fig. 11–20. Rigid and spring frog arrangements. (Source: *Railway and Track Structures Cyclopedia*, pp. 379 and 382.)

2. Manganese steel insert crossings. Manganese steel cast inserts are fitted into rolled rails to form the wing and points of frogs.
3. Solid manganese steel crossings in which each frog is a single solid casting.
4. Double slip switches with movable point crossings.

The choice of crossing type is dependent on speeds of operation, the angle of intersection, and the degree of curvature of the track.

Figure 11–21 shows typical tie layouts at railroad crossings.

11–27. Railroad Grade Intersections. In the design of a highway that intersects a railroad at grade, consideration must be given to approach grades, sight distance, drainage, volume of vehicular traffic, and the frequency of regular train movements at the particular intersection. The particular type of surfacing and kind of construction at railroad crossings at grade will depend upon the class of railroad and kind of roadway improvement.

All railroad intersections at grade require proper advance warning signs. At crossings on heavily traveled highways where conditions justify, automatic devices should be installed. Recommended standards for railroad-highway grade-crossing protection have been adopted by the Association of American Railroads.

The use of grade separations at railroad crossings is recommended at all mainline railroads that consist of two or more tracks and at all single-line tracks when regular train movements consist of six or more trains per day. Other considerations for separating railroad and highway traffic are the elements of delay and safety.

Railroad grade-separation structures may consist of an overpass on which the highway is carried over the railroad or an underpass which carries the highway under the railroad. The selection of the type of structure will depend in large part upon the topographical conditions and a consideration of initial cost. Drainage problems at underpasses can be serious. Pumping of surface and subsurface water may have to be carried on a large part of the time and the failure of power facilities sometimes causes flood conditions at the underpass with the resultant stoppage of traffic.

There are approximately 225,000 highway grade crossings. Twenty per cent of these have special protective devices such as gates or flashing lights. Eight per cent, however, have signs only. The financial and human toll of inadequate grade separations can be inferred from the annual statistics of grade-crossing accidents. Normally, each year there are between 1,500–1,800 deaths and 3,500–4,000 injuries involving approximately $300 million in economic losses. Thirty thousand crossings have traffic volumes which warrant immediate improvement. The expected

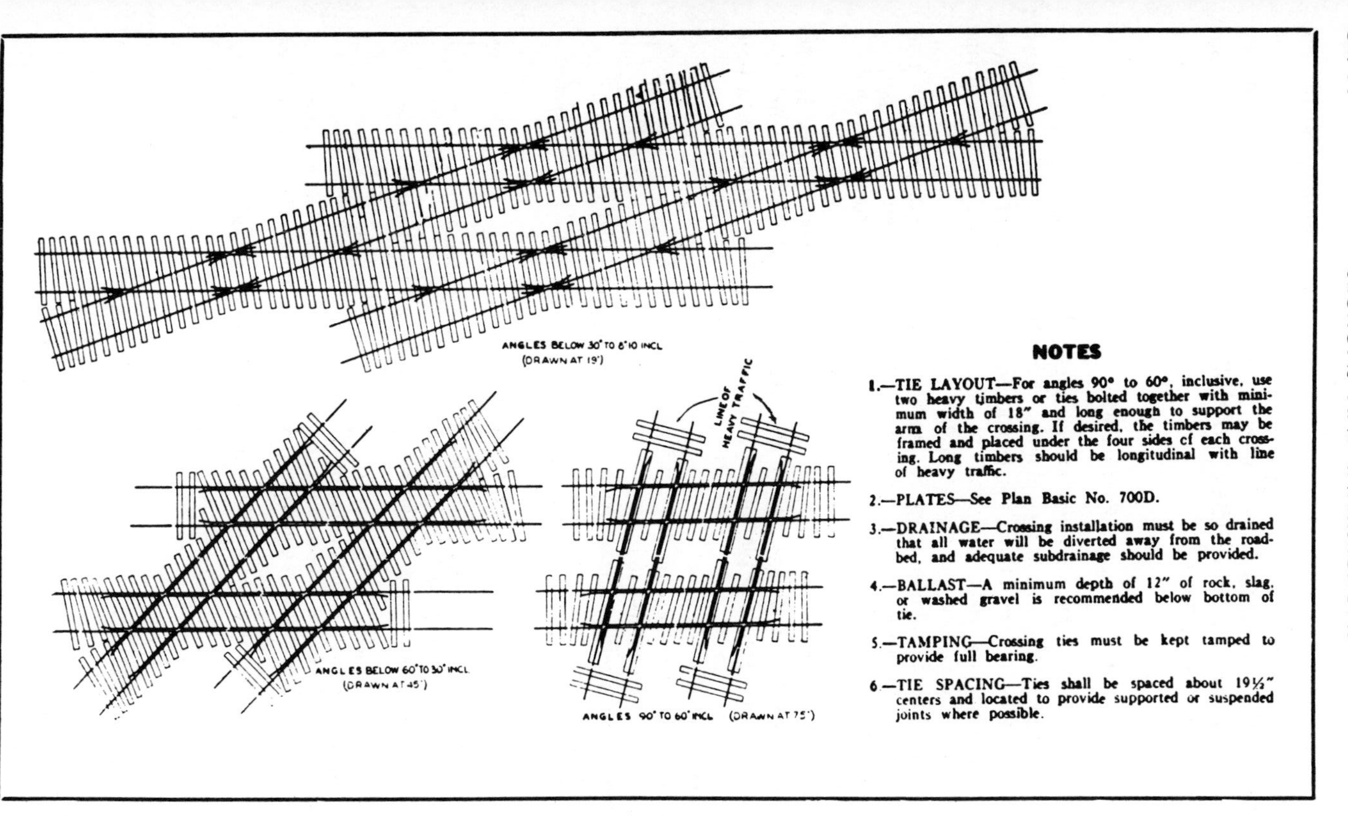

Fig. 11–21. A.R.E.A. tie layouts at railroad crossings. (Source: *Railway and Track Structures Cyclopedia*, p. 400.)

net benefits from improvement are anticipated to amount to $6 billion, while improvement costs would amount to $580 million. During a period from 1963–1967 the total expenditures on federally aided rail-highway grade-crossing projects were $955,124,095, of which the federal contribution amounted to $807,704,653.

PROBLEMS

1. Two intersecting highways cross at an angle of 85°. Make a sketch of a cloverleaf grade separation for complete traffic flow.
2. What would be the result on the geometric design of (1) a primary and (2) of a secondary road system if truck and tractor-trailer combinations are permitted to be larger than the present design vehicle?
3. Discuss the different philosophies that form the basis for the radically different design criteria for a local residential subdivision street and an expressway.
4. How does the United States railroad gage and other appurtenances compare with that of European countries?
5. Suppose a subdivision developer has a 3,000 foot by 1,000 foot rectangular piece of property which has streets abutting on two of the adjacent sides. Sketch a street layout and plat of the residential lots using the design criteria given below:

   Minimum lot size—⅓ acre
   Maximum lot size—½ acre
   Minimum lot frontage—100 feet
   Minimum set-back of houses from right-of-way line—50 feet

   Assume the land is relatively flat.

REFERENCES

1. Ritter, Leo J., Jr., and Paquette, Radnor J., Highway Engineering, Third Edition, The Ronald Press, New York (1967).
2. Urqhart, L. C. (ed.), *Civil Engineering Handbook,* McGraw-Hill, New York (1959).
3. *A Policy on Geometric Design of Rural Highways,* American Association of State Highway Officials, Washington, 1965.
4. *Manual of Recommended Practice,* American Railway Engineering Association.
5. *Railway Track & Structures Cyclopedia,* Simmons-Boardman Corporation, New York, 1955.
6. "Recommended Practices for Subdivision Streets," *Traffic Engineering,* January, 1967.

12

Design of Streets, Highways, and Railways: Drainage and Earthwork Operations

Two important aspects of railway and roadway design will be discussed in this chapter: the provision of adequate drainage and earthwork operations.

DESIGN OF DRAINAGE STRUCTURES AND FACILITIES

In this chapter, the discussion of street, highway, and railway drainage will be limited to a discussion of surface drainage, which is the process of controlling and removing excess water from the traveled way. The specialized problem of dealing with subsurface drainage will not be covered but is discussed in Reference 1.

Three major topics will be discussed in this section:

1. estimation of runoff;
2. hydraulic design of culverts and drainage ditches;
3. a comparison of alternative drainage systems.

The first of these topics deals with techniques for estimating the quantity of runoff water on the bases of certain rainfall data and land use characteristics. Culvert design essentially encompasses a discussion of non-uniform open channel flow, while the design of drainage ditches involves uniform open channel flow. Approaches used in the design of undergound storm drainage systems such as those used for city streets are given in Section 18–7.

In the sections which deal with hydraulic design, emphasis will be placed on the use of empirical data rather than on fluid flow theory. For the reader who is more interested in theoretical approaches to drainage design, reference should be made to one or more of the numerous textbooks in fluid mechanics which have been published.

12–1. Surface Drainage. Before proceeding with a discussion of the techniques for estimating runoff quantities, it should first be noted that consistent with other design objectives, every effort should be made to remove precipitation from the traveled way as expeditiously as possible. Water which is not removed quickly from a highway or railway nearly always is harmful to the load carrying capability of the pavement system. Furthermore, flood waters serve as a deterrent to free traffic movements and create unnecessary perils for the users of the facility. Uncontrolled water movements may weaken, damage, or even destroy transportation structures and the pavement system. For these reasons, highway designers provide pavement crown and shoulder slopes to expedite the removal of surface water.[1] Similarly, railroad design engineers specify an open-graded ballast material and a sloped subgrade to insure adequate and quick drainage. Well-designed culvert and bridge structures must be provided for railways and highways alike to prevent destructive back waters and roadway overtopping from occurring.

To insure adequate drainage, unpaved side ditches should have a slope of at least 0.5 per cent. Experience has shown that paved ditches with an average slope of about 0.3 per cent will drain satisfactorily.

12–2. Streamflow Records. It may be possible, especially in the case of design of large culverts and bridges, to base estimates of peak flow on historical empirical data. Flood-frequency studies based on statistical analysis of streamflow records have been published by a number of states as well as the U.S. Geological Survey.

> Where suitable streamflow records are not available, useful information may be gained from observations of existing structures and the natural stream. Drainage installations above and below the proposed location may be studied and a design based upon those which have given satisfactory service on other portions of the same stream. Lacking this information, an examination of the natural channel may be made, including the evidences left by flood crests which have occurred in the past, and an estimate made of the quantity of water which has been carried by the stream during flood periods. Measurements may be made and values assigned to the slope, area, wetted perimeter, and roughness coefficient of the flood channel, and the quantity may be estimated by the Chezy formula. The Chezy formula is of the form,
>
> $$Q = Ca\sqrt{rS} \tag{12–1}$$

[1] Cross-section design criteria to insure good drainage are given in Chapter 11.

where:

Q = quantity of flow, cubic feet per second
C = roughness coefficient, varying from 30 to 80, depending on the condition and nature of the channel
a = area of the flow cross-section, square feet
S = slope of the channel, feet per foot
r = hydraulic radius, feet, = a/p, where p = wetted perimeter, the length of the boundary of the cross-section of the flow channel in contact with the water [1]

Streamflow records are, of course, often not available, and a detailed physical examination of the existing channel may not be feasible. These conditions commonly prevail, in fact, for small channels which drain areas of only a few hundred acres. A number of formulas and analytical procedures have been developed for estimating runoff from small drainage areas. Before we describe some of these analytical tools, a brief discussion of the factors which influence the magnitude of surface runoff will be given.

12–3. The Nature of the Problem. It should be noted at the outset that the estimation of surface runoff is not an exact science. When two or more methods are used to estimate runoff, discrepancies in the estimates of 50 per cent or more are not uncommon. Difficulties in the estimation of runoff stem from the fact that there are a large number of factors which affect the answer. Certain of these factors are difficult to quantify and may be subject to change with time. In the final analysis, no matter what technique of estimation is used, the application of sound engineering judgment is required to produce a reliable answer.

12–4. Coefficient of Runoff. Runoff results from precipitation which falls on the various surfaces of the watershed. A part of the precipitation evaporates and some of it may be intercepted by vegetation. A portion of the precipitation may infiltrate the ground or fill depressions in the ground surface. The storm runoff for which roadside ditches and drainage structures must be designed is, then, the precipitation minus the various losses which occur.

These losses and, thus, the runoff are strongly dependent upon the slope, vegetation, soil condition, and land use of the watershed. The designer should remember that certain of these factors, notably vegetation and land use, will not remain constant with time. For example, a drainage structure which is designed to accommodate a watershed used for agricultural purposes may prove totally inadequate if the land is later developed into a residential subdivision.

Most analytical procedures for estimating runoff involve the use of a coefficient or runoff to take into consideration the hydrologic nature of the drainage area. Values of the coefficient of runoff for one method of estimating the runoff quantity are given by Table 12–1. If the drainage area under consideration consists of several land-use types, a coefficient should be chosen for each such subarea. The runoff coefficient for the entire area should then be taken as the weighted average of the coefficients for the individual areas.

12–5. Rainfall Intensity, Duration, and Frequency. Rainfall intensity is the rate at which the rain falls, typically expressed in inches per hour. Because of the capriciousness of weather, it is necessary to discuss rainfall intensity in the context of rainfall frequency and duration.

Rainfall intensity-duration data have been collected and published by the U.S. Weather Bureau for various sections of the United States. Typical rainfall intensity-duration graphs are shown in Fig. 12–1. To utilize such data, the designer is at once confronted with the decision of which curve should be used. In making this decision, the designer must weight the physical (as well as social) damages which might occur from a flood of a given frequency against the additional costs of designing the structure so as to lessen the risk of these damages. Referring to the hypothetical data in Fig. 12–1, for example, the choice of the 50-year curve, instead of the 25-year curve, would mean designing for a more severe storm but at a higher cost. Conversely, the choice of a five-year frequency would result in a less costly system but at the risk of more frequent runoffs which exceed the capacity of the drainage facility. The selection of rainfall frequency is to a large extent a matter of experience and judgment, although agency or departmental policy may dictate this choice.

As can be seen from Fig. 12–1, rainfall intensity varies greatly with the duration of rainfall. The average rainfall intensity for short periods of time is much greater than for long periods. In the design of most railroad and highway drainage facilities, a duration equal to the *time of concentration* is chosen.

12–6. Time of Concentration. The time of concentration is defined as the time required for a particle of water to flow from the most remote point in the drainage area to the outlet end of the drainage structure. In other words, it consists of the time of overland flow plus the time of flow in the drainage system. For most railroad and highway culverts, the time of flow in the drainage system is negligible in comparison to the time of overland flow. Thus, for the purposes of this chapter, the terms time of overland flow and time of concentration may be used synonomously.

Time of concentration varies with the land slope, type of surface, rainfall intensity, the size and shape of the drainage area, and many other factors. A number of empirical studies have been made relating

TABLE 12-1

Coefficients of Runoff To Be Used in the Rational Formula

Use for a Culvert Design

TYPE OF SURFACE	VALUE C
Impervious Surfaces	0.90–0.95
Steep Barren Surfaces	0.80–0.90
Rolling Barren Surfaces	0.60–0.80
Flat Barren Surfaces	0.50–0.70
Rolling Meadow	0.40–0.65
Deciduous Timberland	0.35–0.60
Conifer Timberland	0.25–0.50
Orchard	0.15–0.40
Rolling Farmland	0.15–0.40
Flat Farmland	0.10–0.30

Use for a Storm Sewer Design in an Urban Area

TYPE OF SURFACE	VALUE C
Watertight Surfaces such as roofs and pavements	0.70–0.90
Block Pavements with Open Joints	0.50–0.70
Macadam Pavements	0.25–0.60
Gravel Surfaces	0.15–0.30
Parks, Cultivated lands, lawns, etc., depending on slopes and character of soil	0.05–0.30
Wooded Areas	0.01–0.20

Use for an Airport Drainage Design

TYPE OF SURFACE	VALUE C
Watertight Roof Surfaces	0.75–0.95
Asphalt runway pavements	0.80–0.95
Concrete runway pavements	0.70–0.90
Gravel or Macadam pavements	0.35–0.70
Impervious Soils (heavy)*	0.40–0.65
Impervious Soils with turf*	0.30–0.55
Slightly pervious soils*	0.15–0.40
Slightly pervious soils with turf*	0.10–0.30
Moderately pervious soils*	0.05–0.20
Moderately pervious soils with turf*	0.00–0.10

Source: *Concrete Pipe Handbook*, American Concrete Pipe Association, 1965.

*For Slopes from 1–2 per cent.

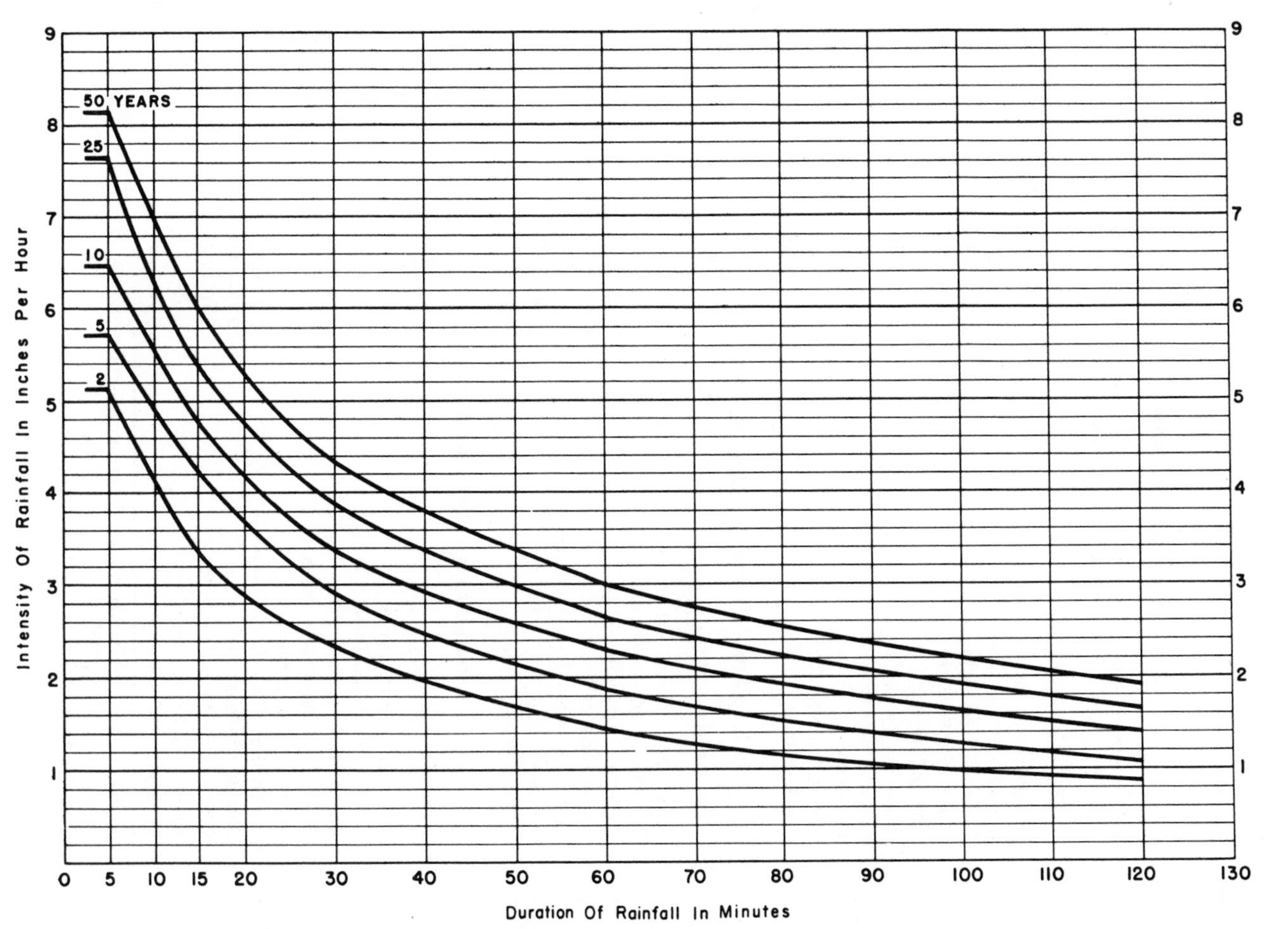

Fig. 12–1. Typical rainfall intensity-duration curves. (Courtesy Federal Aviation Administration.)

time of concentration to the slope and dimensions of the drainage area. An estimate of time of concentration can be obtained from Fig. 12–2 which is based on a study of six watersheds in Tennessee which varied in size from 1.25 to 112 acres.

12–7. The Rational Method. By far, the most popular method for estimating runoff from small drainage areas is the rational method. In

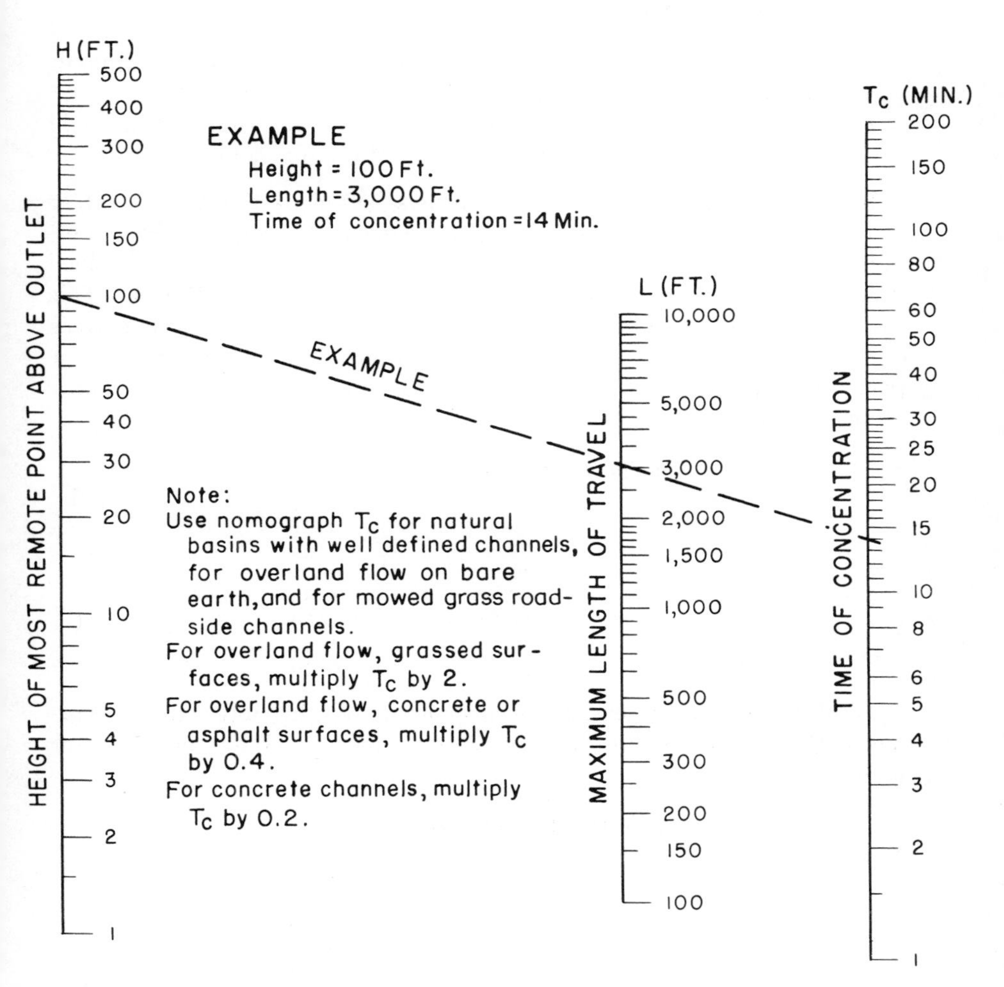

Fig. 12–2. Time of concentration of small drainage basins. (Source: *Design of Roadside Drainage Channels,* 1965.)

the rational formula, the quantity of water falling at a uniform rate is related by simple proportion to the total quantity that appears as runoff.

$$Q = CIA \tag{12-2}$$

where:

Q = runoff, in cubic feet per second.

C = a coefficient representing the ratio of runoff to rainfall. Typical values of C are given in Table 12–1.

I = intensity of rainfall, in inches per hour for the estimated time of concentration.

A = drainage area in acres. The area may be determined from field surveys, topographical maps, or aerial photographs.

The designer should contact the nearest office of the U.S. Weather Bureau for rainfall intensity data. In the event that suitable local rainfall intensity data are not available, approximate data may be obtained from Fig. 12–3, a chart published by the Bureau of Public Roads. This map shows rainfall intensity values in inches per hour for various areas of the contiguous United States for a two-year, 30-minute rainfall. The two-year rainfall intensity for other durations may be obtained by multiplying by the following factors:

Rainfall Duration (Minutes)	*Factor*
5	2.22
10	1.71
15	1.44
20	1.25
40	0.80
60	0.60
90	0.50
120	0.40

To obtain rainfall intensities corresponding to other intervals of recurrence, the values on Fig. 12–3 should be multiplied by the following factors:

Recurrence Interval (Years)	*Factor*
1	0.75
2	1.00
5	1.30
10	1.60
25	1.90
50	2.20

It may be noted that the rational formula is not dimensionally correct. It happens, however, that a one-inch depth of rainfall applied at a uniform rate to an area of one acre during one hour will produce 1.008 cubic feet per second of runoff if there are no losses. Thus, the

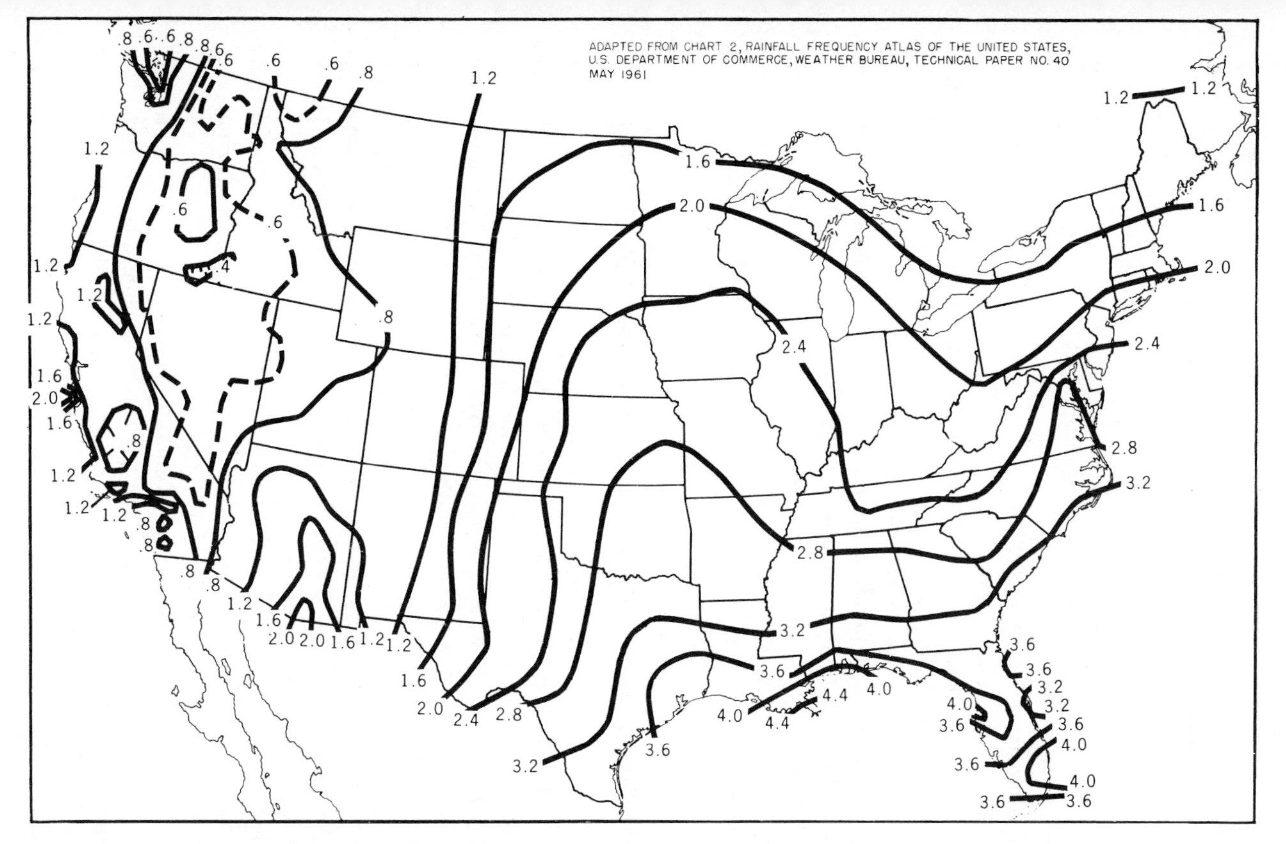

Fig. 12–3. Map of the contiguous United States showing 2-year, 30-minute rainfall intensity. (Source: *Design of Roadside Drainage Channels*, 1965.)

runoff coefficient represents the approximate fraction of the total water which falls that reaches the lowest point in the drainage basin.

12–8. The Burkli-Ziegler Formula. Another runoff formula which has gained wide acceptance is the Burkli-Ziegler formula:

$$Q = AIC \sqrt[4]{\frac{S}{A}} \tag{12–3}$$

where:

Q = quantity of water reaching culvert or screw in cu. ft. per sec.
A = drainage area, in acres
I = average rate of rainfall in inches per hour during the heaviest rainfall (2.75 in. per hr. is commonly used in the Middle West)
S = average of slope of ground, in feet per 1000 ft.
C = a coefficient depending upon the character of the surface drained and determined approximately a follows:
c = .75 for paved streets and built up business blocks
c = .625 for ordinary city streets
c = .30 for village with lawns and macadam streets
c = .25 for farming country

HYDRAULIC DESIGN OF CULVERTS AND DRAINAGE CHANNELS

In the preceding paragraphs we have discussed concepts and design procedures relating to the estimation of the quantity of runoff from a drainage basin. The following sections will deal with principles and techniques for the hydraulic design of culverts and other drainage structures for streets, highways, and railways. Some of the applicable fundamental principles govern fluid flow in conduits flowing under pressure and in open channels. This material is admittedly sketchy. The reader who wishes a more extensive treatment of this subject is referred to Chapter 3 of Reference 4 where a brief but well written discussion on the hydraulics of culverts is given.

Following a brief review of hydraulic principles, various types of culvert flow will be discussed after which typical design charts for culverts and open channels will be introduced and described.

12–9. Fundamental Principles—Conduits Flowing Full. Two fundamental principles form the basis for the theory of conduits accommodating fluid flow under pressure: the conservation of mass (expressed as the continuity equation), and the conservation of energy (expressed as the Bernoulli equation).[2]

[2] These principles are treated in more depth in Chapter 14 which deals with pipeline transportation.

The continuity equation merely states that the quantity of flow throughout a given flow system is constant.

$$Q = VA \tag{12-4}$$

where:

Q = rate of flow, cubic feet per second
V = average velocity of flow, feet per second
A = cross-sectional area, square feet.

The Bernoulli equation states that the total (pressure, kinetic, and potential) energy at a selected section of a flow system is equal to the energy at some previous section provided allowance is made for any energy added to or taken from the system.

$$Z_1 + \frac{P_1}{w_1} + \frac{V_1^2}{2g} - \Sigma H_L = Z_2 + \frac{P_2}{w_2} + \frac{V_2^2}{2g} \tag{12-5}$$

where:

$\frac{P}{w}$ = pressure energy
$\frac{V^2}{2g}$ = kinetic energy
Z = potential energy
ΣH_L = summation of energy losses (head losses)

The head losses, ΣH_L, are usually expressed in terms of velocity of flow. These losses are due primarily to friction losses within the conduit and entrance losses.

Friction Losses. Friction losses in a conduit flowing full are most commonly obtained by the Darcy-Weisbach equation:

$$H_f = f \frac{L}{D} \frac{V^2}{2g} \tag{12-6}$$

where L and D refer, respectively, to the length and diameter of the pipe; $V^2/2g$ is the velocity head; and f is the friction factor. The friction factor is a dimensionless measure of pipe resistance which depends on the characteristics of the pipe and the flow. For culverts the friction factor is most commonly given in terms of the coefficient of the Manning equation, n. The relationship between n and f is given by the following equation:

$$f = 185 \frac{n^2}{D^{1/3}} \tag{12-7}$$

It has been shown that the friction factor is not constant for a given type of culvert material but varies also with the Reynolds number. This

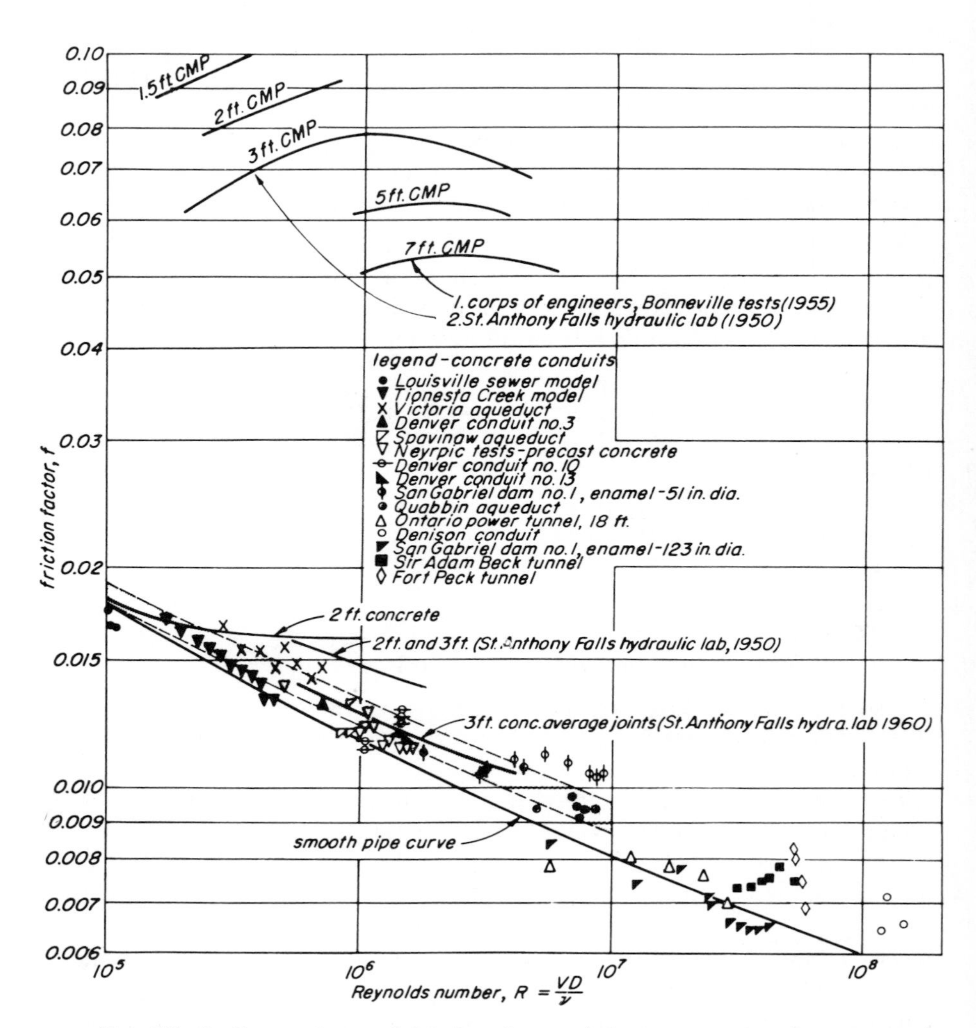

Fig. 12–4. Comparison of friction factor f for concrete and corrugated metal culverts. (Source: *Handbook of Concrete Culvert Pipe Hydraulics*, 1964.)

is shown by Fig. 12–4, which gives friction factors for concrete and corrugated metal culverts under various flow conditions.

Entrance Losses. Another major source of head loss occurs when flow is constricted in a culvert entrance. Entrance losses are due mainly to the expansion of flow following the entrance constriction. These losses may be computed by multiplying the velocity head of the fuel pipe times a constant value called the entrance loss coefficient, K_e.

$$H_L = K_e \frac{V^2}{2g} \tag{12-8}$$

Entrance loss coefficients vary widely with different types of entrance geometry. Typical values of K_e for various types of entrances are given by Table 12–2.

TABLE 12-2

Entrance Loss Coefficients for Submerged Circular Pipe Culverts

Type of Entrance	Entrance Head Loss Coefficient, K_4
Pipe entrance with headwall	
Grooved edge	0.19
Rounded edge (0.15D radius)	0.15
Rounded edge (0.25D radius)	0.10
Square edge (corrugated metal pipe)	0.43
Pipe entrance with headwall and 45° wingwall	
Grooved edge	0.20
Square edge (corrugated metal pipe)	0.35
Headwall with parallel wingwalls spaced 1.25D apart	
Grooved edge	0.30
Square edge (corrugated metal pipe)	0.40
Miter entrance for 2:1 embankment slope	0.62
Projecting entrance	
Grooved edge, thick wall	0.25
Square edge, thick wall	0.46
Sharp edge, thin wall (corrugated metal pipe)	0.92

Source: *Handbook of Concrete Culvert Pipe Hydraulics*, Portland Cement Association, 1964.

12–10. Fundamental Principles—Open Channel Flow. In open channels, the water surface, which is exposed to atmospheric pressure, serves as a flow boundary. Flow in open channels must, therefore, adjust itself so that the pressure at the water surface is equal to the pressure of the atmosphere. Open channel flow exists in conduits flowing part full as well as in open drainage ditches.

The basic laws of continuity of flow and conservation of energy also underlie the analysis of flow in open channels. In this case, however, the term for pressure is eliminated from the Bernoulli equation since the flow occurs under atmospheric pressure, which is assumed to be constant. The equation becomes:

$$Z_1 + d_1 + \frac{V_1^2}{2g} - \Sigma H_L = Z_2 + d_2 + \frac{V_2^2}{2g} \tag{12–9}$$

where:

Z = elevation of the bottom of the channel above a horizontal datum
d = depth of water in the channel
V = average velocity of flow

The energy relationships for open channel flow are shown by Fig. 12–5. Uniform flow occurs when the total energy line shown in the

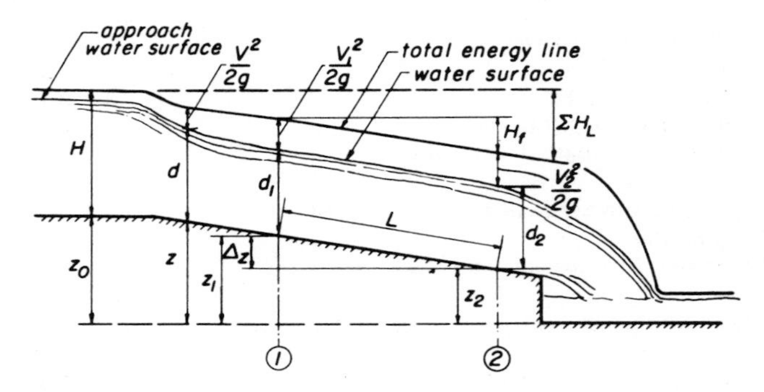

Fig. 12–5. Definition sketch for open channel flow. (Source: *Handbook of Concrete Culvert Pipe Hydraulics,* 1964.)

figure is parallel to the channel slope. Uniform flow is not often attained in highway and railway culverts.

It is instructive to note certain relationships involving the specific energy, which is defined as the sum of the depth (potential energy) and the velocity head (kinetic energy). Specific energy diagrams for a rectangular channel and a circular channel are shown, respectively, as Figs. 12–6 and 12–7.

From the specific energy diagrams, it will be noted that it is possible to have many different discharges at the same energy level. However, the greatest attainable discharge at a given energy level occurs at what is called the critical depth d_c. This flow condition, which might be considered a desirable goal in design, seldom occurs in practice.

12–11. Types of Culvert Flow. The type of flow occurring in a culvert depends upon the total energy available between the inlet and outlet.

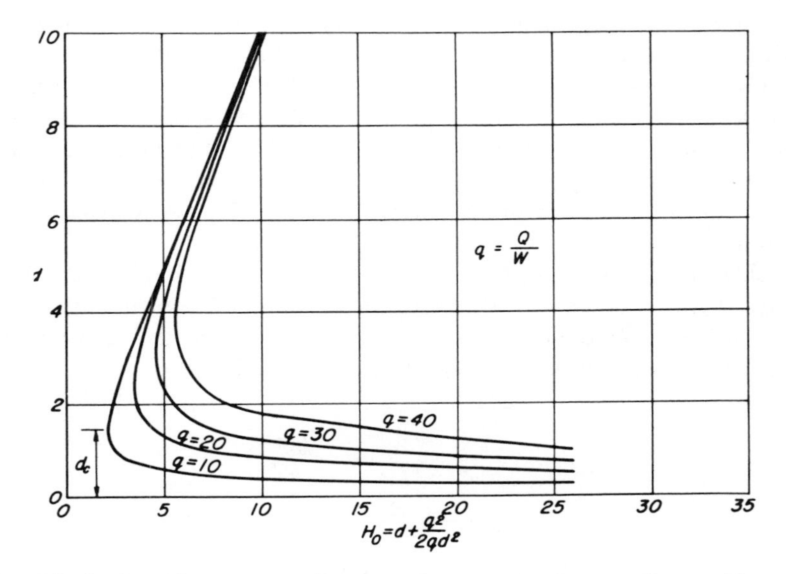

Fig. 12–6. Specific energy diagram for rectangular section. (Source: *Handbook of Concrete Culvert Pipe Hydraulics*, 1964.)

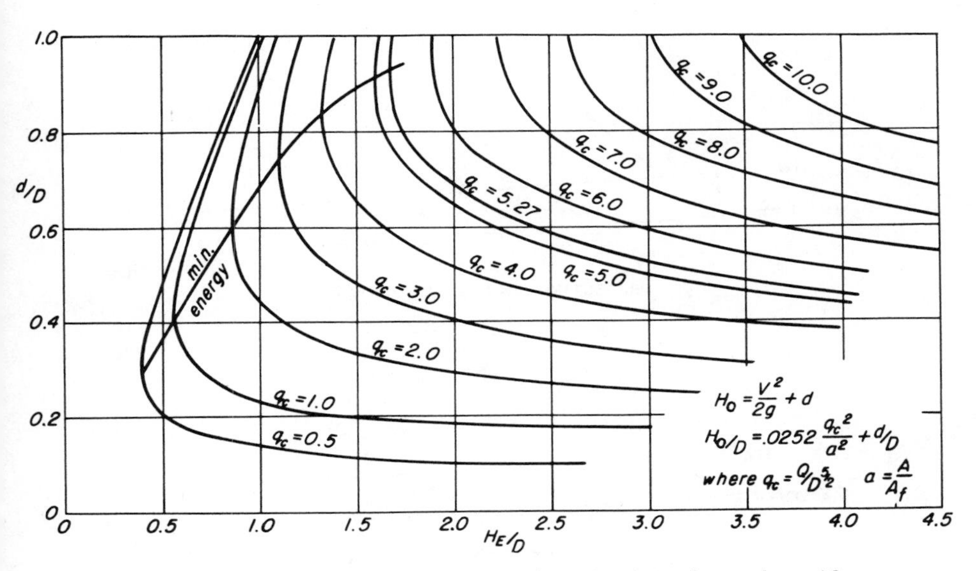

Fig. 12–7. Specific energy diagrams for circular channels. (Source: *Handbook of Concrete Culvert Pipe Hydraulics*, 1964.)

The flow that occurs naturally is that which will completely expend all of the available energy. Energy is thus expended at entrances, in friction, in velocity head, and in depth.

The flow characteristics and capacity of a culvert are determined by the location of the *control section.* A control section in a culvert is similar to a control valve in a pipeline. The control section may be envisioned as the section of the culvert which operates at maximum flow; the other parts of the system have a greater capacity than is actually used.

Laboratory tests and field studies have shown that highway and railway culverts operate with two major types of control: *inlet control* and *outlet control.* Examples of flow with inlet control and outlet control are shown, respectively, by Figs. 12–8 and 12–9.

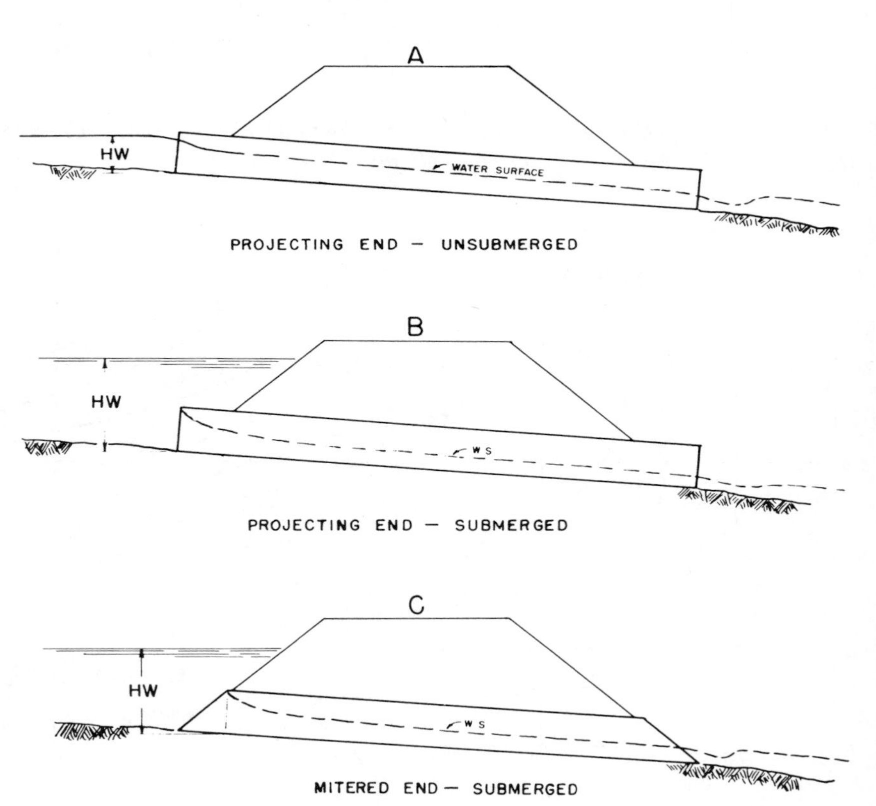

Fig. 12–8. Examples of flow with inlet control. (Source: *Hydraulic Charts for the Selection of Highway Culverts,* December, 1965.)

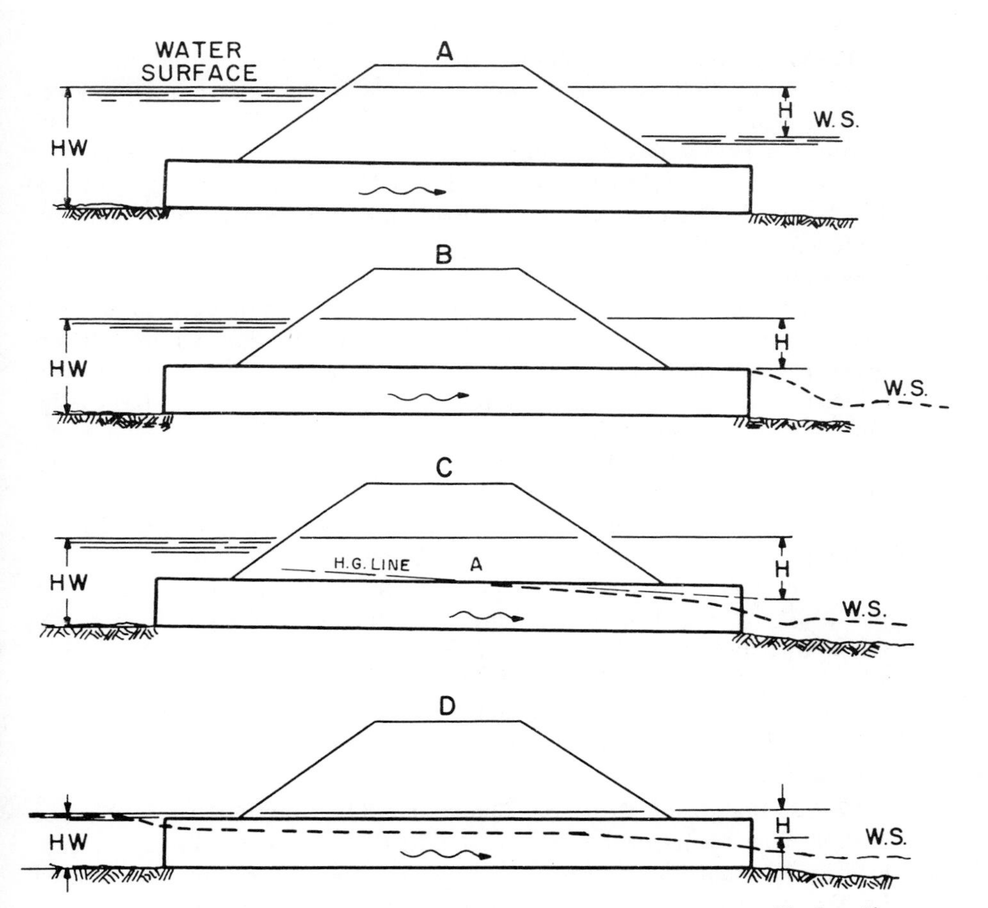

Fig. 12–9. Examples of flow with outlet control. (Source: *Hydraulic Charts for the Selection of Highway Culverts*, December, 1965.)

Under inlet control, the discharge capacity of a culvert depends primarily on the depth of headwater at the entrance and the entrance geometry (barrel shape, cross-sectional area, and type of inlet edge). Inlet control commonly occurs when the slope of the culvert is steep and the outlet is not submerged.

Maximum flow in a culvert operating with outlet control depends on the depth of headwater and entrance geometry and the additional considerations of the elevation of the tailwater in the outlet, the slope, roughness, and length of the culvert. This type of flow most frequently occurs on flat slopes, especially where downstream conditions cause the tailwater depth to be greater than the critical depth.

12–12. Culvert Design Charts. It is possible by involved hydraulic computations to determine the probable type of flow under which a given culvert will operate and to estimate its capacity. These computations may often be avoided by using design charts and nomographs published by the Bureau of Public Roads. From these charts and graphs, the headwater depths for both inlet control and outlet control may be determined for practically all combinations of culvert size, material, entrance geometry, and discharge.

Capacity Charts. Examples of capacity charts published in the Bureau of Public Roads' Hydraulic Engineering Circular Number 10 are given as Fig. 12–10 and Fig. 12–11. The solid line curves, which were plotted from model test data, represent inlet control. The dashed-line curves were computed for culverts of various lengths on relatively flat slopes and operating under outlet control. The dotted line, stepped across the curves, indicates the upper headwater limit for the unrestricted use of the charts. For greater headwaters, more reliable results are obtained by the nomographs described below.

Nomographs. Hydraulic charts in the form of nomographs have been published by the Bureau of Public Roads as Hydraulic Engineering Circular Number 5. Examples of these nomographs are shown as Fig. 12–12 and Fig. 12–13. The inlet control charts are based on laboratory research conducted by the National Bureau of Standards and the U.S. Coast and Geodetic Survey. The nomographs for outlet control were prepared from computations based on fundamental energy relationships.

12–13. Design of Drainage Channels. The simplest type of open channel flow occurs in long channels. In this case, equilibrium is established such that the energy losses due to friction are counterbalanced by the gain in energy due to slope. Discharge of this type, which is known as uniform flow, can be computed by Manning's equation:

$$Q = \frac{1.486}{n} AR^{2/3}S^{1/2} \tag{12–10}$$

where:

n = channel friction factor (see Table 12–3)
A = cross sectional area of flow

$$R = \text{hydraulic radius} = \frac{A}{\text{wetted perimeter}}$$

The Manning equation can be solved for discharge in a given channel if the depth of flow is known. The more common problem of solving for the depth of flow corresponding to a known discharge requires re-

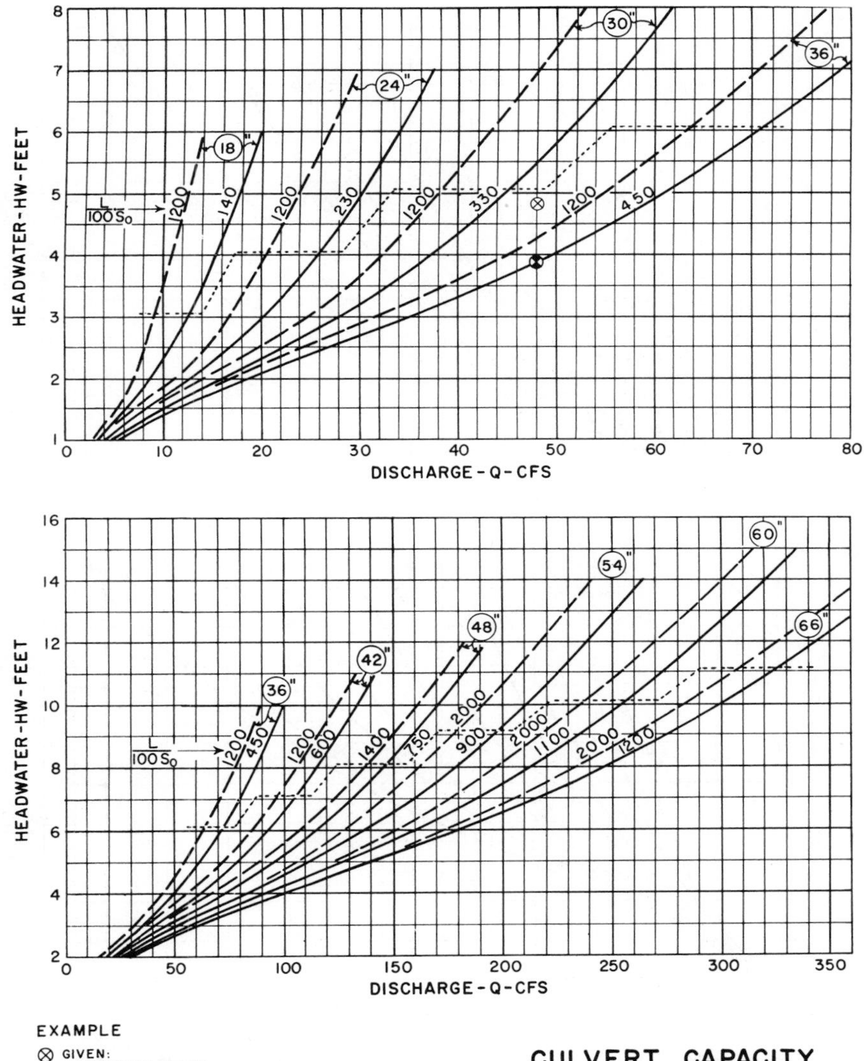

Fig. 12–10. Culvert capacity chart for circular concrete pipes with square-edged entrance. (Source: *Capacity Charts for the Hydraulic Design of Highway Culverts,* March, 1965.)

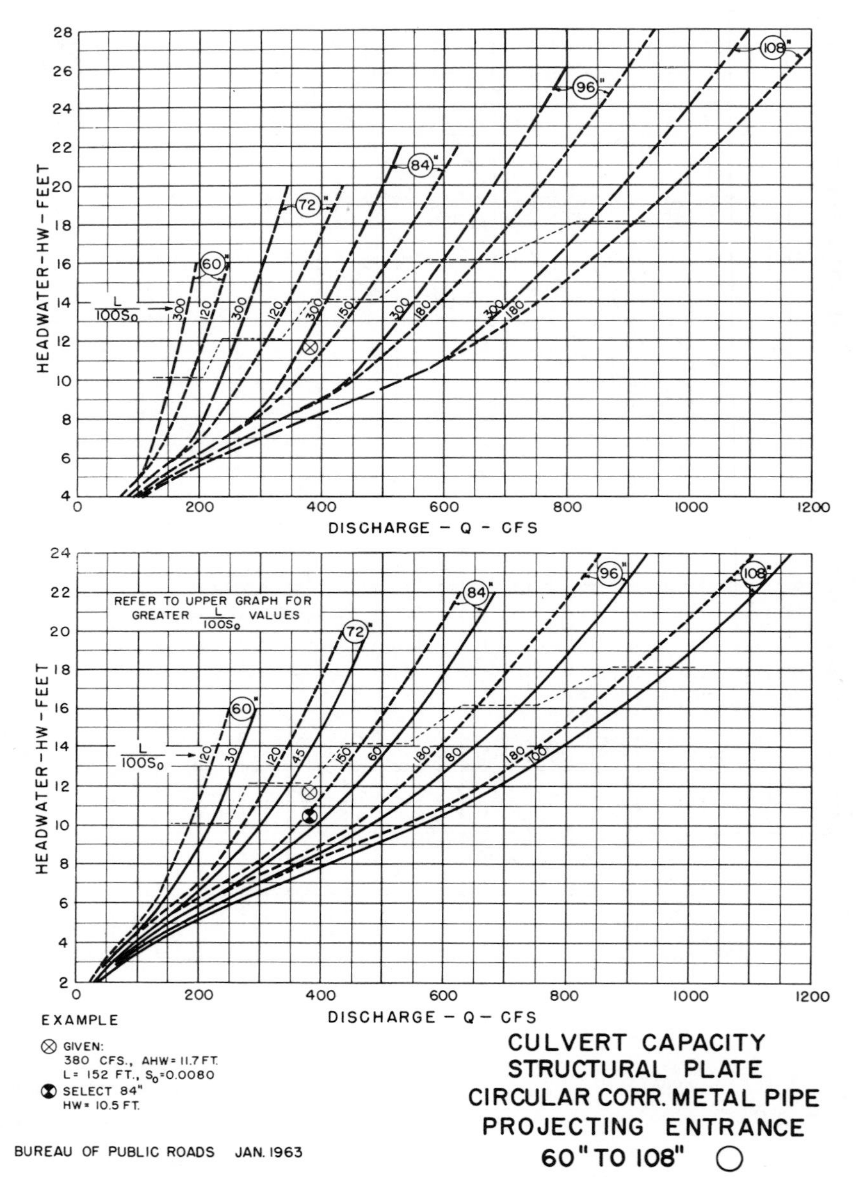

Fig. 12–11. Culvert capacity chart for structural plate circular corrugated metal pipes with projecting entrance. (Source: *Capacity Charts for the Hydraulic Design of Highway Culverts*, March, 1965.)

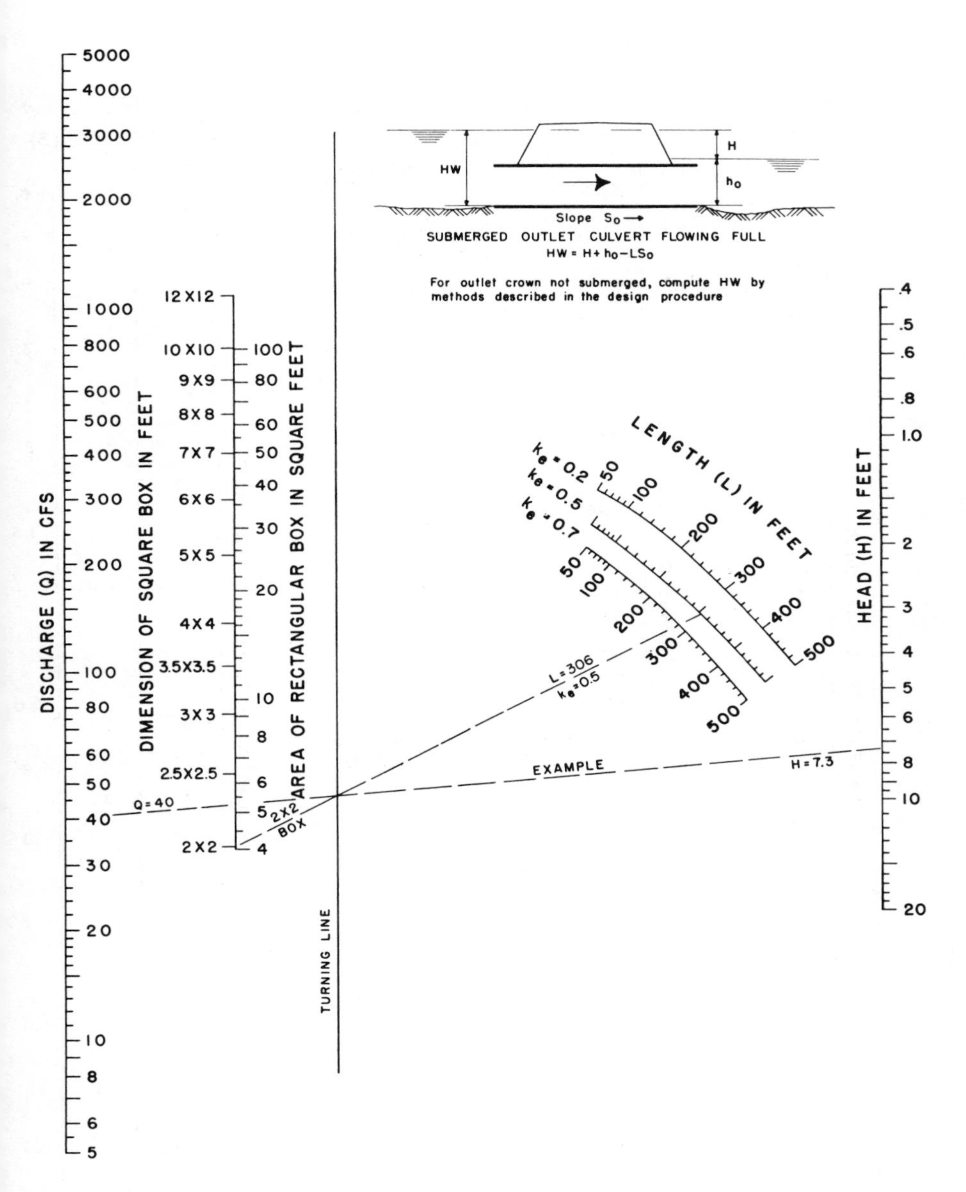

BUREAU OF PUBLIC ROADS JAN. 1963

HEAD FOR
CONCRETE BOX CULVERTS
FLOWING FULL
n = 0.012

Fig. 12–12. Nomograph for head for concrete box culverts flowing full. (Source: *Hydraulic Charts for the Selection of Highway Culverts,* December, 1965.)

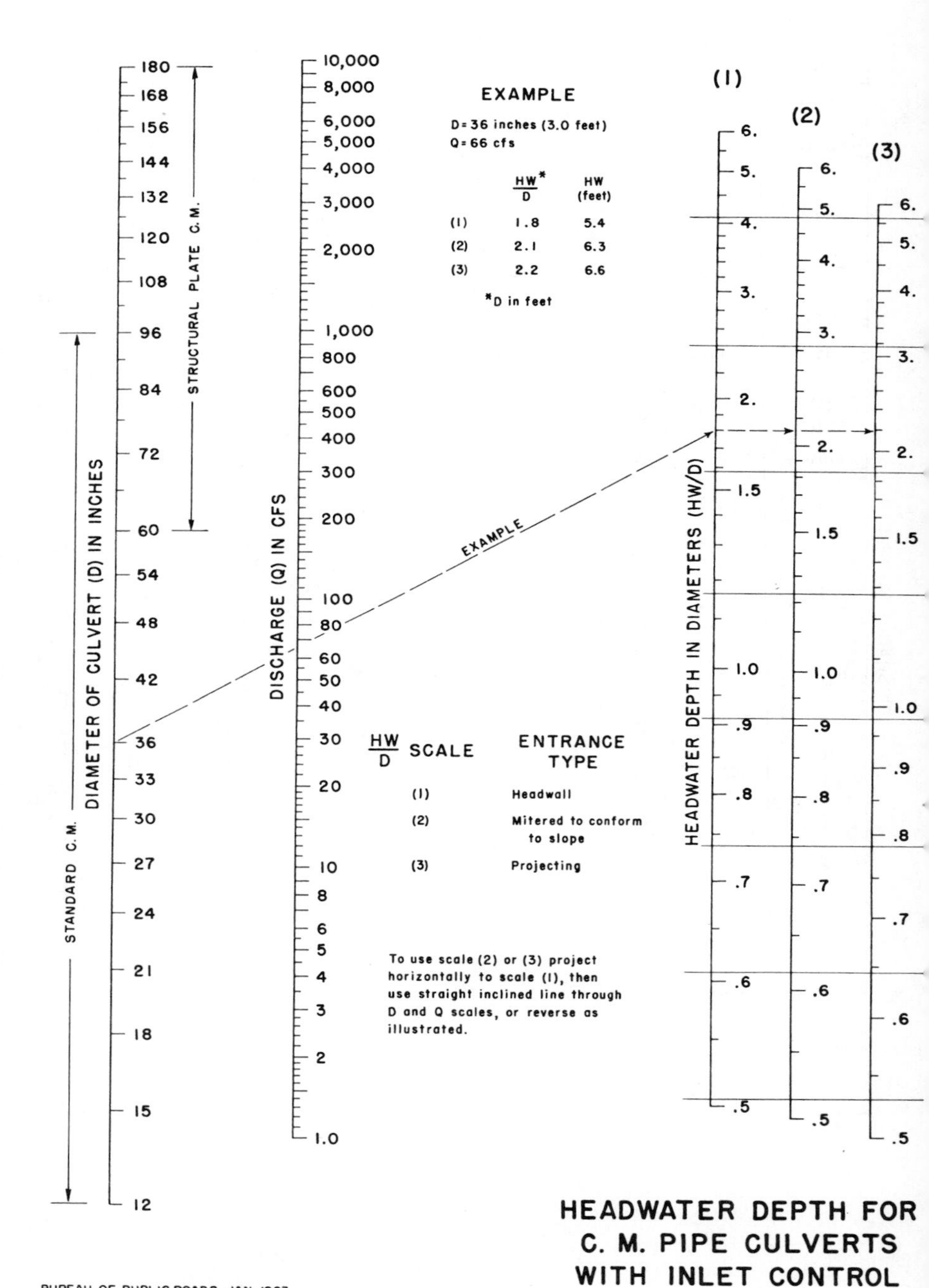

Fig. 12–13. Nomograph for headwater depth for corrugated metal culverts with inlet control. (Source: *Hydraulic Charts for the Selection of Highway Culverts*, December, 1965.)

TABLE 12-3

Values of Manning's Roughness Coefficient (Open Channels)

Type of Lining	Values of *n*
Smooth concrete	0.013
Rough concrete	0.022
Riprap	0.030
Asphalt, smooth texture	0.013
Good stand, any grass—depth of flow more than 6 inches	0.09–0.30
Good stand, any grass—depth of flow less than 6 inches	0.07–0.20
Earth, uniform section, clean	0.016
Earth, fairly uniform section, no vegetation	0.022
Channels not maintained, dense weeds	0.08

Source: Ritter, Leo J., Jr. and Paquette, Radnor J., *Highway Engineering*, Third Edition, The Ronald Press, 1967.

peated trials. Charts have been published, however, providing a direct solution to the Manning equation for various sizes of rectangular, triangular, circular, and trapezoidal cross-sections [7].

The Manning equation most commonly applies to the underground storm drainage systems (discussed in Section 18–7) and to the design of open drainage channels.

An example of one of the channel charts available in the literature is shown as Fig. 12–14. This chart is applicable to a trapezoidal channel with 2:1 side slopes and a constant two-foot bottom width.

> Depths and velocities shown in the chart apply accurately only to channels in which uniform flow at normal depth has been established by sufficient length of uniform channel on a constant slope when the flow is not affected by backwater.
>
> Depth of uniform flow for a given discharge in a given size of channel on a given slope and with $n = 0.030$ may be determined directly from the chart by entering on the Q-scale and reading normal depth at the appropriate slope line (or an interpolated slope). Normal velocity may be read on the V-scale opposite this same point. This procedure may be reversed to determine discharge at a given depth of flow.
>
> For channel roughness other than $n = 0.030$, compute the quantity Q times n and use the $Q.n$- and $V.n$-scales for all readings, except those which involve values of critical depth or critical velocity. Critical depth for a given value of Q is read by interpolation from the depth lines at the point where the Q-ordinate and the critical curve intersect, regardless of channel roughness. Critical velocity is the reading on the V-scale at

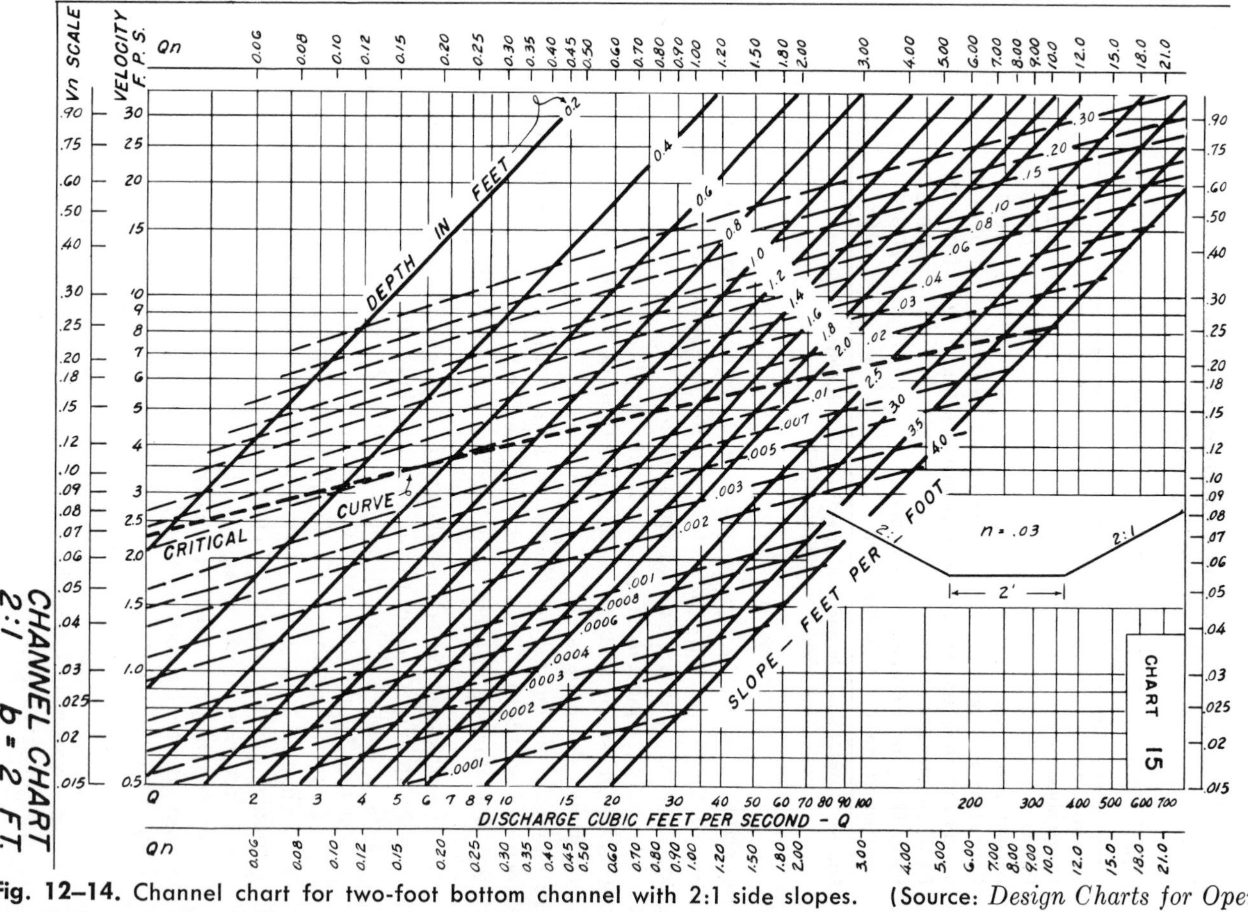

Fig. 12–14. Channel chart for two-foot bottom channel with 2:1 side slopes. (Source: *Design Charts for Open Channel Flow*, August, 1961.)

this same point. Where $n = 0.030$, the critical slope is read at the critical depth point. Critical slope varies with n; therefore, in order to determine the critical slope for values of n other than 0.030, it is first necessary to determine the critical depth. Critical slope is then read by interpolation from the slope lines at the intersection of this depth with the $Q.n$-ordinate [1].

Suppose, for example, one wishes to determine the depth and velocity of flow in a trapezoidal channel ($n = 0.030$) with 2:1 side slopes and a two-foot bottom width given a flow of 100 cubic feet per second and a slope of 4.0 per cent. The chart is entered at $Q = 100$ cubic feet per second and a line is projected vertically until it intersects the slope line $S_o = 0.04$. At the point of intersection, the normal depth $d_n = 1.6$ feet and the corresponding normal velocity $V_n = 10$ feet per second is read.

To find the critical depth, velocity, and slope for these conditions, the line $Q = 100$ cubic feet per second is projected upward to its intersection with the critical curve. At the point of intersection, the following values are read:

Critical depth $d_c = 2.3$ feet
Critical velocity $V_c = 6.6$ feet per second
Critical slope $S_c = 0.014$

The normal depth is less than the critical depth indicating that the flow is rapid and not affected by backwater conditions.

Suppose $n = 0.012$ and other conditions are the same as given above. In this case, the product $Q.n$ is determined and the $Q.n$ scale is used.

$$Q.n = 100(0.012) = 1.2$$

Entering the $Q.n$ scale at a value of 1.2 and proceeding as before yields the following values:

$$d_n = 1.2 \text{ feet}$$

$$V_n = \frac{0.24}{0.012} = 20 \text{ feet per second}$$

$$d_c = 1.5 \text{ feet}$$

$$V_c = \frac{0.17}{0.012} = 14 \text{ feet per second}$$

$$S_c = 0.016$$

Erosion Control. The designer should remember that erosion of open channels is not only aesthetically offensive, but it may also increase roadway maintenance costs and, in its most drastic form, may create roadway hazards. There are several steps that can be taken to minimize the probability and lessen the effects of erosion. In the first place, erosive channel velocities should be avoided whenever possible. The magnitude of stream velocity which causes erosion varies, of course, with the type of channel lining. Unlined ditches in certain fine-grained soils may be eroded by streams flowing as slow as 1.5 to 2.0 feet per second, while other untreated soils may resist erosive forces caused by streams flowing up to 6.0 feet per second. The American Society of Civil Engineers has published helpful maximum allowable velocities for various types of channel linings [1].

Another step that the designer can take to control erosion is to seed, sod, or pave the channel lining. Generally, seeding operations can be successfully applied to ditches which have mild slopes and low quantities of flow. Sodding can generally be successfully used for moderate slopes and velocities. In cases in which velocities greater than about 6.0 feet per second are expected, the designer must resort to a special treatment such as soil cement or paving with asphalt or portland cement concrete.

Culvert Selection. While a variety of materials has been utilized, the large majority of culverts in common use fall into two classes:

1. reinforced concrete culverts
2. corrugated metal culverts

Reinforced concrete culverts fall into two general classes: pipes and box culverts. Concrete pipes come in a variety of diameters ranging from 12 to 108 inches. Box culverts are usually formed and poured in place with rectangular or square cross section.

Concrete culverts are durable and able to withstand large stresses imposed by heavy wheel loads. In many localities, concrete culverts cost less than comparable sizes of corrugated metal culverts. Possessing a smaller roughness coefficient, concrete culverts are more efficient hydraulically than corrugated metal culverts. On the other hand, concrete culverts are heavy and not easily installed, particularly in cases involving steep slopes and large sizes.

Corrugated metal culverts are made of galvanized steel of varying thicknesses. These culverts are manufactured in diameters from 8 to 96 inches. Corrugations, typically 2⅔ inches from crest to crest and ½ inch deep, are formed in the sheet metal for added strength.

Corrugated metal culverts are easy to handle and install, even in large sizes and steep slopes. Where very large pipe sizes are required, corrugated metal culverts may be formed from heavy plates which are bolted together at the field site.

Corrugated metal culverts may be purchased as pipe arches, the vertical dimension (rise) being about 0.6 the horizontal dimension (span).

Arch culverts are commonly and advantageously used in low fill areas where the headroom is limited. Although corrugated metal culverts are subject to deterioration in certain severe exposure conditions, this may be largely overcome by the use of a pipe in which the invert has been covered with a heavy bituminous mixture.

EARTHWORK OPERATIONS

12–14. Introduction. Practically all highway and railroad construction jobs involve a considerable amount of earthwork. Earthwork operations in general are those construction processes which involve the soil or earth in its natural form, which precede the building of the pavement structure itself. These processes may include everything that pertains to the grading and drainage structures, which will also include clearing and grubbing, roadway and borrow excavation, the formation of embankments and the finishing operations for the preparation of the highway or runway pavement or railroad ballast. Any or all of these construction processes may be performed on a given project and they may overlap to some extent.

Clearing is the removal of trees, shrubs, brush, etc., from within designated areas, while grubbing refers to the removal of roots, stumps, and similar obstacles to a nominal depth below the existing ground surface. Frequently, clearing and grubbing comprise a single contract item and may include the removal of topsoil to a shallow depth. Excavation refers to the removal of earth from its natural resting place to a different place for the highway or railroad foundation. Embankments required in construction are usually formed in relatively thin layers or lifts of soil and compacted to a high degree of density. Such embankments are called rolled earth embankments or rolled earth fills. Hydraulic fills are also sometimes required in construction.

Finishing operations include such items or trimming and finishing of slopes and the fine grading operations required to bring the grade to the desired final elevation.

Broadly speaking, earthwork operations may include all the operations involved in bringing the foundation to the point where the surface or ballast is to be applied.

12–15. Earthwork Equipment. In modern practice earthwork operations are accomplished largely by the use of highly efficient and versatile machines. Machines have been developed which are capable of performing every form of earthwork operation efficiently and economically. These include tractors, bulldozers, scraper units, shovel crane units for the excavation and moving of earth, and the sheepsfoot, pneumatic-tired and steel wheel rollers for compacting the earth as well as vibratory compactors and pneumatic tampers. Motor graders, trucks, and other specialized equipment may be used for earthwork operations.

A more detailed discussion of earth moving equipment may be found in Reference 1.

12–16. Excavation. Excavation may be classified into two types: (1) rock excavation and (2) common excavation. Rock excavation will include boulders which may be over 0.5 cubic yards or more in volume and all hard rock which has to be removed by blasting. In some cases the requirement relative to the removal by blasting is not specified and a phrase such as solid well-defined ledges of rock is substituted in the definition for rock excavation. Even though a contractor may choose to rip some materials, this operation may still be called rock excavation by some agencies.

Common excavation consists of all excavation not included in the rock definition.

Borderline cases frequently arise in which there may be some question as to the proper classification of a portion of the work. To eliminate this condition most agencies use the term unclassified excavation to describe the excavation of all materials, regardless of their nature. Many feel that the unclassified term tends to put the risk of uncertainty on the bidder, which results in higher prices than when rock and common excavation units are used.

If there is not sufficient material within the cross-section of construction for completion of the grade, additional material may be obtained from borrow pits which may be called borrow excavation. Sometimes additional excavation may be obtained by changing the cross-section by widening. This is called side borrow. All excavation is paid for on a cubic yard basis, measured in place before excavation occurs.

12–17. Construction of Embankments. Embankments are used in construction when it is required to maintain a grade for the roadway, runway, or railway. Usually the grade is built up in a fill or embankment section from material in a cut or excavation section and borrow excavation if needed and is termed a rolled-earth embankment.

Rolled-earth embankments are constructed in relatively thin layers of loose soil. Each layer is rolled to a satisfactory degree of density before the next layer is placed and the fill or embankment is thus built up to the desired height by the formation of successful layers or lifts. Most agencies at the present time require layers to be from 6 to 12 inches thick before compaction begins, when normal soils are encountered. Specifications may permit an increase in layer thickness where large rocks are used in the lower portion of a fill, up to a maximum thickness of 24 inches.

The layers are required to be formed by spreading the material to uniform thickness before compaction is permitted. End dumping from trucks without spreading is definitely not permitted. The only exception to this rule may be when the embankment foundation is such that it cannot support the weight of the spreading and the compacting equipment. In such cases end dumping may be permitted until sufficient thickness can support the equipment.

A close relationship between the compaction procedure and the type of soils used for embankment purposes has to be carefully evaluated. The treatment of the various types of soils is not considered in this text.

12–18. Control of Compaction. Practically all soils exhibit a similar relationship between moisture content and density (dry unit weight) when subjected to dynamic compaction. Practically every soil has an optimum moisture content at which the soil attains maximum density under a given compactive effort. In the laboratory this relationship is usually performed under the Standard Proctor or the Standard AASHO Method (T99). Briefly stated, this procedure uses the soil that passes the No. 4 sieve which is placed in a 4-inch diameter mold having a volume of $\frac{1}{30}$ of a cubic foot. The soil is placed in three layers of about equal thickness and each layer is subjected to 25 blows from a hammer weighing 5.5 pounds, having a striking face 2 inches in diameter and falling through a distance of 12 inches (12,375 foot-pounds per cubic foot).

Due to the use of heavier compaction equipment in recent years and in order to correlate more effectively laboratory procedures with field conditions, the procedure was modified and is now known as the Modified Proctor or Modified AASHO (T180) compactor. Under the modified procedure the same mold is employed using 25 blows from a 10 pound hammer dropping a distance of 18 inches on 5 equal layers (56,250 foot-pounds per cubic foot).

Regardless of which method is used, the optimum moisture and maximum density are usually found in the laboratory by a series of determinations and the results are plotted. Figure 12–15 shows the moisture-density relationship for a typical soil under dynamic compaction. The zero air voids curve shown in Fig. 12–15 represents the theoretical density which this soil would attain at each moisture content if all the void spaces were filled with water, namely if the soil were completely saturated.

After the laboratory density has been determined, recommendations are made for field conditions. A large majority of agencies require compaction to a certain percentage of the maximum density, as determined in the laboratory. This may range from 90 to 100 per cent.

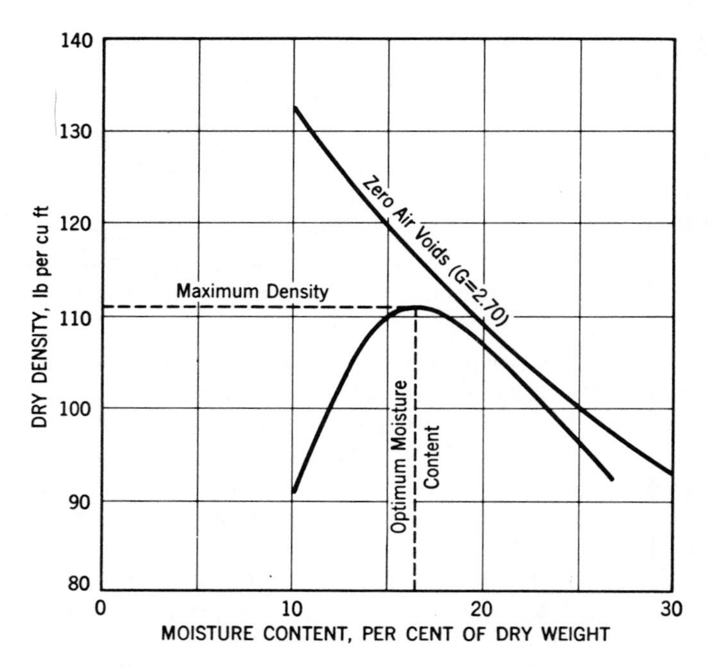

Fig. 12–15. Moisture-density relationships for a typical soil under dynamic compaction. (Source: Leo J. Ritter and Radnor J. Paquette, *Highway Engineering,* Third Edition, The Ronald Press, 1967.)

In order to determine if compaction meets laboratory requirements, field testing is required. Three methods have been widely used, which are designated as the sand, the balloon, and the heavy oil methods. Some use of nuclear devices to measure in place density and moisture content are being used. The details of these tests will not be discussed here.

12–19. Special Embankment Foundations. In swampy areas, particularly where peat and other highly organic soils are encountered, special treatment may be required in the construction of embankments to prevent failure of the embankment. These methods in general use may be classified as (1) gravity subsidence, (2) partial or total excavation, (3) blasting, (4) jettying, and (5) vertical sand drains.

In gravity subsidence a fill may simply be placed on the surface of an unsatisfactory foundation soil and allowed to settle at will, with no special treatment of the underlying soil. A temporary surface may be placed and, as settlement occurs, the temporary surface may be replaced.

Under the partial or total excavation, the undesirable soil may be partially or completely removed and backfilled with suitable material. Total excavation of the undesirable material is expensive, but has its advantages in that the pavement surface may be placed immediately. Under partial excavation, gravity subsidence takes place and the temporary surface has to be replaced.

When blasting is used in swampy areas, a fill is usually placed first by the end-filling method. When the fill is complete, dynamite charges are used and so timed that the swampy material is thrust sideways and the fill material settles in place.

In jettying, the fill is made in a similar way as for blasting. The process of jettying involves the pumping of water into the underlying soil in order to liquefy it and thus aid in its displacement by the weight of the fill or embankment.

Vertical sand drains have been used increasingly in recent years. Vertical sand drains generally consist of circular holes or shafts from 18 to 24 inches in diameter, which are spaced from 6 to 20 feet apart on centers beneath the embankment section and are carried beneath the layer of compressible soil. The holes are backfilled with suitable granular material. A sand blanket from 3 to 5 feet thick is placed on top of the drains of the embankment. The embankment is then constructed by normal methods on top of the sand blanket. The weight of the embankment forces the water out of the sand drains and consolidation takes place.

12–20. Computing Earthwork Quantities. The amount of earthwork on a project is usually one of the most important features in its design. Earthwork includes the excavation of material and any hauling required for completing the embankment. Payment for earthwork is based on excavated quantities only and generally includes the cost of hauling. An additional item of payment called overhaul is often used to provide for hauling of the excavated material beyond a freehaul distance. As stated previously, the excavated material may be obtained from within the construction limits or from borrow pits.

In order to determine earth excavation and embankment requirements before construction, the grade line for the proposed highway or railway will have to be determined and cross-sections will have to be made of the original ground. Earthwork quantities are then determined by placing templates of the proposed grade over the original ground. The areas in cut and the areas in fill are determined and the volumes between the sections are computed. Figure 12–16 shows template sections and

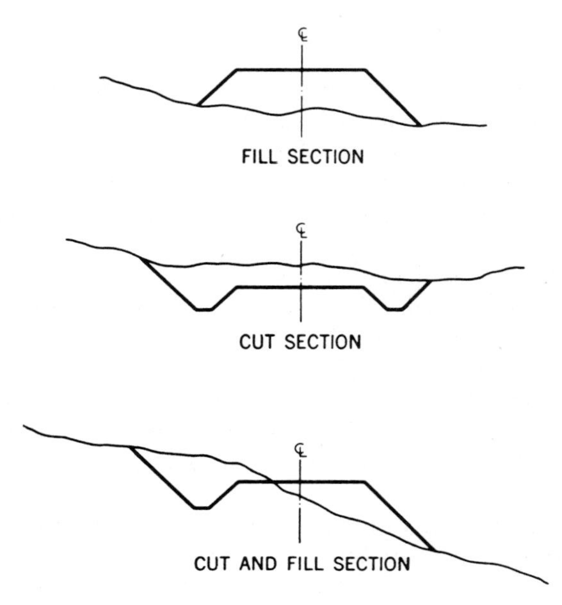

Fig. 12–16. Original ground line and template sections. (Source: Leo J. Ritter and Radnor J. Paquette, *Highway Engineering,* Third Edition, The Ronald Press, 1967.)

original ground in cut and fill. The terms cut and fill are used for areas of the section and the terms excavation and embankment generally refer to volumes.

Cross-sections are plotted on standard cross-section paper to any convenient scale. A scale of 1 inch equals 5 feet vertically and horizontally is common practice. Each cross-section should show the location or station of the original ground section and template section, the elevation of the proposed grade at that station, and the areas of cut and fill for each section. The computed volumes of excavation and embankment may be placed on the sheet between two successive cross-sections to facilitate the tabulation of earthwork quantities.

The areas of cut and fill may be measured by the use of a planimeter, a computation method using coordinates, or some other suitable method.

Volumes may be computed by the average end area method or by the prismoidal formula. The average end area method is based on right prisms and the computed volumes are slightly in excess of those computed by the prismoidal formula. This error is small when the sections do not change rapidly; however, when sharp curves are used, a prismoidal correction should be applied. The average end area method is generally used by a majority of agencies. The formula for the average end area method is as follows:

$$V = \frac{\frac{1}{2}L(A_1 + A_2)}{27} \qquad (12\text{–}11)$$

where:

V = volume in cubic yards
$A_1 + A_2$ = area of end sections in square feet
L = distance between end sections in feet

When a section changes from a cut section to a fill section, a point is reached where zero cut and zero fill occur. At this point, it will be necessary to take additional cross-sections so that the proper volumes may be computed. A typical volume sheet for computing excavation and embankment quantities is shown in Fig. 12–17. Manual methods described above are being rapidly supplanted by methods based on electronic computers. In typical cases and for earthwork quantities at the design state, the computer input includes the center line data, the cross-section data, template data, and other pertinent information. The computer produces a tabulation of the cut and fill volumes at each station and the difference between them adjusted for shrinkage and swell, the cumulative volumes of cut and fill, mass diagram coordinates, and slope stake coordinates.

12–21. Shrinkage and Swell. When the freshly excavated material is hauled to an embankment, the material increases in volume; however, during the construction process of compacting the embankment, the volume decreases below that of its original volume. This is known as shrinkage. In estimating earthwork quantities, this factor must be taken into consideration. The amount of shrinkage varies with the type of soil, the depth of the fill, and the amount of compactive effort. An allowance of 10 to 15 per cent is frequently made for high fills with from 20 to 25 per cent for shallow fills. The shrinkage may be as high as 40 or 50 per cent for some soils. This generally also allows for shrinkage due to loss of material during the hauling process and loss of material at the toe of the slope.

When rock is excavated and placed in the embankment, the material will occupy a larger volume. This increase is called swell and may amount to 30 per cent or more. The amount of swell is not important when small amounts of loose rock or boulders are placed in the embankment.

12–22. The Mass Diagram. A mass diagram is a graphical representation of the manner in which earth is moved. It does not show any earth which is used to balance cuts and fills without end hauling such as casting or blade work, but reflects only the excess or deficiency after such work is completed. In other words, a mass diagram is a continuous curve showing the accumulation of the algebraic sum of the yardage from some initial station to any succeeding station. A mass diagram will show the location of balance points, the direction of haul, and the amount of earth taken or hauled to any location. It also indicates if borrow or waste must be made. It is a valuable aid in the supervision of grading operations and is helpful in determining the amount of overhaul and the most economical distribution of the material.

The overhaul distance may be defined as the length of haul beyond a certain distance known as freehaul. This freehaul distance may be as low as 500 feet and as long as 3,000 feet or more. Some agencies do not consider any overhaul which means that the cost of excavation and hauling of the material is included in the cost of earth excavation. The overhaul distance is found from the mass diagram by determining the distance from the center of mass of the excavated material to the center of mass of the embankment. This distance may be measured in stations or in miles. Thus, a cubic yard station is the hauling of one cubic yard of excavation one station beyond the freehaul distance and one cubic yard mile is the hauling of one cubic yard of material a distance of 1 mile beyond the freehaul limits. The dividing line between cubic yard stations and cubic yard miles will vary according to the specifications. Some agencies may state that excavation hauled up to ½ mile beyond the freehaul limits shall be measured in yard stations and any excavation hauled beyond the freehaul limits over ½ mile shall be in yard miles. Thus with a freehaul distance of 1000 feet, the dividing line between yard stations and yard miles will be a distance of 3,640 feet. In practice these are called cubic yard miles, but actually they are cubic yard ½ miles.

Several methods for determining overhaul are in use. The graphical method, the method of movements, and the planimeter method are a few. Various computer programs have been developed to perform these calculations, which result in a large saving of time. The graphical method will be illustrated here to give some idea as to the approach to the problem and theory involved. For a more detailed explanation of the other methods reference is made to Chapter 8 of *Highway Engineering* by Ritter and Paquette [1].

12–23. Graphical Method for Determining Overhaul. A mass diagram is plotted to scale from the earthwork volumes computed for the project similar to that indicated in Fig. 12–17. The scales used are generally 1 inch equals 500 feet horizontally and 1 inch equals 500 cubic yards vertically; however, other convenient scales may be used. Such a curve may be as indicated in Fig. 12–18. The balance points and direction

VOLUME SHEET

SHEET No. 1	PROJECT No. 00				COUNTY Madison			COMPUTED BY J. P.			DATE		
	End Area, Square Feet				Volume in Cubic Yards						Bal. Volumes Diff. bet. 5 and 10		M. O.
Station	1 Total Cut	2 Fill	3 Loss	4 Unsuit. Material Waste	5 Total Cut	6 Fill	7 Shr % 20	8 Loss	9 Unsuit. Material Waste	10 Fill + 7, 8 & 9	11 Cut +	12 Fill −	13
286	60	35											0
287	54	47			211	152	30			182	29		+29
288	48	50			189	179	36			215		26	+3
289	24	65			133	213	43			256		123	−120
290	12	43			67	200	40			240		173	−293
291	0	56			22	183	37			220		198	−491
292	6	42			11	181	36			217		206	−697
293	7	63			24	195	39			234		210	−907
294	13	97			37	296	59			355		318	−1225
295	28	110			76	383	75			458		382	−1607
296	43	148			131	478	96			574		443	−2050
297	87	160			278	571	114			685		407	−2457
298	109	89			363	461	92			553		190	−2647
299	156	50			490	257	52			309	181		−2466
300	112	37			496	161	32			193	303		−2163
301	83	20			342	106	21			127	215		−1948
302	68	9			268	54	9			63	205		−1743
303	56	5			230	26	5			31	199		−1544
304	54	0			273	9	2			11	192		−1352
305	42	0			178	0	0			0	178		−1174
				Total	3749	4105					4923		

Rock Exc.	Earth Exc.	Borrow	Fill	Shrkg 20%	Loss	Waste		Overhaul	Borrow O'haul

Fig. 12–17. Volume sheet and computations.

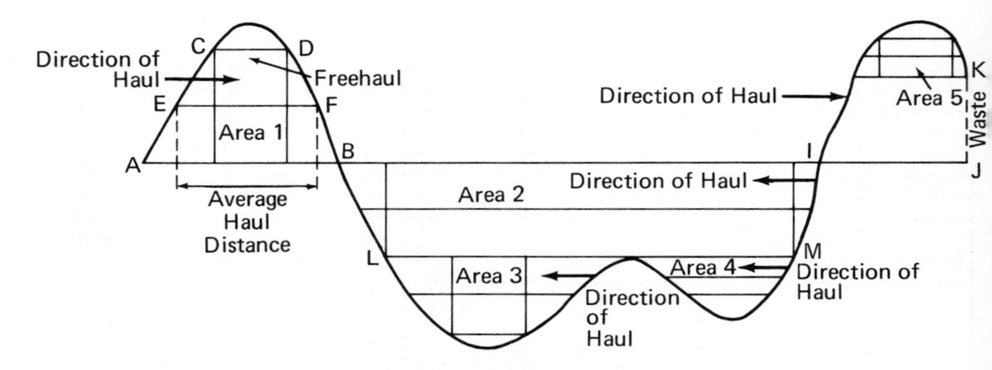

Fig. 12–18. A mass diagram.

of haul are as indicated. To determine the overhaul for each balance, a line equal to the freehaul distance, 1,000 feet in this case, is drawn parallel to the base line AV. The shaded area indicates freehaul and is eliminated from further consideration. The next step is to drop perpendiculars from C and D to the base line. These perpendiculars are bisected and extended to intersect the mass diagram curve at E and F. The distance EF is the average haul distance. The vertical distance CG or DH indicates the number of cubic yards. To compute the overhaul for this balance, it will be the average distance EF in stations less the freehaul distance in stations multiplied by the volume in cubic yards. This is done for each balance. Where the curve changes within a balance the area has to be divided as indicated by the line LM and the process repeated. The summation of each area will give the total overhaul for the project. KJ indicates that so many cubic yards of material has to be wasted. If this were below the base line AJ, it would indicate that borrow would have to be made.

This method gives the average haul and it has its limitations. If the slopes of the curve are more or less uniform, a minimum of error occurs. If the slopes are irregular, the method of moments or some other method should be used.

12–24. Length of Economical Haul. Where it is necessary to haul material long distances it is sometimes more economical to waste material excavated from the roadway or railway section and borrow material from a borrow pit from within the freehaul distance. The length of economical haul can be determined by equating the cost of the excavation plus the cost of overhaul to the cost of excavation in the road or railway plus the cost of excavation from the borrow pit.

If h equals the length of haul in stations beyond the freehaul distance, e equals the cost of excavation, and o equals the cost of overhaul, then to move 1 cubic yard of material from cut to fill, the cost will be $e + ho$ and the cost to excavate from cut, waste the material, borrow, and place 1 cubic yard in the fill will equal $2e$. Assuming that the cost of the roadway excavation is equal to the cost of borrow excavation, then:

$$e + ho = 2e$$

and:

$$h = \frac{e}{o} \text{ stations}$$

12–25. Preparation of Final Plans. Earth excavation quantities, overhaul, and borrow requirements, are generally the first items that are determined when plans are being made and then other details are prepared. These will include items such as the drainage structures, surfacing items, fine grading and finishing operations, seeding or sodding and other items that may be necessary for the completion of the project.

When the plans are completed a summary is made of all items of work, estimated quantities, and units of measurement. From these items preliminary estimates of cost can be made. Specifications and a proposal are prepared. When this is done the project is ready to be advertised for bids and contracts for construction.

PROBLEMS

1. A highway culvert is to be designed for a 630-acre drainage area in Nashville, Tennessee. The height of the most remote point above the outlet is 50 feet and the maximum length of water travel is 1,800 feet. The drainage area consists of rolling farmland with relatively impervious soil. Using a two-year recurrence interval, determine the runoff in cubic feet per second.
2. Using the data in Problem 1, what maximum runoff would be expected in a 25-year recurrence interval?
3. Given an allowable headwater depth of 8 feet and a runoff of 265 cubic feet per second, determine the required size of circular corrugated metal pipe and the actual headwater depth. The slope for the location is one per cent and the length of the culvert is 120 feet.
4. Solve Problem 3 using an allowable headwater depth of 14 feet. Assume a headwall is to be provided and that the pipe will operate with inlet control.
5. Determine the depth and velocity of flow in a trapezoidal channel ($n = 0.03$) with 2:1 side slopes and a two-foot bottom width given a flow of 150 CFS and a slope of 0.5 per cent.
6. Solve Problem 5 using a value of $n = 0.012$.
7. Discuss the advantages from side borrow over that from borrow pits.

8. Assuming a freehaul distance of 1,500 feet, the cost of excavation $0.65 per cubic yard, and the price of overhaul is $0.10 per cubic yard station, what is the limit of economical haul?
9. On a construction project, the total road excavation amounted to 50,600 cubic yards. The contractor also obtained dirt from three borrow pits as follows: (1) 5,600 cubic yards, (2) 2,500 cubic yards, and (3) 1,450 cubic yards. The volumes in the fill section before and after construction amounted to 40,900 cubic yards. What was the shrinkage factor?
10. Consult your local highway office and report what new techniques are being used for computing earth work.

REFERENCES

1. RITTER, LEO J., JR., and PAQUETTE, RADNOR J., *Highway Engineering,* Third Edition, The Ronald Press, New York (1967).
2. *Concrete Pipe Handbook,* American Concrete Pipe Association, 1965.
3. *Design of Roadside Drainage Channels,* Hydraulic Design Series No. 4, Bureau of Public Roads, 1965.
4. *Handbook of Concrete Culvert Pipe Hydraulics,* Portland Cement Association, 1964.
5. *Hydraulic Charts for the Selection of Highway Culverts,* Hydraulic Engineering Circular No. 5, Bureau of Public Roads, December, 1965.
6. *Capacity Charts for the Hydraulic Design of Highway Culverts,* Hydraulic Engineering Circular No. 10, Bureau of Public Roads, March, 1965.
7. *Design Charts for Open Channel Flow,* Hydraulic Design Series No. 3, Bureau of Public Roads, August, 1961.

13

Design of Land Transportation Terminals

Terminal facilities are needed for all modes of transportation whether by land, sea, or air. The moving of people and commodities has to have an origin and a destination. The origin of one may be the destination of another. Regardless of the mode of transportation the terminal facilities should be considered as a part of the physical plant of the transportation system.

The motor vehicle, which includes passenger cars, taxis, trucks, and buses, will have to have terminal facilities for loading and unloading purposes. Passenger cars need parking space either at the curb, parking lot, garage, or shopping center. Taxis will need taxi cab stands and trucks will need loading zones or other special terminal facilities, while buses need specially designed and well-located terminal buildings.

By the same token, railroads have to have terminal facilities for the loading and unloading of passengers and commodities. Passenger terminals should be located where walking can be kept at a minimum while classification yards for the making up of freight trains should be located outside of the urban area because of the large area required.

Air and water transportation terminals also need well-designed facilities. While water transportation terminals have been fairly well stabilized, the phenomenal increase in air traffic has required the expansion of present facilities to the point where additional airports are required.

Land transportation facilities will be discussed in this chapter, while air terminals will be discussed in Chapter 17 and water transportation terminals are covered in Chapter 21.

13–1. The Parking Problem. The parking problem today is one of the most important confronting most urban areas. In the larger cities the lack of proper parking facilities is resulting in decentralization of business with the resulting loss of land values in the central shopping district. The parking problem usually begins not at the point of destination but at the point of origin, when the motorists decide where and how to go. The increase in population, especially in urban areas, has increased the motor vehicle ownership and usage, complicating traffic conditions. Also from an economic standpoint, there is a definite relationship between the volume of business and the ability to accommodate vehicles in a designated area.

In order to provide adequate terminal facilities, a parking survey is necessary to determine the parking habits and requirements of motorists and the relation of these factors to other users of existing parking facilities. The information secured in a parking survey should include (1) the location, kind, and capacity of existing parking facilities, (2) the amount of parking space needed to serve present demands, (3) the approximate location of possible additional parking facilities, and (4) the legal, administration, financial, and economic aspects of parking facilities. With this information the design of terminal facilities may proceed.

13–2. Types of Parking. Parking may be classified into two broad divisions, first on-street parking facilities and second, off-street parking facilities. On-street or curb parking can be further classified as unrestricted curb parking where the motorist can park where spaces are available for as long as he wants. Restricted curb parking may be controlled by police through enforcing posted restrictions on signs or by meter-controlled restrictions which are also enforced by police.

On-street parking is convenient for the individual fortunate to find a space; however, curb parking is a deterrent to moving traffic and accidents do occur during parking maneuvers. To improve traffic flow in areas of concentrated land use such as in the central business districts, on-street or curb parking is rapidly being eliminated.

Off-street parking facilities are usually of two basic types, namely, surface lots and multifloor structures or garages. These structures or garages may be further classified as being of the ramp type or the mechanical type.

From an administrative point of view these off-street facilities may be privately owned and operated, public owned, privately operated, or public owned and operated.

All types of off-street facilities have been operated successfully and the type selected will depend on local conditions. Each city will have to decide which type will best serve its needs.

13–3. The Design Vehicle. Parking facilities have to be provided for all makes of vehicles. The size, dimension turning radius, and overall weight of vehicles are the main factors in determining parking standards. The design vehicle, therefore, will be one with dimensions at least as great as these of the majority of cars being manufactured. This information is usually obtained from the Automobile Manufacturers Association. The critical dimensions of the design vehicle are shown in Fig. 13–1. The dimensions of the design vehicle should be flexible enough to be adapted to future changes in automobile dimensions.

13–4. Curb Parking Design. Parking at the curb may be convenient and desirable but the amount of space available within a block may be limited due to driveways, alleys, fire hydrants, loading zones, and other legal restrictions. Also, curb parking prohibits the normal flow of traffic and, for this reason, curb parking in the central business district is usually prohibited. Where curb parking is permitted, parallel or angle parking may be used.

Figure 13–2 shows the geometric requirements for parallel and angle parking. It can be readily seen that angle parking accommodates more vehicles per foot of curb space than parallel parking but as the angle increases, the roadway width needed for the parking maneuver is also increased. Statistics also show that the accident rate for angle parking is higher than parallel parking at the same location. The present trend is to eliminate angle parking as much as possible.

The space needed at the curb for loading zones, bus stops, and taxi stands is determined by the size and number of vehicles using these facilities.

13–5. Parking Lots. Parking lots now in operation consist of small parking areas handling as few as 10 cars to large lots handling 8,000 or more parked cars. These include the corner gas station, vacant lots, and the large areas of shopping centers.

Parking in these areas may be of two types, namely, attendant parking and self-parking. Some motorists prefer attendant parking which means that the attendant can park in a smaller space and very often he can double- or even triple-park vehicles. With attendant parking, capacity of these lots can be increased from 30 to 50 per cent.

Self-parking, however, is usually preferred by the majority of parkers. The individual can park and lock his car and has access to it at all times. Self-parking requires more space for the parking stalls and aisles. Figure 13–3 shows the stall and aisle dimensions required for several different angles of parking. This applies not only to parking lots but to garages as well. As indicated, the parking unit varies with the parking angle. Although angle parking is more readily acceptable to parkers, greater space economy is achieved by 90-degree parking.

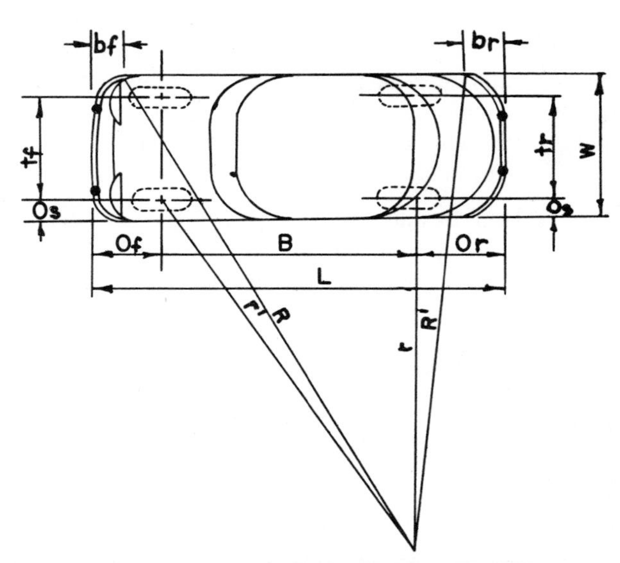

DIMENSIONS OF DESIGN VEHICLE

SYMBOL	DESCRIPTION	VALUE IN FEE
L	OVERALL LENGTH	17.60
W	OVERALL WIDTH	6.70
B	WHEEL BASE	9.80
O_f	FRONT OVERHANG (FRONT AXLE TO BUMPER)	2.90
O_r	REAR OVERHANG (REAR AXLE TO BUMPER)	4.80
O_s	SIDE OVERHANG	0.80
t_r	REAR TREAD (CENTER TO CENTER OF TIRES)	5.10
t_f	FRONT TREAD (CENTER OF CENTER OF TIRES)	5.10
r	MIN. TURNING RADIUS (INSIDE REAR WHEEL)	12.70
r'	MIN. TURNING RADIUS (INSIDE FRONT WHEEL)	16.00
R	MIN. TURNING RADIUS (OUTSIDE POINT, FRONT BUMPER)	22.30
R'	MIN. TURNING RADIUS (OUTSIDE POINT, REAR BUMPER)	19.20
b_f	BUMPER DEPTH, FROM EXTREME OUTSIDE POINT, FRONT	1.00
b_r	BUMPER DEPTH FROM EXTREME OUTSIDE POINT, REAR	0.67
C	MIN. CLEARANCE OF VERTICAL OBSTRUCTION (PUBLIC PARKING)	1.00
C	MIN. CLEARANCE OF VERTICAL OBSTRUCTION (CAPTIVE PARKING)	0.50

NOTE

DESIGN DATA OBTAINED FROM COMPOSITE OF 1962 TO 1966 AUTOMOBLES AS PUBLISHED IN "AUTOMOBILE MANUFACTURERS ASSOCIATION ENGINEERING NOTES."

Fig. 13–1. Dimensions of the Design Vehicle, Parking Design Standards. C. L. Lefler, Traffic Engineer, City of Santa Barbara, California, 1968.

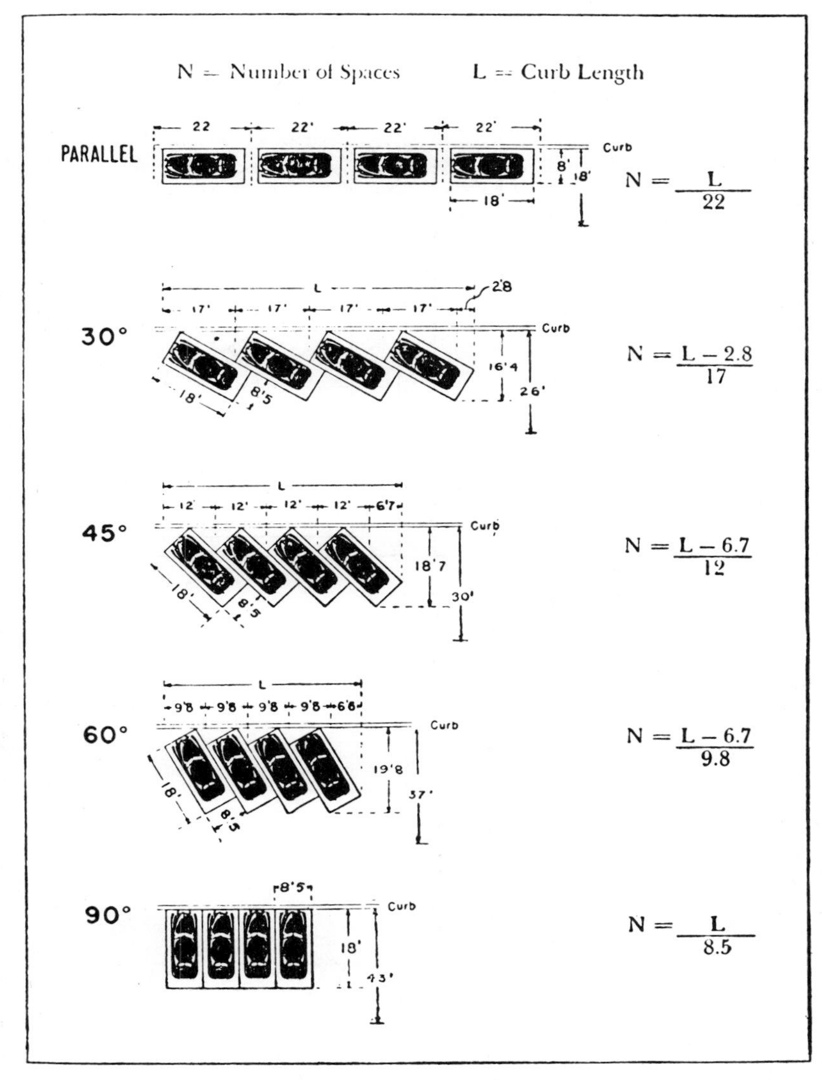

Fig. 13–2. Curb parking—Parking Eno Foundation 1957.

Minimum stall widths should be 8.5 feet for angle parking, 9 feet for 90 degree parking, and approximately 10 feet for parallel parking. The length of a stall should be 18 feet.

The herringbone pattern illustrated in Fig. 13–4 permits economies where space limitations prevent 90-degree parking. This usually calls for one-way aisles as indicated.

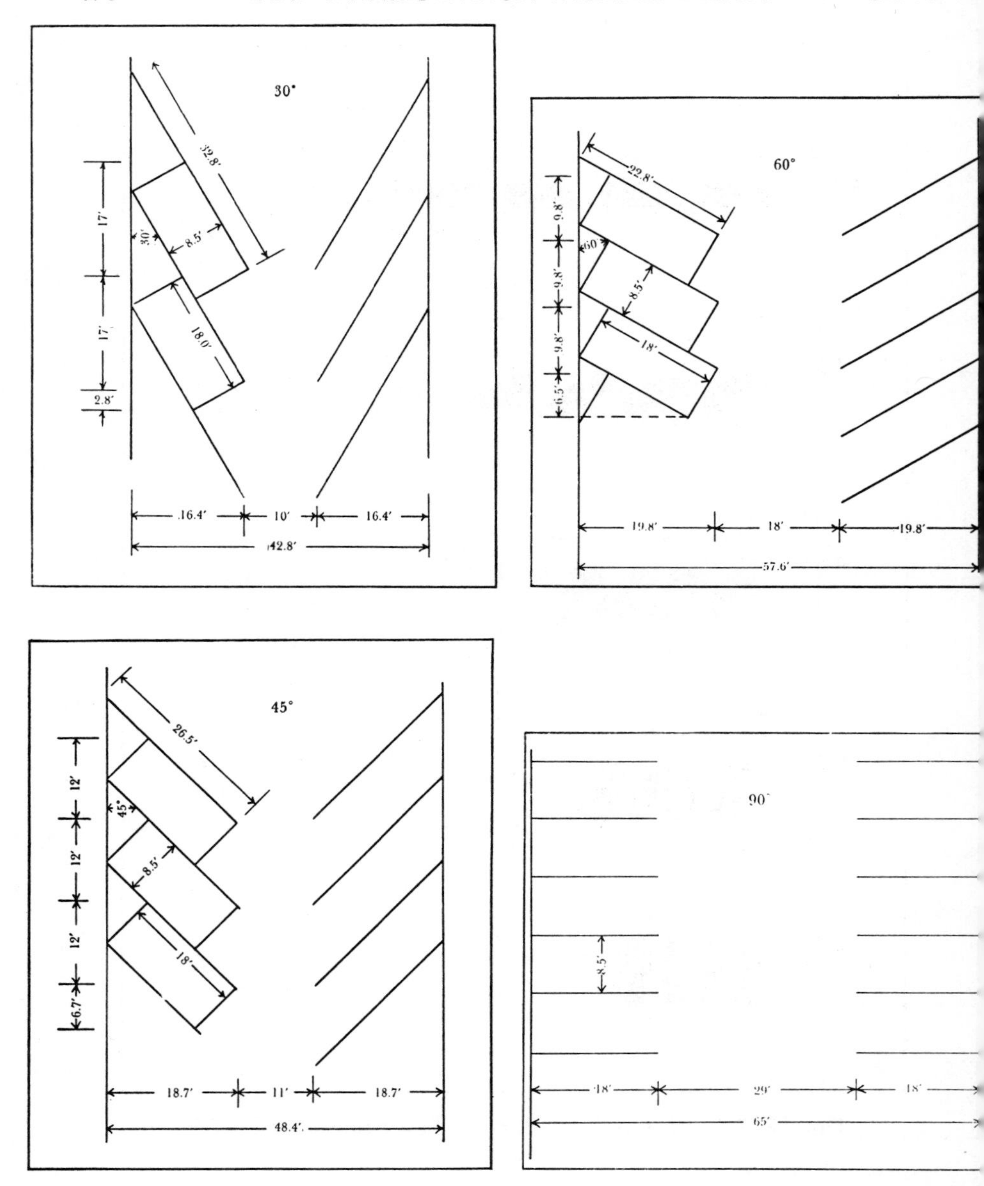

Fig. 13–3. Space requirements for lot or garage parking. C. L. Lefler, Traffic Engineer, City of Santa Barbara, California, 1968.

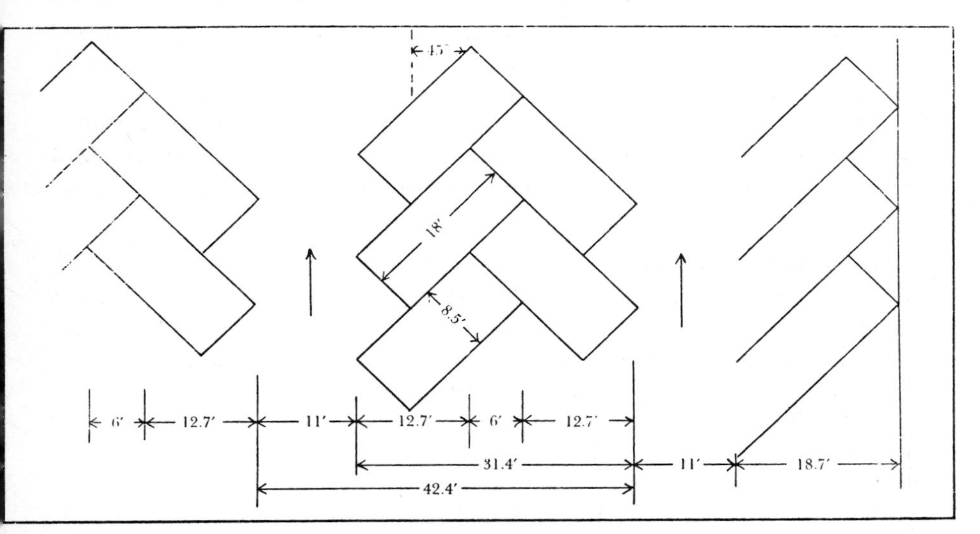

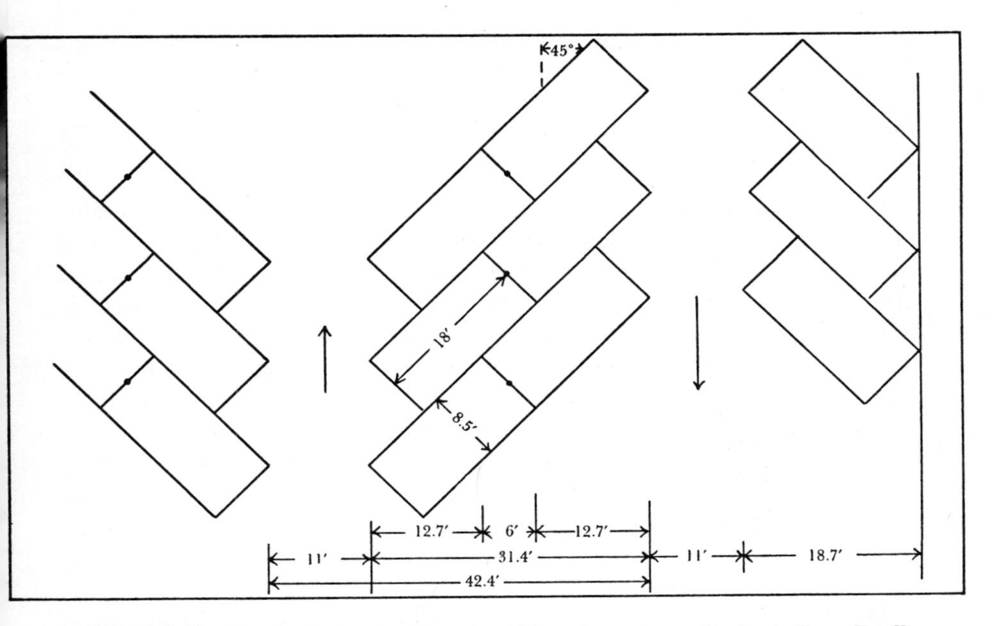

Fig. 13–4. Herringbone pattern parking layouts. C. L. Lefler, Traffic Engineer, City of Santa Barbara, California.

Entrances and exits should be large enough for easy access to and from the street and should be so located as to prevent blocking of or interfering with traffic on the adjacent street.

The layout of a parking lot will depend upon its location within a block, the location of alleys, as well as its size and shape. Figure 13–5 shows the wide variation of layouts that can be obtained for a self-parking lot of a similar size.

13–6. Parking Garage. As motor vehicle registrations increased and parking demands intensified, off-street garages were built to augment parking lots. Early designs were rather makeshift by today's standards. At the present time much progress has been made in the planning, location, and design of modern parking garages.

Older designs required attendant parking while most of today's garages are designed for self-parking. Self-parking reduces manpower requirements, reduces peak hour congestion at entrance and exit points, and lowers damage claims occurring from attendant parking.

Parking may be provided below or above ground. Many underground facilities are developed on sites of public owned property permitting the surface to be used for other purposes. In the construction of today's high-rise apartments and office buildings parking facilities are incorporated as an integral part of the structure. Parking facilities are usually of the ramp type.

13–7. Ramp Garages. A large majority of garages now constructed are of the ramp type. The term "ramp" refers to the sloping surface that connects the floors. In multiple-story garages the individual ramps are placed one above the other to conserve floor space and simplify construction. Cars traveling the ramps follow a circular or an elliptical pattern which may be of one-way or two-way operations. The one-way ramp is more common and provides less delay and lower accident rates.

The floors of the garage may be parallel, connected by ramps, or the floors may be sloping. This permits parking in the ramp surface. The same aisle may be used for the upward and downward movement of vehicles. Some designs use spiral ramps to facilitate more rapid descent. Figure 13–6 illustrates a few of the ramp systems in current use.

The desirable location of parking garages should be away from intersections, preferably at midblock locations to prevent congestion on the street. The design of the receiving area will be dependent on whether attendant or self-parking is to be used. When self-parking is used, cars can flow freely with only enough time to receive a parking ticket from an attendant or a machine. If attendant parking is used, a storage reservoir is needed. The reservoir space required will be dependent upon

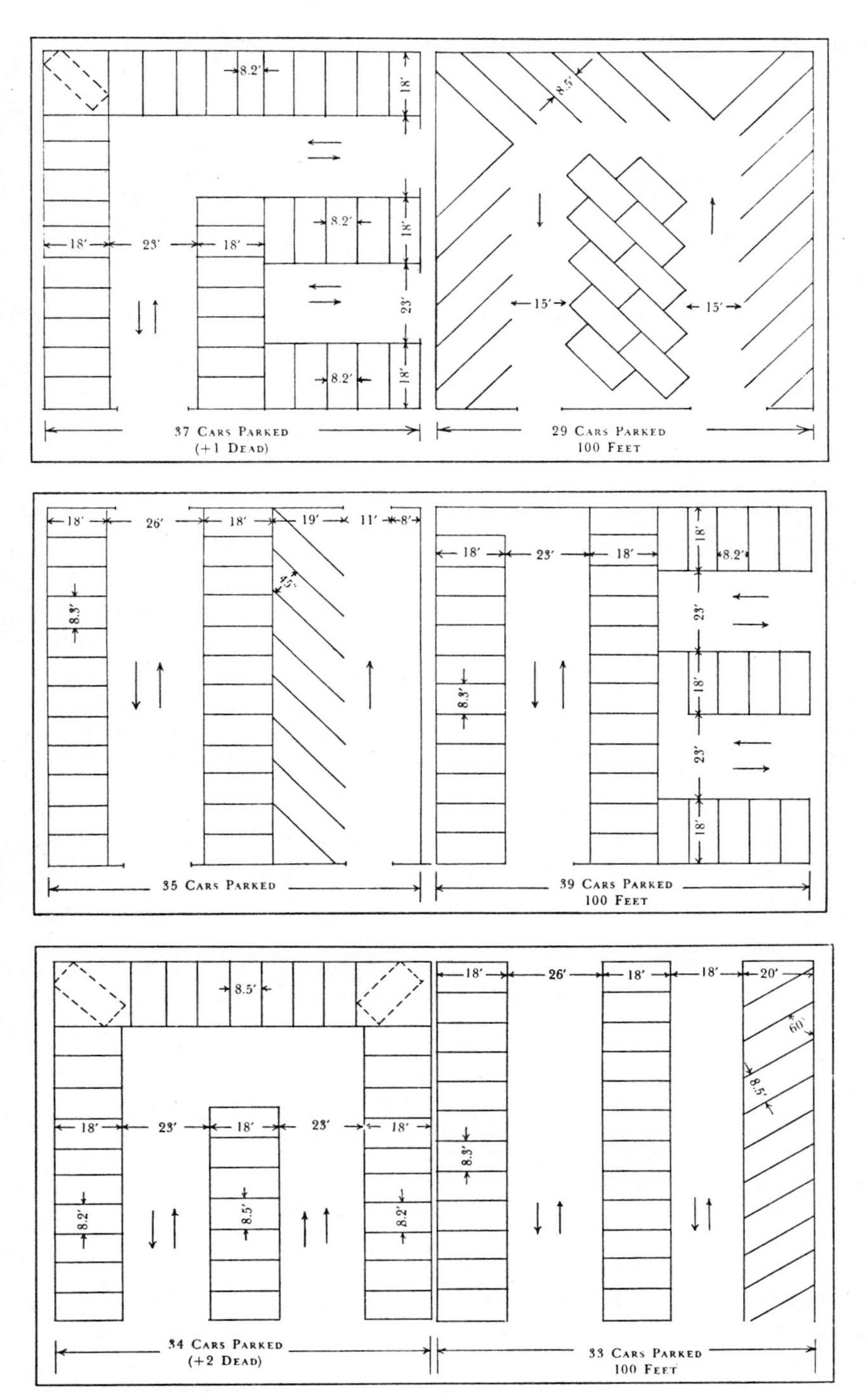

Fig. 13–5. Parking lot arrangements. Parking Guide for Cities, U.S. Printing Office, 1956.

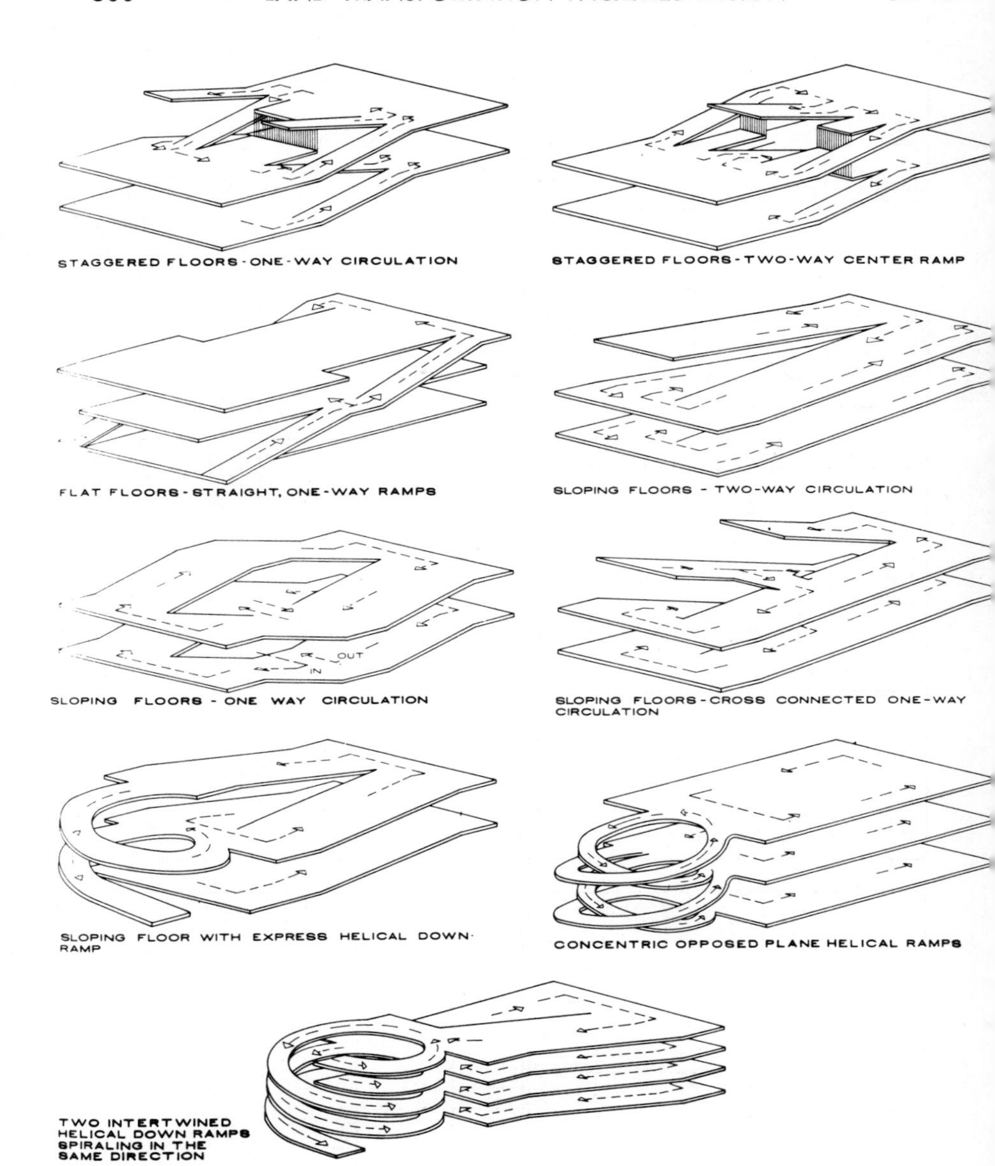

Fig. 13–6. Ramp systems. The American Institute of Architects, New York, N.Y.

the passenger unloading time required for the parking of each vehicle. Figure 13–7 gives the reservoir space required for various vehicle arrival rates.

Design standards for parking garages are determined from the size, dimension, and weight of the vehicle and modern building codes are constantly being revised to accomodate the necessary live load required for multideck and roof-level parking garages.

The unit parking dimensions, which consist of two parking stalls and an aisle, are used for the layout. The parking unit will vary with the angle of parking. While 90-degree parking is more economical, angle parking may be preferred by the users. Suggested parking unit for 45-, 60-, and 90-degree parking are given in Table 13–1.

Other design features include a floor-to-ceiling height of approximately 7.5 feet, which requires a 9.5 to 10.0 foot floor-to-floor height. Ramp

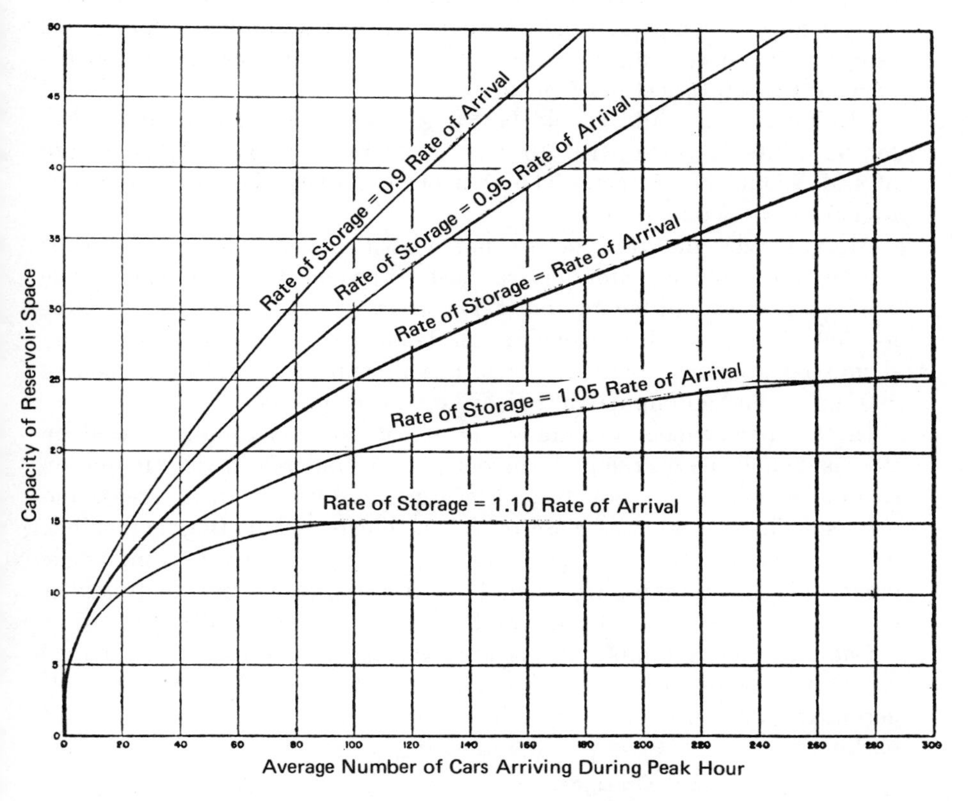

Fig. 13–7. Reservoir spaces required for various vehicle arrival rates, overloaded less than 1 per cent of time. From *Traffic Design of Parking Garages,* The Eno Foundation, 1957.

TABLE 13–1

Suggested Unit Parking Dimensions

Angle of Parking	Direction of Aisles	Unit Parking Dimension
45	One-way	48 to 53 feet
60	One-way	57 to 60 feet
90	Two-way	62 to 65 feet

Source: *Parking in the City Center.* Wilbur Smith & Associates, New Haven, Conn., 1965.

grades of from 5 to 8 per cent may be used on sloping floor garages where ramps provide direct access to stalls. Grades of 10 to 13 per cent should not be exceeded where interfloor ramps are used. A minimum width of ramp for straight ramp is 9 feet and 16 feet for curved ramps. The maximum superelevation of curved ramps should be between 0.1 and 0.15 feet per foot.

Minimum stall width should be 8.5 feet for angle parking and 9 feet for 90-degree parking with 10 feet for parallel parking. The length of stall should be 18 feet. The area per car space should be from 300 to 400 square feet.

Most garages of this type are not heated or ventilated as the walls are left open, but elevator service must be provided. As a general rule one elevator should be provided for every 200 spaces, two elevators for 500 spaces, and three elevators for 1,000 spaces. One outbound lane should be provided for 200 spaces, one inbound lane for every 300 to 500 spaces, and a minimum of two exits per each garage.

In the arrangement of ramps, the down ramp is usually placed on the inside and the up ramp on the outside. Counterclockwise circulation should be used where possible. Elevators, customer waiting areas, and exits should be located as close to the customers' destination as possible. All elevators, stairs, and door openings should be away from traffic lanes. Garage illumination should be from three to five foot-candles in parking areas and 30 foot-candles in cashier and waiting areas.

Column spacing limits parking arrangements. The use of prestressed concrete, post-tensioned construction, and other construction methods permit column free areas which permit better layouts and ease of parking.

13–8. Mechanical Garages. While ramp garages have increased considerably in recent years, some use is being made of mechanical garages. These require less space than ramp garages and have the advantages of being placed on small lots in high-cost land areas. While the initial cost is high due to the mechanical equipment, less operating cost can be obtained due to low manpower requirements. Mechanical garages are usually unable to accommodate heavy peak-hour incoming or outgoing movements.

Cars may be loaded by driving the car onto the elevator, having the elevator control near the drivers' seat so the operator does not have to get out of the car, or by moving the car by means of a dolly. These dollys can push the car in position on the elevator, lift the car by its undercarriage, or lift the car by its wheels. Dolly systems which lift the car permit the car to be locked.

A recently installed mechanical garage in New York City is entirely electronically controlled and requires one cashier-attendant for the eight-level 270-car garage. The location of the stall is indicated on the parking ticket by inserting a key in the control equipment.

13–9. Shopping Centers. Shopping centers in urban and suburban areas have grown appreciably during the last decade and this growth will apparently continue. A shopping center as defined by the Urban Land Institute is "a group of commercial establishments, planned, developed, owned, and managed as a unit with off-street parking provided on the property and related in location, size, and type of shops to the trade area that the unit serves—generally in an outlying or suburban territory."

Shopping centers may consist of neighborhood centers with supermarkets and variety stores as the largest tenants; community centers of intermediate size with a junior department store as the most important tenant; or the largest type of shopping center designed as a Regional Shopping Center. These centers may provide all goods and services found in the Central Business District. Community centers can serve as many as 100,000 people while regional shopping centers can serve from 100,000 to 1,000,000 people residing within 30 minutes driving time from the center.

Parking demand at shopping centers will vary as to the time of day, day of the week, seasons of the year, holidays, and special events. Some studies indicate a relationship between the dollar sales volume and parking demand, others indicate a relationship between merchandising area and parking spaces. The Lenox Square Shopping Center, one of several in the Atlanta, Georgia area provides three parking spaces for each 1,000 square feet of merchandising area. Lenox Square has 6000 available parking spaces.

The layout for the parking area will have to provide adequate entrances and exits, adequate lanes to parking spaces, proper lighting, and, in large areas, a marker system so that the shopper can find his car. A car space should be large enough for all vehicles with easy access into and out of the vehicle. The configuration of the lot will depend upon the location and topography of the area.

13–10. Industrial Plants. Parking at industrial plants poses many problems. At many small industrial sites where off-street employee parking is inadequate, use is made of the adjacent highways and streets, very often spilling over into residential areas. At large industrial sites improperly designed facilities cause conflicts between pedestrians and vehicles resulting in unsafe and inefficient use of the land.

To plan parking facilities properly for industrial plants it will be necessary to make surveys to determine (1) the curb and off-street parking spaces available, (2) the parking demand which will include the number of employees, visitors, and number of shifts, (3) non-driver employees using other modes of transportation, (4) the physical characteristics of the area, and (5) other special studies pertaining to entrances and exits, the volume of traffic on approachiing highways, peak hour usage, and zoning requirements.

Many municipal zoning requirements indicate the number of parking spaces required which is related to the floor space of the industry. The village of Skokie, Illinois, requires one parking space per 300 square feet of office space, and one parking space per 600–800 square feet of manufacturing space, 1500 square feet of storage space, or every two employees, whichever results in the greater number of spaces.

Table 13–2 shows the off-street requirements for manufacturing and processing plants in North Carolina of various sizes as compared to a national survey of the American Society of Planning Officials.

Parking at industrial plants differs from that at shopping centers or commercial parking lots in that large volumes arrive and depart within a relatively short period of time. Entrances and exits should be so

TABLE 13–2

Off-Street Requirements for Manufacturing and Processing Plants

City Size	Name Required	Number of Cities Requiring One Space Per Number of Employees Shown				
		1	2	3	4	Other
2,500–7,500	6	2	1	1	1	1
7,500–15,000	5	0	1	1	1	1
15,000–25,000	2	0	0	2	3	1
25,000–50,000	0	0	2	1	3	0
50,000+	0	2	2	0	4	1
Total	13		6	5	12	4
ASPO Survey	1	0	8	5	1	5

designed that easy access can be provided with a minimum of delay. Many configurations may be obtained for a given layout. Figure 13–8 shows a stall and aisle arrangement for a small industrial plant.

13–11. Bus Terminals. Bus terminals are usually located in the downtown area close to or in the central business district. The number of loading and unloading slots will depend on the bus headway of incoming buses, the number of passengers to be loaded or unloaded, and the time required for the bus to maneuver into or out of the slot. Figure 13–11 shows a bus terminal layout using a 45-degree angle.

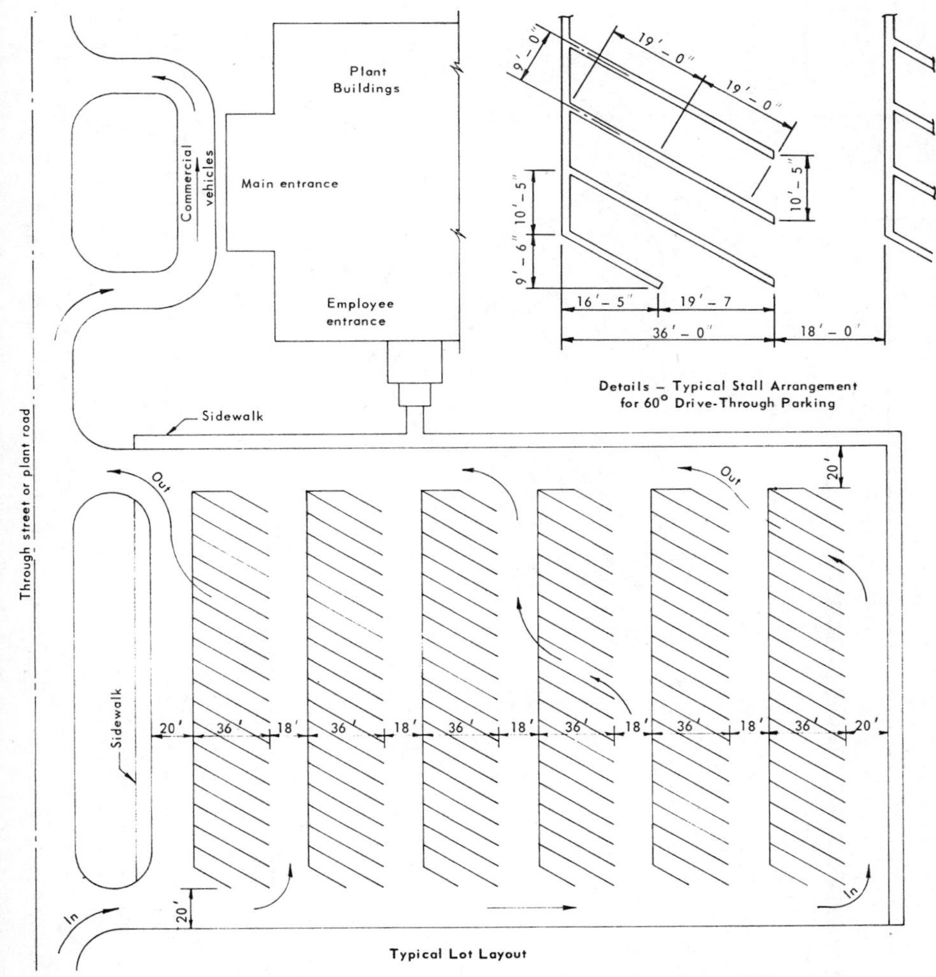

Fig. 13–8. Drive-through-lot layout. (Source: Parking Facilities for Industrial Plants Informational Report of the Institute of Traffic Engineers.)

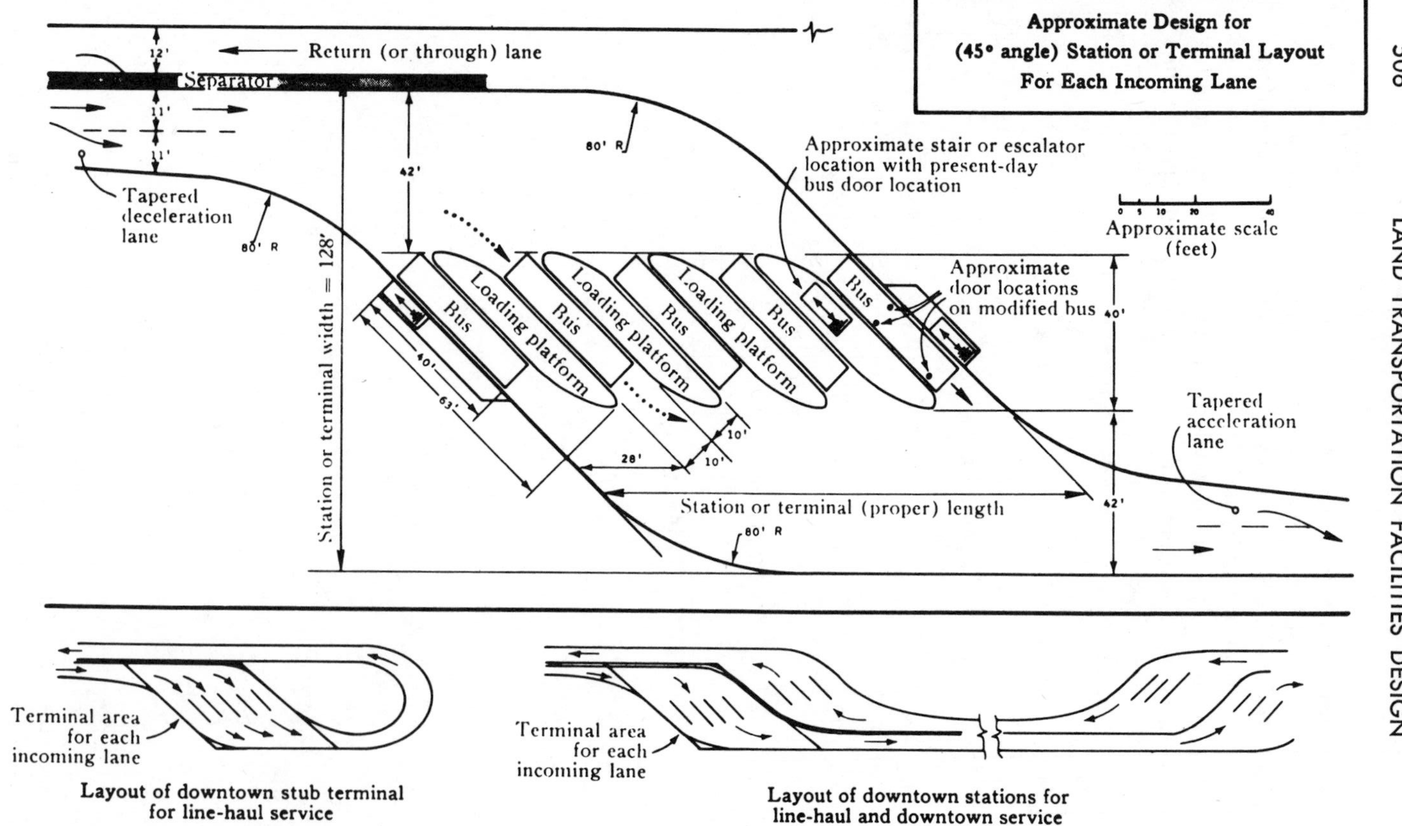

Fig. 13–9. Downtown bus terminal design. (Source: Meyer, Kain, and Wohl, *The Urban Transportation Problem*, Harvard University Press, Cambridge, Massachusetts, 1965.)

TABLE 13-3

Apron Space Required for Single-Maneuver Positioning of Tractor-Trailer Combinations

Overall Length of Tractor-Trailer ft.	Width of Stall ft.	Apron Space ft.
	10	46
35	12	43
	14	39
	10	48
40	12	44
	14	42
	10	57
45	12	49
	14	48

Source: Fruehauf Trailer Co., published in *Architectural Record*, October, 1947.

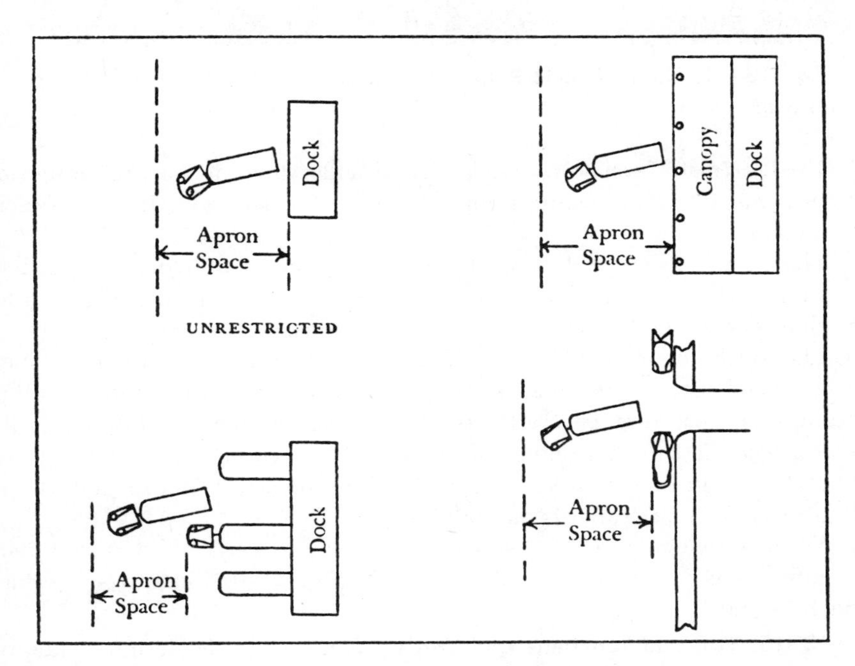

Fig. 13–10. Load dock layouts. (Source: *Parking,* The Eno Foundation.)

Fig. 13–11. Union Pacific Railroad Company, Bailey Yard, North Platte, Nebraska.

13–12. Truck Terminals. Off-street truckloading docks are designed to provide terminal facilities for all types of trucks and tractor-trailer combinations.

The area needed will depend upon the number of trucks using the terminal, the arrival rate, storage and space for maneuvering of the various trucks, and tractor-trailer combinations.

The stalls, doors, and storage lanes should be at least 12 feet wide. Dock height will vary from 48 to 50 inches for most heavy duty units. Ramps or jacks are generally used to adjust for normal differences in truck beds and dock heights. Driveways to docking areas should provide for a turning radius of 45 to 50 feet for small trucks and 70 to 90 feet for tractor-trailer combinations.

Apron space required for single maneuver positions of tractor-trailer combinations is given in Table 13–3 and Fig. 13–10 illustrates loading dock layouts.

13–13. Railroad Terminals and Yards. Railroads need terminal facilities for the loading and unloading of passengers and freight. These facilities are usually provided at or near the central business district. In fact, the growth of many cities developed around these terminals. In some cities joint terminal buildings were constructed to be used by several different railroads for the distribution of passengers and freight. Due to the decrease in passenger traffic many stations have been abandoned, not only in rural areas but in urban areas as well. Freight terminals will still be needed as this is the most important business of the railroads.

Railroad yards have to be provided for shops and facilities for the maintenance of the rolling stock and for the storage of idle cars. The most important need, however, is for the making up of trains for the distribution of freight to various parts of the country. This is accomplished by switching cars to their respective trains. This is generally known as hump switching. The cars are pushed to the top of the hump and they roll toward the center by the force of gravity to their proper place, the speed being controlled by the proper per cent of grades. Formerly, brakemen were used to apply the brakes as they approached the train to be made up. Today, however, car retarders are used on the wheels to control the speed, and the whole operation is operated from a centrally located control tower.

Figure 13–11 shows a classification yard at the Union Pacific Railroad Company's Bailey Yard.

PROBLEMS

1. A corner lot 120 × 120 feet is available for a parking lot. The frontage facing the streets is the only available space for entrance and exit. Draw a plan layout for self-parking.
2. Many new classification yards are being constructed. Write a brief report on one near your locality.
3. Discuss the problem of truckers in the collection and distribution of merchandise in the Central Business District.

REFERENCES

1. American Railway Engineering Association.
2. BURRAGE, R. H., and MOGREN, E. C., *Parking,* The Eno Foundation for Highway Traffic Control, Saugatuck, Conn., 1957.
3. CLEVELAND, D. E., and MUELLER, E. A., *Traffic Characteristics at Regional Shopping Centers,* Bureau of Highway Traffic, Yale University Press, New Haven, Conn. (1967).
4. LEFLER, C. L., *Parking Design Standards,* City of Santa Barbara, Calif., 1968.
5. MEYER, J. R., KAIN, J. F., WOHL, M., *The Urban Transportation Problem,* Harvard University Press, Cambridge, Mass. (1965).
6. *Parking Facilities for Industrial Plants,* Institute of Traffic Engineers, Washington, D.C., 1969.
7. Parking Guide for Cities, Bureau of Public Roads, U.S. Government Printing Office, Washington, D.C., 1965.
8. RICKER, E. R., *Traffic Design of Parking Garages,* The Eno Foundation for Highway Traffic Control, Saugatuck, Conn., 1957.
9. SMITH, WILBUR AND ASSOCIATES, *Parking in the City Center,* New Haven, Conn., 1965.
10. SMITH, WILBUR AND ASSOCIATES, *Transportation and Parking for Tomorrow's Cities,* New Haven, Conn., 1966.

14

Pipeline Transportation

14–1. Introduction. In 1865, Samuel Van Syckel built a two-inch pipeline to transport oil from a northwestern Pennsylvania oil field to a railroad terminal six miles away. This seemingly inauspicious event, probably more than any other, ushered in the era of modern day pipeline transportation. Prior to this time, oil was transported by barrels in horsedrawn wagons at a cost of about $0.30 per barrel-mile. In the decade that followed, longer and larger pipelines were constructed and, in 1878, Byron D. Benson completed the first long distance pipeline, a six-inch line extending 108 miles from Coryville to Williamsport, Pennsylvania. Today, pipelines transport about 18 per cent of the total intercity freight (ton-miles) hauled in the U.S. in a more than 200,000-mile system of pipelines that interlace the country. In this modern system, a gallon of petroleum may be transported more than 1,000 miles at a cost of less than a penny.

14–2. The Nature of Pipeline Transportation. Because pipelines are virtually noiseless and unseen, it is easy to overlook their contribution to the progress and well-being of the nation. Cities depend on pipelines for safe and adequate water supply and for removal of human and industrial wastes. By pipelines, shorelines are reshaped into new land on which rise factories, homes, and centers of shopping and recreation. Serious research is currently underway which could advance solids pipeline flow technology to make it possible to transport packages or capsules of industrial raw materials, grains, and even people through pipes.

The pipeline system of today is predominately utilized for the transportation of petroleum products and natural gas. The vital importance of pipelines is understood when it is realized that approximately 75 per cent of the energy needs of the nation are supplied by petroleum and about 80 per cent of petroleum products are transported by pipelines.

In the United States, the large majority of commercial pipelines are owned by petroleum refining companies. Notable exceptions to this statement are the Pipeline Division of the Southern Pacific Railroad, the Little Inch Division of the Texas Eastern Transmission Company, and the Mid-America Pipeline Company.

Interstate oil pipelines have been declared to be common carriers and as such are subject to regulation by the Interstate Commerce Commission. Many pipelines also operate on an intrastate level and are, therefore, subject to state regulation in addition to that of the ICC.

TECHNOLOGY OF FLUID FLOW IN PIPELINES

The theory of fluid flow in pipelines is based on two fundamental principles: *the conservation of mass* and *the conservation of energy.*

14–3. The Principle of Conservation of Mass. The principle of the conservation of mass states that matter can be neither created nor destroyed. This law leads to the conclusion that the mass of fluid passing one section of pipe per unit of time must simultaneously equal the mass per unit of time passing every other section. This principle is expressed by the *equation of continuity:*

$$G = A_1V_1w_1 = A_2V_2w_2 \qquad (14\text{–}1)$$

where:

G = the weight rate of flow typically expressed in pounds per second
A = the cross-sectional area of flow, square feet
V = average velocity of flow, feet per second
w = the specific weight of the fluid, pounds per cubic feet

The subscripts refer to two arbitrarily selected sections along the pipeline.

In the flow of liquids, incompressibility can be assumed ($w_1 = w_2$) and the equation of continuity may be expressed in terms of Q, the rate of flow with typical units of cubic feet per second.

$$Q = A_1V_1 = A_2V_2 \qquad (14\text{–}2)$$

14–4. The Principle of Conservation of Energy. The principle of conservation of energy states:

> If any system undergoes a process during which energy is added to or removed from it (in the form of work or heat) none of the energy added is destroyed within the system and none of the energy removed is created within the system [1].

In the case of fluid flow in pipelines, one may express the total energy possessed by the fluid at any given flow section as a sum of the energy due to molecular agitation (internal energy), velocity, pressure, and height. By taking into account any energy added to or taken from the system, the conservation of energy principle makes it possible to draw conclusions about the energy possessed by the fluid at other sections along the pipeline.

If one considers one pound of fluid which flows between two sections in a pipeline system, the conservation of energy principle may be expressed by the following general equation:

$$I_1 + \frac{p_1}{w_1} + \frac{V_1^2}{2g} + Z_1 + 778E_H + E_M = I_2 + \frac{p_2}{w_2} + \frac{V_2^2}{2g} + Z_2 \quad (14\text{–}3)$$

where:

I = internal energy

$\frac{p}{w}$ = pressure energy, the ratio of pressure to the specific weight of the fluid

$\frac{V_1^2}{2g}$ = kinetic energy, obtained from the general kinetic equation, one-half the product of mass and velocity squared

Z = potential energy, the height above a known datum

E_H = heat energy added to the fluid in British Thermal Units

E_M = the mechanical energy added to the fluid

g = 32.2 ft./sec.^{2}, acceleration due to gravity

Each term in Equation (14–3) has the units of foot-pounds per pound or simply feet of *head* and the subscripts refer to two arbitrarily selected sections along the pipeline.

In most pipeline applications, it can be assumed that no heat is added to or taken from the system ($E_H = 0$) and that there is no significant change in internal energy ($I_1 = I_2$). If, in addition, no mechanical energy is added to the fluid ($E_M = 0$) and there is no change in the specific weight of the fluid ($w_1 = w_2$), Equation (14–3) reduces to the familiar Bernoulli equation:

$$\frac{p_1}{w} + \frac{V_1^2}{2g} + Z_1 = \frac{p_2}{w} + \frac{V_2^2}{2g} + Z_2 \quad (14\text{–}4)$$

Equations (14–3) and (14–4) apply to an ideal system. In a real fluid system, fluid friction will result in a dissipation of energy which must be accounted for by the addition of another term, h_L, to allow for these *head losses*.

$$\frac{p_1}{w_1} + \frac{V_1^2}{2g} + Z_1 = \frac{p_2}{w_2} + \frac{V_2^2}{2g} + Z_2 + h_L \quad (14\text{–}5)$$

Equation (14–5) may be represented graphically as shown in Fig. 14–1. It should be noted that for the usually assumed conditions of uniform flow, the loss in head is reflected in a loss in pressure and drop in the total energy line and the *hydraulic grade line*.

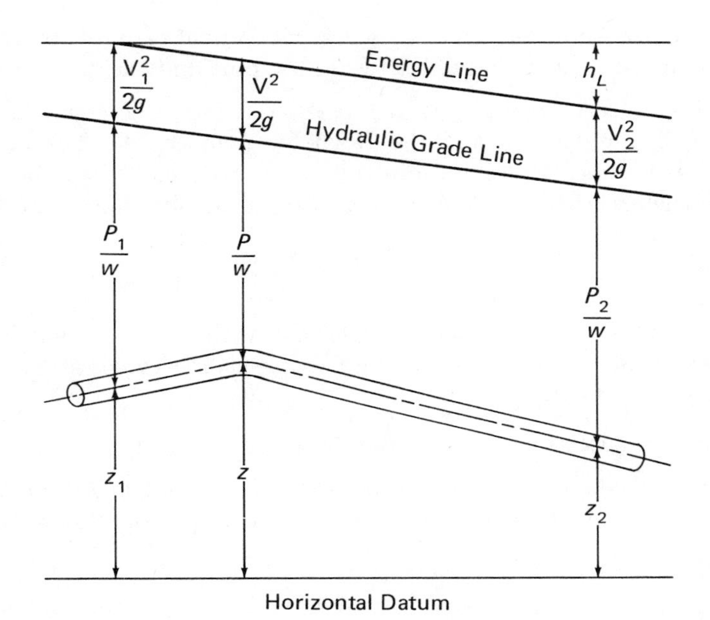

Fig. 14–1. Graphical representation of energy relationships in a pipeline system.

For uniform flow in long pipes of constant cross-section ($V_1 = V_2$), it can be seen from Equation (14–5) that:

$$h_L = \frac{p_1}{w_1} + Z_1 - \frac{p_2}{w_2} - Z_2 \tag{14–6}$$

Under these conditions the head loss is the decrease in *piezometric head* between Sections 1 and 2.

14–5. The Darcy-Weisbach Equation. The head loss in a pipeline was shown by Henri P. G. Darcy and Julius Weisbach to depend on the pipe length (L) and diameter (D), and the velocity of flow (V). In independent nineteenth century research, these scientists developed the following relationship:[1]

$$h_L = f \frac{L}{D} \frac{V^2}{2g} \tag{14–7}$$

[1] This relationship is sometimes expressed as $f \frac{4L}{D} \frac{V^2}{2g}$, the Fanning equation. Values of f used in the Fanning equation are one-fourth the Darcy-Weisbach friction coefficients.

where:

f = the *friction factor*, a dimensionless measure of pipe resistance which depends on the characteristics of the pipe and the flow.

Researchers have shown that the friction factor is a function of the relative pipe roughness and the Reynolds number, R. The Reynolds number is a dimensionless ratio involving the average fluid velocity (V), pipe diameter (D), and the viscosity (μ) and density (ρ) of the fluid:

$$R = \frac{VD\rho}{\mu} = \frac{VDw}{\mu g} \tag{14-8}$$

The variation of the friction coefficient with Reynolds number and pipe roughness is shown by an engineering chart developed by Moody [2]. See Fig. 14–2.

EXAMPLE 14–1. LIQUID PIPELINES PROBLEM

A 22-inch pipeline is being designed to transport 200,000 barrels per day of gasoline (specific weight = 42.1 pounds/cubic foot) between

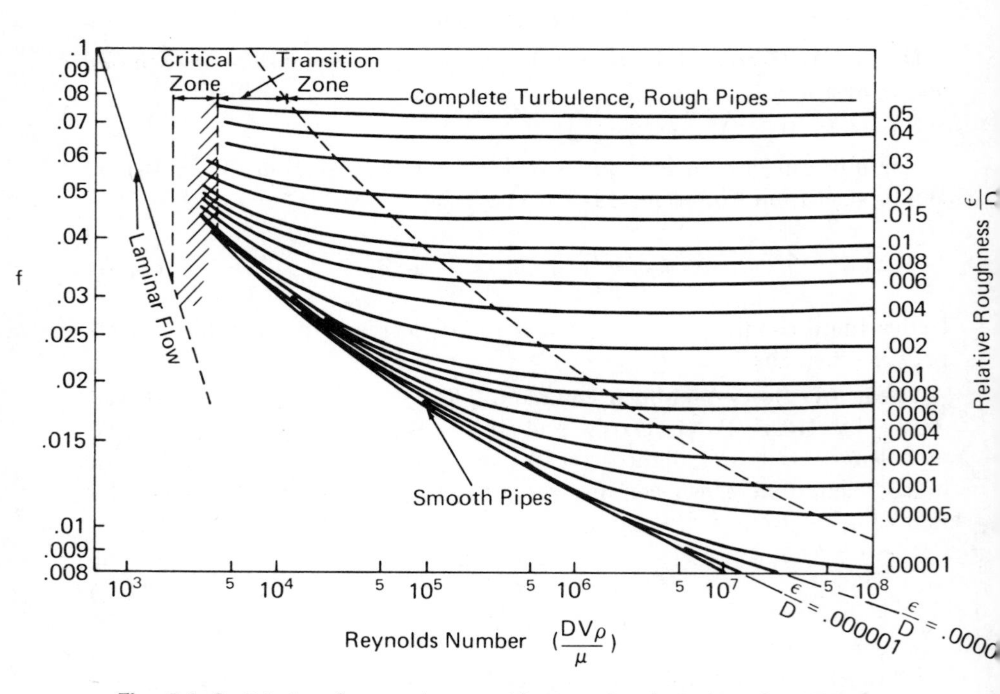

Fig. 14–2. Friction factor chart. (Source: Lewis F. Moody, "Friction Factors for Pipe Flow," *Transactions*, ASME, November, 1944.)

points X and Y. (See Fig. 14–3.) The total length of the pipeline is to be 360 miles. The pipe friction factor is 0.02, and the allowable working pressure is 800 psi. Determine the number of pumping stations required and the distance between adjacent stations.

$$Q = \frac{200{,}000 \text{ bbls./day} \times 42.0 \text{ gal./bbl.}}{(24 \times 3600) \text{ sec./day} \times 7.48 \text{ gal./cu. ft.}} = 13.0 \text{ cfs}$$

$$Q = AV = \frac{\pi}{4}\left(\frac{22}{12}\right)^2 V = 13.0 \text{ cfs}$$

$$V = 4.93 \text{ feet/sec.}$$

$$h_L = f\frac{L}{D}\frac{V^2}{2g} = 0.02\,\frac{5280}{22/12}\,\frac{(4.93)^2}{2g} = 21.8 \text{ feet/mile}$$

This head loss is equivalent to 6.35 psi/mile.

Total pressure drop = 6.35 psi/mile × 360 miles = 2286 psi

Difference in elevation between X and Y is 520 feet which is equivalent to 152 psi. The pumps need to supply only 2286 — 152 = 2134 psi.

$$\text{Number of stations required} = \frac{2134}{800} = 2.67 \text{ (say 3 stations)}$$

As a matter of practice, the pressure is not allowed to drop below about 25 psi. The hydraulic grade line, therefore, in this example will drop 775 psi between stations.

$$\text{Distance between } X \text{ and } A = \frac{775 - (110 \times 42.1/144)}{6.35} = 117.00 \text{ miles}$$

Distances between other stations are found in a similar way. These values are shown in Fig. 14–3.

14–6. Flow in Gas Pipelines. Although the flow of gas in pipelines obeys the same basic laws outlined earlier in this chapter, design computations basically differ from those for liquid pipeline flow. These differences result from the need to consider the thermodynamic effects inherent to compressible fluid flow.

In the following paragraphs, general equations of gas flow will be introduced along with the results of recent research evaluating frictional resistance at the large Reynolds numbers typically experienced for flow in commercial gas lines.

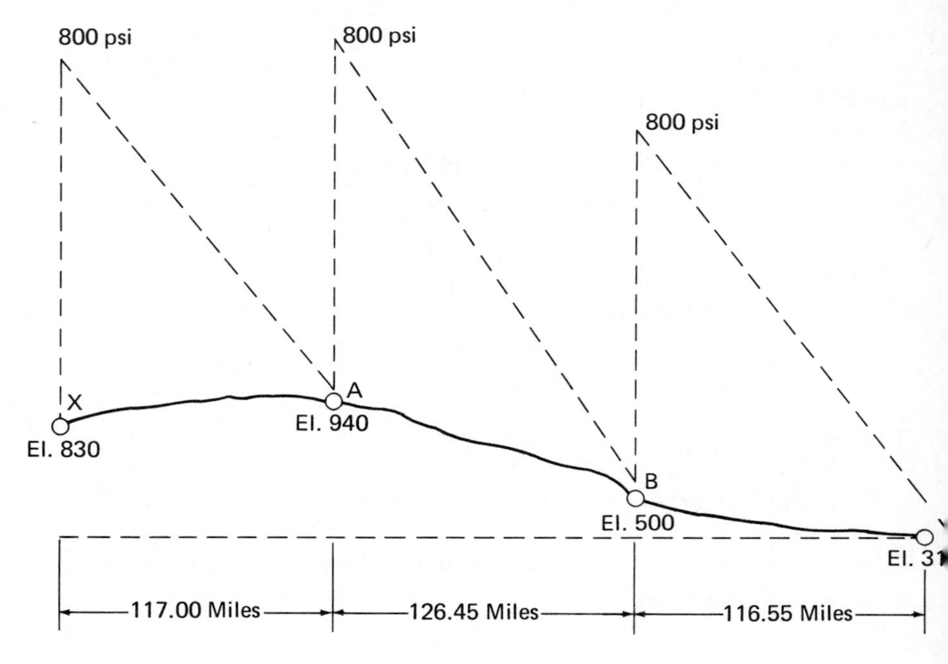

Fig. 14–3. Hydraulic gradient profile for Example 14–1.

In its most general form the equation for flow or *throughput* for a natural gas pipeline is:

$$Q_b = 38.77 \left(\frac{T_b}{P_b}\right) \sqrt{\frac{1}{f}} \left[\frac{p_1^2 - p_2^2 - 0.0375G\,\Delta h P_{\text{avg}}^2/TZ}{SLTZ}\right]^{0.500} D^{2.5} \tag{14–9}$$

This equation is based on the assumptions that the flow is steady state and isothermal and that the kinetic energy change in the length of pipe is negligible. Nomenclature and typical units for this equation are given in Table 14–1.

If the terminal points of the pipeline are at the same elevation, the general equation reduces to:

$$Q_b = 38.77 \left(\frac{T_b}{P_b}\right) \sqrt{\frac{1}{f}} \left[\frac{p_1^2 - p_2^2}{SLTZ}\right]^{0.500} D^{2.5} \tag{14–10}$$

The most troublesome part of these equations is the evaluation of the term, $\sqrt{\frac{1}{f}}$, which is called the *transmission factor*. A number of empirical

TABLE 14-1

Nomenclature and Symbols for Gas Flow Equations

Symbol	Value	Units
T_b	Absolute temperature at reference or base condition	°R
P_b	Absolute pressure at reference or base condition	psia
f	Friction factor	none
p_1	Pressure at inlet of line section	psia
p_2	Pressure at outlet of line section	psia
S	Specific gravity of gas, Dry air–1.000	none
L	Length of section	miles
T	Temperature of flowing gas	°R
Z	Compressibility factor at base conditions	none
D	Inside diameter of pipeline	inches
Δh	Elevation change	feet
R	Reynolds number	none
k	Average height of roughness peak	micro-inches
V	Average velocity	feet per second
ρ	Density of gas	lb-sec^2/feet^4
μ	Viscosity of gas	lb-sec/feet^2
Q_b	Volume of gas flow	cubic feet/day

equations have been proposed over the years for the computation of the transmission factor. These equations which were developed under various field conditions have yielded a confusing variety of flow equations[2] which are found in pipeline industry journals.

Research has shown that at high Reynolds numbers the transmission factor curves tend to level off indicating that under these conditions the transmission factor does not depend on the Reynolds number. At very high Reynolds numbers, the transmission factor is better correlated in terms of a ratio of the wall roughness to the pipe diameter, k/D.

Early research indicated that laminar flow is rare in gas pipelines and that two major types of turbulent flow may occur. The type of flow which develops at low to moderate Reynolds numbers was termed

[2] A discussion of three of the most popular of these equations, the Panhandle A, the New Panhandle, and the Weymouth is given by Bukacek [10].

smooth-pipe flow because of the absence of wall effects. For this type of flow, the flow resistance is a function only of the Reynolds number.

The type of flow existing at very high Reynolds numbers is usually termed *fully turbulent flow* and, in this case, the flow resistance depends on the pipe diameter and the wall roughness and not on the rate of flow.

For turbulent flow near *smooth* boundaries, the transmission factor can be expressed by the following equation:

$$\sqrt{1/f_{SP}} = 4 \log (R/\sqrt{1/f}) - 0.6 \tag{14-11}$$

For turbulent flow near *rough* boundaries, the following equation yields an accurate value for the transmission factor:

$$\sqrt{1/f_{RP}} = 4 \log 3.7D/k \tag{14-12}$$

Recent research of the Institute of Gas Technology [3] has indicated that truly smooth pipe flow is rarely achieved in practice because of the influence of pipe fittings and bends. To allow for this finding, IGT classed gas pipeline flow into two categories: *partially turbulent flow* and *fully turbulent flow*.

Computation of the transmission factor for *partially turbulent flow* can be made by the use of the following equation:

$$\sqrt{1/f_{PT}} = F_f 4 \log (R/1.4 \sqrt{1/f_{SP}}) \tag{14-13}$$

Use of Equation (14–13) requires the following steps:

1. Compute the *bend index* which is defined as the total degrees of bend in a pipe section divided by the length of the section in miles.
2. Determine the *drag factor*, F_f. (See Fig. 14–4.)
3. Select the value of $\sqrt{1/f_{SP}}$ from Table 14–2 or compute it by Equation 14–11.
4. Compute $\sqrt{1/f_{PT}}$ by Equation 14–13.

The transmission factor for *fully turbulent flow* is given by:

$$\sqrt{1/f_{FT}} = 4 \log 3.7D/k_e \tag{14-14}$$

This equation is identical to Equation 14–13 except that the roughness is expressed as the operating or effective roughness of the interior surface rather than the absolute roughness of the pipe wall alone.

It is not always clear which flow regime will exist for a given set of conditions. The Reynolds number for the point of transition from

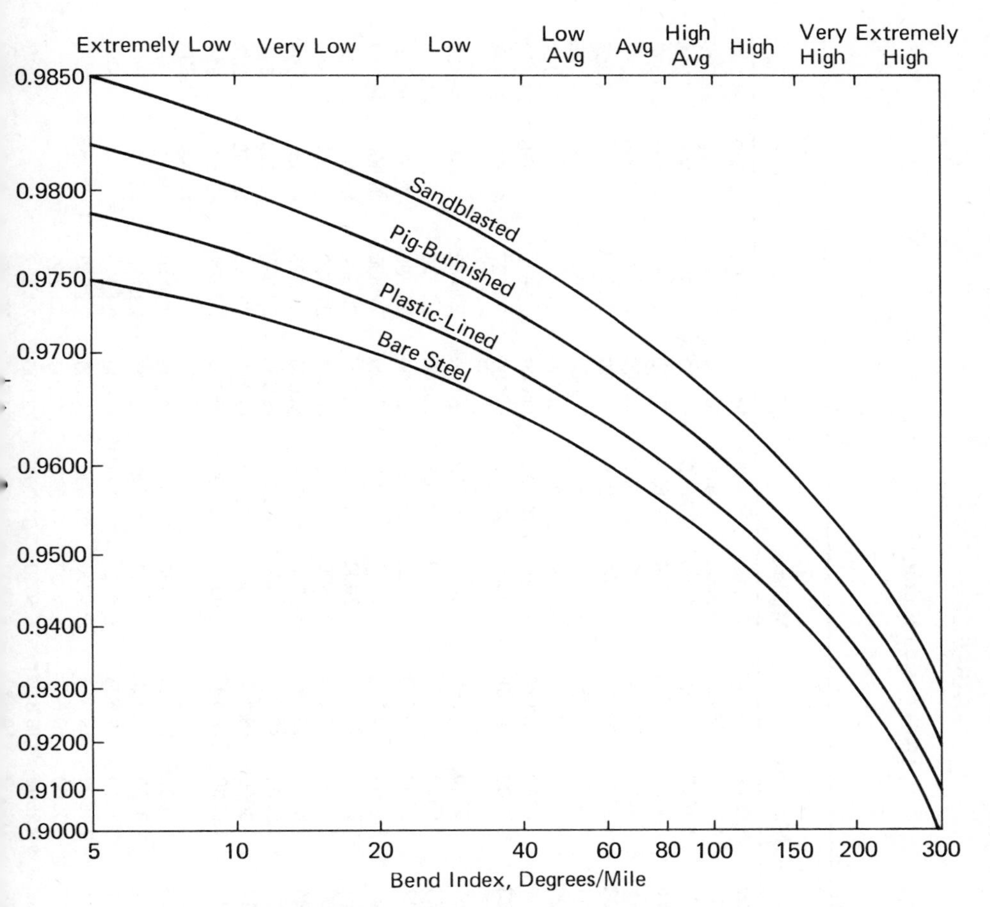

Fig. 14–4. Drag factor as a function of bend index. (Source: Arthur E. Uhl, "Steady Flow in Gas Pipe Lines," *Pipe Line Industry,* 1966–67.)

partially turbulent flow to fully turbulent flow is given by:

$$R = 5.65 \left(\frac{3.7D}{k_e}\right)^{1/F_f} \log \frac{3.7D}{k_e} \tag{14–15}$$

If the Reynolds number for the given conditions is less than that computed by Equation 14–15, partially turbulent flow will exist; otherwise, the flow will be fully turbulent.

The equations which have been introduced make it possible to solve all but the most involved pipeline flow problems. However, use of these

TABLE 14-2

Transmission Factor, F_t

$$F_t = \sqrt{1/f}$$

N_{Re}, Millions	F_t	N_{Re}, Millions	F_t	N_{Re}, Millions	F_t	N_{Re}, Millions	F_t	N_{Re}, Millions	F_t	N_{Re}, Millions	F_t
.10000	14.727	.40000	16.897	1.0000	18.346	4.0000	20.556	10.000	22.028	40.000	24.268
.10500	14.803	.40500	16.917	1.0500	18.423	4.0500	20.576	10.500	22.107	40.500	24.288
.11000	14.876	.41000	16.936	1.1000	18,497	4.1000	20.596	11.000	22.182	41.000	24.308
.11500	14.945	.41500	16.955	1.1500	18.568	4.1500	20.615	11.500	22.253	41.500	24.328
.12000	15.011	.42000	16.974	1.2000	18.635	4.2000	20.635	12.000	22.322	42.000	24.347
.12500	15.075	.42500	16.993	1.2500	18.700	4.2500	20.654	12.500	22.388	42.500	24.366
.13000	15.136	.43000	17.011	1.3000	18.763	4.3000	20.672	13.000	22.451	43.000	24.385
.13500	15.195	.43500	17.029	1.3500	18.823	4.3500	20.691	13.500	22.512	43.500	24.404
.14000	15.251	.44000	17.047	1.4000	18.880	4.4000	20.709	14.000	22.570	44.000	24.423
.14500	15.306	.44500	17.083	1.4500	18.936	4.4500	20.727	14.500	22.627	44.500	24.441
.15000	15.359	.45000	17.065	1.5000	18.990	4.5000	20.745	15.000	22.682	45.000	24.459
.15500	15.410	.45500	17.100	1.5500	19.042	4.5500	20.763	15.500	22.735	45.500	24.477
.16000	15.460	.46000	17.117	1.6000	19.093	4.6000	20.780	16.000	22.786	46.000	24.495
.16500	15.508	.46500	17.134	1.6500	19.142	4.6500	20.798	16.500	22.835	46.500	24.512
.17000	15.554	.47000	17.151	1.7000	19.190	4.7000	20.815	17.000	22.884	47.000	24.530
.17500	15.600	.47500	17.168	1.7500	19.236	4.7500	20.832	17.500	22.930	47.500	24.547
.18000	15.644	.48000	17.184	1.8000	19.281	4.8000	20.849	18.000	22.976	48.000	24.564
.18500	15.687	.48500	17.201	1.8500	19.324	4.8500	20.865	18.500	23.020	48.500	24.581
.19000	15.728	.49000	17.217	1.9000	19.367	4.9000	20.882	19.000	23.063	49.000	24.597
.19500	15.769	.49500	17.233	1.9500	19.408	4.9500	20.898	19.500	23.105	49.500	24.614
										50.000	24.630
.20000	15.809	.50000	17.249	2.0000	19.449	5.0000	20.914	20.000	23.146		
.20500	15.847	.51000	17.280	2.0500	19.488	5.1000	20.946	20.500	23.186		
.21000	15.885	.52000	17.311	2.1000	19.526	5.2000	20.977	21.000	23.225		
.21500	15.922	.53000	17.341	2.1500	19.564	5.3000	21.008	21.500	23.263		
.22000	15.958	.54000	17.370	2.2000	19.601	5.4000	21.038	22.000	23.300		
.22500	15.993	.55000	17.399	2.2500	19.636	5.5000	21.067	22.500	23.337		
.23000	[illegible]	.56000	[illegible]	2.3000	[illegible]	5.6000	[illegible]	[illegible]	[illegible]		

.24000	16.094	.58000	17.483	2.4000	19.739	5.8000	21.152	24.000	23.441
.24500	16.126	.59000	17.510	2.4500	19.772	5.9000	21.180	24.500	23.474
.25000	16.158	.60000	17.537	2.5000	19.805	6.0000	21.207	25.000	23.507
.25500	16.189	.61000	17.563	2.5500	19.836	6.1000	21.233	25.500	23.539
.26000	16.220	.62000	17.589	2.6000	19.867	6.2000	21.259	26.000	23.570
.26500	16.250	.63000	17.614	2.6500	19.898	6.3000	21.285	26.500	23.601
.27000	16.279	.64000	17.639	2.7000	19.928	6.4000	21.310	27.000	23.631
.27500	16.308	.65000	17.663	2.7500	19.957	6.5000	21.335	27.500	23.661
.28000	16.336	.66000	17.687	2.8000	19.986	6.6000	21.360	28.000	23.690
.28500	16.364	.67000	17.711	2.8500	20.014	6.7000	21.384	28.500	23.719
.29000	16.391	.68000	17.735	2.9000	20.042	6.8000	21.408	29.000	23.747
.29500	16.418	.69000	17.758	2.9500	20.069	6.9000	21.431	29.500	23.775
.30000	16.444	.70000	17.781	3.0000	20.096	7.0000	21.454	30.000	23.802
.30500	16.470	.71000	17.803	3.0500	20.122	7.1000	21.477	30.500	23.829
.31000	16.496	.72000	17.825	3.1000	20.148	7.2000	21.500	31.000	23.855
.31500	16.521	.73000	17.847	3.1500	20.174	7.3000	21.522	31.500	23.881
.32000	16.546	.74000	17.869	3.2000	20.199	7.4000	21.544	32.000	23.907
.32500	16.570	.75000	17.890	3.2500	20.224	7.5000	21.565	32.500	23.932
.33000	16.594	.76000	17.911	3.3000	20.248	7.6000	21.587	33.000	23.956
.33500	16.618	.77000	17.931	3.3500	20.273	7.7000	21.608	33.500	23.981
.34000	16.641	.78000	17.952	3.4000	20.296	7.8000	21.628	34.000	24.005
.34500	16.664	.79000	17.972	3.4500	20.320	7.9000	21.649	34.500	24.028
.35000	16.687	.80000	17.992	3.5000	20.343	8.0000	21.669	35.000	24.052
.35500	16.709	.82000	18.031	3.5500	20.365	8.2000	21.709	35.500	24.075
.36000	16.731	.84000	18.069	3.6000	20.388	8.4000	21.747	36.000	24.097
.36500	16.753	.86000	18.107	3.6500	20.410	8.6000	21.785	36.500	24.120
.37000	16.774	.88000	18.143	3.7000	20.432	8.8000	21.822	37.000	24.142
.37500	16.795	.90000	18.179	3.7500	20.453	9.0000	21.858	37.500	24.163
.38000	16.816	.92000	18.214	3.8000	20.474	9.2000	21.894	38.000	24.185
.38500	16.837	.94000	18.248	3.8500	20.495	9.4000	21.928	38.500	24.206
.39000	16.857	.96000	18.281	3.9000	20.516	9.6000	21.962	39.000	24.227
.39500	16.877	.98000	18.314	3.9500	20.536	9.8000	21.996	39.500	24.248

Source: Arthur E. Uhl, "Steady Flow in Gas Pipe Lines," *Pipe Line Industry.*

equations is cumbersome at best and may involve trial and error solutions in certain instances. If repetitive flow calculations are required, a great saving in time may be realized by referring to the American Gas Association's publication TR-10 [4]. In this manual, various terms in the general flow equation are tabulated in factorial form providing a rapid means for the solution of simple flow problems associated with planning and preliminary design work. For more complicated design problems, electronic computer programs are used by the various pipeline companies.

EXAMPLE 14–2. GAS PIPELINES PROBLEM

Natural gas (specific gravity = 0.62) is to be transported from point A to point B, a distance of 150 miles, at a rate of 100 million cubic feet per day. The pipeline, which is to be constructed of bare steel, is to have a bend index of 60 degrees per mile and an effective roughness of 0.3 mil. The inlet pressure is 700 psia and the outlet pressure is 100 psia. Determine the size of pipe required assuming a base pressure of 14.73 psia and a base temperature and flowing temperature of 520° absolute. The specific weight of air at 60°F is 0.0763 pounds per cubic foot. Assume $\mu_{60°F} = 2.5 \times 10^{-7}$ and use a compressibility factor of 0.891. Since the Reynolds number depends on the rate of flow and, thus, on the pipe size, the solution must be done by trial.

A first estimate of pipe size can be made by use of the Weymouth equation, which is applicable to fully turbulent flow.

$$Q_b = 433.5 \frac{T_b}{P_b} D^{8/3} \left[\frac{p_1^2 - p_2^2}{STL} \right]^{1/2}$$

$$100{,}000{,}000 = 433.5 \frac{520}{14.73} D^{8/3} \left[\frac{(700)^2 - (100)^2}{(0.62)(520)(150)} \right]^{1/2} \qquad (14\text{–}16)$$

$$D = 17.6'' \text{ (say 18'')}$$

The weight rate of flow,

$$G = \frac{100{,}000{,}000 \times 0.0763 \times 0.62}{3600 \times 24}$$

$$G = 54.63 \text{ pounds/second}$$

Combining Equations (14–1) and (14–8),

$$R = \frac{GD}{\mu A g} = \frac{54.63 \times 1.5}{2.5 \times 10^{-7} \times (\pi/4)(1.5)^2 \times 32.2} = 5.76 \times 10^6$$

From Fig. 14–4, the Drag Factor $F_f = 0.9600$.

The Reynolds number for the point of transition from partially turbulent flow to fully turbulent flow is given by Equation (14–15):

$$R = 5.65\left(\frac{3.7 \times 18}{0.0003}\right)^{1/0.96} \log \frac{3.7 \times 18}{0.0003}$$

$$R = 11.32 \times 10^6$$

The flow is partially turbulent.
From Table 14–2,

$$\sqrt{1/f_{SP}} = 21.141$$

By equation 14–13,

$$\sqrt{1/f_{PT}} = (0.96)4 \log \frac{5.76 \times 10^6}{1.4 \times 21.141}$$

$$\sqrt{1/f_{PT}} = 20.3$$

Using the general flow equation,

$$100{,}000{,}000 = 38.77\,\frac{520}{14.73}\,(20.3)\left[\frac{(700)^2 - (100)^2}{0.62 \times 520 \times 150 \times 0.891}\right]^{1/2} D^{2.5}$$

$$D = 16''$$

Since this value differs from the assumed, the computations should now be repeated using an assumed value of $D = 16''$. When this is done for this example, however, the answer is not significantly changed.

14–7. Two-Phase Flow. All of the discussion thus far has been confined to single-phase flow, that is, flow of either gas or a liquid. There are situations in which it is economically advantageous to transport simultaneously both a gas and a liquid in a single pipeline. This is called *two-phase flow.* An example of two-phase flow involves certain offshore oil operations where it is extremely expensive to separate the liquid and gas phases in deep water. As the technology advances, two-phase flow is likely to become commonplace in situations such as this.

Two-phase flow may occur over a wide range of liquid to gas ratios and manifests a variety of flow patterns. (See Fig. 14–5).

Two-phase flow technology is much more complex than single-phase flow. However, research accomplished to date suggests that the two-phase design problem is not insurmountable. Indeed, Flanigan [5] has published a design technique which should yield satisfactory results under a wide range of flow conditions. Publication of the details of this method is beyond the scope of this chapter and the interested student is referred to papers by Flanigan [5] and Huitt and Marusov [6].

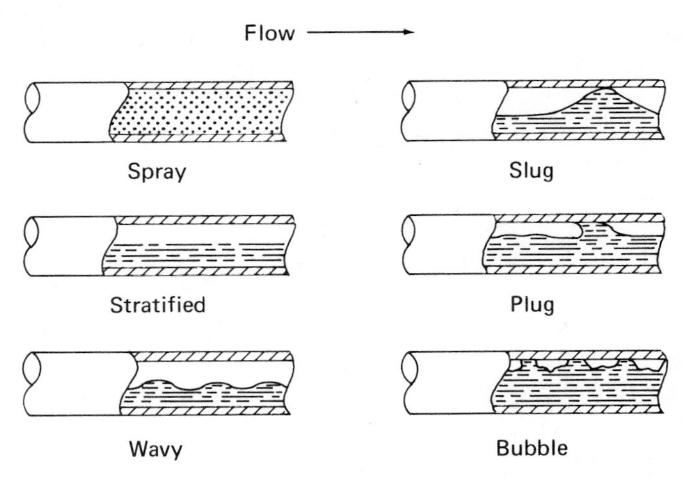

Fig. 14–5. Two-phase flow patterns. (Source: J. L. Huitt and N. Marusov, "Where We Stand in Flow Technology," *Pipe Line Industry,* May, 1964.)

Experimental studies have revealed that two-phase flow results in large drops in pressure far beyond that indicated by single-phase technology. Flanigan [5] suggests that there are two components of pressure drop in two-phase flow:

1. The pressure drop due to friction which increases with increase in gas flow rate. This is the only component in horizontal lines.
2. The pressure drop due to the head of liquid in inclined lines. Practically all of this pressure drop occurs in the uphill section of the pipeline.

Flanigan describes a technique for computing these components which, when added, will yield a satisfactory estimate of the total pressure drop in the transportation of natural gas and condensate. Further research is required to determine whether the Flanigan method applies to other gas-liquid mixtures.

14–8. Transportation of Solids by Pipelines. Another form of two-phase pipeline flow involves the transportation of solids in a liquid or gaseous medium. Although solids pipelines have been used for more than a century, most of activity in this field has occurred since World War II.

For the most part, solids pipeline transportation has been confined to manufacturing, mining, and construction activities. Solids pipelines have most often been used for loading and unloading operations, in-plant movement of materials, and transportation of construction materials for short distances. Pipeline lengths in excess of ten miles have been the exception.

The longest known solids pipeline constructed to date is the 108-mile pipeline which transports coal from Cadiz, Ohio mines to a power plant in East Lake, Ohio.

In order for solids to be transported successfully and economically by pipelines, the following conditions must generally be satisfied [7]:

1. The solid material should not react in any undesirable way with the carrying fluid or become otherwise contaminated within the pipeline system.
2. Attrition during transport should either be beneficial to or its effect be negligible on subsequent operations.
3. The top particle size should be such that it can be handled in commercially available pumps, pipes, and preparation equipment.
4. The solid material should mix easily with, and separate easily from, the carrying fluid at the feeder and discharge terminals, respectively.
5. The solid material should not be corrosive or become so in the carrying fluid.

Researchers have demonstrated that solids pipeline flow may occur in at least four flow regimes:

1. *Flow as a homogeneous suspension.* For velocities experienced in most pipeline applications, this regime involves suspension of particles of diameter of less than 30 microns. Provided the flow is turbulent, the suspension flows like a homogeneous fluid. In this case, head losses may be computed as previously described using density and viscosity values corresponding to the suspension.
2. *Flow as a heterogeneous suspension.* In a heterogeneous suspension, the concentration of particles in a vertical plane is not uniform. This type of flow involves transportation of particles slightly larger than those which flow in a homogeneous suspension. In heterogeneous transport, the solid particles travel with a velocity slightly less than that of the liquid.
3. *Flow by saltation.* In solids pipeline flow involving relatively large particles and low velocities, particles collect at the bottom of the pipe and form a stationary bed. There is virtually no movement of solids in this case. At higher velocities, movement of particles at the interface occurs. This phenomenon is known as saltation.

4. *Flow with a moving bed.* At high-liquid velocities, particles of large diameter may slide forward along the bottom of the pipe in a single mass or as a moving bed. This movement occurs at a much lower velocity than that of the liquid.

Which of these flow regimes occurs in a given instance will depend on the mean velocity, particle size, pipe diameter, and specific gravity of the material. The effect of velocity and particle size on the flow regime is illustrated by Fig. 14–6. The location of boundaries separating the

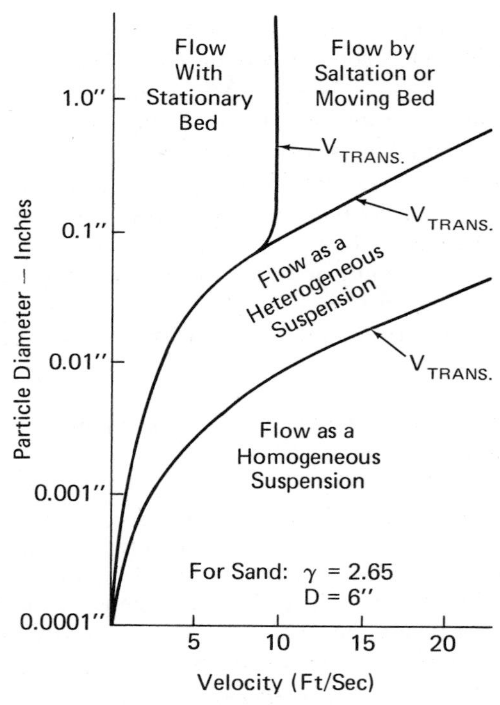

Fig. 14–6. Classifications of flow regimes. (Source: Colorado School of Mines Research Foundation, *The Transportation of Solids in Steel Pipelines,* **1963.)**

various regimes for a given case will, of course, depend on the nature of the solids-liquid mixture.

Basically, the problem of designing a solids pipeline is to find a flow velocity which will prevent settling, yet minimize the friction loss. At the same time, the per cent of solids and diameter of pipe must be sufficiently large to transport the required dry tons per day.

A survey of the Hydraulic Institute [8] revealed that the minimum or critical carrying velocity and friction losses in a pipeline transporting fluid-solid mixtures are dependent on the following factors:

1. The density and/or per cent solids by weight of mixture.
2. The specific gravity of solids relative to the specific gravity of transporting liquid.
3. The gradient of the pipeline.
4. The presence of slimes and/or colloidal material.

It was further reported that friction losses were influenced by interior pipe surface (type of pipe) and the gradation (screen analysis) of the solids being transported.

The Colorado School of Mines Research Foundation has published a book [7] which includes a summary of the results of various research studies of solids pipeline flow. The theories that have been developed in these studies are expressed as equations which apply to most slurry flow conditions encountered in practice. These equations should generally be adequate for the preliminary design of a solids pipeline system. However, large scale pilot testing will usually be required to evaluate adequately the effect of various conditions on flow velocity and friction losses in solids pipeline flow.

OPERATION OF PETROLEUM PIPELINE SYSTEMS

In the following paragraphs, the journey of a petroleum product is traced from the oil well to its ultimate destination. This information, which was abstracted from a publication of the Petroleum Extension Service, University of Texas [9], should provide the reader with a more complete understanding of the problems and practices involved in the operation of pipeline systems.

14-9. Transportation of Crude Oil from Well Head to Tank Battery. The average oil well produces only about 20 barrels per day while a typical 20-inch pipeline is capable of transporting about 200,000 barrels per day. This means that a single pipeline is capable of carrying the output of up to 10,000 wells. Thus, before great quantities of crude oil can be transported in high capacity lines to the refinery, it must first be concentrated in sufficient quantities.

In the interests of simplicity and efficiency, shipments of oil are made in "batches" consisting of several thousand barrels. The oil is usually moved in *gathering lines* from the well head to temporary storage tanks and accumulated in pipeline company tanks until the entire shipment is ready for delivery to the main pipeline. Prior to this time, some preliminary treatment is required to separate gas from the oil and to remove water and other contaminants.

While temporarily stored in the pipeline tank battery, the oil is measured (gauged) and its specific gravity and water content are determined. In modern installation, these measurements may be made automatically as the oil flows through special pipeline sections instrumented with electronic measuring and recording devices.

14–10. Transportation of Crude Oil to Refinery. Efficient operation of a pipeline system requires careful planning and detailed scheduling of oil movements. Every effort is made to operate the pipeline at capacity and at the same time minimize the use of storage tanks.

The typical oil pipeline has a small number of shippers, each of which is continually receiving deliveries and adding oil to the line. The oil remains the property of the shippers and, since the line must remain full, the shipper always has a *balance* of oil in the line. A scheduler canvasses the shippers during the last week in the month to determine the customers' shipping needs and desires. From this information, the monthly barrel-per-day line pumping rates are established.

Once the schedule is prepared, it is important that it is executed according to plan. This is the duty of the dispatcher. The dispatcher's post is manned 24 hours per day. By means of modern and varied communications equipment, he is able to monitor line pressures and rates of flow, receipts and deliveries, and to detect any leaks and breaks in the line that may occur.

Changes in the custody of oil shipments are recorded and controlled by means of a *run ticket* which is signed by the receiver and deliveryman. This document describes the oil in quality and quantity based on readings of meters, gauges, and other such devices, and legally accounts for receipts and deliveries.

14–11. Transportation from Refinery to Market. From the refinery, petroleum products must be transported to areas of high market demand. This final movement is made in *products pipelines.* In these lines, different petroleum products from different shippers flow in adjacent batches.

Products handled in these pipelines include various grades of gasolines, turbine fuel, burning or heating kerosene, and diesel fuels. Figure 14–7 shows a schematic representation of the typical cycle of batches. In certain cases, contiguous batches are segregated by means of rubber spheres called batch separators. Even when batch separators are not used, very little commingling occurs.

Products are shipped to delivery terminals and stored in *break-out tanks* and later relayed to points on the stub lines and to shippers' tanks through smaller lines.

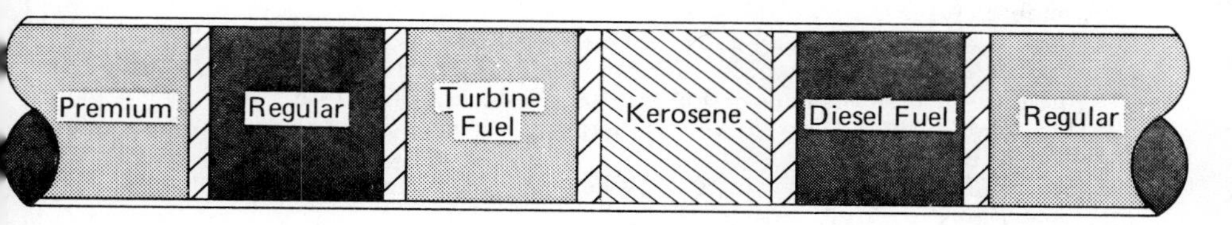

1. Premium—Regular Commingling Can Be Split Both Ways With No Loss or Degradation.
2. Turbine Fuel Commingling Usually Must Be Cut Out and Blended to Regular.
3. Turbine Fuel—Kerosine Commingling Can Be Cut Directly to Turbine Fuel Within Certain Limits.
4. Kerosine—Diesel Fuel Commingling Can Be Cut Directly to Diesel Fuel Within Certain Limits.
5. The Kerosine Buffer Batch Between Diesel Fuel and Turbine Fuel Is Cut Both Ways.

Fig. 14–7. Typical cycle of batches in products pipelines. (Source: Petroleum Extension Service, The University of Texas, *Introduction to the Oil Pipeline Industry,* May, 1966.

At the delivery terminals, the "interface" between adjacent batches is allowed to pass, and stocks are drawn from the center of the batch. If the adjacent batches are similar, the commingled material will be used to upgrade the lower quality product. When two dissimilar products mix, the material is placed in special tanks where it is reblended under strict laboratory procedures.

PROBLEMS

1. An 18-inch pipeline is being designed to transport 93,000 barrels per day of crude oil (specific weight = 55 pounds/cubic foot) from point A to point B, a distance of 290 miles. The pipe friction factor is 0.016, and the allowable working pressure is 650 psi. Point B is 820 feet higher than A. Determine the number of pumping stations required and the distance between adjacent stations. It may be assumed that the ground slope between points A and B is constant.
2. Solve Problem 1 assuming that the oil is to be pumped from point B to point A.
3. Natural gas (specific gravity = 0.62) is to be transported from point X to point Y, a distance of 185 miles, at a rate of 125 million cubic feet per day. The pipeline is to be constructed of bare steel and will have an effective roughness of 0.3 mils. It will have a bend index of 20 degrees per mile. The inlet pressure is 750 psia and the outlet pressure 100 psia. Determine the size of pipe required assuming a base pressure of 14.73 psia and a base temperature and flowing temperature of 520° absolute. Assume $M\mu_{60°F} = 2.5 \times 10^{-7}$ and use a compressibility factor of 0.890.

REFERENCES

1. STOEVER, HERMAN J., *Engineering Thermodynamics,* John Wiley and Sons, New York (1951).

2. MOODY, LEWIS F., "Friction Factors for Pipe Flow," *Transactions,* ASME, November, 1944.
3. UHL, ARTHUR E., "Steady Flow in Gas Pipe Lines," 5 Parts, *Pipe Line Industry,* August, 1966, September, 1966, January, 1967, March, 1967, and April, 1967.
4. Institute of Gas Technology Publication TR-10, Chicago, 1966.
5. FLANIGAN, ORIN, "Two Phase Gathering Systems," *The Oil and Gas Journal,* March 10, 1958.
6. HUITT, J. L., and MARUSOV, N., "Where We Stand in Flow Technology," *Pipe Line Industry,* May, 1964.
7. Colorado School of Mines Research Foundation, *The Transportation of Solids in Steel Pipelines,* 1963.
8. Hydraulic Institute, "Correlation of Data Derived From Two Surveys On Velocities and Friction Losses Involved in Pumping of Fluid-Solid Mixtures," New York, 1959.
9. Petroleum Extension Service, The University of Texas, *Introduction to the Oil Pipeline Industry,* May, 1966.
10. BUKACEK, RICHARD F., "Flow Equations For Natural Gas Pipelines," *Journal of the Pipeline Division,* American Society of Civil Engineers, June, 1958.

15

Belt Conveyor Systems

15–1. Introduction. Belt conveyors provide a highly specialized mode of transportation, being usually limited to short distances and restricted in the type of cargo handled. Great volumes of bulk materials can be transported over rough terrain over grades up to 32 per cent, and for distances of 10 miles or more. Feasibility studies have been made in the Great Lakes region to transport iron ore from a lake port to an Ohio river terminal with a return load of coal to the lake port for a distance of 100 miles.

Modern large scale use of belt conveyors is found in the moving industry and in heavy construction. In the construction of large dams, such as Coulee, Shasta and others, use of belt conveyor systems were found to be most desirable.

15–2. Uses and Limitations. The belt conveyor system is best used for transporting large quantities of pulverized, granular, or lumpy materials for relatively short distances such as coal, ore, grain, sand, and other construction materials. It has the advantage by using steeper grades, the grades, however, being limited to the angle at which slippage will occur for the transported materials. Certain highly abrasive materials may prove costly to transport because of excessive wear on the belt, and high temperature materials may restrict its use. Modern technology, however, has developed belts highly resistant to abrasion and high temperature. Hot material belts permit the use of materials from 275°F to 350°F under controlled conditions. Table 15–1 gives the unit weight of various bulk materials and the maximum grade permitted.

15–3. Elements of the Belt Conveyor System. A short discussion will be given of the various items of a belt conveyor system. These include the belt, the idlers, the take-ups, the tripper, the drive, and belt cleaners.

The Belt. The most common belt is the so-called rubber belt which consists of a cotton duck "carcass," or canvas-like center which gives

TABLE 15-1

Unit Weight of Various Materials

Material	Unit Weight, pcf	Max. Incline Deg	Max. Incline Per Cent
Bauxite, aluminum ore	55–58	17	30
Bauxite, crushed, dry	75–85	20	36
Cement, clinker	88–100	18	32
Clay, wet	95–105		
Coal, bituminous	47–52	18	32
Concrete mix, wet	115–125	12	21
Dolomite, crushed	90–110	17	30
Earth, dry loam	70–80	20	36
Earth, wet loam	104–112	15	27
Granite, broken	96	20	36
Gravel, washed and screened	85	20	36
Gypsum, broken	80–100	17	30
Limestone, broken	95–100	17	30
Ores, sulfides and oxides, broken	125–160	17	30
Sand and gravel, dry	90–105	20	36
Sand and gravel, wet	115–125	12	21
Trap rock, broken	105–110	17	30
Wood chips, dry	15–32		

Source: Hennes and Eske, *Fundamentals of Transportation Engineering,* 2nd Edition, McGraw-Hill, 1969.

TABLE 15-2

Recommended Belt Speeds for Various Materials

Material to be Conveyed	Belt Widths, in. 12	18	24	30	36	42	48	54	60
Light, free-flowing material	300	400	500	600	600	600	700	700	700
Coal and similar lump material	300	400	500	550	550	600	600	650	650
Heavy ore and abrasive material	300	350	450	550	550	550	600	600	600

Source: W. Staniar (ed.), *Plant Engineering Handbook*, McGraw-Hill, 1959.

strength and body, and rubber covers to protect the fibers. The belt is commonly made of several plies of fabric cemented together by rubber. The top cover is usually heavier and made of a high-grade rubber to resist impact. To provide more strength, cord belts have been used in the core overlain by duck material fabric, to give transverse strength (These have to be spliced by vulcanizing). Steel wires have also been imbedded in belts to give more strength. Various synthetic fibers such as rayon, dacron, and glass have also been used. Table 15–2 gives the maximum recommended speeds and Table 15–3 gives the capacity for various belt widths and speeds.

The Idlers. The belt is supported by cylindrical rollers called idlers 4 to 7 inches in diameter, the length of which depends on the width of the belt. Three of these are usually used across the belt forming a flat-bottomed trough. The typical trough angle is about 20°. Spacing of the idlers, which is typically 2½–5½ feet on centers, depends on the belt and commodity being transported. If the spacing is too much, the sag causes shock as the belt meets the rolls, shortening the life of the belt. The idlers usually contain dust proof bearings lubricated by the manufacturer. The idlers are commonly steel tubing, but may be covered with rubber.

Idlers are usually closely spaced at loading points to reduce belt wear and may be of a special design to cushion the shock.

Idlers are commonly spaced at twice the spacing on the return since they only carry the empty belt. Figure 15–1 shows some belt idlers.

The Take-ups. Because of increases (changes) in belt length due to temperature change and belt load, a take-up device is required to reduce and control belt slack. These are usually automatic devices made up of weights and pulleys. Figure 15–2 shows some take-up arrangements.

The Tripper. A tripper is a device which causes the material to be discharged into a chute. The belt is doubled back over two pulleys so that the inertia of the load causes the material to go into the chute where it is discharged to either side of the belt. The tripper may be fixed or movable. For light nonabrasive materials a scraping device may be used to unload the belt. This may be a diagonal member or a V-shaped plow.

The Drive. Power is usually transferred to the belt through a flat-faced cast-iron or steel pulley for large tensions, and the pulley may be "lagged" with diagonal grooves to provide for more efficient transfer of loads. Most belts are driven by a single squirrel-cage induction-type motor.

Belt Cleaners. It is essential that the surface of the belt be kept clean, especially if the material is damp or sticky. Several types of cleaners are used which may be in the form of a scraper, a rotary brush, a Robins Cleaner, or a Hudson Cleaner. The scraper is the most commonly used and, very often, the least effective. This consists of a flat piece of rubber or stainless steel pressed against the moving belt. The rotary brush operates at high speed and rotates in the opposite direction of the belt movement. The Robins Cleaner consists of a rubber helix or spiral gear-like cylinder which rotates against the moving belt. The Hudson Cleaner has a series of rubber scrapers which press against

TABLE 15-3

Belt Capacity

Width of belt, in.	Wt per cu ft of Material, lb	Capacity, Short Tons (2,000 lb) per Hour — Belt Speed, fpm												Cross Section of load, sq ft
		50	100	150	200	250	300	350	400	450	500	550	600	
18	50	17	34	51	68	85	102	119	136					0.227
	75	26	51	77	102	128	153	179	204					
	100	34	68	102	136	170	204	238	272					
	125	43	85	128	170	212	255	298	340					
	150	51	102	153	204	255	306	357	408					
24	50	31	63	94	125	156	187	219	250	281	312			0.417
	75	47	94	141	187	234	281	328	375	422	468			
	100	63	125	188	250	312	375	438	500	563	625			
	125	78	156	234	312	390	468	546	624	702	780			
	150	94	188	281	375	468	562	656	750	843	936			
30	50	50	100	150	200	250	300	350	400	450	500			0.667
	75	75	150	225	300	375	450	525	600	675	750			
	100	100	200	300	400	500	600	700	800	900	1,000			
	125	125	250	375	500	625	750	875	1,000	1,125	1,250			
	150	150	300	450	600	750	900	1,050	1,200	1,350	1,500			
36	50	73	145	218	290	362	435	507	580	652	725	797	870	0.967
	75	109	217	326	434	542	651	760	868	976	1,085	1,195	1,300	
	100	145	290	435	580	725	870	1,015	1,160	1,305	1,450	1,595	1,740	
	125	181	362	543	724	905	1,085	1,270	1,450	1,630	1,810	1,990	2,170	
	150	218	435	653	870	1,090	1,310	1,520	1,740	1,960	2,180	2,390	2,644	

42	50	102	203	304	406	508	609	710	812	913	1,016	1,120	1,220	1.353
	75	152	304	456	608	760	912	1,065	1,218	1,370	1,520	1,670	1,825	
	100	203	406	609	812	1,015	1,218	1,420	1,625	1,825	2,030	2,230	2,440	
	125	254	508	761	1,015	1,270	1,525	1,780	2,030	2,280	2,540	2,790	3,040	
	150	304	609	915	1,220	1,520	1,830	2,130	2,440	2,740	3,040	3,350	3,650	
48	50	138	275	413	550	688	825	963	1,100	1,240	1,375	1,510	1,650	1.833
	75	206	413	620	825	1,030	1,240	1,445	1,650	1,860	2,060	2,270	2,480	
	100	275	550	825	1,100	1,375	1,650	1,925	2,200	2,480	2,750	3,020	3,300	
	125	344	688	1,030	1,375	1,720	2,060	2,410	2,750	3,100	3,440	3,780	4,130	
	150	413	825	1,240	1,650	2,060	2,480	2,890	3,300	3,710	4,125	4,540	4,950	
54	50	179	358	537	716	895	1,075	1,250	1,430	1,610	1,790	1,970	2,150	2.383
	75	268	536	805	1,070	1,340	1,610	1,875	2,140	2,410	2,680	2,950	3,220	
	100	358	715	1,072	1,430	1,790	2,140	2,500	2,860	3,229	3,580	3,930	4,290	
	125	447	894	1,340	1,790	2,240	2,680	3,130	3,580	4,020	4,470	4,910	5,360	
	150	536	1,072	1,610	2,140	2,680	3,220	3,760	4,290	4,830	5,360	5,900	6,440	
60	50	225	450	675	900	1,125	1,350	1,575	1,800	2,020	2,250	2,480	2,700	3.000
	75	337	675	1,010	1,350	1,690	2,020	2,360	2,700	3,040	3,370	3,710	4,050	
	100	450	900	1,350	1,800	2,250	2,700	3,150	3,600	4,050	4,500	4,950	5,400	
	125	562	1,125	1,690	2,250	2,810	3,370	3,940	4,500	5,060	5,620	6,180	6,750	
	150	675	1,350	2,020	2,700	3,380	4,050	4,720	5,400	6,070	6,750	7,420	8,100	
72	50	270	540	810	1,080	1,350	1,620	1,890	2,160	2,430	2,700	2,920	3,240	4.000
	100	540	1,080	1,620	2,160	2,700	3,240	3,780	4,320	4,860	5,400	5,940	6,480	
	150	810	1,620	2,430	3,040	4,050	4,860	5,670	6,480	7,300	8,100	8,900	9,720	

Source: W. Staniar (ed.), *Plant Engineering Handbook*; McGraw-Hill, 1959.

Note: Cubic feet per hour for each 100 fpm belt speed = 6,000 X load cross section in square feet.

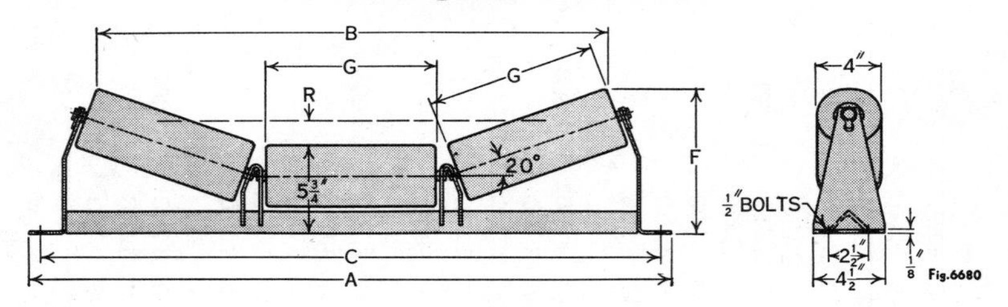

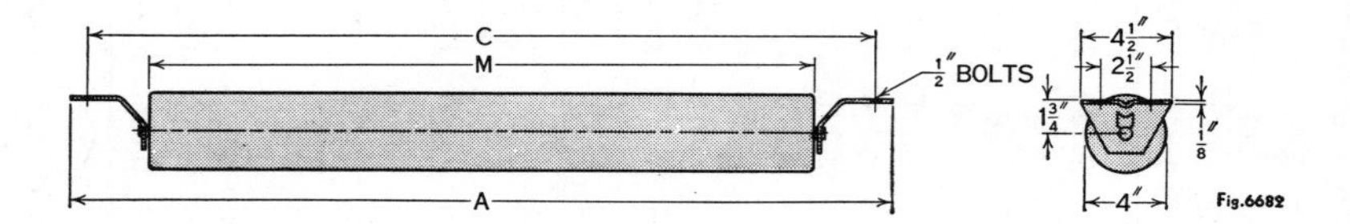

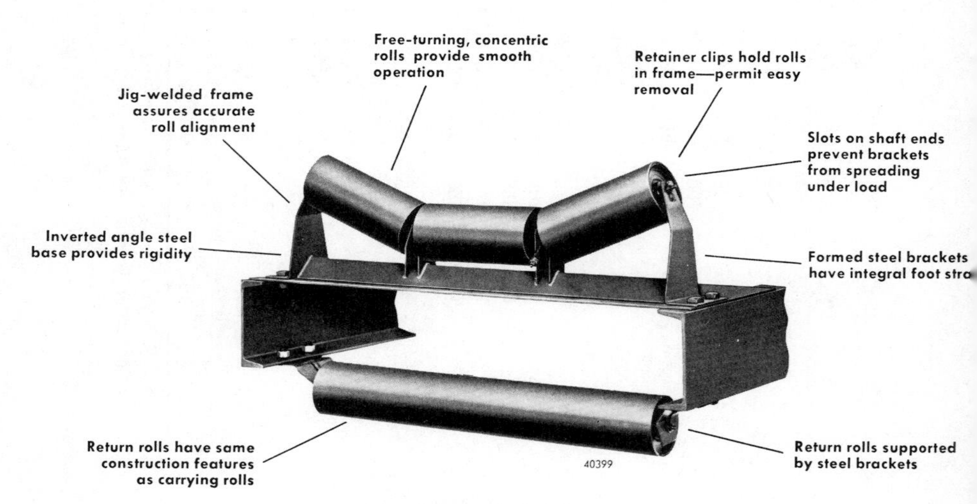

Fig. 15–1. Belt idlers.

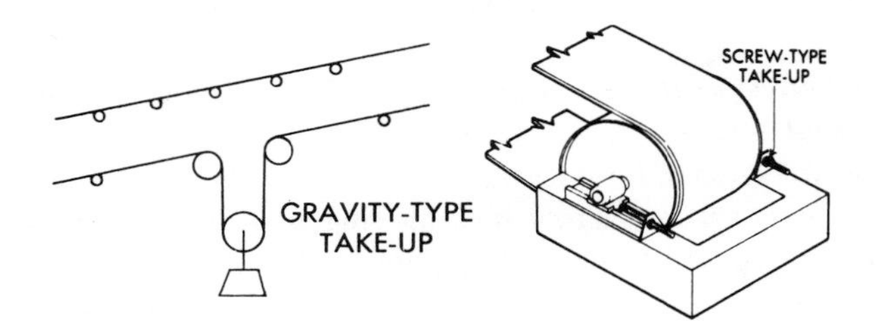

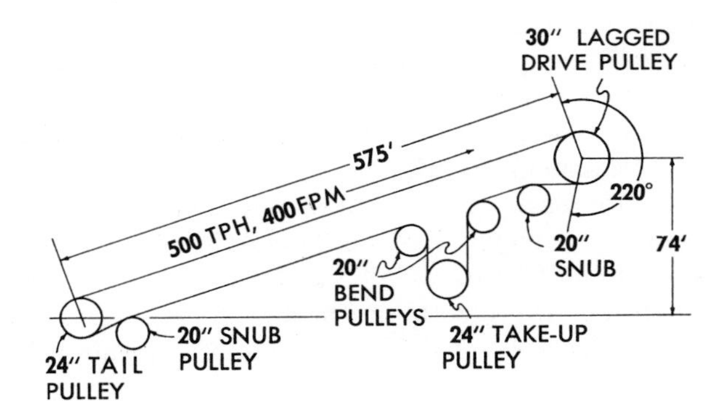

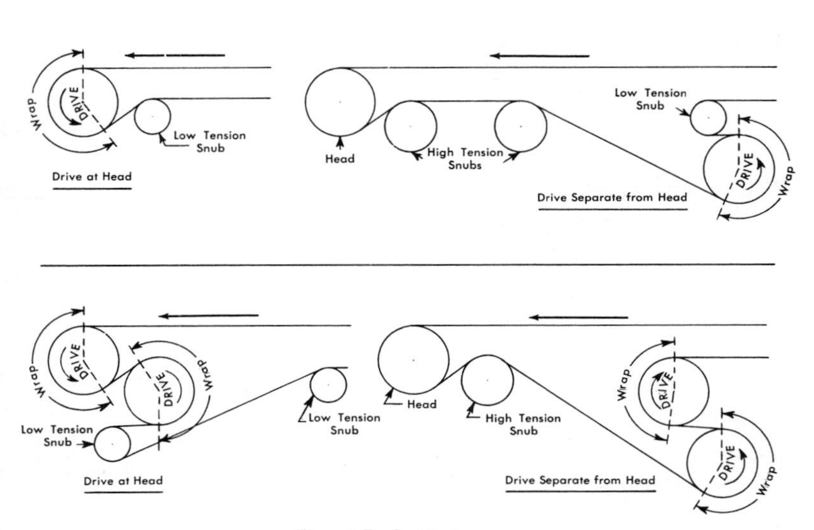

Fig. 15–2. Take-ups.

the belt, rotating in the opposite direction. This is an improved type of scraper.

15–4. Belt Conveyor Design Technology. The designer begins with the requirement that some specific tonnage per hour of some commodity be moved between specific locations. One of the first considerations to be made is between a narrow belt moving at a high speed or a wide belt moving at a slower speed. The choice of speed will depend somewhat on the type of material; also, the maximum lump size will influence the choice of the belt width.

The choice of belt thickness and fabric is determined from manufacturer's recommendations based on the amount of power to be transmitted. The amount of power transmitted is the sum of (1) the power required to move the empty belt which will vary with the length and width of belt, (2) the power required to move the load horizontally which depends on the quantity of material in tons per hour and the length, and (3) the power required for lifting the material which depends on the quantity and amount of lift.

The following approach to the final selection of a belt is taken from the B. F. Goodrich Conveyor Belt Selector Manual (Manual No. IPC-264-3) and in a simplified form follows three steps: (1) determine the belt effective tension, (2) determine the belt operating tension and required horsepower from the total effective tension, and (3) select the correct belt.

15–5. Effective Belt Tension. Maximum operating tension is developed in the belt at rated speed and peak capacity. Tension requirements are found as follows:

Frictional resistance or effective tension necessary to move empty belt is:

$$T_x = F_x L_c G \tag{15–1}$$

Frictional resistance or effective tension necessary to move the material load horizontally is:

$$T_y = F_y L_c Q \tag{15–2}$$

Effective tension to lift or lower the load is:

$$T_z = HQ \tag{15–3}$$

The total effective tension requirement is equal to the sum of these or:

$$T_E = T_x + T_y + T_z \tag{15–4}$$

TABLE 15-4

Weight of Moving Parts, G, for Use in Equation 15-1

This value may be taken from the following table if it is not possible to calculate. It is best to get the details and calculate for long center installations.

Belt Width in.	Light Duty Conveyors 4" Idlers	Regular Duty 5" Idlers	Regular Duty 6" Idlers Belts up to 6 ply	Heavy Duty 6" or 7" Idlers Belts 7 to 10 ply
14	12	14	..	..
16	14	15	20	..
18	15	17	22	..
20	16	18	25	..
24	19	24	30	32
30	25	31	38	45
36	29	37	47	58
42	34	43	55	71
48	..	48	64	84
54	..	..	72	97
60	..	..	81	110
72	..	..	97	135

If the value of "G" is to be calculated, the weight of the belt, weights of all pulleys and shafts turned by the belt, and weights of moving parts of carrying and return idlers must be known.
Source: B. F. Goodrich Co., *Conveyor Belt Selection Manual IPC-264-3*

where:

G = weight of conveyor belting, idlers, pulleys, expressed in pounds per foot of conveyor length (see Table 15–4)

F = coefficient of friction of rolling parts

F_x = value of F to be used to move the empty belt which is .03 for ideal conditions

F_y = value of F to be used to move the load horizontally which is .04 for normal loading and good equipment

L = conveyor length in feet, measured along belt between center of terminal pulleys

L_c = adjusted center length to use in horsepower formulas

$$(L_c = 0.55L + 115)$$

H = vertical distance in feet between loading and discharge points
Q = material weight per foot

$$Q = \frac{33.3C}{S}$$

where:
C = loading rate, short tons per hour
S = belt speed F.P.M.

15–6. Belt Operating Tension and Required Horsepower. In frictional drive systems additional tension must be induced in the belt to prevent slippage on the drive pulley. This added tension is also the tension in the belt on the slack or leaving side of the pulley. It is referred to in the following formulas as T_2 and is estimated by:

$$T_2 = KT_E \qquad (15\text{–}5)$$

where:
T_2 = slack side tension
K = drive factor based on coefficient of friction, wrap, and type of takeup (see Table 15–5).

On those units equipped with a gravity takeup located near the drive and when the total takeup weight is known:

$$T_2 = \frac{\text{Total takeup weight}}{2}$$

The tight side tension (operating tension) in the belt, T_1 is equal to the sum of the effective tension, T_E, and the slack-side tension, T_2, or:

$$T_1 = T_E + T_2 \qquad (15\text{–}6)$$

The horsepower required at the drive pulley is:

$$HP = \frac{T_E S}{33000} \qquad (15\text{–}7)$$

where:
S = speed

The effective tension is the difference between the tensions on the tight side and the slack side of the driving pulley, or:

$$T_e = T_1 - T_2 \qquad (15\text{–}8)$$

As an alternative to Equation (15–5), the following formula may be used with Equation (15–6) to determine the slack-side tension:

$$\frac{T_1}{T_2} = e^{f\alpha} \qquad (15\text{–}9)$$

TABLE 15-5

Drive Factor, K, for Use in Equation 15-5

Angle of Belt Wrap at Drive	Type of Drive	Screw Take-up		Gravity Weighted or Flexible Takeup	
		Bare Pulley	Lagged Pulley	Bare Pulley	Lagged Pulley
150°	Plain	1.5	1.0	1.08	.67
160°	Plain	1.4	.9	.99	.60
170°	Plain	1.3	.9	.91	.55
180°	Plain	1.2	.8	.84	.50
190°	Snubbed	1.1	.7	.77	.45
200°	Snubbed	1.0	.7	.72	.42
210°	Snubbed	1.0	.7	.67	.38
220°	Snubbed	.9	.6	.62	.35
230°	Snubbed	.9	.6	.58	.32
240°	Snubbed	.8	.6	.54	.30
340°	Tandem or Dual	.5	.4	.29	.143
360°	Tandem or Dual	.5	.4	.26	.125
380°	Tandem or Dual	.5	.3	.23	.108
400°	Tandem or Dual	.5	.3	.21	.095
420°	Tandem or Dual	.4	.3	.19	.084
440°	Tandem or Dual	—	—	.17	.074
460°	Tandem or Dual	—	—	.15	.064
480°	Tandem or Dual	—	—	.14	.056

Source: B. F. Goodrich Co., *Conveyor Belt Selection Manual IPC-264-3.*

where:

T_1 = maximum tension
T_2 = Slack side tension
f = coefficient of belt friction (0.30–0.35)
α = angle of contact in radians
e = base of natural logarithms (2.718)

15–7. Selecting the Correct Belt. Generally speaking, long belts are desired because they reduce the frequency of load impact, bending around terminal pulleys and the scraping action of belt cleaners.

Belt selection charts are made up for the various types of belt construction fibers used and are rated according to their tension in pounds per inch, troughability, impact, load support, flexibility, and comparative cost. The type of splice used (vulcanized or mechanical) determines

TABLE 15-6

Belt Selection Chart

1	2	3	4	5	6	7
Construction	Tension Rating lb./in. 196 Required	Troughability	Impact	Load Support	Flexibility	Comparat Cost
8 ply 32 oz.	216					
Nyfil 35	216	X	X	X		
6 ply 36 oz.	198	X	X	X		
Nyfil 43	198	X	X	X		
Nycord 43	198	X	X	X	X	
Rayon 43	198	X	X	X	X	
5 ply 42 oz.	200	X	X	X	X	
Nyfil 50	200	X	X	X	X	X
Nycord 50	200	X	X	X	X	X
Nyfil 60	225	X	X	X		
Nycord 60	225	X	X	X	X	
Rayon 60	225	X	X	X		
4 ply Nyfil 70	220	X				
Nycord 70	220	X				
Rayon 70	220	X	X	X	X	

Source: B. F. Goodrich Co., *Conveyor Belt Selection Manual IPC-264-3.*

the maximum belt operating tension. Vulcanized splices are more efficient and durable than the mechanical type and when used may result in lower cost belt and less frequent replacement of splices. Table 15–6 shows a belt selection chart for various materials.

15–8. Passenger Conveyor Belt Systems. Belt conveyors have been used for moving passengers at railroad stations and airports. The need for moving sidewalks may be an advantage where the air passenger has to cover great distances from the main lobby to the stall of the plane. Some possibility exists for moving of passengers by means of small belt-carried cars or compartments which move continuously over established routes. The use of escalators in a large number of buildings has grown considerably. Escalators in one sense may be considered as a belt conveyor system.

REFERENCES

1. Hudson, Wilbur G., *Conveyors and Related Equipment,* Third Edition, John Wiley and Sons, New York (1954).
2. Goodrich, B. F., *Conveyor Belt Selection Manual IPC-124-3.*
3. Hennes, Robert G., and Ekse, Martin, *Fundamentals of Transportation Engineering,* Second Edition, McGraw-Hill Book Co., New York (1969).
4. Link-Belt, Belt Conveyor Idlers, Book 2816, 1961.

IV

DESIGN OF AIR TRANSPORTATION FACILITIES

16

Airport Planning and Layout

16–1. Introduction. It happened at midmorning on a windy beach at Kitty Hawk, North Carolina on December 17, 1903. A fragile looking two-winged craft with a man at the controls was propelled on a little trolley along a wooden rail. It rose from the rail, surged forward into the wind, and settled back into the sand. Man's first flight in a heavier-than-air craft had lasted 12 seconds and covered a distance of 120 feet. On that morning, Wilbur and Orville Wright were to make three additional flights, the longest lasting for almost a minute. The air age had begun, and the world would never be the same.

Man's ancient aspirations to fly had been, in Wilbur Wright's words, "handed down to us by our ancestors who, in their gruelling travels across trackless lands in prehistoric times, looked enviously on the birds soaring freely through space, at full speed, above all obstacles, on the infinite highway of the air" [1].

The Wrights' success had been preceded by centuries in which inaccurate and often ludicrous theories of flight were proposed, and abortive and sometimes disastrous attempts were made to leap and glide and soar.

A groundswell of interest in air transportation developed in 1783 when Joseph and Etienne Montgolfier demonstrated in Annonay, France that man could travel by balloons filled with hot air. Later that same year, a French physicist, J. A. C. Charles, made a successful flight in a balloon filled with hydrogen.

Still later, when it became possible to steer these *Montgolfières and Charlières,* they came to be known as dirigibles. Strong interest in transportation by the slow and cumbersome dirigibles continued until the 1930's, when several spectacular disasters occurred to the lighter-than-air crafts. These dramatic tragedies, along with progress in heavier-than-airflight technology, resulted in the assignment of the dirigible to very limited and specialized uses.

The Wright brothers' flight came at a time when there was arduous and seemingly frenetic activity by other aerial pioneers. The conviction that man was on the brink of successful heavier-than-air flight encouraged widespread study and experiment.

Fig. 16–1. The first heavier-than-air flight, December 17, 1903. (Courtesy The Library of Congress.)

The Englishman, George Cayley, has been called the father of aerial navigation. His experiments during the first half of the nineteenth century, with small and full scale gliders, demonstrated the feasibility of flight in heavier-than-air craft. In 1866, another Englishman, F. H. Wenham, in a report to the first meeting of the Aeronautical Society of Great Britain, did much to advance the state of knowledge of aerodynamics and the design of wings.

A German, Otto Lilienthal, made over 2,000 successful glides during the last decade of the nineteenth century, some of which covered several hundred feet. In 1894, Octave Chanute, a successful civil engineer in the United States, who was spurred by Lilienthal's successful experiments, published a historical account of man's attempts to fly. He later designed gliders on his own, using his knowledge of bridge building to improve their structural design. Chanute developed a close friendship

Fig. 16–2. The maiden flight of the Lockheed C-5 Galaxy, the world's largest airplane, June 30, 1968. (Courtesy Lockheed-Georgia Company.)

with Orville and Wilbur Wright and was a source of help and encouragement during their years of experiments at Kitty Hawk.

The airplane made a spectacular, if not decisive contribution to the outcome of World War I, and after the war, more serious attention was given to airplanes as an effective means of transporting people and goods.

Acceptance of air transportation was increased by Charles Lindbergh's dramatic solo flight from New York to Paris in 1927. Air transportation was nearing the threshold of maturity at the advent of World War II, and few can dispute its vital contribution to the war effort.

16–2. Growth of Aviation Activity. The growth of aviation activity since World War II has been unprecedented and extraordinary, and future aviation growth threatens to become unmanageable. In 1966, air transportation accounted for 66 per cent of common carrier passenger miles, as compared to only 13 per cent in 1950 [2]. During 1956–1966, air passenger traffic (in terms of passenger-miles) more than tripled. During that decade, an even higher rate of growth was experienced by operations in general aviation aircraft. Trends in the factors which are accepted as indicators of air travel activity (population, wealth, education, etc.) indicate that growth trends recently experienced in air transportation will be maintained and more likely will be surpassed.

16–3. Nature of the Problem. In the pages that follow, some of the problems and techniques of planning and designing an airport will be described. The remainder of this chapter will be devoted to a discussion of airport planning, site selection, runway orientation, obstruction clearance standards, and typical runway configurations. Chapter 17 will be concerned with the planning and design of the terminal area, while Chapter 18 will relate more specific design criteria and procedures for the runway and taxiway system.

Although these topics will be discussed separately, it should be recognized that there is an interaction between the various components of the problem, and that goals and requirements of one component of the problem will often conflict with the needs of another. The overall task is an iterative one which seeks an optimum airport plan and design which best satisfies the various needs, constraints, and controls of the system.

16–4. Airport Demand. Before engineering plans and designs for a new airport or improvements to an existing facility are made, a study should be made to determine the extent of future needs for airport facilities. Although vital, this study is socio-economic rather than engineering in nature and is based on fundamental principles of marketing. It includes forecasts of annual peak-day and peak-hour volumes of passengers and aircraft, types of aircraft, type of forecast usage (i.e., business, commercial, passenger and freight, pleasure, etc.), as well as factors concerning size of community to be served and economic and population growth trends.

While the detailed techniques for making these forecasts are beyond the scope of this text, the following general observations and relationships are given as matter of interest.

Experience has shown that a community's aviation activity is primarily responsive to: (1) population and population density, and (2) the economic character of the community.

Studies have also shown that a given traveler's propensity and ability to travel by air are closely related to:

1. *Income.* Higher income groups are much more likely to travel by air than those in lower economic classes.
2. *Education.* Persons with a college education or with some college training tend to travel more by air than those who have never been to college.
3. *Occupation.* The majority of air travelers are in managerial, professional, technical, or official occupations.
4. *Length of trip.* As one might expect, air transportation is especially attractive to travelers making long trips, and as the trip length increases, a larger percentage of travels go by air.
5. *Purpose of trip.* Business travel accounts for about 60 per cent of air revenue passenger-miles, the remaining portion consisting of personal travel for vacations and the like.

Finally, it is noted that the preparation of forecasts of aviation demand for a given community can be greatly facilitated by national forecasts by the Federal Aviation Administration (FAA) since a given community's share of nationwide air travel tends to remain relatively constant. Reference can also be made to the *National Airport System Plan,* a publication prepared and maintained by the FAA which lists recommendations for future airport needs to promote the development of an adequate national system of airports. Airport development projects must first be included in the *National Airport System Plan* before being considered for Federal aid airport funds [3].

16–5. Selection of Airport Site. Perhaps the single most important aspect of the planning and design of airports is the selection of an airport site. Mistakes made in this phase of the airport development program can result in the failure or early obsolescence of the facility.

Several contemporary trends have complicated the problem of selection of airport site:

1. Urban sprawl has occurred around most U.S. cities which has been accompanied by increasing scarcity of land and rising land costs.
2. Faster and larger aircraft have appeared requiring longer runways and more aircraft service space along with increased automobile parking and circulation space.
3. The requirements and desires of the public regarding air passenger services have become more elaborate and sophisticated.

Guidelines regarding procedures for making a study of alternate airport sites, along with a discussion of the major factors influencing site selection have been promulgated by the FAA [4]. While the critical investigation and location of airport sites is a responsibility of the airport sponsor[1], FAA endorsement of the site is required if Federal aid is contemplated. An airspace review by the FAA is required in any event.

After airport needs have been established, the FAA recommends that the following procedure be followed in selecting an airport site.

Desk study of area. Before a field investigation is made, a great deal can be learned from a desk study which includes:

1. A review of existing comprehensive land use plans and other community and area plans.
2. An analysis of available wind data to determine the desired runway orientation. (This is discussed in Section 16–6.)

[1] The airport sponsor is usually a state, city, or other local body, although it may be a private organization or individual.

3. A study of USC&GS quadrangle sheets, road maps, and aeronautical charts to select feasible sites for further evaluation.
4. A study of general land costs in the areas of interest.

In this study, special attention should be given to the location of other airports and land transportation facilities, obstructions, topographic features, and atmospheric peculiarities.

Physical inspections. Actually, two physical field inspections should be made of the potential sites: preliminary and final.

1. The preliminary inspection should be made jointly by sponsor representatives of all existing airports and potential sites. After these inspections, sketches of the various sites should be made, as well as an overall sketch or small scale map showing all of the sites under consideration. If possible, an aerial inspection of the various sites should be made and aerial photographs should be taken.
2. On the final inspection, those sites selected during the preliminary inspection are visited. If Federal-aid funds are involved, an FAA airports representative should accompany the inspection group.

Evaluation and recommendations. The FAA recommends that a final report be prepared including a rough cost estimate for each site. The report should list the sites in order of preference and indicate the advantages and disadvantages of each site. In cases involving Federal aid, this report is submitted to the FAA for endorsement, and a site mutually agreeable to the sponsor and the FAA is chosen. Written endorsement by the FAA does not imply a commitment of Federal funds, but it is a necessary step in obtaining funds under the Federal-aid airports program.

If no Federal funds are requested, the FAA should be contacted for further information regarding the initiation of a request for airspace review.

There are at least ten factors which should be considered when analyzing potential airport sites:

1. Convenience to users
2. Availability of land and land costs
3. Design and layout of the airport
4. Airspace obstructions
5. Engineering factors
6. Social factors
7. Availability of utilities
8. Atmospheric conditions
9. Hazards due to birds
10. Coordination with other airports

Convenience to Users. If it is to be successful, an airport must be conveniently located to those who use it. From this viewpoint, the airport ideally would be located near the center of most cities. The obvious problems of air obstructions and land costs rule out this possibility, and most cities have found it necessary to locate the airport several miles from the city center.

Urban sprawl and increasing scarcity and costs of land have resulted in airports being located farther and farther from the city center.[2] At the same time, air speeds have increased with the result that an increasing percentage of air passengers spend more time in the ground transportation portion of the trip than in the air.[3] Yet it is known that the amount of airport use is very sensitive to the ratio of the ground travel time to the total journey time, and as this ratio increases, the air traffic can be expected to decrease precipitously.

In the view of the airport user, travel time is a more important measure of convenience than distance. Thus, a relatively remote potential airport site should not be ruled out if it is conveniently located to a major highway or other surface transportation facility.

Availability of Land and Land Costs. Vast acreages are required for major airports, and it is not uncommon for new airports in large cities to require more than 10,000 acres. Airport planners in Chicago concluded that a desirable size for a new airport there was 11,000 to 13,000 acres [6]. A new airport for Atlanta will require 14,000 acres and an additional area nearly as large as the airport itself will be rendered unsuitable for any uses other than industry, certain recreational activities, and limited recreational development [7]. Los Angeles officials anticipate that by 1980 in the L. A. hub area more than 3,300 acres will be needed for apron space to handle and tie down general aviation aircraft alone [8].

Since desirable airport site land is also in demand for other purposes, land costs are high, and real estate can be expected to appreciate with the planning and development of a new airport facility. In Chicago, four potential airport sites were considered. Cost estimates for the land cost (including appreciation), leveling the site, and relocation of utilities, roads, and railroads ranged from $188.0 million to $443.0 million [5].

[2] An Arthur D. Little study of eleven cities indicated that in the case of the old airport, the average central city to airport distance was 7.0 miles. For these cities, the average distance from the central city to a proposed or new airport was 15.8 miles. The most remote new airport was Dulles International Airport, located 27.0 miles from downtown Washington, D.C. [5].

[3] The ratio of ground travel time to total trip time decreases with increases in the total length of trip. A study of the fifty most heavily traveled city-to-city routes indicated that trips in which the airport-to-airport mileage was 250 miles or less, 51 to 65 per cent of the total trip time was spent in ground travel. For trips in which the airport-to-airport mileage was 1,000 miles or more, this percentage ranged from 22 to 32 per cent [5].

It is important that sufficient land be acquired for future expansion. Failure to do so could mean that a convenient and otherwise desirable airport site would have to be abandoned due to limitations in aircraft operations from an unexpandable runway or inadequate space for aircraft or passenger handling.

The requirements for large acreages, the need for a convenient location, and the high land costs may lead to novel approaches to the problem of site selection. Serious consideration is being given in Chicago, for example, to placing a new airport in a circular dike in Lake Michigan approximately 8.5 miles offshore from the city [6]. It has similarly been suggested that an airport be constructed in Boston harbor [9].

Design and Layout of the Airport. In considering alternate potential airport sites, the basic layout and design should essentially be constant. One should avoid making major departures from the desired layout and design to fit a particular site. In this connection, one consideration is especially important. Runways should be oriented so as to take advantage of prevailing winds, and variations in runway alignment from optimum orientation more than ±10 degrees should normally not be made. (See Section 16–6 for FAA standards on runway orientation.)

Airspace and Obstruction. To meet essential needs for in-flight safety, two requirements must be met:

1. Adjacent airports must be located so that traffic using one in no way interferes with traffic using the other. An airspace analysis should be made to insure that this requirement is met. It is desirable that the assistance of the FAA be sought in conducting this analysis, especially when an airport is to be located in a highly developed terminal complex.
2. Physical objects such as towers, poles, buildings, mountain ranges, etc. must not penetrate navigable airspace. Criteria on "Objects Affecting Navigable Airspace," given in Federal Aviation Regulation, Part 77, should be consulted prior to beginning the site selection process. (See Section 16–7.)

Engineering Factors. An airport site should have fairly level topography and be free of mountains, hills, etc. Further, the terrain should have sufficient slope that adequate drainage can be provided. Areas requiring extensive rock excavation should be avoided as should sites containing peat, muck, and otherwise undesirable foundation materials.

An adequate supply of aggregates and other construction materials should be located within a reasonable distance of the site.

A desirable airport site will be relatively free of timber, although a border of timber along the airport periphery may suppress undesirable noise.

Social Factors. One of the most difficult social problems associated with airport location is that of noise. With the advent of the jet aircraft engine, airport noise has worsened, and despite efforts by industry and government groups, the development of a quiet aircraft engine does not seem likely within the near future.

Airports are not good neighbors, and some control in the development of land surrounding an airport should be exercised. In selecting an airport site, proximity to residential areas, schools, and churches should be avoided, and the runways should be oriented so that these land uses do not fall in the immediate approach-departure paths.

Availability of Utilities. With rare exceptions (e.g., the Dulles International Airport, Washington, D.C.), airports must depend upon existing utilities. The site should be accessible to water, electrical service, telephones, gas lines, etc., and these utilities should be of the proper type and size.

Atmospheric Conditions. Peculiar atmospheric conditions such as fog, smoke, snow, or glare may rule out the use of some potential airport sites.

Hazards Due to Birds. Aircraft impact with birds and bird ingestion into turbine engines have caused numerous air disasters. Airports should not be situated near bird habitats or natural preserves and feeding grounds. At certain potential sites, special work such as filling of ponds and closing of dumps may be required to insure that birds will not present a hazard to aircraft flights.

Coordination with Other Airports. Studies of aviation activity in heavily populated metropolitan areas indicate that more than one major airport will be required to meet future air travel needs. Atlanta planners are considering the construction of a second major airport despite proposed $2.73 million improvements to its present major airport. A Los Angeles study indicated that the ground transportation and parking capacities for the International Airport would be exceeded by 1975 [8]. A satellite system of airports is, therefore, being planned to meet the needs of the Los Angeles area. In cases such as these, it is clear that individual airport requirements must be determined in relation to the needs of the entire metropolitan area and each airport must be considered as a part of a total system.

On the other hand, air services for small cities often may be most beneficially provided by locating the airport between two or more adjacent communities.

16–6. Runway Orientation. Because of the obvious advantages of landing and taking-off into the wind, runways are oriented in the direction of prevailing winds. Aircraft may not safely maneuver on a runway when the wind contains a large component at right angle to the direction of travel. The point at which this component (called the cross wind) becomes excessive will depend upon the size and operating characteristics of the aircraft. Certain large planes can safely operate with cross winds up to 30 or 35 mph, but small aircraft tend to have difficulty in cross winds greater than 15 mph.

According to FAA standards, runways should be oriented so that aircraft may be landed at least 95 per cent of the time with cross-wind components not exceeding 15 mph.

Wind Rose Method. A graphical procedure utilizing a wind rose is typically used to determine the "best" runway orientation insofar as prevailing winds are concerned. (See Fig. 16–3.)

If available, wind data for a potential airport site are obtained from the weather bureau. If suitable weather records are not available accurate wind data for the area should be collected. (Another alternative would be to form a composite wind record from nearby wind-recording stations.) The wind data are arranged according to velocity, direction, and frequency of occurrence as shown by Table 16–1. This table indicates the percentage of time wind velocities within a certain range and

TABLE 16–1

Typical Wind Data

Wind Direction	Percentage of Winds			
	4–15 mph	15–31 mph	31–47 mph	Total
N	4.8	1.3	0.1	6.2
NNE	3.7	0.8	..	4.5
NE	1.5	0.1	..	1.6
ENE	2.3	0.3	..	2.6
E	2.4	0.4	..	2.8
ESE	5.0	1.1	..	6.1
SE	6.4	3.2	0.1	9.7
SSE	7.3	7.7	0.3	15.3
S	4.4	2.2	0.1	6.7
SSW	2.6	0.9	..	3.5
SW	1.6	0.1	..	1.7
WSW	3.1	0.4	..	3.5
W	1.9	0.3	..	2.2
WNW	5.8	2.6	0.2	8.6
NW	4.8	2.4	0.2	7.4
NNW	7.8	4.9	0.3	13.0
Calms		0–4 mph		4.6
Total				100.0%

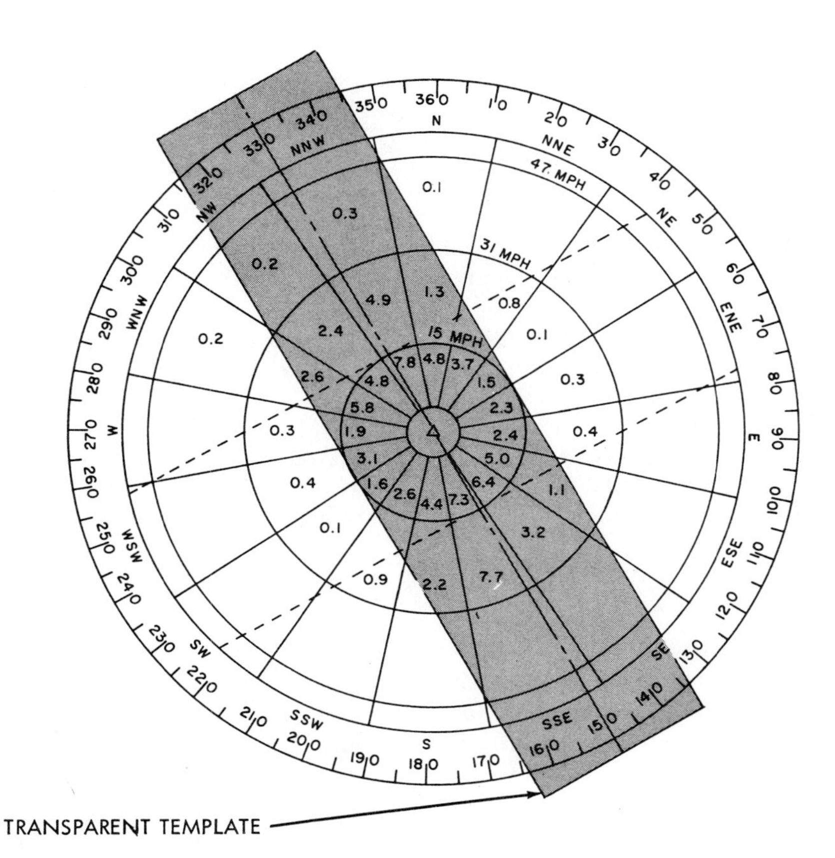

LOCATION

△ 4.6% CALMS, 0–3 M.P.H.

Fig. 16–3. A typical wind rose.

from a given direction can be expected. For example, the table indicates that for the hypothetical site, northerly winds in the 4–15 mph range can be expected 4.8 per cent of the time.

These data are plotted on the wind rose by placing the percentages in the appropriate segment of the graph. On the wind rose, the circles represent wind velocity in miles per hour, and the radial lines indicate wind direction. The data from Table 16–1 have been properly plotted on Fig. 16–3.

The wind rose procedure makes use of a transparent template on which three parallel lines have been plotted. The middle line represents the runway centerline and the distance between it and each of the outside lines is equal to the allowable cross-wind component (e.g., 15 mph).

The following steps are necessary to determine the "best" runway orientation and to determine the percentage of time that orientation conforms to the cross-wind standards:

1. Place the template on the wind rose so that the middle line passes through the center of the wind rose.
2. Using the center of the wind rose as a pivot, rotate the template until the sum of the percentages between the outside lines is a maximum. When the template strip covers only a fraction of a segment, a corresponding fractional part of the percentage shown should be used.
3. Read the true bearing for the runway on the outer scale of the wind rose beneath the centerline of the template. In the example the best orientation is 150°–330° or S 30° E, true.
4. The sum of percentages between the outside lines indicates the percentage of time that a runway with the proposed orientation will conform with cross-wind standards.

16–7. Objects Affecting Navigable Airspace. Part 77 of the Federal Aviation Regulations [11] establishes standards for determining obstructions in navigable airspace, sets forth the notice requirements of certain proposed construction or alteration, provides for aeronautical safety studies of obstructions, and provides for public hearings on the hazardous effect of proposed construction or alteration.

Because of its importance and relevance to the planning and design of airports, Subpart C of this publication, which sets for FAA standards regarding air navigation obstructions, is here quoted.

SUBPART C—OBSTRUCTION STANDARDS

§ 77.21 Scope.

(a) This subpart establishes standards for determining obstructions to air navigation. It applies to existing and proposed man-made objects, objects of natural growth, and terrain. [The standards apply to the use of navigable airspace by aircraft and to existing air navigation facilities, such as an air navigation aid, airport, Federal airway, instrument approach procedure, approved off-airway route, control zone, or transition area. Additionally, they apply to a planned facility or use, or a change in an existing facility or use, if a proposal therefore is on file with the FAA or the Department of Defense on the date the notice required by § 77.13(a) is filed.]

(b) Minimum obstruction clearance altitudes are considered in place of minimum en route altitudes in applying the standards of this subpart to objects whenever planning information available at the time of filing of the notice required by § 77.13(a) indicates a need to lower the minimum en route altitude of a segment of a Federal airway, and that need may be filled by an additional VOR, DME, or other air navigation aid.

[(c) The standards in this subpart apply to the effect of construction or alteration proposals upon an airport if, at the time of filing of the notice required by § 77.13(a), that airport is—

[(1) Available for public use and is listed in the Airport Directory of the current Airman's Information Manual or in either the Alaska or Pacific Airman's Guide and Chart Supplement; or,

[(2) A planned or proposed airport or an airport under construction, that is the subject of a notice or proposal on file with the Federal Aviation Administration, and, except for military airports, it is clearly indicated that that airport will be available for public use; or,

[(3) An airport that is operated by an armed force of the United States.

§ 77.23 Standards for determining obstructions.

(a) An existing object, including a mobile object, is, and a future object would be, an obstruction to air navigation if it is of greater height than any of the following heights or surfaces:

(1) A height of 500 feet above ground level at the site of the construction or alteration.

(2) A height that is 200 feet above ground level or above the established airport elevation, whichever is higher, within three statute miles of the established reference point of an airport, excluding heliports, with its longest runway more than 3,200 feet in length, and that height increases in the proportion of 100 feet for each additional statute mile of distance from the airport up to a maximum of 500 feet.

(See Fig. 16–4.)

(3) A height that is 100 feet above ground level or 100 feet above the elevation of the approach end of the runway, whichever is higher, within an instrument approach area and within three statute miles of the runway end, and that height increases in the proportion of 25 feet for each additional statute mile of distance outward from the runway end up to a maximum of 250 feet and continuing at that height to a distance of ten statute miles from the runway end.

(See Fig. 16–5.)

(4) A height which would increase an instrument approach minimum flight altitude.

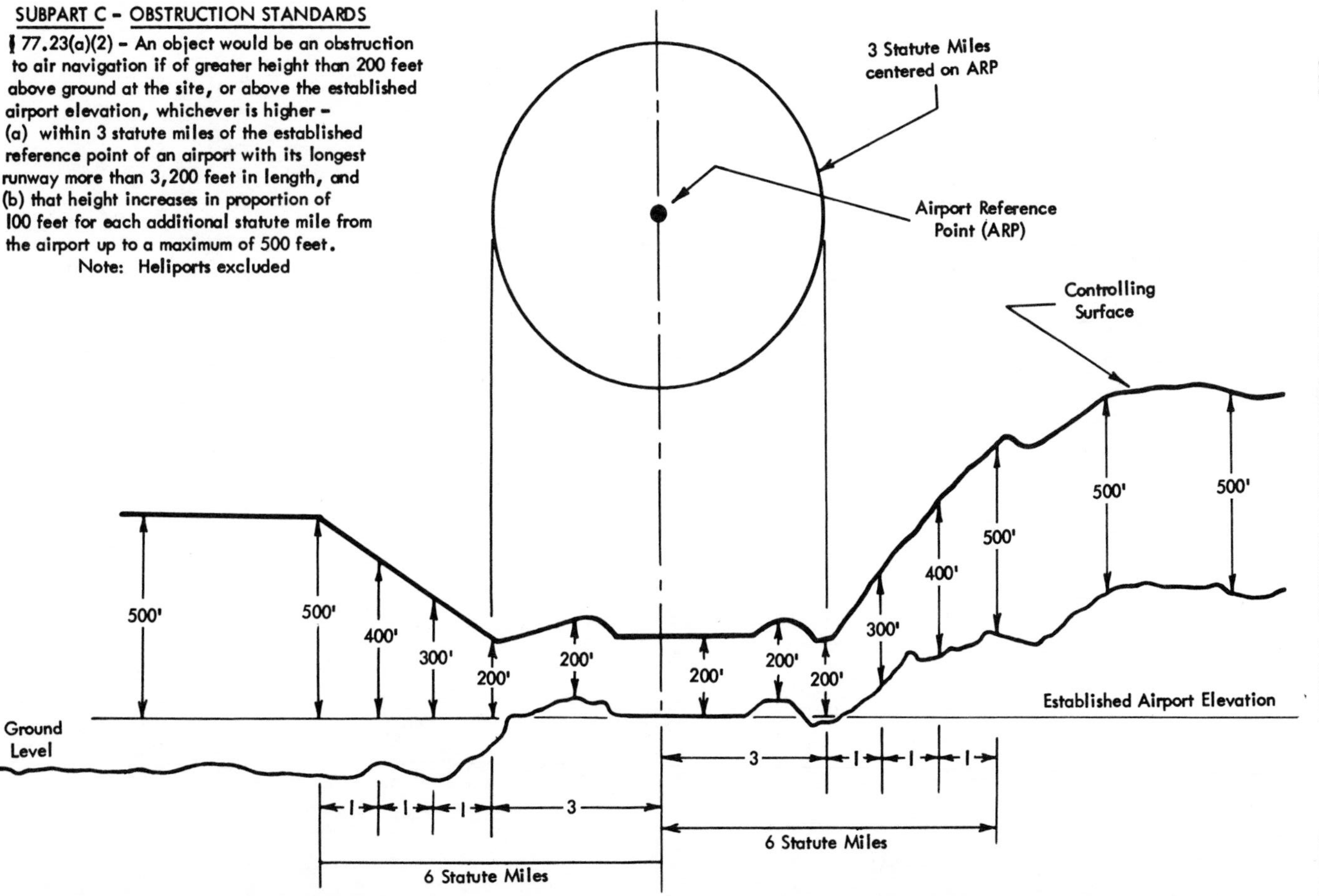

Fig. 16–4. Obstruction standards in the vicinity of airports. (Courtesy Federal Aviation Administration.)

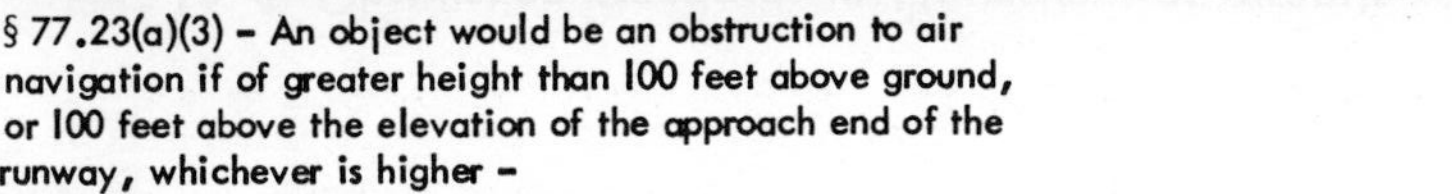
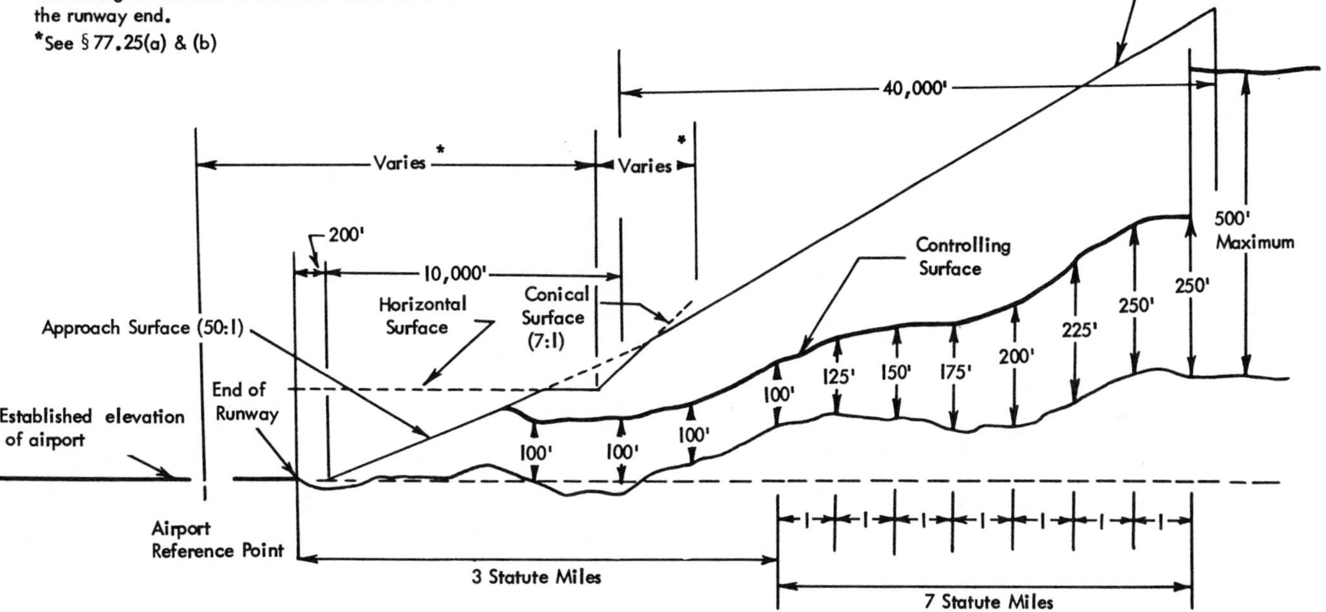

Fig. 16–5. Obstruction standards within instrument approach areas. (Courtesy Federal Aviation Administration.)

(5) A height in or under a Federal airway, transition area, or control zone, or within five statute miles of the course of an approved off-airway route, that is either 200 feet above ground level or 1,451 feet below the established minimum flight altitude, whichever is higher.

(6) An imaginary surface that begins at an altitude of 500 feet below the minimum en route altitude of each Federal airway or approved off-airway route and extends from the lateral boundaries of that airway and from a distance of five statute miles horizontally on both sides from the course of that route. For a distance of 25 statute miles along the airway or route from the nearest electronic air navigation aid upon which the airway or route is based, the imaginary surface extends outward and upward at a slope of 50 to 1 to five statute miles horizontal distance from the boundaries of each airway and ten statute miles horizontal distance on both sides from the course of each route. At greater distance than 25 statute miles along the airway or route from the nearest such aid, the imaginary surface begins at the same height and distance in relation to each airway and route but extends outward the five statute miles distance on a horizontal plane.

(7) An imaginary surface that begins at an altitude of 500 feet below the minimum altitude established for any initial approach, transition or procedure turn of any instrument approach procedure, or for any holding procedure, and extends outward and upward from the boundary of the area involved, including any buffer zone, at a slope of 50 to 1 for five statute miles horizontal distance.

[(8) The surface of a takeoff and landing area of an airport or any imaginary surface established under §77.25, §77.27, §77.28, or §77.29. However, no part of the takeoff and landing area itself will be considered an obstruction. Each airport imaginary surface that is established for a civil airport is based on runway lengths corrected in accordance with the current FAA airport design standards to no gradient and standard conditions of temperature and elevation.]

(b) Except for traverse ways on or near an airport with an operative ground traffic control service furnished by an air traffic control tower or by the airport management and coordinated with the air traffic control service, the standards of paragraph (a) of this section apply to traverse ways used or to be used for the passage of mobile objects only after the heights of those traverse ways are increased 17 feet for an interstate highway, 15 feet for any other highway, 25 feet for a railroad, and, for any other traverse way, an amount equal to the height of the highest unshielded mobile object that would normally traverse it.

§ 77.25 Civil airport imaginary surfaces related to airport reference points.

The following civil airport imaginary surfaces are established with relation to the airport reference point which is fixed at the approximate center of the airport takeoff and landing area and is given the established airport elevation. The size of each such surface is based on the corrected length of the longest runway of the airport. For the purposes of this Part, a runway is the area designated for the landing and takeoff of aircraft.

(a) Horizontal surface—a circular plane, 150 feet above the established airport elevation, with a radius from the airport reference point of:

[(1) 5,000 feet, for an airport with its longest runway no more than 3,200 feet in length and for all airports constructed to "VFR Airports" standards or designated as "Utility Airports."]

(2) 7,000 feet, for an airport with a runway more than 3,200, but not more than 6,000 feet in length.

(3) 11,500 feet, for an airport with a runway more than 6,000, but not more than 7,500 feet in length.

(4) 13,000 feet, for an airport with a runway more than 7,500 feet in length.

(b) Conical surface—a surface extending from the periphery of the horizontal surface outward and upward at a slope of 20 to 1 for the horizontal distances, and to the elevations, above the airport elevation, of:

[(1) 3,000 feet, to an elevation of 300 feet, for an airport with its longest runway no more than 3,200 feet in length and for all airports constructed to "VFR Airports" standards or designated as "Utility Airports."]

(2) 5,000 feet, to an elevation of 400 feet, for an airport with a runway more than 3,200, but not more than 6,000 feet in length.

(3) 7,000 feet, to an elevation of 500 feet, for an airport with a runway more than 6,000 feet in length.

(See Fig. 16–6.)

§ 77.27 Civil airport imaginary surfaces related to runways.

[The following civil airport imaginary surfaces are established for runways based upon their corrected lengths, whether the airport is constructed to "VFR Airports" standards or designated as a "Utility Airport," and whether the runway is an ILS runway, i.e., one equipped with a precision landing aid such as ILS, ground-controlled approach (GCA), or precision approach radar (PAR).

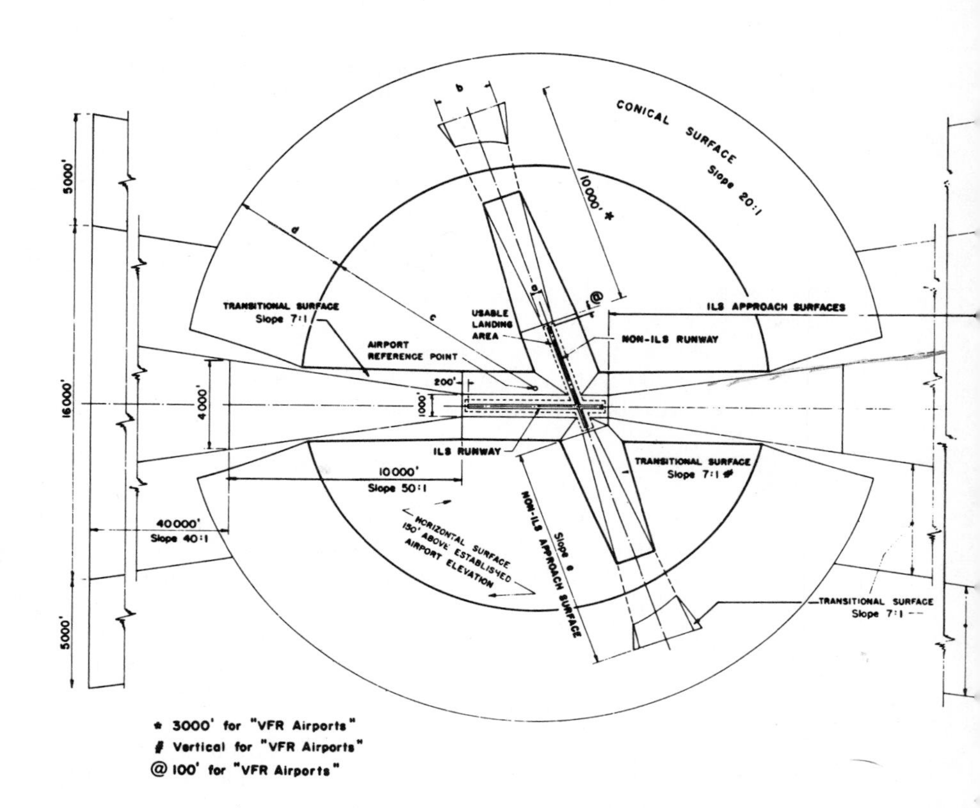

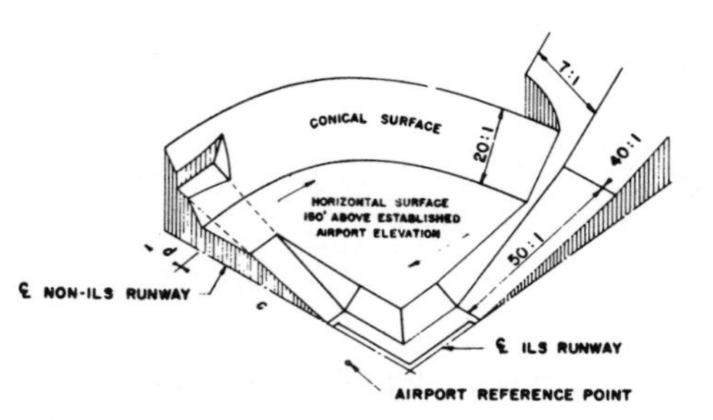

Longest runway	Distance in feet				Sl
	a	b	c	d	
VFR type	200	500	5000	3000'	2
3200' & less	250	2250	5000	3000	2
Over 3200' to 4200'	400	2400	7000	5000	4
Over 4200' to 6000'	500	2500	7000	5000	4
Over 6000' to 7500'	500	2500	11500	7000	4
Over 7500'	500	2500	13000	7000	4

Fig. 16–6. Civil airport imaginary surfaces related to airport reference points. (Courtesy Federal Aviation Administration.)

[(a) Primary surface—a surface longitudinally centered on a runway and extending in length 100 feet beyond each end of a runway of an airport constructed to "VFR Airports" standards and 200 feet beyond each end of a runway of any other airport, including runways of airports designated as "Utility Airports." The elevation of any point on the longitudinal profile of a primary surface, including the extensions, coincides with the elevation of the centerline of the runway, or extension, as appropriate. The width of a primary surface is 200 feet for runways of airports constructed to "VFR Airports" standards. For other airports the width is:

[(1) 250 feet, for non-ILS runways 3,200 feet or less in length, including runways of airports designated as "Utility Airports."]

(2) 400 feet, for non-ILS runways more than 3,200, but not more than 4,200 feet in length.

(3) 500 feet, for non-ILS runways more than 4,200 feet in length.

(4) 1,000 feet, for ILS runways.

(b) ILS approach surface—a surface longitudinally centered on the extended centerline of an ILS runway, beginning at each end of the primary surface and extending outward and upward at a slope of 50 to 1 for a horizontal distance of 10,000 feet and at a slope of 40 to 1 for an additional 40,000 feet. This surface is the width of the primary surface at the beginning and expands uniformly to a width of 16,000 feet at a distance of 50,000 feet from the end of the primary surface.

(c) Non-ILS approach surface—a surface longitudinally centered on the extended centerline of the runway, beginning at each end of the primary surface, with slopes and dimensions as follows:

(1) Airports constructed to "VFR Airports" standards—the surface begins 200 feet wide at each end of the primary surface and extends outward and upward at a slope of 20 to 1, expanding to a width of 500 feet at a horizontal distance of 3,000 feet.

(2) Airports not constructed to "VFR Airports" standards—

[(i) Runways of 3,200 feet or less in length, including runways of airports designated as "Utility Airports"—the surface begins 250 feet wide at each end of the primary surface and extends outward and upward at a slope of 20 to 1, expanding to a width of 2,250 feet at a horizontal distance of 10,000 feet.]

(ii) Runways more than 3,200, but not more than 4,200 feet in length—the surface begins 400 feet wide at each end of the primary surface and extends outward and upward at a slope of 40 to 1, expanding to a width of 2,400 feet at a horizontal distance of 10,000 feet.

(iii) Runways more than 4,200 feet in length—the surface begins 500 feet wide [at each end of the primary surface] and extends outward and upward at a slope of 40 to 1, expanding to a width of 2,500 feet at a horizontal distance of 10,000 feet.

(d) Transitional surfaces—these surfaces apply only at airports constructed to other than "VFR Airports" standards. They extend outward and upward at right angles to the runway centerline at a slope of 7 to 1 from the edges of the primary and the approach surfaces until they intersect the horizontal or conical surface, except that transitional surfaces for those portions of ILS approach surfaces that project through and beyond the limits of the conical surface, extend a distance of 5,000 feet measured horizontally from the edges of those portions of the approach surfaces and at right angles to the runway centerline.

(e) Vertical surfaces—these apply only at airports constructed to "VFR Airports" standards. They extend upward from the edges of the primary surfaces and the approach surfaces until they intersect with the horizontal surfaces.

§ 77.28 Military airport imaginary surfaces.

(a) *Related to airport reference points.* These surfaces apply to all military airports where the length of the longest runway is over 5,000 feet. At all other military airports, the appropriate provisions of § 77.25 apply.

(1) *Inner horizontal surface*—a plane is oval in shape at a height of 150 feet above the established airfield elevation. The plane is constructed by scribing an arc with a radius of 7,500 feet about the centerline at the end of each runway and interconnecting these arcs with tangents.

(2) *Conical surface*—a surface extending from the periphery of the inner horizontal surface outward and upward at a slope of 20 to 1 for a horizontal distance of 7,000 feet to a height of 500 feet above the established airfield elevation.

(3) *Outer horizontal surface*—a plane, located 500 feet above the established airfield elevation, extending outward from the outer periphery of the conical surface for a horizontal distance of 30,000 feet.

(b) *Related to runways.* These surfaces apply to all military airports where the length of the longest runway is over 5,000 feet. At all other military airports, the appropriate provisions of § 77.27 apply.

(1) *Primary surface*—a surface located on the ground or water longitudinally centered on each runway with the same length as the runway. The width of the primary surface for runways longer than 5,000 feet is 2,000 feet. However, at established bases where substantial construction has taken

place in accordance with a previous lateral clearance criteria, the 2,000-foot width may be reduced to the former criteria.

(2) *Clear zone surface*—a surface located on the ground or water at each end of the primary surface, with a length of 1,000 feet and the same width as the primary surface.

(3) *Approach clearance surface*—an inclined plane, symmetrical about the runway centerline extended, beginning 200 feet beyond each end of the primary surface at the centerline elevation of the runway end and extending for 50,000 feet. The slope of the approach clearance surface is 50 to 1 along the runway centerline extended until it reaches an elevation of 500 feet above the established airport elevation. It then continues horizontally at this elevation to a point 50,000 feet from the point of beginning. The width of this surface at the runway end is the same as the primary surface, it flares uniformly, and the width at 50,000 is 16,000 feet.

(4) *Transitional surfaces*—these surfaces connect the primary surfaces, the first 200 feet of the clear zone surfaces, and the approach clearance surfaces to the inner horizontal surface, conical surface, outer horizontal surface or other transitional surfaces. The slope of the transitional surface is 7 to 1 outward and upward at right angles to the runway centerline.

(See Fig. 16–7.)

§ 77.29 Airport imaginary surfaces for heliports.

(a) *Heliport primary surface.* The area of the primary surface coincides in size and shape with the designated takeoff and landing area of a heliport. This surface is a horizontal plane at the elevation of the established heliport elevation.

(b) *Heliport approach surface.* The approach surface begins at each end of the heliport primary surface, with the same width as the primary surface, and extends outward and upward at a slope of 8 to 1 to the minimum en route elevation where its width is 500 feet.

(c) *Heliport transitional surfaces.* These surfaces extend outward and upward from the lateral boundaries of the heliport primary surface and from the approach surfaces at a slope of 2 to 1 for a distance of 250 feet measured horizontally from the centerline of the primary and approach surfaces.

(See Fig. 16–8.)

16–8. Runway Configurations. Inherent in the layout and design of an airport is the need to arrange a given runway efficiently in relation to other runways and service facilities such as the terminal building, aprons, hangers, and other airport buildings. The importance of efficient runway configuration lies in its controlling influence on runway capacity.

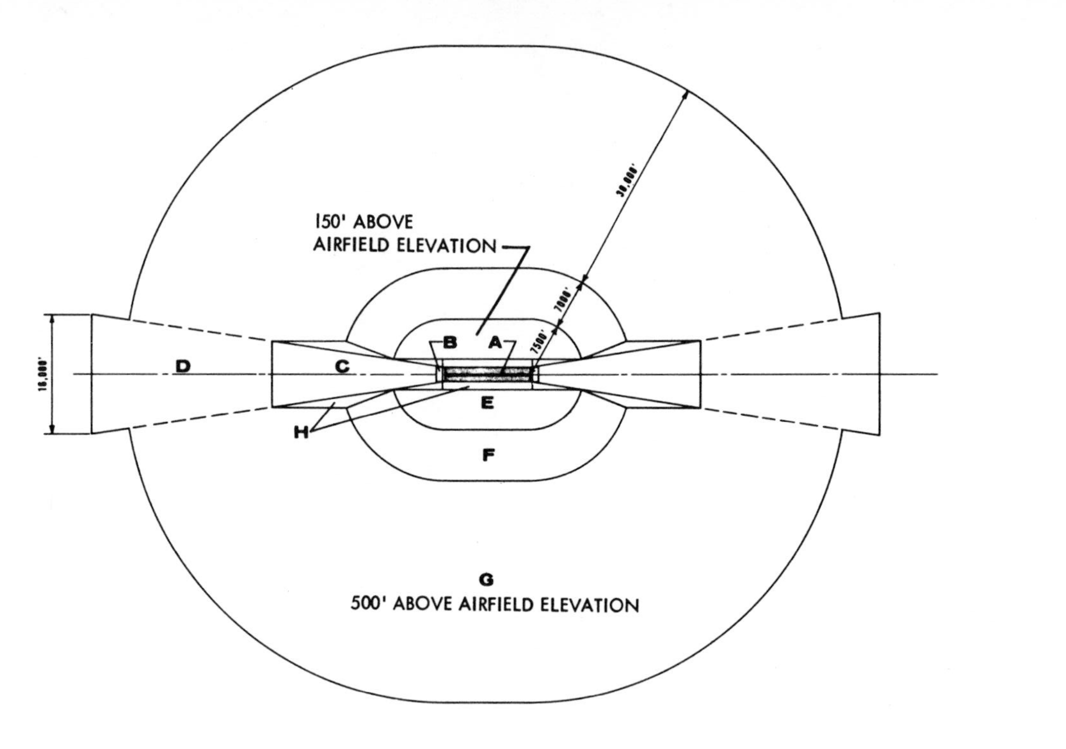

LEGEND

A Primary Surface
B Clear Zone Surface
C Approach–Departure Clearance Surface (Glide Angle)
D Approach–Departure Clearance Surface (Horizontal)
E Inner Horizontal Surface
F Conical Surface
G Outer Horizontal Surface
H Transitional Surface

Note: These surfaces apply to all military airports when the length of the longest runway is over 5,000 feet. At all other military airports, the appropriate provisions of § 77.25 and § 77.27 apply.

Fig. 16–7. Military airport imaginary surfaces. (Courtesy Federal Aviation Administration.)

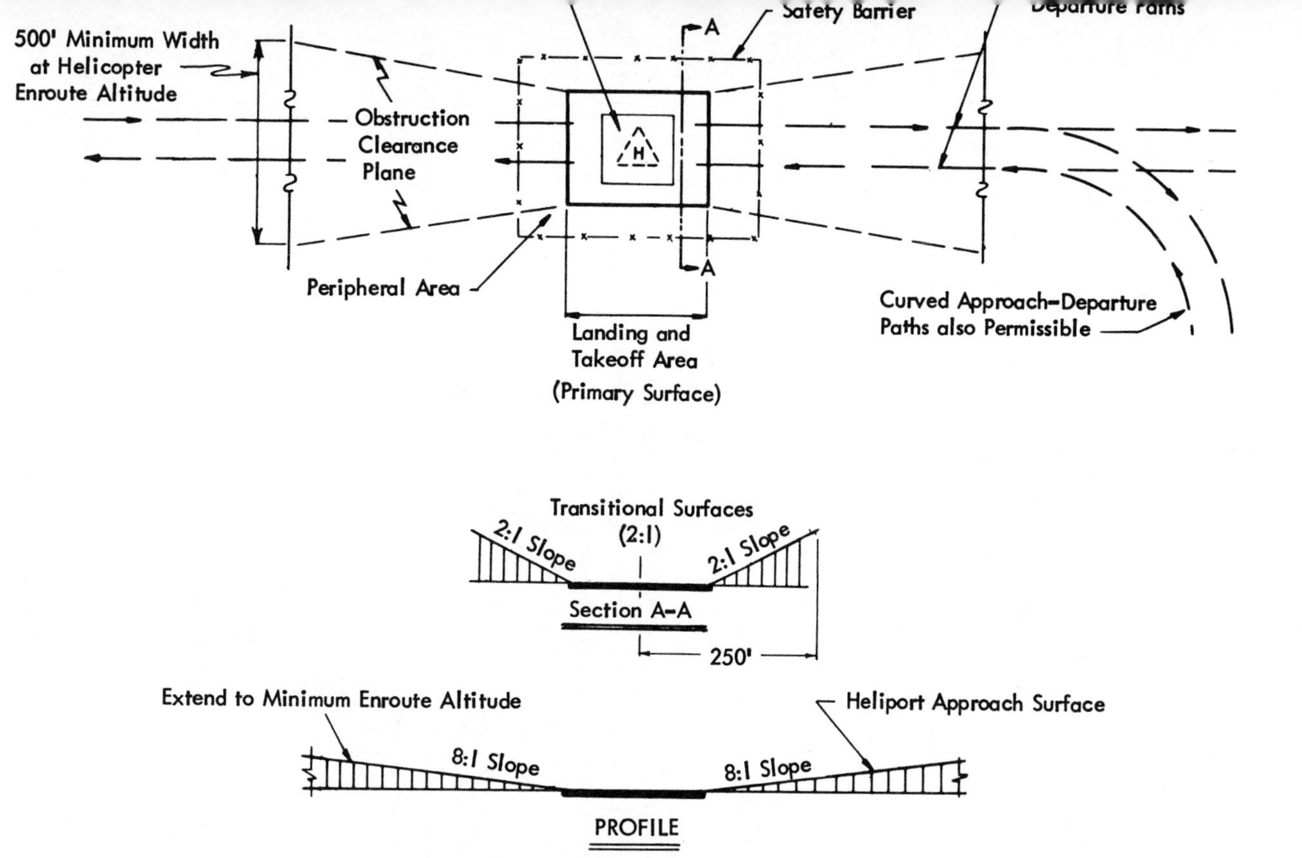

Fig. 16–8. Airport imaginary surfaces for heliports. (Courtesy Federal Aviation Administration.)

The capacity of a runway system is defined as the maximum number of operations (i.e., landings or take-offs) per hour that the system can reasonably be expected to accommodate. The ability of a runway to accommodate aircraft is customarily described in terms of "practical" capacities, which are based upon an assumption of a "reasonable" or "acceptable" average level of delay. Thus, during extremely busy periods, a runway system may accommodate a larger number of aircraft per time period than its practical capacity, but during these hopefully rare periods, unusually large delays will occur.

The capacity of a runway system depends to a considerable extent on the types and relative percentages of various groups of aircraft (termed aircraft *mix*). Generally, the greater the percentage of large aircraft, the smaller the runway capacity.

Runway capacity also varies widely with weather conditions, and it is necessary to distinguish between operations in VFR and IFR conditions. Landings and take-offs at an airport may be made under visual flight rules (VFR) or instrument flight rules (IFR). It will be recalled from Section 5–25 that VFR operations are made in good weather conditions and the aircraft is operated by visual reference to the ground. IFR operations are made in periods of inclement weather and poor visibility, and under these conditions, positive traffic control is maintained by radar and other electronic devices. Runway capacity under IFR conditions is normally less than under VFR conditions.

The FAA [12] has published approximate practical hourly and practical annual capacities for various runway configurations. Some relatively simple runway configurations are shown by Table 16–2 along with practical annual capacities (PANCAP) and practical hourly capacities. The following aircraft mixes relate to the four numbers shown for each configuration given in Table 16–2:

Number	*%A*	*%B*	*%C*	*%D*
1	0	0	10	90
2	0	30	30	40
3	20	40	20	20
4	60	20	20	0

Categories of aircraft for determination of airport capacities are generally as follows:

Type A—4-engine jet and larger
Type B—2- and 3-engine jet, 4-engine piston, and turbo prop
Type C—executive jet and transport type twin-engine piston
Type D—light twin-engine piston and single-engine piston.

It should be noted that the data given by Table 16–2 are based on average traffic and airport conditions. Procedures for making adjust-

TABLE 16-2

Airport Capacities for Long-Range Planning Purpose (Courtesy Federal Aviation Administration)

Runway Configuration				Practical Hourly Cap.	
Layout	Description	Mix	Pancap	IFR	VFR
A 	Single Runway (arrivals = departures)	1	215,000	53	99
		2	195,000	52	76
		3	180,000	44	54
		4	170,000	42	45
B 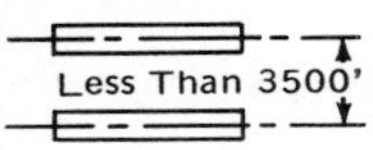 (Less Than 3500')	Close parallels (IFR dependent)	1	385,000	64	198
		2	330,000	63	152
		3	295,000	55	108
		4	280,000	54	90
C 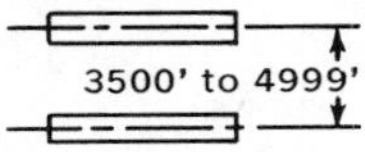 (3500' to 4999')	Independent IFR approach/departure parallels	1	425,000	79	198
		2	390,000	79	152
		3	355,000	79	108
		4	330,000	74	90
D 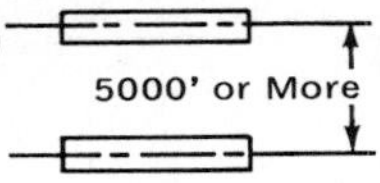 (5000' or More)	Independent IFR arrivals and departures	1	430,000	106	198
		2	390,000	104	152
		3	360,000	88	108
		4	340,000	84	90
K_1 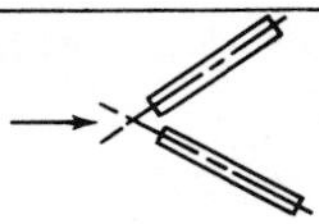	Open V, dependent, operations away from intersection	1	420,000	71	198
		2	335,000	70	136
		3	300,000	63	94
		4	295,000	60	84
K_2 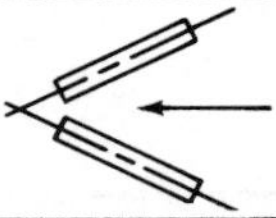	Open V, dependent, operations toward intersection	1	235,000	57	108
		2	220,000	56	86
		3	215,000	50	66
		4	200,000	50	53
L_1 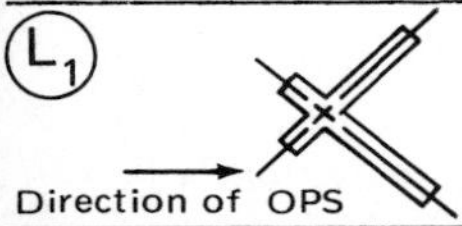 (Direction of OPS)	Two intersecting at near threshold	1	375,000	71	175
		2	310,000	70	125
		3	275,000	63	83
		4	255,000	60	69
L_3 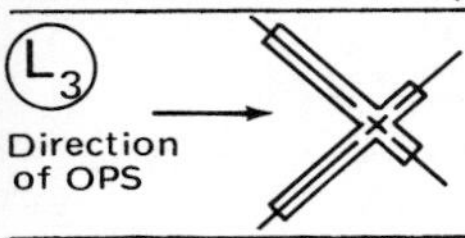 (Direction of OPS)	Two intersecting at far threshold	1	220,000	55	99*
		2	195,000	54	76*
		3	180,000	46	54
		4	175,000	42*	57
J 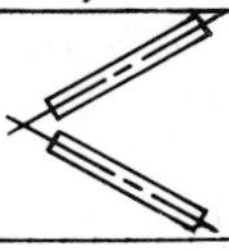	Widely Spaced Open V with independent operations	1	425,000	79	198*
		2	340,000	79	136
		3	310,000	76	94
		4	310,000	74	84
O_1 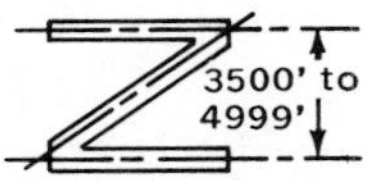 (3500' to 4999')	"Z" configuration and parallel with both intersecting	1	465,000	87	217
		2	430,000	87	167
		3	390,000	87	118
		4	365,000	81	99

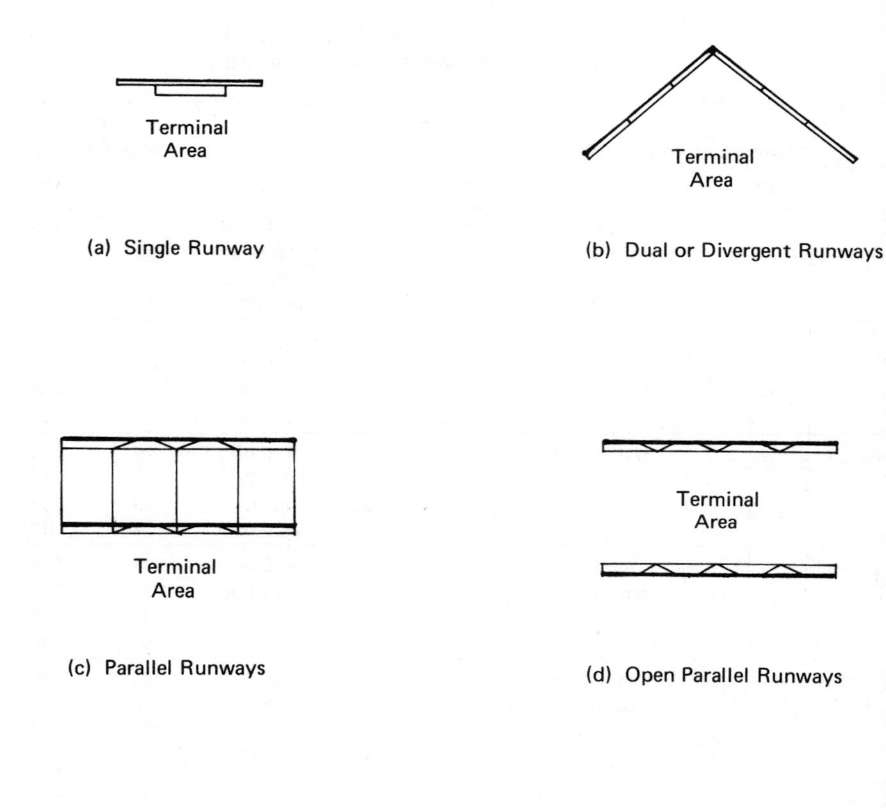

(a) Single Runway

(b) Dual or Divergent Runways

(c) Parallel Runways

(d) Open Parallel Runways

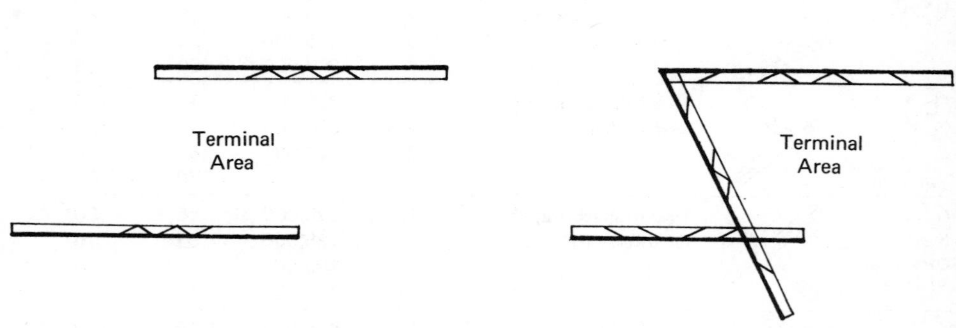

(e) Tandem Parallel Runways

(f) Three Runway System

Fig. 16–9. Typical airport configurations.

ments to the capacity values for local variations in traffic, airport layout, and weather are given by Reference 12. A more detailed approach to the estimation of airport capacities is given by FAA Advisory Circular AC 150/5060-1A [13], which is recommended for use in short range planning.

Typical runway configurations are shown schematically by Fig. 16–9. The simplest runway configuration is a single runway system, which has a practical capacity of 45 to 99 operations per hour in VFR conditions and 42 to 53 operations per hour under IFR conditions.

Frequently, a second runway is added to take advantage of a wider range of wind direction. This system has a higher capacity than a single runway system provided winds are not strong. In conditions of high winds and poor visibility, this system operates as a single runway.

An airport's runway capacity can be increased by adding a second parallel runway. With a separation of about 700 feet, simultaneous landings in the same direction[4] may be made on the parallel runways under VFR conditions with a resulting capacity of 90 to 198 operations per hour. With a runway separation of at least 3,500 feet, simultaneous operation may be made under IFR conditions, and the IFR capacity of this system is 74 to 106 operations per hour.

The parallel runway system shown in Fig. 16–9 (c) has the disadvantage of requiring aircraft using the outboard runway to taxi across the runway adjacent to the terminal area. This disadvantage may be overcome by placing the terminal facilities between the two runways as shown by Fig. 16–9 (d).

In a location where there are prevailing winds from one direction a large percentage of the time, the parallel runways may be staggered or placed in tandem as shown by Fig. 16–9(e). This makes it possible to reduce taxiway distances by using one runway exclusively for take-off operations and the other runway for landings. This configuration, however, requires a great deal of land.

Large airports may require three or more runways. The best configuration for a multiple runway system will depend on the minimum spacing required for safety, prevailing wind directions, topographic features of the airport site, shape and amount of available space, and the space requirements for aprons, the terminal and other buildings.

PROBLEMS

1. Using the wind data given below, construct a wind rose and indicate what would be the best orientation for a runway based on these prevailing winds.

[4] Simultaneous VFR operations in opposite directions on parallel runways require 1400 foot distance between runway centerlines during daylight hours and 2800 foot distance during periods of darkness [14].

Wind Direction	*Percentage of Winds* 15–31 mph	31–47 mph
N	2.0	—
N NE	0.2	0.2
NE	1.0	0.5
E NE	4.3	0.1
E	1.3	—
E SE	1.3	0.3
SE	0.7	0.4
S SE	3.1	0.1
S	5.4	0.2
S SW	4.2	0.5
SW	1.3	—
W SW	0.4	0.1
W	3.1	0.5
W NW	1.8	0.7
NW	0.6	0.4
N NW	2.5	0.8
%Winds 4–15 mph ≐ 58.2		
% Winds 0–4 mph = 3.8		

2. Obtain for an airport near you wind data similar to these given in Problem 1. Construct a wind rose and determine optimum runway orientation based on prevailing winds. How does this orientation compare with existing airport's runway(s)?

3. A general aviation airport is being planned to serve a city of 35,000 people. The non-instrument runway is to be 3,600 feet long. Indicate whether the following objects will be considered obstructions to air navigation by the FAA.

a. A 220-foot radio tower which is not in the landing approach, located 3.5 miles from the airport reference point. The ground elevation at the tower is 25 feet higher than the established airport elevation.

b. A planned 75-foot high office building within the landing approach one-half mile from the end of the runway.

4. A 5,500 foot military runway is to be constructed parallel to and two miles away from a hilly ridge which is 300 feet above the airfield elevation. Does this ridge constitute a hazard to navigation?

5. A heliport is to be located atop a 30-story building. A 38-story building which is located one-quarter of a mile from the heliport lies in the approach path. The top of the latter building extends 80 feet above the top of the touchdown area. Does the 38 story building constitute a hazard to air navigation?

REFERENCES

1. Josephy, Alvin M., Jr., (ed.), *The American Heritage History of Flight,* American Heritage Publishing Co., (1962).
2. Schriever, B. A., and Seifert, W. W., *Air Transportation 1975 and Beyond,* M.I.T. Press, Cambridge, Mass. (1968).

3. *Public Law 91-258, Airport and Airway Development Act of 1970.*
4. *Airport Site Selection,* Department of Transportation, Federal Aviation Administration, AC 150/5060-2, July 19, 1967.
5. Deem, Warren H., and Reed, John S., *Airport Land Needs,* Arthur D. Little, Inc., Communication Service Corporation, Washington, D.C.
6. Pikarsky, Melton, and Corey, John B. W., "An Airport in Lake Michigan for Chicago?" *Civil Engineering,* Vol. 38, No. 9, September, 1968.
7. *Initial Site Selection for the Second Atlanta Air Carrier Airport,* December, 1968, R. Dixon Speas Associates, Manhassett, New York.
8. Fox, Francis T., "The Satellite Airport System and the Community," *Civil Engineering,* Vol. 38, No. 7, July, 1968.
9. "Jet Airport in Boston Harbor?" *Civil Engineering,* Vol. 38, No. 9, September, 1968.
10. *Airport Design,* Airport Engineering Branch, FAA, 1961.
11. *Federal Aviation Regulations, Part 77, Objects Affecting Navigable Airspace,* including changes 1–7, February 1, 1969.
12. FAA Advisory Circular AC 150/5060-3A, "Airport Capacity Criteria Used in Long-Range Planning," December 24, 1969.
13. FAA Advisory Circular AC 150/5060-1A, "Airport Capacity Criteria Used in Preparing the National Airport Plan," July 8, 1968.
14. FAA Advisory Circular AC 150/5330-2A, "Runway/Taxiway Widths and Clearances for Airline Airports," July 25, 1968.

17

The Airport Terminal Area

17–1. Introduction. This chapter will present planning procedures and design criteria for the airport terminal area. For purposes of this chapter, the airport terminal area generally consists of that portion of the airport other than the landing and take-off areas. It includes the automobile parking lots, aircraft parking aprons, the terminal building, and facilities for interterminal and intraterminal transportation.

The importance of a well-conceived airport terminal area design can be seen by considering the numerous and varied component movements that a typical airline passenger makes (see Fig. 17–1). A passenger leaves his home or rented room and travels to the airport by automobile or one of a variety of public travel modes. From the automobile parking lot or vehicle-unloading platform, the passenger and his baggage move to the ticket and passenger service counter by walking, moving sidewalks, or other means. From the gate-loading position, passengers usually walk the short distance to the plane. Additional travel time is involved as the aircraft taxis to the runway holding apron where it waits for control tower clearance to take off. This procedure is essentially reversed at the destination end of the trip. For the relatively small percentage of airline passengers who make enroute stops or transfers, the movements are more numerous and complex.

Each component movement in a typical trip involves possibilities of congestion and delay. It follows that each service facility within the terminal area must be carefully planned and designed to accommodate peak-hour traffic volumes if unacceptable delays are to be avoided. It is also apparent that a balanced, integrated layout and design of the airport terminal area is required to provide a smooth uninterrupted flow of people, baggage, and freight. This design must be sufficiently flexible to allow for oderly expansion of service areas without prohibitive costs.

Terminal facilities vary widely in size, design, and layout depending primarily upon the airport type and size and the volume of air traffic.

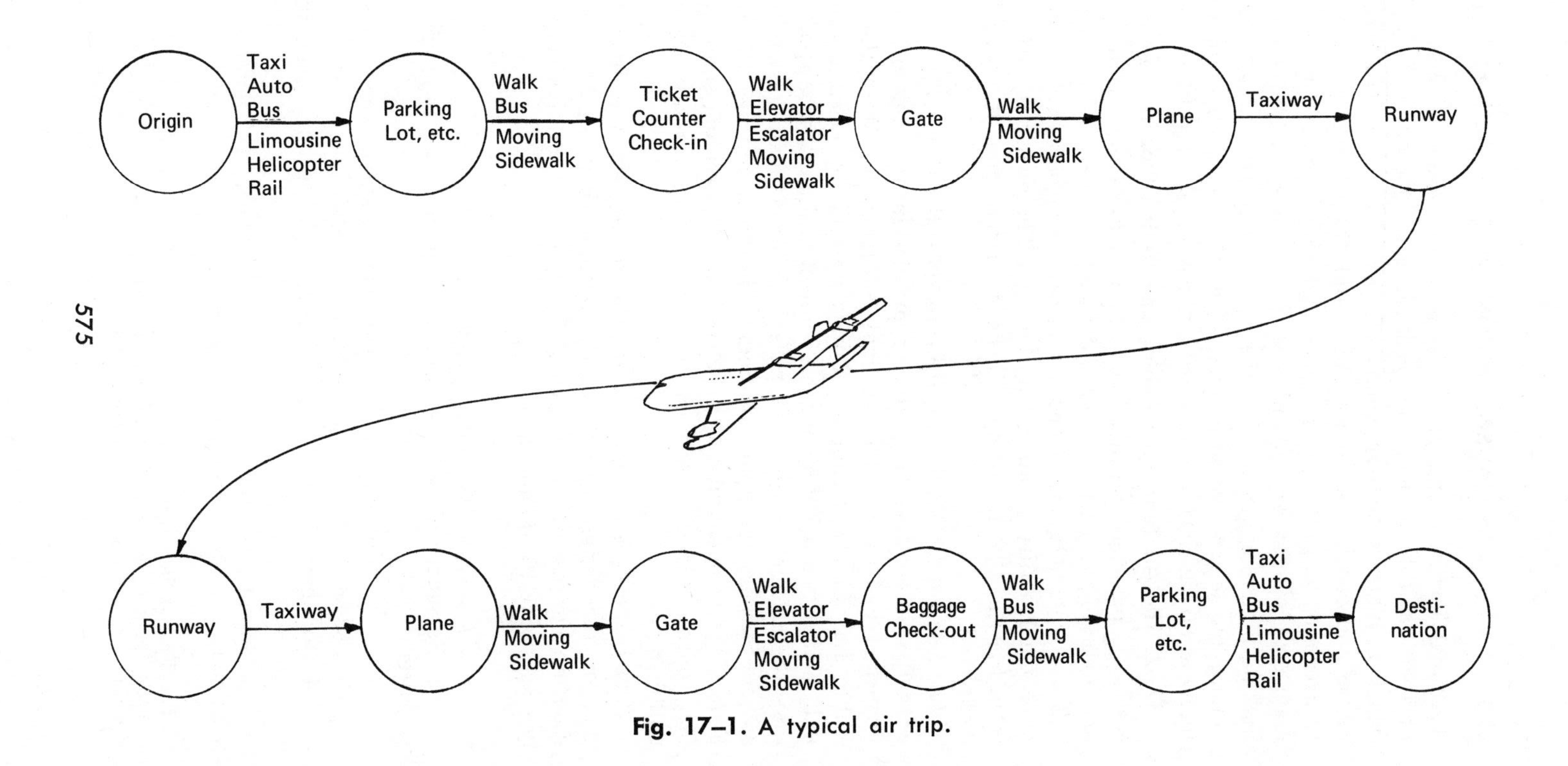

Fig. 17–1. A typical air trip.

Terminal facilities at a utility airport,[1] for example, may consist only of two small buildings: (1) a fixed-based operators' (FBO) building which provides space for commercial activities such as aircraft maintenance and repair, air charter, etc., and (2) an administration building to accommodate pilots, passengers, and visitors and to house the airport manager's office. At the other end of the spectrum, there are cities making plans for airport facilities to accommodate more than 20 million annual passengers by providing elaborate terminal complexes with as many as 80 parking positions for commercial jet aircraft.

17–2. The Air Terminal Planning and Design Process. The design process for a passenger terminal at a commercial carrier airport is a complex one involving at least four organizations:

The Airport Owner. Commercial carrier airports are typically owned by a municipality or airport authority. The owner is particularly concerned with the financing of the terminal and its operation when completed.

The Federal Government. Prior to 1961, the Federal government took an active interest in the planning and design of terminal buildings and furnished Federal aid for terminal construction. With the passage of Public Law 87-255, terminal buildings were no longer eligible for Federal aid. There is a belief among government and industry airport specialists that the Congress will soon pass legislation again providing Federal aid for airport terminal buildings. In the meantime, the role of the Federal government in this area consists of providing sight line criteria for control tower personnel and requesting aircraft clearances to buildings.

The Airlines. As prime tenants, the airlines exert an important influence on the design features of the terminal building and its environs. Each airline provides estimates of the following terminal needs, usually in stages of 5, 10, 15, and 20 years:

1. Estimate of enplaning traffic
2. Passenger aircraft scheduled departures (by type and weight of aircraft)
3. Space requirements for ticket counters, baggage claims and services, operations, and maintenance and supply
4. Number, type, and size of aircraft parking positions at the terminal
5. Special requirements such as telephones, TV flight information, pneumatic tubes, and passenger transportation systems

[1] The FAA defines a utility airport as one which serves general aviation aircraft of 12,500 pounds or less. Information and criteria for planning and development of utility airports is given by Reference 1.

6. Building and space requirements for air freight
7. Maintenance facilities
8. Fixed ramp facilities such as heated air and air conditioning for aircraft, water, and electric power
9. Fuel requirements
10. Automobile parking requirements for employees

Concessionaires. The airport terminal building is a commercial venture of considerable magnitude and consumer services and rentals may produce over half of the terminal revenues. Although most of the concessions are leased after the building is designed, design information on the size and location of restaurants, shops, and other tenants is required.

Consulting architectural and engineering firms are usually employed by the airport owner to develop a terminal design to satisfy the varied requirements and wishes of the airport management, the Federal government, the airlines, and other tenants of the airport.

17–3. Air Terminal Layout Concepts. Foremost in the air terminal designer's mind are two objectives: (1) to minimize walking distances from the automobile parking lots to the terminal and from the terminal to the aircraft, and (2) to minimize passenger, aircraft, and automobile congestion.

Increasing space needs for both automobiles and aircraft have taxed the designer's ingenuity in recent years. This has been especially true, of course, at large airports. The predominant preference of passengers, visitors, and well-wishers to travel to the airport by automobile largely explains the growing demand for automobile parking spaces. The use of larger aircraft has complicated the designer's task in at least three ways. In the first place, parking and service area required for these aircraft has increased making it more difficult to park large numbers of these planes within a reasonable walking distance from the terminal building. Secondly, the sudden deposition of "batches" of several hundred passengers results in surges of terminal passenger traffic, and space and logistic needs for these peak flows must be provided. Finally, the taller aircraft with higher tails and greater floor heights require higher vertical clearances and more flexibility in the heights of loading platforms and bridges.

To satisfy the objective of minimizing passenger walking distances, various physical layout schemes have evolved over the years. When this has failed, designers have turned to the use of moving belts and special vehicles to provide transportation within the terminal area.

17–4. Air Terminal Layout Schemes. Four air terminal layout schemes that have been used are shown in Fig. 17–2. The oldest and simplest

(a) Frontal System

Terminal Building

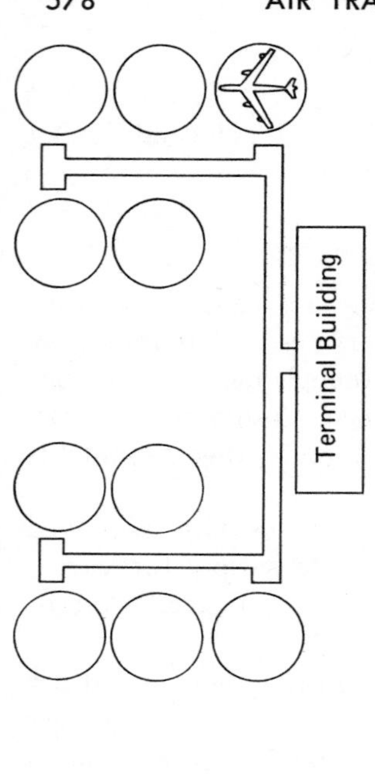

(b) Terminal With Fingers

(c) Terminal With Satellite Structures

Terminal Building

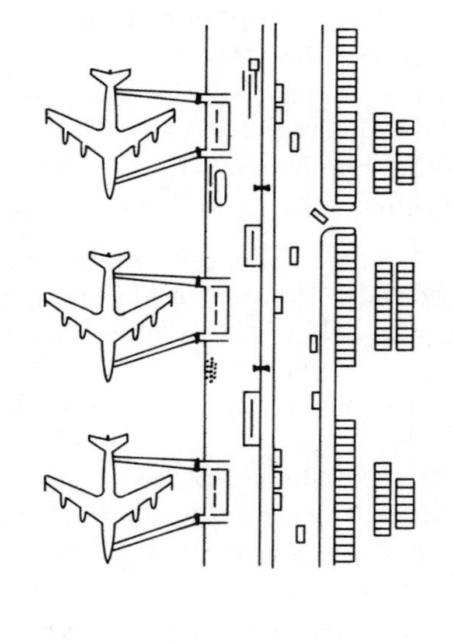

(d) Gate Arrival Scheme

Fig. 17–2. Terminal layout schemes.

concept is the *frontal system,* shown in Fig. 17–2(a), in which the aircraft park parallel to the terminal building. This system is adequate for small airports where the number of aircraft served is small and where there are few or no flights by commercial air carriers.

A second layout scheme, shown by Fig. 17–2(b), involves the use of concourses or *fingers* which extend from the terminal building to the aircraft parking areas. These fingers may simply be fenced walkways but commonly are enclosed structures which are heated and air conditioned throughout. Enclosed concourses protect passengers from the elements, aircraft noise, and propeller and jet blast.

In the mid-1950's, airlines began to process the passengers at the gate by accepting their tickets at a "holding area" within the terminal concourse structure. Later, when it became the practice to increase these areas in size and to provide comfortable furnishings, they came to be known as "departure lounges."

Commenting on the departure lounge loading concept, Thompson [2] states: "Aside from the increase in the number of aircraft gates required for the increased schedules, no other single development had such an impact on terminal design (and, incidentally, cost) as this did."

A logical extension of the departure lounge concept has been the recent development of *satellite* enplaning structures, sometimes called flight stations. (See Fig. 17–2(c).) These satellite structures are self-contained areas with departure lounges, toilets, concessions, and limited food and beverage areas. These structures typically serve 5 to 10 loading gates and are connected to the terminal building with a concourse not more than 500 feet in length. A sketch of a satellite enplaning structure is shown by Fig. 17–3.

Figure 17–2(d) shows the *gate arrival scheme* proposed for the Kansas City Midcontinent International Airport [3]. This concept employs a very long terminal building which essentially is a concourse or finger with aircraft loading positions on one side and the terminal roadway, curb, and public entrances on the other. Because of its length, special ground transportation facilities must be provided for passengers moving from one point in the terminal to another. Kansas City proposes the use of shuttle buses for this purpose.

With the gate arrival scheme, convenient automobile parking is provided and abundant curb lengths are available for discharging and picking up passengers. The principal disadvantage of the gate arrival scheme is that it causes some loss of cross-utilization of space and facilities. That is, more rest rooms, concessions, etc., would be required than in a centralized terminal design.

To accommodate growing air traffic volumes, several large airports are turning to the *unit terminal approach* in which each airline has

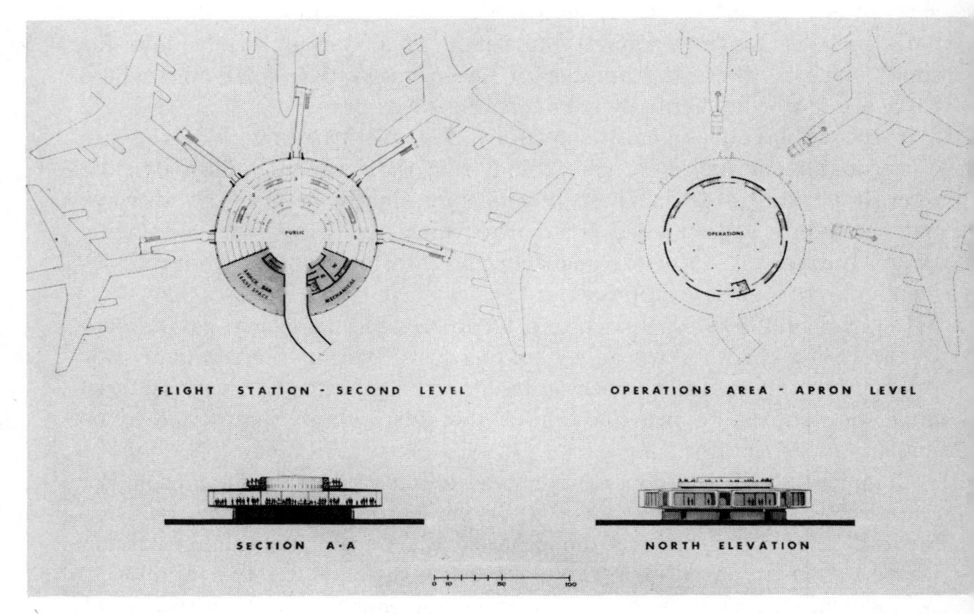

Fig. 17–3. A satellite enplaning structure. (Courtesy City of Houston.)

its own separate terminal building. Four unit terminals are proposed for the Houston Intercontinental Airport [4], for example, each of which will be constructed with four satellite enplaning structures. Each satellite structure will house five gate positions. (See Fig. 17–4.)

Four unit terminals are planned for the Kansas City MCI airport, each employing the gate arrival terminal layout scheme. Each terminal building will have fifteen aircraft loading positions, a coffee shop and cocktail lounge, barber shop, five rest rooms, and two standup snack bars. Three gound transportation centers will be provided in each building. A layout plan for the Kansas City airport is shown by Fig. 17–5.

17–5. Intraterminal and Interterminal Transportation. Despite efforts to minimize walking distance by thoughtful air terminal layout and design, distance remains a problem at existing airports. A study [5] of nine major airports[2] revealed that for originating passengers the average maximum walking distance to the nearest gate was 565 feet, while the minimum walking distance to the farthest gate was 1,342 feet. Passengers who transferred from one airline to another were confronted with an average walking distance of 4,091 feet.

[2] Chicago (O'Hare), New York (John F. Kennedy), Los Angeles International, Atlanta, San Francisco International, Dallas, Miami International, Philadelphia International, and Detroit Metropolitan Airports.

Fig. 17–4. The proposed Houston Intercontinental Airport with four unit terminals, each with four satellite flight stations. (Courtesy City of Houston.)

To overcome the problem of excessive walking distances, two basic approaches have been used: (1) moving sidewalks, and (2) vehicle systems.

Moving sidewalks have been used at the Dallas, Geneva, and Amsterdam airports, and their use is proposed for other airports. More extensive use is also planned at Dallas.

Moving sidewalks are generally limited to distances of not more than 900 feet [6]. Their principal usefulness, therefore, lies in closing the walking distance between ticket counters and aircraft loading areas.

There are two types of moving sidewalks: (1) belt conveyors, and (2) horizontal escalators using pallets similar to the treads used on escalators. To prevent mishaps, hand rails must be provided which move at the same speed of the sidewalk.

A wide variety of vehicular systems have been developed or proposed for transportation within the terminal area. Conventional buses are used at Atlanta to transport passengers from a remote parking lot. Specially designed airport buses are used in Tokyo and at several European airports to transport passengers from the terminal to the aircraft.

Perhaps the most dramatic example of special vehicles for interterminal transportation is the mobile lounge which was developed for the

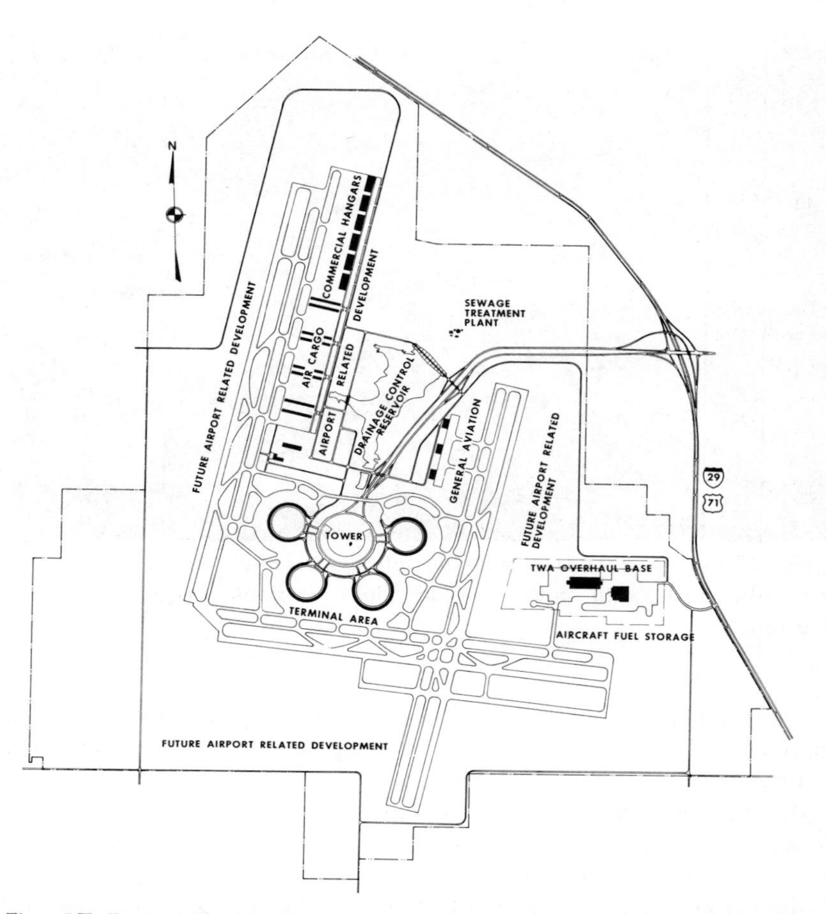

Fig. 17–5. Unit terminals at Kansas City utilizing "gate arrival" air Company.)

Dulles International Airport in Washington, D.C. (See Fig. 17–6.) Mobile lounges are essentially holding areas on wheels which transport passengers from the terminal building to airplanes. These vehicles are about 50 feet long, 15 feet wide, and have a seating capacity of 60 people. They can be driven from either end and cost about $235,000 each.

A similar vehicle has been designed by the Budd Company called the Plane-Mate. The body of this vehicle can be raised or lowered to match with passenger loadings ramps and aircraft floors of various heights.

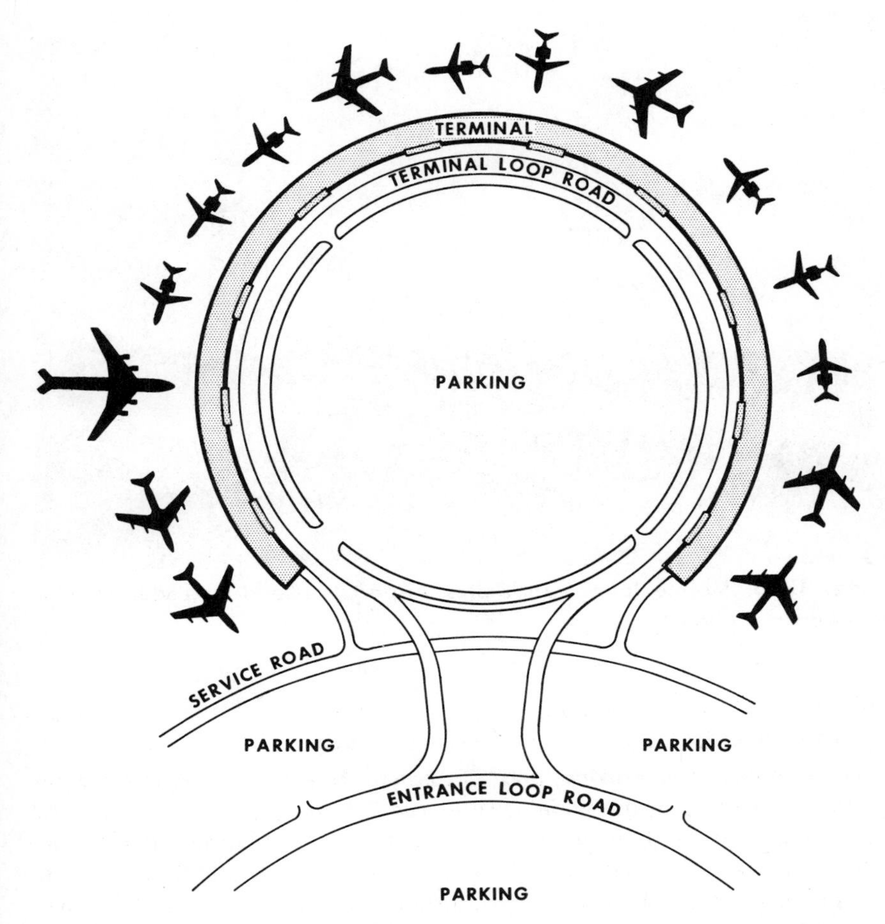

terminal layout scheme. (Courtesy, Burns and McDonnell Engineering

A monorail is planned for Dallas' Love Field to transport passengers from a remote parking lot.

For the Tampa Airport, the Westinghouse Electric Corporation has designed a fully automatic system for shuttling passengers approximately 1,000 feet between the land-side terminal building and four satellite enplaning structures. The system utilizes two rubber-tired vehicles for each enplaning structure, each capable of transporting about 100 standing passengers. (See Fig. 17–7.)

The unit terminals at Houston shown by Fig. 17–4 will be connected by an underground battery-operated, electronically guided train. There

Fig. 17–6. Mobile lounge at Dulles Airport. (Courtesy Federal Aviation Administration.)

will be four trains consisting of the power unit and three passenger cars capable of transporting eight passengers each.

17–6. Automotive Parking and Circulation Needs. Increasing airline activity and the growing popularity of the automobile have combined to create unprecedented demand for parking and circulation facilities at most airports. In small and medium-sized cities, more than half of the air passengers travel to and from the airport by passenger car. Although this percentage may be only 20 to 25 per cent for major airports, the total volume of passenger car traffic to these airports taxes the ingenuity of the airport designer in his efforts to provide adequate and convenient parking and circulation needs.

Ideally, the airline passenger, who is usually carrying baggage, should be provided a parking space within 300 to 400 feet of the terminal building. A maximum walking distance from parking lot to the terminal building of 1,000 feet is recommended by the FAA [7].

The classic parking plan has been a ground-level parking lot adjacent to the airport passenger terminal. At large airports, designers have found it impossible to provide sufficient parking on a single level, and the trend is toward the provision of multilevel parking structures. The design of these structures is similar to the design of a downtown parking garage (described in Chapter 13), the principal differences being the need

Fig. 17–7. An automatic system for transporting passengers to and from satellite enplaning structures at Tampa Airport. (Courtesy The Westinghouse Corporation.)

to provide shorter walking distances and larger stalls for passengers carrying baggage. Parking garages at airports also tend to be larger than downtown garages, and certain airports are now considering ultimate garage capacities of 5,000 to 10,000 spaces.

Most new airport parking facilities provide 1.5 to 2.0 parking spaces per peak-hour passenger. Typical parking stalls are 8.5 to 9.0 feet in width and 18 feet wide. A popular design features angle parking (60 degrees) on each side of a central one-way aisle 22 feet in width. Thus, about 275 square feet of net parking area, including the aisle, is required. When space needs of baggage drop-off and pick-up areas, sidewalks, elevators, and stairs are included, approximately 340 square feet per stall may be required.

Airport parking demand may be divided into several categories:

1. Passengers
2. Visitors bringing passengers and well-wishers
3. Employees
4. Business callers
5. Rental cars, taxis, limousines, etc.

Since the parking characteristics of parkers in these various categories differ in time of occurrence and duration, separate parking analyses should be made for each category.

Preferably, parking spaces for short time parkers should be located nearest the terminal, and many airports charge higher fees for close-in parking spaces. It is the practice for most large airports to separate parking spaces for employees some distance from the terminal building. Shuttle buses may be required to transport employees to their destinations.

If possible, the airport should have direct connections to a controlled access highway system. Within the airport, vehicular circulation is generally counterclockwise and one-way. This permits passengers to be loaded and discharged safely from the right side of the vehicle. At-grade intersections should be avoided in the circulation system, and traffic should be separated by destination at the earliest possible point. The use of overhead or tunnel crossings should be considered to prevent mixing of pedestrian and vehicular traffic.

17–7. Terminal Apron Space Requirements. The term "apron" or "ramp" refers to an area for the parking or holding of aircraft. In terms of operational efficiency of the airport, the terminal apron, which is situated adjacent to the terminal building, is most important.

There are three primary factors which determine the space requirements for a passenger terminal apron:

1. Size of gate positions
2. Number of gate positions
3. Aircraft parking configuration

Size of Gate Positions. The size of gate positions is principally determined by the size and maneuverability of aircraft, but it is also influenced by desirable wing-tip clearances and the manner in which the aircraft is moved into the gate position and serviced.

The FAA promulgates information on the principal dimensions of various aircraft in use. (See References 8 and 9.) An example of the type of aircraft data published by the FAA is given by Fig. 17–8. The FAA has also published some guidelines on aircraft wing-tip clearances [10].

DOUGLAS DC-9-40 SERIES

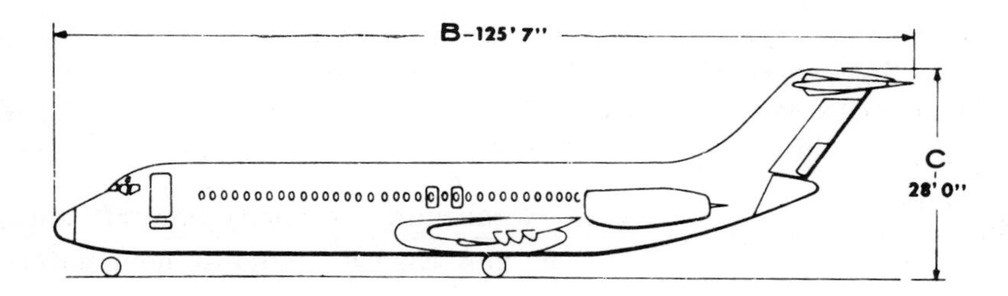

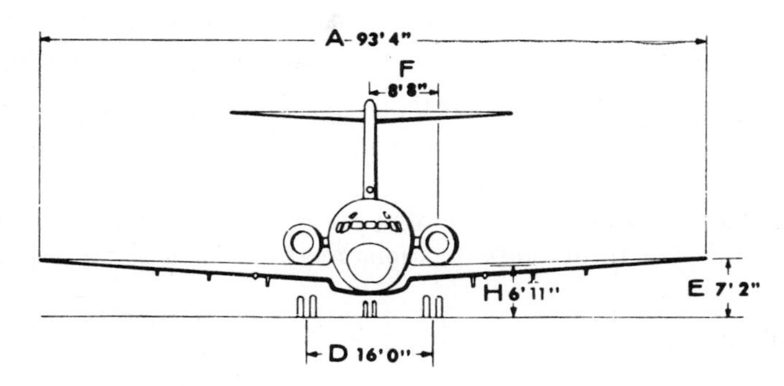

Fig. 17–8. Typical aircraft data. (Courtesy Federal Aviation Administration.)

The amount of space required for maneuvering and servicing the aircraft will vary depending upon airline operational procedures. The airport engineer should, therefore, consult with the various airlines on this matter during the early phase of the design process.

Number of Gate Positions. The number of gate positions required depends on (1) the peak volume of aircraft to be served, and (2) how long each aircraft occupies a gate position.

Gate occupancy time will depend on:

1. Type of aircraft
2. Number of deplaning and enplaning passengers
3. Amount of baggage
4. Magnitude and nature of other services required
5. Efficiency of apron personnel

A 1961 study [11] of nine airports indicated that there is wide variation in the time and magnitude of peak gate occupancy, but each airport had a definite pattern which remained relatively constant over the years. It was further reported that there was a wide range of productivity per gate for various airports.

For an existing airport, surveys should be made to measure and analyze the time required by aircraft at a gate position.

In order to achieve a balanced airport design, the number of gate positions should be related to the capacity of the runway as follows:

$$\text{Number of gate positions} = \frac{1}{2} \times \frac{\text{total runway capacity (aircraft/hour)}}{\text{gate capacity (aircraft/hour/gate)}}$$

Stated another way:

$$\text{Number of gate positions} = \frac{1}{2} \times \frac{\text{total runway capacity (aircraft/hour)}}{\text{60/average gate occupancy time (minutes)}}$$

The factor ½ in these relationships takes into consideration the fact that each parked aircraft represents two operations, a landing and a take-off.

Stafford and Stafford [11] proposed the following formula for forecasting the number of future gate positions required for a given airport:

$$\text{Future gates} = \left[(\text{present gates} - 2) \times \frac{\text{Future passengers}}{\text{Present passengers}} \right] + 2$$

This equation is applicable to any group of aircraft which have mutual use of gates, and separate calculations should be made for each such group. For example, if four groups of five gates are considered separated for traffic, and the future/present traffic ratio is three, for each group:

$$\text{Future gates} = (5 - 2) \times 3 + 2 = 11$$

Thus, the total gates required for four groups will be 44.

Stafford and Stafford recommend that in actual practice approximately 15 per cent be added to the number of gates computed by the above formula to allow for contingencies of operations such as early arrivals or delayed departures.

Aircraft Parking Configuration. Parking configuration refers to the orientation of aircraft in relation to the adjacent building when the aircraft is parked. There are a variety of parking configurations that may be used. The aircraft may be nosed-in, nosed-out, parked parallel, or at some angle to the building or concourse. The parallel parking system, in which the longitudinal axis of the aircraft is parallel to the adjacent building has been extensively used and it provides the best configuration for passenger flow. At major airports, the trend is toward the use of a nose-in parking configuration with the aircraft being pushed away from the loading bridge or gangplank by a tractor after it is loaded. This parking configuration is especially suitable for use with circular satellite enplaning structures. In this case, the aircraft parks in the wedge. This permits flexibility in the provision of space for various sizes of aircraft since the amount of available space can be varied by moving the aircraft along the radial bisector.

17–8. The Terminal Building. A well-designed teminal building is a vital element in the successful operation of an air carrier airport. It must provide for the smooth and efficient transfer of passengers and their baggage between surface transportation vehicles and aircraft.

The terminal building must provide ordered space and facilities for a variety of functions relating to air passenger service, air carrier operations, and operation and maintenance of the airport.

Required Air Passenger Service Functions. The terminal building usually houses the following air passenger service functions:

1. Ticket sales
2. Baggage checking and claiming
3. Provision of flight information
4. Handling and processing of mail and light cargo
5. Restroom services
6. Security protection
7. Passenger boarding and deplaning
8. Waiting and resting

Additional Facilities for Convenience of Passengers. Most moderate-sized to large airports provide numerous and varied facilities for the convenience of passengers. These include:

1. Restaurants and coffee shops
2. Gift shops
3. Bank
4. Insurance sales
5. Bar
6. Newsstand
7. Telephones
8. Car rental agencies
9. Storage lockers
10. Medical services
11. Barber shop
12. Hotel and motel accommodations

Air Carrier Operations. Consideration must also be given to space requirements in the terminal building for air carrier operations including:

1. Operations room for pilots and freight crews
2. Ground crew ready rooms
3. Air crew ready rooms
4. Communications center

Airport Operations and Maintenance. Finally, space may be required in the terminal building for the functions listed below relating to the operation and maintenance of the airport. Certain of these functions may be housed in separate buildings.

1. Air traffic control
2. Ground traffic control
3. Airport administration
4. FAA and other governmental administrative functions
5. Airport maintenance
6. Fire protection
7. Employee cafeterias
8. Utilities

In view of the rapidly expanding and changing nature of air travel activity, it is especially important that terminal buildings be planned and designed to allow easy expansion and change. Generally, a rectangular configuration rather than an odd-shaped building is preferred.

The design should be such that it will not be necessary to relocate kitchen facilities, toilets, and other such costly installations should expansion be required. Nonbearing partitions should be used whenever possible to allow for re-allocation of space to meet changing requirements. In short, the terminal building design should be expansible and flexible, and long range plans should be made to change it and add to it as traffic and economic conditions dictate.

The terminal building design should provide for separation of service areas to prevent passenger and baggage congestion. Specifically, the lobby and waiting room activities should be separated from baggage handling activities.

The design should provide for ease of circulation of enplaning and deplaning passengers. Enplaning passengers should be able to move directly and smoothly to the ticket counter, thence through the waiting room area to the aircraft loading gate. Deplaning passengers should be able to follow a direct route from the aircraft to the baggage claim area and thence to the passenger loading platform. These movements are illustrated schematically by Fig. 17–9.

In the design of high-capacity terminal buildings, consideration should be given to the use of two- or three-level circulation systems. By providing two or more levels in the terminal building, a vertical separation of passenger and baggage flow can be realized. A multiple-level building also makes it easier to separate arriving and departing passengers.

Space Requirements for the Terminal Building. Before one can obtain a reliable estimate of space requirements for the various functions of the terminal building, it is necessary to estimate the typical peak-hour passengers.[3] In the case of general aviation airports, typical peak-hour passengers have been shown to depend on the number of hourly aircraft operations [1]. (See Fig. 17–10.) At the larger air carrier airports, the number of peak hour-passengers can be obtained from Fig. 17–11, which relates typical peak-hour passengers to total annual passengers. More precise estimates of peak-hour passengers can be obtained by making actual counts of passengers, employees, visitors, and customers using existing terminal facilities during busy periods. Passenger forecasts should allow for anticipated growth during a ten-year design period.

The FAA has published design recommendations to aid those planning terminal buildings [7]. In this publication, space requirements for the various activities in the terminal building may be estimated from graphs relating space needs to typical peak hour passengers. Examples of these graphs are shown as Figs. 17–12, 17–13, 17–14, and 17–15, which give

[3] Typical peak-hour passengers are defined as the total of the highest number of passengers enplaning and deplaning during the busiest hour of a busy day of a typical week.

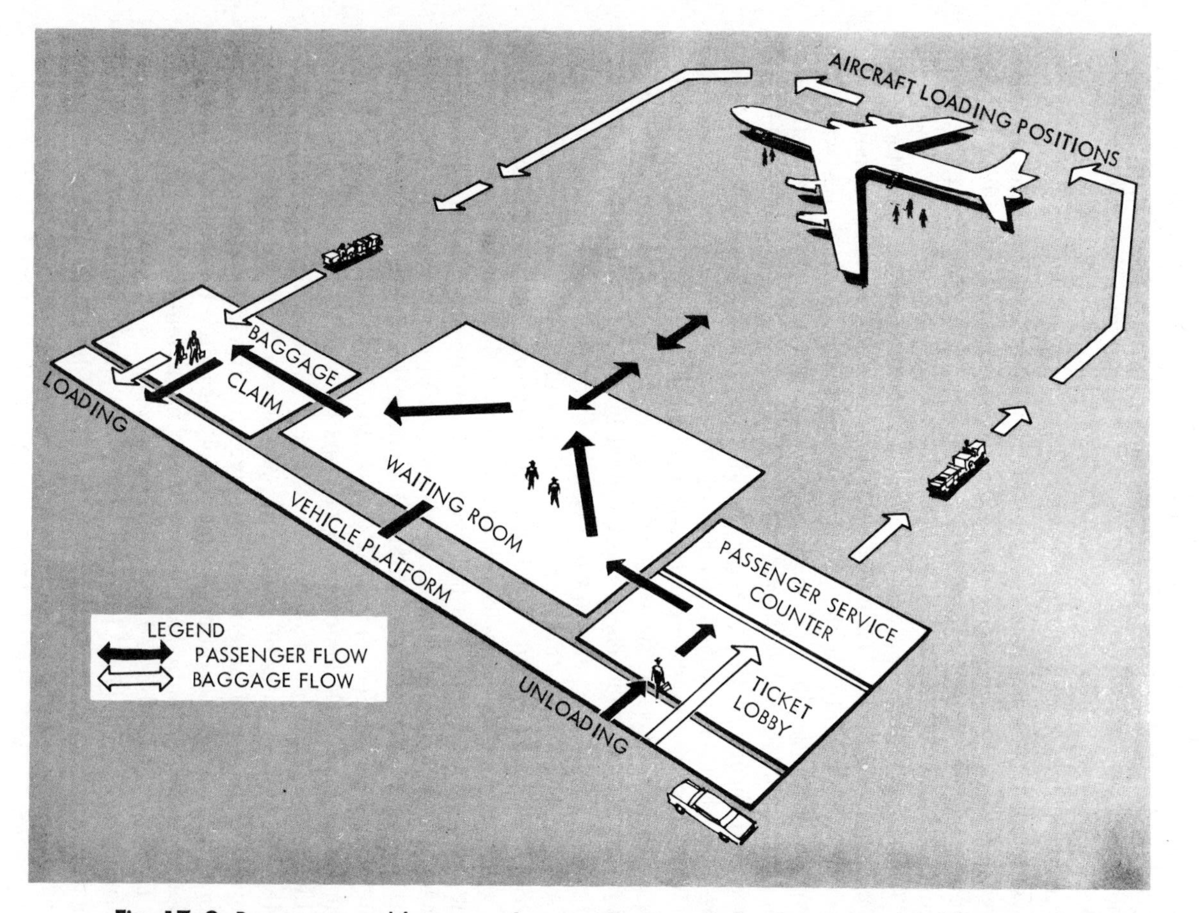

Fig. 17–9. Passenger and baggage flow. (Courtesy Federal Aviation Administration.)

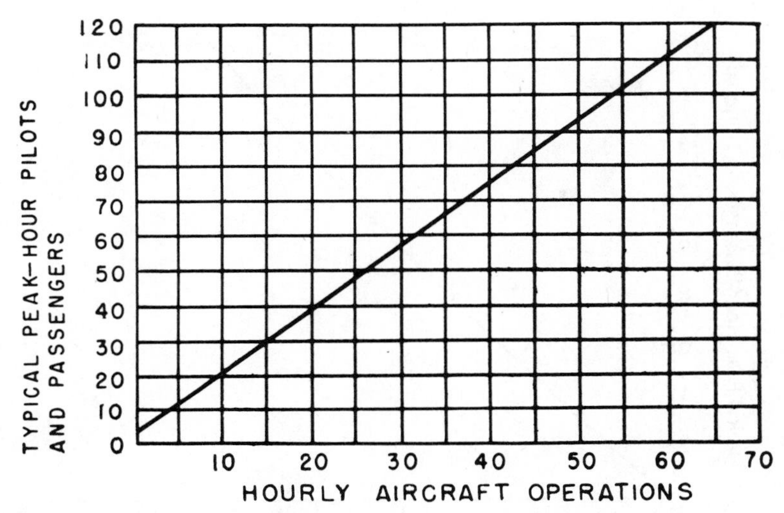

Fig. 17–10. The relationship between typical peak-hour pilots and passengers and hourly aircraft operations at general aviation airports. (Source: *Utility Airports,* Department of Transportation, Federal Aviation Administration, November, 1968.)

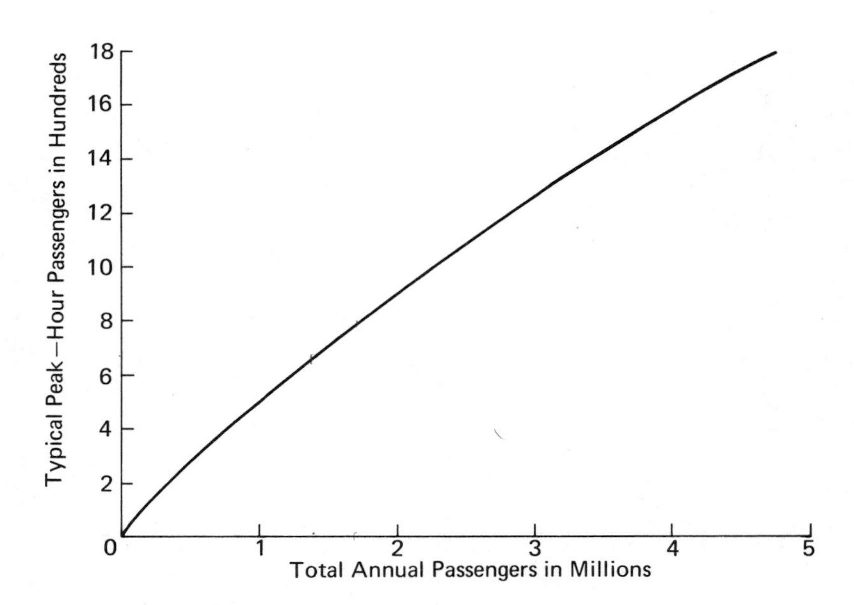

Fig. 17–11. Typical peak-hour passengers related to annual passengers. (Source: *Airport Terminal Buildings,* Federal Aviation Agency, September, 1960.)

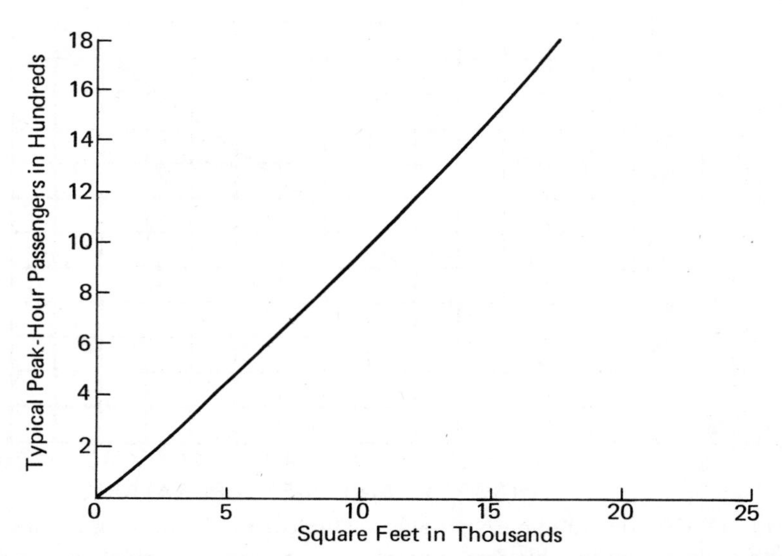

Fig. 17–12. Ticket lobby area. (Source: *Airport Terminal Buildings,* Federal Aviation Agency, September, 1960.)

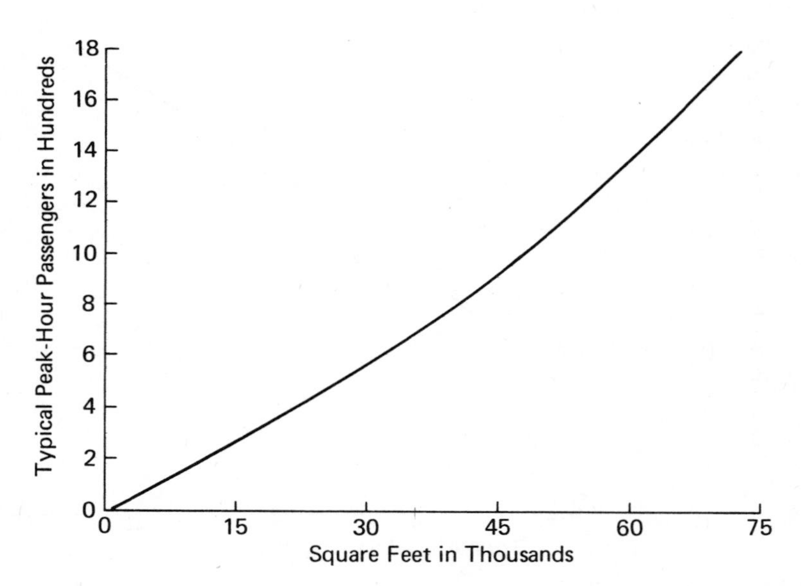

Fig. 17–13. Airline operations space. (Source: *Airport Terminal Buildings,* Federal Aviation Agency, September, 1960.)

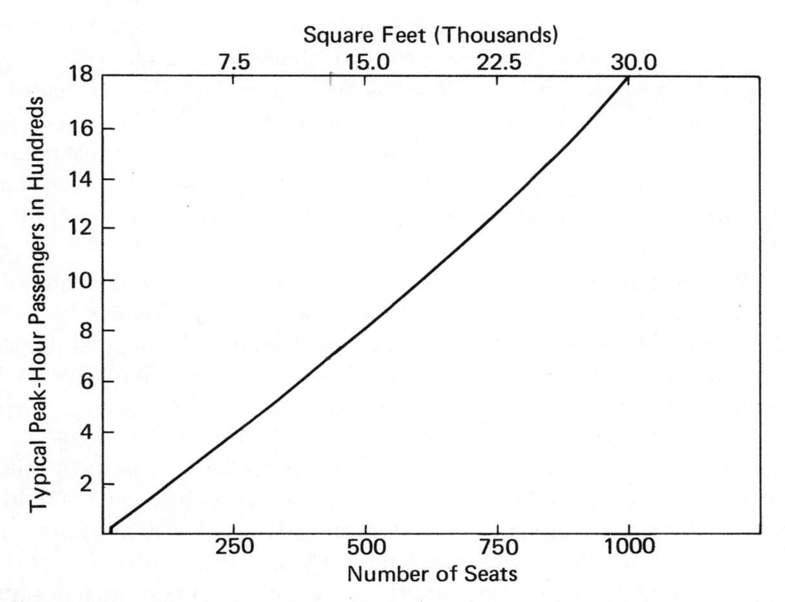

Fig. 17–14. Waiting area. (Source: *Airport Terminal Buildings,* Federal Aviation Agency, September, 1960.)

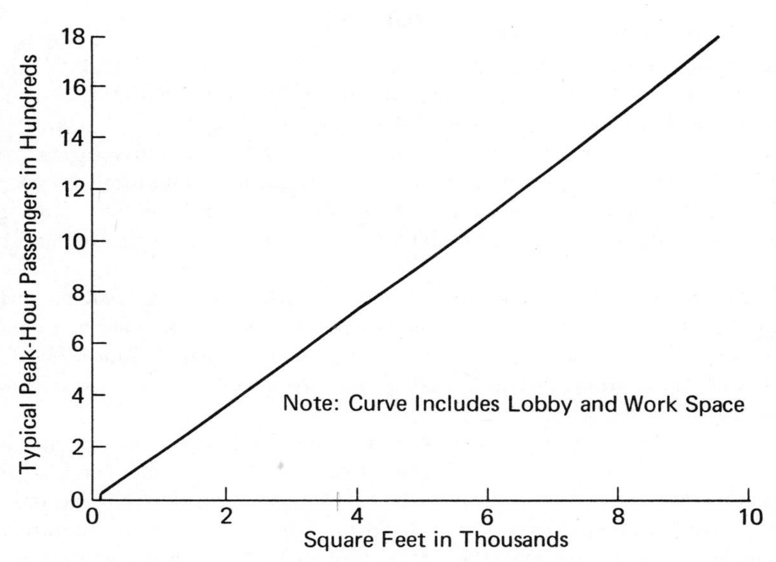

Fig. 17–15. Baggage claim area. (Source: *Airport Terminal Buildings,* Federal Aviation Agency, September, 1960.)

estimates of space requirements, respectively, for the ticket lobby area, airline operations space, waiting area, and baggage claim area. Similar graphs are given in that publication for the linear feet of passenger service counter, and the areas required for public eating facilities, news, novelties and gift stands, and public restrooms. While the graphs are useful for planning purposes, it is noted that more detailed study may be required for design purposes to allow for local variations from the norm.

17–9. Conclusion. An airport terminal area is a complex and delicately balanced microcosm. Inevitable changes in passenger loads, actions by the airlines and airport management, and technological developments tend to upset this balance, sometimes resulting in undesirable and near-intolerable passenger and aircraft congestion and delays. Airport passenger loads typically fluctuate in the short run and increase in the long run. Airline companies react to heavier traffic by instituting innovations in passenger and baggage handling procedures and facilities. The introduction of larger and faster aircraft creates a need for more and larger waiting rooms and baggage pick-up areas and may overload auto parking and circulation facilities. Airport planners and designers accept an awesome challenge in attempting to provide terminal facilities which resist obsolescence in the face of constant change.

PROBLEMS

1. A study for a proposed general aviation airport indicates that 25 aircraft operations can be expected during a typical peak hour. Estimate the number of automobile parking spaces that should be provided.
2. An air carrier airport which presently serves 1.2 million passengers expects a 40 per cent increase within 10 years. A total of 1,000 automobile parking spaces is presently provided in a 600-foot-square parking lot. How many additional spaces should be provided to accommodate the estimated additional parkers?
3. Using the data given in Problem 2, estimate the maximum walking distance to the terminal building if all of the parking spaces are provided in a one-level parking lot. Assume that all of the spaces are added on the most distant side of the parking lot and that it is not possible to increase the width of the lot.
4. A proposed airport runway system is expected to have a capacity of 80 operations per hour under VFR conditions and 35 operations per hour under IFR conditions. A study of airline activities at an existing airport gave an average gate occupancy time of 28 minutes. Based on this information, how many gate positions should be provided at the new airport?
5. Four airlines which serve a certain airport gave the following estimates for the expected increase in passenger traffic within a 10-year design period.

Airline	*Expected Increase*
A	90%
B	70%
C	80%
D	50%

Airlines A and B presently have exclusive use of 10 gates each, and the smaller airlines C and D share eight gates. Estimate the total number of gates which will be required to accommodate the future traffic.

6. An airport terminal building is to be designed to accommodate 1.4 million passengers per year. Estimate the space required for the following activities:

Ticket lobby
Airline operations
Waiting room
Baggage claim

REFERENCES

1. *Utility Airports,* Department of Transportation, Federal Aviation Administration, AC 150/5300-4A, November, 1968.
2. THOMPSON, ARNOLD W., "Evolution and Future of Airport Passenger Terminals," *Journal of the Aero-Space Transport Division, Proceedings of the American Society of Civil Engineers,* October, 1964.
3. RUNYAN, EDWIN JOE, "Gate Arrival Terminal for Kansas City MCI Airport," *Transportation Engineering Journal,* Proceedings of the American Society of Civil Engineers, May, 1969.
4. "Houston Intercontinental Airport Brings People and Planes Together in a Straight-Line Series of Terminals," *Architectural Record,* August, 1968.
5. SCHRIEVER, BERNARD A., and SIEFERT, WILLIAM W., *Air Transportation 1975 and Beyond, A Systems Approach,* The M.I.T. Press, Cambridge, Mass. (1968).
6. GABRIELSEN, WILLIAM C., "People Movers: An Appraisal of Secondary Transportation Systems and a Guide to Their Selection, *Architectural Record,* August, 1968.
7. *Airport Terminal Buildings,* Federal Aviation Agency, September, 1960.
8. *Aircraft Data,* Federal Aviation Administration, AC 150/5325-5A, January 12, 1968.
9. *Is Your Airport Ready for the Boeing 747,* Federal Aviation Administration, AC 150/5325-7, January 15, 1968.
10. Department of Transportation, Federal Aviation Administration Advisory Circular AC 150/5330-2A, July 26, 1968.
11. STAFFORD, PAUL H., and STAFFORD, D. LARRY, "Space Criteria for Aircraft Aprons," *Transportation Engineering Journal, Proceedings of the American Society of Civil Engineers,* May, 1969.

18

Airport Design Standards and Procedures

18–1. Introduction. In many respects, this chapter is the most important of the three chapters which deal with air transportation. It presents specific design standards and procedures which are required for the preparation of plans and specifications for an airport. Topics covered in this chapter will include runway lengths, geometric design of the runway system, earthwork, drainage, paving, and lighting and marking.

It should be remembered that the design standards given in this chapter are recommended standards rather than absolute requirements. Developed by the FAA for all parts of the nation, the standards are based on broad considerations. Local conditions and requirements may justify deviation from a particular standard in order to secure an advantage relating to another design feature. In such a case, the designer should be prepared to justify his decision to deviate from an accepted engineering design standard. In any event, the designer would be well advised to check with the nearest office of the FAA.

18–2. Runway Length. One of the most important design features for an airport is runway length. Its importance stems from its dominant influence on air safety, and size and cost of the airport.

Design runway length is most influenced by the performance requirements of the aircraft using the airport, especially when operated with its maximum landing and take-off loads. Variations in required runway length are caused by:

1. Elevation of the airport
2. Average maximum air temperature at the airport
3. Runway gradient

The FAA publication *Utility Airports* [1] provides a family of curves, reproduced as Fig. 18–1, which give recommended runway lengths for three utility airport groups.

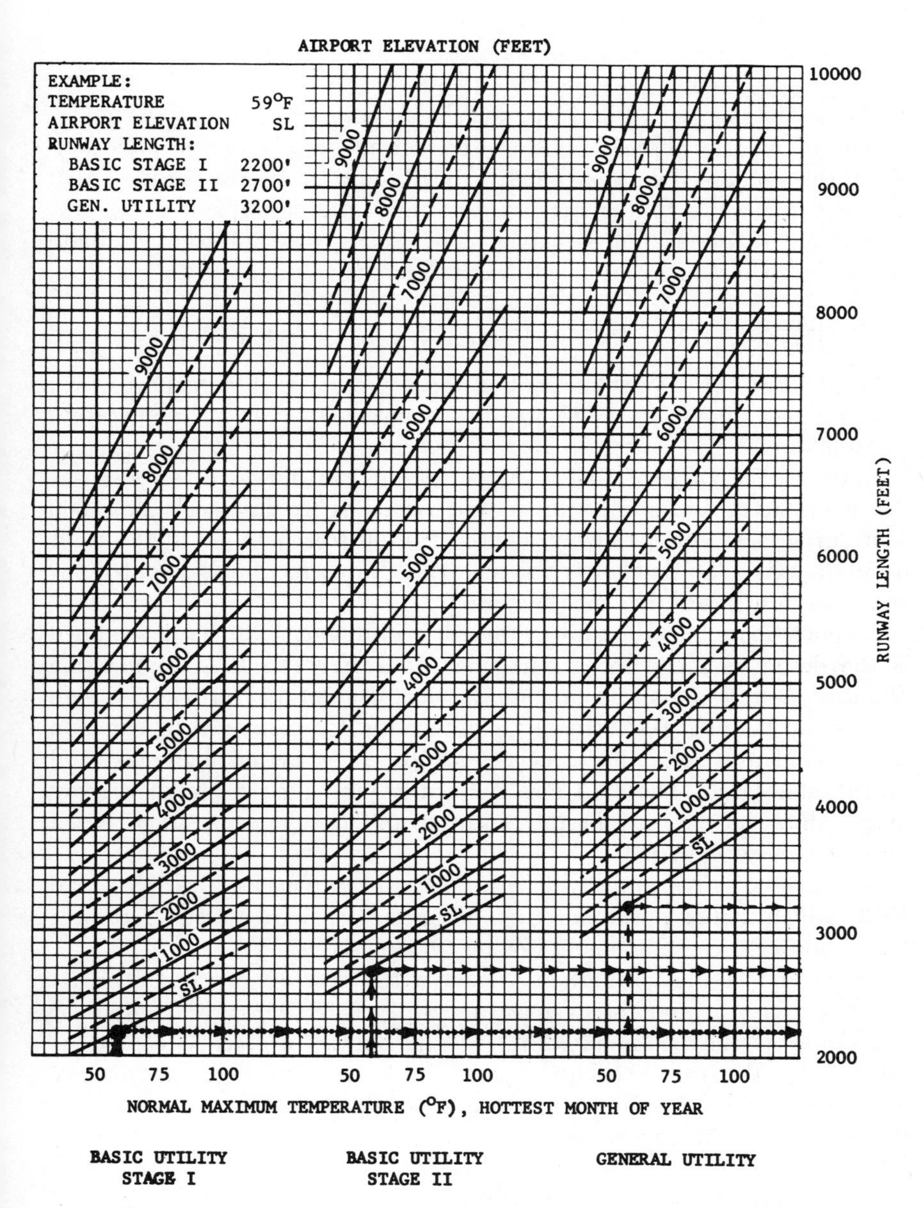

Fig. 18–1. Runway length curves for utility airports. (Source: *Utility Airports,* Federal Aviation Administration, November, 1968.)

1. Basic Utility—Stage I.
2. Basic Utility—Stage II.
3. General Utility.

To use the curves in Fig. 18–1, one should enter the appropriate family of curves on the abscissa axis at the normal maximum temperature.[1] From this point, a line is extended vertically until it intersects the slanted line corresponding to the airport elevation, interpolating if necessary. The point of intersection is extended horizontally to the right ordinate where the required runway length can be read.

Example 18–1. Use of Utility Airport Design Curves

What length of runway is required for a General Utility Airport which is 5,000 feet above sea level and has a normal maximum temperature of 80°F?

Entering the General Utility Airport family of curves at an abscissa value of 80°F, and projecting a line vertically to intersect the 5,000 foot curve, a required runway length of 6,100 feet is found at the right ordinate axis. This value is the answer. No correction is required for runway gradient or other such factors.

A much more precise runway length requirement can be determined if the design is made for a particular aircraft. The FAA publication *Runway Length Requirements for Airport Design* [2] provides design curves for landing and take-off requirements for airplanes in common use in civil aviation. These curves are based on actual flight test and operational data. The FAA also frequently promulgates similar runway length data for aircraft in the research and development stage based on preliminary data [3].

Examples of FAA runway length curves are given as Figs. 18–2 and 18–3, which indicate the aircraft performance characteristics, respectively, for a Beech D18S and a Boeing 707-300 series.

These and similar performance curves for take-off are based on an *effective runway gradient* of zero per cent. Effective runway gradient is defined as the maximum difference in runway centerline elevations divided by the runway length. The FAA specifies that the runway lengths for take-off should be increased by the following rates for each one per cent of effective runway gradient:

1. For piston and turboprop powered airplanes—20%
2. For turbojet powered airplanes—10%

[1] The normal maximum temperature is defined as the arithmetical average of the daily highest temperature during the hottest month. This information is usually available from the nearest U.S. Weather Bureau office.

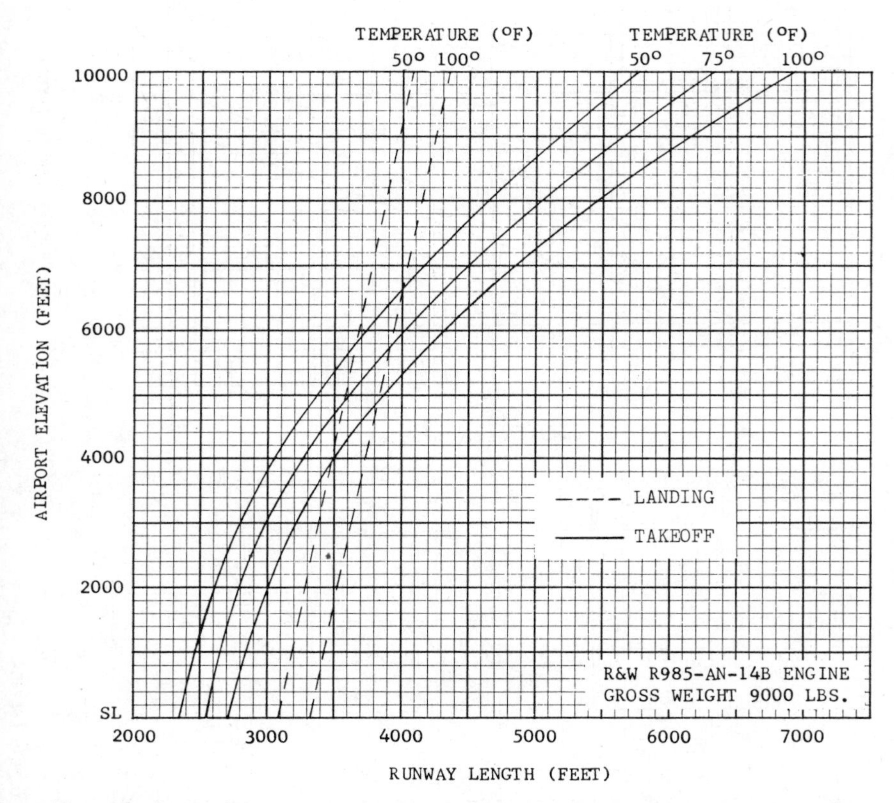

Fig. 18–2. Performance curve for Beech D18S aircraft. (Source: *Runway Length Design Requirements for Airport Design,* Federal Aviation Administration, November 8, 1967.)

Example 18–2. Runway Length Requirement for a Beech D18S

What length of runway is required for a Beech D18S if the airport elevation is 7,000 feet and the normal maximum temperature is 75°F? The maximum difference in elevation between the highest and lowest point on the runway centerline is 40 feet.

By reference to Fig. 18–2, it is seen that a runway length of 4,500 feet is required. This value should be increased by an amount of 20 per cent for each per cent of effective grade.

$$\text{Effective Grade} = \frac{40 \times 100\%}{\text{Design Runway Length}}$$

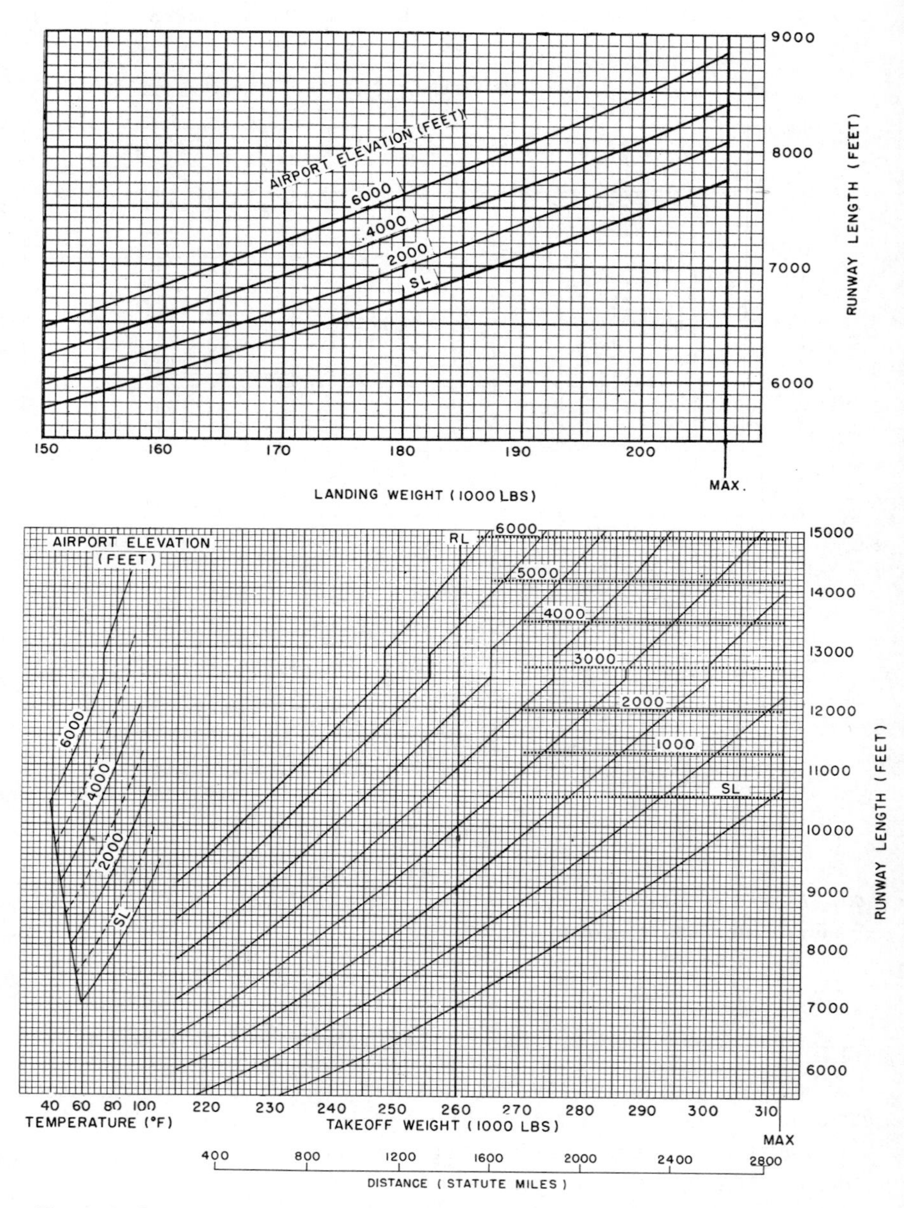

Fig. 18–3a. Landing performance curve for Boeing 707-300 Series aircraft. (Source: *Runway Length Design Requirements for Airport Design*, Federal Aviation Administration, November 8, 1967.)

Fig. 18–3b. Takeoff performance curve for Boeing 707-300 Series aircraft. (Source: *Runway Length Design Requirements for Airport Design*, Federal Aviation Administration, November 8, 1967.)

$$\text{Design Runway Length} = 4500 + \left(\frac{40 \times 100\%}{\text{Design Runway Length}}\right)(0.20 \times 4500)$$

This expression can be solved directly or by trial to yield a design runway length of 5,193 feet, which is rounded to 5,200 feet.

Example 18–3. Runway Length Requirement for a Boeing 707-300 Series Aircraft

What length of runway is required for a Boeing 707-300 Series aircraft, given the following design conditions?

1. Normal maximum temperature—85°F
2. Airport elevation—3,000 feet
3. Flight distance—1,600 miles
4. Maximum landing weight—207,000 pounds
5. Effective runway gradient—0.4 per cent

Runway Length Required for Landing. Figure 18–3(a) is entered on the abscissa axis at the maximum landing weight (207,000 pounds) and this point is projected vertically to intersect with the 3,000 foot airport elevation line (by interpolation). This point of intersection is extended horizontally to the right ordinate scale where a runway length required for landing of 8,250 feet is read.

Runway Length Required for Take-off. The following steps are required to determine the runway length required for take-off from Fig. 18–3(b):

1. Enter the temperature scale on the abscissa axis at the given temperature (85°F).
2. Project this point vertically to the intersection with the slanted line corresponding to the airport elevation (3,000 feet).
3. Extend this point of intersection horizontally to the right until it coincides with the reference line (RL).
4. Then, proceed up and to the right or down and to the left parallel to the slanted lines to the intersection of the elevation limit line (in this case 3,000 feet), or until reaching a point directly above the aircraft's take-off weight or distance (e.g., 1,600 miles), whichever occurs first.
5. Project this point horizontally to the right and read the required runway length for take-off at the right ordinate scale. In this example, a length of 11,000 feet is required for take-off.

6. Increase this runway length for effective gradient (0.4 per cent).

$$\text{Runway Length (Take-Off)} = 11{,}000 + (11{,}000 \times 0.10 \times 0.4) = 11{,}440 \text{ feet}$$

$$\text{Design Runway Length} = 11{,}440 \text{ feet (largest of two values)}$$

18–3. Longitudinal Grade Design for Runways. In the interests of safe and efficient aircraft operations, longitudinal runway grades should be as flat as possible, and grade changes should be avoided. A maximum longitudinal grade of 1½ per cent is generally specified for air carrier airports, and a 2 per cent maximum is recommended for utility airports [1].

Where grade changes are necessary, the recommended maximum changes in grade and minimum lengths of vertical curves given by Fig. 18–4 should be used. Careful thought should be given to what the future runway gradient requirements may be, and the runway gradient design should provide for future needs as well as present.

To minimize any hazard associated with objects on the runway, sight distance should preferably be provided for the full length of the runway. For runway and landing strips equal to or less than 3,400 feet, the FAA specifies that any two points 10 feet above the runway centerline must be mutually visible for a distance of one-half the runway length. Where the landing strip is longer than 3,400 feet, sight distance is measured from points five feet above the runway centerline. Otherwise, the sight distance criteria for the longer landing strips is identical to that given above.

Vertical curves should be provided when there is a change in grade as great as 0.4 per cent. The length of vertical curve should be at least 300 feet for each one per cent of grade change at utility airports and 1,000 feet for each one per cent of grade change at air carrier airports. A minimum vertical curve length of 1,000 feet is specified for runways which will serve the Boeing 747.

18–4. Runway and Taxiway Cross-Section. The dimensional requirements for runways and taxiways vary widely depending on the airport size and the type and size of aircraft served. The runway, a paved load bearing roadway, is typically 50 feet wide at small airports and 150 feet wide at large airports. Shoulders are provided along each side of the runway as a safety measure should an aircraft lose control and veer from the runway. Shoulders, which are typically stabilized earth with grass cover, vary in width from 25 feet at the smallest airports to 175 feet at the largest airports. The runway with adjacent shoulders is called the *landing strip*.

The taxiway structural pavement is typically 20 to 40 feet wide at utility airports and 50 to 100 feet at air carrier airports. In the latter

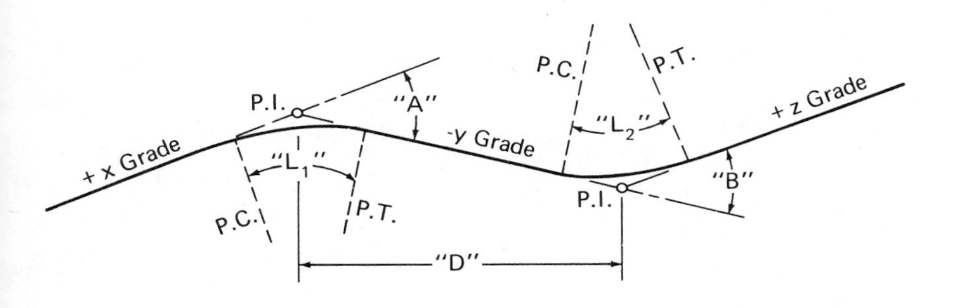

	Airports with the longest landing strip equal to or less than 3400 feet	*Airports with the longest landing strip longer than 3400 feet
Maximum grade change such as "A" or "B" (in percent) not to exceed	2 percent	1 1/2 percent
Length of Vertical curve "L_1" or "L_2" (in feet) for each one percent grade change	300 feet	1000 feet
Distance (in feet) between points of intersection for Vertical curves or "D" Note: "A" and "B" in percent	250 ("A" + "B") feet	1000 ("A" + "B") feet

*For airports that will serve only small aircraft operating under visual flight rules the criterion for airports with longest strip equal to or less than 3400 feet may be used.

Fig. 18–4. Vertical curves for runways. (Source: *Airport Surface Areas Gradient Standards,* Federal Aviation Administration, May 12, 1966.)

case, an additional 15 feet of pavement width should be provided on taxiway curves. In the interests of safety, taxiway center lines are located 150 to 600 feet from runway center lines. Shoulders are also provided for taxiways.

A typical runway and taxiway cross section is shown as Fig. 18–5. Table 18–1 gives recommended dimensional standards for utility airports,

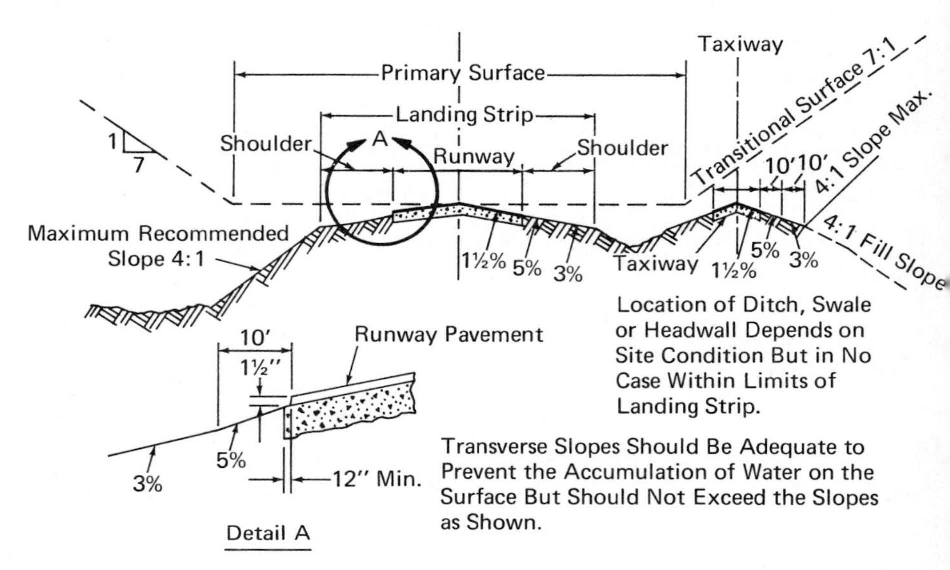

Fig. 18–5. Typical runway and taxiway cross section. (Source: *Utility Airports,* Federal Aviation Administration, November, 1968.)

TABLE 18-1

Recommended Dimensional Standards for Utility Airports

	Airport Group		
	Basic Utility Stage I	**Basic Utility Stage II**	**General Utility**
Landing Strip Width	100′	120′	150′
Runway Width	50′	60′	75′
Taxiway Width	20′	30′	40′
Taxiway Centerline to:			
Fixed or movable obstruction	50′	50′	50′
Airplane tie-down area	75′	75′	75′
Runway Centerline to:			
Taxiway centerline	150′	150′	200′
Property line (non-taxiway side)	200′	200′	250′
Building restriction line	250′	250′	300′

Source: *Utility Airports,* FAA, November, 1968.

while similar standards for air carrier airports are given by Tables 18–2a and 18–2b.

To utilize the data in Table 18–2a, it is first necessary to convert the planned runway length to standard sea level conditions by removing the effects of elevation, temperature, and runway gradient. The *basic runway length* is determined by dividing the planned runway length by the product of three factors which represent local elevation, temperature, and gradient conditions.

$$\text{Basic Runway Length} = \frac{\text{Planned Runway Length}}{F_e \times F_t \times F_g}$$

The elevation factor, F_e, allows for a seven per cent change in runway length for each one thousand feet of airport elevation. Thus:

$$F_e = (0.07 \times E + 1)$$

where

E = airport elevation, thousands of feet

The temperature factor, F_t, applies a rate of one-half per cent for each one degree of difference between the normal maximum temperature, T (°F), at the airport and the standard sea level temperature (59°F), corrected for airport elevation. This factor is computed by the formula:

$$F_t = 0.005[T - (59 - 3.566E)] + 1$$

The gradient factor, F_g, allows for a rate of 20 per cent (10 per cent for turbojet aircraft) for each one per cent of effective gradient, G. Thus:

$$\text{For turbojets, } F_g = (0.10G + 1)$$
$$\text{For other aircraft, } F_g = (0.20G + 1)$$

EXAMPLE 18–4. UTILIZATION OF TABLE 18–2

A certain airport, elevation 7,500 feet, has a planned runway length of 8,600 feet. The normal maximum temperature is 85°F and the effective gradient is 0.6 per cent. Which column of data should be used in Table 18–2? (Assume no turbojet aircraft will be used.)

$$F_e = (0.07 \times 7.5 + 1) = 1.525$$
$$F_t = 0.005[85 - (59 - 3.566 \times 7.5)] + 1 = 1.264$$
$$F_g = 0.20 \times 0.6 + 1 = 1.120$$

$$\text{Basic Runway Length} = \frac{8{,}600}{1.525 \times 1.264 \times 1.120} = 4{,}000 \text{ feet.}$$

This would indicate that Column 1 of Table 18–2a should be used. However, if the future airport plan indicated that the planned basic runway

TABLE 18–2a

Recommended Dimensional Standards for Airline Reports—Runways

	Basic Runway Lengths	
	1 **3200–4200 Feet**	**2** **More Than 4200 Feet**
Landing Strip Width	400′ Min.	500′
Runway Width	100′	150′
Runway centerline to:		
Obstacle	200′	250′
Taxiway centerline	250′	400′
Aircraft parking or tiedown area	425′	650′
Property line	500′	750′
Building line, non-precision runway	500′	750′
Building line, precision runway	750′	750′

Source: *Runway/Taxiway Widths and Clearances for Airline Airports*, FAA, July 25, 1968.
*Exceptions are made for certain navigational, meteorological, and visual aids.

TABLE 18–2b

Recommended Dimensional Standards for Airline Airports—Taxiways

		Dimensional Criteria (Feet) Airplane Taxiway Design Group			
Design Item	**Symbol**	I[a]	II[b]	III[c]	IV[d]
1. Taxiway Structural Pavement Width on Tangents	W_T	50	75	100	125
2. Taxiway Structural Pavement Width on Curves	W_C	65	90	115	140
3. Taxiway Shoulder Width	—	20	25	35	40
4. Safety Area Width	—	110	150	220	310
5. Taxiway and Apron Taxiway Obstacle Free Area Width	—	210	270	360	470
6. Terminal Taxilane Obstacle Free Area Width	—	160	210	290	390
7. Separation Distance from Taxiway C_L to Taxiway C_L	S_T	200	300	300	400
8. Separation Distance from Taxiway C_L to Runway C_L	S_R	400	400	600	1,000
9. Radius of Taxiway C_L Curves	R	100	150	150	200

Source: *Airport Design Standards—Airports Served by Air Carriers—Taxiways*, FAA Advisory Circular AC 150/5335-1A, May 15, 1970.
[a]For airplane models B-727-100, B-737-100, B-737-200, BAC-1-11, CV-58C, DC-9-10, DC-9-30, F-27, SE-210, DC-9-40.
[b]For airplane models DC-7, DC-8-50, B-707-100, B-707-300, DC-8-63, B-727-200.
[c]For airplane models DC-10-10, L-1011, B-747, L-500.
[d]For future aircraft.

length exceeds 4,200, the runway and taxiways clearances in Column 2 should be used in the initial construction.

18–5. Taxiways and Turnarounds. Taxiways are used to facilitate the movement of aircraft to and from the runways. Where air traffic warrants, the usual procedure is to provide a taxiway parallel to the runway centerline for the entire length of the runway. This makes it possible for landing aircraft to exit the runway more quickly and decreases delays to other aircraft waiting to use the runway.

At smaller airports, air traffic may not be sufficient to justify the construction of a parallel taxiway. In this case, taxiing is done on the runway itself and turnarounds should be constructed at the ends of the runway. A typical taxiway turnaround is shown as Fig. 18–6. A parallel taxiway is warranted if the annual operations total 50,000 or if there are as many as 20 itinerant operations during peak hours [1]. Where less traffic is anticipated, partial parallel taxiways may be suitable.

The design of the taxiway system will be determined by the volume of air traffic, the runway configuration, and the location of the terminal building and other ground facilities. The following general guidelines should be helpful in designing the taxiway system:

1. Taxiway routes should be direct and uncomplicated. Generally, taxiways should follow straight lines, and curves of long radius should be used when curves are required.
2. Whenever possible, taxiways should be designed so as not to cross active runways or other taxiways.
3. A sufficient number of taxiways should be provided in order to avoid congestion and complicated routes between runway exit points and the apron area.

At large and busy airports, the time an average aircraft occupies the runway frequently will determine the capacity of the runway system and the airport as a whole. This indicates that exit taxiways should be conveniently located so that landing aircraft can vacate the runway as soon as possible.

At utility airports, three exit taxiways will generally be sufficient: one at the center and one at each end of the runway.

Two common types of exit taxiways are illustrated by Fig. 18–7. Perpendicular exit taxiways may be used when the design peak hour traffic is less than 30 operations per hour. To expedite the movement of landing aircraft from the runway, most modern air carrier airports provide exit taxiways which are oriented at an angle of about 30° to the runway centerline. This makes it possible for aircraft to leave the

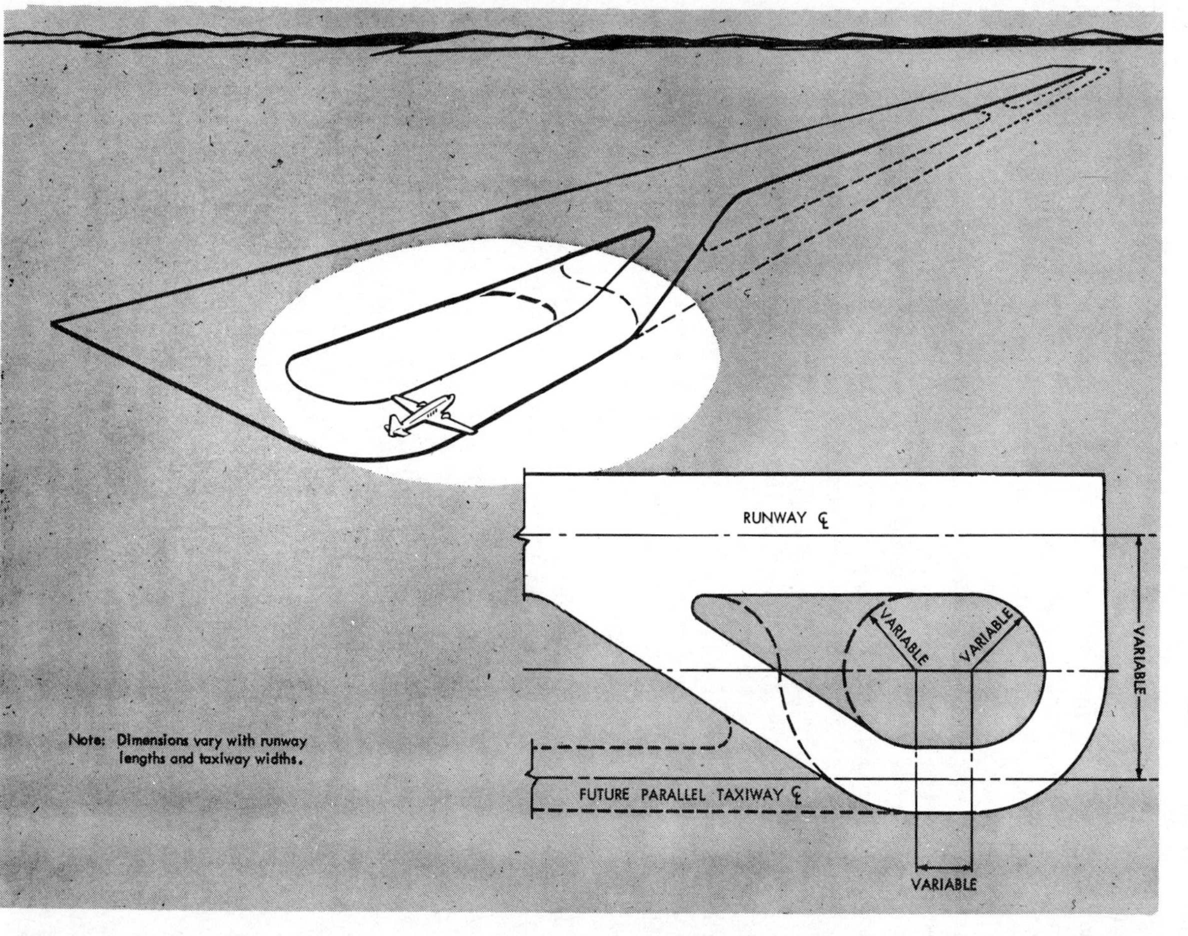

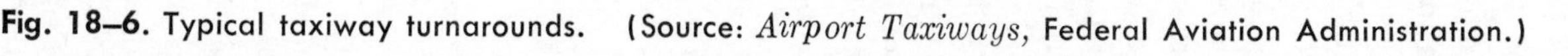

Fig. 18–6. Typical taxiway turnarounds. (Source: *Airport Taxiways*, Federal Aviation Administration.)

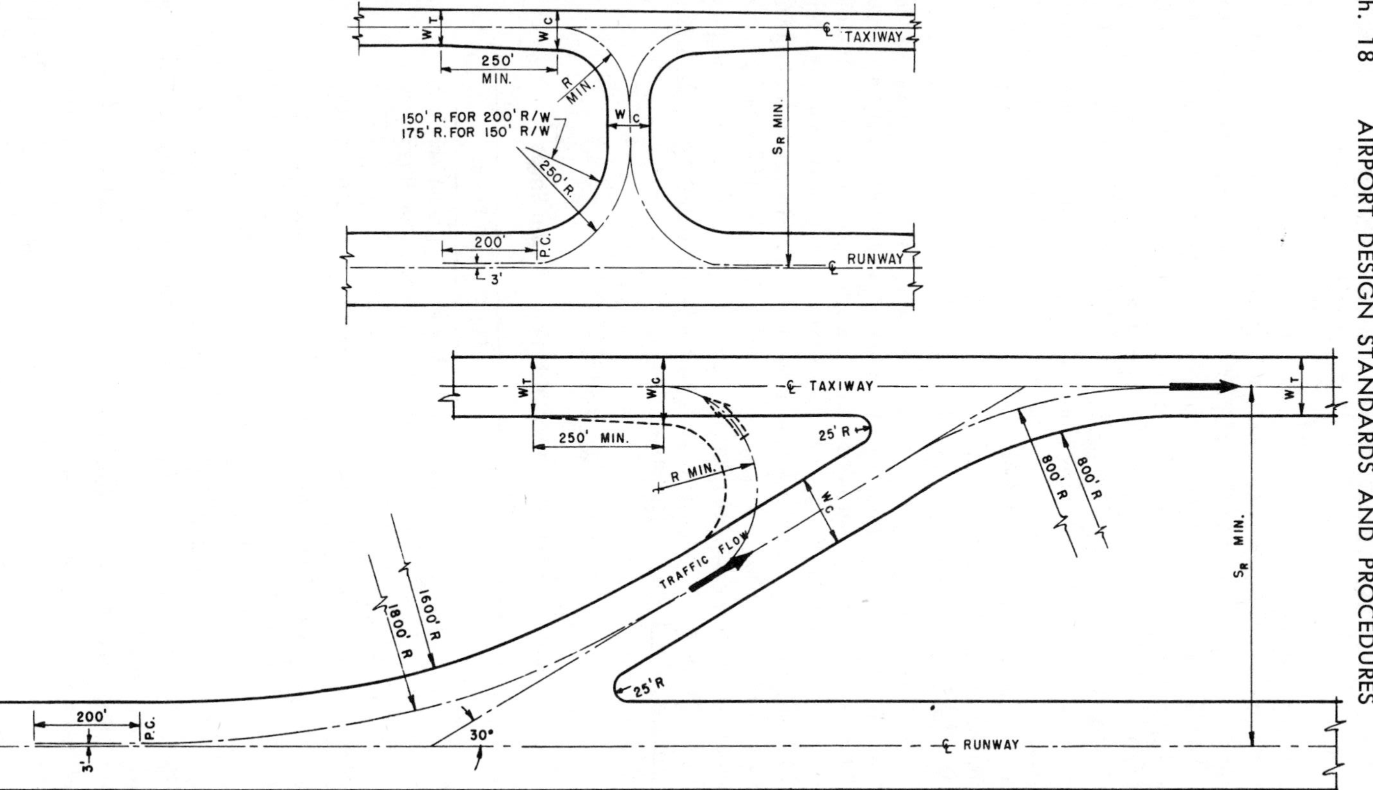

Fig. 18–7. Recommended design for exit taxiways. (Source: *Airport Taxiways*, Federal Aviation Administration.)

runway at speeds up to 60 miles per hour, increasing the efficiency and capacity of the airport system.

The proper location of exit taxiways is dependent on the touchdown point and landing roll of the aircraft, as well as the configurations of the exits. The FAA [6] recommends that the points of curvature of the angled type of exit be located at intervals beginning approximately 3,000 feet from the threshold to approximately 2,000 feet of the stop end of the runway. For the 90 degree type, the P.C.'s of the taxiway exits should be located at intervals beginning about 3,500 feet from the threshold to approximately 2,000 feet from the stop end of the runway. Where the runway length exceeds 7,000 feet, intermediate exits at intervals of approximately 1,500 feet are recommended.

18–6. Holding Aprons. A holding apron is an area provided adjacent to the taxiway near the runway entrance for aircraft to park briefly while cockpit checks and engine runups are made preparatory to take-off. The use of holding aprons reduces interference between departing aircraft and minimizes delays at this portion of the runway system.

In the case of utility airports, the FAA [1] recommends that holding aprons be installed when air activity reaches 30 operations per normal peak hour, 20,000 annual itinerant operations, or a total of 75,000 annual operations. Space to accommodate at least two but not more than four is recommended for small airports.

TABLE 18–3

Factors for Determination of Holding Apron Space

	For Aircraft With:	
	Dual Wheel Undercarriages	Dual Tandem Landing Gears
Length Parallel to Taxiway	1.35 to 1.50	1.60 to 1.75
Width Outside Taxiing Corridor	0.65 to 0.75	0.80 to 0.90

Source: *Is Your Airport Ready for the Boeing 747?*, FAA, January 15, 1968.

To determine the amount of holding apron space required for a given aircraft, the factors given in Table 18–3 should be multiplied by the aircraft wing span [3].

A sketch of a typical holding apron is shown as Fig. 18–8.

18–7. Airport Drainage. A well-designed drainage system is an essential requirement for the efficient and safe operation of an airport. Inadequate drainage facilities not only will result in costly damages due to

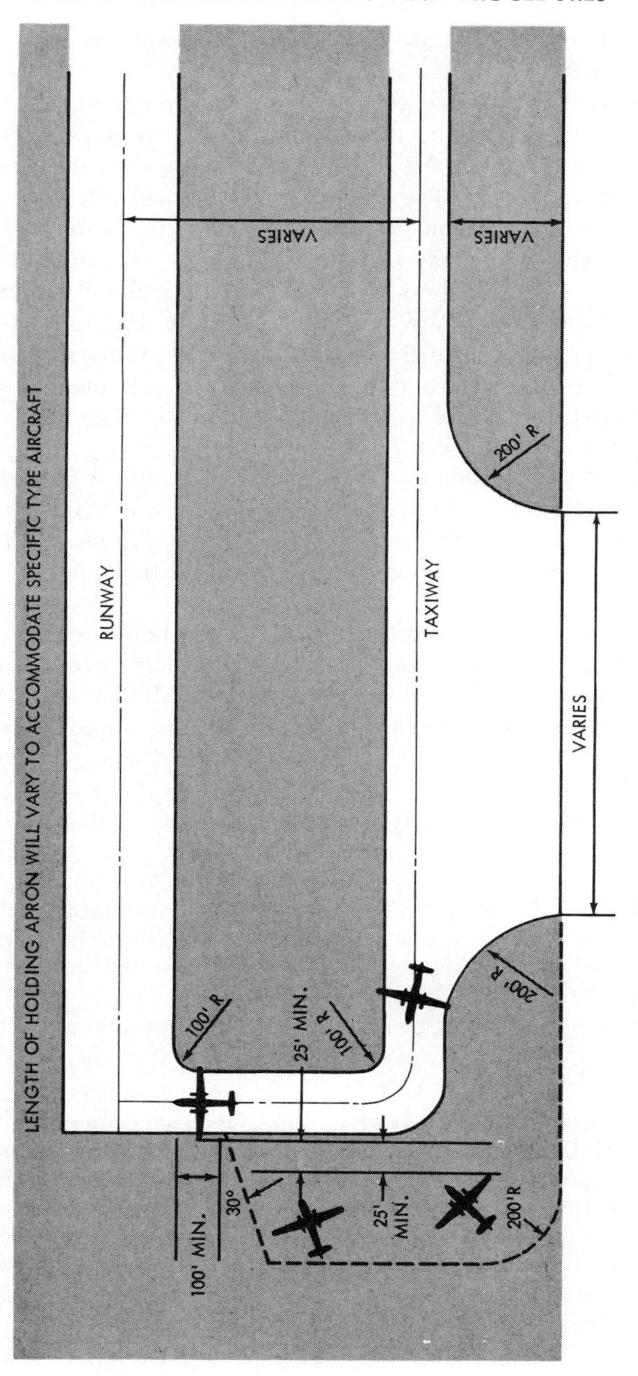

Fig. 18–8. Typical holding apron. (Courtesy Federal Aviation Administration.)

flooding, but also may cause hazards to air operations and even result in the temporary closing of a runway or airport.

The design of a drainage system is based on the fundamental principles of open channel flow given in Chapter 12, and in certain respects, the design procedures for airport drainage are identical to those for railway and highway drainage. The computation of runoff, for example, is accomplished by the Rational Formula, described in Chapter 12. On the other hand, an airport has certain peculiarities regarding its drainage requirements. Characterized by extensive areas and flat slopes and a critical need for the prompt removal of surface and subsurface water, airports are usually provided with an integrated drainage system. This system consists of surface ditches, inlets, and an underground storm drainage system. Typical drainage systems for landing strips and runways are shown in Reference 7.

The underground conduits are designed to operate with open channel flow, and because pipe sections in this system are long, uniform flow can be assumed. The hydraulic design of the channels and conduits, therefore, is usually accomplished by application of the Manning equation.

Storm drain inlets are placed as needed at low points and are typically spaced 300 to 500 feet. Manholes are provided to permit workmen to inspect and maintain the underground system. Manholes are commonly placed at every abrupt change of direction and approximately every 300 to 400 feet on tangents. Typical inlet and manhole designs are given by Reference 7.

The design of a drainage system for an airport involves the following steps:

1. Using the proposed grading plan as a basis, a layout of the drainage system is made. The grading plan, which should show the proposed finished grade by one-foot contour lines, will make it possible to select appropriate locations for drainage ditches and inlets and to determine the tentative layout of the underground pipe system.

2. Drainage structures and pipelines are usually identified by numbers or letters for easy reference in design computations.

3. For each drainage sub-area, the runoff is computed by means of the Rational Formula. (See Chapter 12.) This involves the estimation of a runoff coefficient and a time of concentration (including flow time in the pipe system), and the selection of a design rainfall intensity from an intensity-duration curve similar to Fig. 12–1. In this connection, the FAA recommends a storm frequency of five years [7].

4. Beginning with the uppermost pipe section, the slope and pipe size is selected to carry the design flow. Design charts such as

that shown as Fig. 18–9 are used for this purpose. As the design progresses along the line, each succeeding pipe section carries the water from its surface drainage area plus that contributed through its inlet structure.

Example problems for the actual design of a drainage system for a portion of an airport have been abstracted from the FAA publication *Airport Drainage*[2] and are given below.

EXAMPLE 18–5. DRAINAGE DESIGN WITHOUT PONDING

Suppose it is desired to design an underground drainage system to accommodate the surface flow from the apron and taxiways shown by Fig. 18–10. Inlets and line segments are first numbered and lengths are scaled from the map and recorded as shown by Columns 1, 2, and 3 in Table 18–4.

Columns 4 through 10 record the data required for the calculation of runoff for various subareas in the system. These calculations are made by the Rational Formula which is adequately described in Chapter 12. It is noted that for a given inlet, time of concentration equals the inlet time (Column 4), or time required for water to flow overland from the most remote point in the subarea, plus flow time (Column 5) through the particular pipe segment. Flow time is computed by dividing the pipe length by the velocity of flow (Column 12).

Column 11 shows the accumulated runoff which must be accommodated.

Columns 12 through 16 show data pertaining to the hydraulic design of the system. The slope of a pipe section (Column 14) is based on such factors as topography, amount of cover, depth of excavation, elevation of the discharge basin or channel, and discharge velocity. With the slope and accumulated runoff, the size of pipe required (Column 13), velocity of flow (Column 12), and pipe capacity (Column 15) can be determined by means of a design chart for the Manning equation similar to Fig. 18–9. In this example, concrete pipe was used and a Manning roughness coefficient, n, of 0.015 was assumed.

EXAMPLE 18–6. DRAINAGE DESIGN WITH PONDING

Suppose we wish to drain the area shown in Fig. 18–11 with a single pipe to permit ponding of a short duration between the taxiways. It will be noted that this area is part of that shown in Fig. 18–10, except the contours have been changed to permit drainage by a single inlet. Suppose further that a 21-inch pipe is used to drain the area and that

[2] A helpful publication by this same title has been published by the Portland Cement Association [8].

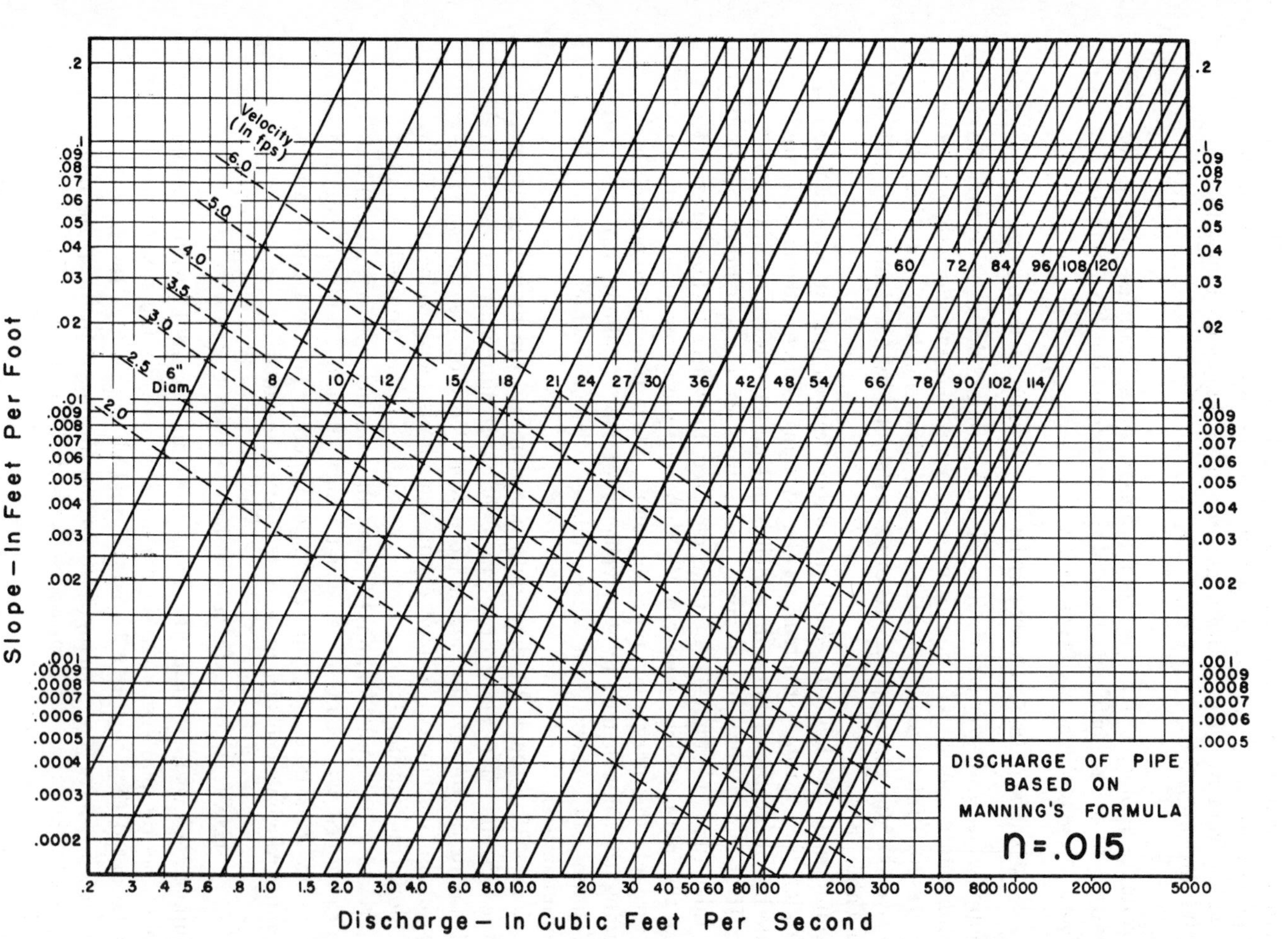

Fig. 18–9. Design chart for uniform flow. (Source: *Airport Drainage*, Federal Aviation Administration.)

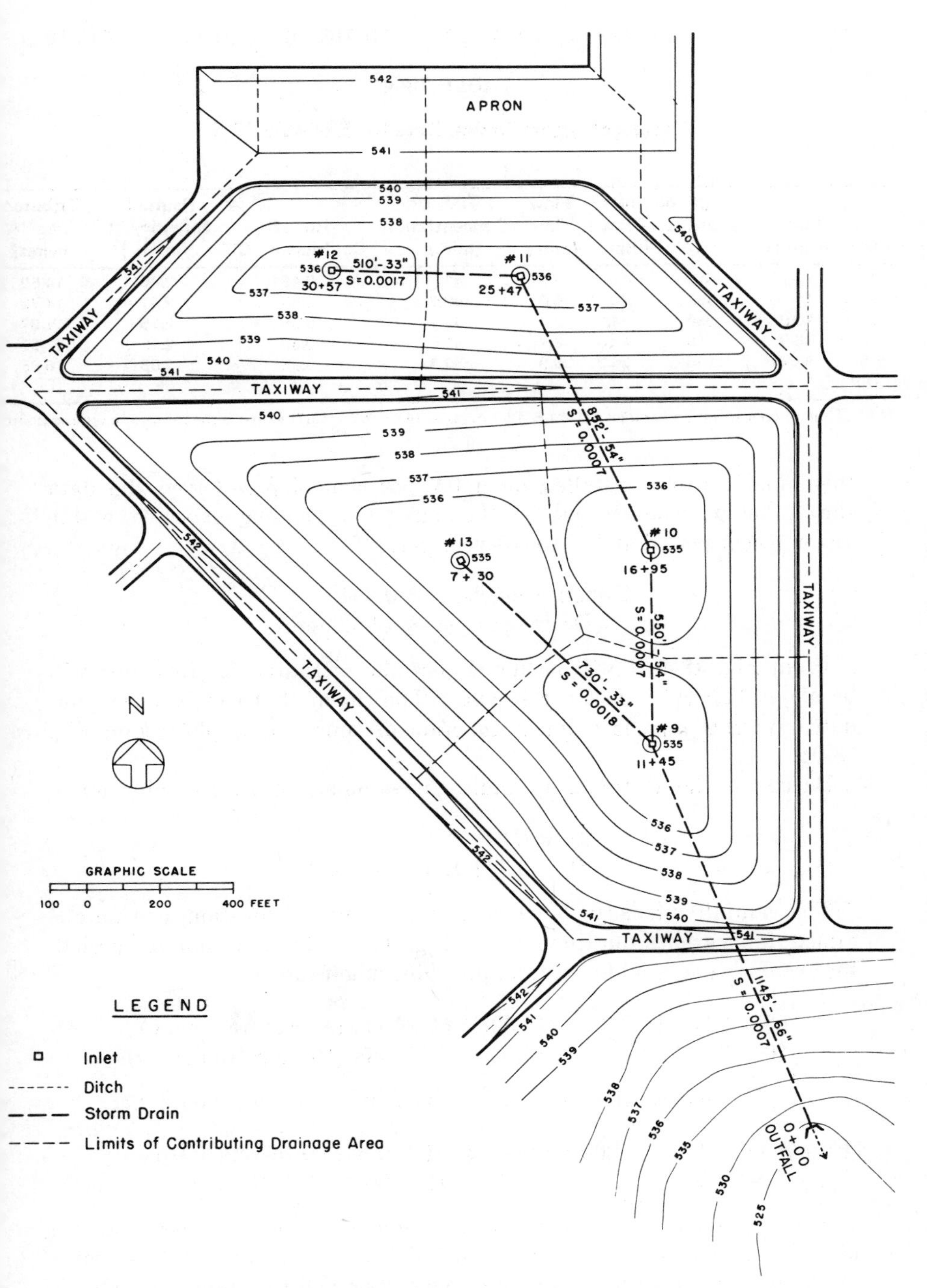

Fig. 18–10. Portion of an airport showing drainage design. (Courtesy, Federal Aviation Administration.)

TABLE 18–4

Drainage System Design Data for Example 18-5

Inlet	Line Segment	Length of Segment (ft.)	Inlet Time (min.)	Flow Time (min.)	Time of Concentration (min.)	Runoff Coefficient "C"	Rainfall Intensity "I" (in./hr.)	Tributary Area "A" (acres)
12	12-11	510	41	2.7	41	0.49	2.40	14.69
11	11-10	852	40	5.0	43.7	0.53	2.31	14.72
10	10-9	550	34.8	3.3	48.7	0.35	2.15	11.97
13	18-9	730	48.6	3.7	48.6	0.35	2.16	21.50
9	9-Out	1145	36.3	5.9	52.3	0.35	2.03	16.05
Out								

Note: Time of concentration for inlet No. 11 is 43.7 minutes (41 + 2.7 = 43.7) which is the most time remote point

this pipe is to be installed on a 0.7 per cent slope. Given the data shown below, what would be the maximum ponding expected within (a) a 5-year period and (b) a 10-year period?

$$\text{Runoff coefficient} = 0.354$$
$$\text{Drainage area} = 49.52 \text{ acres}$$

From Fig. 18–9 it will be noted that the discharge for this pipe will be about 11.5 cubic feet per second. The runoff that can be accommodated by this pipe is a linear function of time and is plotted on Fig. 18–12.

Based on the Rational Formula, the amount of runoff (cfs) is:

$$Q = CIA$$
$$Q = 0.354 \times I \times 49.52$$

The rainfall intensity, I, which is dependent on duration, can be obtained for various durations from Fig. 12–1. In a 30-minute period, for example, one would expect a maximum runoff rate of:

$$Q = 0.354 \times 2.90 \times 49.52 = 50.84 \text{ cfs (5-year frequency)}$$
$$Q = 0.354 \times 3.38 \times 49.52 = 59.25 \text{ cfs (10-year frequency)}$$

Thus, the corresponding runoff values in cubic feet would be:

$$\text{Runoff} = 50.84 \times 1{,}800 \text{ seconds} = 91{,}500 \text{ cu. ft. (5-year frequency)}$$
$$\text{Runoff} = 59.25 \times 1{,}800 \text{ seconds} = 106{,}500 \text{ cu. ft. (10-year frequency)}$$

Similarly, runoff values have been computed for other times and are plotted on Fig. 18–12. It will be noted from the graph that the maximum difference between the pipe capacity line and the cumulative runoff

TABLE 18-4

(Continued)

Runoff "Q" (cfs)	Accumulated Runoff (cfs)	Velocity of Drain (ft./sec.)	Size of Pipe (in.)	Slope of Pipe (ft./ft.)	Capacity of Pipe (cfs)	Invert Elevation	Remarks
17.28	17.28	3.18	33	.0017	18.90	530.65	(n = 0.015)
18.02	35.30	2.84	54	.0007	45.00	528.03	See note below
9.01	44.31	2.84	54	.0007	45.00	527.44	See note below
16.25	16.25	3.27	33	.0018	19.40	530.11	
11.40	71.96	3.24	66	.0007	77.00	526.05	
						525.25	

for this inlet. Likewise time of concentration for inlet No. 10 is 48.8 minutes (41 + 2.7 + 5.0 = 48.7).

curves occurs at a time of approximately 60 minutes, and that maximum ponding values are:

$$P = 76{,}600 \text{ cu. ft. (5-year frequency)}$$
$$P = 102{,}500 \text{ cu. ft. (10-year frequency)}$$

While these values are less than the storage capacity between the inlet and contour 536, Fig. 18–12 indicates that it would require more than three hours for the 21-inch pipe to empty the ponding area even when considering the five-year flood. Since ponding over a long period of time is undesirable from the standpoint of safety and pavement performance, a larger culvert should be used.

18–8. Grading and Earthwork. The proper grading of an airport is required to provide safe and efficient grades for aircraft operations, to maintain good surface drainage, and to control erosion. Airport grading is characterized by wide, flat, and rounded slopes, with smooth transitions provided between graded and ungraded areas. Yet, because of the costs associated with earthwork operations, unnecessary grading should be avoided. Where future expansion of an airport is anticipated, the grading should be consistent with the ultimate proposed grades. Proposed grading operations are usually shown by means of a grading plan which shows original and proposed contour lines. (See Fig. 18–13.)

Grading quantities are usually computed by the average end area formula discussed in Chapter 12.

$$\text{Volume (cubic yards)} = \frac{L}{27} \times \frac{(A_1 + A_2)}{2}$$

However, because of the relatively flat topography and large expanses of areas to be graded at airports, it may be advantageous to consider

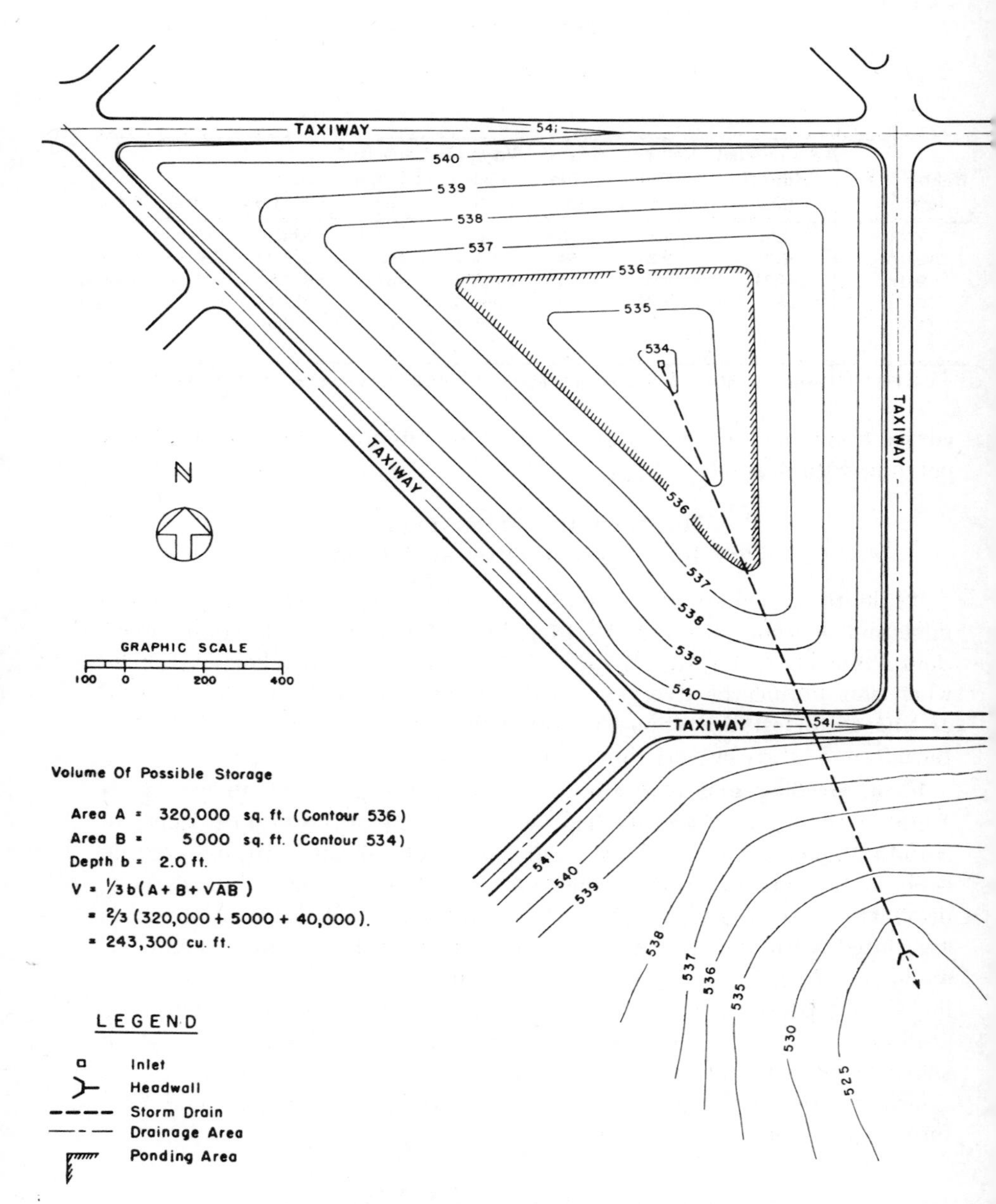

Fig. 18–11. Example of providing for ponding area. (Courtesy, Federal Aviation Administration.)

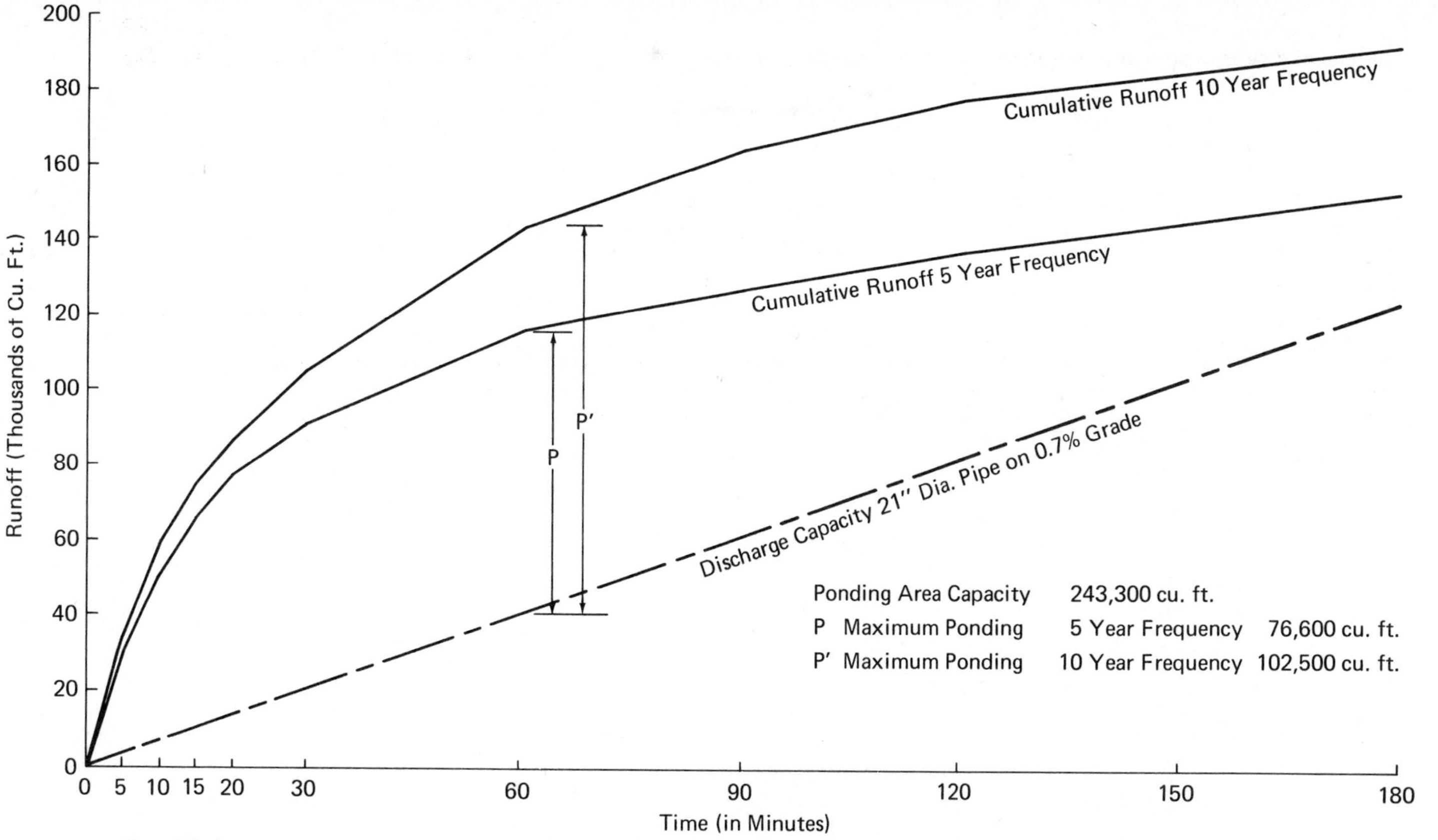

Fig. 18–12. Cumulative runoff for ponding in Fig. 18–11. (Courtesy, Federal Aviation Administration.)

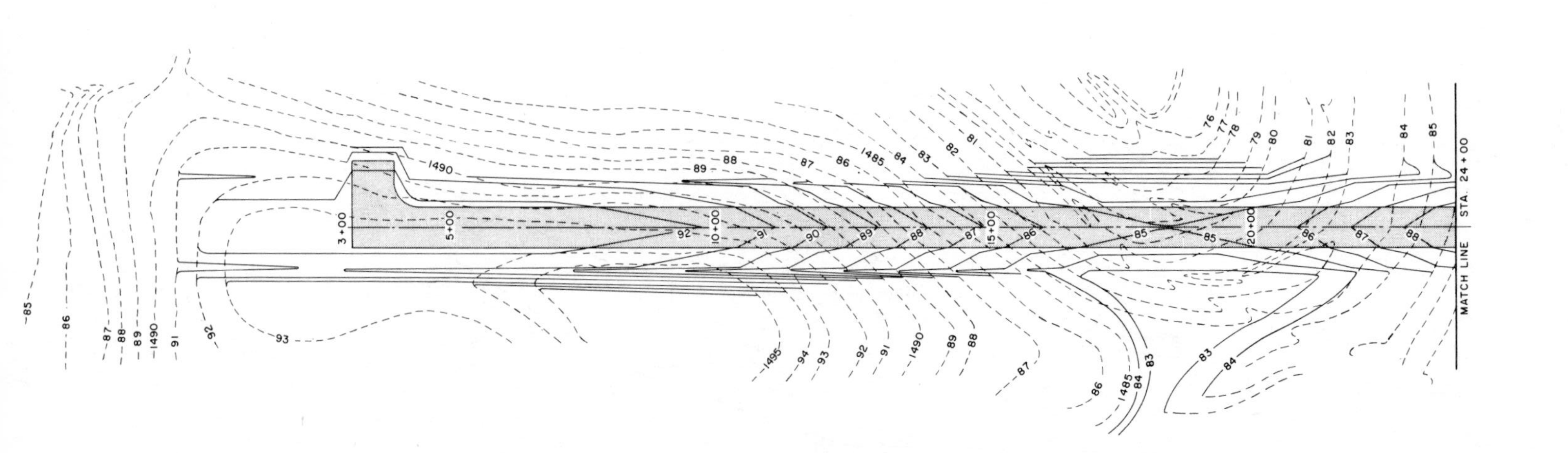

Fig. 18–13. A typical grading plan. (Source: *Utility Airports*, Federal Aviation Administration, November 1968.)

the areas enclosed by the original and final contour lines as end areas, A_1 and A_2. These areas can be measured with a planimeter and the contour interval becomes the length, L, used in the above formula. Where an embankment or excavation section ends between two contours, the vertical distance must be estimated.

AIRPORT MARKING AND LIGHTING

18–9. Visual Aids Requirements. During the major portion of flights while flying at altitude, pilots are assisted by magnetic compasses, gyros, and electronic devices, but presently these instruments are not reliable when the aircraft is within about 200 feet from the ground. Thus, landings and take-offs are accomplished largely "by eye," and during these critical operations various visual aids are placed at an airport to assist the pilot. For operations in the daytime and in good weather, the pilot is aided by airport markings. In inclement weather and at night, he depends on airport lighting.

In addition to visual aids to help a pilot locate and identify an airport or runway, he especially needs assistance in properly approaching the runway and landing the aircraft. Walter and Roggenveen [9] have pointed out that, while landing, an airplane is a moving coordinate system which is approaching a stationary coordinate system, the runway. These coordinate systems are shown by Fig. 18–14. The aircraft may not only move about each of the three axes, but it may also rotate about them. The pilot must, therefore, rotate, orient, and translate the aircraft so that it coincides with the coordinate system of the runway.

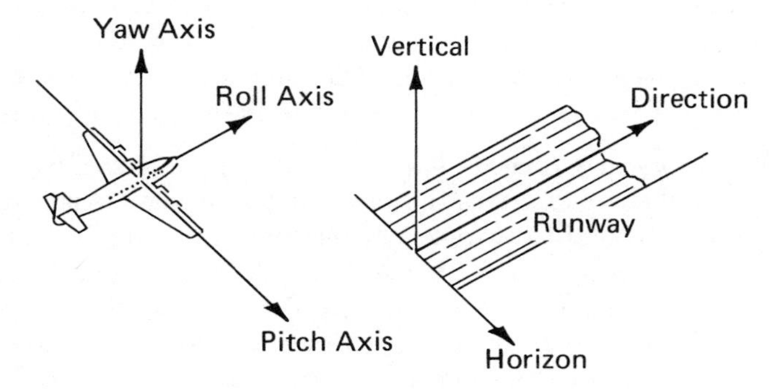

Fig. 18–14. Runway and airplane coordinate systems. (Source: Walter and Roggeveen, "Airport Approach, Runway and Taxiway Lighting Systems," *Journal of the Air Transport Division,* June, 1958.

To make a safe landing, a pilot must make correct judgments regarding:

1. alignment—whether the plane is headed straight for the runway
2. roll—weather the aircraft is properly banked in relation to the ground surface
3. the height of the aircraft above the runway, and
4. its distance from the end of the runway

In periods of good visibility, these judgments can be made by reference to familiar objects on the ground such as trees and buildings. When the visibility is restricted due to inclement weather or darkness, the pilot requires visual aids in the form of airport markings and lights.

18–10. Hangar and Strip Markers. At small airports which do not have a paved runway, the landing strip may be identified by strip markers such as that shown by Fig. 18–15. For additional locational guidance, the name of the airport and an arrow showing true north may be provided as a hangar marker.

18–11. The Segmented Circle Marker System. Another visual aid commonly used at small airports is the segmented circle. It consists of a series of pointed markers arranged in the form of a circle of 50-foot radius. The segments of the circle are typically three feet in horizontal width and 6 to 12 feet in length. Typical details are shown by Fig. 18–16.

The segmented circle helps a pilot to identify an airport and provides a standard location for various signal devices. A wind cone or sock is usually placed at the center of the segmented circle. If the airport has more than one runway, a landing direction indicator may be provided at the center of the circle in the form of an arrow or tee. To indicate the landing pattern and orientation of landing strips, indicators may be placed at the perifery of the segmented circle as shown by Fig. 18–16.

18–12. Runway and Taxiway Marking. Runway and taxiway marking consists of numbers and stripes that are painted on the pavement. Each end of a runway is marked with a number nearest one-tenth the magnetic azimuth of the runway centerline measured clockwise from the magnetic north. For example, a runway oriented N 10°E would be numbered 1 on the south end and 19 on the north end. Additional information is needed when two or more parallel runways are used and the designations L,C,R,LC, and RC are used to identify, respectively, left, center, right, left-center, and right-center runways. The numbers and letters are about 60 feet tall and 20 feet wide.

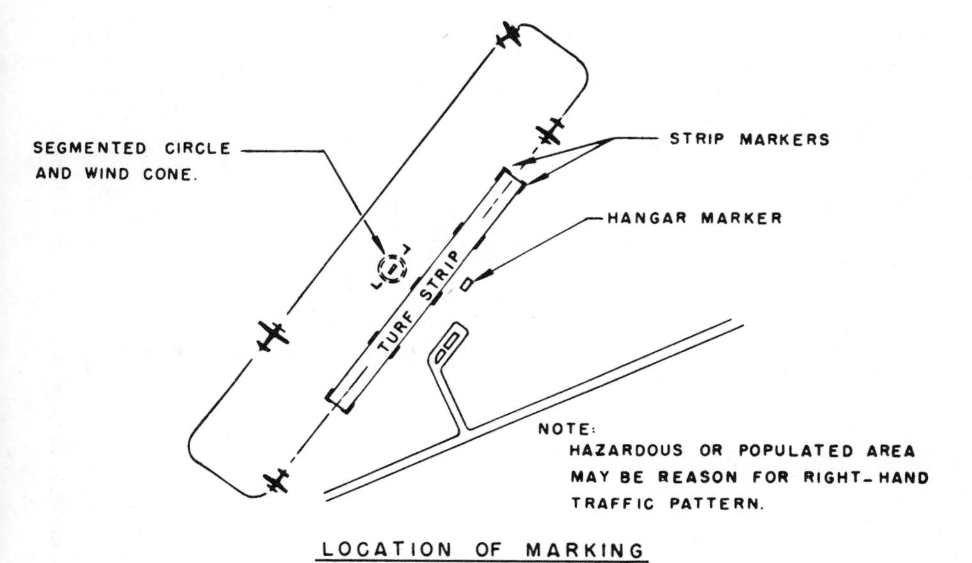

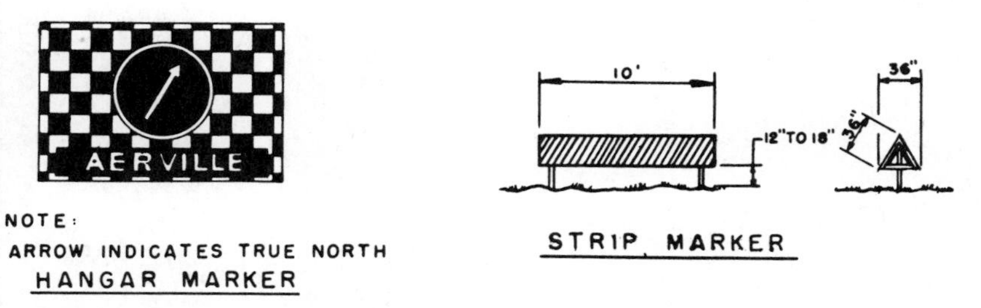

Fig. 18–15. Hangar and strip markers for utility airports. (Source: *Utility Airports,* Federal Aviation Administration, November, 1968.)

For purposes of runway marking, the FAA groups runways into three classes : (1) basic runway, (2) instrument runway, and (3) all-weather runway. Recommended markings for these runway classes are shown by Fig. 18–17. In addition to runway numbers, the following runway markings may be used:

1. a dashed centerline stripe
2. threshold markers
3. side stripes

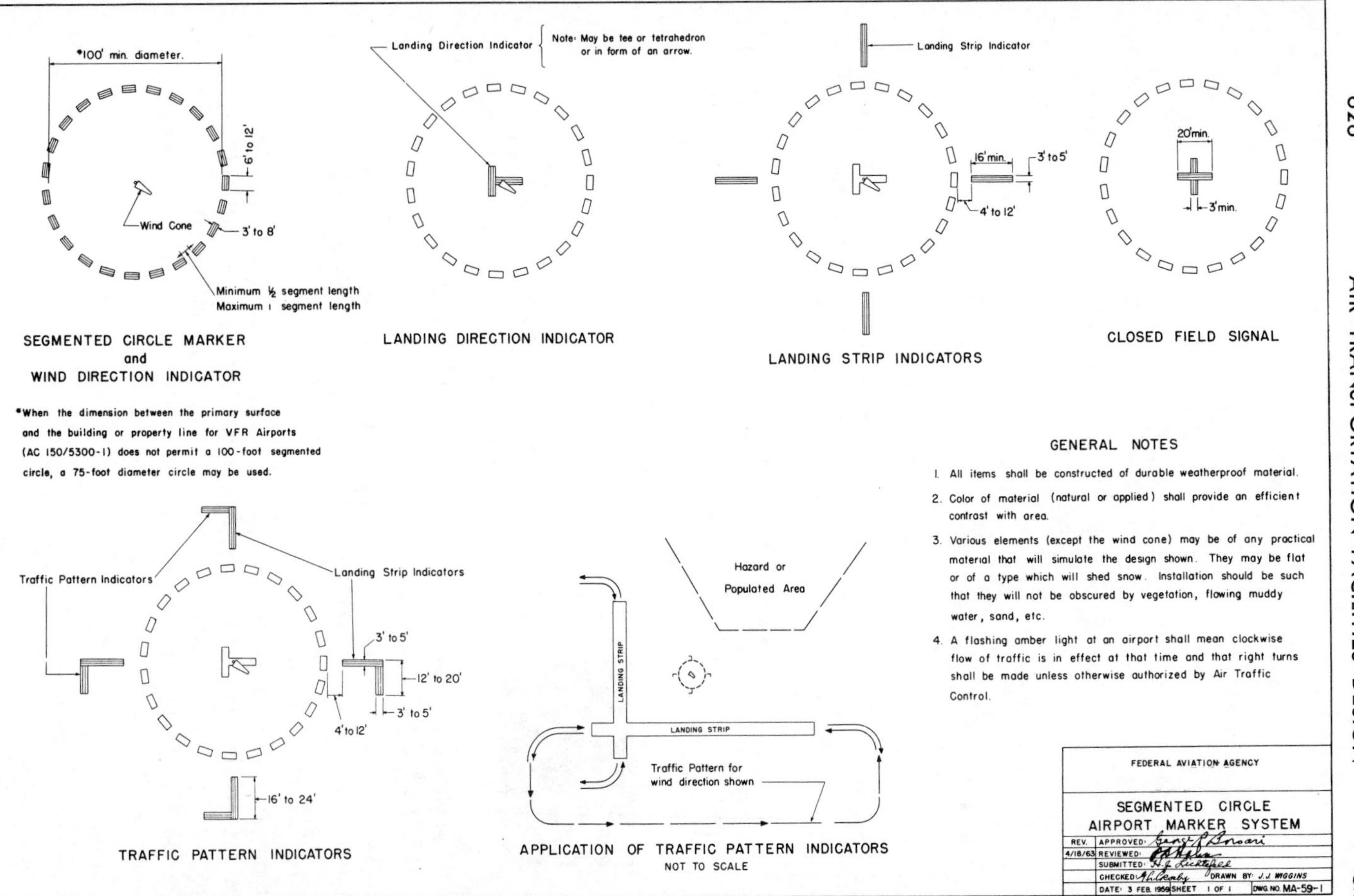

Fig. 18–16. Segmented circle airport marker system. (Courtesy, Federal Aviation Administration.)

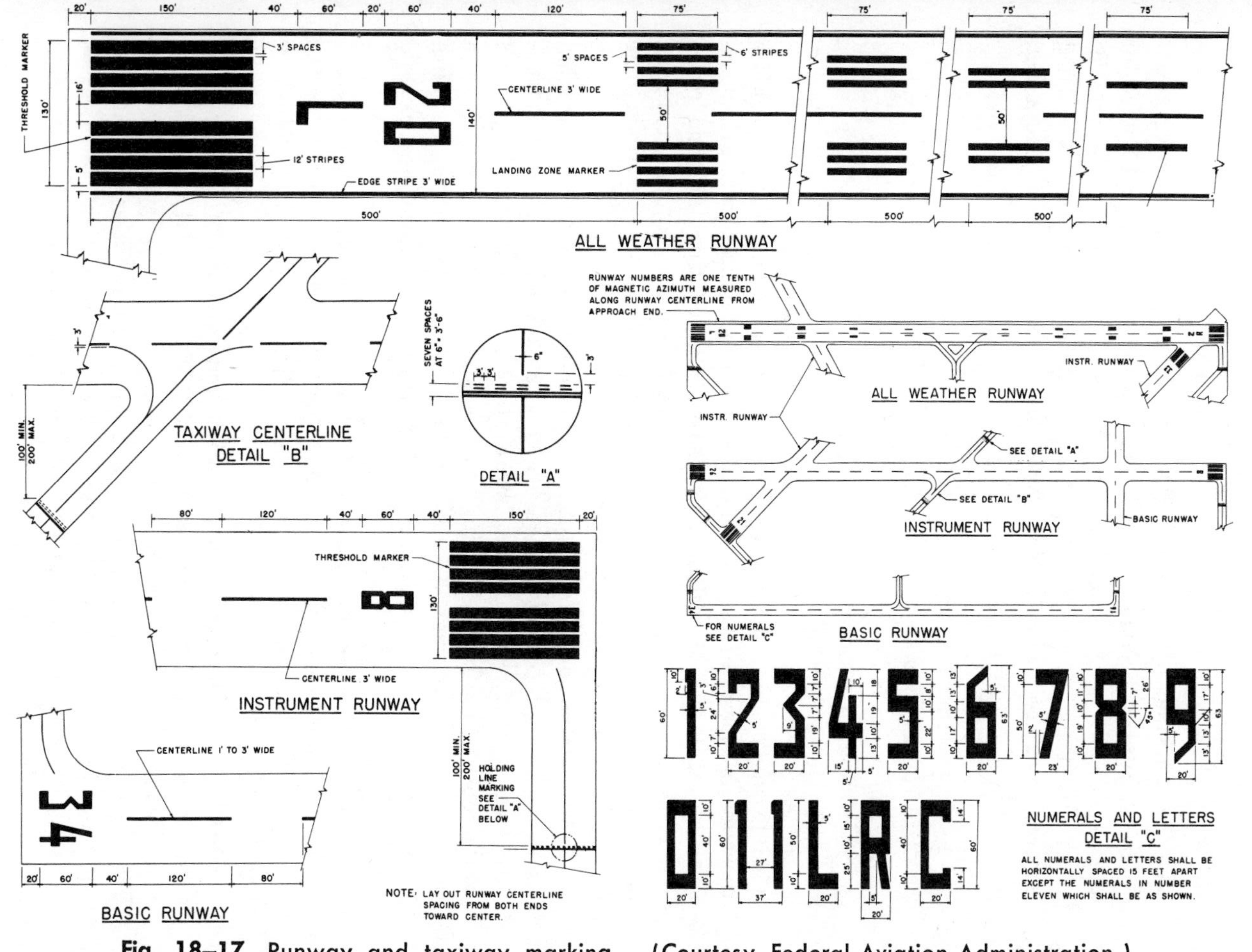

Fig. 18–17. Runway and taxiway marking. (Courtesy, Federal Aviation Administration.)

4. landing zone markers to indicate distance from the end of the runway
5. taxiway centerline stripes

Runway markings are typically white and taxiway markings are yellow.

18–13. Airport Lighting. The lighting which should be provided for a given airport depends on the airport size, the nature and volume of air traffic at night and during periods of inclement weather, and local meteorological conditions. Five types of airport lighting are described in the following paragraphs:

1. Obstruction lighting
2. Airport beacons
3. Approach lighting
4. Runway lighting
5. Taxiway lighting

This listing is not intended to be an exhaustive one, but it includes those major classes of airport lighting utilized to facilitate aircraft operations at night and in inclement weather. Illumination for wind cones and wind tees, ceiling light projectors, and lighting for aprons, hangars, and auto parking lots, though important, will not be described here due to limitations in space.

18–14. Obstruction Lighting. In the interests of air safety, obstruction lights must be placed on towers, bridges, smokestacks, and other structures which may constitute a hazard to air navigation. Single and double obstruction lights, flashing beacons, and rotating beacons are used to warn airmen during darkness or other periods of limited visibility of the presence of obstructions. A standard color, aviation red, is used for these lights.

The number, type, and placement of obstruction lights on a given structure will depend principally on its height. FAA standards for the lighting of obstructions are given in the publication *Obstruction Marking and Lighting* [10].

18–15. Airport Beacons. The location and presence of an airport at night is indicated by an airport beacon. While limited use has been made of a 10-inch "junior" beacon, a 36-inch beacon is typically used. The standard airport beacon rotates at a speed of 6 RPM and is equipped with an optical system which projects two beams of light 180 degrees apart. One light beam is green and the other is clear. A split beam is used to distinguish military airports.

18–16. Approach Lighting. Approach lights provide guidance to pilots during the few seconds it takes to travel from the approach area to the runway threshold. A pilot must be able to immediately identify and easily interpret approach lights. These lights provide the pilot information regarding the alignment, roll, and height of his aircraft and its distance from the runway threshold.

The basic needs for approach lights have been objectively and thoughtfully determined on the basis of the glide angle, visual range, cockpit cutoff angle, and landing speed of the aircraft (taken to be 1.3 times the stalling speed). It is agreed, for example, that approach lights should extend at least 1,400 feet from the runway threshold for nonprecision approaches and about 3,000 feet for precision approaches. However, the FAA has only recently standardized approach lighting requirements, and a confusing variety of approach lighting systems remain in use. As late as 1958, Walter and Roggeveen [9] reported that there were eleven major systems used in the U.S. and in 1961, Finch [11] described approach lighting conditions as "chaotic."

The FAA [12] recommends eight standard approach lighting systems. These systems can be grouped into three categories:

1. Medium Intensity Systems
2. Simplified Short Systems
3. Standard 3,000 Feet High Intensity Systems

All of the recommended FAA approach lighting configurations have a series of light bars installed perpendicular to the extended runway centerline at specified spacings. These light bars are composed of five lamps each. All of the recommended FAA configurations also feature a wide light bar installed at a distance of 1,000 feet from the runway threshold. Sketches of FAA approach lighting systems are shown as Figs. 18–18, 18–19, 18–20, and 18–21. Except as noted, approach lights are white.

Medium Intensity Systems. Medium Intensity Approach Lighting Systems are economy systems which utilize 150 watt lamps. These systems are recommended for utility airports. There are three recommended medium intensity approach configurations:

MALS—Medium Intensity Approach Lighting System. (See Fig. 18–18.)

MALSF—Medium Intensity Approach Light System with Sequenced Flashers. This is the same as MALS except it is equipped with three sequenced flasher lights. This system would be used where approach area identification problems exist.

MALSR—Medium Intensity Lighting with Runway Alignment Indicator Lights. This system is similar to MALS except that eight

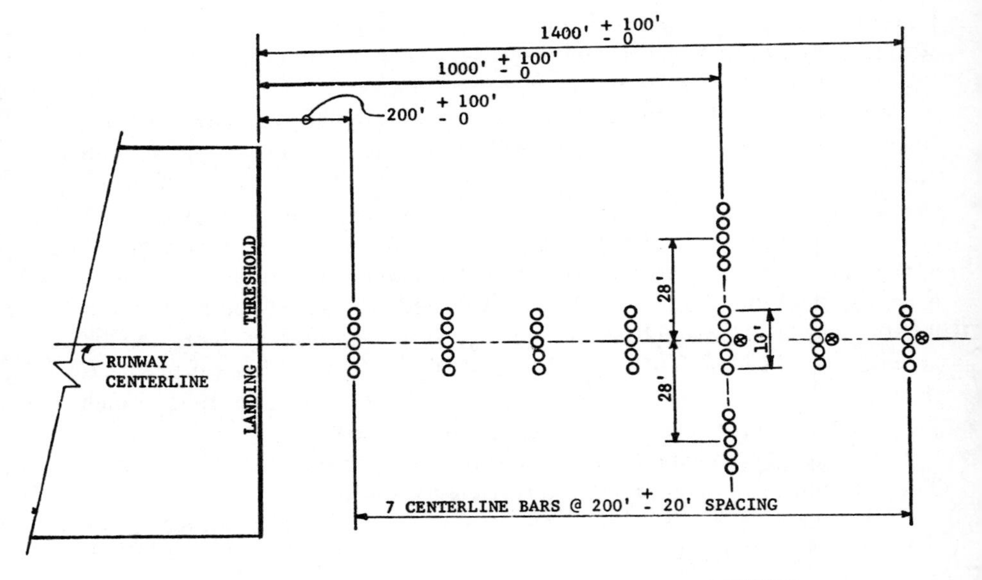

Fig. 18–18. MALS and MALSF approach lighting configurations. (Source: *Visual Guidance Lighting Systems,* Federal Aviation Administration, May 29, 1969.)

flashing lights are installed along the extended runway centerline at 200 foot spacings extending the total length to 3,000 feet.

Simplified Short Systems. Simplified Short Approach Lighting Systems are high intensity systems (utilizing 300 watt lamps) which may easily be upgraded to Standard 3,000 Foot High Intensity Systems.

SSALS—Simplified Short Approach Lighting System. (See Fig. 18–19.)

SSALF—Simplified Short Approach Lighting System with Sequenced Flashers. This is the same as SSALS except it is equipped with three sequenced flashers. It is recommended for use where approach area identification problems exist.

SSALR—Simplified Short Approach Lighting System with Runway Alignment Indicator Lights. SSALR is an economy-type system which can be used for precision approaches. It is similar to SSALS

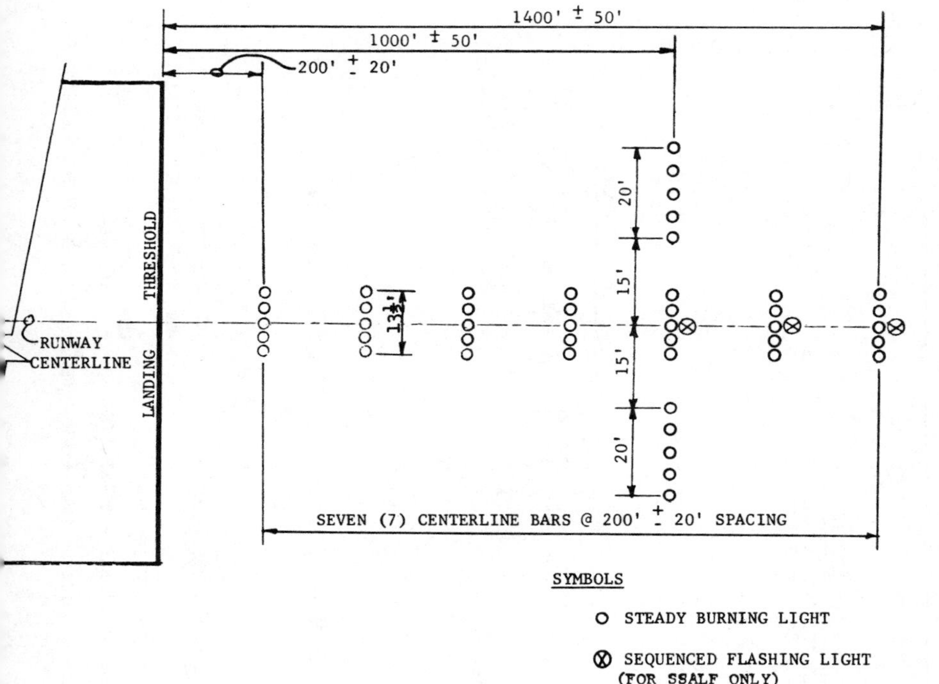

Fig. 18–19. SSALS and SSALF approach lighting configurations. (Source: *Visual Guidance Lighting Systems,* **Federal Aviation Administration, May 29, 1969.)**

except that eight flashing lights are installed along the extended runway centerline at 200 foot spacings extending the total length to 3,000 feet.

Standard 3,000 Foot High Intensity Systems. These systems are the most elaborate and expensive approach lighting systems recommended by the FAA. In addition to steady burning lights, these configurations feature a system of sequenced flashing lights. One such light is installed at each centerline bar starting 1,000 feet from the threshold and extending outward to the end of the system. The sequence flashing lights appear as a ball of light travelling at a speed of approximately 4,100 miles per hour [12].

There are two Standard 3,000 Foot High Intensity Systems, ALSF-1 and ALSF-2. Configurations for these systems are shown in Figs. 18–20 and 18–21. The ALSF-2 configuration has two lines of lights embedded

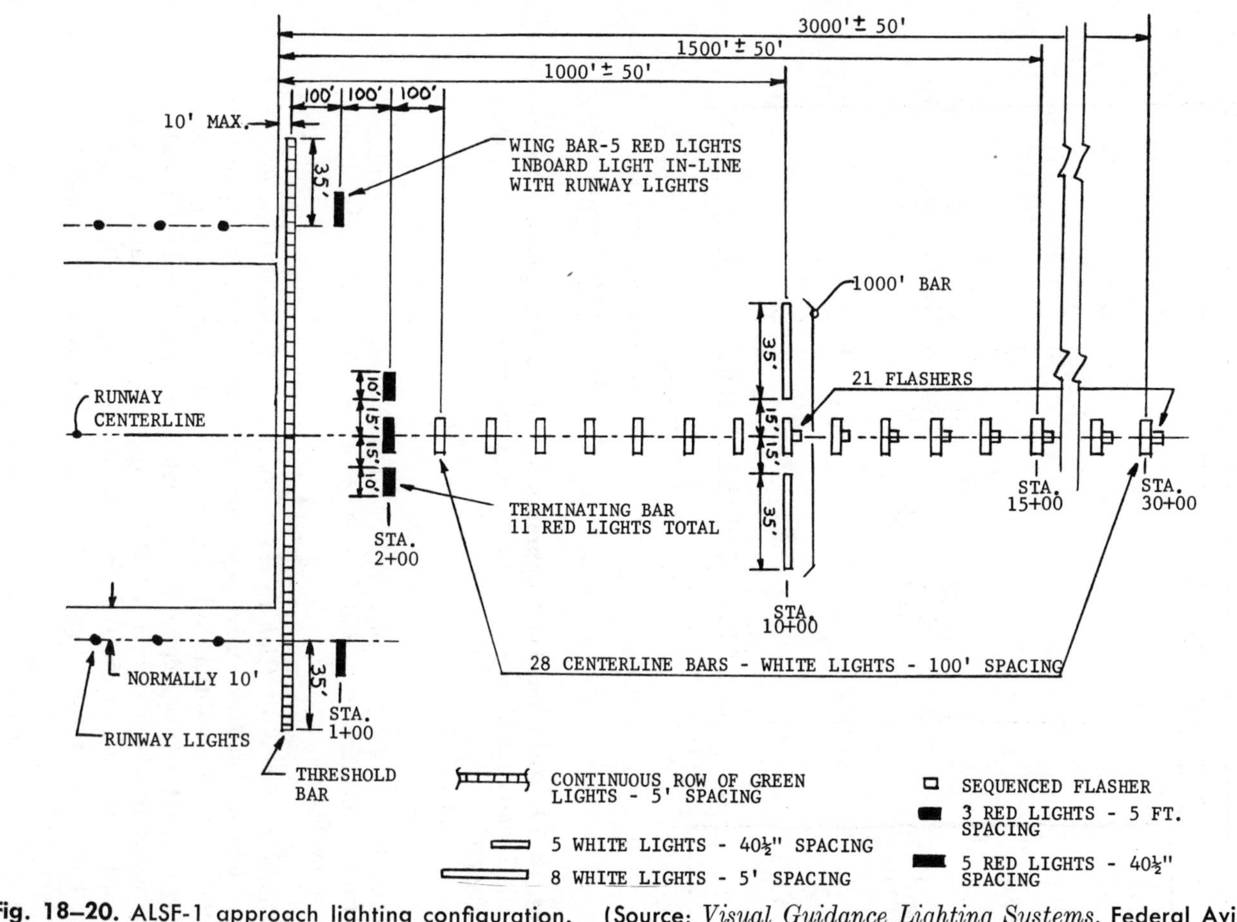

Fig. 18–20. ALSF-1 approach lighting configuration. (Source: *Visual Guidance Lighting Systems,* Federal Aviation Administration, May 29, 1969.)

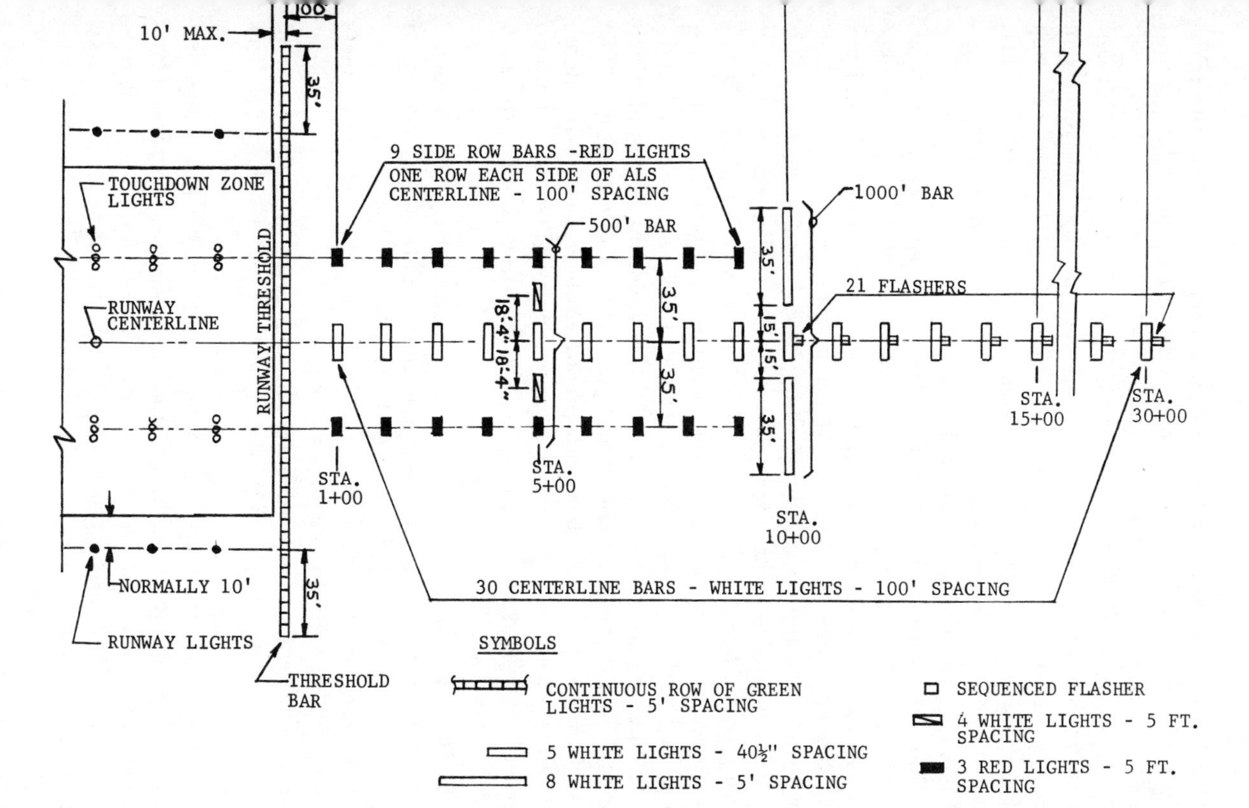

Fig. 18–21. ALSF-2 approach lighting configuration. (Source: *Visual Guidance Lighting Systems*, Federal Aviation Administration, May 29, 1969.)

in the pavement to provide guidance in the central portion of the runway in the touchdown zone.

18–17. Other Visual Aids for Aircraft Approaches. There are two additional visual aids systems used to facilitate easier and safer aircraft approaches: (1) VASI, (2) REIL.

The Visual Approach Slope Indicator System (VASI) basically consists of light bars placed on each side of the runway, 600 feet from the runway end (downwind bars), and a second set of bars on each side of the runway, 1,300 feet from the runway end (upwind bars). There are three basic VASI configurations, the 12-box, 4-box, and 2-box. The 12-box configuration is shown by Fig. 18–22.

Each light box in the VASI system projects a split beam of light, the upper segment being white and the lower red. When a pilot makes an approach, he sees white lights if the approach is too high and red lights if the approach is too low. When a proper approach is made, the downwind bars appear white and the upwind bars appear red. This system is primarily intended for use in VFR weather conditions.

Runway End Identifier Lights (REIL) are sometimes placed at the ends of runways to provide rapid and positive identification of the approach end of a runway. The system consists of two synchronized flashing lights, one on each end of the runway threshold, the beams of which are aimed 10–15 degrees outside a line parallel to the centerline. The REIL system is used where there is a preponderance of confusing lights from off-airport sources such as motels, automobile lights, etc. It would not normally be used if sequenced flashers are used in the approach lighting system.

18–18. Runway Lighting. Runway lighting includes: (1) threshold lights and (2) runway edge lights. Threshold lights, which consist of a line of green lights extending across the width of the runway, identify to the pilot the runway end and help him decide whether to complete his landing or to execute a missed approach.

Runway edge lights consist of lines of lights installed not more than 10 feet from the pavement edge and spaced not more than 200 feet on centers.

The FAA specifies two types of runway lighting systems: a medium intensity runway lighting system [13] and a high-intensity runway lighting system [14]. The high-intensity system is used for instrument runways, on runways that are approved for straight-in approach procedures, and on non-instrument runways where there is traffic congestion and poor visibility conditions.

Runway edge lights are aviation white except that aviation yellow is used in the last 2,000 feet of instrument runway to indicate the caution zone.

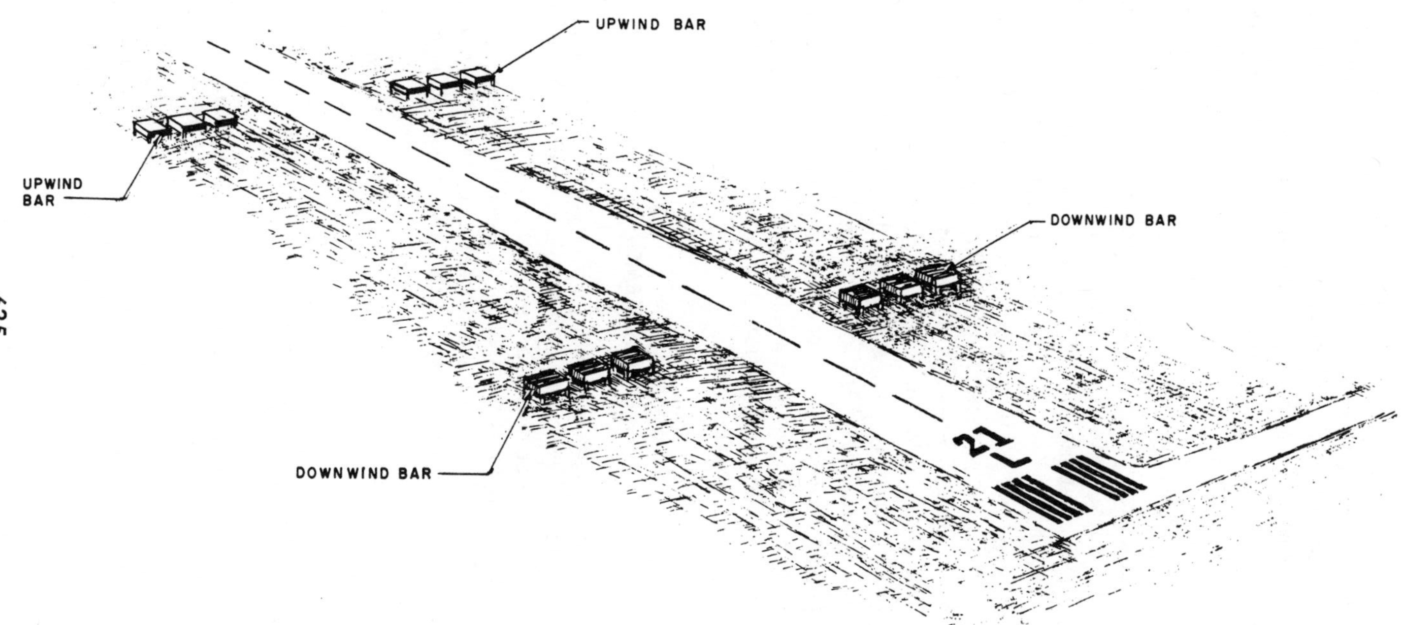

Fig. 18–22. Typical visual approach slope indicator facility installation. (Courtesy, Federal Aviation Administration.)

18–19. Taxiway Lighting. Taxiway lighting provides guidance for pilots for maneuvering along the system of taxiways which connect the runways and the terminal and hangar aprons. The conventional taxiway lighting consists of omnidirectional blue lights located on each side of the taxiway pavement. These lights are offset not more than 10 feet from the pavement edge and spaced longitudinally not more than 200 feet apart. Much shorter spacings are required on short curves and at intersections.

Taxiway centerline lights may be installed instead of taxiway edge lights in new construction, and may supplement taxiway edge lights where operations occur in low visibility or where taxiing confusion exists.

The taxiway centerline lighting system consists of single semiflush lights (with less than 3/8 inch protrusions above the pavement surface) inset in the taxiway pavement along the centerline. These lights are steady burning and have a standard color of aviation green.

Detailed information on the design, installation, testing, maintenance, and inspection of taxiway edge and centerline systems are given, respectively, by References 15 and 16.

18–20. Airport Paving. It is regretted that space limitations preclude the inclusion of a discussion of the subject of airport paving. The reader is referred to the FAA publication *Airport Paving* [17] for a complete discussion of this important topic.

PROBLEMS

1. What length of runway is required for a general utility airport which is 6,200 feet above sea level and has a normal maximum temperature of 52°F? The effective runway gradient is 1.3 per cent.
2. What length of runway is required for a Beech D18S aircraft if the airport elevation is 2,500 feet and the normal maximum temperature is 75°F? The maximum difference between the highest and lowest point on the runway is six feet.
3. What length of runway is required for a Boeing 707-300 series aircraft given the following design condition:

Normal maximum temperature	64°F
Airport elevation	3,800 feet
Flight distance	2,100 miles
Maximum landing weight	190,000 pounds
Effective runway gradient	0.65%

4. Profile grade data for a proposed airport runway are given below. Does the proposed longitudinal grade design conform with the requirements of the FAA? The first vertical curve is 1,600 feet long and the second is 1,200 feet long.

Station	*Grade*	*Comment*
0 + 00		Begin runway
	−0.85%	
30 + 50		P.I. Station No. 1
	+0.65%	
54 + 50		P.I. Station No. 2
	−0.50%	
96 + 00		End runway

5. A proposed runway is to be 6,000 feet long. What are the recommended widths of the runway and taxiway? The runway is at an elevation of 3,200 feet and has a normal maximum temperature of 48°F. The effective gradient is 1.6 per cent. It may be assumed that turbojets will not use this runway.

6. Obtain rainfall data for your locality from the U.S. Weather Bureau or some other source. Prepare a rainfall intensity-duration curve similar to Fig. 18–10 (5-year curve). Using the other data given in Example 18–5, design a drainage system for the apron and taxiways shown by Fig. 18–10.

7. Design a drainage system for the apron and taxiways shown by Fig. 18–10 given the runoff values shown below. Other required data are given by Table 18–5.

Inlet	*Runoff "Q"*
12	22.40
11	26.91
10	11.69
13	18.27
9	15.02

REFERENCES

1. *Utility Airports,* Federal Aviation Administration, Advisory Circular AC 150/5300-4A, November, 1968.
2. *Runway Length Design Requirements for Airport Design,* Federal Aviation Administration, Advisory Circular AC 150/5325-4, through change 8, November 8, 1967.
3. *Is Your Airport Ready for the Boeing 747,* Federal Aviation Administration, Advisory Circular AC 150/5325-7, January 15, 1968.
4. *Airport Surface Areas Gradient Standards,* Federal Aviation Administration, Advisory Circular AC 150/5325-2A, May 12, 1966.
5. *Runway/Taxiway Widths and Clearances for Airline Airports,* Federal Aviation Administration Advisory Circular, AC 150/5330-2A, July 25, 1968.
6. *Airport Design Standards—Airports Served by Air Carriers—Taxiways,* Federal Aviation Administration, Advisory Circular AC 150/5335-1A, May 15, 1970.
7. *Airport Drainage,* Federal Aviation Administration, Advisory Circular AC 150/5320-5A, 1966.
8. *Airport Drainage,* Portland Cement Association, 1966.
9. Walter, C. Edward, and Roggeveen, Vincent J., "Airport Approach, Runway and Taxiway Lighting Systems," *Journal of the Air Transport Division,* American Society of Civil Engineers, June, 1958.
10. *Obstruction Marking and Lighting,* Federal Aviation Administration, Advisory Circular AC 70/7460-1, February, 1968.

11. Finch, D. M., "Recent Developments in Airport Lighting," *Journal of the Air Transport Division,* American Society of Civil Engineers, August, 1961.
12. *Visual Guidance Lighting Systems,* Federal Aviation Administration, Handbook No. 6850.2, May 29, 1969.
13. *Medium Intensity Runway Lighting System,* Federal Aviation Administration, Advisory Circular AC 150/5340-16A, December 19, 1967.
14. *High Intensity Runway Lighting System,* Federal Aviation Administration, Advisory Circular AC 150/5340-13A, April 14, 1967.
15. *Taxiway Edge Lighting System,* Federal Aviation Administration, Advisory Circular AC 150/5340-15A, November 1, 1967.
16. *Taxiway Centerline Lighting System,* Federal Aviation Administration, Advisory Circular AC 150/5340-19, November 14, 1968.
17. *Airport Paving,* Federal Aviation Administration, Advisory Circular AC 150/5320-6A, September 12, 1968.

V

DESIGN OF WATER TRANSPORTATION FACILITIES

19

Introduction to Water Transportation

19–1. Introduction. Since ancient times, water transportation has broadened the horizons of man and profoundly influenced the growth and development of civilization. Historians report that as early as 6,000 B.C. the Egyptians had ships with masts and sails, and galleys were used on the Nile River as early as 3,000 B.C. During the reign of King Solomon (circa 961–922 B.C.), Phoenician galleys sailed from Biblical Tyre and Sidon bringing copper from Cyprus, papyrus from Egypt, and ivory, gold, and slaves from Africa. These large, low, and typically one-decked vessels were propelled by both sails and oars. Galleys, which were often manned by slaves, were used by Rome in her war with Carthage (140 A.D.). These vessels continued to be used throughout the Middle Ages, especially in the Mediterranean Sea.

In early America:

> River boats were nothing more than huge rafts called *flatboats* or broad-horns. Generally, they were flat-bottomed and boxlike, covered from stem to stern. The flatboat was a one-way vessel, dependent entirely upon currents for propulsion with only occasional guidance from its handlers. At the end of its downstream run, it was usually broken up and the lumber sold.
>
> The *keelboat* began to appear on the rivers at about the turn of the 19th Century. It was a long, narrow vessel with graceful lines, sturdily built to withstand many trips both downstream and upstream. The keelboat could carry as much as 80 tons of freight. It was floated downstream under careful guidance, and cordelled upstream. Cordelling took two forms: the crew walked along the bank and pulled the keelboat with ropes; or they literally pushed it upstream with iron-tipped poles which extented to the bottom of the river. One historian has estimated that there were as many as 500 keelboats operating on the Ohio River and its tributaries by 1819.

> The *steamboat* was invented in 1807; and in 1811 the river steamer *New Orleans* was launched in Pittsburgh and went into operation between there and New Orleans. By 1835 New Orleans was posting the arrival of over 1,000 steamboats per year. Records indicate that by 1852 the public landing at Cincinnati was recording the arrival of steamboats at an annual rate of 8,000, about one per hour. The tonnage handled by steamboats on the rivers of the United States at the height of the packet boat era, just before the War Between the States, is reported to have exceeded the tonnage handled by all the vessels of the British Empire [1].

Although the feasibility of steamboat travel in the open sea was demonstrated by Col. John Stevens in 1809, sailing vessels continued to dominate ocean transportation until shortly before the Civil War.

Scheduled ocean travel was initiated on January 5, 1818 when the *James Monroe,* a 424-ton ship, sailed from New York to Liverpool. This packet or liner service, which gave uncommon services to passengers and fast movement of freight, was highly successful. Larger, faster, and more expensive ships were built, additional lines were organized, and the packet service was extended.

The first half of the nineteenth century was a prosperous period for shipping companies. The quest to sail "at a fast clip" resulted in the design and building of over 400 clipper ships during the period 1846–1855. These vessels traveled at speeds which rival some of the faster commercial ships in service today. Despite the speed and classic beauty of these vessels, the clipper ships were soon replaced by steam ships, and following the Civil War, there were sharp decrease[1] in shipping under the American flag. This decline in U.S. influence in ocean shipping is felt to the present day.

19–2. Recent Growth in Water Transportation. In recent years, substantial increases in water transportation activity have been noted. These increases are attributed to growing world population, the development of new products and new sources of raw materials, and general industrial growth, especially in the petroleum industry. Accompanying the increases in water transportation tonnage, larger ships have been built and innovative storage and loading facilities have been developed. These changes have created an ever increasing need for the most modern and efficient port facilities.

Data furnished by the Maritime Administration, U.S. Department of Commerce, indicate that there has been substantial growth in the

[1] For all practical purposes the river fleet was destroyed during the Civil War and was not rebuilt. There was little progress in inland water transportation from the end of the Civil War until about 1920 [1].

number and gross tonnage[2] of the merchant ships of the world in recent years:

Year	*Ships*	*Gross Tons*
1948	12,643	71,549,000
1958	16,966	112,314,000
1968	19,361	184,242,000

Gross tonnage has grown at a much greater rate than the number of ships, reflecting the trend to larger ships.

Similar growth has been experienced in inland water transportation as indicated by the data furnished by the American Waterways Operators, Inc. [1, 2]:

Year	*Towing Vessels*	*Barges*	*Capacity (net tons)*
1948	4,127	11,689	8,740,569
1958	4,169	15,221	13,771,757
1968	4,240	18,416	20,940,261

Today, nearly 99 per cent of overseas freight tonnage is transported by ships and approximately 16 per cent of the freight (ton-miles) within the continental U.S. moves on the inland waterway system.

19–3. The Nature of Water Transportation. By its nature, water transportation is most suitable for bulky and heavy commodities which have to be moved long distances and for which time of transport is not a critical factor.

Some of the most common classes of cargo transported by ocean-going ships are:

1. General cargo
2. Bulk cargo
3. Heavy machinery
4. Motor vehicles
5. Wood and wood products
6. Livestock
7. Perishable foods

General cargo, the most important class, refers to a wide variety of packaged goods including such materials as:

1. Cutlery, hardware, and implements
2. Electrical goods
3. Fabrics
4. Shoes, leather, and other leather goods
5. Pottery and glass

[2] Gross tonnage is a cubic measurement: one gross ton equals 100 cubic feet of storage space. The data given are for ocean going ships of 1,000 gross tons and over.

Bulk cargo includes chemicals, dyes, grains, and crude and refined petroleum.

Heavy machinery is transported by ship to practically every nation in the world. For the most part, the machinery is exported from the United States, Switzerland, Italy, Germany, Denmark, and Canada.

More than 540,000 automobiles, trucks, and buses were exported by the United States in 1968 while 1.7 million vehicles, mostly automobiles, were imported. These vehicles were transported by ships.

Special storage and hangling facilities are required for lumber, livestock, perishable foods, and many other commodities transported by ocean-going ships.

TABLE 19-1

Principal Commodities Transported on the U.S. Inland Waterways (Exclusive of the Great Lakes)

Commodity	Net Tons (1967)	Per Cent
Bituminous Coal and Lignite	115,570,662	23.1
Other Petroleum and Coal Products	71,764,673	14.3
Sand and Gravel	58,650,911	11.7
Crude Petroleum	50,677,470	10.1
Gasoline, Jet Fuel, and Kerosene	50,492,198	10.1
Marine Shells, unmanufactured	23,362,780	4.7
Rafted Logs	18,290,048	3.7
Chemicals and Chemical Products	17,551,703	3.5
Grain and Grain Products	17,397,743	3.5
Iron, Steel Products, and Scrap	7,126,418	1.4
Liquid Chemicals	6,792,734	1.4
Soybeans	4,719,270	0.9
Nonmetallic Minerals, excluding fuels	3,725,137	0.7
Limestone Flux and Calcareous Stone	3,673,194	0.7
Building Cement	3,548,977	0.7
Pulpwood, Lumber, and Lumber Products	2,936,387	0.6
Iron Ore and Concentrates	2,781,355	0.6
Fresh Fish and Shellfish	2,459,108	0.5
Clay, ceramic	2,100,363	0.4
Fertilizer and Fertilizer materials	1,935,023	0.4
Crude Products	1,926,267	0.4
Paper and Paper Products	1,636,675	0.3
Sugar	929,166	0.2
Molasses, inedible	856,877	0.2
Bauxite, aluminum ores and concentrates	668,133	0.1
Other Commodities	29,339,461	5.8
Grand Total	500,912,733	100.0

Source: *1967 Inland Waterborne Commerce Statistics*, The American Waterways Operators, Inc., April, 1969.

The large majority of the traffic moving over the inland waterways is made up of bulk commodities as Table 19–1 indicates.

Freight movement on the Great Lakes consists predominantly of iron ore, steel, coal, and grain.

One of the most important developments in water transportation to occur in recent years has been the shipment of certain types of freight in large sealed boxes called containers. (See Fig. 19–1). The use of containers promises to speed dramatically the handling of freight and decrease water transportation costs.

In the paragraphs that follow, coastal environmental conditions which complicate the engineering design and shorten the useful life of port and harbor facilities will be discussed.

Fig. 19–1. A modern container ship. (Courtesy United States Lines).

Due to space limitations, it will not be possible to discuss a number of topics which relate to water transportation. Specifically, the important topics of river hydraulics, beach erosion, and the design of locks, dams, and canals will not be covered. For information on these topics, the reader may refer to other textbooks or to the publications listed at the end of this chapter.

DESIGN FOR THE COASTAL ENVIRONMENT

19–4. The Coastal Environment. The design of durable port and harbor facilities is one of the most challenging problems which face the engineer. The environment of the seacoast is harsh and corrosive, and water transportation facilities must be designed to withstand the various destructive biological, physico-chemical, and mechanical actions which are inherent to the coastal environment. Enormous forces of winds, waves, and currents are imposed on port and harbor structures. Wood structures must withstand the forces of decay and the attack of termites and other biological life. Concrete structures must be designed and constructed to highest engineering standards to prevent rusting of the reinforcement and spalling of the concrete. Without protection, steel structures corrode and do not last long in the coastal environment.

19–5. Wind. Wind is the approximate horizontal movement of air masses across the earth's surface. Winds result from changes in the temperature of the atmosphere and the corresponding changes in air density. Wind exerts a pressure against objects in its path which depends on the wind velocity. The equation for the calculation of wind pressure on a structure is:

$$p = Kv^2 \qquad (19\text{–}1)$$

where p is expressed in pounds per square foot, wind velocity is expressed in miles per hour, and K is a factor which depends principally on the shape of the structure. The values of K most frequently used range from 0.0025 to about 0.0040.

Considerable judgment is required in computing wind forces on coastal structures and port facilities. It should be remembered that loading equipment will not generally be used when winds exceed about 15 miles per hour, and ships will not usually remain alongside a wharf during a severe storm.

19–6. Waves. While the design of buildings and other structures must accommodate wind loads, our principal interest in wind in coastal design lies in its contribution to the formation of waves. Waves may be caused by earthquakes, tides, and man-made disturbances such as explosions and moving vessels; however, the waves of principal interest in coastal design are those formed by winds.

When wind moves across a still body of water, it exerts a tangential force on the water surface which results in the formation of small ripples. These irregularities tend to produce changes in the air stream near the water surface. Pressure differentials in the air stream are formed which cause the water surface to undulate. As the wind continues, this process is repeated and the waves grow.

The form and size of water waves have been the subject of considerable scientific observation and research. There is now general acceptance that the surface of deep-water waves is approximately trochoidal in form. According to the trochoidal theory, wave movement can be described by assuming that individual particles are not translated, but rotate in a vertical plane about a horizontal axis. This is consistent with the observed tendency of a floating object in deep water to rise and fall and oscillate but not be translated by the waves.

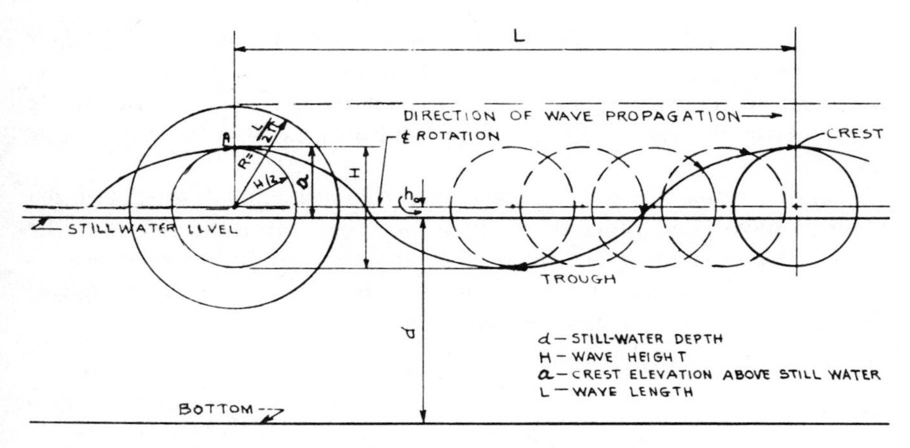

Fig. 19–2. Deep water wave characteristics. (Source: Alonzo Quinn, *Design and Construction of Ports and Marine Structures*, McGraw-Hill Book Co., 1961. By permission of the publisher.)

Figure 19–2 shows the surface of an ideal deep-water wave as a trochoid formed by the rotating particles of water. The trochoid is described by a point on a circle which rotates and rolls in a larger concentric circle. The center of the circle moves along a line which lies above the still water level. The amount this line is elevated above the still water level depends on the wave steepness. The diameter of the smaller circle is equal to the wave height and the circumference of the larger circle equals the wave length. The line thus described represents the wave surface. Particles beneath the wave surface also tend to rotate similarly in circular paths, but the radii decrease rapidly and roughly exponentially with depth.

The speed of the wave form, or wave velocity, in feet per second,

$$V = \frac{L}{T} \tag{19–2}$$

where:

L = wave length, or distance between consecutive crests, feet
T = wave period, or time for wave to travel one wave length, seconds

The speed of waves in deep water is approximately equal to the velocity acquired by a body falling freely through a height equal to one-half the radius of a circle the circumference of which equals the length of the wave. Thus:

$$V = \sqrt{2g\left(\frac{1}{2}\right)\left(\frac{L}{2\pi}\right)} = \sqrt{\frac{gL}{2\pi}} \tag{19–3}$$

In this equation, g is acceleration due to gravity.

In shallow water (i.e., when the water depth is less than one-half the wave length), the paths of water particles are influenced by the frictional forces of the sea bed. This causes the orbit of the water particles to become approximately elliptical with the major axis horizontal.

Shallow water waves are exceedingly complex and at least a dozen theories have been proposed to describe them. It is agreed that the velocity of shallow water waves is a function of the water depth.

The following equation should give suitable estimates of wave velocities in depths less than one-tenth of the wave length:

$$V = \sqrt{gD} \tag{19–4}$$

where:

D = water depth, feet.

A general equation for wave velocity in shallow water has been proposed by G. B. Airy:

$$V = \sqrt{\frac{gL}{2\pi}\tanh\left(\frac{D}{R}\right)}$$

or:

$$V = \sqrt{\frac{gL}{2\pi}\tanh\left(\frac{2\pi D}{L}\right)} \tag{19–5}$$

It is noted that Equation (19–5) generalizes to Equation (19–3) when D become greater than L.

As a comparison of shallow and deep water equations indicate, the velocity of the wave decreases as the wave moves into shallow water.

When waves approach the shore at an oblique angle, the portion of the wave nearest the shore slows down with the result that the wave swings around and tends to become parallel to the shore. At the same time, the wave lengths decrease as the wave period remains constant. This phenomenon is known as *wave refraction*. The U.S. Navy Hydrographic Office has published a graphical procedure for determining the direction of waves and the lengths of refracted waves [3].

When waves move into shallower depths, as along the coast, the orbits of the particles become distorted due to the friction exerted by the bottom. This causes the major axis of the elliptical path to tilt shoreward from the horizontal, and the wave gradually transforms from a purely oscillatory wave to a wave of translation. (See Fig. 19–3.) It

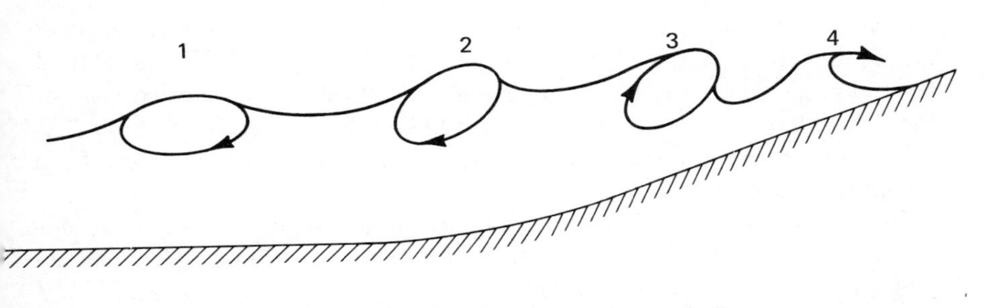

Fig. 19–3. Development of a wave of translation.

is at this point that waves are capable of exerting great forces against bulkheads, breakwaters, and other coastal structures. Techniques for estimating these forces will be described in Chapter 20.

Wave heights have been found to vary with wind velocity and fetch, the straight-line stretch of open water available for wave growth. Several empirical equations have been proposed for the estimation of maximum wave height. Two of the earliest such equations were proposed by Thomas Stevenson in 1864:

$$H_{\max} = 1.5F + 2.5 - \sqrt[4]{F} \text{ for } F < 30 \text{ nautical miles} \quad (19\text{–}6)$$

$$H_{\max} = 1.5\sqrt{F} \text{ for } 30 < F < 900 \text{ nautical miles} \quad (19\text{–}7)$$

where:

$H_{\max}$ = maximum wave height, feet
F = Fetch, nautical miles (1 nautical mile = 6080 feet).

Because of the difficulties inherent in measuring maximum wave heights, observers customarily refer instead to the significant wave height. Significant wave height is defined as the average height of the highest one-third of the waves for a stated interval. The maximum wave height, which should be used for design, is equal to approximately 1.87 times the significant height.

Wilson [4] fitted a curve to deep water wave height data from fourteen sources and reported the following dimensionless relationship between the significant wave height and fetch and wind velocity:

$$\frac{gH}{U^2} = 0.26 \tanh\left[\frac{1}{10^2}\left(\frac{gF}{U^2}\right)^{1/2}\right] \tag{19–8}$$

In Equation (19–8) g is acceleration due to gravity.

Curves for forecasting deep water waves are given by Figs. 19–4 and 19–5. With these curves, one may estimate the significant wave height and the significant wave period knowing the wind speed and either the duration of wind or the fetch length. These graphs should be entered at the left side at the value of wind speed, U, and the U-line should be followed from the left side across the page to its intersection of the fetch line or the duration line, whichever is intersected first. At this point, the significant wave height and period may be read from the respected curves.

In shallow water areas, smaller wave heights and shorters wave periods will be experienced. Empirical curves for the estimation of wave heights and lengths in shallow waters have been developed by Thijsse and Schijf [5] and published by the U.S. Army Corps of Engineers [6]. These are shown by Fig. 19–6. Use of these curves is demonstrated by Example 19–1.

Example 19–1

Given a fetch of 10 miles, a wind speed of 66 feet per second, and a mean water depth of 20 feet, determine the wave height and wave length.

$$\frac{gD}{U^2} = \frac{32.2(20)}{(66)^2} = 0.148$$

$$\frac{gF}{U^2} = \frac{32.2(52{,}800)}{(66)^2} = 3.9 \times 10^2$$

From Fig. 19–6:

$$\frac{gH}{U^2} = 3.5 \times 10^{-2} \quad \text{and} \quad \frac{Lg}{2\pi U^2} = 7 \times 10^{-2}$$

$$H = \frac{(66)^2(3.5 \times 10^{-2})}{32.2} = 4.7 \text{ feet}$$

$$L = \frac{2(66)^2(7 \times 10^{-2})}{32.2} = 59.6 \text{ feet}$$

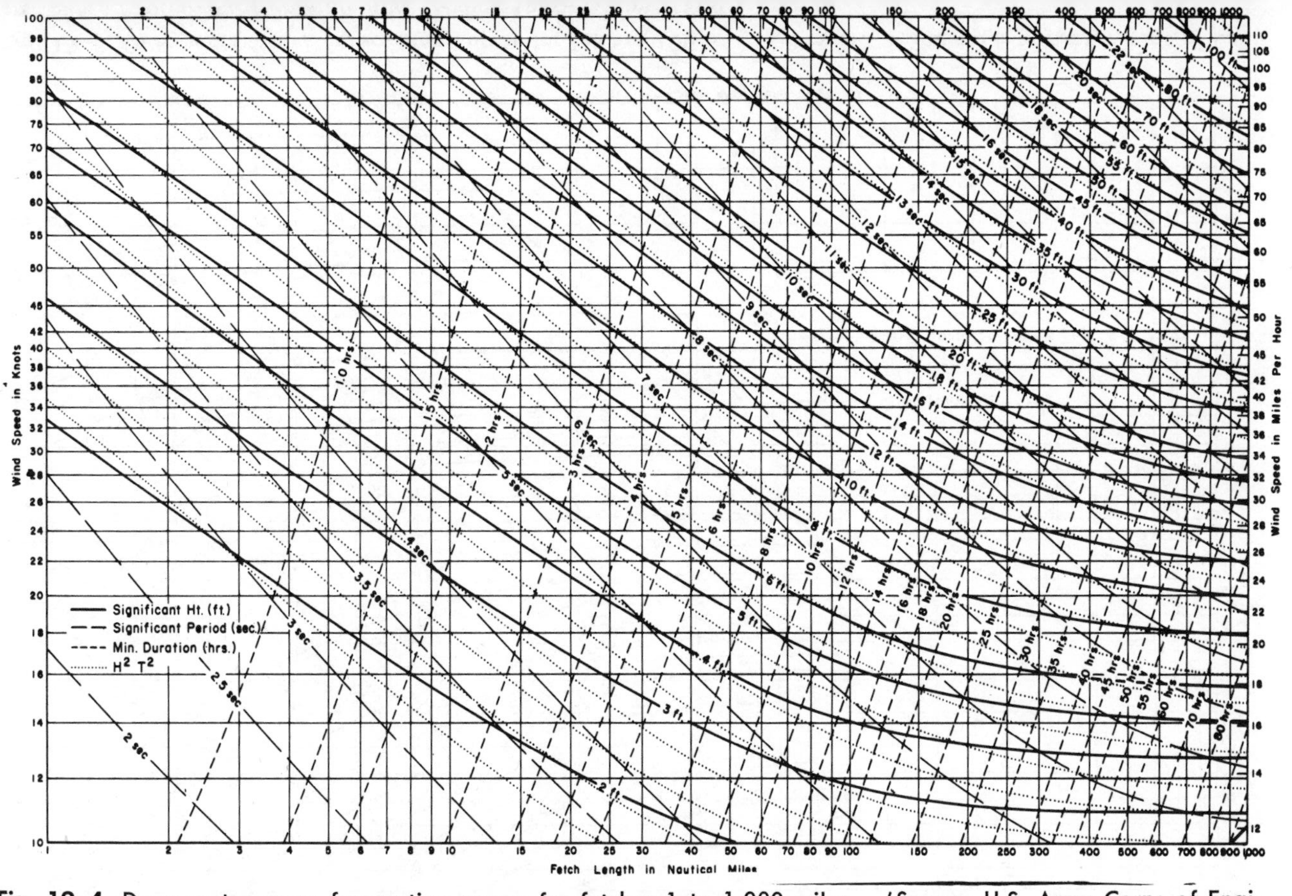

Fig. 19–4. Deep water wave forecasting curves for fetches 1 to 1,000 miles. (Source: U.S. Army Corps of Engineers, *Shore Protection, Planning, and Design*, Third Edition, 1966.)

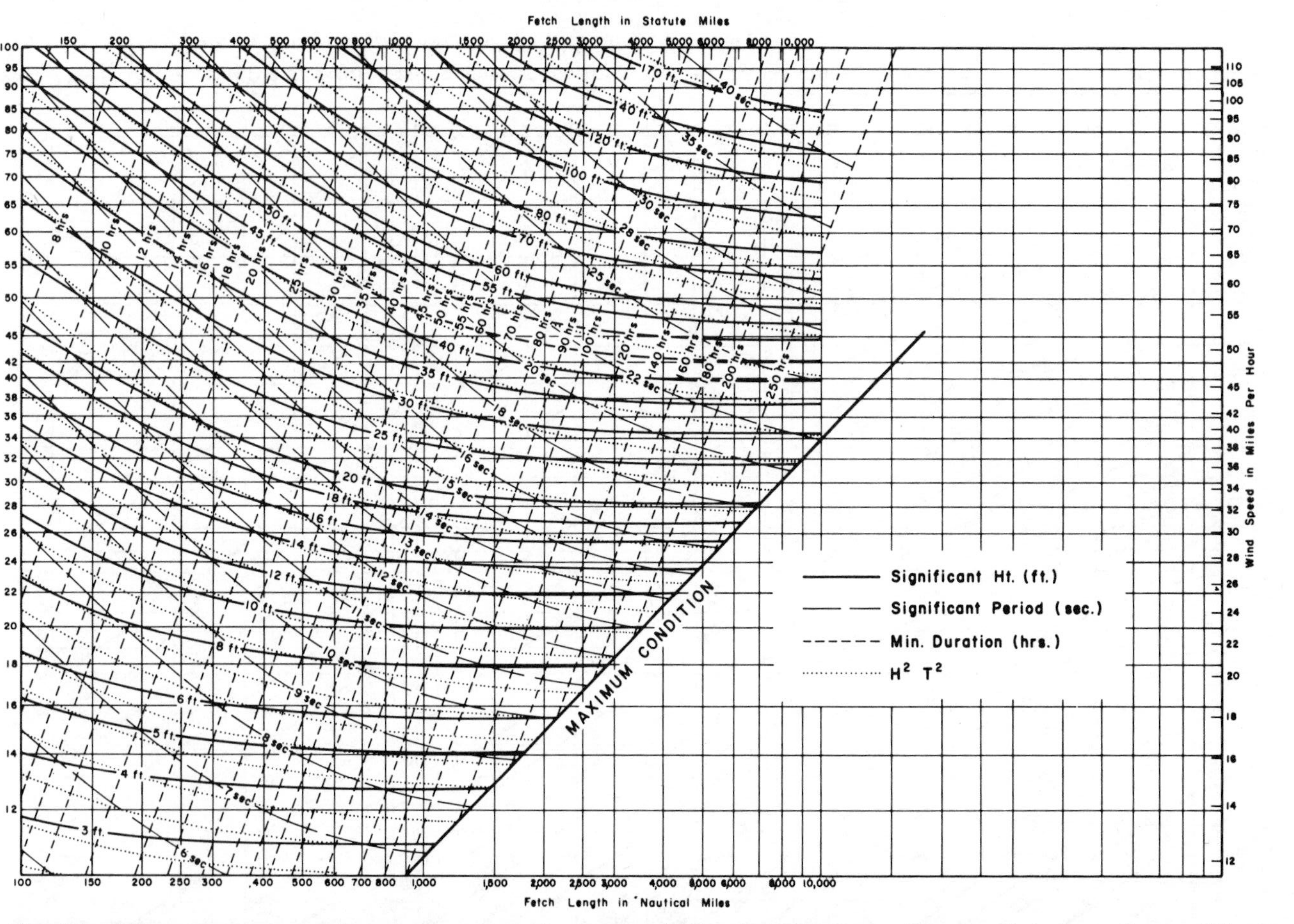

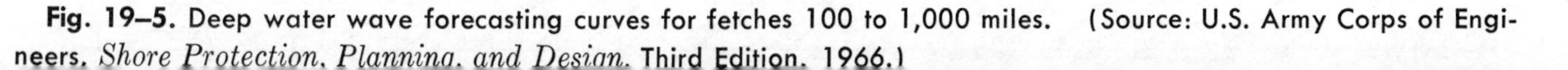

Fig. 19–5. Deep water wave forecasting curves for fetches 100 to 1,000 miles. (Source: U.S. Army Corps of Engineers. *Shore Protection, Planning, and Design.* Third Edition. 1966.)

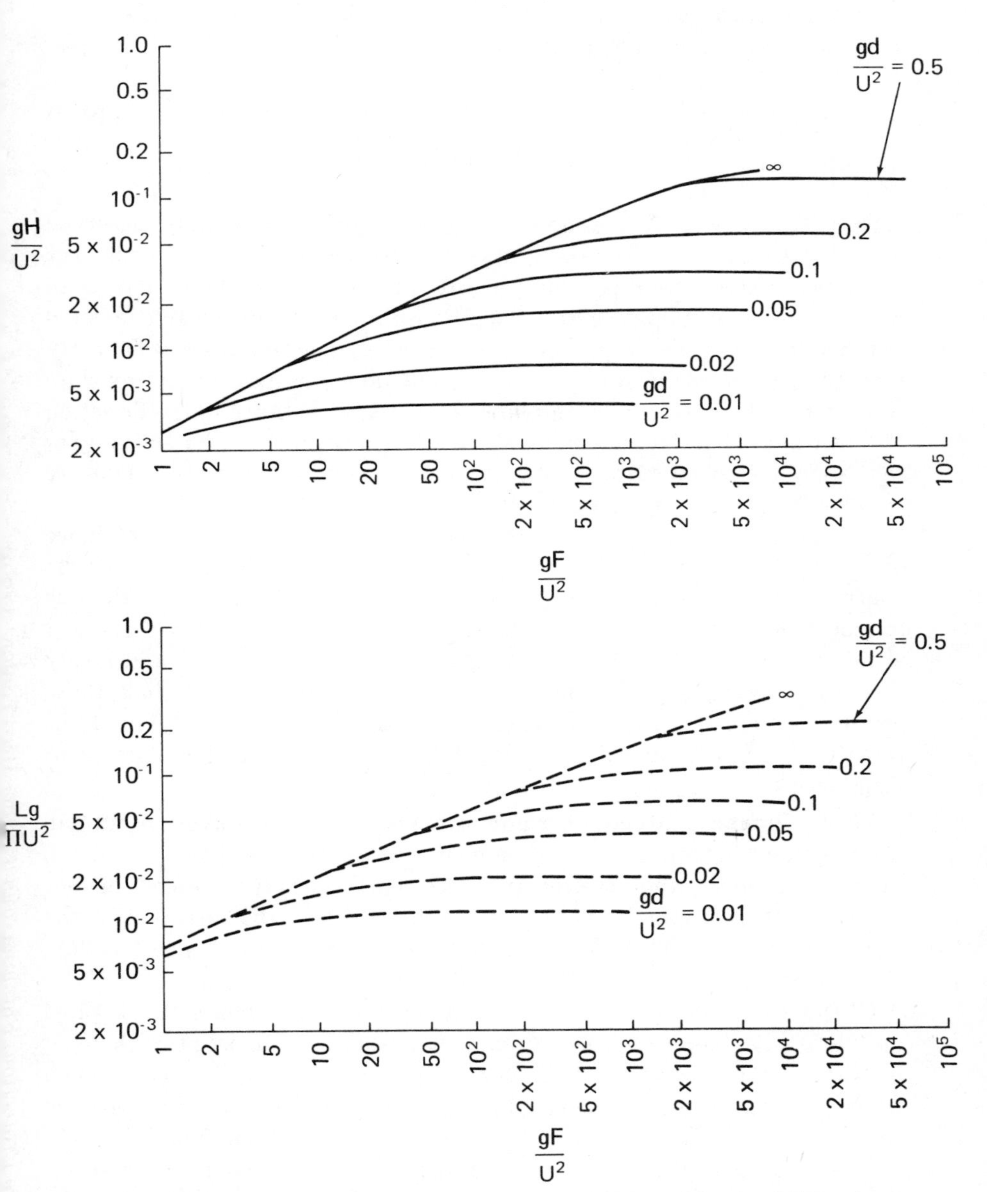

Fig. 19–6. Growth of waves in a limited depth. (Source: U.S. Army Corps of Engineers, *Shore Protection, Planning, and Design,* Third Edition, 1966.)

By combining Equations (19–2) and (19–5), the period of shallow water waves may be calculated:

$$T = \sqrt{\frac{2\pi L}{g \tanh \dfrac{2\pi D}{L}}} \tag{19–9}$$

Only rarely will the designer be able to refer to reliably observed wave height data for the design location. In the usual case, the engineer will find it necessary to use empirical relationships to estimate maximum wave height. In applying these relationships, it should be remembered that fetch is that distance over which storms are active for a sufficiently long time to generate waves. For inland lakes, it may be appropriate to take this distance as the shoreline-to-shoreline distance in the direction of the wind. For larger bodies of water, winds may not be active over the shoreline-to-shoreline distance and reference must be made to weather data for an estimate of fetch.

The estimation of wave heights is not an exact science, and the choice of wind speeds and fetch lengths is not a simple one. One must be concerned with moving storm systems with winds which vary with time in magnitude and direction. Allowances must be made for reefs and shoals which would tend to inhibit wave growth. Bays and inlets may accentuate wave growth and result in waves much higher than those estimated from empirical equations and forecasting curves. For additional information on this complex subject, the reader should refer to Reference 6.

19–7. Currents. Much of water movement, as in waves, occurs in form rather than in the translation of water particles. Since water is viscious, however, the rotating particles which constitute waves do not return to their original position but rather drift in the direction of the wave movement. This flow, which occurs at a velocity smaller than the velocity of the wave itself, is called a current.

Offshore currents may result from hydraulic head when water is piled up along the coast because of mass translation associated with tides or waves. An example of this type of current is the Gulf Stream which flows northeasterly at a speed of about four miles per hour between the southern tip of the Florida peninsula and the Bahama Islands.

The engineer's principal interest in currents lies in his efforts to stabilize erodible shoreline and to maintain navigable inlets.

19–8. Tide. Some knowledge is required by the designer of port and harbor facilities of the nature and effects of tides. The tide is the alternate rising and falling of the surface of the oceans, gulfs, bays, and coastal rivers caused by the gravitational attraction of the moon and sun. In most places, such as on the Atlantic coast of the United States, the tide ebbs and flows twice in each lunar day (24 hours and 50 minutes). Larger than usual tides called *spring tides* occur when the sun and moon act in combination as when there is a new moon or full moon. Smaller than usual tides, called neap tides, occur when the moon is at first or third quarter.

In addition to the effects of the moon and sun, the magnitude and nature of a tide at a given location and time will be influenced by:

1. geographical location
2. physical character of the coastlines
3. atmospheric pressure
4. currents

At certain inland and landlocked seas, such as the Mediterranean Sea and the Gulf of Mexico, the tides are practically negligible. At other places, such as the Bay of Fundy, local topographical peculiarities contribute to tides as high as 100 feet.

Tidal charts and tables are published for the various ports of the world by the U.S. Coast and Geodetic Survey, the British Admiralty, and other organizations.

DETERIORATION AND TREATMENT OF MARINE STRUCTURES

Port and harbor structures must withstand some of the most destructive environmental conditions found in the world. This is especially true of piles and other elements of substructures which are subject to attack by marine life as well as the corrosive effects of salt and sea. Generally, these destructive effects are most pronounced in a sea water environment and some of the destructive agents, such as certain species of marine borers, are not found in fresh waters.

In the following paragraphs, the causes of deterioration and methods of protection of wood, concrete, and steel waterfront structures will be discussed. This material is quoted from the ASCE Manuals of Engineering Practice, number 17 and 27 [7, 8].

19–9. Deterioration and Treatment of Wood Structures. In water front structures timber is damaged by three forms of attack: decay, insects (termites and wharf borers), and marine borers. The two former attack above, and the latter below, water level. Abrasion of fender piles and piles and timber in ferry slips also causes damage.

Decay. All forms of decay are the result of the action of certain low forms of plant life called fungi. Certain substances in wood form the food of the fungi and, as these substances are consumed, the wood is disintegrated. The species of fungi that cause decay require food, air, the right amount of moisture, and a favorable temperature. Timber that is continuously below water level does not decay. Poisoning the food supply by impregnating the timber with a suitable preservative is the easiest and surest method of preventing decay.

Termites. Where they are active, termites are probably more destructive to timber than any other land organism, except fungi. Sometimes called "white ants," they are not true ants, but resemble them somewhat in appearance and method of life. There are two types of termites, the "subterranean" and the "dry wood." The former type requires moisture and therefore must have access to the ground at all times. The latter type flies and does not need contact with the ground. The subterranean termites are widely distributed in the United States and do an enormous amount of damage to infested structures. The "dry wood" termite is not found much north of Norfolk, Va., on the east coast, and San Francisco, Calif., on the west coast.

Termites are not immigrants. There are 42 species native to the United States and many more in the tropics. Their chief food is cellulose which they obtain from dead wood. They are one of the few groups of insects of which this is true. There is no species of timber known to the Committee which is immune to the attack of termites. The heartwood of some trees is resistant. These trees are: Teak (*Tectona grandis*) and sal (*Shorea robusta*) of India; cypress pine (*Callistris robusta*) and camphor wood (*Cinnamon camphora*) of the Orient. Redwood, heart cypress, and heart long leaf pine, containing a large amount of pitch, are not as readily attacked as other native timbers. Because there are so few termite-resistant woods known, it is necessary, in order to secure immunity from attack, to use ordinary commercial timbers properly treated, or to so insulate all wood from the ground that the termites cannot attack it. Insulation will not protect against attack by "dry wood" termites.

Wharf Borers. These animals do considerable damage, but they are not of as much commercial importance as others of the organisms attacking timber. The damage is done by the young of a winged beetle. The beetle lays its eggs in the cracks and crevices of the timber and the larvae or worms that destroy the timber are hatched from these eggs. They do not work below water level and seem to prefer timber that is not far above high water, or timber that is wet by salt spray at times.

Marine Borers. There are two main divisions of these very destructive animals: The Molluscan group is related to the oyster and clam; and, the Crustacean group is related to the lobster and crab. Their methods of attack on timber are entirely different. The first group enters the timber through a minute hole and as the animals grow they destroy the interior of the timber. The second group destroys the outside of the timber. The attack of the molluscan borers can only be found by the most careful inspection of the surface, or by cutting into it. The attack of the second group is easily seen and measured by surface inspection. The rate of destruction in heavy attacks is several times more rapid by the molluscans than by the crustaceans.

Molluscan Borers. The molluscan group is divided into two general groups, the *Teridinidae* and the *Pholadidae,* with very different physical characteristics. Both groups are classified, biologically, into several genera and many species. The teredo or shipworm group has grayish, slimy, wormlike bodies with the shells used for boring on the head. The burrow is lined with a smooth nacreous lining. The size of the mature animals of the common species varies with the species, ranging from ⅜ in. in diameter and 5 or 6 in. in length to 1 in. in diameter and 4 to 5 ft long. A species, thus far only identified in some of the Pacific Islands, may be more than 3 in. in diameter and 3.5 ft in length. In areas of heavy attack, an unprotected pile may be totally destroyed, so far as its bearing value is concerned, in 6 to 8 months. Animals of the teredine group have been found in harbors of continental North America only in salt or brackish water; but in Australasia, India, and in some parts of South America they have been found in fresh water.

The pholad group also uses its shells for boring but the body of the animal is enclosed by the shells and there is no lining in the burrows. Some species of this group bore in concrete as well as in soft rock and mud. The entrance holes made by this group are somewhat larger than those made by the teredo group, but are still small and hard to find by surface inspection. Although some specimens of this group have been found in temperate waters, they have not, so far, done much damage to structures except in tropical and semitropical waters.

Crustacean Borers. There are three important genera of crustacean borers: *Limnoria, Chelura,* and *Sphaeroma. Limnoria lignorum,* the most widely distributed species of this group, resembles a wood louse in appearance and has a body from ⅛ to ¼ in. in length with a width of about one-third the length. The head bears a pair of eyes and two pairs of short feelers or antennae, and the mouth has a pair of strong, horny-tipped mandibles with which the boring is done. The body has seven pairs of legs ending in sharp, hooked claws so that it can move around freely and cling to the timber. It uses its gill plates for swimming. *Limnoria* destroys timber by gnawing interlacing branching burrows on the surface. As many as 400 animals per square inch have been counted on timber under heavy attack. They are found from the Arctic Circle to the Tropics, in salt or brakish water, and in clean or polluted water. The greatest intensity of attack is generally found near the mud line or at half tide; but it may be anywhere between these limits, or it may be uniformly distributed.

Sphaeroma is very much like *Limnoria,* except that it is larger. Some specimens are ½ in. in length and ¼ in. in width. It is not so widely distributed or found in as great numbers as *Limnoria,* but is capable of causing considerable damage. It is generally found in Southern waters, either salt or fresh; but sometimes it appears in the temperate zone.

Chelura is allied to the ordinary sand hopper. It is slightly larger than *Limnoria* and is found in the same localities. When in great numbers, it seems to drive out *Limnoria.* The joints in the body, the antennae, and legs are heavily feathered with long hairs and it has a spike projecting from the middle of its back. This genus is destructive in European waters and many Pacific Island harbors. It had not been found in important numbers on the continental United States until 1935 when it appeared in enormous numbers in Boston, Mass. and several other New England harbors where it has shown itself to be more destructive than *Limnoria.*

Protection Methods

Many methods of protection for timber have been tried since the beginnings of history. Few have been proved to have permanent value. A brief statement of these methods follows:

Bark. To be of any value as a protective agent the bark must be unbroken so that the borers cannot have access to the timber beneath. This makes necessary extreme care in driving and the covering of all knots and other openings by sheet metal which will last as long as the bark. Piles so protected have longer life in infested water than peeled piles. This type of protection is useful only for temporary structures.

Charring and Tarring. This method has been used since the time of the Phoenicians, but has only temporary value.

Pile Coatings. A pile coating is defined as "Protection that is an integral part of the wood surface, becoming so by a slight penetration, which through the content of a substance chemically harmful or disagreeable to the marine wood borers prevents them from entering the timber." Practically, this means a paint. It is not effective except for a short time because the toxic element is small in quantity, is leached out quickly, and because the coating is soon destroyed by abrasion. This method is useful for the protection of a temporary structure through one season, if an efficient material be used.

Pile Armors. Under this heading are classified those methods of protecting piles which depend wholly on some mechanical method of preventing the borer from coming in contact with the timber. The following methods and materials have been used:

(a) Steel, Iron, and Zinc Sheathing.—These sheet metals do not have a sufficiently long life in sea water to justify the expense of placing.

(b) Muntz Metal.—This metal, alloy of copper and zinc, has given good service at times, but unless the alloy is homogeneous, electrolytic corrosion will soon destroy it. Service records indicate that the material is unreliable. In some cases long service has been obtained.

(c) Copper Sheathing.—Copper resists corrosion and where it can be protected from damage, the use of this material results in an effective and long-lived protection. The initial cost is relatively high, the sheets are easily damaged by abrasion, and the metal is of sufficient value to invite theft.

(d) Fabric Sheathing.—There are a number of patented methods of wrapping piles in fabric of various kinds, soaked in various materials. This sheating is effective as long as it is unbroken, but it is not practicable to handle and drive piles without creating holes in the sheathing.

(e) Vitrified Pipe Casings.—This type of protection is effective as long as the pipe is unbroken, but either wave action or drift is likely to break the pipe and then it becomes worthless. If used, casings should extend from a point below any possibility of scour to a point above high water. After placing, pipe should be filled with sand or lean concrete made with fine aggregate.

(f) Concrete Casings.—Like other methods this one is efficient as long as there is no breakage or deterioration sufficient to give borers access to the wood. These casings may be either cast in place or precast and either type may be used for protecting piles in place, that have been attacked.

(g) Cast Iron Casings.—This is one of the most durable materials known when immersed in sea water. Pipe should be filled with sand or lean concrete to minimize breakage. Cast iron casings are expensive, but their cost is sometimes justified because of their long life.

Impregnation with Toxics. This form of protection is the most generally used. It usually results in the lowest annual cost for the structure if treatment and preservative are the best. A great advantage of this type of protection is that it also protects from decay, termites, and wharf borers, as well as marine borers. The best and most reliable preservative so far known is coal tar creosote, used in sufficient quantity and carefully applied.

19–10. Deterioration and Protection of Concrete Structures. *Plain or reinforced-concrete piles entirely embedded* in earth generally may be considered permanent. The elevation of the water table does not, in general, affect their durability. There is the possibility that in isolated areas concrete piles may be damaged by the percolation of ground water charged with destructive acids, alkalies, or chemical salts. Ground waters move rapidly through sandy soils, and hence corrosive effects would be more pronounced, while in clays the movement is so slow that the action would be unimportant. Destructive chemicals in the ground water may be due to: (*a*) Wastes from manufacturing plants, leaky sewers, leaching from storage piles of soft coal or cinder fill containing sulfuric acid and other destructive compounds; (*b*) sodium and magnesium sulfates leached from the ground itself; or (*c*) organic acids resulting from the decay of vegetable matter.

Reinforced-concrete piles above the ground surface, like other reinforced-concrete construction, are subject to attack by weathering and any destructive elements carried in the air. In damp seacoast climates, the exposure may be severe because of moisture penetrating permeable concrete and reaching the reinforcement, rusting it and spalling the sides and edges of the pile. In general, where conditions permit or favor corrosion of steel, the reinforced pile is threatened, particularly if the concrete is defective or if there is insufficient cover over the reinforcement.

Reinforced-concrete piles in waterfront structures are called upon to meet unfavorable exposure conditions when they extend from the harbor bottom into water and air. Such piles are subject to:

1. Abrasion by floating objects or scouring sand.
2. Attack by *Pholads* (rock-boring mollusks). These mollusks have been found in concrete and masonry structures in such widely scattered locations as Los Angeles (Calif.) harbor, Panama Canal Zone, and Plymouth, England. The damage thus far appears to have been done to concrete of unquestionably poor quality and not to have been of great economic importance; but it must be considered.
3. Chemical action of polluted waters on concrete. The waters of some western streams and lakes are highly destructive to concrete; so also are some of the salts in sea water. Special cements recently have been developed in an attempt to produce concretes which will be durable when exposed to sea water. Most of the claims for durability are not as yet substantiated by long-time exposure.
4. Frost action on porous concrete.
5. Destructive action caused by rusting of the reinforcement and spalling of the concrete. This is the most serious weakness of reinforced-concrete piles when used in waterfront structures. It is particularly serious when the structures are located in tidal waters where alternate wetting and drying of the concrete due to the rise and fall of the tide—especially if combined with alternate freezing and thawing—accelerates the destructive action. Rough water also promotes this destructive action by keeping the piles soaked with spray to a high elevation on a windy day and allowing them to dry

out on calm days. Experience has shown that the quality of the aggregates, the composition of the cement, the cover over the steel and the workmanship in mixing and placing the concrete are controlling factors in attempting to attain the desired permanence of the piles. Protection to the pile is afforded by oil coating in locations where oil floats on the surface of the water, as near oil docks.

Spalling of precast reinforced-concrete piles may be minimized by:

1. Deep embedment of the reinforcement;
2. Use of dense, rich concrete;
3. Jackets of wood or metal from below to above the tidal range or continued oil coating; and
4. Careful handling of piles to minimize stresses and to avoid cracking during placement.

19–11. Deterioration and Protection of Steel Structures. *Steel piles entirely embedded in relatively impervious earth* generally may be considered permanent. Entire embedment usually may be assumed as extending downward from a level about 2 ft below the ground surface. Atmospheric oxygen is blanketed off by the surrounding soil, thereby inhibiting progressive corrosion. Variations in ground-water level do not have any effect on the durability of steel piles under this condition, except as the upper layers are previous to air and water.

Occasionally entirely embedded steel piles may be subjected to destructive action if the surrounding earth or ground water contains corrosive compounds, from sources such as coal piles, "alkali" soils, cinder fill, or wastes from mines and manufacturing plants. Cinder or ash fill which has been in place for a long time, subjected to the leaching action of ground and surface waters, is not necessarily deleterious. Tests and analyses of the ground materials should be made in questionable cases.

Steel piles protruding into the air from the ground are subject to some rusting at the ground line and for a short distance below, due to the action of surface water and the presence of air and organic matter in the top soil. Protection of the piles at this point can be achieved by a renewable concrete encasement or a suitable coal-tar coating, extending from about 2 ft below to a short distance above ground level. Above ground level, atmospheric corrosion can be prevented by painting, the same as structural steel.

Steel piles protruding from the ground into open water, as in trestle bridges or waterfront structures, are subjected to varying degrees of deterioration depending, in general, on whether the water is fresh or salt. In the case of fresh-water exposure, there is usually very little deterioration, although the action of salt water subjects exposed steel piles to more severe deterioration and unfavorable performance, as more fully discussed later.

Steel piles in fresh water in most cases do not require protection. Where there is pollution from industrial wastes, the piles may be protected above the mud line by the application of a suitable coal-tar coating before driving. Such a coating is especially desirable at the water surface where deterioration of the steel, although not very active, is relatively greater than in the totally immersed parts of the piles. Renewable concrete encasement of the piles from about 2 ft below to a short distance above water level also will prove effective. The parts of the piles above water level can be treated as an ordinary steel structure and maintained in good condition by painting.

Steel Piles in Sea Water. A greater length of life can be expected from steel piling if in protected waters than if subjected to wave action in the open ocean. The various parts are affected as follows:

(*a*) *That portion of a pile below the bottom* is generally not subject to rapid deterioration.

(*b*) *At and immediately above the bottom,* deterioration may be accelerated rapidly by the abrasive action of water-borne sand agitated by waves and currents. This condition is usually present only in shallow waters where wave and tidal action is most active. Sometimes destructive organic substances consisting of decayed marine animal or vegetable matter deposited on the bottom may cause accelerated deterioration in a narrow zone at the mud line. Tests and analyses of the ground should be made in questionable cases. Under either of these conditions it is desirable to protect the piling by some form of renewable or replaceable encasement in the zone of accelerated deterioration.

(*c*) *Between the bottom and low-tide level,* corrosion is usually more active in the upper part where the oxygen content of the water is greatest. In some waters there are certain types of marine shellfish which attach themselves to steel piles below water level and may have a deleterious effect by causing excessive pitting. Suitable coal-tar coating and synthetic resin or zinc chromate paint will protect against such action as long as they last. Encasement applied to the piles before or after driving should be effective.

(*d*) *Between low and high-tide levels,* corrosion may be extremely active in the region of alternate wetting and drying of the piles. The presence of oil on the water surface in many harbors results in a protective coating forming on the steel piles. This decidedly retards corrosive action.

(*e*) *Above high-tide level,* corrosion also may be severe, especially if waves subject this zone to the action of spray which tends to build up an accumulation of salt. Deterioration is also relatively greater in locations having a high temperature and a humid atmosphere. Protective coatings or encasement with concrete will lengthen the life of steel piles under this condition and also under that covered in paragraph (*d*).

Steel piles should be suitably protected against corrosive action from immediately below low water to above the spray line.

Corrosion of steel piles caused by electrolysis is rare. It may be a factor under certain conditions. Steel piers or trestles carrying pipe lines which may have picked up stray currents, or carrying direct-current power lines, may transmit enough current to the piles to cause electrolysis. Insulating the source of the current from the piles will prevent this condition. In buildings such as power houses, or in cities where direct current is used, electrolysis may occur, and care should be taken to provide proper insulation to prevent direct currents from reaching steel piles. The superstructure also should be properly grounded.

Under the most common conditions, where the tops of steel piles are embedded in concrete footings and thereby insulated from the rest of the structure, electrolysis is not a factor. Local electrolytic action and corrosion may be set up in salt water where the steel pile forms one pole of a battery, with its other pole in some dissimilar metal in the water close by.

Copper-bearing steel with a minimum of 0.20% copper affords greater resistance against atmospheric corrosion than plain carbon steel. For both materials, when completely immersed in either fresh or salt water or within the tidal range of sea water, the resistance against deterioration is about the same. Care must be taken to guard against the use of copper-bearing steel in contact with plain carbon steel or wrought iron, and thus avoid harmful local electrolytic corrosive action between dissimilar metals, especially in the presence of sea water.

Rate of loss of metal is not constant. If loss due to corrosion is plotted against time of exposure, the result will probably not be a straight line, the loss being heaviest in the first years of exposure. Many factors, however, will modify this, such as shock, which removes the rust scale periodically.

PROBLEMS

1. Compute the wave velocity and period of a wave which is 250 feet long. The average depth of water is 175 feet.
2. Estimate the velocity of a 250 foot long wave in water 15 feet deep.
3. Estimate the velocity of a wave 480 foot long and 18 feet high in water 120 feet deep.

4. Compute the maximum deep water wave height for a fetch of 45 nautical miles and a wind speed of 38 knots:
   (a) using the Stephenson equation
   (b) using graph.
5. Compute the maximum deep water wave height for a fetch of 400 statute miles and a wind speed of 15 miles per hour:
   (a) using the Stephenson equation
   (b) using graph.
6. Determine the wave height and wave length for a fetch of 8 miles, a mean water depth of 10 feet, and a wind speed of 25 miles per hour.
7. Given, a fetch of 30 nautical miles and a wind speed of 26 knots:
   (a) estimate the maximum wave height.
   (b) what would be the significant wave height?
   (c) estimate the significant wave period.
   (d) what minimum duration of a 26 knot wind would cause this wave?
   (e) estimate the speed of the significant wave.

REFERENCES

1. Carr, Braxton B., "Inland Water Transportation Resources," *U.S. Transportation, Resources, Performance and Problems,* National Academy of Sciences, National Research Council, Publication 841-S, August, 1960.
2. *1967 Inland Waterborne Commerce Statistics,* The American Waterways Operators, Inc., April, 1969.
3. U.S. Navy Hydrographic Office, *Breakers and Surf,* November, 1944.
4. Wilson, Basil W., Trans. ASCE, Vol. 128, Part IV, Paper #3416, 1963.
5. Thijsse, J. Th., and Schijf, J. B., "Penetration of Waves and Swells into Harbors," *Proceedings, XVII International Navigation Congress,* Lisbon, 1949.
6. U.S. Army Corps of Engineers, *Shore Protection, Planning and Design,* Technical Report No. 4, Third Edition, 1966.
7. American Society of Civil Engineers, *Timber Piles and Construction Timbers,* Manual of Engineering Practice No. 17, 1939.
8. American Society of Civil Engineers, *Pile Foundations and Pile Structures,* Manual of Engineering Practice No. 27, 1946.

OTHER REFERENCES

Cornick, Henry F., *Dock and Harbour Engineering,* Volume 2, Charles Griffin and Company, Ltd., London (1959).

Mitchell, C. Bradford, "Pride of the Seas," *American Heritage,* December, 1967.

Quinn, Alonzo, *Design and Construction of Ports and Marine Structures,* McGraw-Hill Book Company, New York (1961).

20

Planning and Design of Harbors

20–1. Introduction. A *harbor* is a partially enclosed area of water which serves as a place of refuge for ships. The term port refers to a portion of a harbor which serves as a base for commercial activities. Harbor and coastal structures are the means by which protection from waves and winds is provided and the erosion of beaches and coastlines is controlled. Port facilities make it possible for ships to obtain fuel and supplies, to be repaired, and to transfer passengers and cargo.

This chapter will deal with the planning and design of coastal and harbor structures, while Chapter 21 will be concerned with the planning and design of port facilities.

20–2. Classes of Harbors. Harbors may be classified into one of several categories according to function and protective features. There are many *natural harbors* in the world where protection from storms is provided by natural topographical features. Natural harbors may be found in bays, inlets, and estuaries and may be shielded by offshore islands, peninsulas, or reefs. Other natural harbors are protected by virtue of the fact that they are located in river channels some distance from the sea. Several famous world harbors are located as much as 50 miles inland.

Where sufficient protection from storms has not been provided by nature, it may be provided by the construction of breakwaters and jetties. Harbors thus formed are known as *artificial harbors.* Planning considerations and design criteria for breakwaters and jetties will be discussed in Section 20–4.

Some degree of protection from storms is provided by *harbors of refuge* and *roadsteads.* These areas are easily accessible but generally offer less protection than does a harbor. The term "harbors of refuge" refers to convenient protected anchorage areas which are usually found along established sea routes and dangerous coasts. A roadstead is a tract of water which is protected from heavy seas by a bank, shoal, or breakwater. Port facilities are not provided in harbors of refuge or roadsteads although these facilities may be provided within a nearby harbor.

Additional classifications of harbors include: *commercial harbors* which provide protection for ports engaged in foreign or coastwise trade; and *military harbors* within which the dominant activity is the accommodation of naval vessels.

20–3. Desirable Features of a Harbor Site. There are at least five desirable features of a harbor site:

1. Sufficient depth
2. Secure anchorage
3. Adequate anchorage area
4. Narrow channel entrance in relation to harbor size
5. Protection against wave action

The depth of harbor and approach channel should be sufficient to permit fully loaded ships to navigate safely at the lowest low water. Obviously, the harbor depth required depends principally on the draft[1] of the ships using the harbor.

The harbor and channel depth below the lowest low water should be at least the maximum draft anticipated plus an additional five feet, approximately. This additional depth is to allow for the tendency of a ship to "surge" when in motion and to provide a clearance of at least three feet below the ship's keel as a factor of safety.

A summary of the average drafts of various types of ships in the world's ocean fleet is given in Chapter 4, along with average drafts of typical inland waterway vessels. Generally, the average draft of ocean-going ships varies from 22 to 31 feet depending on ship type, while the draft of the largest commercial oceangoing ship (a tanker) is 57 feet. Few if any ocean ports have channel depths to accommodate the largest tankers and special provisions must be made to unload these vessels offshore to smaller tankers or by means of pipelines.

Generally, ocean ports maintain harbor and channel depths of 35 to 40 feet; however, the trend to larger ships indicates that modern ocean harbors may require depths in excess of 40 feet.

On the inland waterway system, a nine-foot operating depth is considered standard. Channel depths on the inland system vary a great deal, however, and about 26 per cent of the inland waterway channel miles has a depth of less than six feet while approximately 18 per cent is 14 feet and over [1].

[1] Draft is the vertical distance between the waterline and the keel. Unless otherwise noted, when the word draft is used here it will refer to the full load draft.

Conceivably, the selection of a harbor site may be influenced by the soil conditions along the bottom of the anchorage area. Generally, firm cohesive materials provide good anchorage, while light sandy bottoms are poor anchorage areas. Other factors being equal, more anchorage area will be required when poor bottom soil conditions previal. This follows from the fact that maximum resistance to ship movement occurs when the anchor cable is as nearly horizontal as possible.

The shape and extent of anchorage area is principally dependent on five factors:

1. the maximum number of ships to be served
2. the sizes of the ships
3. the method of mooring
4. maneuverability requirements
5. topographic conditions at the proposed site

Because objectionable waves will be generated in large harbors, artificial harbors should be built as small as possible consistent with the needs for convenient and safe maneuvering and mooring.

Space requirements for ships at anchor varies a great deal depending on the method of mooring. A ship which is secured by a single anchor will occupy a circular area with a radius of the ship length plus approximately three times the water depth. Thus, a 600-foot long ship anchored by a single anchor in a harbor 50 feet deep will require about 40 acres of anchorage area. Considerably less area is required for a ship which is secured by two anchors. Such a ship will occupy a circular area with a diameter little more than the ship length, or about 6.5 acres for a 600-foot ship. However, when a clear area is maintained for the anchor cable, a rectangular area 1000 to 1200 feet long by about 500 to 600 feet wide is required for a typical merchant freighter. Thus, a merchant freighter secured by two anchors requires a total of about 16 acres of harbor space.

The minimum turning radius for a ship is equal to about twice the ship length. Thus a typical ocean going freighter will require an additional 30 to 35 acres for maneuvering if a full size turning basin is provided.

As one might expect, harbors of the world vary a great deal in area, the smallest fishing harbors being less than 10 acres and the largest harbors being more than 10 square miles.

To minimize wave action within the harbor, the harbor entrance should be as narrow as possible provided it meets the requirements for safe and expeditious navigation and provided it does not cause excessive tidal currents. Currents in excess of 4 to 5 feet per second will adversely affect navigation and may cause scour of breakwaters and other protective works.

The required entrance width will naturally be influenced by the size of the harbor and the ships that use it. As a rule of thumb, the width of entrance should be roughly equal to the length of the largest ship using it. While this guideline should be helpful for planning purposes, the entrance width and location used for design purposes should be determined by model tests.

COASTAL STRUCTURES

The term "coastal structures" used here will be used in the broad sense to include:

1. off-shore structures (breakwaters) to lessen wave heights and velocities,
2. structures which are built at an angle to the shore such as jetties and groins to control littoral drift, and
3. structures built at or near the shoreline to protect the shore from the erosive forces of waves. In this category are seawalls, bulkheads, and revetments.

20–4. Breakwaters and Jetties. Breakwaters are massive structures built generally parallel to the shoreline to protect a shore area or to develop an artificial harbor. A jetty is a structure built roughly perpendicular to the shore extending some distance seaward for the purpose of maintaining an entrance channel and protecting it from waves and excessive or otherwise undesirable currents. Jetties are usually built in pairs, one on each side of a channel or the mouth of a river. Structurally, breakwaters and jetties are similar; however, the design standards for jetties may be slightly lower than those for breakwaters. This results from the fact that jetties are not subject to direct wave attack to as great an extent as are breakwaters.

To avoid redundance, the remaining discussion of this section will focus principally on the planning and design of breakwaters, and little else will be said about jetties. However, one especially important feature relating to the design of jetties should be mentioned, that of hydraulic design. Because of the possibility of erosion and accretion due to changes to the velocity and direction of channel currents, the determination of distance between jetties must be carefully made. In making this decision, the designer should study the magnitude and direction of existing tidal currents and the effect that the construction of jetties might have on these currents. Important inputs to this decision will, of course, be existing topographical features of the area and the channel width needed for navigation. Unfortunately, few analytical tools are available to help the designer with this problem, and he must rely on engineering judgment and, whenever feasible, model studies.

While a wide variety of breakwaters have been built, those successfully employed generally fall into two classes:

1. mound breakwaters
2. wall breakwaters.

By far the most popular type of breakwater is the rubble mound breakwater. As the name suggests, this type of breakwater consists of a mound of large stones extending in a line from the shore or lying parallel to and some distance from the shore. Typically, this type of breakwater is constructed of stones ranging in weight from 500 pounds to more than 16 tons each. The smaller stones are used to construct the core, while the largest sizes, being most resistant to displacement, serve as armor stones which comprise the outer layer of the mound. Commonly, the largest armor stones are used on the seaward side of the breakwater.

Where adequate quantities of armor stone are not available in suitable sizes, precast concrete armor units may be used. Various shapes of these units have been used, including tetrapods, quadripods, and tribars. See Fig. 20–1 and Table 20–1. These shapes are patented and a royalty charge must be paid for their use. A rubble mound breakwater utilizing an armor layer of tetrapods is shown as Fig. 20–2.

In certain instances, a relatively impervious material is used in the core of the breakwater. When a sand-clay or shale is used for the base and core material, the breakwater is classed as a solid fill structure. The use of a fine-grained material in the core may be used because of economy or in order to prevent the effect of waves passing through the structure. The voids of the upper portion of the core of the breakwater may also be filled with hot asphaltic concrete or portland cement concrete in order to improve the stability or imperviousness of the structure.

Vertical wall breakwaters constitute a second major class of breakwaters. This class of breakwater differs in concept and design from rubble mound breakwaters. The designer of a vertical wall breakwater must be concerned with the ability of the total structure to remain stable under the attack of waves, whereas in the case of rubble mound breakwaters, he must focus his attention on the stability of the individual stones.

The principal advantage of vertical wall breakwaters is that less rock is required than is needed for rubble mound construction. These break-

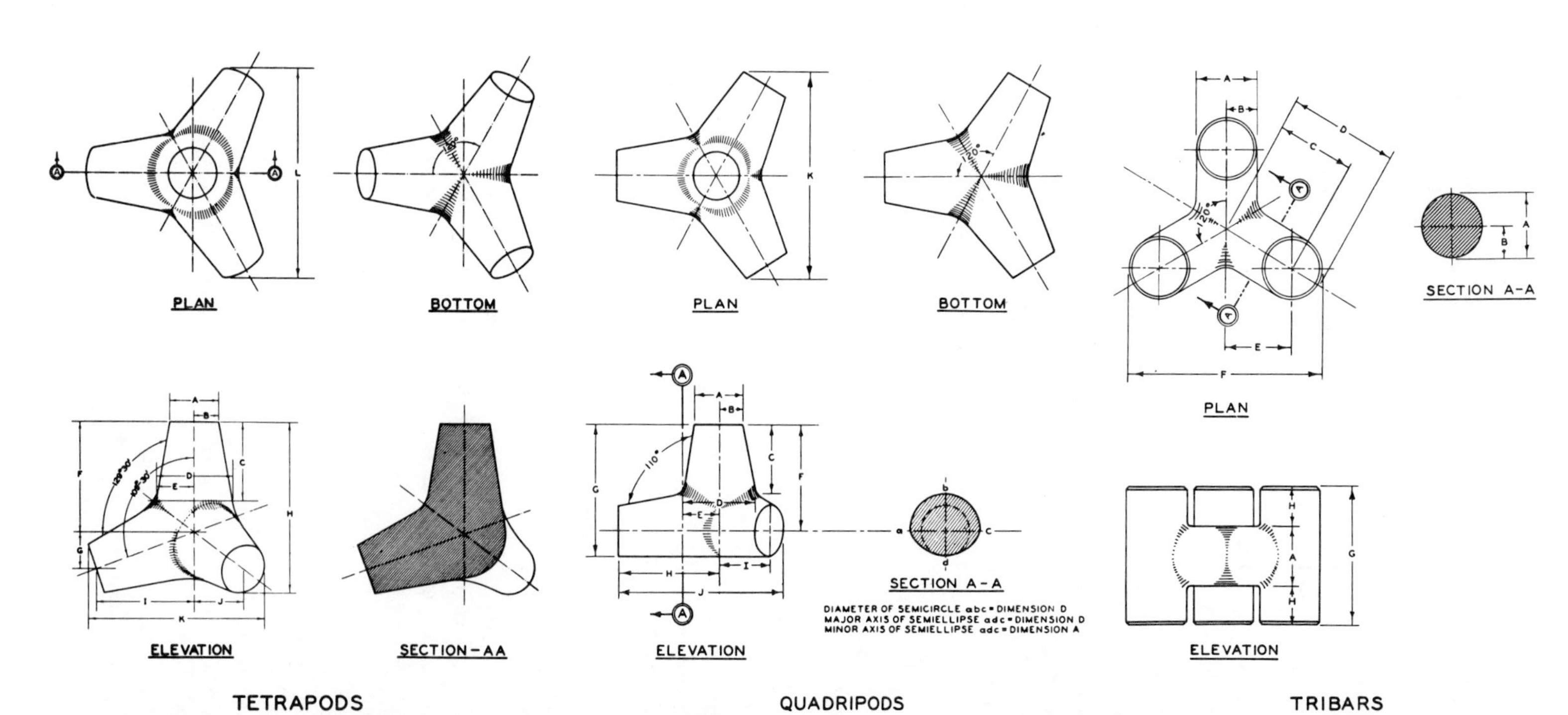

Fig. 20–1. Typical concrete armor units. (Source: *Wave Forces on Rubber-Mound Breakwaters and Jetties.*)

TABLE 20-1

Design Information for Concrete Armor Units

	Tetrapods			Quadripods			Tribars		
	Small	Medium	Large	Small	Medium	Large	Small	Medium	Large
Volume of Armor Unit (cu. ft.)	7.14	214.29	571.43	7.14	214.29	571.43	7.14	214.29	571.43
Weight of Armor Unit (Tons)*	0.53	16.02	42.71	0.53	16.02	42.71	0.53	16.02	42.71
Average Thickness of Two Layers Placed Pell-mell (feet)	4.01	12.45	17.26	3.66	11.37	15.77	3.85	11.97	16.60
Number of Armor Units Per 1000 Square Feet (2 Layers, Pell-mell)	280.18	29.02	15.18	261.05	27.04	14.15	161.34	16.71	8.74
Dimension (Feet)									
A	0.89	2.76	3.83	0.93	2.88	4.01	1.08	3.35	4.64
B	0.44	1.38	1.91	0.46	1.44	2.00	0.54	1.67	2.32
C	1.40	4.36	6.05	1.28	3.98	5.52	1.82	5.67	7.86
D	1.38	4.30	5.96	1.38	4.28	5.94	1.29	4.00	5.54
E	0.69	2.15	2.98	0.69	2.14	2.97	1.11	3.44	4.77
F	1.89	5.88	8.16	1.97	6.12	8.49	3.29	10.22	14.17
G	0.63	1.96	2.72	2.43	7.57	10.49	2.15	6.69	9.28
H	2.94	9.14	12.68	1.97	6.12	8.49	0.54	1.67	2.32
I	1.78	5.54	7.69	0.99	3.06	4.25			
J	0.89	2.77	3.84	3.36	10.43	14.47			
K	3.21	9.97	13.83	3.88	12.05	16.70			
L	3.54	10.98	15.23						

Source: *Wave Forces on Rubble-Mound Breakwaters and Jetties*, U.S. Army Corps of Engineers, September, 1961.
*Assuming specific weight = 149.5 lbs./cu. ft.

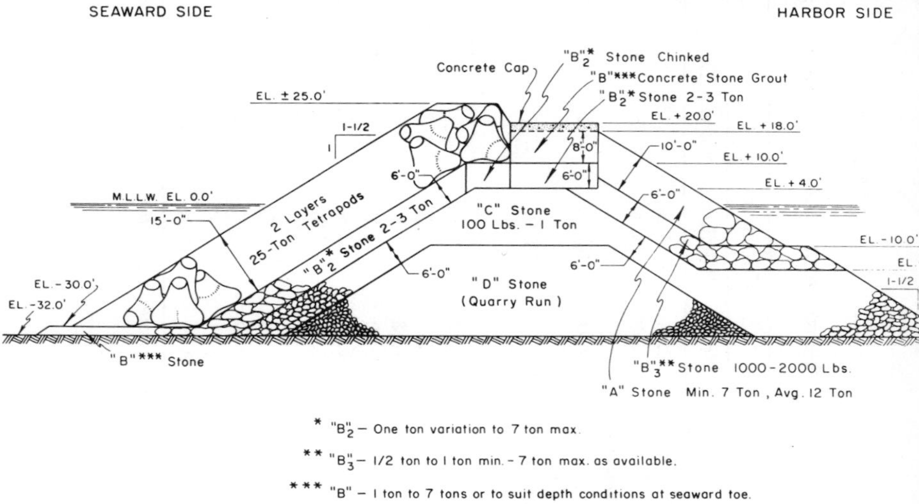

Fig. 20–2. Tetrapod-rubble-mound breakwater at Crescent City, California. (Courtesy U.S. Army Corps of Engineers.)

waters being less massive provide more usable harbor area and make it possible to have a narrower harbor entrance. On the other hand, rubble mound breakwaters can be constructed on foundations which would be unsuitable for the support of a vertical wall breakwater. Waves tend to break and be dissipated on the slopes of rubble mound breakwaters, and consequently these structures do not have to be constructed to heights as great as those required for vertical wall breakwaters.

Vertical wall breakwaters include the following types:

1. timber or pre-cast concrete cribs filled with large stones
2. concrete caissons filled with stone or sand
3. sheet piling breakwaters.

Timber or pre-cast concrete cribs consist of large box-like compartments of open construction which are placed on a prepared foundation and then filled with stone. Concrete caissons are massive water-tight boxes which are floated into position, settled on a prepared foundation and filled with stone, earth, or sand. Usually, these caissons are then covered with a concrete slab.

The simplest sheet piling breakwater consists of a single row of piling which may or may not be strengthened by vertical piles. Another type of sheet piling breakwater consists of double walls of sheet piling connected by tie bars with the space between the walls filled with stone or sand. Finally, vertical wall breakwaters may be built of cellular sheet pile structures which are filled with earth, stone, or sand to provide stability. A photograph of such a breakwater is shown as Fig. 20–3.

One can find, of course, breakwaters which are neither mound type nor vertical wall type structures but which are composite structures containing features common to both broad classifications. Small vertical walls are often superimposed on top of rubble mound breakwaters, and it is not uncommon to find rubble mound foundations which support massive vertical wall breakwaters.

Several novel concepts in breakwater design have been recently studied, including floating breakwaters, hydraulic breakwaters, and bubble breakwaters. These proposals appear to have very limited usefulness.

Laboratory research on floating breakwaters constructed with cylinders or pontoons has suggested that these devices may be useful for temporary harbors for seasonal or transient marine activities or for the development of marinas for recreational purposes [3].

Hydraulic breakwaters consist of a series of horizontal water jets which generate a surface current in opposition to incident waves. These breakwaters do not appear to be useful in a conventional sense, being inefficient and costly for either deep water or shallow water waves [4].

Bubble breakwaters consist of a curtain of air bubbles released from a line of jets on the sea bed in an attempt to create a surface current which would cause oncoming waves to break. Studies have shown that

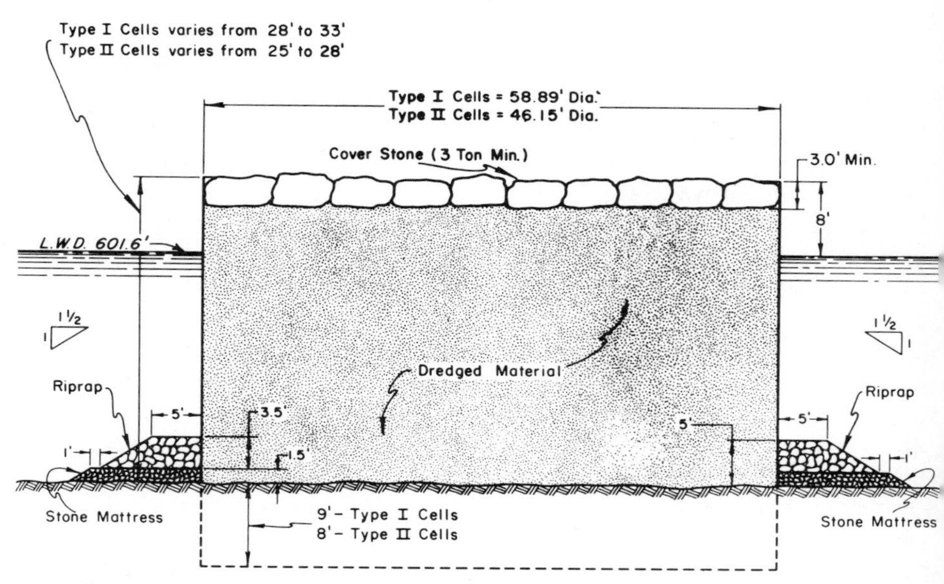

Fig. 20–3. Cellular steel sheet-pile breakwater at Grand Marais Harbor, Michigan. (Courtesy U.S. Army Corps of Engineers.)

the quantity of air required to control effectively ocean waves is astronomical, and the cost of operating a full-scale bubble breakwater system is prohibitive [5].

In the final analysis, the selection of breakwater type will depend on its purpose, foundation conditions, wave forces, availability of materials, and costs.

Breakwaters may or may not be connected to the shore and may be constructed as a single unit or as a series of relatively short structures. The latter approach may provide good protection from waves without the formation of undesirable sand shoals between the breakwater and the shore.

The height of a breakwater will depend principally on maximum tide elevation, wave height, and breakwater type. Waves tend to break on mound structures, and the amount of wave runup[2] will vary with the angle of breakwater slope, wave height and length, and the smoothness and permeability of the face of the structure. Figure 20–4 may be used for guidance in determining the required height of mound breakwaters.

A wave will rise higher on a vertical wall breakwater, and these structures should be built to an elevation equal to or greater than about 1.5 times the maximum wave height plus the elevation of maximum tide.

As a rule of thumb, the minimum width of the top of a mound breakwater should be approximately equal to the height of the maximum wave [6]. The width of the top and bottom of a vertical wall breakwater should be determined by an analysis of the various forces on the structure. These dimensions should be sufficient to prevent the structure from overturning, allowing a suitable factor of safety.

20–5. Seawalls, Revetments, and Bulkheads. Seawalls, revetments, and bulkheads are structures which are placed parallel to the shoreline to separate the land from the water. While these structures have this general purpose in common, there are significant differences in specific function and design. Seawalls are massive structures which are usually placed along otherwise unprotected coasts to resist the force of waves. Seawalls are gravity structures which are subjected to the forces of waves on the seaward side and the active earth pressure on the shoreward side.

A revetment is also used to protect the shore from the erosive action of waves. It is essentially a protective pavement which is supported by an earth slope. A bulkhead is not intended to resist heavy waves but simply to serve as a retaining wall to prevent existing earth or fill from sliding into the sea.

[2] Wave runup is the vertical height above stillwater level to which water will rise on the face of the structure.

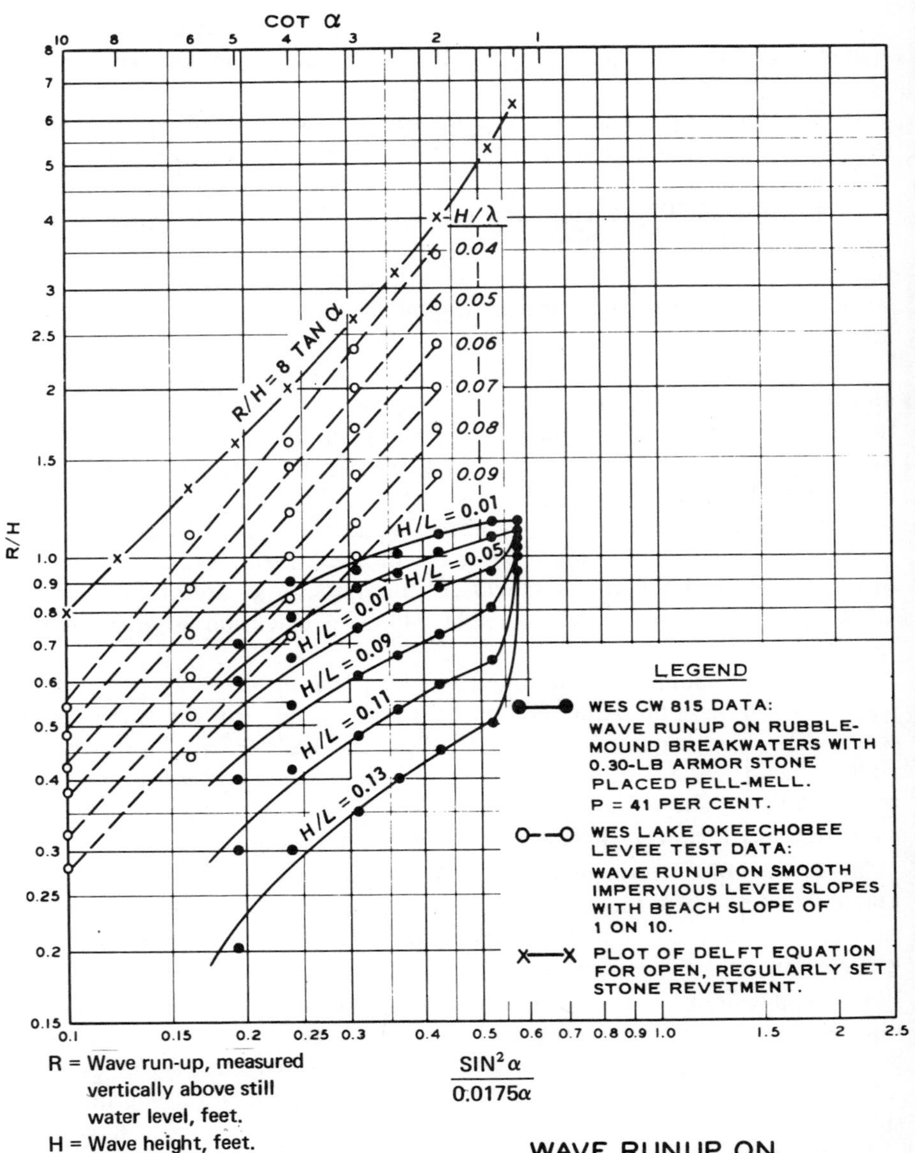

Fig. 20–4. Wave runup on rubble-mound breakwaters and smooth impervious slopes. (Source: *Wave Forces on Rubble-Mound Breakwaters and Jetties*, U.S. Army Engineer Waterways Experiment Station, Corps of Engineers, September, 1961.)

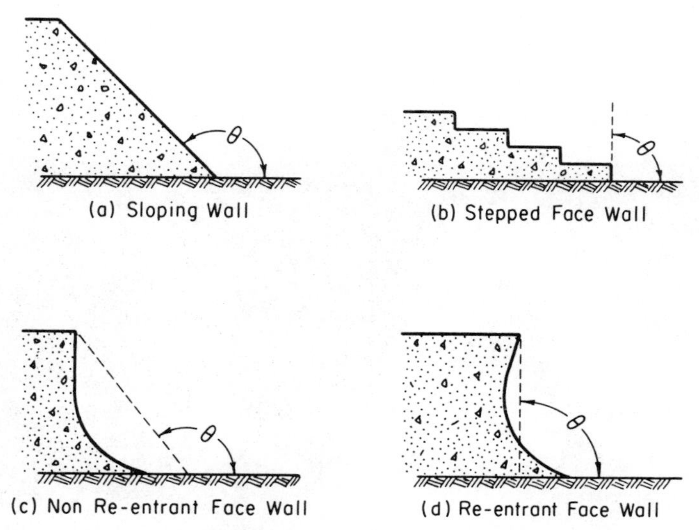

Fig. 20–5. Typical sea walls.

Typical structural types of seawalls are shown by Fig. 20–5. These include a sloping wall, stepped-face wall, and a curved wall. The latter wall face may either be the non re-entrant type, which is essentially a vertical wall, or a re-entrant type, which turns the wave back upon itself.

Sloping or vertical-faced seawalls offer the least resistance to wave overtopping. The amount of wave overtopping can be substantially decreased by the use of an armor block facing such as tetrapods [7]. The stepped-face sea wall is used under moderate wave conditions. It, too, may experience objectionable wave overtopping when subjected to heavy seas and high winds.

Under the most severe wave conditions, massive curved sea walls are the most frequently used. For this type of structure, the use of a sheet pile cut-off wall at the toe of the seawall is recommended to reduce scour and undermining of the base. As a further precautionary measure to prevent scour, large rocks may be piled at the toe of the structure. Both of these features may be seen in Fig. 20–6.

There are two broad classes of revetments, rigid and flexible. The rigid type of revetment consists of a series of cast-in-place concrete slabs. In essence, this type of revetment is a small sloping sea wall.[3]

[3]The distinction between seawalls, revetments, and bulkheads is not sharp. A sloping faced seawall in one locality may be called a revetment in another. Similarly, a vertical wall structure may be termed a bulkhead by some observers and a seawall by others.

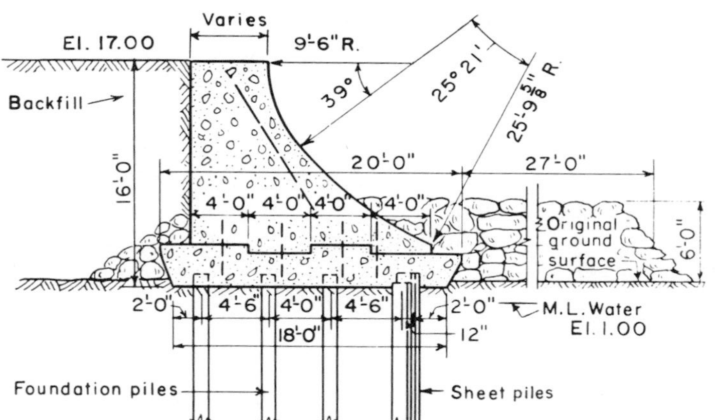

Fig. 20–6. Concrete curved-face seawall at Galveston, Texas. (Courtesy U.S. Army Corps of Engineers.)

Flexible or articulated armor type revetments are constructed of rip-rap or interlocking concrete blocks which cover the shore slope. Typical rip-rap revetments have armor stones which weight one-half to 3 tons. Concrete blocks used for revetments are commonly 1.5 to 4.0 feet square and 2 to 12 inches thick. Sketches of typical flexible revetments are shown as Fig. 20–7.

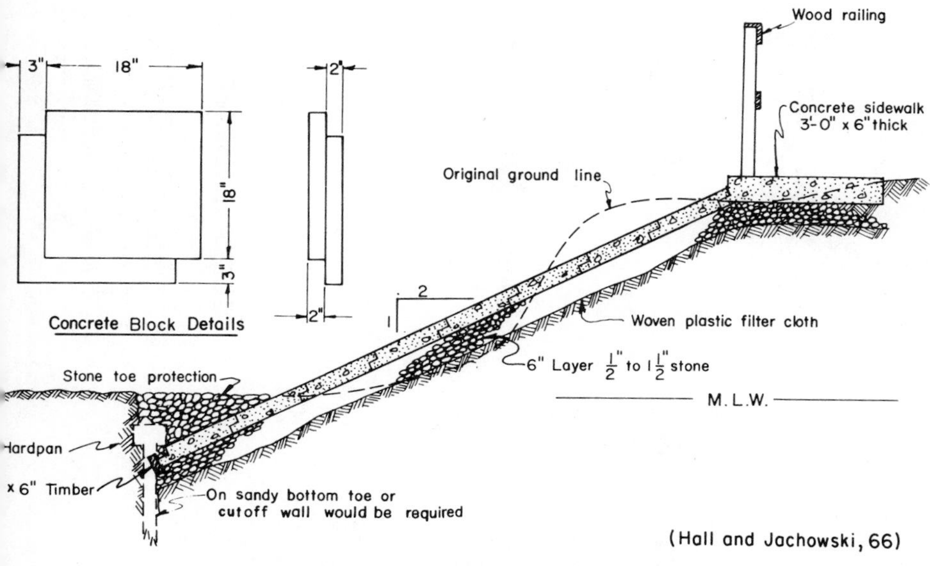

INTERLOCKING CONCRETE-BLOCK REVETMENT
(BENEDICT, MARYLAND)(MODIFIED)

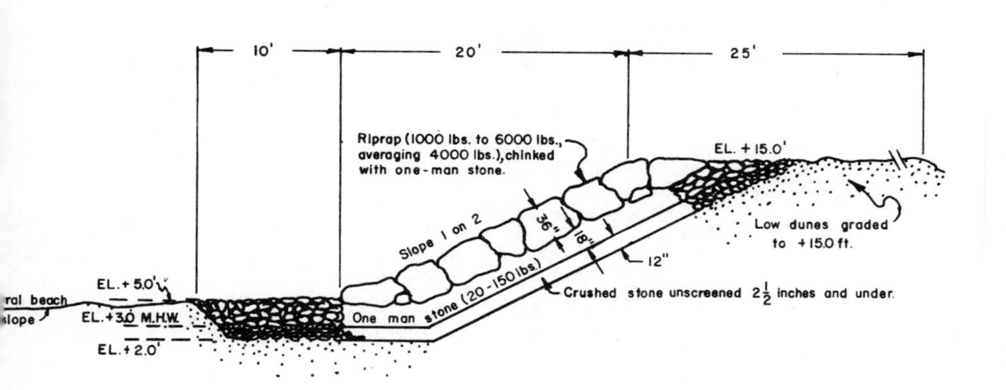

RIPRAP REVETMENT
(FORT STORY, VIRGINIA)

Fig. 20–7. Typical flexible revetments. (Courtesy U.S. Army Corps of Engineers.)

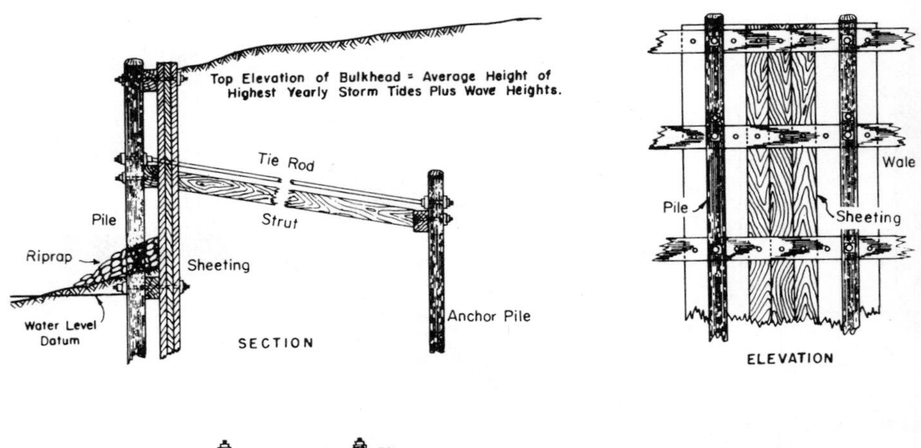

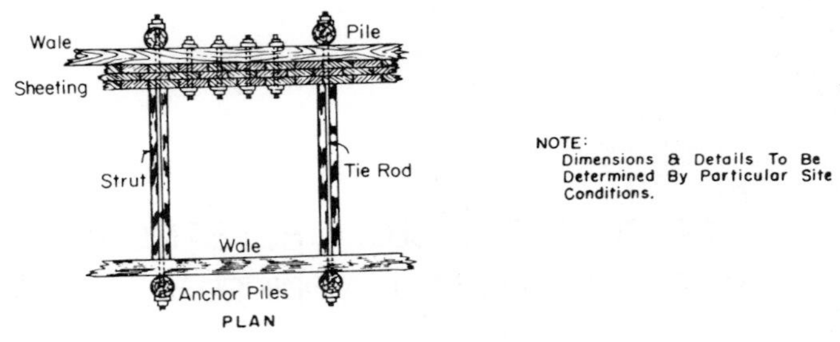

NOTE:
Dimensions & Details To Be Determined By Particular Site Conditions.

TIMBER SHEET PILE BULKHEAD

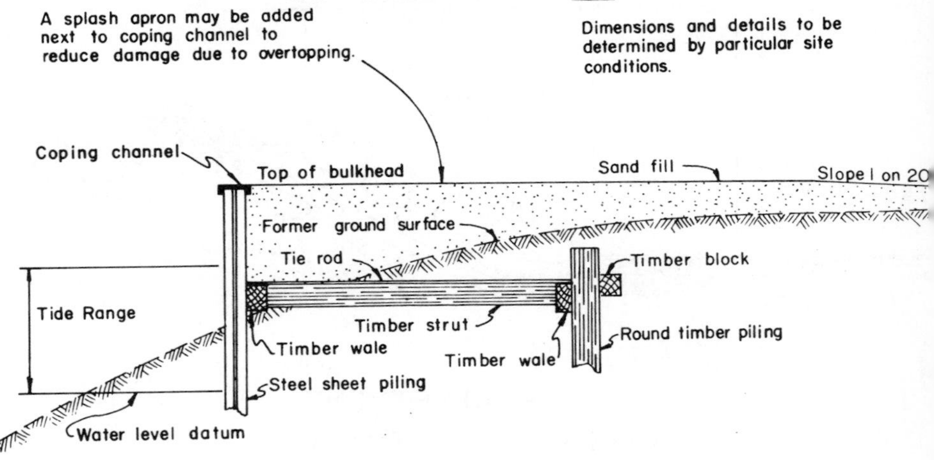

STEEL SHEET PILE BULKHEAD

Fig. 20–8. Typical bulkheads. (Courtesy U.S. Army Corps of Engineers.)

Two common types of bulkheads are illustrated by Fig. 20–8. A bulkhead is supported in a cantilever fashion by the soil into which it is driven. Additional support for a bulkhead may be provided by tie rods connected to vertical piles, which are driven some distance shoreward.

The choices of the location and length of seawalls, revetments, and bulkheads are usually straightforward depending as they do on local circumstances. The location of these structures with relation to the shoreline will generally coincide with the line of defense against further erosion and encroachment of the sea. The length of these structures depends on how much shoreline is to be separated from the water or protected from the sea.

A critical factor in the design of seawalls, revetments, and bulkheads is the determination of the height of the structure. This determination will hinge on a choice between two basic approaches to the problem. One approach is to design the structure so as to prevent wave overtopping which might damage the structure, flood facilities on the landward side, and possibly endanger human lives. This approach is based on an estimation of the wave runup, which, when added to the still water elevation, establishes the minimum elevation of the crest of the structure. A second design approach is to recognize that it may not be feasible to construct the protective facility high enough to insure that no wave overtopping will ever occur. This approach attempts to estimate the volume of water which will pass over the top of the structure under the most critical wave conditions and to attempt to provide appropriate facilities to expel the overtopped water.

Unfortunately, little empirical research has been carried out to establish definitive relationships between either wave runup or overtopping discharge and the various characteristics of the wave and structure which influence these values. Practically all of the research in these areas has been conducted using small-scale models. It has been found that there is a scale effect whereby the predicted values of wave runup and overtopping volume from small-scale tests are smaller than those observed in practice. Pending the completion of more definitive research, the designer may be guided by the work of Saville [8, 9] in the estimation of wave runup and that of Tsuruta and Goda [10] in the estimation of overtopping discharge. The latter work describes an investigation of overtopping discharge due to irregular waves, taking into account the facts that waves in nature are not of uniform height and that there may be interference from preceding waves.

20–6. Groins. The erosion of beach areas represents one of the most complex and difficult problems facing the coastal engineer. This erosion results from the effects of breaking waves, especially when the waves approach the shoreline at an oblique angle. Waves approaching a shoreline obliquely create a current which generally parallels the shoreline. This current, which is called a *longshore current,* sweeps along the sandy particles which constitute the beach bottom. The material which moves along the shore under the influence of waves and the longshore current is called *littoral drift.*

It is not always understood that man-made structures may stabilize one beach area while creating additional problems of beach erosion in adjoining areas. Impermeable seawalls, bulkheads, or revetments which are constructed along the foreshore often upset the delicate natural regime and cause or increase erosion immediately in front of the structure and along the unprotected coastline downdrift from the structure.

The most common approach to the control of beach erosion is to build a groin or a system of groins. A groin is a structure which is constructed approximately perpendicular to the shore in order to retard erosion of an existing beach or to build up the beach by trapping littoral drift. The groin serves as a partial dam which causes material to accumulate on the updrift or windward side. The decrease in the supply of material on the downdrift side causes the downdrift shore to recede. The shoreline changes which follow the construction of a properly designed groin system are shown by Fig. 20–9.

There are two broad classes of groins: permeable and impermeable. Permeable groins permit the passage of appreciable quantities of littoral

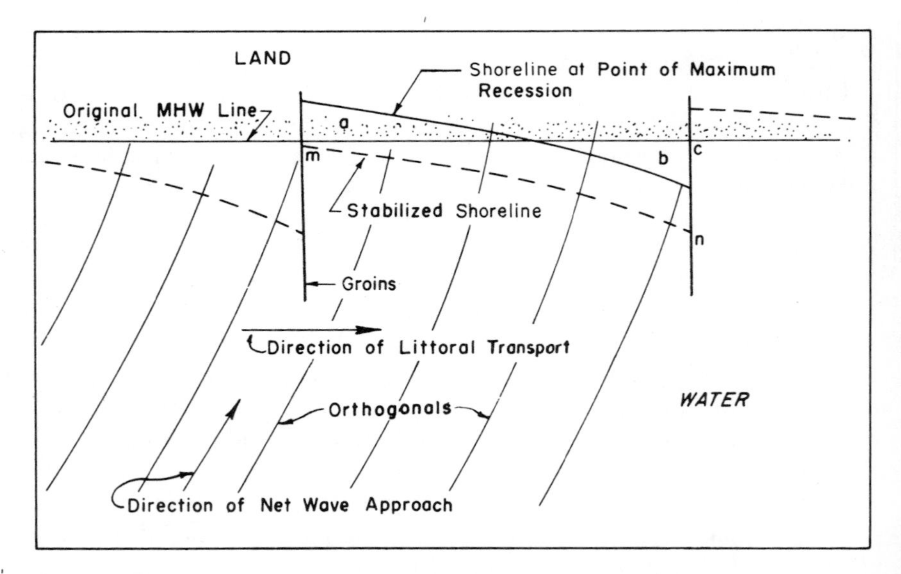

Fig. 20–9. Groin system operation. (Courtesy U.S. Army Corps of Engineers.)

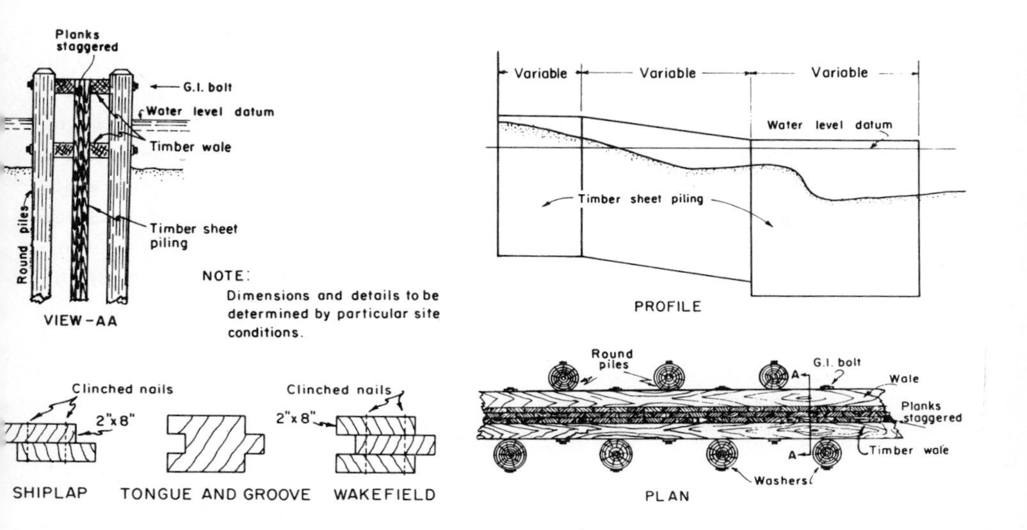

TIMBER SHEET-PILE GROIN

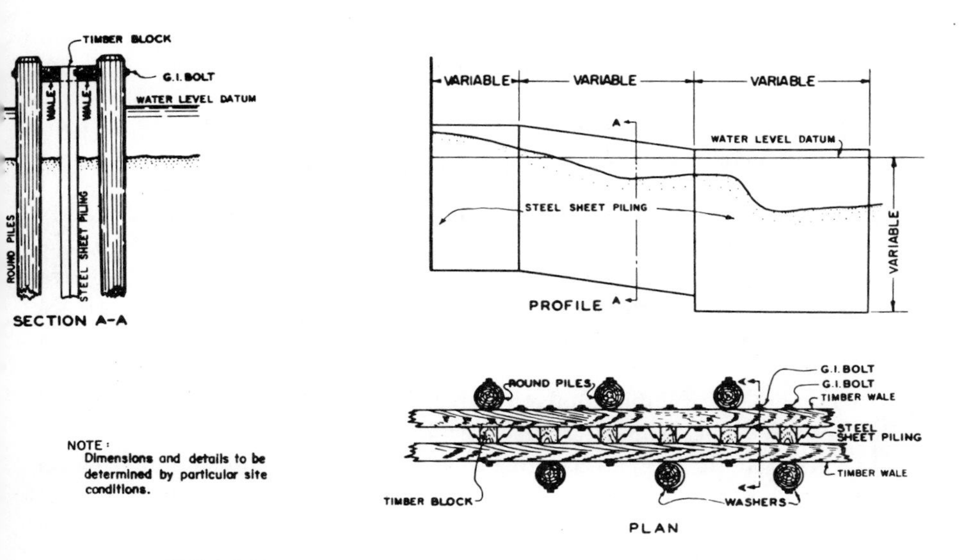

TIMBER - STEEL SHEET-PILE GROIN

Fig. 20–10. Typical timber groins. (Courtesy U.S. Army Corps of Engineers.)

drift through the structure. Impermeable groins, the most common type, serve as a virtual barrier to the passage of littoral drift.

A wide variety of groins has been constructed, utilizing timber, steel, concrete, and stone. Two examples of timber groins are shown by Fig. 20–10. It will be noted that in the design shown by Fig. 20–10(b) steel sheet piling provides the impermeable barrier. Cellular steel sheet-pile groins have also been successfully employed, as have prestressed concrete sheet-pile groins and rubble mound groins. The latter type is usually constructed with a core of fine quarry run stone to prevent the passage of sand through the structure.

The selection of the type of groin will depend to a large degree on the following factors:

1. availability of materials
2. foundation conditions
3. presence or absence of marine borers
4. topography of the beach and uplands.

The hydraulic behavior of a system of groins is exceedingly complex and its performance will be influenced by:

1. the specific weight, shape, and size of the particles which constitute the littoral drift;
2. the height, period, and angle of attack of approaching waves;
3. the range of the tide and the magnitude and direction of tidal currents;
4. the design features of the groin system, including the groin orientation, length, spacing, and crown elevation.

Groins are usually built in a straight line and oriented approximately normal to the shoreline. There appears to be little advantage to the use of curved structures or groins of the "T" or "L" head types. These types tend to be more expensive to build and will normally experience more scour at the end of the structure than will be experienced with the straight groin.

There are no reliable analytical techniques for the determination of the desirable length of a groin. The length will depend on the nature and extent of the prevailing erosion and the desired shape and location of the stabilized shoreline. The total length, which typically is on the order of 100 to 150 feet, is comprised of three sections, illustrated by Fig. 20–11:

1. the horizontal shore section which extends from the landward end to the desired location of the crest of the berm;

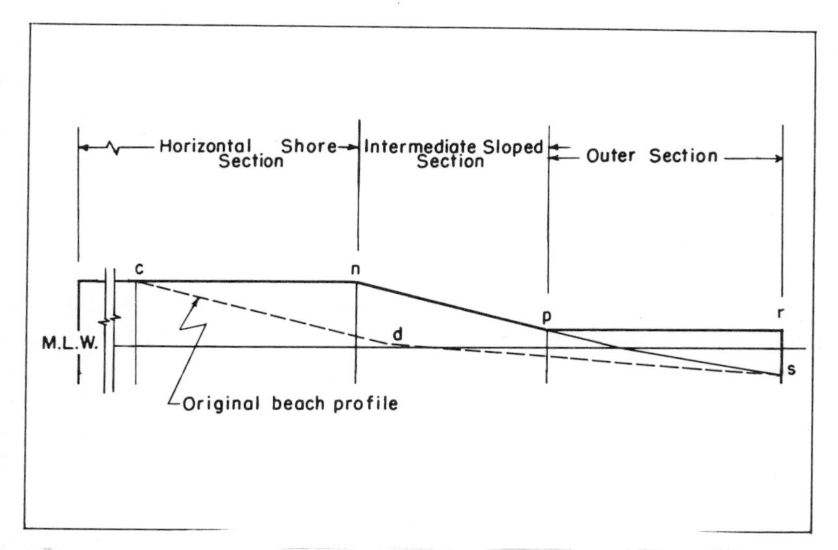

Fig. 20–11. Typical groin profile. (Courtesy U.S. Army Corps of Engineers.)

2. the intermediate sloped section, which extends between the horizontal shore section and the outer section, roughly paralleling the slope of the foreshore;
3. the outer section, which is horizontal.

The spacing of groins in a groin system depends on the groin length and the desired alignment and location of the stabilized shoreline. As a rule of thumb, the U.S. Army Corps of Engineers [9] recommends that the spacing between groins should be equal to two or three times their length from the berm crest to the seaward end.

The elevation of the crest of a groin will determine to some extent the amount of sand trapped by the groin. In cases where it is desirable to maintain a supply of sand on the leeward side of the groin, it may be built to a low height, allowing certain waves to overtop the structure. If no passage of sand beyond the groin is desired, the elevation of the crest should be such that storm waves will not overtop the structure.

WAVE FORCES

Attempts by engineers to control the powerful and relentless forces of waves have often met with disappointment and failure. Rarely is an engineer called upon to design a structure to withstand forces of the magnitude of those imposed by ocean waves during a storm. Furthermore, wave phenomena are exceedingly complex and reliable analytical equations for the estimate of wave forces are not generally available. Yet, by means of empirical data collected over a period of many years, the design of coastal and harbor structures can be approached with confidence by the application of sound engineering principles and procedures.

The magnitude of wave forces on coastal and harbor structures will vary a great deal depending on whether or not the waves break, the shape and slope of the face of the structure, and its roughness and permeability.

20–7. Vertical Walls Subjected to Non-Breaking Waves. Coastal structures located in protected areas or deep water may be subjected to non-breaking waves. In the case of vertical walls subjected to non-breaking waves, a satisfactory estimate of wave forces may be obtained by the Sainflou method. This method, which was proposed by the Frenchman George Sainflou in 1928, is based on the assumption that pressures due to non-breaking waves are essentially hydrostatic.

Observations have shown that when a non-breaking wave strikes a vertical wall the reflected wave augments the next oncoming wave resulting in the formation of a standing wave or *clapotis*. The height of a clapotis is approximately twice the height of the original wave.

The orbit center of the standing wave lies a distance h_o above the stillwater level. This distance may be computed by the following equation:

$$h_o = \frac{\pi H^2}{L} \coth\left(\frac{2\pi d}{L}\right) \tag{20–1}$$

where:

H = wave height, feet
L = wave length, feet
D = stillwater depth, feet.

Assuming the same stillwater level exists on both sides of the wall, the pressures on the wall are shown by Fig. 20–12. When the wave is in the crest position the resultant pressures are landward as shown by Fig. 20–12(a). When the trough of the clapotis is at the face of the wall, the net pressures are directed seaward as shown by Fig. 20–12(b).

The pressure P_1 is given by the equation:

$$P_1 = \frac{wH}{\cosh\left(\frac{2\pi d}{L}\right)} \tag{20–2}$$

where:

w = Specific weight of water, $\frac{\text{pounds}}{\text{cubic foot}}$

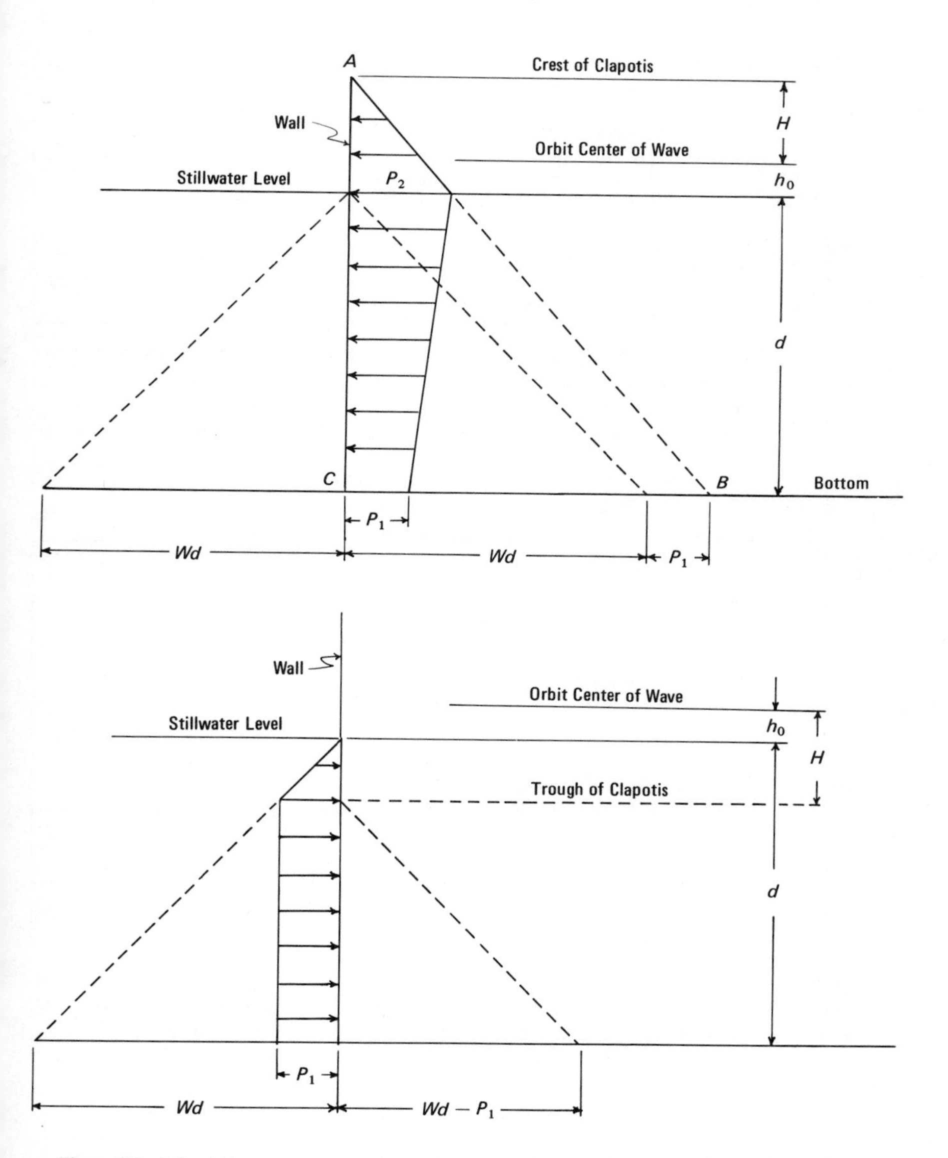

Fig. 20–12. Wave pressures on vertical walls, according to Sainflou method.

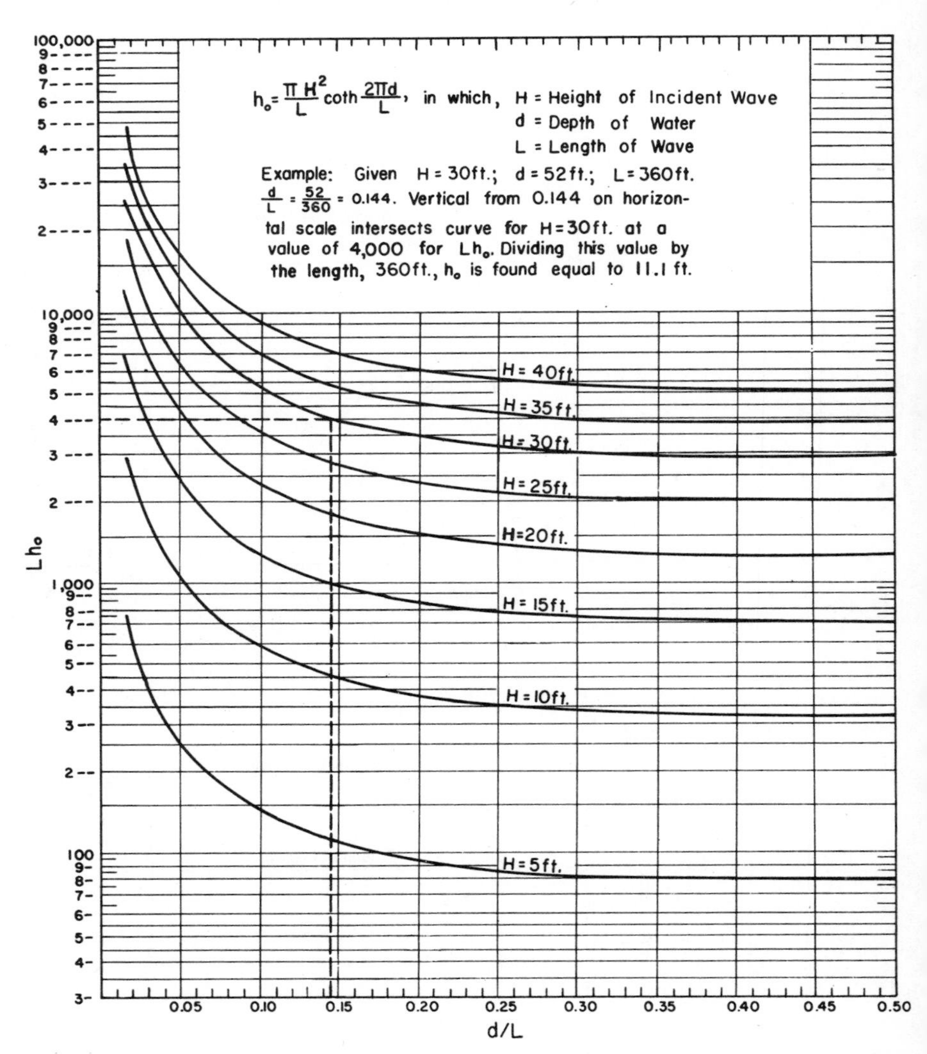

Fig. 20–13. Determination of value of h_o in Sainflou's formula. (Source: U.S. Army Corps of Engineers, *Shore Protection, Planning, and Design,* 3rd ed., 1966.)

Graphical solutions for h_o and P_1 may be determined, respectively, by Figs. 20–13 and 20–14.

By simple proportion, it can be seen that:

$$P_2 = (wd + P_1)\left(\frac{H + h_o}{H + h_o + d}\right) \tag{20–3}$$

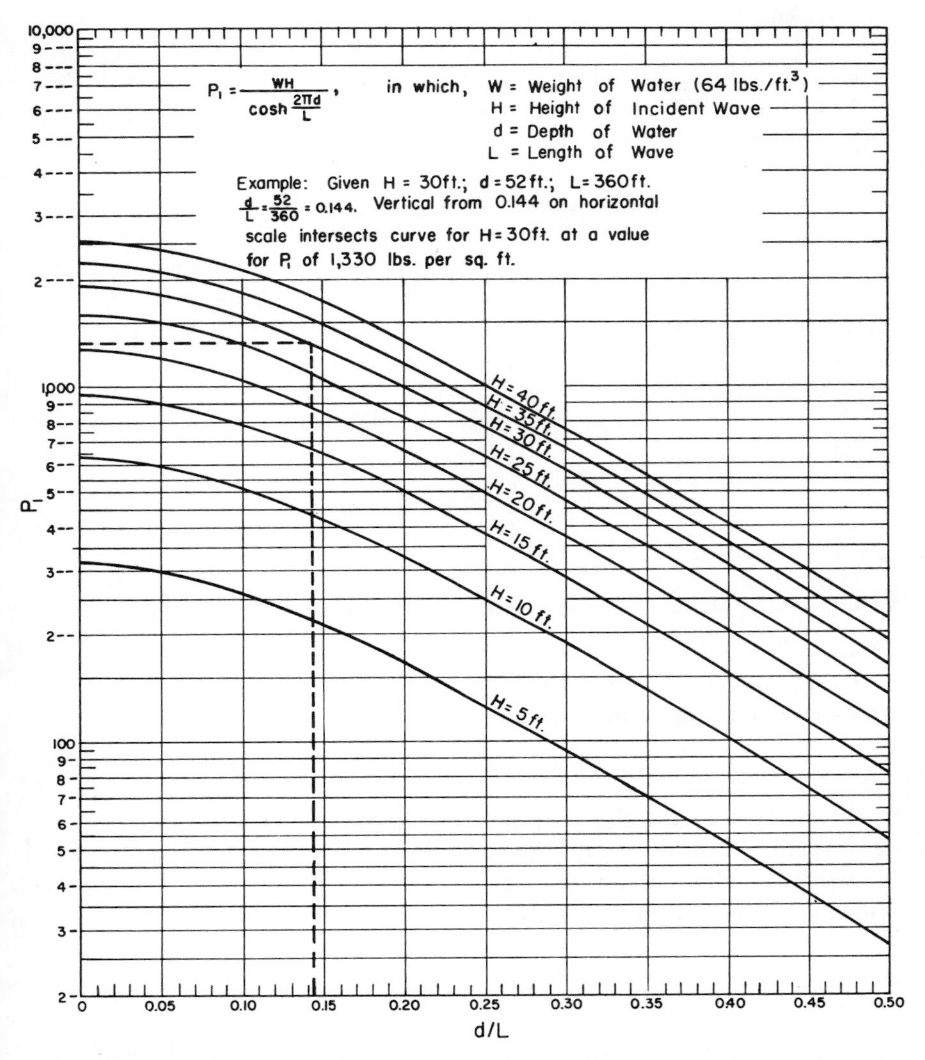

Fig. 20–14. Determination of value of P_1 in Sainflou's formula. (Source: U.S. Army Corps of Engineers, *Shore Protection, Planning, and Design,* 3rd ed., 1966.)

It is noted that if there is no water on the landward side of the wall, the resultant pressure with the crest at A would be that shown by the triangle ACB, Fig. 20–12(a).

20–8. Forces Due to Breaking Waves. Bulkheads, seawalls, and vertical-wall breakwaters are often located so as to be exposed to the force of breaking waves. Research has shown that these waves are much more complex than non-breaking waves. Model studies have indicated that structures exposed to breaking waves must withstand both hydrostatic and dynamic pressures. The dynamic pressure is typically intense and of short duration.

The method commonly used to evaluate forces due to breaking waves was developed by R. R. Minikin [11]. The Minikin method was originally developed to analyze forces on a composite breakwater consisting of a concrete superstructure supported by a rubble mound substructure. According to this method, the dynamic pressure is assumed to be a maximum at stillwater level, decreasing parabolically to zero at a distance of one-half the wave height above and below the stillwater level. (See Fig. 20–15.)

The magnitude of the maximum dynamic pressure in pounds per square foot is given by:

$$P_m = \pi g w \left(\frac{H}{L}\right) \frac{d}{D} (D + d) \qquad (20\text{–}4)$$

where:

d = depth of water at the toe of the vertical wall, feet.

Other values in Equation (20–4) are as previously defined.

The dynamic force per linear foot of structure is obtained from the area of the force diagram in Fig. 20–15(a):

$$\text{Dynamic force per linear foot} = \tfrac{1}{3} P_m H \qquad (20\text{–}5)$$

In addition to the dynamic pressure, the Minikin method recognizes that there is a hydrostatic pressure acting shoreward due to the height of the wave above stillwater level. (See Fig. 20–15(b).) The magnitude of this force at the stillwater level is:

$$P_s = \frac{wH}{2} \qquad (20\text{–}6)$$

Assuming that hydrostatic pressures exist on both sides of the wall, the net hydrostatic force per linear foot of structure is:

$$\text{Hydrostatic force per linear foot} = P_s d + \frac{P_s(H/2)}{2} \qquad (20\text{–}7)$$

It follows that the resultant unit wave force per linear foot of structure is:

$$R = \frac{P_m H}{3} + P_s d + \frac{P_s H}{4} \qquad (20\text{–}8)$$

An adaptation of the Minikin method has been used to calculate forces on caissons or walls which have no substructure. In this case, the values of D and L in Equation (20–4) refer, respectively, to the water depth and wave length measured one wave-length seaward from the structure. This is illustrated by Fig. 20–16. Reference [9] indicates that this adaptation of the Minikin method yields reasonably reliable results provided the bottom slope is at least 1:15. When the actual slope is flatter than 1:15, it is recommended that pressure derived for a 1:15 slope be used.

The preceding equations for dynamic wave pressures are applicable to vertical wall faces and should give approximate values for stepped face structures. Walls which slope backwards will experience smaller dynamic forces, and Minikin gives the following equation for the horizontal dynamic wave pressure:

$$P'_m = P_m \sin^2 \theta \tag{20–9}$$

where:

θ is the angle between the wall face and the horizontal

20–9. Wave Forces on Rubble-Mound Structures. The stability of rubble-mound structures depends on the ability of individual armor units which comprise the armor layer to resist displacement. Thus, the designer's task is to determine the size of individual armor units required to withstand the attack of storm waves. Because of the complexity of the problem, it is necessary to rely on empirical equations which have been derived by means of extensive laboratory model tests as well as observation of wall failures.

The following equation developed by the U.S. Army Corps of Engineers [2, 9] is considered to be the most reliable guide for the estimation of the weight of each armor unit:

$$W = \frac{w_r H^3}{K_D(S_r - 1)^3 \cot \alpha} \tag{20-10}$$

where:

W = weight of armor unit in primary cover layer, lbs.
w_r = unit weight (saturated surface dry) of armor unit, lbs/ft^3
H = design wave height at the structure site
S_r = specific gravity of armor unit, relative to the water in which structure is situated $\left(S_r = \frac{w_r}{w_w}\right)$.
w_w = unit weight of water, freshwater = 62.4 lbs/ft^3, sea water = 64.0 lbs/ft^3.
α = angle of breakwater slope measured from horizontal, degrees.
K_D = an empirical constant. (See Table 20–2.)

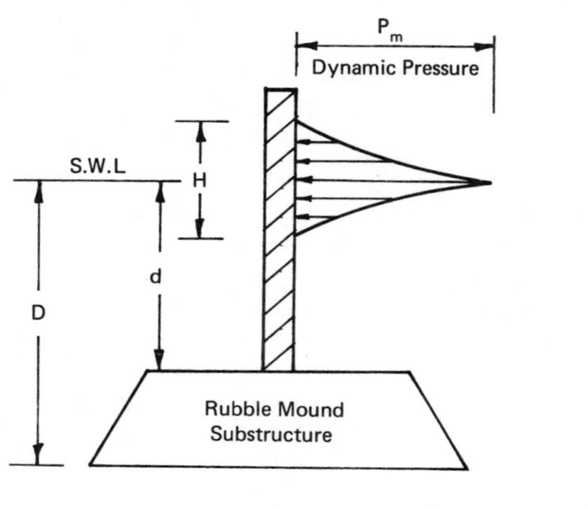

(a) Dynamic Pressure Diagram.

(b) Hydrostatic Pressure Diagra

Fig. 20–15. Minikin wave pressure diagram.

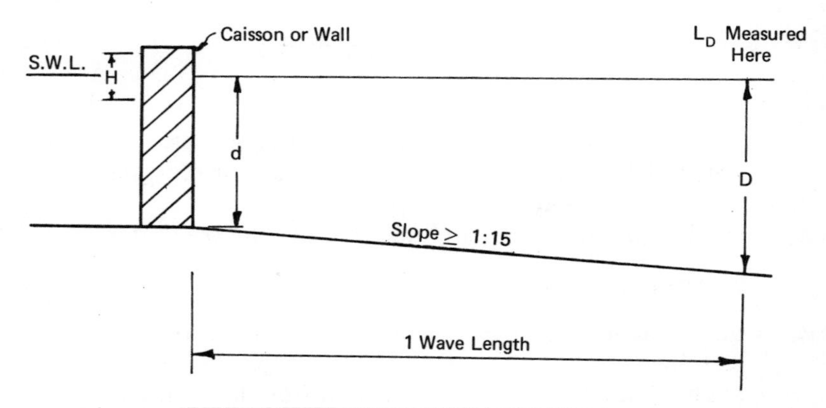

Fig. 20–16. Parameters for Adaptation of Minikin Method.

TABLE 20-2

K_D Values for Use in Determining Armor Unit Weight No-Damage Criteria

			Structure Trunk		Structure Head	
Armor Units	n[a]	Placement	Breaking Wave[b]	Nonbreaking Wave[c]	Breaking Wave[b]	Nonbreaking Wave[c]
Smooth rounded Quarrystone	2	Random	2.5	2.6	2.0	2.4
Smooth rounded Quarrystone	>3	Random	3.0	3.2	—	2.9
Rough angular Quarrystone	1	Random[d]	2.3	2.9	2.0	2.3
Rough angular Quarrystone	2	Random	3.0	3.5	2.7	2.9
Rough angular Quarrystone	>3	Random	4.0	4.3	—	3.8
Rough angular Quarrystone	2	Special[e]	5.0	5.5	3.5	4.5
Modified Cube	2	Random	7.0	7.5	—	5.0
Tetrapod	2	Random	7.5	8.5	5.0	6.5
Quadripod	2	Random	7.5	8.5	5.0	6.5
Tribar	2	Random	8.5	10.0	5.0	7.5
Tribar	1	Uniform	12.0	15.0	7.5	9.5

Source: U.S. Army Corps of Engineers, *Shore Protection, Planning, and Design*, 1966.
[a]n is the number of units comprising the thickness of the armor layer.
[b]Minor-overtopping criteria.
[c]No-overtopping criteria.
[d]The use of single layer of quarrystone is not recommended except under special conditions and when it is used, the stone should be carefully placed.
[e]Refers to special placement with long axis of stone placed normal to structure face.

This equation directly accounts for variations due to the wave height, slope of the structure face, and the unit weights of the water and the armor unit.

The empirical coefficient, K_D, is used to allow for the effect of the following variables:

1. shape and roughness of the armor unit
2. number of units comprising the thickness of the armor layer
3. permeability of the structure as affected by the placement of the armor units

4. whether the structure is subjected to breaking or non-breaking waves
5. whether the armor unit is to cover the structures trunk or head.

Experience has shown that the head of a breakwater or jetty is more likely to sustain extensive damage than is the trunk. The provision of different values of K_D for the head and trunk allows for this fact.

It is noted that the values given in Table 20–2 provide little or no factor of safety.

PROBLEMS

1. Estimate the wave runup on a rubble-mound breakwater with a 3:1 side slope if the wave height is 18 feet and the wave length is 360 feet.
2. Using the wave data in Problem 1, what wave runup would be expected for an open regularly set stone revetment with a 3:1 side slope?
3. Using the wave data in Problem 1, what wave runup would be expected for a smooth impervious levee with a 3:1 side slope and a 10:1 beach slope?
4. According to the Sainflou method, estimate the shoreward pressure on a vertical wall at the stillwater level and at the bottom due to a non-breaking wave which is 22 feet in height and 280 feet in length. Compute these pressures using appropriate equations, then check your results using Figs. 20–13 and 20–14. The stillwater depth is 50 feet.
5. A composite breakwater is subjected to the following wave conditions:
   Wave height, H = 30 feet
   Wave length, L = 320 feet
   Still water depth to top of substructure d = 40 feet
   Depth to bottom, D = 60 feet
   Determine: (a) maximum dynamic pressure and the dynamic force per linear foot on the breakwater; (b) hydrostatic force per linear foot.
6. Two layers of tetrapods are to be randomly placed on a breakwater which will be exposed to 15-foot high breaking waves. The breakwater has side slopes of 2.5 horizontal to 1.0 vertical. Determine the minimum weight of armor unit required to resist displacement for: (a) trunk of breakwater; (b) head of breakwater.

REFERENCES

1. The American Waterways Operators, Inc., "How Far Does the Shipper's Dollar Move a Ton of Freight," March, 1969.
2. *Wave Forces on Rubble-Mound Breakwaters and Jetties,* U.S. Army Engineer Waterways Experiment Station, Corps of Engineers, September, 1961.
3. BREBNER, A., and OFUYA, A. O., "Floating Breakwaters," *Proceedings of Eleventh Conference on Coastal Engineering,* American Society of Civil Engineers, September, 1968.
4. NECE, R. E.; RICHEY, E. P.; and RAO, V. S., "Dissipation of Deep Water Waves by Hydraulic Breakwaters," *Proceedings of Eleventh Conference on Coastal Engineering,* American Society of Civil Engineers, September, 1968.

5. BULSON, P. S., "The Theory and Design of Bubble Breakwaters," *Proceedings of Eleventh Conference on Coastal Engineering,* American Society of Civil Engineers, September, 1968.
6. QUINN, ALONZO D., *Design and Construction of Ports and Marine Structures,* McGraw-Hill Book Co., New York (1961).
7. SHIRAISHI, NAOFUMI; NUMATA, ATSUSHI; and ENDO, TAIJI, "On the Effect of Armour Block Facing on the Quantity of Wave Overtopping," *Proceedings of Eleventh Conference on Coastal Engineering,* American Society of Civil Engineers, September, 1968.
8. SAVILLE, T., JR., "Wave Run-up on Shore Structures," *Proceedings,* American Society of Civil Engineers, Vol. 82, No. WW2, April, 1956.
9. U.S. Army Corps of Engineers, *Shore Protection, Planning and Design,* Technical Report No. 4, Third Edition, 1966.
10. TSURUTA, SENRI, and GODA, YOSHIMI, "Expected Discharge of Irregular Wave Overtopping," *Proceedings of Eleventh Conference on Coastal Engineering,* American Society of Civil Engineers, September, 1968.
11. MINIKIN, R. R., *Wind, Waves, and Maritime Structures,* Second Edition, Charles Griffin and Co., Ltd., London (1963).

21

Planning and Design of Port Facilities

21–1. Introduction. Increasing world population, the industrialization of nations, and the reshaping of national boundaries have caused substantial increases in the volume of ocean commerce. The needs of increased world trade have resulted in increases in the number and size of the merchant fleet, which has created a growing need for the construction of new ports and the enlargement and modernization of existing port facilities.

In planning and designing port improvements, the engineer must choose between alternate port layout schemes and substructure designs. He must obtain reliable forecasts of the number and time distribution of ship movements in order to anticipate the number of required berths. Adequate dimensions for channels and berths must be provided to allow safe and expeditious ship movements and berthing. Sufficient apron space must also be provided for the loading and unloading of ships and for the taking on of fuel and supplies.

General cargo terminals require properly designed transit sheds for sorting and temporary storage of packaged freight. Long-term storage facilities, both open and covered, are usually required and special facilities may be needed for the handling and storage of chemicals, grains, and other materials in bulk.

Every attempt should be made in planning and designing a port to anticipate innovations and improvements in cargo handling technology in order that prohibitively expensive alterations will not be required at a later date.

21–2. General Layout and Design Considerations. In the early planning stages of the development of a port facility, certain basic decisions must be made regarding the general arrangement and layout of the facility. One such decision regards the choice of type of wharf.

A *wharf* is a structure built on the shore of a river, canal, or bay so that vessels may lie alongside and receive and discharge cargo and passengers. A wharf built generally parallel to the shoreline is called a *marginal wharf* or *quay*. A wharf built at an angle to the shore is called a *finger pier* or simply a *pier*. The berthing and maneuvering space between adjacent piers is called a *slip*.

Marginal wharves provide for easier berthing of ships and are, therefore, generally preferred over finger piers by steamship and stevedoring companies. Other advantages claimed for marginal wharves are:

1. The cost per berth may be lower.
2. Operational needs of ship owners, truckers, and stevedoring companies are better satisfied.
3. The costs of channel maintenance are less.
4. There is less hazard from waterfront fires and explosions.
5. The continuous line of wharf apron facilitates emergency movements of cargo between adjacent buildings or ships along the waterfront.

Piers, which generally favor rail operations, are preferred by railroad companies. The pier-type layout usually provides more berths per unit length of waterfront than does the marginal wharf arrangement.

The finger pier or slip type of construction may be obtained by extending the piers outward from the shoreline or by reshaping the shoreline by dredging the area between adjacent piers.

The choice of type of wharf layout will depend to a large extent on local conditions. At a port which has ample harbor area for ship maneuvering, but which has a scarcity of waterfront land, constructing piers from the shore may prove advantageous. At a location where there is a scarcity of water area for ship movement and berthing, consideration should be given to dredging of slips. In the latter case, space must be provided for dumping the spoilage that is removed. Such excavation, of course, also removes from service land which otherwise could be devoted to on-shore port operations and storage.

21–3. Pier and Wharf Substructures. Basically, there are two broad classes of wharf substructures: (1) *solid fill type,* and (2) the *open type*. The solid fill type, illustrated by Fig. 21–1, consists of a vertical wall which is backfilled by earth which supports a paved deck. The wall is commonly a cantilevered, anchored steel sheet-pile bulkhead. Gravity structures such as cellular steel bulkheads and cribs of timber or concrete have also been successfully employed.

In the open-type construction, the wharf superstructure is supported by timber, concrete, or steel piles. In this type of substructure, trans-

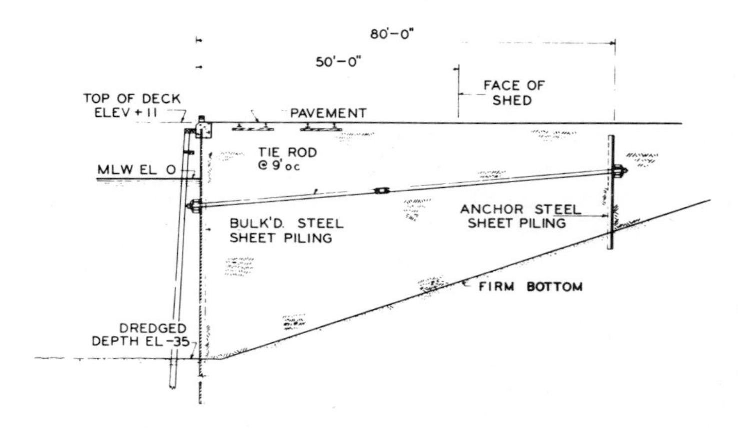

Fig. 21–1. Solid fill type of wharf construction. (Courtesy The American Association of Port Authorities.)

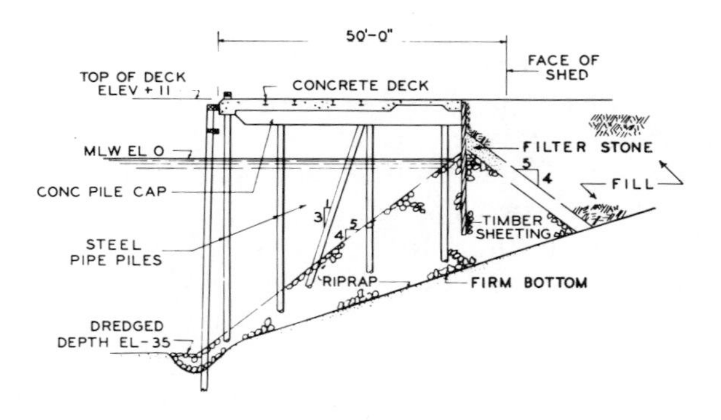

Fig. 21–2. High-level open-type of wharf construction. (Courtesy The American Association of Port Authorities.)

verse rows of bearing piles are driven and capped with concrete girders. Longitudinal beams may also be provided to support unusually heavy concentrated loads. Alternatively, where heavy concentrated loads are not expected, the piles are closely spaced and capped with a flat slab. A typical open-type of wharf substructure is shown by Fig. 21–2.

The principal advantage of the solid fill type of wharf substructure is that its great mass provides adequate resistance to the impact of mooring ships. Solid fill substructures are inexpensive (except in deep water), stable, and require little maintenance. Because this type of substructure serves as a barrier to currents and tides, it is used principally to support marginal wharves.

The open type of wharf substructure is more economical in deep water locations and where a high-level superstructure is required. Since this type of substructure offers little restriction to water movements, it can be used to support piers in rivers and coastal areas alike.

Open substructures supported by timber piles are subject to decay and may be attacked by marine borers. Considerable risk of wharf fires is associated with the use of timber piles. These objections to the open type of substructure may, of course, be largely overcome by using concrete piles. Steel piles are usually encased in concrete above the low water line to inhibit corrosion.

A variation of the open-type wharf substructure utilizes a *relieving platform* on which fill is superimposed, capped by a paved deck. This type of design offers the advantages of high resistance to impact and economy of construction. High load concentrations are spread by the fill, lessening or eliminating the need for massive longitudinal girders. The relieving platform type of design is less subject to deterioration and decay than is the conventional open-type design, especially if the platform is located below the elevation of mean low water. A typical open-type wharf substructure with a relieving platform is shown by Fig. 21–3.

21–4. Fender Systems. During the mooring process, a great deal of damage can be done to the wharf and ship unless some sort of protective device is provided to absorb the energy of the moving vessel. Such

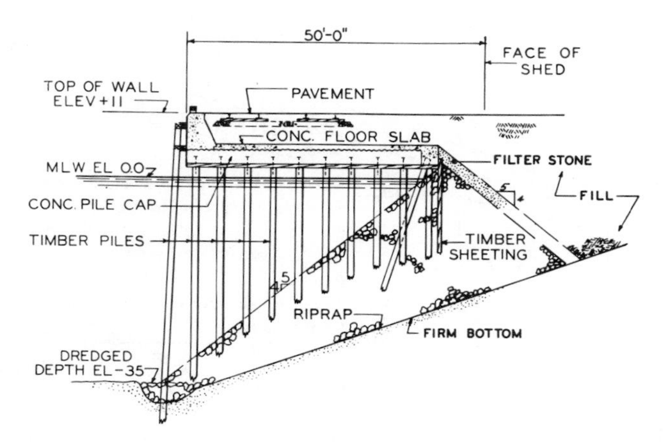

Fig. 21–3. Open-type wharf substructure with concrete relieving platform. (Courtesy The American Association of Port Authorities.)

a device is also needed to lessen the effect of the bumping and rubbing of the ship against the wharf while the ship is secured. Protective installations which meet these needs are called *fenders*. A wide variety of fenders has been employed, including:

1. pile fenders
2. timber hung systems
3. rubber fenders
4. gravity type fender systems
5. hydraulic fenders.

One of the simplest fender systems involves a row of vertical wood piles which are driven on a slight batter and secured to the top edge of the wharf. By this system, impact is absorbed by deflection and by compression of the wood. A timber pile fender may be seen in Fig. 21–3. A floating log called a "camel" is often placed between the ship and the pile fenders to distribute impact loads along the fender system and keep the ship away from the face of the dock. The energy absorption of timber piles depends on the pile diameter and length and the type of wood. (See Fig. 21–4.) It should be remembered that the energy absorption capabilities of wood piles, which are rather limited at best, decrease sharply with deterioration and wear.

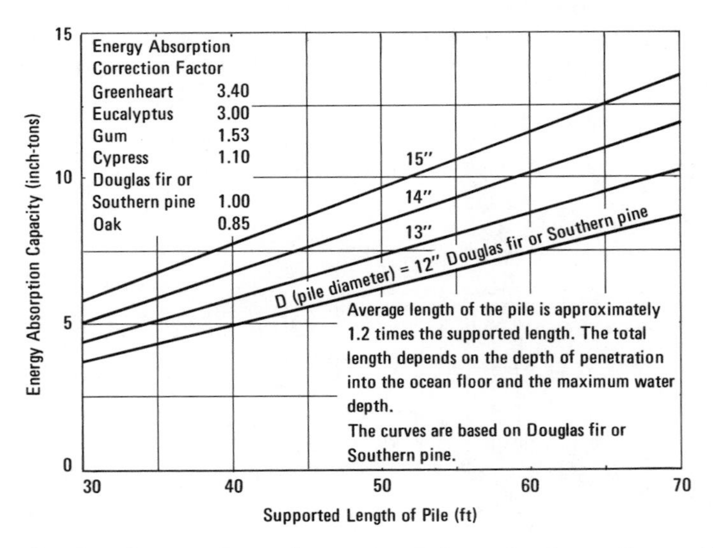

Fig. 21–4. Energy-absorption of timber fender piles. (Source: Theodore T. Lee, "Design Criteria Recommended for Marine Fender Systems," *Proceedings of Eleventh Conference on Coastal Engineering,* September, 1968.)

Pile fenders constructed of steel and concrete have also been used, but generally these systems do not perform as satisfactorily as timber piles.

In locations where the water is calm and the tidal range is small, a timber-hung fender system may be used. In this system, vertical wood members are secured to the face of the dock and terminate near the water surface. Typically, horizontal wood members are attached between the vertical members and the face of the dock. Timber-hung systems have a low energy-absorption capacity which depends entirely on the compression of the wood.

Rubber has been used extensively and effectively in fender systems and in a variety of ways. Cylindrical or rectangular rubber blocks are sometimes used in wood fender systems to improve energy-absorption capability. These blocks, which are placed behind horizontal members or vertical piles, absorb energy by compression.

Several patented rubber fender devices have been employed, including Raykin fender buffers and Lord fenders. Raykin fender buffers consist of layers of rubber cemented to steel plates and formed in a "V-shape." (See Fig. 21–5.) Energy is absorbed by this device as the rubber layers

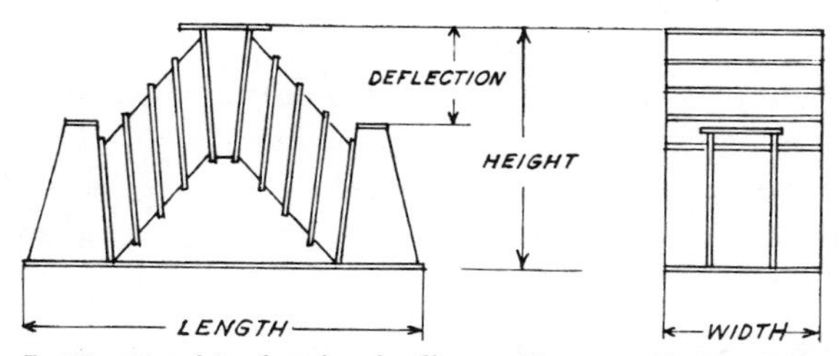

Fig. 21–5. The Raykin fender buffer. (Source: A. D. Quinn, *Design and Construction of Ports and Marine Structures,* 2nd ed., McGraw-Hill Book Co., 1972.)

distort in response to the shearing forces imposed by a mooring vessel. Lord fenders consist of an arch-shaped rubber block bonded between two steel plates. Impact energy is absorbed by the bending and compression of an arch-shaped rubber column.

Hollow rubber cylinders have also been used as fenders. These vary in size from about 5–18 inches in outside diameter, and the inside diameter is typically one-half the oustide diameter. These cylinders are draped along the face of a wharf, suspended by a heavy chain. These fenders are suitable to protect a solid and deep wall such as the face of a relieving platform-type of substructure.

The energy-absorption characteristics of rubber fenders are described in graphs and tables available from the rubber companies which sell these products.

Gravity-type fenders are made of large concrete blocks or cylinders which are suspended from the edge of the wharf deck. When a ship impacts the fender system, these heavy objects are lifted a short distance absorbing the energy of impact. One such fender uses concrete-filled steel tubes which measure two to three feet in diameter and 20 to 25 feet in length. These cylinders, which weigh about 15 tons, are suspended vertically along the side of the wharf and are hinged in such a way so as to be lifted when a lateral impact force is applied. Wood rubbing strips are usually attached to the seaward side of these cylinders.

A new concept in fender design was recently proposed [4] involving the use of clusters of water-filled plastic cells. A floating fender system, this device has been named a "hi-dro cushion camel" by its designers. The cylindrical cells are completely enclosed except for an air vent hole and a pressure regulating orifice. When impacted, water is released through the orifice causing the impulse period to be lengthened and the forces to be reduced. The cell clusters are sandwiched between two structural diaphragms as shown by Fig. 21–6. Tests of this device at the Treasure Island U.S. Naval Station indicated that the hi-dro cushion camels may effectively absorb the energy of a 100 ton vessel berthing at a speed of 6 feet per second or a 1,000 ton vessel berthing at two feet per second.

In the preceding paragraphs, a brief summary description of some of the more popular fender systems has been given. Table 21–1 lists the principal advantages and disadvantages of these systems.

In selecting a type of fender system, the designer must consider a variety of factors, but the most important factors are:

1. mass of ships to be berthed
2. speed of berthing (normal to the dock)
3. environmental conditions at the port.

The speed of approach normal to the dock taken for design purposes varies from about 0.1 feet per second to 1.25 feet per second, depending principally on the ship size, exposure of the wharf to wind, waves, tides and currents, and the availability and type of docking assistance. Some guidance on selection of docking speed has been published by Lee [2].

The kinetic energy of a docking ship is given by the equation:

$$E = \frac{1}{2} Mv^2 = \frac{1}{2} \frac{W}{g} v^2 \tag{21–1}$$

where:

M = mass of ship
v = berthing velocity normal to face of the dock
W = displacement of the ship
g = acceleration due to gravity

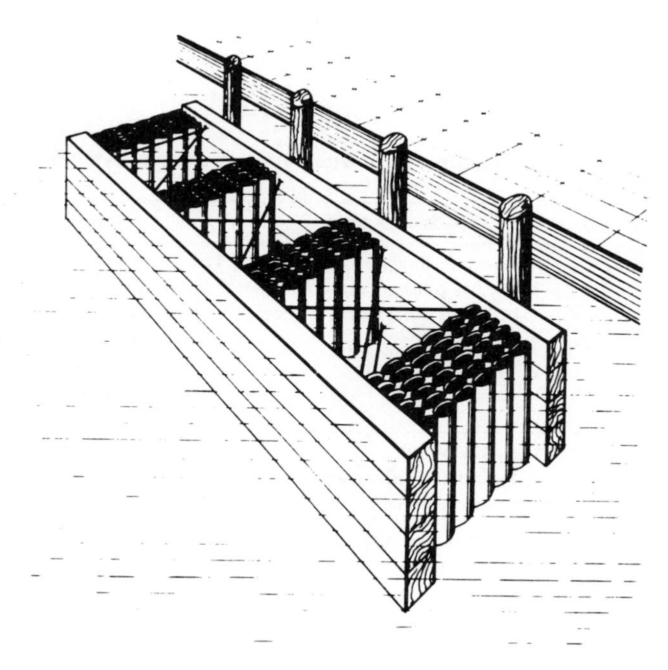

Fig. 21–6. The Hi-Dro Cushion floating fender system. (Courtesy American Society of Civil Engineers.)

For ships docking at moderate to high speeds, the kinetic energy is increased due to the mass of water moving alongside the ship. To allow for this effect, the mass value used in Equation (21–1) should be increased by about 60 per cent.

For planning purposes, it may be assumed that one-half of the kinetic energy is to be absorbed by the fender system.

The load-deflection characteristcis of various fender systems is shown by Fig. 21–7.

21–5. Estimation of Required Number of Berths. The planning of the size of a port begins with an appraisal of the volume of present and future commerce and types of shipping. This essential information allows the port planner to estimate the number, type, and sizes of the ships to be accommodated. The forecasting of the anticipated volume and type of freight is an area of concern of economists and planners, and engineers are seldom involved in this type of activity. While the detailed procedures for making such a study do not fall within the purview of this chapter, a brief general discussion of certain aspects of this problem will be given.

TABLE 21-1

Comparison of Various Types of Fender Systems

Fender System	Advantages	Disadvantages
Standard Pile, timber	1. Low initial cost. 2. Timber piles are abundant in U.S. and most world regions.	1. Energy absorption capacity is limited. It declines as result of biodeterioration. 2. Susceptible to mechanical damage and biological deterioration. 3. High maintenance cost if damage and deterioration is significant.
Standard Pile, steel	1. High strength. 2. Feasible for difficult seafloor conditions.	1. Vulnerability to corrosion. 2. High cost.
Standard Pile, reinforced concrete	1. Insignificant effects of biodeterioration.	1. Energy-absorption capacity is very limited. 2. Corrosion of steel reinforcement through cracks.
Standard Pile, prestressed concrete	1. Resistance to natural and biological deterioration. 2. Better energy-absorption characteristics than reinforced concrete piles.	1. Limited strain-energy capacity, if rubber buffers are not provided.
Timber Hung System	1. Very low initial cost. 2. Less biodeterioration hazard.	1. Low energy-absorption capacity. 2. Unsuitability for locations with significant tide and current effects.
Rubber Fender Systems: Rubber-in-compression	1. Simplicity and adaptability. 2. Effectiveness at reasonable cost.	1. High concentrated loading may result; frictional force may be developed if rubber fenders contact ship hull directly. 2. Higher initial cost than standard pile system without resilient units.

Source: Theodore T. Lee, "Design Criteria Recommended for Marine Fender Systems," *Proceedings of Eleventh Conference on Coastal Engineering,* American Society of Civil Engineers, September, 1968.

TABLE 21-1

(Continued)

Rubber Fender Systems: Rubber-in-shear	1. Capable of cushioning berthing impact from lateral, longitudinal, and vertical directions. 2. Most suitable for dock-corner protection. 3. High energy-absorbing capacity for serving large ships of relatively uniform size. 4. Favorable initial cost for very heavy duty piers.	1. "Raykin" buffers tend to be too stiff for small vessels and for moored ships subject to wave and surge action. 2. Steel plates are subject to corrosion. 3. Bond between steel plate and rubber is a problem. 4. High initial cost for general cargo berths.
Rubber Fender Systems: Lord Flexible Fender	1. High energy-absorption and low terminal-load characteristics.	1. Possible destruction of bond between steel plates and rubber. 2. Possible fatigue problems.
Rubber Fender Systems: Rubber-in-torsion	1. Capable of resisting impact load from all directions.	1. Possible destruction of the bond between steel casting and rubber. 2. Possible fatigue problems.
Gravity-Type Fender Systems	1. Smooth resistance to impacts induced by moored ships under severe wave and swell action. 2. High energy absorption and low terminal load can be achieved through long travel for locations where excessive distance between ship and dock is not a problem.	1. Heavy berthing structure is required. 2. Heavy equipment is required for installation and replacement. 3. High initial and maintenance costs. 4. Excessive distance between dock and ship caused by the gravity fender is undesirable for general cargo piers and wharves.
Hydraulic and Hydraulic-pneumatic Fender	1. Favorable energy-absorption characteristics for both berthing and moored ships.	1. High initial and maintenance cost.

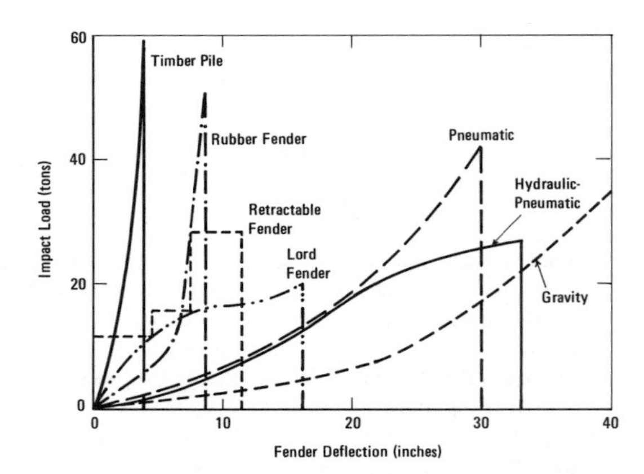

Fig. 21–7. Load-deflection characteristics of various fender systems. (Source: Theodore T. Lee, "Design Criteria Recommended for Marine Fender Systems," *Proceedings of Eleventh Conference on Coastal Engineering,* September, 1968.)

The commerce to be shipped through a proposed port will depend most of all on the nature and size of its tributary area or *hinterland.* The hinterland is generally defined as that area within which the overall cost of freight movements through the port in question is equal to or less than corresponding costs via competing ports, based on existing rates and charges [5].

The forecasting of freight movements for a proposed port would normally include an evaluation of existing and future levels of activity in manufacturing, mining, oil drilling, agriculture, and forestry within the hinterland. Significant changes in berthing, cargo handling, and storage capabilities of competing ports should also be carefully considered. When expansion of existing port facilities is contemplated, the current level of port activity and its efficiency in handling current traffic are considerations of foremost concern.

Basically, a marine terminal must be able to perform three functions:

1. unload and load a ship's cargo with efficiency and dispatch
2. provide adequate temporary and long-term storage for incoming and outgoing cargo
3. provide rail and/or highway connections for movement of freight into and out of the port area.

While the practical capacity of a port may be limited by any one of these functions, the first is generally the controlling factor. Thus, the operating capacity is essentially the product of the cargo-handling rate (tons/day/occupied berth) and the number and extent of utilization of berths.

The rate of loading and discharging of cargo depends on:

1. types of cargo
2. vessel type and size (especially, number of hatches)
3. availability and size of stevedore gangs
4. degree of mechanization and methods of cargo handling.

At a typical general cargo berth, the average loading or unloading rate is about 25–30 tons per hatch per hour or 200–240 tons per hatch per eight-hour day. Obviously, the cargo handling rate will depend on the number of hatches being worked simultaneously and the length of working day. For planning purposes, an average of 40 stevedore gang-hours per working day may be used [5]. This may consist of five gangs working eight hours each, four gangs working 10 hours each, or any similar combination. On this basis, the maximum loading rate for an occupied berth is 1,000–1,200 tons/berth/day. In using these rates to estimate annual berth capacities, it should be remembered that a given berth will not be occupied 100 per cent of the time during the year.

The estimation of the number of berths required must be made in the face of fluctuations in demand. In cold climates, allowance must be made for the closing of the port in winter months due to the formation of ice. Consideration may also have to be given to seasonal fluctuations in transportation of certain products.

Independent studies by Plumlee [6], Nicolaou [7], and Fratar, et al. [8] have shown that ships arrive at a public seaport in accordance with a random pattern and that the Poisson probability distribution may be used to predict the number of days on which a particular number of ships will be present.

According to the Poisson law:

$$F_n = \frac{T(\bar{n})^n e^{-\bar{n}}}{n!} \tag{21–2}$$

F_n = the number of units of time that n ships are present during T time units.

$\bar{n}$ = the average number of ships present.

e = the Naperian logarithmic base, 2.71828.

The Poisson equation makes it possible to calculate the distribution of ship arrivals if only the average number of ships present during the period in question is known. For example, if it is known that an average number of ships present at a certain port is 8.0, the number of hours during a one-year period (8,760 hours) that 10 ships will be present is:

$$F_{10} = \frac{8760(8)^{10}e^{-8}}{10!} = 872$$

This and similar values are plotted in Fig. 21–8 which shows the Poisson distribution for $\bar{n} = 8$ and a period of one year. The calculations indicate that at a nine-berth port, there are 872 hours during the year in which one ship will be waiting for a berth. Similarly, at a port with 11 berths, there will be 872 hours during the year in which one berth will be vacant.

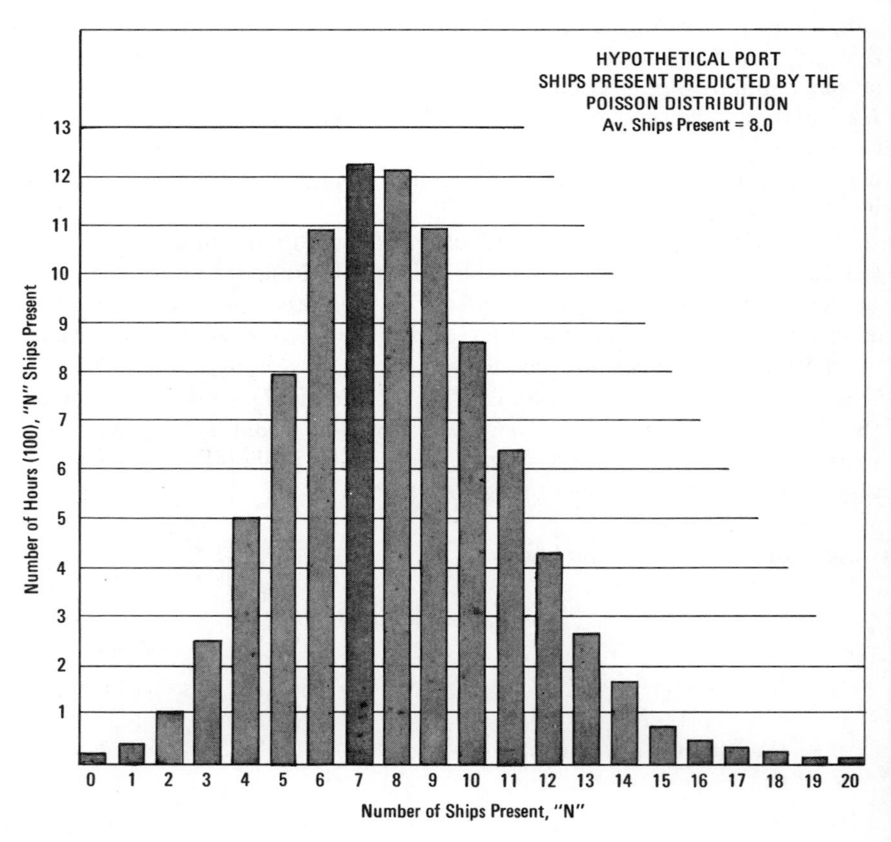

Fig. 21–8. Ship distribution at a hypothetical port. (Courtesy American Society of Civil Engineers.)

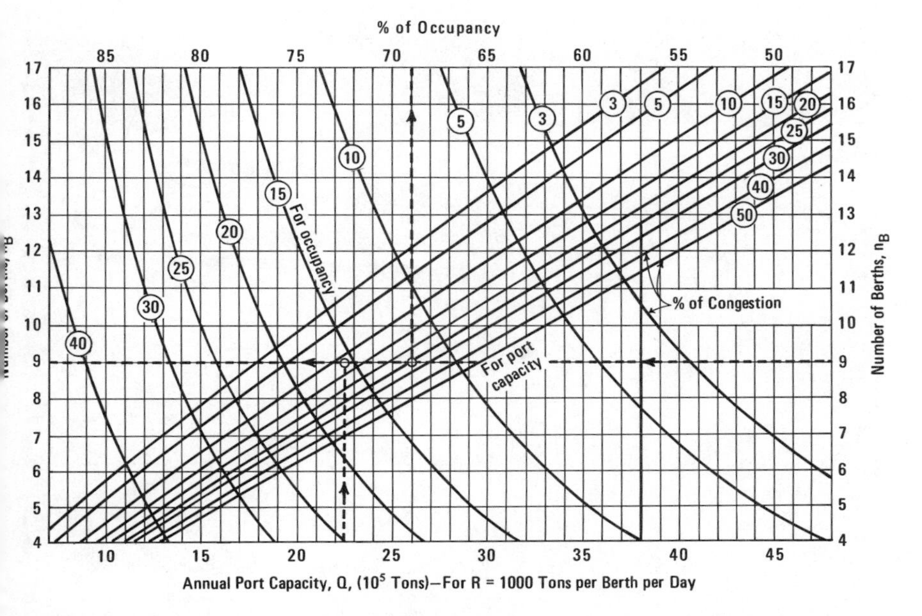

Fig. 21–9. Relationship between annual port capacity, congestion, and berth occupancy. (Source: S. N. Nicolaou, "Berth Planning by Evaluation of Congestion and Cost," *Proceedings,* American Society of Civil Engineers, November, 1967.)

Since there are costs involved in idle berths as well as waiting ships, the optimum number of berths is a compromise, best determined by minimization of the sum of the annual costs of vacant berths and waiting ships. Using this approach, Nicolaou [7] has developed a useful graph, shown as Fig. 21–9, which relates annual port capacity to the number of berths.

It will be noted that two parameters are shown on Fig. 21–9: *per cent of congestion and per cent of occupancy. Per cent of congestion* is defined as the per cent of time for which the number of ships in port exceeds the number of berths available. *Per cent of occupancy* is defined as the per cent of time that the total number of berths available at the port, N_B, is occupied.

Nicolaou [7] showed that per cent of congestion is related to average cost of a ship waiting for a berth, C_S, and the average cost of an idle berth, C_B, by the following inequality:

$$\% \text{ Congestion} < 100\left(1 - \frac{C_S}{C_B + C_S}\right) \tag{21-3}$$

Fig. 21–10. Port facilities at Port Everglades, Florida. (Courtesy The American Association of Port Authorities.)

The annual port capacity, Q, is given by the following equation:

$$Q = N_B R T \left(\frac{\% \text{ occupancy}}{100} \right) \tag{21-4}$$

where:

N_B = number of berths
R = annual average cargo handling rate per berth, tons/day
T = time period, usually 365 days

The use of Fig. 21–9 is illustrated by a numerical example.

EXAMPLE 21–2

Given: T = 365 days; $Q = 1.8 \times 10^6$ tons; R = 800 tons per day; C_B = \$250; and C_S = \$900.

$$Q' = Q\left(\frac{1000}{R}\right) = (1.8 \times 10^6)\left(\frac{1000}{800}\right) = 2.25 \times 10^6 \text{ tons}$$

By Equation (21–3),

$$\% \text{ Congestion} < 100\left(1 - \frac{900}{250 + 900}\right) = 21.7\%$$

Entering Fig. 21–9 with a value of $Q' = 2.25 \times 10^6$ tons and for a value of % Congestion = 21.7, the chart gives a value of $N_B = 9$ and % of Congestion = 12.0. Re-entering the chart with a value of $N_B = 9$ and intersecting the per cent of congestion curves as shown, we find that:

$$\% \text{ Occupancy} = 69.0$$

from which

$$Q = 9(800)(365)\left(\frac{69.0}{100}\right) = 1.813 \times 10^6 \text{ tons}$$

Figure 21–9 should give suitable planning estimates of the number of required berths for the range of values shown. More complete analyses, which are beyond the scope of this text, may be made by the use of queuing theory [9] or by computer simulation.

21–6. Berth and Slip Dimensions. The space required alongside a wharf for the berthing depends on:

1. size and type of ships served
2. wharf configuration
3. mooring procedures.

Terminals which accommodate tankers and other bulk cargo ships vary widely with regard to berth and mooring space requirements. There are no generally accepted standards for berth and slip dimensions for these facilities, as these space needs will depend not only on the factors listed above but also on the cargo handling procedures and equipment. In contrast, terminals which serve ocean-going general cargo ships are conventional and typical dimensions may be useful in the planning of these facilities.

From the standpoint of ease of mooring, the choices of wharf configuration in order of desirability are:

1. marginal wharf
2. two-berth pier
3. four-berth pier.

The required berth length is equal to the length of a ship, plus a small clearance between adjacent ships and space for the ships' lines. Thus, a typical berth length for ocean-going general cargo ships is about

600 feet. Further allowance may be required for space for mooring and docking.

In planning a marginal wharf to serve ocean-going general cargo ships, about 700 feet should be allowed for a one-berth facility and a minimum of 600 feet per berth is required for multiple berth facilities. A desirable pier length for a two-berth pier is 700 feet and for a four-berth pier, about 1,300 feet.

For a wharf configuration utilizing two-berth piers, a slip width of at least 250 feet is needed. This dimension is roughly the sum of the beams of two ships and the length of a tugboat. When four-berth piers are used, a minimum slip width of about 325 feet is required. This dimension is based on the beams of two ships (moored) plus the beam of another ship (mooring) plus the length of a tugboat.

In summary, the minimum and desirable pier length and slip width for a two-berth pier and a four-berth pier are:

	Pier Length		*Slip Width*	
	Minimum	*Desirable*	*Minimum*	*Desirable*
two-berth piers	625′	700′	250′	300′
four-berth piers	1200′	1300′	325′	375′

21–7. Transit Sheds and Apron. At a typical general cargo terminal, temporary dry storage space is provided in a large building called a *transit shed.* In addition to providing short-term storage, the transit shed may also provide space for customs activities and for port administration and security. The transit shed should not be used for long-term storage, and when long-term storage facilities are required, warehouses and open storage areas are usually located shoreward of the transit shed.

Experience has shown that for a typical dry-cargo ship, a transit shed which has an area of 85,000 to 90,000 square feet is desired. This value provides adequate space to accommodate a single average-sized dry cargo vessel and includes an allowance of about 40 per cent for aisles and other non-storage areas. The 90,000 square foot requirement is based on a net average storage height of about 12 feet. It allows for the discharging and loading of the entire contents of an average cargo vessel. Proportionately smaller transit sheds may be used at small terminals where ships are expected to discharge and take on only partial loads.

A common length of transit shed is 500 to 550 feet. The larger value provides a sheltered storage area adjacent to the entire length of a large merchant cargo vessel. A minimum transit shed width of 165 feet is recommended for marginal wharf terminals [1]. Where a single shed

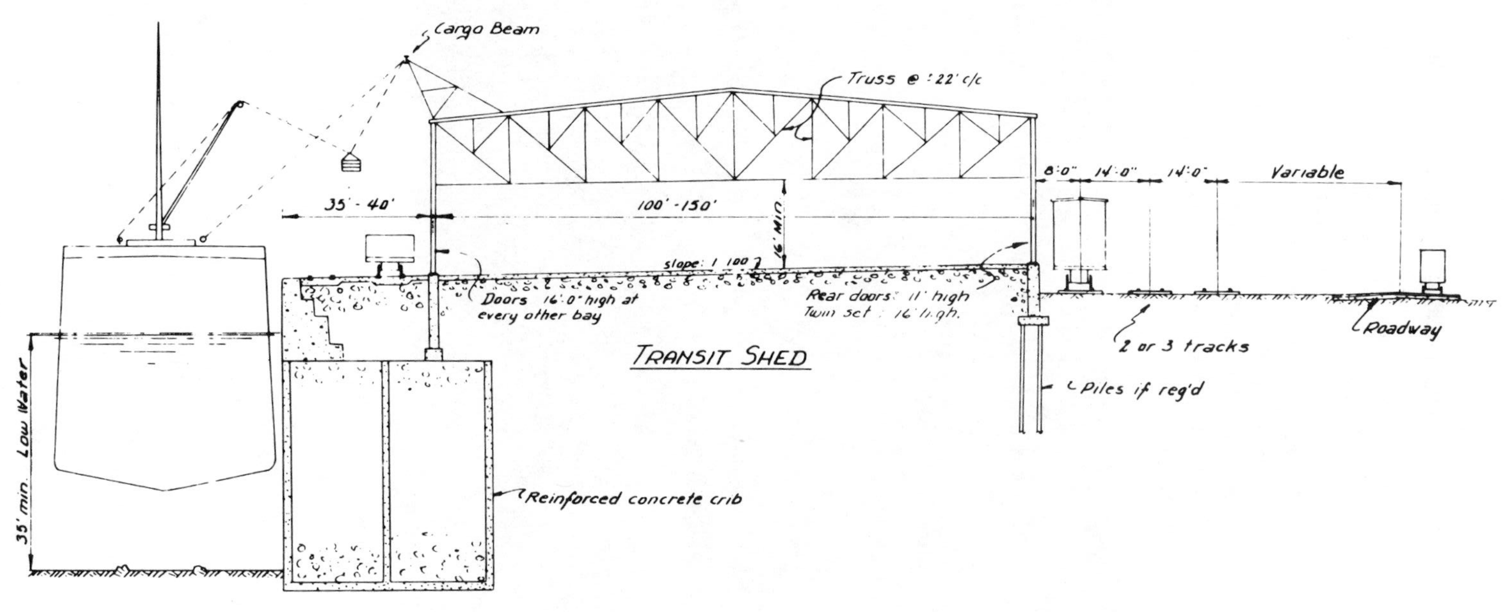

Fig. 21–11. Typical cross-section of wharf with transit shed with solid fill type wharf construction. (Courtesy The American Association of Port Authorities.)

serves two berths, one on each side of a pier, approximately twice this value would be required to provide the specified 90,000 square feet per berth.

A 20-foot wide covered platform along the rear (shoreward) side of the transit shed is recommended to facilitate the transfer of cargo between trucks and railroad cars and the transit shed. Door openings are usually provided in alternate bays along both the apron side and platform side of the transit shed.

Fig. 21–12. View of port apron. (Courtesy The American Association of Port Authorities.)

The portion of the wharf or pier which lies between the waterfront and the transit shed is called the *apron*. This uncovered space is needed for mooring, and for the loading, unloading, and movement of cargo into the transit shed. Along the waterfront edge of the apron, space must be reserved for bollards, cleats, and other mooring devices. Connections for electric power, and telephone and water service must be provided. These connections are usually housed in service boxes built into the deck of the apron along the waterfront edge. At least two service boxes per berth are recommended.

When railroad service is desired along the apron, the rails are constructed flush with the apron deck and the tracks are constructed on 13.5 foot center-to-center spacing. Rail supported gantry cranes may be installed along the apron, in which case 3 to 5 feet of apron width must be allocated for each crane rail.

The required total apron width will vary from about 20 feet to more than 60 feet depending on the facilities provided. Table 21–2 gives ranges of recommended apron widths for various design conditions.

TABLE 21–2

Ranges of Recommended Apron Widths for Various Design Conditions

Design Condition	Range of Recommended Apron Width
Basic Width (work and mooring space)	20 ft. to 30 ft.
Basic Width with 1 Railroad Track	30 ft. to 38 ft.
Basic Width with 2 Railroad Tracks	38 ft. to 50 ft.
Basic Width with 3 Railroad Tracks	52 ft. to 60 ft.

Note: Add 3 to 5 feet to the above values when a half portal gantry crane is to be provided and 6 to 10 feet when a full portal crane is to be used.

21–8. Container Ports. One of the most significant developments in water transportation to occur in recent years has been the increased usage of containers. Nielsen [10] suggested that the coming of containerization has made perhaps the greatest impact on the shipping industry since the invention of the steam engine.

Containers are simply boxes, typically 8 by 8 by 20 feet, although several lengths are used. The usage of containers eliminates much of the handling of small units of cargo at a port facility. A container is typically loaded and sealed at the place of origin and it remains sealed until it arrives at its destination.

Containers arrive at the port facility by truck or by rail. Upon arrival, the containers are weighed, logged in, and temporarily stored in an assigned location in a marshalling area. The containers are later moved from the marshalling area to wharfside and transferred to a ship by means of heavy container cranes. This procedure is reversed

Fig. 21–13. Modern container terminal at Elizabeth, New Jersey. (Courtesy The Port of New York Authority.)

at the destination end of the trip. The principal elements of a large container port may be seen in Fig. 21–13.

At least five advantages may be listed for a container port:

1. The berth capacity is great, often being five times as high as the capacity of a traditional general cargo berth.
2. Overall transit time is less.
3. Container ports offer greater safety for waterfront employees.
4. There is less damage to cargo.
5. Less pilferage occurs at a container port.

The principal disadvantage of a container port is the large amount of land required for the marshalling area. The need to provide large and expensive container cranes is also a disadvantage of container ports.

Planning considerations and design criteria for a container port have been described in an article by Tozzoli and Wilson [11] which is reprinted in part in the following paragraphs. It should be remembered that the facilities and criteria discussed apply to a major port. Although the terminal components described will be similar for a smaller container port, the sizes will be proportionately reduced.

Most of the criteria normally used by port planners in designing conventional breakbulk port facilities are of no real value when applied to full containership berths. Even now true full containerization, as part of an intermodal system, is in its infancy, so criteria developed today for container ports must be kept continuously under review to adapt to changing transport practices.

Since containers are meant to be easily interchangeable between the three basic modes of transportation—ship, truck, and rail—facilities should be located convenient to all railroads serving the area, with easy highway access in all directions. The container berths themselves should be along a quay rather than a finger pier because of the large amounts of supporting area required per berth.

The components that make up a typical container terminal are ship berth, container cranes, marshalling area, container packing shed, entry facilities, container inspection garage, and equipment storage. Most can be seen in the accompanying photo.

Since the tendency in containerships is to make them larger and longer, a standard container berth is likely to be longer than the present length of general cargo berths. If the container berths are placed along a quay, a certain amount of flexibility in berth length is provided for the port. At the Elizabeth Port Authority Marine Terminal, we started with a standard berth length of 600 ft, later increased to 640 ft. Even now, particularly for containerships that have stern ramp roll-on/roll-off capabilities, a standard berth length is considered to be 750 ft.

Most container vessels planned or under construction, do not have shipboard cranes to handle the containers. Thus, container cranes on shore will usually be a requirement. (See photo.) Normally two cranes working simultaneously will unload and load a containership. In some instances as many as four cranes have been used simultaneously on a single ship.

Marshalling areas—two major types

Of the two most prevalent marshalling area designs, one has all containers stored one high on truck-trailer chassis. Each space is completely accessible at all times. This, of course, gives complete flexibility within the yard. As the containers arrive, they are merely assigned locations in the yard and remain in this position until taken aboard ship by yard tractors. How-

Fig. 21–14. The loading of a containership by means of a land crane. (Courtesy The Port of New York Authority.)

ever, using this method requires more marshalling area space than any other.

In the second system, containers are not stored on chassis; they are placed by a straddle carrier which can "block-store" containers two high. Obviously, the more block storage, the less flexible the system becomes, but the less marshalling area required. Each container port has some of each kind of marshalling area, but the percentages vary.

The real question is whether there is a point at which certain amounts of block storage can be utilized without delaying the loading and unloading of the ship and without incurring excessive yard costs for rehandling containers.

At the New York City area's port it has been concluded that with a nominal standard berth of 750 ft, 14 acres of marshalling area should be provided. On this basis the nearest public road behind the berths and marshalling area is set at 850 ft from the stringpiece. This distance provides sufficient space to store containers for the largest services. This is particularly true when the terminal consists of more than one berth since, when activities at one of the berths is low, the excess capacity in its marshalling area can be used for the adjacent active berth.

Less-than-container loads

A less-than-container load (LTCL) packing shed normally is provided. It need not be contiguous to the marshalling area, and definitely should not assume the normal location of a transit shed. Any structures near the stringpiece tend to encumber the steady movement of containers to and from the cranes during loading and unloading operations. The size of the packing shed varies widely from line to line and from trade route to trade route. The shed's general configuration should resemble a typical truck terminal—basically designed to have delivery trucks arrive at one side of the building, and the cargo be moved from these trucks directly into waiting containers on the opposite side of the building with a minimum flooring of cargo. Thus it tends to be long and narrow with emphasis on the number of truck and container doors necessary. As an example one of the major packing sheds at Elizabeth is 1,000 ft long and 100 ft wide with a total of 162 truck doors. . . .

The truck entrance to the terminal usually consists of two or three entry truck lanes with a corresponding number of departure lanes. Each lane is normally provided with a truck scale to weigh the containers in or out. At this entry-exit point, a building will usually be located to handle the necessary paper work: here also, positions in the marshalling yard are assigned to incoming containers.

Approach public roads to the terminal should be generous. Container operations generate substantial truck traffic, peaking on days that ships are in port. This peaking necessitates truck holding lines waiting to enter the terminal. At Elizabeth, a six-lane road is provided at the berths.

Located somewhere near this building and adjacent to the marshalling area is a small garage for the physical inspection of arriving or departing containers. This is a requirement because the responsibility for the containers changes hands as they arrive or leave the terminal. In addition, a maintenance garage is normally provided for whatever stevedoring devices are used to handle the containers in the marshalling yard.

It appears that direct transfer of containers from rail to ship is not practical. In such a transfer the string of cars would have to be continuously moved during the loading and unloading operation to bring the containers under the crane. Instead, it is more likely that the containers will be removed off-site and brought to the marshalling area on rubber tires. Thus tracks will not be required at the stringpiece although they should be provided to the packing shed.

PROBLEMS

1. Ships arrive at a port in accordance with a Poisson probability distribution. On the average there are five ships in port.
(a) Determine the percentage of time during the year that there will be 0, 1, 2, 3, 4, 5, 6, 7, and 8 ships present.
(b) Plot a histogram of these data similar to Fig. 21–8.
(c) What percentage of time during the year will there be more than seven ships present?
(d) What is the probability that there will be fewer than five ships present?

2. Suppose that on the average it costs $1,100 per day for a ship to wait for a berth. Using the arrival data given in Problem 1, what would be the annual costs of waiting at a port which has eight berths?

3. Suppose that the average cost of an idle berth is $300 per day. Using the arrival data given in Problem 1, what would be the annual costs of idle berth time at a port which has eight berths?

4. Two million tons of cargo is to be handled at a port which has a daily cargo handling rate of 900 tons per berth. The average cost of a waiting ship is $1,200 and the average cost of an idle berth is $350.
(a) How many berths should be provided?
(b) What will be the per cent of congestion?
(c) What will be the per cent of occupancy?
(d) Determine the annual capacity of the port.

REFERENCES

1. The American Association of Port Authorities, *Port Design and Construction,* Washington, D.C., 1964.
2. Lee, Theodore T., "Design Criteria Recommended for Marine Fender Systems," *Proceedings of Eleventh Conference on Coastal Engineering,* American Society of Civil Engineers, September, 1968.
3. Newman, Frank H., "Developing Port Facilities on Houston's Ship Channel," *Proceedings,* American Society of Civil Engineers, Vol. 82, No. WW3, May, 1956.
4. Ford, D. B.; Young, B. O.; and Walker, G. W., "Hi-Dro Cushion Camel—A new Floating Fender Concept," *Proceedings of Eleventh Conference on Coastal Engineering,* American Society of Civil Engineers, September, 1968.
5. Brant, Austin E., Jr., "The Port of Chicago," *Proceedings,* American Society of Civil Engineers, Vol. 84, No. WW4, September, 1958.
6. Plumlee, Carl H., "Optimum Size Seaport," *Proceedings,* American Society of Civil Engineers, Vol. 92, No. WW3, August, 1966.
7. Nicolaou, S. N., "Berth Planning by Evaluation of Congestion and Cost," *Proceedings,* American Society of Civil Engineers, Vol. 93, No. WW4, November, 1967.
8. Fratar, T. J.; Goodman, A. S.; and Brant, A. E., Jr., "Prediction of Maximum Practical Berth Occupancy," *Proceedings,* American Society of Civil Engineers, Vol. 86, No. WW2, June, 1960.

9. Jones, John Hugh, and Blunden, W. R., "Ship Turn-Around Time at the Port of Bangkok," *Proceedings,* American Society of Civil Engineers, Vol. 94, No. WW2, May, 1968.
10. Nielsen, E. F., "Containerport Engineering For The Port of Oakland," *Civil Engineering* magazine, Vol. 39, No. 1, January, 1969.
11. Tozzoli, Anthony J., and Wilson, John S., "Planning and Construction of the Elizabeth, N.J., Port Authority Marine Terminal," *Civil Engineering* magazine, Vol. 39, No. 1, January, 1969.

VI

CONCLUSION

22

Future Developments in Transportation

Attempting to predict the future can be a fruitless exercise. The normal activity of the engineering planner is not so much to predict future conditions but rather to mold the near future into desirable patterns. At some point, however, it is necessary to evaluate the trends of technology to attempt to place the present in its proper perspective at a point in time. Current technology can then be seen within its true context of the continuum of technological change. The subject of transportation has undoubtedly generated more futuristic ideas than any other field of human enterprise. The fantasies of one era are the realities of another. The three-month wagon trip of the early nineteenth century from New York to California was cut to six days by the introduction of the railroad. Early airplanes further slashed this time to a matter of ten hours. Jet transports now make the trip in five hours. With the introduction of supersonic and hypersonic aircraft the journey could be made in a matter of minutes. The proposals for future transportation methods are legion. In this chapter only a few have been selected for discussion to provide an overview of the type of thinking that is seeking to innovate the transportation industry.

LAND TRANSPORTATION—URBAN AREAS

22–1. Rapid Rail Transit. Conventional rapid rail transit systems, sometimes called "duorail" systems, have been in existence since the late nineteenth century. There is, therefore, nothing new about such vehicles. During the 1960's and early 1970's, however, there was a considerable revival in interest in the construction of rail rapid transit systems, after it became apparent that the undesirable environmental impact of highway-oriented urban transportation systems had been underestimated. No dramatic changes in technology are anticipated in the next two decades. Changes are more likely to come from attempts to raise the level of service provided by existing systems, and from the design of new systems to similarly high standards of service.

Totally automatic train operations are currently possible and have been installed in such new networks as the BART system in San Francisco and the Victoria Line in London. By means of automated systems, operating delays can be minimized with headways as low as 90 seconds during peak flows. Automatic train control with designed back-up and fail safe systems provides safer operating conditions than manned control [1].

Short range improvements in existing and future rail systems can be expected from a variety of minor technological developments. Higher operating speeds will be available with the development of ligher rail cars and better traction equipment. Improved smoother braking systems can be anticipated from the development of new hydraulic, dynamic, magnetic, and air brake systems. Better suspension systems in coordination with more reliable rubber-tired wheels will give more acceptable riding qualities. Maintenance procedures under the pressure of modernization will improve the reliability of the systems. Improved methods of cleaning, especially of the ferritic dust which is associated with rail transit, will improve the image of this form of urban mass transport.

Many of the existing rail transit systems are operating with facilities and equipment fifty years old or more. Considerable attention to raising the aesthetic level of the mode will result in the attraction of increased ridership [2]. The degree of attention that new systems will pay to aesthetics is emphasized by Fig. 22–1 which shows designs of elements of proposed systems. The current low status of many urban rail transit systems has been blamed on management structures which in the past have been more production-oriented (i.e., concerned with the capacity to move individuals) than market-oriented (concerned with the attraction of ridership) [3]. Future systems will be designed to provide a minimum of environmental pollution, including visual and noise pollution. These are known to be aesthetically undesirable and, therefore, repulsive to potential riders [4].

The form of development of urban rail transit systems is probably presaged by the BART system in San Francisco and the Transit Expressway line planned for Pittsburgh. The BART system is an attractively designed regional system. Its station areas are designed to serve as collection points for low density residential areas. These stations provide for connections to bus systems operating in the residential areas, and provide parking lots and dropping-off areas for "park-and-ride" and "kiss-and-ride" patrons. The use of the private automobile as a feeder to the rail transit system has proved popular both in the modern Cleveland and Toronto lines.

Rail transit can be expected to function along corridors also serviced by freeways. Conversion of some freeway corridors to rapid transit lines has been recommended where densities increase to a point that transit operations are feasible [5]. Future planners can anticipate the phased progression of a transportation corridor from traditional freeway, to dual-mode bus systems, then low density transit systems, and, finally, rail transit systems capable of moving high volumes of traffic.

Monorail systems have captured the imagination of the layman as the modern key to the solution of the urban transportation problem. In fact, monorail systems were feasible in the last century. The Wuppertal monorail in Germany has been in operation since 1901 [6]. Other than as exhibitions at World's Fairs (Seattle and New York, for example) there has been little commercial development of the systems. Switching problems, the inherent sway of the vehicles, and the hard ride characteristic of many systems have inhibited further development of monorails. An exception to this may be the French SAFEGE system, which has developed a vehicle capable of operation at speeds up to 100 mph in all weather conditions with completely automatic control. The system is feasible for the low to medium urban densities. These have in the past been unservicable by conventional systems which require reasonably high densities [7]. Other advantages of the monorail system include the low cost of the supporting structure and the minimal amount of right-of-way required for operation. In built-up areas the systems can be cantilevered over street rights-of-way and in areas of extreme congestion can be supported in tunnel installations. Figure 22–2 shows the attractive design of a modern monorail system.

In 1964, at the cost of $60 million, Tokyo completed an 8.2-mile monorail connecting the airport to the downtown area. Despite a capacity of 71,000, the line carries only 7,500 passengers daily, losing $5 million annually. Even with satisfactory operational characteristics such as a smooth, safe, and rapid ride, this line has failed because of poor location. Monorail systems appear, however, to have very limited future use since by technological adaptation, conventional support systems can develop rapid transit systems with small attractive vehicles capable of serving low density areas. The Pittsburgh Transit Expressway is an example of what future low-to-medium volume systems could resemble.

22–2. Dual Mode Bus Systems. Dual mode bus systems are already in the testing stage. The "Hy-Rail" experiment in Philadelphia indicated the potential attractiveness of a bus vehicle which has the flexibil-

Fig. 22–1 (a). Car Interior. (Source: Bay Area Rapid Transit District.)
Fig. 22–1 (b). Station Concourse. (Source: Westinghouse Corporation.)

Fig. 22–1 (c). Aerial Transit Structure. (Source: Bay Area Rapid Transit District.)

ity and adaptability of a conventional bus for operation in suburban residential areas, yet can be operated on a fixed guide system of rails over exclusive right-of-way for the line-haul portion of the trip to downtown areas. The bus, which is indistinguishable in operational characteristics from a conventional bus in the suburban areas, is fitted with flanged steel wheels which can be lowered to engage on rails. Traction is provided by the rear rubber-tired wheels. This concept has been tested in both Australia and the United States.

Other more advanced dual-mode bus systems have been suggested. Vehicles would operate on the street system using electric batteries or conventional gasoline motors. When using the guideway, power is picked up from a center-guide or power rail. At the same time the electrical batteries can be recharged. British researchers have developed mechanical systems capable of automatically guiding dual-mode buses along exclusive rights-of-way at high operating speeds.

22–3. Dual Mode Automobile Systems. Two very similar dual mode automobiles are in the prototype and preliminary planning states—the *StaRRcar* and the *Urbmobile* [8].

Fig. 22–2. Proposed Modern Monorail System. (Source: General Electric Co.)

The Alden Self Transit Rail and Road car is a dual-mode automobile which in the most advanced concept is community owned. For suspension and propulsion it relies on its own rubber-tired wheels at all times. When operating under manual control off the guideway, the vehicle is propelled by its own electric motor. While on the guideway the vehicle picks up electric power to drive the motor and recharge the batteries. Anyone using the StaRR car system drives the vehicles from home to the guideway and enters the CBD under the control of the guidance system. On leaving the vehicle in the central area it is reassigned on demand to another operator. For the return trip the first available vehicle is taken and driven first under control, then independently to the driver's home. For an additional charge the vehicle is kept overnight. The driver operates the vehicle with an automatic billing device and is charged for the duration and type of usage. Figure 22–3 shows a model of the StaRRcar.

The *Urbmobile,* a Volkswagen-sized vehicle conceived at the Cornell Aeronautical Laboratory, operates with two sets of wheels. For manual control, the vehicle uses conventional rubber tires. On the same axles, inside the rubber-tired wheel, steel flanged wheels are positioned to en-

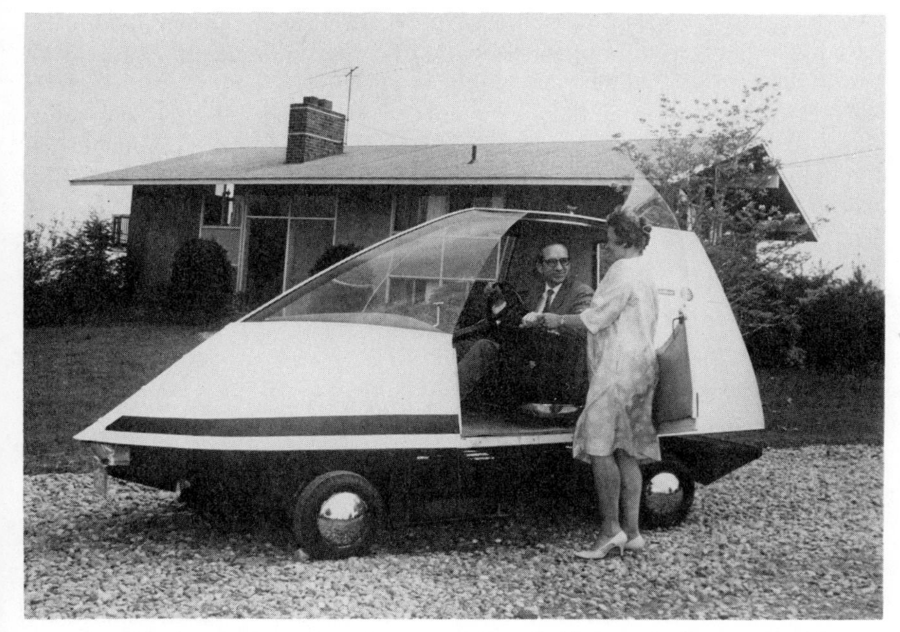

Fig. 22–3. StaRR Car. (Source: Alden Self Transit Corporation.)

gage the rails used on the guideway. While in operation on the guideway the batteries which drive the vehicle in the unguided mode are recharged.

The dual mode automobiles offer some significant advantages. Being electrically powered they minimize pollution. They provide portal-to-portal service and retain much of the privacy of the individually owned automobile. Their size and the fact that they can be operated on an automated guideway would permit high traffic densities and volumes simultaneously (it is quite possible that single guideways will accommodate 20,000 persons per hour in comparison with current freeway lanes carrying only 2,000 vehicles per hour under ideal conditions). A large additional advantage of this type of vehicle is that under the type of public ownership indicated, there would be no need for the extensive amount of vehicle storage currently needed in downtown areas for privately owned vehicles. Vehicles could be summoned into the CBD according to the demand, giving the area high accessibility by the auto mode, without the destruction of high density amenities by extensive surface parking lots.

22–4. Bus Systems. About 75 per cent of the nation's mass transit is supplied by motor bus transport. The developments in this mode, therefore, have a strong influence on the transit usage that can be anticipated. Experts would appear to agree that the fundamental characteristics of bus transit vehicles are unlikely to undergo radical changes over the next five years and that, even after this period, changes will occur slowly, reflecting the huge current investment in capital equipment [9]. Excluding the dual mode bus, which has been previously discussed, several improvements which could appear over the next few years could be [20]:

1. articulated buses
2. double deck buses
3. dial-a-bus systems

The *articulated bus* has a greater vehicle capacity than conventional buses, without a decrease in passenger safety and comfort. These buses can carry 75 seated persons compared with the conventional load of 55. As a result, the labor costs per passenger mile accruing chiefly from driver costs are significantly decreased. Increased use of articulated vehicles can be anticipated as expressway and busway bus systems are introduced and broadened. *Double deck buses* have been in use in many countries, especially those supplied by British vehicle manufacturers. With improved steering systems, and fare collection devices that require only one driver-operator, these vehicles could be expected to come into service in the United States. Their chief advantage is a higher passenger capacity within standard wheel base lengths.

The most innovative form of bus service that is likely to be introduced over the next few years is *dial-a-bus* service [1]. Dial-a-bus service is a demand-actuated system. The concept is relatively simple. Bus transit is supplied in a form closely resembling taxi service and in this way attempts to solve some of the basic flaws in conventional mass transit. Public transport systems usually provide inadequate and infrequent levels of service at the collection and distribution ends of the trip. The density of trips generated in various areas does not permit scheduled linkages between desired origins and destinations. By the use of radio linkage and computer-based scheduling, the dial-a-bus system provides a dynamic response to the changing demand patterns for service. In principle, a central computer maintains a constant monitor on the location of buses, their current passenger load, and their destinations. A potential passenger calls on his telephone requesting a bus. Based on a preprogrammed optimal selection procedure, the computer selects which bus will make the pickup, at the same time updating its memory to reflect the new routing. Systems concept studies indicate that with only a slight increase in fare above conventional transit costs, a service similar in some ways to taxi service can be provided by public transit systems. Since the system is ideally suited to low densities of origins and destinations, it can be anticipated to attract off-peak transit riders who currently use autos when faced with very high headways of off-peak scheduled systems. In this way off-peak riding will actively contribute to the overall costs of community transit service. The dial--a-bus concept is entirely feasible with existing technology and can be expected to be in operation in urban areas during the 1970's.

22–5. High-Density Personal Movements. High-density pedestrian movements in central areas are likely to be accommodated by a variety of movement devices. It is likely that these devices will connect with line-haul rapid transit systems and serve as the downtown distribution elements of integrated transportation systems. The following discussion is limited by space considerations to merely a few types of vehicles that could satisfy this type of movement.

People-movers are now used on a limited basis at Disneyland in Anaheim. The constantly moving cars, assembled in four-car trains, run over a fixed route, driven by revolving wheels in the roadway itself. Speeds vary along the route from 1 to 7 mph. Access and egress to the non-stop car trains is from a transfer table, synchronized to revolve at the same peripheral speed as the conveyor [8]. This loading system is similar to a circular loading platform for the Telecanape operated in 1964 at Swiss Exhibition in Lausanne [6].

A similar type of personal movement vehicle was the *Carveyor,* used in the General Motors and Ford pavilions at the New York World's Fair. The accelerating and decelerating devices in this case were special belt conveyors. The system was also non-stop, with access and egress from the cars being provided over a transitional moving belt. This system used individual cars, which were able to negotiate gentle curves.

Guide-o-matic trains are small electrical battery powered and electronically guided locomotive-hauled industrial and mass transit trains. The guidance system follows a buried cable accepting electronic input for control. This system has been used to connect passenger movements within the Houston Airport.

A more futuristic movement device is the *Bouladon Integrator.* This system is essentially a horizontal elevator. Figure 22–4 shows a sketch of this concept. Access to the system is by a multiple escalator composed of several compartments. As the escalator moves forward and upward there is a sideways component of acceleration until the sideways velocity is equal to the line-haul speed of the system. At this speed the compartments fall into line behind each other, forming a continuous train of compartments. Transfer can be made to a moving belt or train, and the integrator then reverses its acceleration process. This device is in the concept stage only.

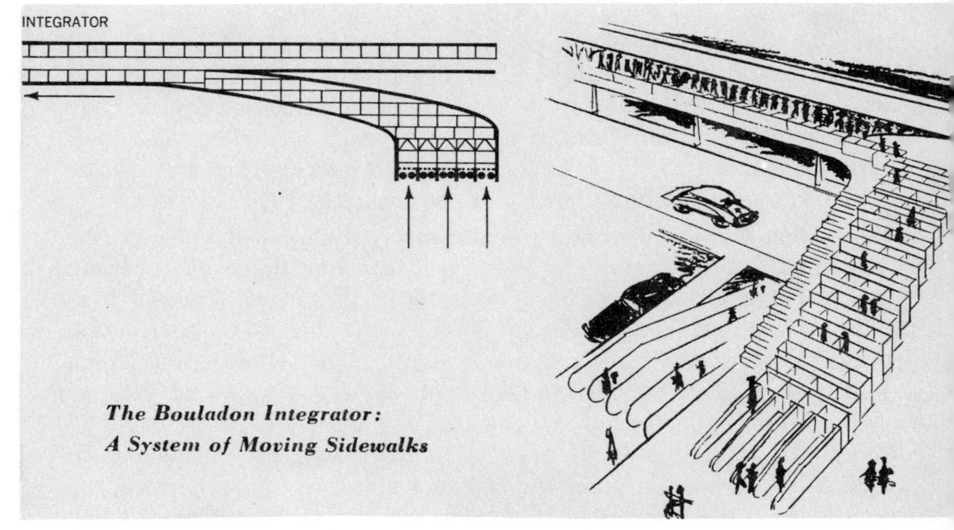

Fig. 22–4. Bouladon Integrator. (Source: *Transportation in the World of the Future,* Hellman, p. 107.)

22–6. Improved Automotive Systems. The large amount of capital that has been invested in the automotive transport system (in the form of vehicles, roads, manufacturing capacity, and manpower skills), insures that a great deal of the technological research and development will take place in relation to private and commercial automobiles. The following developments are likely to take place over the next 20 years:

New Power Plants. Considerable research energy will be expended on the creation of new power plants. Electrical vehicles, using new high-energy-density batteries will be commonplace. The costs of these batteries will be brought down to levels more competitive with internal combustion engines. It is likely that the high pollution gasoline engine will be severely restricted in the face of public demand for a cleaner environment. Other power plants such as the external combustion engine will attain a substantial share of the automobile market.

Highway Electronic Systems. The use of electronic devices to increase the efficiency of road operations will become widespread. Some of the following areas already have limited acceptance; in the next 20 years their acceptance will substantially widen:

1. Traffic surveillance systems to identify and locate vehicles and to provide aid to stranded vehicles.
2. Traffic control systems ranging from computerized traffic signals, ramp and merge metering to fully automated highways (see Section 22–9).

3. Improved commercial traffic control. Monitoring devices will provide an improved data base for optimal scheduling, routing, and maintenance of commercial fleets.
4. Regulatory monitoring. Vehicles will be automatically identified, weighed, and checked for regulation violations. [10]

Safety Design Improvements. Findings of the extensive safety research in the late sixties and early seventies will be translated into safer vehicle and roadway designs. Significant improvements in the high highway accident rates can be expected within ten years [11].

Environmental Design. The full impact of the 1970 Highway Act will take place gradually. Future highway systems will be designed with a heavy emphasis on environmental design. Extensive use can be predicted for the *joint development* concept where multiple governmental programs are used to provide transportation facilities over shared rights-of-way [12]. While technically possible now, there is a need to develop procedures for interagency relationships which will permit the completion of joint development designs.

LAND TRANSPORTATION—INTERURBAN AND SUBURBAN AREAS

22–7. High-Speed Tracked Vehicles. The decline of the U.S. passenger train has been discussed in previous chapters. There appears to be a legitimate question concerning the status of the passenger train, and the viability of rail passenger transport. Rail passenger trains currently have low operating speeds, especially when considered in comparison with their chief competitor, the jet airplane. Yet, in the face of increasing congestion of the national air hubs, there appears to be a definite place for rail passenger service in short-haul intercity service. In spite of significantly lower current line-haul speeds, the center-city-to-center-city speeds of the passenger train could be made highly competitive with the air-road transport combination on intercity movements up to 300 miles long. The great speed advantage of the jet aircraft is lost by take-off and landing delays at the airfield combined with slow, long airport-to-city-center trips by road. While long-haul passenger movements cannot be attracted back to the railroads, it appears entirely feasible that short-haul intercity movements can be served by railroads if levels of service can be raised economically to a competitive level.

The Japanese Tokaido line has proved that short-haul trips can be diverted from the air mode to the rails. This high-speed line operating between Tokyo and Osaka has an annual ridership of 12 million. Since it went into operation, the parallel air services have lost such a proportion of their former patronage that their services have been curtailed. Even with two intermediate stops, the 300-mile run between the two principal Japanese cities is made in three hours averaging 120 mph.

The Japanese experience, coupled with the general western European experience of increasing rail passenger ridership, has led the U.S. Federal government to support the development of high-speed tracked vehicles which could relieve the growing congestion of the airways by diverting a sizable proportion of the short-haul movements. Two types of vehicle are under extensive testing:

1. high-speed rail vehicles
2. high-speed air cushion tracked vehicles.

Included in high-speed rail vehicles are the *Metroliner* and the *Turbo-train*. The Penn Central Metroliner as put in service in the Northeast Corridor is a high-speed conventional electrified train as shown in Fig. 22–5. In order to put this demonstration project into service, it was

Fig. 22–5. The Metroliner. (Source: U.S. Department of Transportation, Office of High Speed Ground Transportation.)

necessary to upgrade the existing track with over 400 miles of welded rail and 180 miles of overhead electrification wire. The train, which is manufactured by the Budd Company, is capable of top speeds in the area of 150 mph. In practice, top speeds will be limited to 110 mph, with average operating speeds being in the region of 90 mph.

The Turbotrain, manufactured by Pullman Standard after a design by the United Aircraft Company, is capable of top speeds of 160 mph. Speeds are maintained between 110–120 mph for operational purposes. Even at these speeds, in the initial service periods, it was found that existing roadbed and signal systems were overtaxed. Because of its self-contained power plant, the Turbotrain, unlike the Metroliner, has no need of route electrification modifications, costing an average of $400,000 per mile. An innovation on the Turbotrain designs is the use of a pendulous suspension system which permits inward banking of the vehicle on curves. The effect is a lessened centrifugal force on the passenger and a subsequent increase in riding comfort. Figure 22–6 shows a three-car version of a Turbotrain.

Fig. 22–6. The Turbotrain. (Source: U.S. Department of Transportation, Office of High Speed Ground Transportation.)

In 1970, the Department of Transportation underwrote the design costs of high speed *Air Cushion Tracked Vehicle (ACTV)*. This vehicle is planned to service the Los Angeles International Airport by late 1972. The Grumman ACTV, shown in Fig. 22–7, will be designed to attain top speeds of 300 mph. Operational speeds of 150 mph are anticipated over a 16-mile track. The development of a satisfactory tracked air cushion vehicle will permit a breakthrough in high speed ground transportation, providing a surface vehicle with speeds which are realistically competitive with air transport.

To date the main thrust for ACTV systems has been provided by research in Britain and France. The Ministry of Technology in the United Kingdom has provided $5.6 million to develop a *Hovertrain* over

Fig. 22–7. An Air Cushion Tracked Vehicle (Grumman). (Source: U.S. Department of Transportation, Office of High Speed Ground Transportation.)

an 18-mile track. The vehicle will be driven by a linear electric motor at 300 mph and carry 16 passengers. The French are currently working with the *Aerotrain*. This turbo-prop propelled vehicle has been tested with a half-scale model at 125 mph; with rocket boosters speeds of 215 mph were attained.

It seems most likely that the independent research being carried out will result in high-speed ground transportation capable of speeds of 300 mph. At these velocities, *mechanical* support systems such as conventional suspensions and wheels become dynamically unstable. Two other forms of support are possible; *fluid* and *magnetic*. Experience with ACTV systems would indicate that fluid support systems using air as the fluid are technologically feasible. At very low temperatures, metals become superconductive and are capable of conducting the very large currents required to generate the magnetic forces to support vehicles. Alloys such as niobium-titanium have been suggested for superconductors at cryogenic temperatures. Various conceptual schemes have been advanced for the support of vehicles by pontoons containing supercooled alloys generating substantial magnetic fields [8].

The experience gained in high-speed rail transportation in the 1970's will almost certainly lead to the development of high-speed conventional rail and ACTV systems for short-haul interurban trips that will become increasingly difficult to handle with air transport systems.

22–8. Tube Transport. At very high speeds, ground transport vehicles need to operate in a controlled environment which precludes the possibility of collision with debris and eliminates the hazards of ice, snow, and other severe weather conditions. Of the many tube systems which have been proposed in the past, only two different concepts will be discussed: The Edwards gravity vacuum tube and Foa's turboprop tube vehicle.

The Edwards Tube concept works on the principle of a pneumatic catapult [13]. In conventional tube travel systems, air is displaced by the vehicle passing through the tube. Some air bypasses the cars, the remainder is forced ahead by a piston-like action. Since a significant amount of energy is used in the displacement of large volumes of air, the system could be made more efficient by evacuating air ahead of the vehicle. Not only is resistance to movement decreased, but the differential pressure between the front and back of the vehicle can be used as a propellant force. Edwards suggests a 200 mph system where air is sucked out ahead of the vehicles and additional acceleration is provided by tunnel inclines. Because of the lack of air resistance, the vehicles have no need of streamlining, as shown in Fig. 22–8. Braking can be provided by the introduction of air ahead of the vehicle and by conventional braking systems. Braking at stations is assisted by

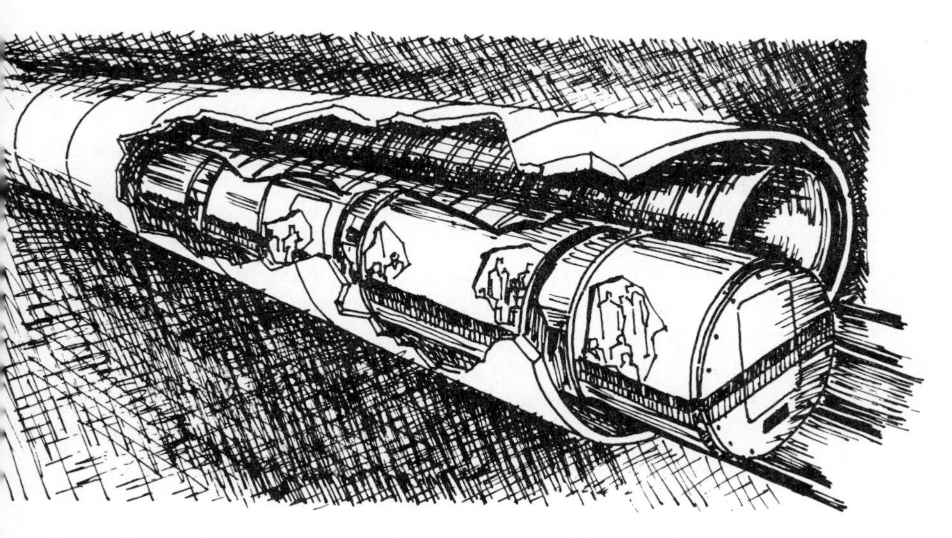

Fig. 22–8. The Edwards Tube. (Source: *Metropolitan,* March-April 1969, Bobit Publishing Co.)

inclined alignments. The suspension system recommended by the designer is conventional steel wheels on rails. By a judicious selection of tube grades and differential pressures, it is felt that very high speeds can be attained with barely perceptible acceleration and deceleration rates. The standard of ride that could be attained by such a system would make it highly acceptable.

A somewhat different concept has been proposed by Dr. J. V. Foa in conjunction with work carried out at the Rennselaer Polytechnic [14]. Recognizing that a large volume of air must be transferred from in front of the vehicle to the back, Foa proposes to use this air as input to a large engine system similar to a jet engine. The ingested air is heated in the engine and expelled to give propulsion. Vehicle speeds in the region of 350 to 400 mph are anticipated. This concept has been called "tube flight." At very high speeds, the air passing around the sides provides a multidirectional suspension system, and the wingless vehicle effectively flies through the tube with equal pressures on all sides. Any tendency for the vehicle to strike the tube sides is corrected by pressure increases which occur as the air gap between vehicle and tunnel lining decreases.

Tube transport is probably far in the future. Based on current technology and present rate of advances it seems unlikely that this type of transport will come into being in the next 50 years. However, the large capacity of tube systems and the high velocities that seem attainable would indicate that future high-speed ground transport systems will develop in this direction.

22–9. Electronic Highways. During the decade of the seventies the first tentative steps will be taken towards electronically automated highways. The Experimental Route Guidance System (ERGS) is an example of the type of preliminary system which will precede fully automated systems. The system depends on a vehicle transceiver which communicates with computer linked roadside transceivers; the driver is given navigation information, providing him with the necessary information to find his destination. In use, the driver inputs his final destination into his vehicle receiver. As he proceeds on his way, the vehicle passes over detection devices which trigger the car transmitter. Information concerning the final destination is sent to a roadside computer which computes routing information. This is transmitted back to the car, giving the driver turning directions. If the driver misses a turn, at the next intersection, new corrected information is transmitted. In this way directions are continually given until the destination is reached [15].

The experience that has been gained in the past in electronic highway surveillance, computerized traffic control, and controlled ramp merging will lead to the eventual development of the fully automatic highway. A

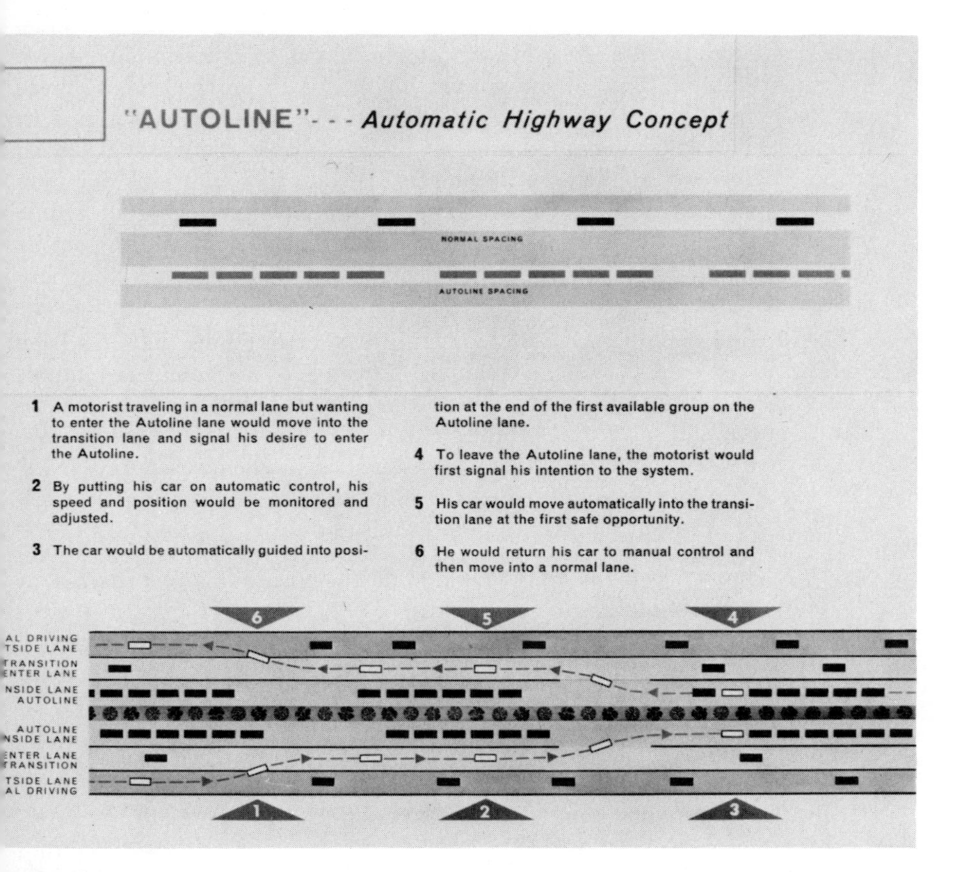

Fig. 22–9. An Automated Highway Concept. (Source: General Motors.)

schematic version of the automatic highway is shown in Fig. 22–9. The G.M. Autoline concept proposes the following operational characteristics for the future system. The roadway at merging sections is composed of three lanes: an outside lane for normal driving, a middle transition lane, and a fully automatic inner lane. When merging, the motorist would move from the normal driving lane into the transition lane and at that time would place his car under automatic control. His speed would be automatically monitored and adjusted to permit a smooth merge into the automatic traffic stream. To leave the Autoline, the process is reversed. On signaling a desire to leave the automated lane, the car is first moved into the transition lane under full automatic control. Once in the transition lane, the car can be returned to full manual control. Fully automatic highways offer the possibility of high-volume, high-speed movements since headways can be cut to a minimum. G.M. indicates that one lane of automated highway will carry 9,000 vehicles per hour at 70 mph; this is a sixfold increase in vehicular volume in comparison with conventional freeway lanes. Other systems have been proposed that would increase automatic speeds up to 150 mph for long distance movements [16]. The timing for the introduction of electronic highways is in doubt. Some experts indicate that systems will be available in the 1970's. Other less optimistic seers indicate that they are at least 25 years away.

22–10. Rail Freight. One of the significant innovations that has been underway for some years and can be expected to be completed imminently is a nationwide information system which will provide instant identification and location of freight cars. *Automated Car Identification* (ACI) has been used by some individual railroads for several years. The extension of these individual systems into a nationwide network will provide several benefits:

1. A shipper will be able to pinpoint the location and progress of individual shipments.
2. The system will provide information that will readily identify locations where additional cars are needed, and locations where there are surplus cars.
3. By optimizing car movements based on identification of needs and surpluses, car utilization can be expected to increase by up to 10 percent. This is equivalent to an increase of rolling stock by the same amount.
4. Maintenance procedures can be integrated with the information system. More optimal maintenance schedules will decrease maintenance costs. [20]

The concept of the ACI system is essentially simple. A scanning device at yards and other recording points scans multicolored service stripes on the freight car which serve to identify it individually. The scanning device is capable of reading the identification at speeds up to 80 mph in the most inclement weather. The scanned information is transmitted to the railroad's own computer which stores the input in an information data bank. By interconnections of the individual railroad's computers with a central computer system in Washington, information concerning individual freight cars is available to all railroads on computer demand. This interconnecting system is known as Tele Rail Automatic Information Network (TRAIN). Delays in the implementation of the system have not been due to technological difficulties, but rather from the high cost of the computers for small railroads and incompatibility of existing computer systems. As these problems are resolved the system will be available on a nationwide basis.

Other areas of innovations in rail freight transportation will come from widespread extensions of existing technology and practices such as greater use of bulk cargo unit trains, container unit trains, piggy back service, and the use of jumbo freight cars.

22–11. Short-Haul Air Transportation—STOL and VTOL. Very rapid changes in air transportation can be expected in the next fifteen years. It can be anticipated that conventional air transportation patterns will change. Air hubs in the last ten years have continued to suffer increasing congestion with rapid increases of air passenger travel. As traffic increased, airports became less desirable neighbors and with their increased space demands moved further from the city centers. As a result, although speeds of air passenger transports have rapidly increased, as shown in Chapter 4, the overall city-center-to-city-center travel time has remained relatively constant due to increasing airport access time. Economic studies indicate that new systems which could serve the city centers on a direct basis can be competitive with the currently congested outlying airports [18]. These systems, capable of operating from sites of restricted area in the central cities, are commonly called Vertical or Short Take-off and Landing (V/STOL) systems.

Vertical Take-off and Landing (VTOL) Aircraft. Vertical take-off and landing aircraft have been available for many years in the form of the helicopter. The helicopter in its present form has extensive shortcomings. The speed of rotation of the motor must be mechanically limited to prevent rotor failure. Additionally, the rotor causes considerable vibration within the aircraft. Maintenance hours on helicopters can amount to twenty times the flying time. Perhaps the most serious drawback is the fact that passenger seat-mile costs are 10 to 15 times higher than for conventional aircraft, due principally to the high lift power required for vertical take-off which is not necessary in flight. Experts indicate that the helicopter can be regarded only as an interim vehicle because of its limited economic range, 50 miles.

Advanced forms of VTOL aircraft are currently in the design stage. These include [18]:

> *Fan in Wing Aircraft.* Lift fans are buried in the wing roots, taking power from concentrically mounted gas generators. Two engines provide cruise power which is additionally deflected to aid in the hover stage. Tail, wing tip, and nose fans are provided for control in the hover stage. Speeds of 550 mph appear possible.
>
> *Tilt Wing Aircraft.* Interconnected turboshaft engines supply the power fan lift with the wing tilted into the vertical position. As

the wing tilts forward the engines supply cruise power for speeds at 300 mph or more.

Compound Helicopter. The rotor is powered conventionally for vertical take-off. For cruising at high speeds, power is transferred to horizontal turboprop or jet thrust. Stub wings provide sufficient lift at cruise speeds of 275–300 mph.

Folding Tilt Rotor. Lift is supplied by wing-tip rotors during the hover and transition phase. In conventional flight the rotors are folded rearward into wing tip nacelles. Convertible fan engines provide shaft power for the rotor drive and convert to give fan thrust for conventional flight. Speeds up to 500 mph are possible.

It is unlikely that rotorless VTOL aircraft will be generally available until 1980 or even 1985. Until that time STOL aircraft are likely to be developed as interim vehicles.

Short Take-off and Landing (STOL) Aircraft. These aircraft appear to have the more immediate advantage with respect to available technology. STOL systems will almost certainly be available within 7 to 15 years. STOL aircraft operate on very short landing strips. In 1970, the FAA set forward STOL airport design criteria, setting runway lengths at 1,500 ft. STOL aircraft can be of the *high lift* or *high acceleration* type.

High Lift STOL. These aircraft use relatively simple high lift devices such as externally blown flaps. By changing the speed of air passage over the wing surfaces, extraordinarily high lift can be achieved.

High Acceleration STOL. By the use of auxiliary engines additional power is supplied for acceleration and lift at take-off and for deceleration after landing. During cruise flight the auxiliary engines are unused.

STOL systems are likely to be phased out with the introduction of nonrotor VTOL aircraft due to the need for extensive costly real estate in central urban areas.

Many problems face the development of V/STOL systems [17]. Among the most important problems which must be solved are:

1. The ability to control site and aircraft induced wind shears in built up areas.

2. The safe provision of ground services such as fuel in areas close to high-density urban centers.

3. Prevention of hazards from aircraft and other objects falling from the elevated STOL sites.

4. Environmental deterioration from noise, air, and aesthetic pollution.

22–12. Long-Haul Supersonic Aircraft. The U.S. aerospace manufacturing industry proposes commercial supersonic operation with the Boeing 2707. This aircraft will fly at 1,800 mph or approximately three times the speed of sound with a range of 4,000 miles. The 286-foot long plane will carry 298 passengers. Originally to be a swing wing aircraft, a delta type fixed wing design was finally chosen. The Boeing 2707 will not, if manufactured, be the first operational commercial supersonic aircraft. The British-French "Concorde" and the Russian TU-144 are anticipated to enter commercial operations in 1973.

The "Concorde" is a smaller, slower aircraft than the proposed U.S. SST. Since the aircraft is not constructed of the lightweight heat resistant titanium metal, the design speed at 1,400 mph is somewhat lower than that of the Boeing. The passenger capacity of the 193-foot aircraft is 135 persons [19].

In characteristics, the TU-144 is similar to the Concorde in size, speed, and passenger capacity. It is not anticipated that the Russian supersonic will enter into the world aviation market to any greater degree than previously designed Soviet aircraft.

The expense of the production of supersonic aircraft is expected to be high. Concorde project costs will exceed $2 billion. This amounts to four times the original estimate [9], and is financed in its entirety by the British and French governments. The entry of the United States into the SST field appears to hinge on governmental support of development costs.

The economic future of the SST has raised many questions. Market projections by the aviation industry have indicated that the introduction of these high-speed aircraft will generate large increases in intercontinental traffic. The world market for SST transports in the 1970's and 1980's is estimated to be at least 400 vehicles for international traffic and 100 for domestic flights. Some analysts question the long-term economics of the SST vehicles. Unforeseen environmental problems, such as high noise levels, sonic boom, and stratospheric pollution may significantly diminish the anticipated utility of these aircraft. Both the Concorde and the proposed Boeing 2707 will generate high take-off and landing noise levels. In transcontinental operations, the sonic boom creates a sufficiently disturbing effect that several European countries have banned supersonic operations in their national airspace.

Additional concern is voiced over the effect of the discharge of the upper atmosphere of fine particulate and gaseous matter. Ecologists have warned that excessive pollution will cause climatic changes which would be virtually irreversible.

The rapid advances in space technology in the 1960's have led aircraft engineers to consider the design of even faster passenger aircraft. This

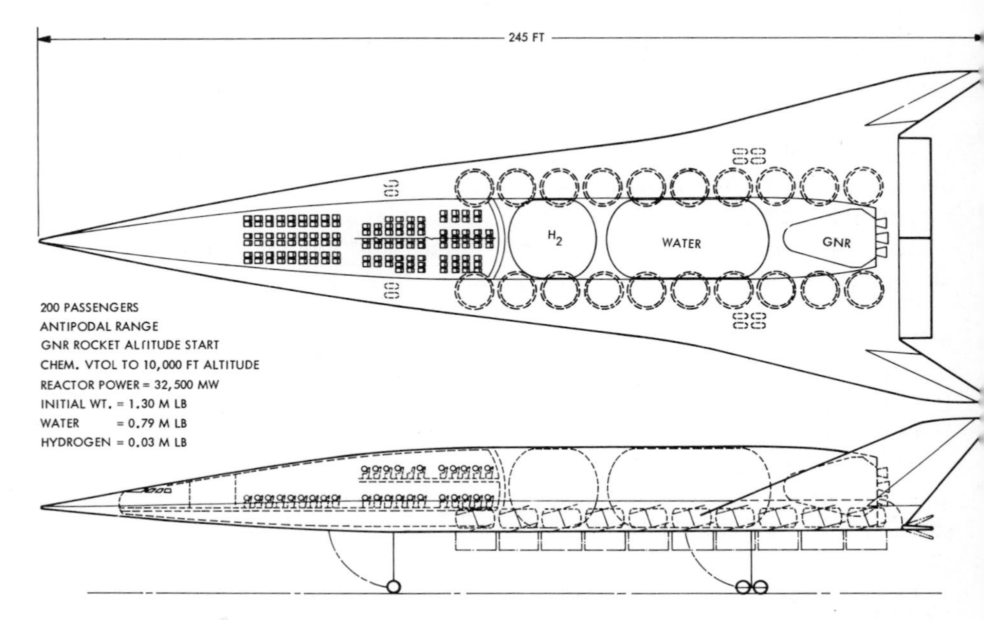

Fig. 22–10. The Hypersonic Transport (HST). (Source: Lockheed Missiles and Space Co.)

generation of aircraft has been called the *hypersonic transport* (HST). Figure 22–10 shows a conceptualization of the appearance of this plane. These aircraft would travel at speeds close to 4,000 mph. Cruising altitudes would be on the very fringes of space at 100,000 feet or more. Such a concept is currently uneconomic. Technological advances coupled with changes in transportation demand and personal income structures could make this concept a reality within 25 years.

WATER TRANSPORTATION

22–13. Water Transportation. Very dramatic changes have been underway for some years in the area of water transportation. Technological advances have permitted the introduction of new transportation concepts in the last few years which can be expected to bring about trends of continuing change. Freight movements are entering the era of *modular freight transport*. The first indications of the change came about with the rapid expansion of container ships and containerization in the 1960's. By 1970, the U.S. maritime fleet contained approximately 70 containerships. Additionally, each major shipping line has adapted at least some of its ships to a container carrying capacity. Often this capacity is as tie-down space above decks. The great advantage of containerships, in the face of heavy capitalization costs, lies in their ability to be rapidly unloaded. One containership holding 1,000 standard $8' \times 8' \times 20'$ containers can be unloaded and reloaded in one day. By comparison the same cargo volume would require five days turnaround time for conventional handling [8]. The rapid growth and widespread acceptance of the containerization process has spawned radical ideas for the integration of land, inland water, and ocean transport. These concepts have been executed in the form of *LASH* and *Seabea* vessels. These innovative ships have extended the concept of containerization by moving modular freight to the ship rather than vice versa.

Lighter aboard ships (*LASH*) vessels are under construction and will be introduced into the Pacific trade during the early seventies. These vessels are 814 feet long with a 100 foot beam. Two shipboard traveling cranes can load or discharge 49 lighters and 356 standard 20′ containers. The modular concept can be seen to be complete. Inland waterway traffic can move in containers aboard lighters. The previously loaded lighter is hoisted directly aboard the LASH vessel which is specially designed to accept the lighter as a loading module. Loading and unloading can be expedited, saving both time and damages. The *Seabea* vessel of the Lykes Line, shown in Fig. 22–11, makes use of the same general

Fig. 22–11. A Seabea Vessel. (Source: Lykes Lines.)

rationale. This ship can load 38 barges containing up to 1,500 containers. Loading in this case is achieved by a hydraulic elevator system with barges floated in and out at the loading and unloading stage.

Hydrofoil ships have generated considerable interest in the past. They could be used extensively in coastal waters in the future if the problems of excessive vibration at speed, and foil damage from floating debris can be overcome. Speeds of 75 mph are possible from this type of vessels. Hydrofoils actually "fly" in water. The movement of water over the submerged foil supplies a sufficient upward thrust to lift the hull of the ship clear of the water, cutting water resistance on the ship to a fraction of its value otherwise. The high density of water relative to air means that only a small area of water foil is required in comparison with the large wings required in air. Ship builders anticipate that large hydrofoil vessels will soon be able to ply the Atlantic in 36 hours, offering increased competition to the air mode in the form of cheaper transportation without a significant lowering of the level of service. British and Russian engineers continue to work extensively on hydrofoil development. Both these nations have inland and coastal waterways where hydrofoil vessels could be of benefit.

Surface effect ships (SES) are similar to land-based air cushion vehicles (ACV). Currently, the SEN-4 vehicle shuttles 600 persons across the English Channel. Vessels of this type are currently in use also in the U.S.S.R. and the U.S. on a smaller scale. Somewhat noisy and difficult to control except in isolated conditions, these machines will continue to be developed for limited transportation solutions. They are unlikely to change transportation patterns significantly except in a local area.

Nuclear powered cargo ships were ushered into existence with the launching of the N.S. Savannah in 1961. The lack of fuel storage space requirement and high power capacities of the nuclear vessel make these vessels ideal cargo ships for operational speeds in excess of 30 knots. At the present time cost of construction is the chief block to development of nuclear fleets. Other problems occur relative to the special training required for crews. Some nations have presented difficulties in granting permission to nuclear ships for docking privileges within their territorial area. Improvements in nuclear technology are likely to render nuclear power plants and the ships that carry them more economic. At this time, which is likely to be within the next 20 years, nuclear powered ships will become commonplace.

22–14. Summary. Change is not dependent simply on the availability of technology. It is dependent rather on the whole structure of national and international priorities. The decision to invest in the development of certain technological areas is intermixed with social, political, and economic decisions in both the governmental and business spheres. Technology changes rapidly in time of war under the pressure of national survival. In peace time, energies are diverted into fields of more social consequence. The concepts which have been discussed in the preceeding sections of this chapter are all technologically feasible at the present time. Undoubtedly, each of the concepts that has been mentioned will be built and used somewhere for some period of time. Its eventual utility and the degree of universal acceptance depends on the social, economic, and political structure of the world into which it is introduced, and whether it is in fact "an idea whose time has come."

REFERENCES

1. *Tomorrow's Transportation: New Systems for the Urban Future,* U.S. Department of Housing and Urban Development, Washington, D.C., 1968.
2. *New Concepts in Urban Transportation Systems,* Special Issue, Journal of the Franklin Institute, Lancaster, Pennsylvania, November 1968.
3. Schneider, L., *Marketing Mass Transit,* Harvard University Press, Cambridge, Mass. (1965).
4. Chermayeff, P., "Orientation and Transit Systems," *Highway Research Record No. 251,* Highway Research Board, Washington, D.C., 1968.
5. Leish, Jack, "Transportation Systems in the Future Development of the Metropolitan Areas," *Highway Research Record No. 239,* Highway Research Board, Washington, D.C., 1969.
6. Richards, Brian, *New Movement in Cities,* Studio Vista, London (1966).
7. Morris, Robert, "Transportation Planning for New Towns," *Highway Research Record No. 293,* Highway Research Board, Washington, D.C., 1969.
8. Hellman, Hal, *Transportation in the World of the Future,* M. Evans and Company, New York (1968).
9. Hill, F. Norman, "What's Ahead for Transit," *Traffic Engineering,* February, 1968.
10. Goldsmith, A., and Cleven, G. W., "Highway Electronic Systems—Today and Tomorrow," *IEEE Transactions on Vehicle Technology,* Volume VT-19, February, 1970.
11. Bartelsmeyer, R. R., "Highways in the 70's," *Traffic Engineering,* May, 1970.
12. Ashford, Norman, "Joint Development and Urban Highways," *The Professional Engineer,* January, 1970.
13. Edwards, L. K., "High Speed Tube Transportation," *Scientific American,* August, 1965.
14. Foa, J. V., *An Introduction to Project Tube Flight,* Rensselaer Polytechnic, September, 1966.
15. Volpe, John A., "Transportation of the Future," *Limestone,* National Limestone Institute, Washington, D.C., Fall, 1969.
16. Vehicular Technology, *IEEE Transactions,* February, 1970.
17. Ashford, Norman, and Covault, D. O., "Areas of Research and Evaluation of V/STOL Transport Systems," Annual Meeting of the American Society of Civil Engineers, Boston, July, 1970.
18. *Study of Aircraft in Short Haul Transportation Systems,* The Boeing Company, Renton, Washington, August, 1967.
19. *Airports for the Future,* Institution of Civil Engineers, London, 1967.
20. *Transport Technological Trends,* Transportation Association of America, Washington, D.C., January, 1970.

Index